4th International Conference on Mechatronics and Computer Technology Engineering (MCTE 2021)

Journal of Physics: Conference Series Volume 2113

Xi’an, China
15-17 October 2021

ISBN: 978-1-7138-5084-7
ISSN: 1742-6588

Printed from e-media with permission by:

Curran Associates, Inc.
57 Morehouse Lane
Red Hook, NY 12571

Some format issues inherent in the e-media version may also appear in this print version.

Institute of Physics
Dirac House, Temple Back
Bristol BS1 6BE UK

Phone: 44 1 17 929 7481
Fax: 44 1 17 920 0979

techtracking@iop.org

Additional copies of this publication are available from:

Curran Associates, Inc.
57 Morehouse Lane
Red Hook, NY 12571 USA
Phone: 845-758-0400
Fax: 845-758-2633
Email: curran@proceedings.com
Web: www.proceedings.com

TABLE OF CONTENTS

Preface

Peer review declaration

MECHANICAL POWER CONTROL AND SENSOR DESIGN

ELECTRICAL CIRCUIT DEVICE AND SMART GRID DETECTION

COMPUTER TECHNOLOGY AND DATA MODEL ALGORITHM IDENTIFICATION

Preface

2021 4th International Conference on Mechatronics and Computer Technology Engineering (MCTE 2021) has been held successfully on October 15-17, 2021. Due to the COVID-19 pandemic around the world and with the strict travelling rules in China, it is still difficult to take international travel for our attendees abroad. Therefore, MCTE 2021 was held both in physical (Xi'an, China) and online (Zoom).

The conference mainly focuses on mechatronics, computer technology engineering, computer science and other research fields, aiming at providing a platform for experts, scholars, engineers and technical researchers engaged in mechatronics, computer technology engineering and computer science research to share scientific research achievements and cutting-edge technologies, understand academic development trends, broaden research ideas, strengthen academic research and discussion, and promote cooperation in industrialization of academic achievements. Experts, scholars, business people and other relevant personnel from universities and research institutions at home and abroad are cordially invited to attend and exchange.

During the conference, the conference model was divided into three sessions, including oral presentations, keynote speeches, and online Q&A discussion. In the first part, some scholars, whose submissions were selected as the excellent papers, were given about 5-10 minutes to perform their oral presentations one by one. Then in the second part, keynote speakers were each allocated 30-45 minutes to hold their speeches. There were 90 individuals who attended this hybrid conference.

In the second part, we invited six professors as our keynote speakers. Prof. Wanmi Chen, from Shanghai University. His mainly research focus on Machine Vision, Machine Learning, Dynamic Role Assignment, Decision Optimisation, Motion Control. And then we invited Assoc. Prof. ZAIN ANWAR ALI, Sir Syed Pakistan UE&T, Karachi, Pakistan. His research area including Bio-Inspired Optimization Algorithms Control Theory and Control Engineering Network-based Systems, Robotics, Autonomous Unmanned System i.e.(Aerial and Underwater Vehicles), Industrial Controls & Automation, Programming Logic Controller PLC. Assoc. Prof. Zeashan Hameed Khan, Air University, E-9, Islamabad, Pakistan. He mainly researches on Smart Manufacturing, Industry 4.0, Renewable Energy, Automation and Control, Biomedical Sensors and Systems etc. Prof. Wei Chen, Nanjing Institute of Technology. He devoted in Intelligent Robotic Control Technology, Image Acquisition and Processing, Multi-sensor Information Fusion. Prof. Na Li, from Xidian University. She made a wonderful speech: *Structural design of new antenna system based on electromechanical coupling theory.* And then we invited Assoc. Prof. Wei Wei, Xi'an University of Technology, China as our finale keynote speaker. He performed a speech: *Gradient-Driven Parking Navigation Using a Continuous Information Potential Field Based on Wireless Sensor Network.* Their insightful speeches had triggered heated discussion in the third session of the conference. Every participant praised this conference for disseminating useful and insightful knowledge.

The proceedings are a compilation of the accepted papers and represent an interesting outcome of the conference. Topics include but are not limited to the following areas: Computing Science & Technology, Intelligent Mechatronics & Applications and other related areas. All the papers have been through rigorous review and process to meet the requirements of international publication standard.

We would like to acknowledge all of those who supported MCTE 2021. The help and contribution of each individual and institution was instrumental in the success of the conference. In particular, we would like to thank the organizing committee for its valuable inputs in shaping the conference program and reviewing the submitted papers.

The Committee of MCTE 2021

Journal of Physics: Conference Series **2113** (2021) 011001 doi:10.1088/1742-6596/2113/1/011001

Committee member

Committee Chair
Assoc. Prof. ZAIN ANWAR ALI, Sir Syed Pakistan UE&T, Karachi, Pakistan
Assoc. Prof. Zeashan Hameed Khan, Air University, E-9, Islamabad, Pakistan

Program Committees
Prof. Xiaofeng Li, Heilongjiang International University, China
Prof. Guobao Xiao, Minjiang University/ College of Computer and Control Engineering, China
Dr. Han Liu, National University of Defense Technology, China

International Technical Program Committees
Prof. Wenbin Gu, Hohai University, China
Prof. Yuliang Liu, Zhejiang Ocean University, China
Prof. Tiejun Cui, Liaoning Technical University, China
Prof. Jingfeng Shao, Xi'an Polytechnic University, China
Prof. Yegang Chen, Yangtze Normal University, China
Prof. Wenli Zhang, Beijing University of Technology, China
Dr. Fugui Li, China Academy of Launch Vehicle Technology, China
Dr. Aiguo Wang, Foshan University, China

Peer review declaration

All papers published in this volume of Journal of Physics: Conference Series have been peer reviewed through processes administered by the Editors. Reviews were conducted by expert referees to the professional and scientific standards expected of a proceedings journal published by IOP Publishing.

- **Type of peer review: Double-blind**
- **Conference submission management system: AI Scholar System**
- **Number of submissions received: 156**
- **Number of submissions sent for review: 151**
- **Number of submissions accepted: 86**
- **Acceptance Rate (Number of Submissions Accepted / Number of Submissions Received X 100): 55.1%**
- **Average number of reviews per paper: 2**
- **Total number of reviewers involved: 50**
- **Any additional info on review process:**
- **Step 1. Each of selected paper will be reviewed by two/three professional experts in the related subject area.**
- **Step 2. Review Reports received from the experts will be judged by one of the editors either Review Reports are logical or not?**
- **Step 3. If not logical, then editor can assign new reviewer or can also judge at his/her own.**
- **Step 4. If logical, then Review Reports will be sent to authors to modify the manuscript accordingly.**
- **Step 5. Authors will be required to revise their papers according to the points raised.**
- **Step 6. Revised version will then be evaluated by the editor for the incorporation of the points raised by the reviewers.**
- **Step 7. Then the editor will send the revised manuscript to the reviewers again for re-evaluation.**
- **Step 8. If the reviewers approve the revise version of the manuscript, then will be accepted for publication.**
- **Contact person for queries:**
  Xuexia Ye
  publication@keoaeic.org
  AEIC Academic Exchange Information Centre

Human-posture-tracking sensor selection for motion recognition

Chongwei Tan

The Second High School attached to Beijing Normal University, Beijing, Beijing, China

battleel@shsbnu.net

Abstract: Very recently special attention has been paid to soft sensors for motion tracking. It is known from the work by Chen et al [2021, Comp Anim Virtual Worlds. 2021; e1993.] that a wearable motion tracking system was developed, in which five sensors were placed around the region of arm and shoulder. In this study, we explore the effect of different sensors on motion recognition and select the sensors with excellent differentiation for different movements.

1. Introduction

Using sensors to capture motion or action is a very important cutting-edge technology in the field of VR or animation [1], especially focusing on flexible sensors. Previous studies focused on the comprehensive effects of all parts (palms, arms, shoulders, and fingers [2]) on motion recognition, while our research focused on the selection of the part which has a higher degree of discrimination for distinguishing different actions.

The movement of the arm mainly depends on the change of the elbow and shoulder joints. In [3], Chen et al developed a wearable garment with some flexible sensors, three of which are on the shoulders and the last one is on the elbows. They tracked the rotation of the joint by monitoring the change in the resistance of the stretchable sensor. For example, when the user bends the elbow joint, the sensor is stretched, and its resistance value increases accordingly. Inspired by this work, here we are interested in which sensor would have better performance for movement (pattern) recognition. We mainly tested two patterns, one is to rotate the arm around the body in a circle, and the other is to raise the arm and then fall vertically to form a semicircle. Besides, we chose to use two distances, Euclidean distance and cosine distance to measure the degree of discrimination between different sensors., and use Principal Component Analysis dimensionality (PCA) reduction to visualize the discrimination ability.

We summarize our key findings as below:

- Compared with cosine distance, Euclidean distance is a more discriminative measure;
- The result of Euclidean distance and the result of PCA dimensionality reduction show a relatively strong association;
- The Euclidean distance of sensor 2 and the results of PCA both show that the two patterns have a high degree of discrimination.

2. Data Specification

2.1 Data Sources
We use the dataset in [3]. The data were collected from a man whose age is 28. He did 7 different patterns of actions according to his preferences. He kept repeating these actions so that one could collect enough data. The collection was about 1.5 hours each time, including the time to rest if he got tired.

2.2 Sensors
The sensor is sewn on the clothes by hand using flat stitch technology. There are four sensors, three around the shoulder joint and the last one at the elbow joint.

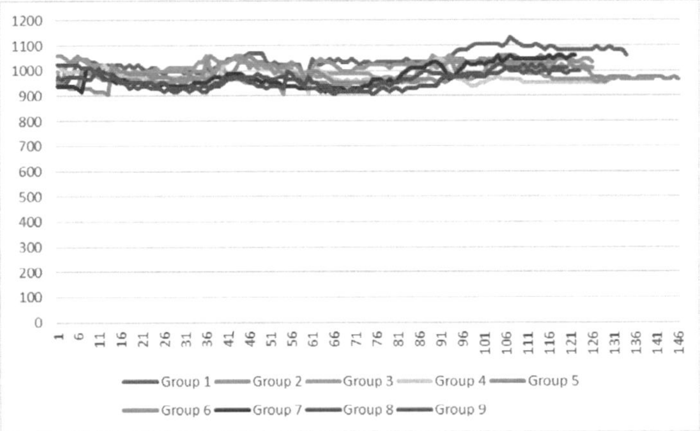

Figure 1. Pattern 1 and the data collected by Sensor 1

3. Data analysis

3.1 Data preprocessing
I drew the data of different groups at the same sensor into a line graph and observe the trend. I mainly completed Pattern 1 and Pattern 2. I found that these data are not the same in length, some are long and some are short, but there are no outliers. Therefore, I simply cut the data so that all groups of data have the same length. It is shown that this preprocessing does not affect the data analysis.

3.2 Calculating distance

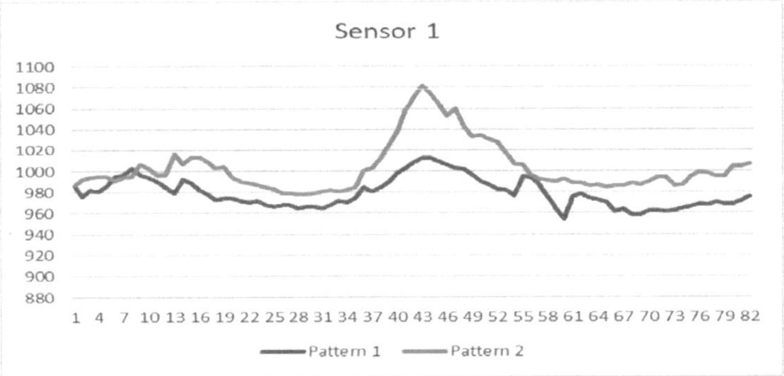

Figure 2. Different data from Sensor 1

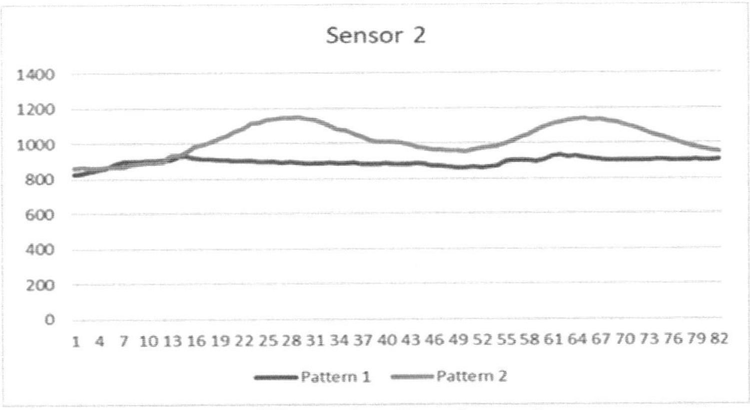

Figure 3. Different data from Sensor 2

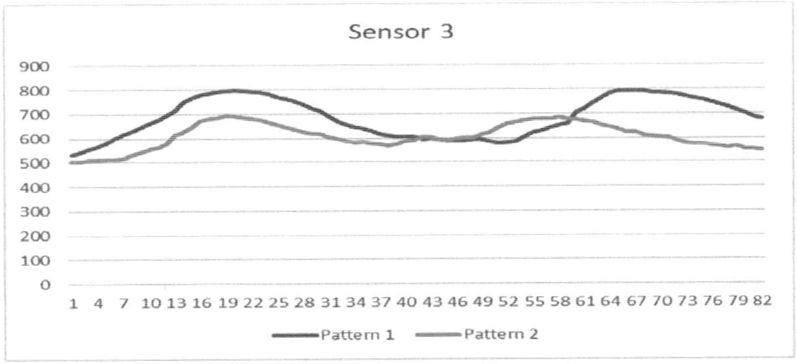

Figure 4. Different data from Sensor 3

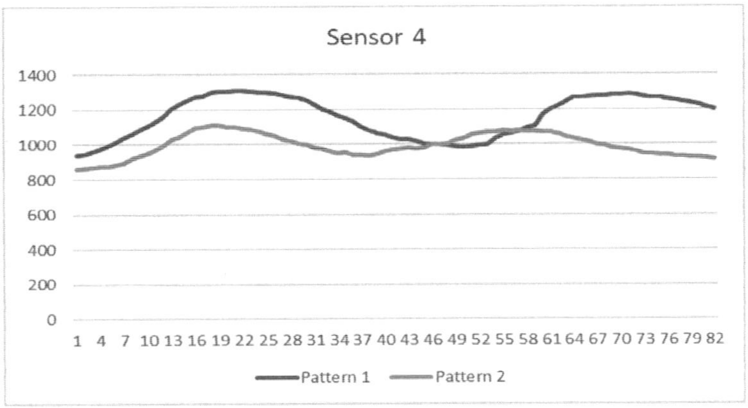

Figure 5. Different data from Sensor 4

I calculated the distance between the two patterns at different sensors, using the cosine distance and the Euclidean distance.

Euclidean distance: In mathematics, the Euclidean distance between two points in Euclidean space is the length of a line segment between the two points. Its formula is:

$$d(x,y) = \sqrt{\sum_{i=1}^{n}(x_i - y_i)^2} \quad (1)$$

Cosine distance (Cosine similarity) is a measure of similarity between two non-zero vectors of an inner product space. It is defined to equal the cosine of the angle between them, which is also the same as the inner product of the same vectors normalized to both have length 1. The cosine of 0° is 1, and it is less than 1 for any angle in the interval (0, π] radians. It is thus a judgment of orientation and not magnitude: two vectors with the same orientation have a cosine similarity of 1, two vectors oriented at 90° relative to each other have a similarity of 0, and two vectors diametrically opposed have a similarity of -1, independent of their magnitude. Its formula is:

$$sin(x,y) = cos\,\theta = \frac{\vec{X}\cdot\vec{Y}}{\|X\|\cdot\|y\|} \quad (2)$$

Euclidean distance reflects the absolute difference in value, while cosine distance reflects the relative difference in direction.

Table 1. Calculations of two distances

	Pattern 1	Pattern 2	pattern 3	Pattern 4
Euclidean distance	259.7172	1342.576	968.3688	1801.898
Cosine distance	0.999982	0.997053	0.994095	0.995003

By taking a look at the mean plots of four patterns, we can see that the cosine distances are very close, so there is no discrimination. It reminds us that it is more convincing to use Euclidean distance to distinguish.

3.3 Principal Component Analysis (PCA)

In the real coordinate space, the principal component [4] of the point set is a sequence of p unit vectors, where the i-th vector is the direction of the straight line orthogonal to the i-1th vector and most suitable for the data. Here, the best fit line is defined as the line that minimizes the average square distance from the point to the straight line. These directions form an orthonormal basis, in which the different dimensions of the data are linearly uncorrelated. Principal component analysis (PCA) is the process of calculating principal components and using them to make fundamental changes to the data. Sometimes only the first few principal components are used and the rest are ignored for exploratory data analysis and prediction models. It is usually used for dimensionality reduction. The method is to project each data point only on the first few principal components to obtain low-dimensional data while keeping the data changes as much as possible. The first principal component can be equivalently defined as the direction that maximizes the variance of the projection data. The i-th principal component can be used as a direction orthogonal to the i-1th principal component to maximize the variance of the projection data.

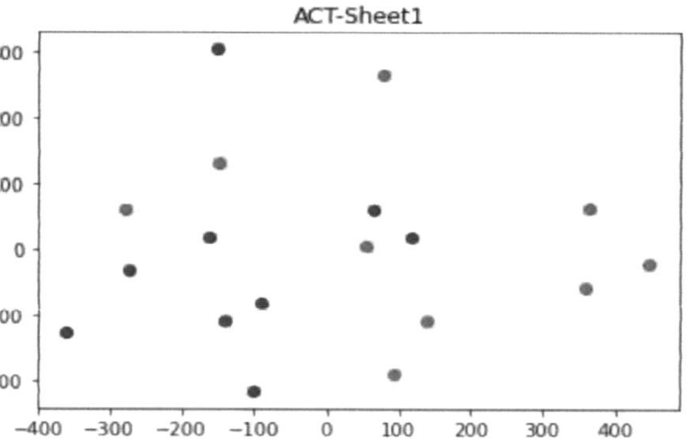

Figure 6. PCA of Sensor 1

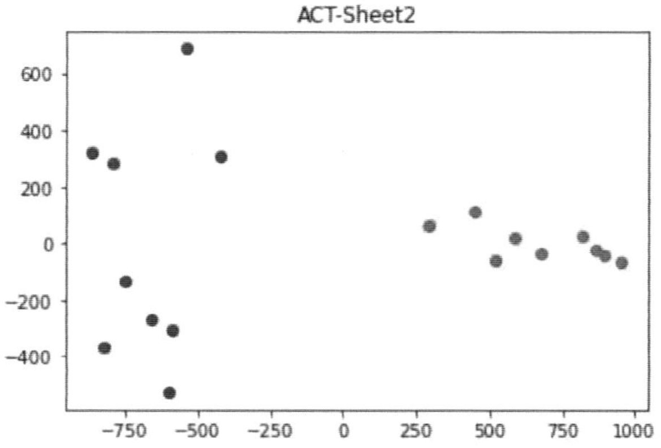

Figure 7. PCA of Sensor 2

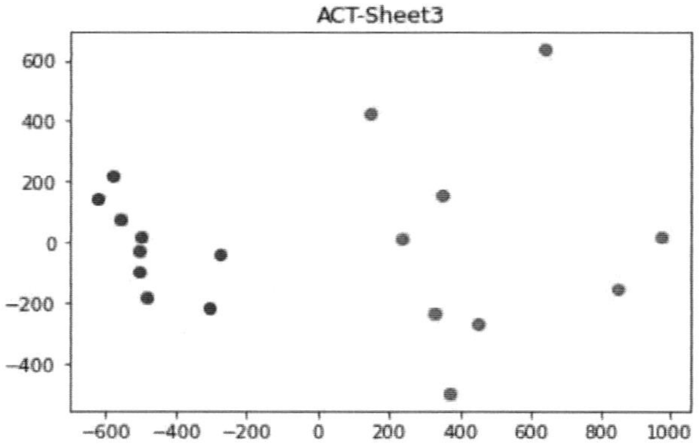

Figure 8. PCA of Sensor 3

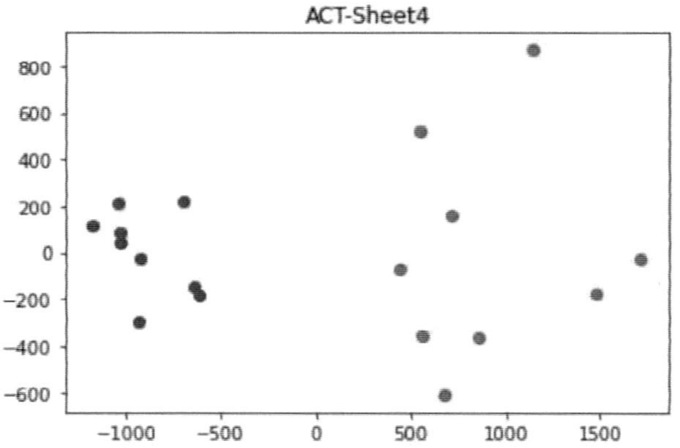

Figure 9. PCA of Sensor 4

4. Conclusions

By comparing the two distances (cosine distance versus Euclidean distance), the discrimination of cosine distance is not as good as the discrimination of Euclidean distance. One possible explanation is that the cosine distance characterizes whether the trends of the two vectors are consistent, while the Euclidean distance pays more attention to the absolute value of the signal. Thus, in this study, we found that the absolute value of the signal is a more important evaluation criterion than the trend.

In future plan, we will improve our work in the following two directions. One is that we only studied two of the seven patterns (movements) in this work. Different conclusions may appear during the comparison of other patterns. Specifically, the performance of different distances may differ in other patterns. The second is we will use more comprehensive data preprocessing methods than this study. Here we simply align the vectors (remove the redundant components), there are some other ways to align different vectors by shrinking or amplifying the signals.

References

[1]. Ruibo L, Qijia S, Siqi W, Christina R, Devin B, and Xia Z. (2019). Reconstructing Human Joint Motion with Computational Fabrics. *Proc. ACM Interact. Mob. Wearable Ubiquitous Technol.* 3, 1, Article 19, 26 pages.

[2]. Oliver G, Shihao W, Daniele P, Otmar H, and Olga S. (2019). Interactive Hand Pose Estimation using a Stretch-Sensing Soft Glove. *ACM Trans. Graph.* 38, 4, Article 41, 15 pages.

[3]. Chen, Z., Chen, X., Ma, Y., Guo, S., Qin, Y., & Liao, M. (2021). Human posture tracking with flexible sensors for motion recognition. *Computer Animation and Virtual Worlds*, e1993.

[4]. Wold S, Kim E, and Paul G. (1987) "Principal component analysis." *Chemometrics and intelligent laboratory systems* 2.1-3 : 37-52.

Decentralized Path Planning for Multi-Objective Robot Swarm System

Zhuokai Wu[1]

[1] Xi'an Gaixin No.1 Middle School, Xi'an, Shaanxi, 710000, China

*Corresponding author's e-mail: alinda@wedo365.com

Abstract. The multi-robot path planning aims to explore a set of non-colliding paths with the shortest sum of lengths for multiple robots. The most popular approach is to artificially decompose the map into discrete small grids before applying heuristic algorithms. To solve the path planning in continuous environments, we propose a decentralized two-stage algorithm to solve the path-planning problem, where the obstacle and inter-robot collisions are both considered. In the first stage, an obstacle- avoidance path-planning problem is mathematically developed by minimizing the travel length of each robot. Specifically, the obstacle-avoidance trajectories are generated by approximating the obstacles as convex-concave constraints. In the second stage, with the given trajectories, we formulate a quadratic programming (QP) problem for velocity control using the control barrier and Lyapunov function (CBF-CLF). In this way, the multi-robot collision avoidance as well as time efficiency are satisfied by adapting the velocities of robots. In sharp contrast to the conventional heuristic methods, path length, smoothness and safety are fully considered by mathematically formulating the optimization problems in continuous environments. Extensive experiments as well as computer simulations are conducted to validate the effectiveness of the proposed path-planning algorithm.

1. Introduction

In the early days, mobile robots were used only to help with production in factories. However, with the development of science and technology, such mobile robots are gradually applied to disaster relief, education, medical care, agricultural production, information and communication fields. When performing a locating pathfinder mission, the robot is equipped with many advanced equipments, the main functions of these devices include mapping of the environment, obstructive detection, control the motion and use navigation techniques to avoid obstacles. The most important function of any navigation for a robot is to find and plan a safe route from start position to terminal point.

The most common approach for path planning is to divide the continuous environment into a discrete gridded space. Considerable investigations have been devoted into this scheme. For instance, the genetic algorithm (GA) [1] and particle swarm optimization (PSO) [2] were developed to find the suitable trajectory with the shortest length in a discrete gridded environment. However, the computational complexity as well as time cost may become extremely high due to the NP-hard nature of this approach [3]. Furthermore, the gridded environments may result in unrealistic trajectories, where sharp turns are inevitable.

To address these challenges, an enhanced genetic algorithm (EGA) was developed in [4], where five genetic operators were designed to improved the performance at each iteration in terms of the predefined objective function. In [5], an informative path-planning problem was derived and resolved using an evolutionary strategy. Unfortunately, the aforementioned algorithms are not suitable for dynamic scenarios since the training process is highly depended. In [6], a decentralized method using velocity control was derived for multi-vehicle traffic networks, where the path-planning problem was only investigated in two-lane scenarios. The control barrier function (CBF) is utilized to guarantee the safe distance among vehicles.

The existing methods for multi-robot path planning can be classified into two categories, namely centralized and decoupled methods. For the centralized approaches, the constraints and trajectory generation for all robots are combined into one optimization problem such that the solutions are near to the global optimum. However, the time complexity may become extremely high due to the combination of multiple constraints and configurations. Therefore, the centralized approaches usually cannot adapt dynamic environments. In contrast, decoupled methods first derive an individual trajectory for each robot and then resolve the possible conflicts of the generated trajectories. A typical decoupled approach was proposed in [7], where the wait-strategy is adopted by predefining the priority of robots. After the trajectories are individually generated, the robot with lower priority is ordered to wait temporarily if two robots are collision-prone. Despite the collisions are avoided, this wait-strategy is nor time-efficient, neither feasible in practice.

In this work, we propose a decoupled approach for multi-robot path planning in which two stages are adopted. In the first stage, obstacle-avoidance trajectories are generated for the individual robots. In contrast to the heuristic algorithms based on the gridded environments, such as A* and Dijkstra [8], the proposed trajectory-generation algorithm mathematically formulates an optimization problem by approximating the obstacles into convex-concave constraints. After obtained the trajectories, we control the velocities of the robots to avoid possible inter-robot collision while considering the time efficiency. The proposed approach is decentralized since each robot only computes its own trajectory and velocity according to the surrounding environments.

In summary, the main contributions of this work can be summarized as follows:

1) A two-stage *decoupled* approach is developed for multi-robot path planning in continuous environments. The proposed algorithm can be implemented in decentralized online operations.

2) As compared to the conventional heuristic algorithms that decompose the spaces into discrete grids, we derive the obstacle-avoidance trajectories by formulating an optimization problem in continuous environments, where the obstacles are circular-approximated.

3) The inter-robot collisions of the planned trajectories are avoided using velocity control by formulating a quadratic programming (QP) problem. In the meanwhile, the travel time of robots is minimized by adopting the control Lyapunov function (CLF). Finally, experiments and simulations are implemented to validate the effectiveness of the proposed approach.

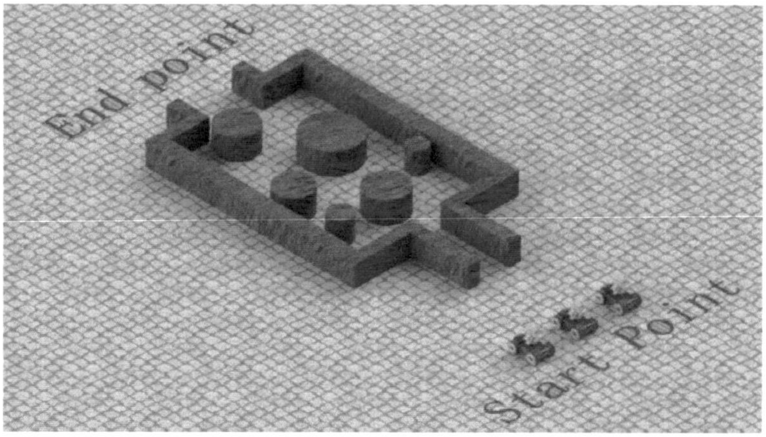

Fig. 1. The diagram of the multi-objective robot swarm system.

The remainder of this paper is organized as follows. Section II provides the brief review of CBF-CLF used in this work. Next, the multi-robot system model and problem statement are presented in Section

III. To solve the path-planning problem, we first generate the obstacle-avoidance trajectories using DCCP in Section IV. Next, the possible collisions of robots are avoided by solving the QP problem derived in Section V. Furthermore, performance of the proposed approach is evaluated in Section VI, and Section VII concludes this work.

Notation: Uppercase boldface and lowercase boldface letters are used to denote matrices and vectors, respectively. $\|A\|$ stands for the ℓ_2 norm of **A** while $\dot{A}$ denotes the derivative of **A**. $\mathbf{L_f A}$ is the Lie derivative of function A to f, $\hat{A}$ the estimator of $\hat{A}$, where $\hat{A}$ can be a matrix, vector, or scalar. A^T stands for the transpose of matrix A. Finally, $\nabla f(A)$ represents the gradient of function $f(A)$.

2. A brief review of CBF-CLF problem formulation

Firstly, we provide an elegant introduction of control barrier functions (CBFs). The complete details can be found in [9]. A continuous function $\Gamma(x)$ is defined as a class **K** function if $\Gamma(x)$ is strictly increasing and $\Gamma(0) = 0$ for $x \leq 0$.

In this paper, we consider a continuous-time affine control system formulated by

$$\dot{x} = f(x) + g(x)u \tag{1}$$

Where $f : R^n \to R^n$ and $g : R^n \to R^{n\times m}$ are both locally Lipschitz. In addition, $x \in R^n$ and $u \in R^m$ are the state andcontrol input, respectively.Barrier functions are used to define the set of all states by

$$C := \{x \in R^n : h(x) \geq 0\} \tag{2}$$

where $h : R^n \to R^n$ is continuously differentiable. If there exist class κ functions β_1, β_2 and γ . A CBF function $B(x)$ for a control system is defined by the following inequalities:

$$\frac{1}{\beta_1(h(x))} \leq B(X) \leq \frac{1}{\beta_2(h(x))} \tag{3}$$

$$\inf_{u\in U}\left[L_f B(x) + L_g B(x)u - \frac{\gamma}{B(x)}\right] \leq 0 \tag{4}$$

for all $x \in C$. L_f and L_g are the Lie derivatives of the function $B(x)$ to f and g, respectively. The Lie derivatives are defined

$$L_f B(x) = \nabla Bf = \frac{\partial B}{\partial x} f(x) \tag{5}$$

where $\frac{\partial B}{\partial x} \in R^{n\times 1}$ and $f(x) \in R^{n\times 1}$ Thus, $L_f B(x)$ is a scalar.For a CBF $B(x)$ and $\gamma > 0$, any Lipschitz continuous controller u with

$$K_{cbf}(x) = \{u \in U : L_f B(x) + L_g B(x)u - \frac{\gamma}{B(x)} \leq 0\} \tag{6}$$

renders set C forward invariant [9] for the affine control system in Equation (1).

A continuously differentiable function $V : R^n \to R^n$ is an exponentially stabilizing control Lyapunov function (CLF) if there exist positive constants c_1, c_2, $c_3 > 0$ such that for any $x \in R^n$, the following inequalities hold:

$$c_1 \|x\|^2 \leq V(X) \leq c_2 \|x\|^2 \tag{7}$$

$$\inf_{u\in U}\left[L_f V(x) + L_g V(x)u + c_3 V(x) \leq 0\right] \tag{8}$$

Given a CLF $V(X)$, any Lipschitz continuous controller u holds

$$K_{clf}(x) = \{u \in U : L_f V(x) + L_g V(x)u + c_3 V(x) \leq 0\} \tag{9}$$

The CLF and CBF can be combined into a single controller through a quadratic programming (QP) problem represented as

$$u(x)^* = \arg\min_{u\in U} \frac{1}{2} u^T H_x(x) u + F(x)^T u \tag{10}$$
$$subject\ \text{to}: L_f V(x) + L_g V(x) u + c_3 V(x) - \boldsymbol{\delta}_0 \leq 0$$
$$L_f B(x) + L_g B(x) u - \boldsymbol{\alpha}(h(x)) \leq 0$$

Where C3≥0 is a constant and α(h(x)) is a class K function. $H_x(\boldsymbol{x})$ is a positive definite and $F(x) \in R^{1\times m}$. In addition,δ_0 is the relaxation factor on the CLF.

3. System overview and problem statement

3. 1 Multi-robot system model

Consider a swarm of Z robots moving in a continuous environment. All the robots are equipped with LiDARs and RGB-D cameras to detect the surrounding situations. The velocity and position of adjacent robots as well as the locations of M obstacles can be measured using the equipped sensors. The obstacles in arbitrary shapes are centered $\{ p_m \}_{m=1,..,M} \in R^2$.

We assume the start and destination locations are given for each robots. The task for each robots is efficiently traveling from the start point to its destination in a continuous environment while avoiding collisions with obstacles and other robots. Furthermore, we assume the robots consist of motion primitives that allow them to deterministically move from $y_{z,n}$ to $y_{z,n+1}$ with velocity $v_{z,n}$, where $\{ y_{z,n} \}_{n=1,..,N} \in R^2$ are the set of sampled points in the trajectory of the z-th robot. The maximal velocity and acceleration of a robot are given as v_{max} and a_{max}, respectively.

3.2 Path-Planning Problem Statement

For the multi-objective path planning, the designated robots are assigned to achieve their own destinations. The objectivesand constraints are summarized as follows:

- **Path Length:** The sum of path lengths is one of the most important metric for the performance of the proposed path-planning algorithm. Let L_z the path length of the *z*-th robot that travels from the start point to its destination. The objective function can be modeled as the minimization of $\sum_{z=1}^{Z} L_z$.
- **Time Efficiency:** The minimal travel time can be achieved if all robots move at v_{max} in the planned trajectory. Therefore,the travel time of a robot is lower bounded by $\frac{L_z}{v_{max}}$. Obviously, the possible collisions among the robots will increase the max travel time.
- **Safety:** The safety constraints can be divided into two categories, namely *static* and *dynamic* collision-avoidance con- straints. Firstly, the *static* collision avoidance refers to the avoidance of static obstacles in the given environment. Next,the inter-robot collisions are considered as *dynamic* collision avoidance. The safety distance must be satisfied with bothof the constraints during the movement of the robots.
- **Smoothness:** A smooth and continuous UGV's trajectory without sharp turns is desired in practice. We will first characterize the smoothness of the generated trajectories. For any three ordered positions $\{\boldsymbol{y}_{z,n-1}, \boldsymbol{y}_{z,n-1}, \boldsymbol{y}_{z,n+1}\}$ with$n = 2, \cdots, N-1$, the turning angle is defined as

$$cos\, \beta_{z,n} = \frac{(y_{z,n+1} - y_{z,n})^T \cdot (y_{z,n} - y_{z,n-1})}{\|y_{z,n+1} - y_{z,n}\| \cdot \|y_{z,n} - y_{z,n-1}\|} \tag{11}$$

We propose to minimize the following metric to ensure the smoothness:

$$S = \sum_{z=1}^{Z} \sum_{n=2}^{N-1} |\beta_{z,n}| \tag{12}$$

In this work, we aim to achieve the aforementioned objectives subject to the constraints for each robot.

4. Path planing for a single robot based on dccp

4.1 Obstacle-Avoidance Path Planning for A Single Robot
In the sequel, we investigate the path-planning problem for a single robot. The inter-robot collisions are neglected temporarily.The goal of path planning is to find the shortest trajectory connecting initial point $\boldsymbol{s}_{\text{int}}$ and destination $\boldsymbol{s}_{\text{des}}$ that avoids Mobstacles, which are approximated as $\tilde{M}$circles using **Algorithm 1**. Apart from obstacles, the available road denoted by Γ is assumed to be known in advance. For instance, the available roads and corridors are usually constrained by the kerbstones or walls.

We assume a continuous trajectory is discretized into N points. The problem is represented as

$$\begin{aligned} &\underset{y_{z,1},\cdots,y_{z,n}}{\text{minimize}}\ L_z \\ &\text{subject to } C1{:}\ y_1 = s_{z,int}, y_N = s_{z,des} \\ &\qquad C2{:}\ \left\| y_{z,n} - y_{z,n-1} \right\| \le \frac{L_z}{N} \\ &\qquad C3{:}\ \left\| y_{z,n} - p_m \right\| \ge r_m, m = 1,\ldots,\tilde{M} \\ &\qquad C4{:}\ y_{z,n} \in T \end{aligned} \tag{13}$$

where L_z is the path length for $z = 1,\ldots, Z$. C_1 provides the start and destination locations while C_2 ensures the smoothnessof the generated trajectory. A larger N will produce a smoother trajectory but the computational complexity is also improved as the increment of variables. In addition, C_3 guarantees the safety distance between the robot and circular-approximated obstacles.

The constraint C_3 is a convex-concave constraint, which cannot be solved by generic solvers. To cope with this issue, we linearize the problem in Equation (13) around a given point $\boldsymbol{x}_n^{(i)}$:

$$\begin{aligned} &\underset{y_{z,1},\cdots,y_{z,n}}{\text{minimize}}\ L_z + \frac{\beta}{2}\left\| y_{z,n} - y_{z,n}^{(i)} \right\|^2 \\ &\text{subject to } C1, C2, C4 \text{ in Equation(13)} \\ &\qquad F(y_{z,n})\text{-}G(y_{z,n}^{(i)})\text{-}\left\langle \nabla G(y_{z,n}), y_{z,n} - y_{z,n}^{(i)} \right\rangle \le 0, N = 1,\ldots,N \end{aligned} \tag{14}$$

where we set $F(y_{z,n}) = r_m$ and $G(y_{z,n}) = \left\| y_{z,n} - p_m \right\|$, The subscript $^{(i)}$ denotes the i-th iteration for the optimization of Equation (13). Furthermore, β is added to the objective function as a regularization term.

Using the sequential convex programming algorithm in [11], the problem in Equation (14) can be iteratively solved via efficient solver, such as DCCP [12]. This problem can be regarded as a difference of two convex (DC) problem, therebythe convergence of Equation (14) is guaranteed. It is worth noting that the obtained trajectories cannot be verified as global optimums due to the non-convex nature.

Once the trajectories for all of the robots are generated, the sum of path lengths is lower bounded by

$$L_{min} = \sum_{z=1}^{Z} L_z^* \tag{15}$$

where L_z^* is the obtained path length from Equation (13). We will validate the effectiveness of the proposed algorithm by comparing the path lengths with other algorithms in Section VI.

5. Collision-aware path planning for multi-robot scenarios

With the given trajectory for all of the robots, the collision-avoidance problem is addressed by velocity control in the sequel. Based on CBF-CLF reviewed in Section II, the time-varying velocities of robots are obtained by solving a quadratic programming (QP) problem. For each robot, it only needs to acquire the location and velocity of the nearest robot. Thus, the operation is decentralized and can be implemented online.

5.1 Leader-follower dynamic model

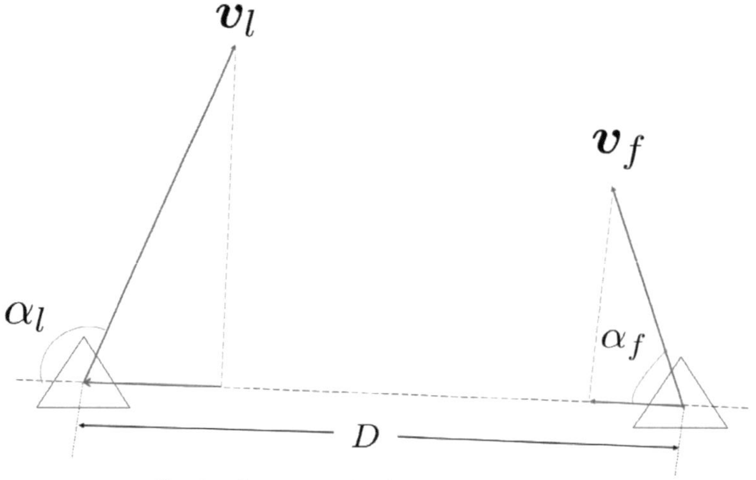

Fig. 2. The LF model for velocity control.

Once the trajectories of robots are generated in Section IV, we now consider the collision-aware path planning by velocity control. In the decentralized dynamic model, we assume one robot is able to obtain the location and velocity of its nearest robot. Since no central coordinator exists, each robot can only control its own velocity. Thus, we propose a leader-follower (LF)dynamic model, where the velocity of the leader robot cannot be changed while the follower robot executes the decentralized algorithm.

Consider a 2-D LF dynamic model that the velocities and locations of the both robots are known at the follower robot. The trajectories are represented by $\boldsymbol{L_l}$ and $\boldsymbol{L_f}$. At the t-th instant, the velocities and locations are given as $[\{v_f(t), \tilde{y}_f(t)\}_{t=1,2,\dots}] \in L_f$ and $[\{v_l(t), \tilde{y}_l(t)\}_{t=1,2,\dots}] \in L_l$ for the follower and leader robot, respectively. As shown in Fig. 2, the velocity components between two robots are denoted by:

$$\tilde{v}_l = \|v_l(t)\| \cos(\alpha_l) \tag{16}$$

$$\tilde{v}_f = \|v_f(t)\| \cos(\alpha_f) \tag{17}$$

Where the α_l and α_f are given as follows

$$\cos(\alpha l) = \frac{(\tilde{y}_l(t+1) - \tilde{y}_l(t))^T (\tilde{y}_l(t) - \tilde{y}_f(t))}{\left\|\tilde{y}_l(t+1) - \tilde{y}_l(t)\right\| \bullet \left\|\tilde{y}_l(t) - \tilde{y}_f(t)\right\|} \tag{18}$$

$$\cos(\alpha_f) = \frac{(\tilde{y}_f(t+1) - \tilde{y}_f(t))^T (\tilde{y}_l(t) - \tilde{y}_f(t))}{\left\|\tilde{y}_f(t+1) - \tilde{y}_f(t)\right\| \bullet \left\|\tilde{y}_l(t) - \tilde{y}_f(t)\right\|} \tag{19}$$

Furthermore, the distance of two robots is denoted by

$$D(t) = \|\tilde{y}_l(t) - \tilde{y}_f(t)\| \tag{20}$$

Given $\boldsymbol{x}(t) = (\tilde{v}_f(t), \tilde{v}_l(t), D(t))$, the dynamics of an LF system can be compactly represented as

$$\dot{x}(t) = \begin{bmatrix} -F_r(v_a(t))/M \\ \alpha_g(t) \\ \tilde{v}_l(t) - \tilde{v}_f(t) \end{bmatrix} + \begin{bmatrix} 1/M \\ 0 \\ 0 \end{bmatrix} u(t) \tag{21}$$

where $F_r(v_a(t)) = f_0 + f_1\tilde{v}_f(t) + f_2\tilde{v}_f(t)^2$ is the aerodynamic drag; the coefficients f_0, f_1, and f_2

Journal of Physics: Conference Series **2113** (2021) 012002 doi:10.1088/1742-6596/2113/1/012002

depend on the practical environments. ***ag(t)*** is the random noise of the measured velocity of the leader robot respect to a Gaussian distribution with zero mean. Next, we consider solving the decentralized collision-aware path planning by formulating a QP problem at the *t-th* instant. For notation simplicity, the state variable $\boldsymbol{x(t)}$ is written as $\boldsymbol{x}$ temporarily.

5.2 Travel Time Minimization

The minimum travel time can be achieved when the robots move with the maximal velocity. We sort the path lengths from the largest to the smallest, *i.e.*, $L_z \geq L_{z+1}$. Thus, the lower bound of the travel time to accomplish the tasks is

$$T_{min} = \frac{L_Z}{\|v_{max}\|} \tag{22}$$

To minimize the travel time, we should select the robot with a larger path length as a leader robot since the follower robot may reduce its velocity to avoid the inter-robot collisions. In this way, the robot with a larger path length can keep a relatively larger velocity.

Given the dynamic model, we give a "soft constraint" for each robot to achieve a desired velocity:

$$v_d = \|v_{max}\| cos(\alpha f) \tag{23}$$

This constraint only works when the safety distance is ensured. The soft constraint can be written in terms of a candidateCLF:

$$V(x) = (\tilde{v}(f) - v(d))^2 \tag{24}$$

For any $c > 0$, the following inequality verifies that $V(\boldsymbol{x})$ is a valid CLF [13]

$$\inf_{u \in U} \left[L_f V(x) + L_g V(x) u + c_3 V(x) \right] \leq 0 \tag{25}$$

5.3 CBF-CLF QP formulation

Now we consider the constraints for inter-robot collision avoidances. We express the safety distance as a "hard constraint"given by

$$D \geq \tilde{v}_f \tag{26}$$

where τ is a constant reaction time for the follower robot. The hard constraint ensures the safety distance of two robots. We consider the function $h(x) = D - \tau\tilde{v}(f)$ and a candidate CBF $B(x) = -\log(\frac{h(x)}{1+h(x)})$, Recalling Equation (6),we set $\gamma = 1$

$$\inf_{u \in U} \left[L_f B(x) + L_g B(x) u - \frac{1}{B(x)} \right] \leq 0 \tag{27}$$

which validates $B(\boldsymbol{x})$ is a CBF.

As a result, a CLF-CBF QP is formulated by combining the above constraints [14]:

$$\begin{aligned} u^* &= \arg\min \frac{1}{2} u^T H_{acc}(x) u + F_{acc}(x)^T u \\ \textit{subject to} &: \mathrm{C1}:[\mathrm{L_g V(x)},-1]\mathrm{u} \leq \mathrm{b}_{clf} \\ &\mathrm{C2}:[\mathrm{L_g B(x)},0]\mathrm{u} \leq \mathrm{b}_{cbf} \\ &\mathrm{C3}:[1,0]\mathrm{u} \leq \mathrm{b}_{fc} \end{aligned} \tag{28}$$

where

$$
\begin{aligned}
L_g B(x) &= \frac{\tau}{M(1+D-\tau\tilde{v}_f)\left(D-\tau\tilde{v}_f\right)} \\
L_f B(x) &= \frac{-\tau F_r / M + \tilde{v}_f - \tilde{v}_l}{(1+D-\tau\tilde{v}_f)\left(D-\tau\tilde{v}_f\right)} \\
L_g V(x) &= \frac{2(\tilde{v}_f - v_d)}{M} \\
L_f V(x) &= \frac{2F_r(\tilde{v}_f - v_d)}{M}
\end{aligned} \tag{29}
$$

and

$$
\begin{aligned}
b_{clf} &= -L_f V(x) - cV(x) \\
b_{cbf} &= -L_f B(x) - \frac{1}{B(x)} \\
b_{clf} &= a_{fc}
\end{aligned} \tag{30}
$$

The constraint C_1 is generated by CLF and the additive parameter σ makes the CLF constraint "soft". The constraint C_2 guarantees the "hard" safety distance for two robots while a_{fc} is the acceleration limit of a robot in constraint C_3. It is worth noting that the combination of CLF and CBF assures the requirements are not in conflict.

In addition, the cost function related to the control yields the following function in terms of u:

$$
C(u) = \left|(F_r - u)/M\right|^2 = \frac{1}{M^2}(F_r^2 + u^2 - 2uF_r) \tag{31}
$$

which can be converted into the matrix form in Equation (28)

$$
H_{acc} = 2\begin{bmatrix} \frac{1}{M^2} & 0 \\ 0 & p_{sc} \end{bmatrix}, F_{acc} = -2\begin{matrix} Fr/m \\ 0 \end{matrix} \tag{32}
$$

where p_{sc} is the relaxation weight of the parameter σ.

The QP problem in Equation (28) can be solved by an available solver such as the function quadprog in MATLAB.

Algorithm 1 : Decentralized Path Planning Algorithm

Input:

Initial and terminal positions of robots:

$\{y_{z,int}\}_{z=1,\ldots Z}, \{y_{z,des}\}_{z=1,\ldots Z};$

The positions of the circular approximated obstacles:

$\{p_m\}_{m=1,\ldots M};$

Procedures:

1: Generate the obstacle-avoidance trajectories by Equation (13): $\{y_{z,n}\}_{n=1,\ldots,N}$ for z=1,...,Z
2: **for** n = 1,..., N **do**
3: Sort the left travel distances for all the robots;
4: **for** z = 1,..., Z **do**
5: Select the leader robot and the nearest robot as the follower;
6: Calculate the controller u by solving Equation (28);
7: Update the position and velocity by Equation (33) and Equation (34), respectively;
8: **end for**
9: **end for**
10: **Outputs:** The center coordinates and radius of K∗ circles.

5.4 Time-varying control systems

The velocities are updated by

$$
\tilde{v}_f(t+1) = \tilde{v}_f(t) + \frac{u(t) - F_r}{M}\Delta T \tag{33}
$$

where ΔT is a time slot given by

$$\Delta T = \frac{\left\| \tilde{y}_l(t+1) - \tilde{y}_l(t) \right\|}{\tilde{v}_l(t)} \tag{34}$$

Furthermore, we assume the leader and follower robot are the zl-th and zf -th robot, respectively. The sampled points are given by $\tilde{y}_l(n) = y_{zl,n}$ while the initial position of the follower robot is given by $\tilde{y}_f(1) = \tilde{y}_{zf_{\text{int}}}$. Given the time slot according to the leader robot, the sampled trajectories of the follower robot are updated by

$$\tilde{y}_l(t+1) = \tilde{y}_l(t) + v_f(t)\Delta T \tag{35}$$

As the velocity components of the LF model are derived, we can straightforwardly calculate the velocities of the both robots by:

$$v_f(t+1) = \frac{\tilde{v}_f(t+1)}{\cos(\alpha f)} \cdot \frac{\tilde{y}_f(t+1) - \tilde{y}_f(t)}{\left\| \tilde{y}_f(t+1) - \tilde{y}_f(t) \right\|} \tag{36}$$

In each instant, the left travel distances of robots are sorted. The robot with a larger left distance is selected as the leader robot. In this way, the travel time can be minimized according to Section V-B.

The decentralized path planning algorithm is elaborated in **Algorithm 1**, where the DCCP and CBF-CLF are implemented. Specifically, the obstacle-avoidance trajectories for each robots are first generated by temporarily ignoring the possible collisions using DCCP. Next, the inter-robot collisions are considered by formulating a QP problem using CBF-CLF. In the time-varying dynamic system, the possible collision is avoided by controlling the velocities of robots.

6. Performance evaluation

To evaluate the performance of the proposed multi-objective path-planning algorithm, the concept-proof experiments are presented in this section. We use multiple RoboMaster produced by Da-Jiang Innovations (DJI) to accomplish the experiments. The RoboMaster can be controlled via computer with the SDK provided by DJI[1]. Each RoboMaster is equipped with a LiDar and a RGB-D camera such that the state of other robots can be detected. We assume the geographical information of the given environment is acknowledged.

TABLE I PARAMETERS USED IN THE EXPERIMENTS.

M	2	f_0	0.1	f_1	5
f_2	0.25	p_{sc}	10	$v_g(0)$	1
$v_a(0)$	0.1	$v_h(0)$	3	τ	4
δ	0.01	c	10	α_{fc}	20

6. 1 Velocity Control for LF Dynamic Model

We first investigate the velocity control of two robots in the LF dynamic model. The leader robot is observed by the follower robot. After obtained the position and velocity of the leader robot, the follower robot dynamically control its velocity to avoid possible collisions. If no collision exists, the follower robot is prone to move with the maximum velocity.

In Fig. 3, the velocities of two robots are illustrated by exploiting the CBF-CLF-based algorithm derived in Section V.

Fig. 4 reveals the variation of the states and the control input in the LF dynamic system. The control input is reported inthe bottom figure.

Journal of Physics: Conference Series **2113** (2021) 012002 doi:10.1088/1742-6596/2113/1/012002

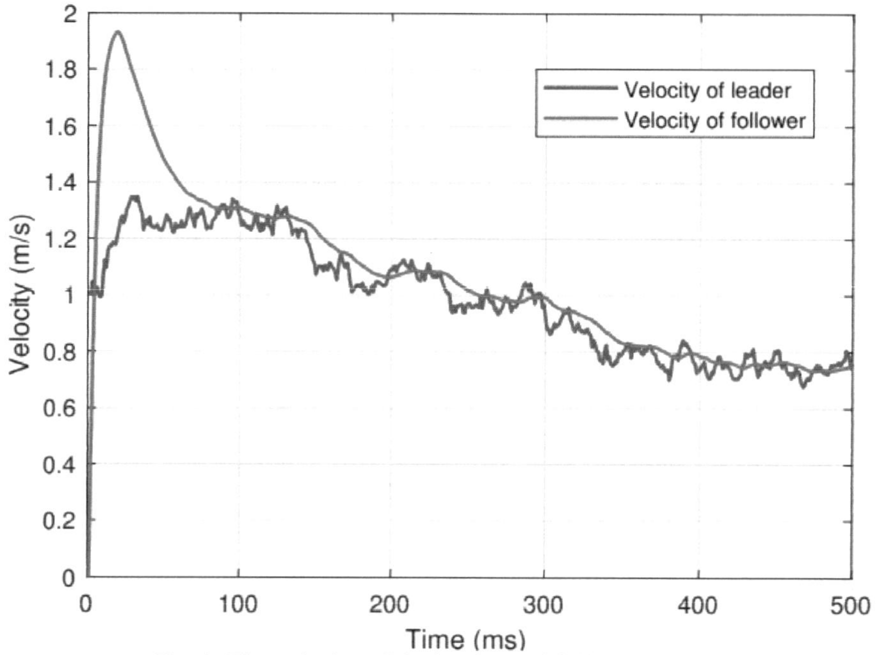

Fig. 3. The velocity of the leader and follower robots.

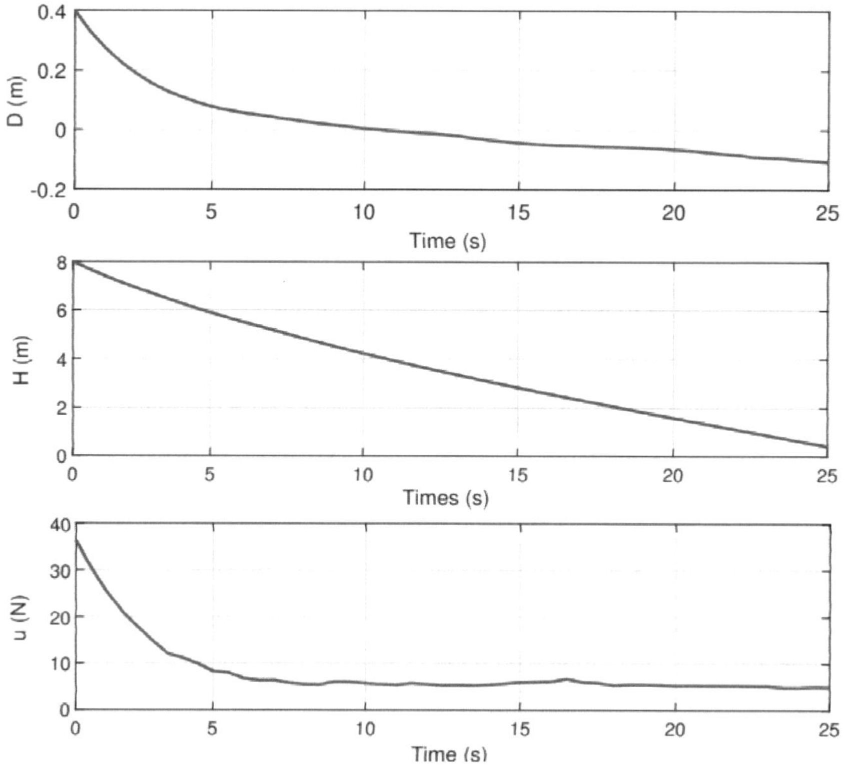

Fig. 4. The distance between the two robots (up), and the controller of two robots (bottom).

6. 2 Path-Planning for the Robot Swarm

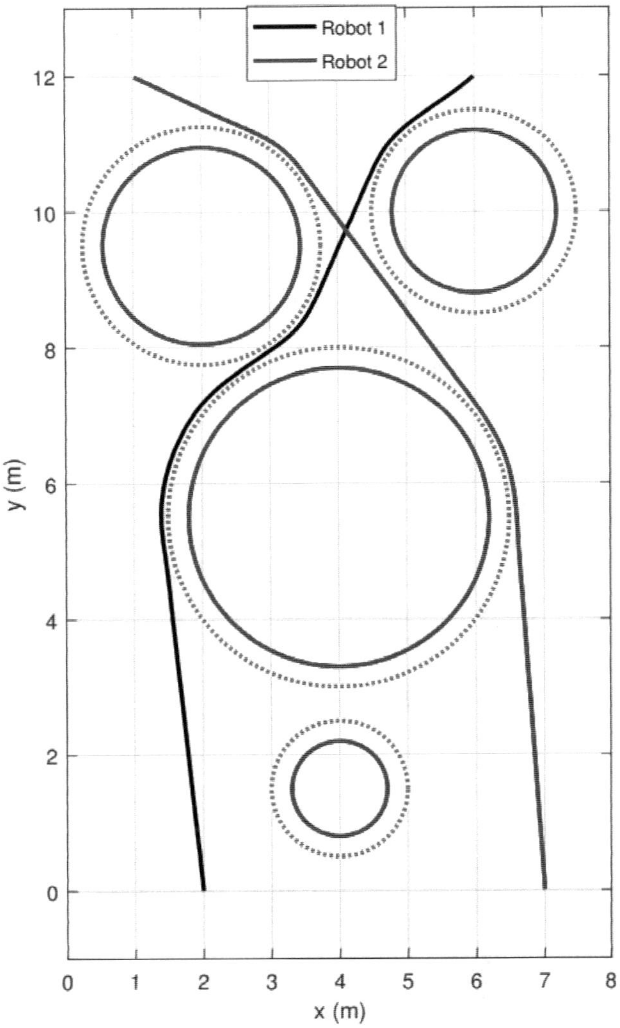

Fig. 5. The trajectory of two robots.

Next, we consider the multi-objective path planning for the robot swarm. Assume that four RoboMasters are assigned tasks to achieve different destinations in a continuous environment, where the obstacles are randomly deployed.

The left side of Fig. 5 shows the trajectories generated using the circular-approximated obstacles corresponding to thepractical environment in the right side.

7. Conclusion and Future Work

In this work, a decentralized path-planning approach is proposed for multi-robot scenarios. Based on DCCP and CBF-CLF, the decoupled problems are separately resolved in two stages. Firstly, the trajectories are generated by neglecting the inter- robot collisions. Next, we formulate a QP problem using CBF-CLF to design the velocity controller for each robot. In thisway, the travel time as well as the trajectory length are minimized. The decentralized algorithm operates on the robots by implementing the LF model. Finally, extensive experiments verify the advantages of the proposed decentralized approach, showing the improvements on efficiency and solution quality.

In our future works, we will consider the dynamic path planning problem for the robots. For instance, one of the most important path-planning problems is the existing of the dynamic obstacles. It requires much more computational resource for the computer equipped on the robot. Moreover, the path-planning problem can be also extended to the 3-D scenarios, such as the unmanned aerial vehicles (UAVs).

References

[1] M. Elhoseny, A. Tharwat, and A. E. Hassanien, "Bezier curve based path planning in a dynamic field using modified genetic algorithm," Journal of Computational Science, vol. 25, pp. 339–350, 2018.
[2] T. Ṭ. Mac, C. Copot, D. T. Tran, and R. De Keyser, "A hierarchical global path planning approach for mobile robots based on multi-objective particle swarm optimization," Applied Soft Computing, vol. 59, pp. 68–76, 2017.
[3] P. Surynek, "An optimization variant of multi-robot path planning is intractable," in Proceedings of the AAAI Conference on Artificial Intelligence, vol. 24, no. 1, 2010.
[4] M. Nazarahari, E. Khanmirza, and S. Doostie, "Multi-objective multi-robot path planning in continuous environment using an enhanced genetic algorithm," Expert Systems with Applications, vol. 115, pp. 106–120, 2019.
[5] G. Hitz, E. Galceran, M.-E` . Garneau, F. Pomerleau, and R. Siegwart, "Adaptive continuous-space informative path planning for online environmental monitoring," Journal of Field Robotics, vol. 34, no. 8, pp. 1427–1449, 2017.
[6] W. Xiao, C. Belta, and C. G. Cassandras, "Decentralized merging control in traffic networks: A control barrier function approach," in Proceedings of the 10th ACM/IEEE International Conference on Cyber-Physical Systems, 2019, pp. 270–279.
[7] S. Zeng, Y. Zhu, and C. Qi, "HTN-based multi-robot path planning," in 2016 Chinese Control and Decision Conference (CCDC), Yinchuan, China, May. 2016, pp. 4719–4723.
[8] A. Candra, M. A. Budiman, and K. Hartanto, "Dijkstra's and a-star in finding the shortest path: a tutorial," in 2020 International Conference on Data Science, Artificial Intelligence, and Business Analytics (DATABIA). IEEE, 2020, pp. 28–32.
[9] A. D. Ames, X. Xu, J. W. Grizzle, and P. Tabuada, "Control barrier function based quadratic programs for safety critical systems," IEEE Transactions on Automatic Control, vol. 62, no. 8, pp. 3861–3876, 2016.
[10] N.-s. P. Hyun, P. A. Vela, and E. I. Verriest, "A new framework for optimal path planning of rectangular robots using a weighted Lp norm," IEEE Robotics and Automation Letters, vol. 2, no. 3, pp. 1460–1465, 2017.
[11] F. Debrouwere, W. Van Loock, G. Pipeleers, Q. T. Dinh, M. Diehl, J. De Schutter, and J. Swevers, "Time-optimal path following for robots with convex–concave constraints using sequential convex programming," IEEE Transactions on Robotics, vol. 29, no. 6, pp. 1485–1495, 2013.
[12] X. Shen, S. Diamond, Y. Gu, and S. Boyd, "Disciplined convex-concave programming," in 2016 IEEE 55th Conference on Decision and Control (CDC), Las Vegas, USA, Dec. 2016, pp. 1009–1014.
[13] A. D. Ames, J. W. Grizzle, and P. Tabuada, "Control barrier function based quadratic programs with application to adaptive cruise control," in 53rd IEEE Conference on Decision and Control, Los Angeles, USA, Dec. 2014, pp. 6271–6278.
[14] K. Galloway, K. Sreenath, A. D. Ames, and J. W. Grizzle, "Torque saturation in bipedal robotic walking through control lyapunov function-based quadratic programs," IEEE Access, vol. 3, pp. 323–332, 2015.

Control strategy of optimal distribution of feet forces for quadruped robots based on virtual model

Xinjie Chen[1], Yuegang Tan[2*], Ruiya Li[3], Yecheng Mei[4]

[1,2,3,4]School of Mechanical and Electrical Engineering, Wuhan University of Technology, Wuhan, 430000, China

*Corresponding author's e-mail: ygtan@whut.edu.cn

Abstract: In order to improve the dynamic walking performance of quadruped robot, a control method of optimizing feet forces distribution based on virtual model is proposed. In the supporting phase, the virtual model control method is applied to solve the virtual force of the torso. Combined with the gravity of the center of mass (CoM), the distribution problem between the virtual force of the CoM and the feet forces of the supporting legs is transformed into a quadratic programming (QP) problem, which is solved by Gurobi to realize the optimal distribution of the feet forces. Similarly, the virtual force of the swinging leg is solved by using the virtual model, and the joints torques of the robot's swinging legs are obtained by combining the inverse dynamics feedforward of the swinging legs. Through the simulation of quadruped robot's trot gait walking by webots and vs2019, it is verified that this method can stabilize the robot's attitude angles and body speeds near the target values. Compared with the feet forces distribution method that abandons the lateral force control, the application of this method makes the fluctuation range of the attitude angles of the robot and the ground reaction forces (GRFs) of the supporting legs smaller. It is proved that this control method can effectively improve the walking stability and robustness of the quadruped robot.

1. Introduction

Quadruped mammals in nature have a strong locomotion ability. The robot of bionic quadruped has good adaptability and movement flexibility in complex terrain because of its discrete contact with the environment and its legs have multiple degrees of freedom [1]. The locomotion control algorithm of robot is the core module that can realize flexible and stable movement. The most famous control algorithm of quadruped robot was proposed by Raibert which was based on spring loaded inverted pendulum (SLIP) [2]. In the robot's movement process, the leg mass is ignored, and the diagonal leg and torso are regarded as a SLIP model. Ignoring the leg mass, the diagonal legs and the torso are regarded as a SLIP model, and the locomotion control of the robot is realized by controlling the forward speed, bounce height and attitude of it. Zero moment point method (ZMP) was first proposed by Vukobratovic, a Yugoslav scholar. It is the most commonly used static stability control method to ensure that the robot is under static stability conditions at every time in the process of motion according to ZMP static stability criterion [3]. The neural network control method based on the response and reflection mechanism of central pattern generator (CPG) is also widely used control method of quadruped robots [4]. The control method based on virtual model (VMC) firstly assumes the existence of virtual components on the control object, such as spring and damper, then analyzes the virtual force on the control object, and establishes the mapping relationship between virtual force and joint force through Jacobian matrix [5-7].

Handling the contact forces between the robot and the ground and reasonably distributing the joint forces is the basis for realizing the flexible and efficient movement of quadruped robot. HyQ [8-9] quadruped robot of IIT and StarlETH [10] quadruped robot of Zurich University of technology used VMC to achieve better plantar interaction and compliance characteristics. In the process of robot locomotion control, how to realize the reasonable force distribution is one of the most important problems. Zhang [11] proposed that abandoned the lateral force artificially, and the plantar force distribution was obtained by applying constraint $f_{Fy} = f_{Hy}$. E Ming Cheng et al. [12] proposed using weighted average vertical force to realize force distribution which was not reasonable in unstructured terrain. In order to realize the locomotion stably and optimal plantar force distribution of quadruped robot during dynamic locomotion, this paper transforms the distribution problem of CoM virtual force and feet forces into a QP problem, the optimal distribution of feet forces is realized by finding the optimal values.

2. Quadruped robot model

The model of quadruped robot is shown in Figure 1. The robot has 12 degrees of freedom, FL represents the front left leg of the quadruped robot (FR: the front right leg, HL: the hind left leg, HR: the hind right leg). Each leg has three active joints: adduction joint θ_1, hip joint θ_2 and knee joint θ_3. The whole length of the body is $2(L + d_1)$, the width is 2W, and the structural parameters of the legs are l_1、l_2、d_2 , The specific values are shown in Table 1.

Table 1 robot structural parameters

Parameter	Value of parameter
length of half body L/mm	150
Width of half body W/mm	50
Height of body H/mm	30
body offset d_1/mm	47
hip offset d_2/mm	62.3
thigh link length l_1/mm	220
shin link length l_2/mm	212
whole body mass m/kg	20

2.1 Forward kinematics model

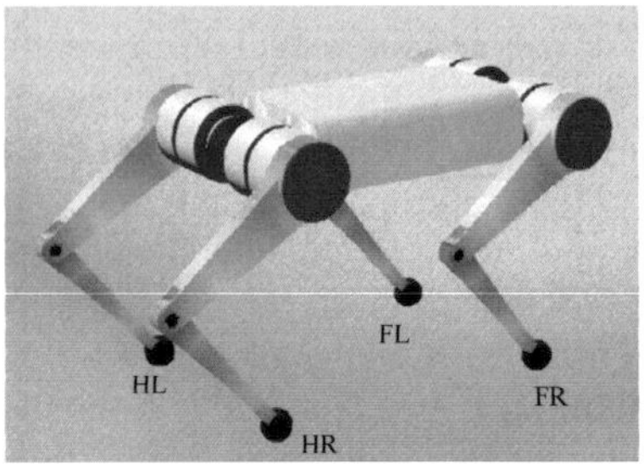

Figure 1. Quadruped simulation model

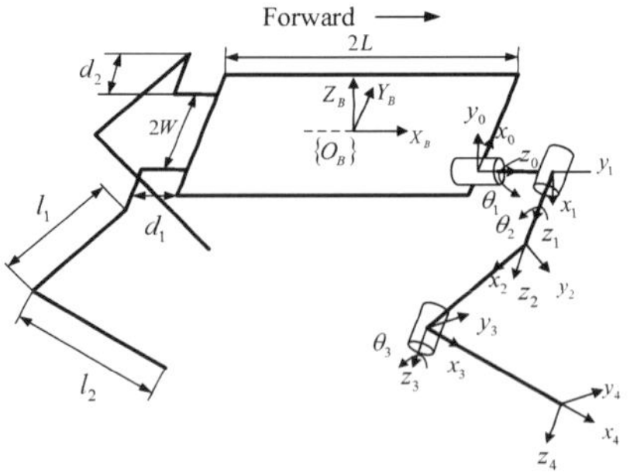

Figure 2. Coordinate system of FR leg

For the sake of description, coordinate system $\{O_B\}$ is established at the CoM of the body. The coordinate systems $\{O_i\}(\mathrm{i} = 0,1,2,3,4)$ in FR leg are established as shown in Figure 2, and coordinate system $\{O_0\}$ is established at the hip and coordinate system $\{O_4\}$ is established at the foot end. The transformation matrix from foot coordinate system to body coordinate system is calculated according to ***D-H*** method as:

$$ {}^{B}_{4}T = {}^{B}_{0}T\,{}^{0}_{4}T = \begin{bmatrix} s_{23} & -c_{23} & 0 & l_2 s_{23} + l_1 s_2 + d_1 + L \\ s_1 c_{23} & -s_1 s_{23} & -c_1 & l_2 s_1 c_{23} + l_1 s_1 c_2 - d_2 c_1 - W \\ -c_1 c_{23} & c_1 s_{23} & -s_1 & -l_2 c_1 c_{23} - l_1 c_1 c_2 - d_2 s_1 \\ 0 & 0 & 0 & 1 \end{bmatrix} \tag{1} $$

$$ s_i = \sin(\theta_i), c_i = \cos(\theta_i), s_{23} = \sin(\theta_2 + \theta_3), c_{23} = \cos(\theta_2 + \theta_3)(i = 1,2,3) $$

The forward kinematics (**FK**) equation can be obtained from the transformation matrix as follows:

$$ P = \begin{pmatrix} p_x \\ p_y \\ p_z \end{pmatrix} = \begin{pmatrix} l_2 s_{23} + l_1 s_2 + d_1 + L \\ l_2 s_1 c_{23} + l_1 s_1 c_2 - d_2 c_1 - W \\ -l_2 c_1 c_{23} - l_1 c_1 c_2 - d_2 s_1 \end{pmatrix} \tag{2} $$

According to Equation (2), the Jacobi matrix can be obtained by differentiating the joint angles with time as follows:

$$ J = \begin{bmatrix} 0 & l_1 c_2 + l_2 c_{23} & l_2 c_{23} \\ l_2 c_1 c_{23} + l_1 c_1 c_2 + d_2 s_1 & -l_1 s_1 s_2 - l_2 s_1 s_{23} & -l_2 s_1 s_{23} \\ l_1 s_1 c_2 + l_2 s_1 c_{23} - d_2 c_1 & l_2 c_1 s_{23} + l_1 c_1 s_2 & l_2 c_1 s_{23} \end{bmatrix} \tag{3} $$

2.2 Inverse dynamic model

The inverse dynamics model is established by Lagrange equation, and its general calculation model is as follows:

$$ \boldsymbol{\tau} = \frac{\mathrm{d}}{\mathrm{dt}} \frac{\partial \boldsymbol{L}}{\partial \dot{\boldsymbol{q}}} - \frac{\partial \boldsymbol{L}}{\partial \boldsymbol{L}} \tag{4} $$

In Equation (4), $L = E_K - E_P$, E_K is the sum of the kinetic energy of the robot leg link, E_P is the sum of the potential energy of the robot leg link; $\boldsymbol{q} = (\theta_1, \theta_2, \theta_3)$ is the joint angles of the robot. The kinetic equation as shown in Equation (5) can be obtained through differential derivation calculation and arrangement, where $\boldsymbol{M}$ represents inertial parameters, $\boldsymbol{V}$ represents Coriolis and centrifugal forces, and $\boldsymbol{G}$ represents gravity.

$$ \boldsymbol{\tau}_{ID} = \boldsymbol{M}(\ddot{\theta}_i) + \boldsymbol{V}(\dot{\theta}_i^2, \dot{\theta}_i \dot{\theta}_j) + \boldsymbol{G} \tag{5} $$

3. Virtual model controller

3.1 Stance phase virtual model

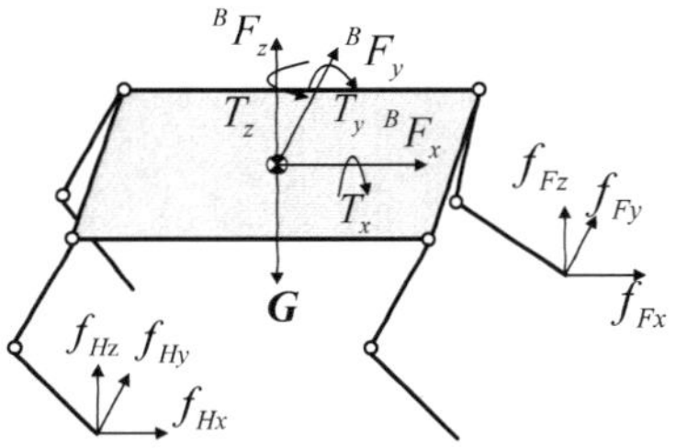

Figure 3. Static forces acting on the robot. The supporting forces and torques of the supporting legs are balanced with the virtual forces and gravity of the CoM of the torso.

Compared to the torso, the mass of the robot leg is very small so that it can be ignored in the virtual model, and the mass of the robot lies in its CoM. Statics analysis is performed on the supporting legs in each gait cycle of the robot in the stable motion of diagonal gait, as shown in Figure 3.

$$\begin{bmatrix} {}^{B}\boldsymbol{F} \\ {}^{B}\boldsymbol{T} \end{bmatrix} = \begin{bmatrix} \boldsymbol{I} & \boldsymbol{I} \\ {}^{B}\boldsymbol{P}_1 \times & {}^{B}\boldsymbol{P}_2 \times \end{bmatrix} \begin{bmatrix} \boldsymbol{F}_f \\ \boldsymbol{F}_h \end{bmatrix} \tag{6}$$

Where $\left({}^{B}\boldsymbol{F}, {}^{B}\boldsymbol{T}\right)^T$ is the virtual force of the CoM, $\boldsymbol{F}_f = \left(f_{Fx}, f_{Fy}, f_{Fz}\right)^T$ and $\boldsymbol{F}_h = \left(f_{Hx}, f_{Hy}, f_{Hz}\right)^T$ are the force of front and rear the support legs, and ${}^{B}\boldsymbol{P}_i$ is the position of the ith supporting leg in the body coordinate system. The relationship between them is shown in Equation (6).

Based on the principle of VMC, the external force on the centroid of the robot's torso is equivalent to a virtual system composed of spring-damping. The virtual spring-damping system is used to follow the desired three-dimensional locomotion and attitude of the robot. As shown in FIG. 4, a virtual force model with six degrees of freedom is established in the CoM.

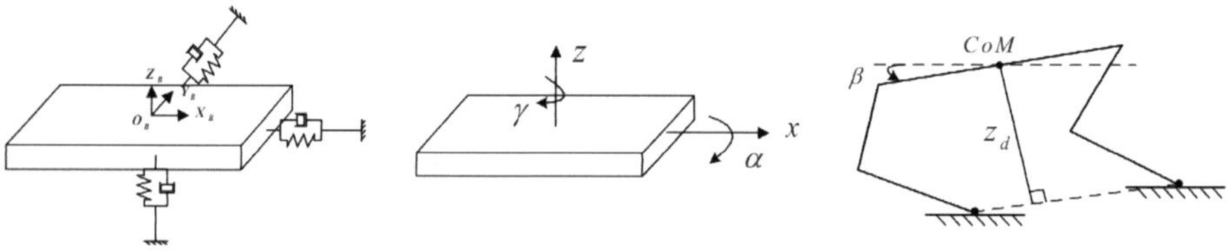

Figure 4. Dimensional spring damping model and attitude control of body

According to the virtual model of each dimension, the virtual force of the CoM can be obtained as:

$$\begin{aligned} {}^{B}F_x &= K_x\left(x_d - x\right) + B_x\left(\dot{x}_d - \dot{x}\right) \\ {}^{B}F_y &= K_y\left(y_d - y\right) + B_y\left(\dot{y}_d - \dot{y}\right) \\ {}^{B}F_z &= K_z\left(z_d - z\right) + B_z\left(\dot{z}_d - \dot{z}\right) \\ {}^{B}T_x &= K_\alpha\left(\alpha_d - \alpha\right) + B_\alpha\left(\dot{\alpha}_d - \dot{\alpha}\right) \\ {}^{B}T_y &= K_\beta\left(\beta_d - \beta\right) + B_\beta\left(\dot{\beta}_d - \dot{\beta}\right) \\ {}^{B}T_z &= K_\gamma\left(\gamma_d - \gamma\right) + B_\gamma\left(\dot{\gamma}_d - \dot{\gamma}\right) \end{aligned} \tag{7}$$

In Equation (7), $\left(\boldsymbol{P}_d, \ \psi_d\right)^T = \left[x_d, y_d, z_d, \alpha_d, \beta_d, \gamma_d\right]^T$ are the expected position and attitude angles of the CoM where $\left(\boldsymbol{P}, \ \psi\right)^T = \left[x, y, z, \alpha, \beta, \gamma\right]^T$ are the actual position and attitude angles. In addition, the robot is also constrained by its own gravity, combining Equation (6) ~ (7) and considering the gravity component, it can be obtained:

$$\underbrace{\begin{bmatrix} {}^{B}F_x - mg\sin\theta \\ {}^{B}F_y \\ {}^{B}F_z - mg\cos\theta \\ T_x \\ T_y \\ T_z \end{bmatrix}}_{\boldsymbol{B}} = \underbrace{\begin{bmatrix} 1 & 0 & 0 & 1 & 0 & 0 \\ 0 & 1 & 0 & 0 & 1 & 0 \\ 0 & 0 & 1 & 0 & 0 & 1 \\ 0 & -z_F & y_F & 0 & -z_H & y_H \\ z_F & 0 & -x_F & z_H & 0 & -x_H \\ -y_F & x_F & 0 & -y_H & x_H & 0 \end{bmatrix}}_{\boldsymbol{A}} \underbrace{\begin{bmatrix} f_{Fx} \\ f_{Fy} \\ f_{Fz} \\ f_{Hx} \\ f_{Hy} \\ f_{Hz} \end{bmatrix}}_{\boldsymbol{x}} \tag{8}$$

Since **rank** (**A**)=5<6, there is no unique solution for the distribution problem between the virtual force of the CoM and the robot's support feet. To solve this problem, this paper transformed the problem of feet forces distribution into a QP problem, so as to realize the aim of optimal distribution of feet forces.

Under the condition that the plantar force distribution satisfies Equation 8, the output forces should be as smaller as possible. In order to prevent the foot from slipping, friction cone constraint is added to make the feet forces not exceed the scope of static friction constraint. At the same time, add a constraint to ensure that the vertical forces are greater than 0.

$$\begin{aligned} \min \quad & \boldsymbol{f} = (\boldsymbol{Ax} - \boldsymbol{B})^T \boldsymbol{S} (\boldsymbol{Ax} - \boldsymbol{B}) + \boldsymbol{x}^T \boldsymbol{Wx} \\ \textbf{s.t.} \quad & |f_{ix}| \le \mu f_{iz}, \quad |f_{iy}| \le \mu f_{iz}, \quad f_{iz} \ge 0 \end{aligned} \tag{9}$$

Where $\boldsymbol{S} \in \boldsymbol{R}^{6\times6}$ is a weight diagonal matrix, and $\boldsymbol{W} \in \boldsymbol{R}^{6\times6}$ denotes the penalty diagonal matrix. They are used to make the output follow the target virtual force and prevent excessive virtual force. μ represents the friction coefficient. Since the forces are interactive, the output force of the robot support foot is equal to the support force of the ground to the support foot, and the direction is opposite. Combined the Jacobian matrix calculated in Equation (3), the joint control torques in stance phase as:

$$\boldsymbol{\tau}_{st} = -\boldsymbol{J}^T \boldsymbol{f} \tag{10}$$

3.2 Support phase virtual model

Similarly, virtual model control is still used to control the swinging legs. As shown in FIG. 5, the forces on the swinging leg is equivalent to a virtual system composed of spring damping in three directions.

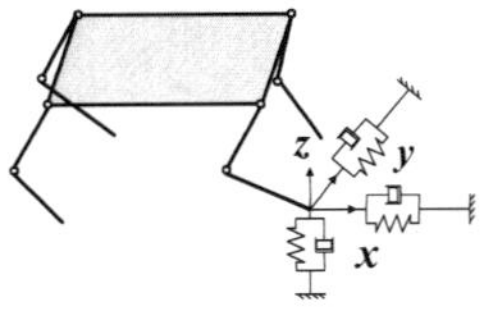

Figure 5. Swing phase virtual model

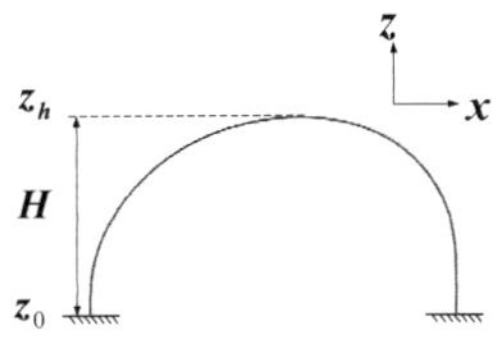

Figure 6. The foot trajectory generated

The virtual force of swinging legs:

$$\begin{bmatrix} f_x^{sw} \\ f_y^{sw} \\ f_z^{sw} \end{bmatrix} = \begin{bmatrix} k_x\left(x_d^{sw} - x^{sw}\right) + k_{vx}\left(\dot{x}_d^{sw} - \dot{x}^{sw}\right) \\ k_y\left(y_d^{sw} - y^{sw}\right) + k_{vy}\left(\dot{y}_d^{sw} - \dot{y}^{sw}\right) \\ k_z\left(z_d^{sw} - z^{sw}\right) + k_{vz}\left(\dot{z}_d^{sw} - \dot{z}^{sw}\right) \end{bmatrix} \tag{11}$$

The virtual forces of swinging legs are mapped to each joint by Jacobi matrix, and the torques can be obtained as follows：

$$\boldsymbol{\tau}_{vmc}^{sw} = \boldsymbol{J}^T \boldsymbol{F}^{sw} \tag{12}$$

Combined with inverse dynamics feedforward torques given in Equation (5), the torques of swinging leg joints can be obtained as follows:

$$\boldsymbol{\tau}_{sw} = \boldsymbol{\tau}_{ID} + \boldsymbol{\tau}_{vmc}^{sw} \tag{13}$$

3.3 Swing trajectory generator

In this paper, the trajectory of the swinging leg is planned in the foot end coordinate system along the x, y and z directions respectively. The trajectory planning in the x direction realizes the forward motion of the robot, while in the y direction realizes the lateral swing motion and in the z direction realizes the vertical step of the swinging leg.

$$x_{sw}(t)=\begin{cases} x_0 & , \quad 0 \le t < a_1T \\ \frac{6L_x}{T_x^5}(t-a_1T)^5-\frac{15L_x}{T_x^4}(t-a_1T)^4+\frac{10L_x}{T_x^3}(t-a_1T)^3, & a_1T \le t < a_2T \\ L_x & , \quad t \ge a_2T \end{cases} \tag{14}$$

$$y_{sw}(t)=\begin{cases} y_0 & , \quad 0 \le t < a_1T \\ \frac{6L_y}{T_y^5}(t-a_1T)^5-\frac{15L_y}{T_y^4}(t-a_1T)^4+\frac{10L_y}{T_y^3}(t-a_1T)^3, & a_1T \le t < a_2T \\ L_y & , \quad t \ge a_2T \end{cases} \tag{15}$$

$$z_{sw}(t)=H\left(\frac{16t^4}{T^4}-\frac{32t^3}{T^3}+\frac{16t^2}{T^2}\right)+z_0 \tag{16}$$

Where T is the time period of the swinging legs, $T_x=T_y=(a_2-a_1)T$, and (x_0, y_0, z_0) is the initial position of the swinging leg, L_x, L_y respectively is the stride length of the swinging leg in the x direction and y direction, and H is the stride height of the swinging leg. The foot trajectory of swinging leg in the *X-Z* plane is shown in FIG. 6.

3.4 Overall control architecture of robot

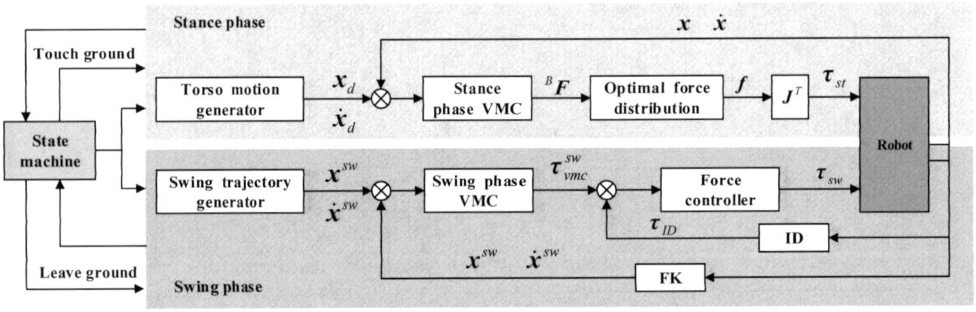

Figure 7. Overview of the control architecture.

As shown in FIG. 7, the overall movement phase of the robot is divided into stance phase and swing phase, and the two phases are switched alternately by the state machine. In the stance phase, the virtual force distribution between of the body centroid and plantar force is transformed into a QP problem, and the real-time optimal force distribution is realized through Gurobi library to find the optimal values. In the swing phase, inverse dynamic feedforward and VMC are combined to achieve smooth trajectory following. FK represents forward kinematics, ID represents inverse dynamics, and force controller is a PD controller.

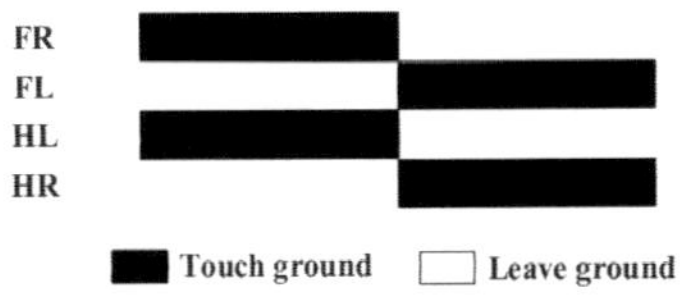

Figure 8. Trot gait sequence diagram. The black squares represent landing state, while the white squares represent leaving ground state.

Gait time sequence refers to the proportion of time in the swing phase and the standing phase for each leg in a complete gait cycle. In nature, the diagonal gait is the most common gait of quadrupeds. When a quadruped robot moves in a trot gait, one diagonal leg touch the ground and the other diagonal leg swing and the stance phase time is equal to the swing phase time. The phase switching is carried out by the state machine to realize the alternating forward movement of the diagonal legs of the robot.

4. Simulation and results

4.1. Walking simulation of quadruped

Walking is an effective method to verify the diagonal gait control algorithm of quadruped robot. Make the target speed of quadruped robot walking from 0.2m /s forward speed, lateral speed and rotation speed to 0.

Table 2 Robot simulation parameters

Parameter	Value of parameter
$K_x, K_y, K_z\ /(N/m)$	0,0,8000
$K_\alpha, K_\beta, K_\gamma\ /(N\cdot m/rad)$	2000,10000,200
$B_x, B_y, B_z\ /(N\cdot s/m)$	200,0.01,800
$B_\alpha, B_\beta, B_\gamma\ /(N\cdot s/m)$	120,50,0
$k_x, k_y, k_z\ /(N/m)$	100,8000,1000
$k_{vx}, k_{vy}, k_{vz}\ /(N\cdot s/m)$	100,100,1000
weight matrix S	diag (2.5,2.5,50,5,5,5)
penalty matrix W	diag(0.1,0.1,0.1,0.1,0.1,0.1)

The robot is controlled by using VMC and optimal feet forces distribution. In Table 2, there are simulation parameters in webots software. After the simulation starts, record the changes of body's speed as well as pitch angle.

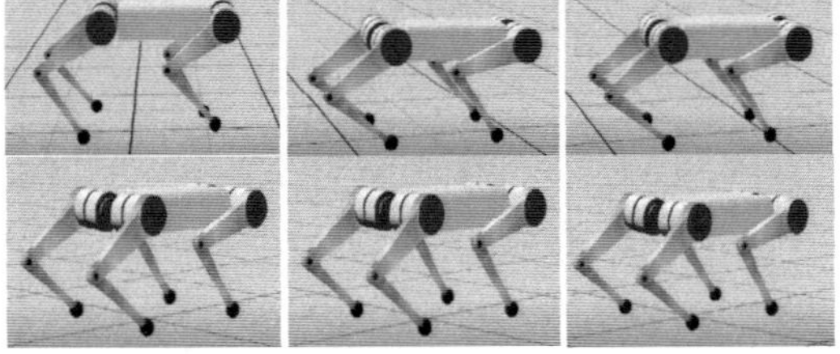

Figure 9. Screenshot of robot walking simulation process in Webots

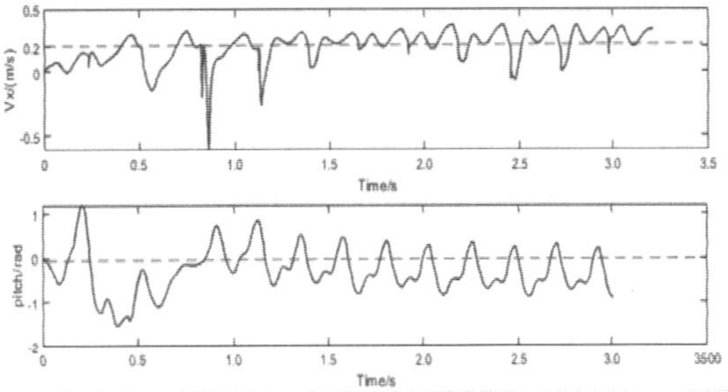

Figure 10. Curve of forward speed and pitch value during walking

The position and posture of the quadruped robot within 0-6 seconds were captured as shown in FIG. 9, which reveals that the robot could move smoothly. Figure 10 shows the target velocity tracking curve of the robot and the change curve of the body's pitch angle. It can be seen from the figure that the forward velocity of the robot fluctuates slightly within the range of 0.2m/s after starting time, which can track the desired velocity well, and the pitch fluctuation is also very small.

4.2 Quadruped status contrast before and after optimization

The above simulation results verify the stability of the proposed method and the feasibility of the optimal foot force distribution algorithm based on virtual model. In order to test the optimization effect of the force distribution method in the locomotion process of quadruped robot, a group of comparative experiments are carried out. Under the same step frequency and step height, compared with the foot force distribution method without lateral force, the advantages of this force distribution method are explained by the foot forces as well as joint torques feedback of the FR leg, and attitude angles of the body.

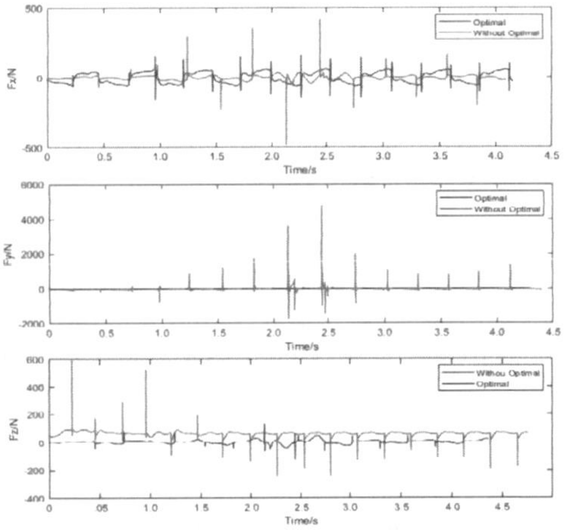

Figure 11. Torque curves of FR Leg

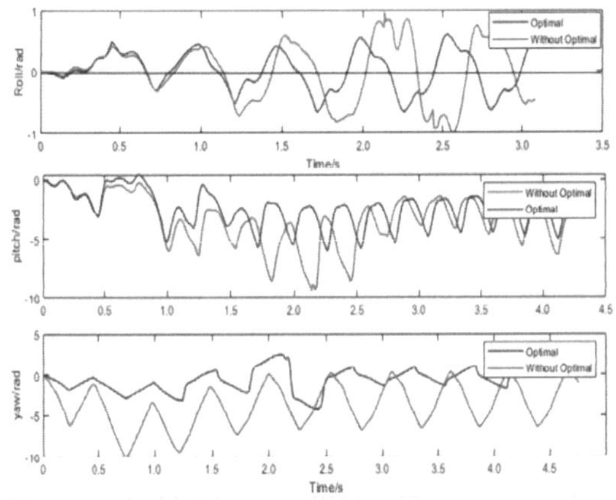

Figure 12. Attitude angle curve

In FIG. 11 and 12, the curves are torques feedback of the FR leg joints and attitude angles, and the blue line is the distribution curve of force control method with optimization while the red line without (the same below). It's obvious that the torque of the robot leg joints as well as fluctuation ranges are significantly reduced after using the optimal feet forces distribution control algorithm. Compared with the control algorithm without optimization, the roll, pitch and yaw and the fluctuation ranges are significantly smaller, which significantly improves the stability of the robot.

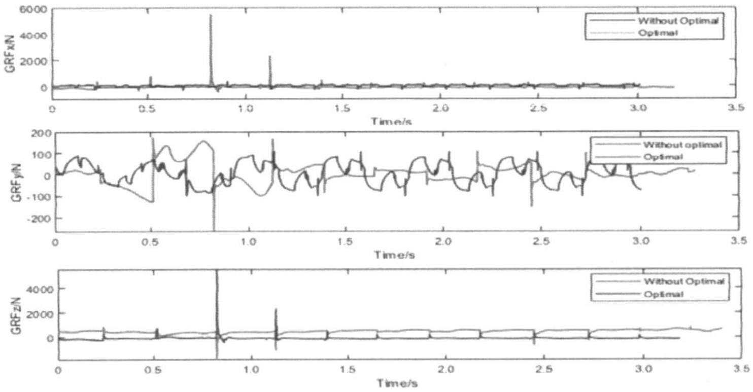

Figure 13. Ground contact force comparison curves of supporting legs

FIG. 13 shows the GRFs change curve of the FR leg. Compared with the control algorithm without optimization, the fluctuation of forces becomes smaller in the X and Y directions, which means that the robot walks more stably. And in Z direction, the GRFs reduced by about 50%, which means that the impact of the ground on the vertical direction is greatly reduced and the robot walks more smoothly.

5. Conclusion

In this paper, an optimal force distribution control method based on virtual model is proposed for the quadruped robot in trot gait. The force distribution problem that cannot be solved directly is transformed into a constrained QP problem, which can be solved through Gurobi C++ library in real time. The optimized force distribution method is applied to the locomotion control of the robot, and the robot can walk stably forward. From the perspective of plantar force impact, joint torques and robot posture change during the movement, the proposed method plays a good role in stabilizing the motion of the quadruped robot and reducing the impact force compared with the force distribution method without lateral force

References

[1] Yang, J.J., Sun, H., Wang, C.H., Chen, X.D. (2019) Review of quadruped robot research. [J]. Navigation Positioning and Timing., 6 (05): 61-73.
[2] Raibert, M., Blankespoor, K., Nelson, G., et al. (2008) BigDog, the rough terrain quadruped robot. [J]. IFAC Proceedings Volumes., 41(2): 10823-10825.
[3] VUKOBRATOVIĆ M, STEPANENKO [J]. (1972) On the stability of anthropomorphic systems. [J]. International journal of humanoid robotics., 15(1-2):1-37.
[4] Zhang, Y., Guo, Z.W., Chen, D.J. Wang, B.R. (2018) Gait control of pneumatic muscle quadruped robot based on Kimura oscillator and virtual model. [J]. Journal of ordnance industry., 39 (07): 1411-1418.
[5] Pratt, J., Chew, C., Torres, A., Dilworth, P., and Pratt, G. (2001) Virtual model control: An intuitive approach for bipedal locomotion. [J]. International Journal of Robotics Research., 20: 129-143.
[6] Desai, R. Geyer, H. and Hodgins, J.K. (2014) Virtual model control for dynamic lateral balance. [C]. IEEE-RAS Int. Conf. Humanoid Robots., Nov. 2014, pp. 856–861
[7] Zhang G T, Rong X W, Hui C, et al. (2016) Torso motion control and toe trajectory generation of a trotting quadruped robot based on virtual model control. [J]. Advanced Robotics., 30(4): 284-297.
[8] Boaventura, T., Buchli, J., Semini, C., et al. (2015) Model-based hydraulic impedance control for dynamic robots. [J]. IEEE Transactions on Robotics, 31(6): 1324-1336.
[9] Winkler, A. W., Mastalli, C., Havoutis, I., et al. (2015) Planning and execution of dynamic whole-body locomotion for a hydraulic quadruped on challenging terrain. In: IEEE International Conference on Robotics and Automation. Washington. 5148-5154.
[10] Gehring, C., Coros, S., Hutter, M., et al. (2013) Control of dynamic gaits for a quadrupedal robot. In: IEEE International Conference on Robotics and Automation. Karlsruhe. 3287-3292.
[11] Zhang, G.T., Rong X.W, Li, Y.B., Chai, H., Li, B. (2016) Gait control method of quadruped robot diagonal trot based on virtual model. [J]. Robot., 38 (01): 64-74
[12] E, M.C., Liu, H., Zhang, X.L., et al. (2014) Compliant gait generation for a quadruped bionic robot walking on rough terrains. [J]. Robot., 36(5): 584-591.

Research on Optical Fiber Sensor Signal Processing Technology Based on Variational Mode Decomposition and Singular Spectrum Analysis

ZheHao DONG[1,a], YanHong DING[2,b], ZhiLi ZHANG[3,c*] and HaiQiang ZHU[4,d]

[1] Electrical Engineering and Automation College, TianJin University of Technology, TianJin city, TianJin province, 300074, China

[2] School of Integrated Circuit Science and Engineering, TianJin University of Technology, TianJin city, TianJin province, 300074, China

[3] Smart Manufacturing Academy, Tianjin Sino-German University of Applied Technology, TianJin city, TianJin province, 300350, China

[4] Electrical Engineering and Automation College, TianJin University of Technology, TianJin city, TianJin province, 300074, China

[a] email: 1904897060@qq.com, [b] email: lucydyh@163.com, [d] email: 2827421712@qq.com

*email: zhangzhili@tsguas.edu.cn

Abstract. To better apply the phase-sensitive optical time-domain reflectometer (Φ-OTDR) in projects with complex environments, given the problem of the low signal-to-noise ratio of the Φ-OTDR signal, a variational modal decomposition (VMD) is proposed. Signal processing algorithm combined with singular spectrum analysis. First, the equalization optimizer (EO) optimizes the VMD. The VMD decomposes the preprocessed signal into a multi-layer eigenmode function (IMF), selects the IMF component according to the correlation coefficient, and after the SSA noise reduction process, the signal Reconstruction to realize the noise reduction processing of the Φ-OTDR signal. Through experiments, it is proved that the relative mean square error (RMSE) of the method in this paper is lower than that of EO-VMD, and the noise reduction performance is better.

1. Introduction

Φ-OTDR has the advantages of real-time feedback, high sensitivity, wide detection range and flexibility. It has good application performance in oil and gas pipelines, water pipelines and perimeter security. But Φ-OTDR has high sensitivity and is easily disturbed by two kinds of noise in engineering[1-4]. One is the noise during system operation, and the other is environmental noise. The former can be filtered out by traditional methods, while the latter is complex random noise, which is difficult to remove, causing useful signals to drown in noise. Therefore, inhibition of the environmental noise in OTDR is the purpose of this paper.

2. The principle of the noise reduction method

2.1. The principle of variational mode decomposition

Variational Modal Decomposition (VMD) is a signal processing method proposed by Dragnmiretskiy K and others in 2014[5]. VMD decomposes the original signal into multiple intrinsic modal components (IMF), and the center frequency and bandwidth of each component have adaptively obtained the optimal solution, thereby realizing the effective separation of modal components (IMF).

The constructed variational model is as follows[6-7]:

$$\min_{\{u_k\},\{\omega_k\}} \{\sum_k \| \partial_t[(\delta(t)+\frac{j}{\pi t})*u_k(t)]e^{-j\omega_k t} \|_2^2\} \qquad s.t. \quad \sum_k u_k = f \tag{1}$$

In formula (1): k is the number of modal layers to be decomposed, $\delta(t)$ is the impulse function, u_k is the kth modal component decomposition, which is subjected to Hilbert transform, Obtain the one-sided spectrum corresponding to k modal components: $\left[\delta(t)+\frac{j}{\pi t}\right]u_k(t)$, ω_k corresponds to the center frequency of each modal component decomposition.

2.2. Principle of equalization optimizer

The balanced detector is a metaheuristic algorithm (simulated natural algorithm) proposed in 2020. Inspired by the dynamic mass balance of the control volume, it simulates the process of entry, departure and generation of the inactive components in the control volume. EO algorithm has the advantages of simple structure, strong adaptability and high flexibility[8]. The algorithm optimization steps are as follows[9]:

1. Initialization: The algorithm performs random initialization within the upper and lower bounds of each optimization variable, as follows:

$$C_i^0 = C_{\min} + rand_i(C_{\max} - C_{\min}), i = 1,2,\cdots, \text{ N} \tag{2}$$

In equation (2), C_{min}, C_{max} are the upper and lower limit vectors of the optimization variable respectively; $rand_i$ is a random vector between [0,1]; N is the population number.

2. Balanced state pool: In order to improve the algorithm's global searchability, the equilibrium state (C_{eq}) will be selected from 5 current optimal candidate solutions. The balanced state pool formed from these candidate solutions is as follows :

$$C_{eq,pool} = \{C_{eq,1}, C_{eq,2}, C_{eq,3}, C_{eq,4}, C_{eq,ave}\} \tag{3}$$

3. Exponential term coefficient F:

$$F = a_1 * sign(r-0.5)[e^{-\lambda t} - 1] \tag{4}$$

In the formula (4), $a_1$is the weight constant coefficient of the global search; both r and λ represent random number vectors.

4. Quality generation rate G: In order to strengthen the local optimization capability of the algorithm, the generation rate is designed as follows:

$$G = G_{CP}(C_{eq} - \lambda C) \ , \ G_{CP} = \begin{cases} 0.5r_i, r_2 \geq 0.5 \\ 0, otherwise \end{cases} \tag{5}$$

In formula (5), G_{cp} is a vector of generating rate control parameters; r_1 and r_2 are random number vectors, and each element value is a random number from 0 to 1.

5. Solution update: For optimization problems, the solution can be updated as:

$$C = C_{eq} + (C - C_{eq})F + G(1-F)/\lambda V \tag{6}$$

2.3. Principles of singular spectrum analysis (SSA)

SSA is a principal component analysis method, which is used to process nonlinear sequences. It has time-frequency domain feature analysis method and does not need to select structural function as a priori information, so it can more flexibly analyze nonlinear and unstable time series, to realize functions such as denoising and feature extraction[10].

The four steps of SSA are embedding, singular value decomposition, grouping and reconstruction[11].

1. Embedding: The signal collected by Φ-OTDR can be regarded as a one-dimensional time series signal sampled at equal intervals. If the window is embedded, the trajectory matrix X can be obtained.

2. Singular value decomposition: singular value decomposition of the trajectory matrix can obtain singular values of window size. Each singular value can be multiplied by its corresponding empirical orthogonal function and principal component to obtain X_i in the $X=\sum_i^n X_i$.

3. Grouping: Divide them into useful signals and noises according to their characteristic values.

4. Reconstruction: Ignore the noise, average the useful signal diagonally, and restore the trajectory matrix to a signal.

3. Experimental verifications

The parameters of the Φ-OTDR system used in the experiment are as follows: the laser wavelength is 1550 nm, the line width is 1.66 kHz, the sampling rate of the acquisition card is 250 MSa/s, the sensing fiber length is 40 km, and the spatial resolution is 0.4m.

3.1. Single-point perturbation algorithm verification

Piezoelectric ceramics are used to disturb the optical fiber at 2.35km. The signals at the disturbance are collected and processed by moving average and difference to remove the system noise. After many experiments, it is found that the sliding average and difference results are the best when the sliding window length is 5 and the step size is 8. The preprocessed signal is used as the original signal. The original signal is collected many times to verify the correlation coefficient criterion. In order to measure the noise reduction performance, this paper selects the relative mean square error (RMSE) as the standard, and the relative mean square error is defined as follows:

$$RMSE = \frac{E[(S-Y)^2]}{E[S^2]} \tag{7}$$

In formula (7), S is the original signal, and Y is the signal after S has been reduced by noise. The smaller the value of RMSE, the better the noise reduction effect. The experiment is shown in figure 1.

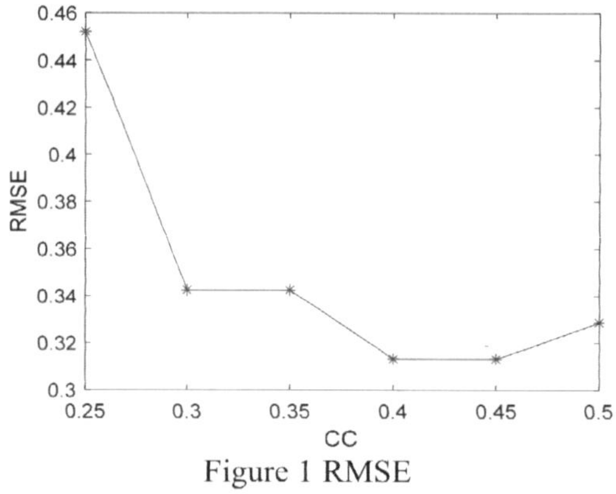

Figure 1 RMSE

It can be seen from the figure that the RMSE of the signal shows a trend of falling first and then rising. When the RMSE is the lowest, the signal noise reduction is the best. Therefore, this paper takes the components with a correlation coefficient lower than 0.4 to perform SSA noise reduction processing. The original signal is noise-reduced using the method in this paper, and compared with the noise reduction effect of EO-VMD, the result is shown in Figure 2.

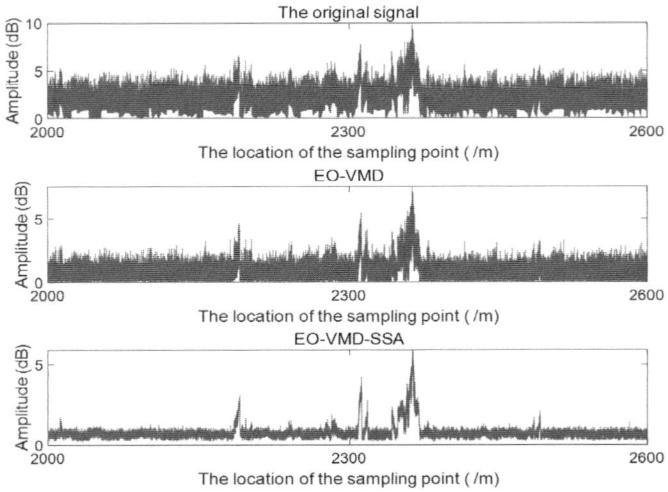

Figure 2 Single point disturbance

After EO-VMD noise reduction, the RMSE of the signal is 0.3251. Although the signal can be decomposed well, the decomposed components are not processed, resulting in poor noise reduction performance. After noise reduction, according to the method in this paper, the RMSE of the signal is 0.1543, which is better than that of EO-VMD. Therefore, this method can decompose the signal properly, reduce the noise power and remove the noise.

3.2. Multi-point disturbance algorithm verification

To verify the noise reduction effect of the abbreviation algorithm in this paper on multi-point disturbance signals, PZT is used for the disturbance at 1.4km and 2.4km of optical fiber, After moving the fiber signal average and difference, noise reduction is carried out by the method in this paper, and compared with EO-VMD. The results are shown in Fig. 3.

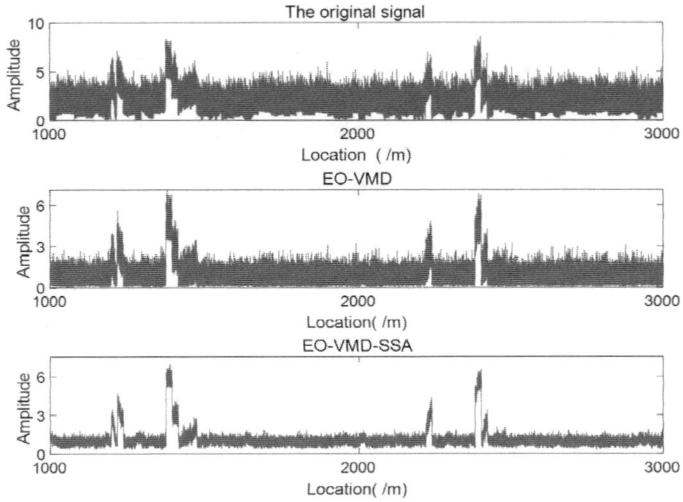

Figure 3 Multi-point disturbance

After EO-VMD noise reduction, the RMSE of the multi-point disturbance signal is 0.3531, and after the noise reduction method in this paper, the RMSE is 0.1324. It can be concluded that the noise reduction in this paper is still effective in the multi-point disturbance.

4. Conclusions

Aiming at the problem of the low signal-to-noise ratio of signals collected by the Φ-OTDR system, a denoising method based on EO-VMD-SSA is proposed. This method uses EO optimized VMD to decompose the signal, then uses SSA to denoise the IMF component with a low correlation coefficient, and then reconstruct it. Through experiments, it is verified that the RMSE of this method after noise reduction is 0.2 higher than that of EO-VMD, which enables the Φ-OTDR system to locate and track interference more accurately.

Acknowledgments

This paper is the innovation team training plan of Tianjin colleges and universities (Project No.: td13-5094), Scientific research plan project of Tianjin Education Commission (Project No.: 2019zd22),

Special project of Tianjin Science and technology plan (Project No.: 19zxjrgx00060),

Tianjin technological innovation guidance project (20ydtpjc00550),

And one of the research achievements of the science and technology cultivation project of Tianjin Sino German University of Applied Technology (zdkt2019-006).

References

[1] Lu Xin & Krebber Katerina.(2021).Characterizing detection noise in phase-sensitive optical time domain reflectometry.. Optics express(12), doi:10.1364/OE.424410.

[2] Zhijie Yu,,Yang Lu,,Xiaoyang Hu & Zhou Meng.(2017).Polarization dependence of the noise of phase measurement based on phase-sensitive OTDR. Journal of Optics(12), doi:10.1088/2040-8986/aa924e.

[3] Arias A,,Shlyagin M G,,Miridonov S V & Manuel Rodolfo Martinez.(2015).Phase-sensitive correlation optical time-domain reflectometer using quantum phase noise of laser light.. Optics express(23), doi:10.1364/OE.23.030347.

[4] He Haijun,,Luo Bin,,Zou Xihua,... & Yan Lianshan.(2018).Enhanced phase-sensitive OTDR system with pulse width modulation Brillouin amplification.. Optics express(18), doi:10.1364/OE.26.023714.

[5] Dragomiretskiy, K.; Zosso, D. Variational mode decomposition. IEEE Trans. Signal Process. 2014,62, 531–544.

[6] Chen Qiang,Zhao Yong & Yan Lixia.(2021).X-ray Pulsar Signal Denoising Based on Variational Mode Decomposition. Entropy(9), doi:10.3390/E23091181.

[7] Nazari Mojtaba & Sakhaei Sayed Mahmoud.(2020).Successive Variational Mode Decomposition. Signal Processing(), doi:10.1016/j.sigpro.2020.107610.

[8]Afshin Faramarzi,,Mohammad Heidarinejad,,Brent Stephens & Seyedali Mirjalili.(2020).Equilibrium optimizer: A novel optimization algorithm. Knowledge-Based Systems(), doi:10.1016/j.knosys.2019.105190.

[9] Abdul hamied Dalia T.,Shaheen Abdullah M.,Salem Waleed A.,Gabr Walaa I. & El sehiemy Ragab A..(2020).Equilibrium optimizer based multi dimensions operation of hybrid AC/DC grids. Alexandria Engineering Journal(prepublish), doi:10.1016/j.aej.2020.08.043.

[10] Xuegeng Mao & Pengjian Shang.(2019).Multivariate singular spectrum analysis for traffic time series. Physica A: Statistical Mechanics and its Applications(), doi:10.1016/j.physa.2019.121063.

[11]Nina Golyandina & Anatoly Zhigljavsky(2013).Singular Spectrum Analysis for Time Series.Springer, Berlin, Heidelberg.

Waterway control assembly for rope core drilling rig and its application method

Dong Fan [1] **and Feifei Lu**[1,2*]

[1] Drilling Technology and Equipment R&D Center ,Xi'an Research Institute of China Coal Technology & Engineering Group ,Xi'an ,shaanxi, 710077,China

[2] Geological Department ,China Coal Research Institute, Beijing, China

[*]Corresponding author's e-mail: lufeifei@cctegxian.com

Abstract. In the process of conventional rope core drilling, the connection and disassembly of waterway occupy a lot of auxiliary time, which seriously affects the drilling efficiency, and even causes drilling or safety accidents. The waterway control assembly and application method for the rope coring rig developed by the author's team can control the waterway flow direction. It does not need to disassemble the water pipe frequently to control the raising speed of the inner pipe and avoid damaging the drilling tool too fast.Introduction.

1. Introduction

In this paper, in view of the existing core drilling machine technology adaptability, low efficiency problems and practical needs, the project team coring drilling tools for scientific research. The integrated control of the fast switching system of the waterway is developed, and the rope coring waterway control method which is safe, efficient and suitable for the construction of near horizontal hole is developed The quick change water system can realize the quick change of common drilling conditions, and the cable-connected water joint with buffering function ensures the safety of equipment and personnel during the construction.

2. Format

The characteristic of wireline coring drilling is that when the core is filled with the inner pipe, the inner pipe is fished up from the drill pipe with a wireline fishing tool without lifting the drill to obtain the core. During drilling, water or mud is usually used as the medium to send water to the hole bottom through the path between the water joint and the mud pump. During normal drilling, since the drill pipe is a rotating part.It is necessary to connect the conventional water joint with the drill pipe, and then connect the rubber hose between the water pipe and the mud pump. In the process of conventional rope core drilling, the connection and disassembly of waterways occupy a lot of auxiliary time, which seriously affects the drilling efficiency. At the same time, due to the need to operate the mud pump and drilling rig at the same time, auxiliary mud pump operators have to be equipped, which not only increases the drilling cost, but also makes it difficult for two people to communicate and cooperate in the noisy drilling yard.It affects the drilling efficiency and even causes drilling or safety accidents[1-3].

3. There are problems in the connection of existing waterways

During drilling, water or mud is usually used as the medium to send water to the hole bottom through the path between the water joint and the mud pump. During normal drilling, since the drill pipe is a rotating part, it is necessary to connect the conventional water joint with the drill pipe, and then connect the rubber hose between the water pipe and the mud pump. When fishing the core (mainly horizontal hole or upward hole). It is necessary to put the fishing device into the coring drill pipe, connect the through-line water joint, and then connect the water and mud pump. The fishing device is quickly lowered by hydraulic power, and then winch to lift the fishing core. When fishing the core, it is necessary to disconnect the through-line water joint from the mud pump to discharge the water in the drill pipe, Especially when drilling the upward hole, the inner pipe assembly is fished out under the action of its own weight. In this process, it is necessary to continuously adjust the size of the drain port to adjust the falling speed of the inner pipe assembly and the speed of winding the steel wire rope with the winch. If the drain port is not adjusted in time or the opening is unreasonable, the steel wire rope is wound too fast, which is not conducive to the neat winding of the steel wire rope, and the inner pipe drops faster. It is easy to damage the drilling tool by directly hard hitting the wireline. During the drilling process, it is necessary to frequently replace the connection between the conventional joint and the wireline water joint, especially the connection between the wireline water joint. Because the wireline water joint cannot be connected with the pipeline rotation, when connecting the wireline water joint and the drill pipe, it is necessary to remove the connecting pipe, tighten the water joint and the drill pipe, and then connect the pipeline again. The operation process is cumbersome and time-consuming, and since the mud pump generally has only one water outlet, it is also necessary to disassemble and connect the connecting pipeline when replacing the conventional water connector and through cable water connector. The waterway replacement process accounts for a large part of the auxiliary time. The efficiency of waterway replacement is an important factor affecting the auxiliary time[4-6].

4. New waterway control assembly

The water channel control assembly for core drilling rig consists of rotatable cable water connector, water channel control valve group, conventional water connector and connecting pipeline. The two connectors A and B in the waterway control valve group are respectively connected with the rotatable cable water connector and the conventional water connector through the connecting pipe. Because the rotatable cable water joint can load and unload the water joint without dismantling the connecting pipeline, the step of screwing and unloading the connecting pipeline is reduced, the auxiliary time is shortened and the labor intensity of workers is reduced. At the same time, because the sealing ring (preferably Y-shaped sealing ring) is used between the rotating sleeve and the mandrel, there is a large gap between the sealing ring and the rotating sleeve before the high-pressure water circuit is connected, the rotating sleeve rotates flexibly, and reliable sealing can be ensured when the high-pressure water circuit is connected.

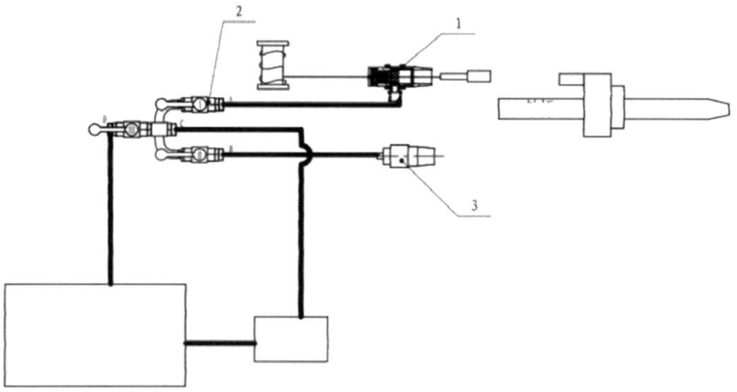

Fig. 1 system composition

The spring installed in the center of the mandrel can avoid hard impact on the through-line water joint during the falling of the inner pipe assembly, so as to achieve the purpose of soft landing, so as to improve the service life of the drilling tool.

The rotatable cable water joint realizes the installation of cable water joint without dismantling the water pipe. The waterway control assembly connects the rotatable through cable water joint, conventional water joint, water inlet pipe and drainage pipe together. By switching the ball valve in the waterway control assembly, the waterway flow direction can be controlled without frequent disassembly of the water pipe. When the inverted hole or horizontal hole is made, the opening of the ball valve III can be adjusted to control the lifting speed of the inner pipe, avoid damaging the drilling tool too fast.

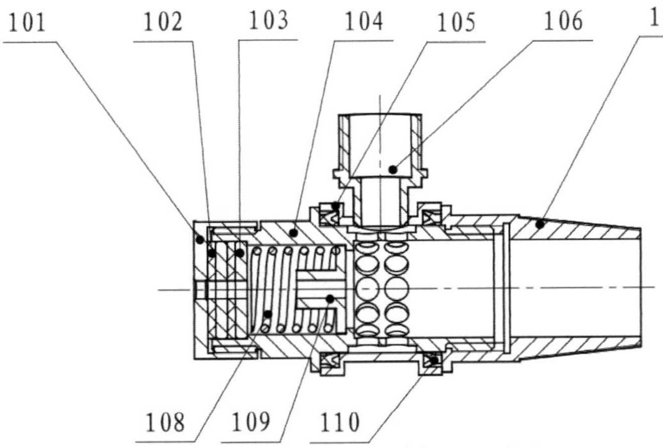

Fig. 2 Rotatable water joint

5. Footnotes

The ZDY1200G drilling rig of Xi'an Academy of Coal Sciences, which is equipped with the waterway control assembly of rope coring drilling, has carried out type test in Cangyuan lead zinc mine, Lincang City, Yunnan Province. A total of 4 boreholes have been constructed, including kz19-1301, with a depth of 512.68m, a core recovery rate of 96.8%, and kz19-1302 - 79 ° rope coring boreholes, with the fastest single day footage of 98.7m, which is the highest single day footage of the company[1]. The drilling rig was used to construct two boreholes in Xunyang mercury antimony mine, Ankang City, Shaanxi Province, with a total footage of 702m. In the process of use, the driller can quickly switch the water system, realize the convenient switching of normal drilling, fish conveying and lifting inner pipe water, save the process of mud pump clutch operation and reduce labor intensity. When lifting the inner pipe, the water path controls the lowering speed of the inner pipe drilling tool through the throttle valve to realize the uniform lifting of the inner pipe and prevent the accelerated falling from damaging the drilling tool.

Fig. 3 Industrial test

Journal of Physics: Conference Series **2113** (2021) 012005 doi:10.1088/1742-6596/2113/1/012005

6. Conclusion

In practical application, the rotatable cable type water can solve the problem that the conventional cable type water cannot rotate when there is a water pipe connected, so the water pipe must be disassembled back and forth. It only needs to replace the water joint as needed, and there is no need to disassemble the connecting water pipe back and forth, which simplifies the operation process, shortens the auxiliary operation time and reduces the labor intensity of workers, The core of the rotatable through cable water joint is provided with a buffer mechanism, which can avoid fishing the inner pipe from falling too fast and damaging the drilling tool due to hard contact when drilling an inclined hole. The waterway control assembly in this patent can easily adjust the water volume and direction of each waterway in drilling to meet the requirements of different drilling procedures. The operation is convenient, and the layout of the waterway control assembly is relatively flexible, It is convenient for the driller to adjust, no additional auxiliary personnel are required, and the staffing is reduced.

References

[1] Fan Dong, Tian Hongliang, Wang Ruize, Chang Jianghua, Lu Feifei, & Zou Zujie et al. (2019). Development of lightweight full hydraulic tunnel core drill. Coal science and technology, v.47; No.541(12), 136-140.

[2] Wang Ruize. (2019). Design and analysis of angle adjustment mechanism of tunnel core drill. Coal mine safety, 50 (02), 115-118

[3] Wei Huanhuan. (2019). Thread strength analysis of high strength rope cored drill pipe joint. Coal mining machinery (3)

[4] Luo Zheng. (2020). Analysis on the development status of rope core construction technology. Building materials and decoration, 000 (002), 217-218

[5] Yang Shaofeng. (2019). Analysis of wireline core drilling technology. China Metal Bulletin (3)

[6] Shi Zhixing. "Analysis of diamond rope core drilling in broken strata." Western Exploration Engineering 022.012 (2010): 36-37

[7] Zhou Yang, et al. "Comprehensive construction method and effect of rope core drilling in broken stratum", China hydraulic society; China Society of hydropower engineering; Water conservancy and hydropower drilling information network; China Water Conservancy Society; China Society of hydropower engineering; Water resources and hydropower drilling information network, 2015

[8] Li Zhimin. "Technology and formula application of small diameter rope core mud." Western Exploration Engineering 31.05 (2019): 55-57

 2113 (2021) 012006 doi:10.1088/1742-6596/2113/1/012006

Design and Experimental Study of Array Eddy Current Sensor for Internal Inspection of Natural Gas Pipeline

Qingxiang Zhou[1], Gangqing Li[1], Kai Hou[1], Feng Cao[1] and Kai Song [2*]

[1]Crrc Qingdao Sifang Co. ,Ltd.

[2]Key Laboratory of Nondestructive Testing, Ministry of Education, Nanchang Hangkong University, Nan-chang, China

[*]Correspondence: kevin.song@crrcgc.cc

Abstract. Array eddy current detection (ECAT) technology has the advantages of fast scanning speed and high detection efficiency, and has a wide range of application prospects. However, the traditional array eddy current sensor has a poor coupling effect with the inner wall of the pipeline and is not suitable for in-pipe inspection. Based on the basic principles of ECAT, a flexible array eddy current sensor made by flexible printed circuit board (FPCB) technology is designed and developed, which can realize 360° detection of defects on the inner wall of natural gas pipelines. The paper uses ANSYS finite element software to establish a simulation model of multi-parameter effects, study the influence of sensor size parameters and detection parameters on eddy current signals, carry out detection experiments on blind hole defects in the inner wall of steel pipes, analyze the response characteristics of defect signals, and verify the feasibility of using the FPCB array eddy current sensor for internal inspection of natural gas pipeline.

1. Introduction

As a new and economical transportation method,pipeline transportation has the advantages of high efficiency, safety and economy, and less damage to the environ-ment compared with other transportation methods. The natural gas pipeline is an im-portant part of the urban gas pipeline network[1], with a small diameter and a large number, which meets the social energy demand. However, in the course of long-term service, the pipeline will gradually age due to chemical corrosion, erosion, wear and other problems, and defects such as cracks and perforations will appear, and the dis-asters caused by pipeline leakage will also increase[2,3], so the directional inspection and maintenance of natural gas pipelines is of great significance for preventing pipe-line leakage and preventing major malignant accidents.

For the internal inspection technology of natural gas pipelines, the commonly used methods include ultrasonic inspection, magnetic flux leakage inspection and eddy current inspection. Ultrasonic testing requires couplant, which has a great influence on slag inclusion in the cavity, and the ultrasonic testing of small-diameter pipes has problems such as curved surface coupling compensation and internal surface diffusion, which reduces the accuracy and reliability of the test results.[4] Magnetic flux leakage detection technology is an important method for pipeline detection. Under the impetus of the medium, a magnetic flux leakage detector is used to record data while detecting the pipeline. The magnetic flux leakage equipment is large in size, long in structure, and large in power consumption. Due to the small diameter of the pipe, it is prone to jamming due to the limitation of the internal space. [5]

Eddy current detection technology is widely used in natural gas pipeline detection due to its high detection efficiency, small sensor size, and high reliability of detection results. Duan Chenggong[6] used the small-diameter pipeline experiment platform to simulate the actual pipeline environment and built a pulsed far-field eddy current de-tection system to realize the detection and display of defects. Zhang Zhihao, Sun Yinjuan[7] and others combined eddy current and ultrasonic testing technology to de-velop a pipeline corrosion detection robot to inspect and record the entire pipeline. Xing Lidong et al.[8] used the new far-field eddy current sensor to rotate at different angles along the pipe circumference, which can effectively obtain the pipe circumfer-ence crack defect information. Wang Xiaofeng et al.[9] designed four sensor models with different structures based on pulsed far-field eddy current technology, which can detect axial cracks in pipelines. Although the above research has achieved many in-structive research results for engineering applications, the conventional eddy current detection technology has a small scanning range and low detection efficiency, and it is easy to cause missed detection. [10]

Eddy Current Array Testing (Eddy Current Array Testing, ECAT) technology has a wide detection range, which can scan a large area of the workpiece surface at one time, and has the same resolution as conventional eddy current point sensors[11], which has a wide range of application prospects. Li Xuechao[12] and others applied array-type pulsed far-field eddy current sensors to achieve quantitative detection of pipeline defects. Han Ning[13] and others designed a rotating magnetic field array sensor that can identify and locate pipeline defects in different orientations. Professor Luo Feilu et al.[14] proposed an oblique-angle array sensor, which can play a very good role in the quantitative measurement of large-area corrosion. Although some ex-ploratory researches with practical application value have been carried out on the sensors designed in the above institutes, the traditional array eddy current sensors are mainly composed of coil-wound or Hall sensors, which have poor coupling effect with the inner wall of the pipeline and low detection accuracy.

In order to complete the 360° omni-directional inspection of the inner wall of the natural gas pipeline and realize the complete coupling with the inner wall of the pipe-line, this paper designs and develops an FPCB array eddy current sensor[15], uses fi-nite element technology to establish a multi-parameter influence three-dimensional simulation model, and optimizes the sensor size parameters. The influence of the de-tection parameters on the eddy current signal is studied, and the feasibility of the FPCB array sensor for detecting the inner wall defects of the natural gas pipeline is verified.

2. The Principle of Array Eddy Current Testing Technology

As shown in Figure 1, it is assumed that the array eddy current sensor is placed above an infinite flat test piece with ideal defects[16]. The length of the test piece is l, the width is w, and the thickness is d. The entire spatial field is set as a linear medium, and the sensor is composed of several close-turned coils of exactly the same size and material. The inner and outer diameters are r_1 and r_2, and the height is h. The current $i(t)$ in the coil is harmonic with time t Change, the total number of turns of the coil unit is W, and the number of turns density can be obtained by formula (1).

$$n_c = \frac{W}{\left[h(r_2 - r_1)\right]} \tag{1}$$

The space occupied by the winding area of the coil unit is V_c, which can be seen from the following formula:

$$\boldsymbol{E} = -\frac{\partial \boldsymbol{A}}{\partial t} - \nabla \varphi \tag{2}$$

In the formula: $\boldsymbol{A}$ is the magnetic vector potential, and φ is the scalar potential. By transforming the formula (2), we can get:

$$\boldsymbol{E} = -\frac{\partial}{\partial t}(\boldsymbol{A} + \int_{-\infty}^{t} \nabla \varphi(t')dt') \tag{3}$$

Let $\boldsymbol{A}^*$ be:

$$\boldsymbol{A}^* = \boldsymbol{A} + \int_{-\infty}^{t} \nabla \varphi(t')dt' \tag{4}$$

Then the electric field intensity $\boldsymbol{E}$ and the magnetic induction intensity $\boldsymbol{B}$ are as follows:

$$\boldsymbol{E} = -\frac{\partial \boldsymbol{A}^*}{\partial t} \tag{5}$$

$$\boldsymbol{B} = \nabla \times \boldsymbol{A} = \nabla \times \left(\boldsymbol{A}^* - \int_{-\infty}^{t} \nabla \varphi(t')dt' \right) = \nabla \times \boldsymbol{A}^* \tag{6}$$

According to formula (6), the flux linkage of the n-th turn of the wire of the coil unit is:

$$\psi_n(t) = \int_{S_n} \boldsymbol{B} \cdot dS = \int_{S_n} \left[\nabla \times \boldsymbol{A}^* \right] \cdot dS = \int_{l_n} \boldsymbol{A}^* \cdot d\boldsymbol{l} \tag{7}$$

In the formula: S_n is the curved surface enclosed by the nth turn of the coil unit, and l_n is the circumference enclosed by the nth turn of the coil unit.

From formula (7), the total flux $\psi(t)$ of the turns of the coil unit can be obtained as:

$$\psi(t) = \frac{1}{i(t)} \sum_{n=1}^{W} \psi_m(t) i_n(t) = \frac{1}{i(t)} \sum_{n=1}^{W} \int_{l_n} \boldsymbol{A}^* \cdot i_n(t) d\boldsymbol{l} \tag{8}$$

In the formula: $i(t)$ is the total current in the coil, $i_n(t)$ is the current in the nth turn of the coil, so $i(t) = i_n(t)$. According to the previous assumptions, the following two substitutions are established:

$$\sum_{m=1}^{W} \phi_{l_m} \Rightarrow \int_{V_c} \tag{9}$$

$$i_m(t)dl \Rightarrow \boldsymbol{J}_c(t)dV \tag{10}$$

Among them, $\boldsymbol{J}_c$ is the current density in the coil unit, so that formula (8) can be rewritten as:

$$\psi(t) = \frac{1}{i(t)} \int_{V_c} \boldsymbol{A} \cdot \boldsymbol{J}_c dV \tag{11}$$

Also, the current density at any point Q on the coil unit is as follows:

$$\boldsymbol{J}_c(t) = n_c(Q) i(t) e(Q) \tag{12}$$

Combining equations (11) and (12), the total flux linkage of the turns of the coil unit can be calculated as follows:

$$\psi(t) = \frac{1}{i(t)} \int_{V_c} n_c(Q) \boldsymbol{A}^*(Q) \cdot e(Q) dV(Q) \tag{13}$$

In the formula: $e(Q)$ is the unit vector in the direction of the current at the point $Q \in V_c, |e| = 1$.

According to the total flux linkage $\psi(t)$ of the turns of the coil unit in equation (13) and Faraday's law of electromagnetic induction, the induced voltage of the coil unit can be calculated as follows:

$$u(t) = \frac{d\psi(t)}{dt} = \int_{V_c} n_c(Q) \frac{\partial \boldsymbol{A}^*}{\partial t} \cdot e(Q) dV(Q) \tag{14}$$

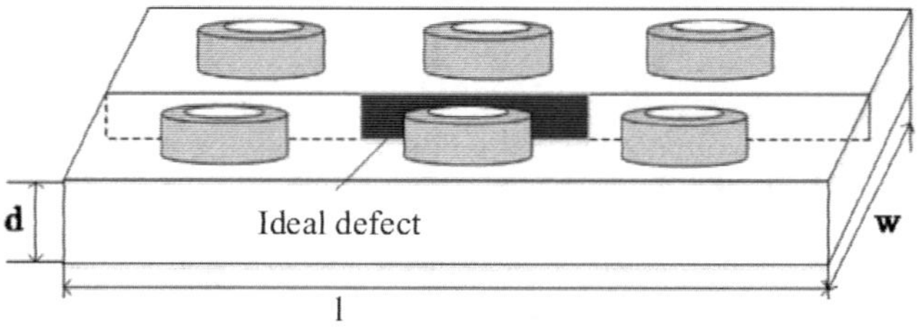

Figure 1. ECAT flat specimen(Ideal defect)

3. Design of FPCB Array Eddy Current Sensor

3.1. Model building

Using ANSYS finite element software to establish a three-dimensional simulation model, as shown in Figure 2. The whole model mainly includes two parts: natural gas pipeline and FPCB array eddy current sensor. In order to enhance the sensitivity of the sensor to defects in the inner wall of the pipeline, the sensor is connected in the form of a bridge, it is composed of a double-layer multi-channel FPCB board, and the coil is a plane spiral structure, which can be approximately composed of several concentric current loops of different diameters, as shown in Figure 3. The sensor is time-sharing switching in time sequence, that is, there is only one channel for detection at a certain time. In order to facilitate experimental research, a single-channel simplified model is adopted. Considering that the inner wall of the natural gas pipeline is a plane relative to the FPCB array eddy current sensor, Use steel plates instead of pipes for testing.

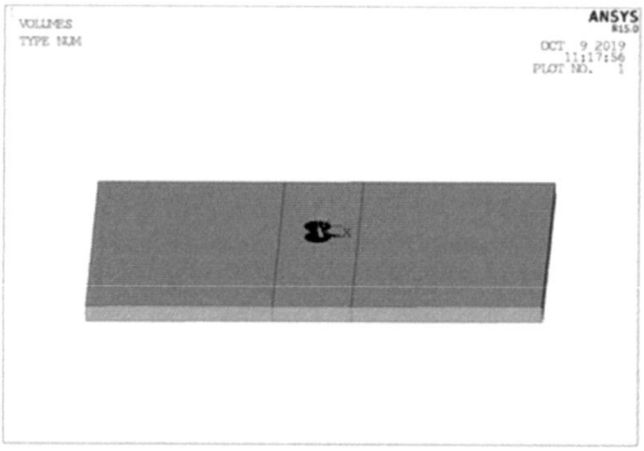

Figure 2. 3D simulation model diagram

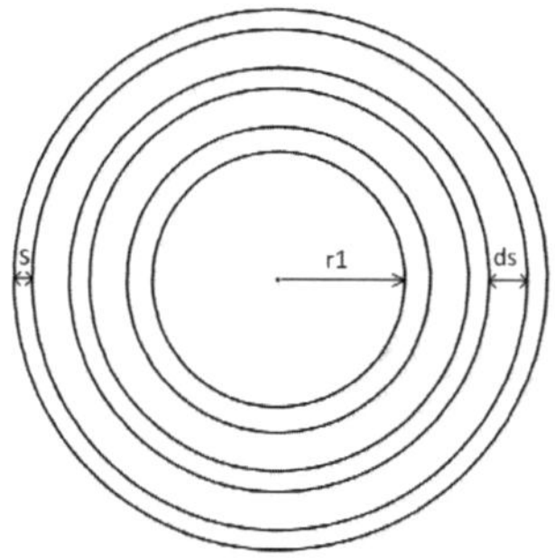

Figure 3. Multi-turn concentric current loop

In the model, the length L of the steel plate is 150mm, the width W is 100mm, the thickness D is 6mm, and the electrical conductivity σ is 36Ms/m. The initial inner diameter r1 of the coil is 8.5mil, the number of turns n is 60 turns, the excitation frequency f is 700kHz, the lift-off height lift is 4mm, and the coil conductivity ρ is 58.8 Ms/m. The overall simulation model parameters are shown in Table 1.

Table 1. Simulation model parameters

Steel plate parameters	**Numerical value**	**Coil parameters**	**Numerical value**
L/mm	150	r_1/mil	8.5
W/mm	100	n	60
D/mm	6	ρ(Ms/m)	58.8
σ(Ms/m)	36	lift /mm	4

Considering the open domain of the electromagnetic field, it is necessary to establish a large enough near-field air area and far-field air area around the model to eliminate the truncation effect. In order to ensure the quality of the grid and improve the calculation accuracy and efficiency, the coils and steel plates with regular shapes are meshed by the mapping division method, and the near and far-field air areas are meshed by the free division method. The model after the mesh is divided As shown in Figure 4.

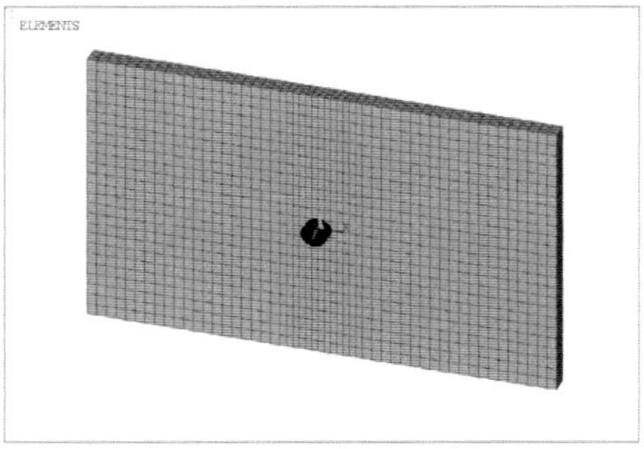

Figure 4. Model meshing

3.2. Parameter optimization

3.2.1. Coil diameter optimization

The wire diameter size directly affects the amount of current passing through the coil, and the wire diameter needs to be optimized first. Set the wire diameters to 3mil, 4mil, 5mil and 6mil respectively, the turn pitch is 4mil, and a blind hole with a diame-ter of 10mm and a depth of 2mm is set at the center of the steel plate. Under the action of a sinusoidal excitation signal of 5V, 600kHz, the sensor selects the optimal wire di-ameter by the induced electromotive force when the sensor passes through the blind hole of the steel plate. The test results are shown in the figure below.

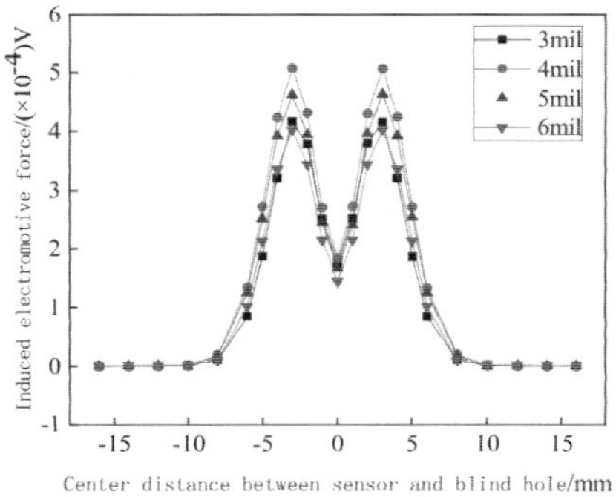

Figure 5. Test results under different wire diameter conditions

From the analysis of Figure 5, it can be seen that the change trend of the sensor induced electromotive force amplitude characteristic curve under different wire diameter conditions is the same. The detection signal increases as the coil approaches the blind hole, and decreases as the coil moves away from the blind hole, reaching a peak value near ±3mm from the center of the blind hole. When the wire diameter is 4mil, the sensor induced electromotive force peak value reaches 5×10-4V, compared with 4mil, when the wire diameter is 3mil, 5mil, and 6mil, the induced electromotive force is 4.16×10-4V, 4.64×10-4V, 4.03×10-4V, attenuated by 21.8%, 9.27% and 25.8% respectively. In order to improve the detection sensitivity of the FPCB array eddy current sensor, the wire diameter should be 4mil.

3.2.2. Coil pitch optimization

Under the wire diameter is 4mil and other parameters remain unchanged, a sim-ulation study is carried out on the turn pitch between the coils. The turn pitch is in-creased from 3 mil to 6 mil in steps of 1 mil and compare the magnitude of the elec-tromotive force induced by the sensor under different turn pitch conditions. The test results are shown in the figure below.

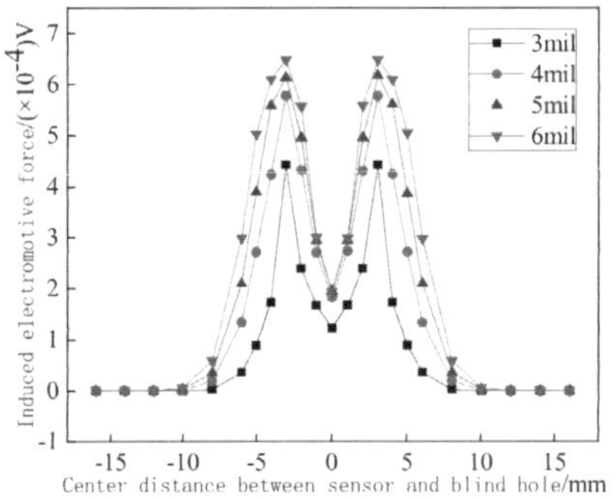

Figure 6. Test results under different turns

It can be seen from Figure 6 that the sensor has a minimum value at the center of the blind hole, and has good symmetry relative to the remaining scanning positions on both sides of the blind hole center. This is because the coil is an axisymmetric structure. The eddy current field is disturbed twice, so the detection signal has double peaks. When the turn pitch is increased from 3mil to 6mil, the peak values of the induced electromotive force are 4.43×10-4V, 5.7831×10-4V, 6.13×10-4V and 6.48×10-4V respectively. Considering that the coil is too large and the sensor volume is too large, which brings difficulties to the subsequent FPC board production. When the turn pitch is 4mil, compared with the 6mil turn pitch, the induced electromotive force is only reduced by 5.4%, and the amount of change between the two is almost Equal, and can detect defects effectively. Therefore, in the case of ensuring the sensitivity of the sensor, the turn pitch is selected to be 4mil for the next step of research.

3.2.3. Coil spacing optimization

The distance between the upper and lower coils of a single channel of the FPCB array eddy current sensor affects the magnitude of the differential voltage signal. In order to study the effect of the coil distance on the detection sensitivity, under the condition of the other parameters unchanged, the simulation and analysis of the sensor's impact on the blind hole under different coil distances The detection sensitivity is shown in Figure 7.

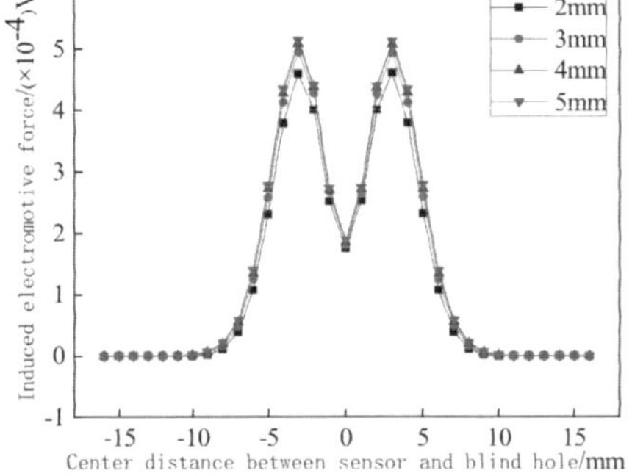

Figure 7. Test results under different center distance conditions

Analysis of Figure 7 shows that when the coil spacing increases from 2mm to 5mm, the peak values of the induced electromotive forceare 4.596×10-4V, 4.951×10-4V, 5.08×10-4V and 0.55×10-4V in sequence. Compared with the coil pitch of 2mm, when the coil pitch is 3mm, 4mm, and 5mm, the peak change amount increases respectively by 0.35×10-4V, 0.49×10-4V and 0.55×10-4V, and in the process of increasing the coil spacing from 3mm to 5mm, the increase in the change amount slows down, that is, as the coil spacing increases, the detection sensitivity of the sensor becomes higher. This is because within a certain range, the farther the two coils are, the greater the differential signal output by the sensor. Therefore, on the premise of ensuring the miniaturization of the sensor, the coil spacing should be increased as much as possible.

3.2.4. The influence of eccentricity on detection signal

Due to the existence of the coil spacing, the curvature of the upper and lower layers of the FPC board is inconsistent, so the centers of the upper and lower coils of each channel cannot be completely overlapped, and there is a certain eccentricity. In order to study the influence of the eccentricity on the detection signal, the offset between the upper and lower coils in a single channel was changed to detect the blind hole shown above. Set the eccentricity to 0.5mm,0.8mm,1mm and1.5mm respectively, and the test results are shown in Figure 8. It can be seen from the figure that when the eccentricity is increased from 0.5mm to 1.5mm, the detection results are almost the same. This is because when the current is constant and the eccentricity changes little, the induced voltage of the two coils in the sensor is almost the same.

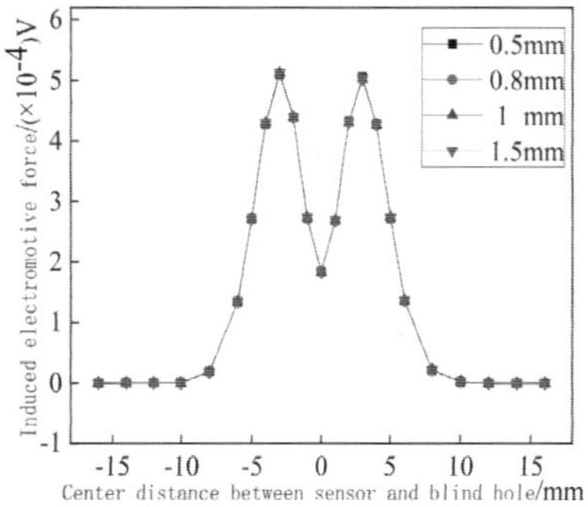

Figure 8. Effect of eccentricity on detection signal

3.2.5. The influence of excitation frequency on detection signal

Since the frequency of the excitation signal affects the sensitivity of the sensor, it is necessary to conduct a simulation study on the excitation frequency. Based on the established sensor simulation model, the excitation signal frequency range is set to 200kHz to 800kHz, and the simulation is performed every 200kHz step, and the maximum voltage when the sensor passes through the blind hole of the steel plate is extracted. The detection result is shown in Figure 9.

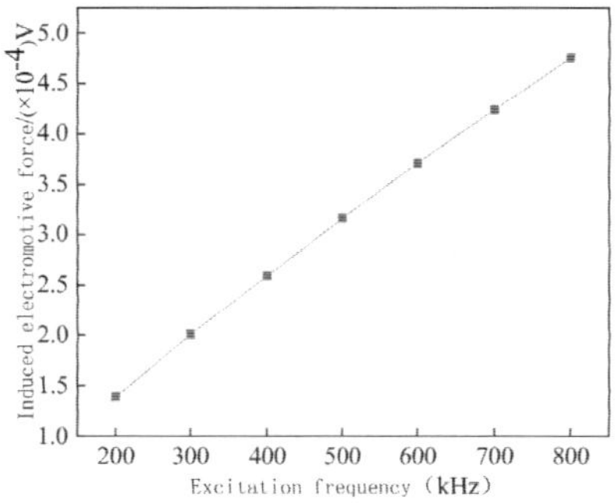

Figure 9. Effect of excitation frequency on the detection signal

It can be seen from Figure 9 that as the excitation frequency increases, the sensor's induced electromotive force amplitude increases linearly. When the excitation frequency is 800kHz, the induced electromotive force is 0.55mV, which is an increase of 175% compared to the excitation frequency of 200kHz. This is because the higher the excitation frequency, the greater the eddy current density on the surface of the steel plate, and the higher the detection sensitivity of the sensor. However, due to the influence of the skin effect, according to $\delta = \sqrt{\frac{1}{\pi f \mu \sigma}}$, the higher the excitation frequency, the lower the depth of eddy current penetration of the steel plate. The eddy current distribution cloud diagram of blind hole defects of steel plate under different excitation frequencies is shown in the figure below.

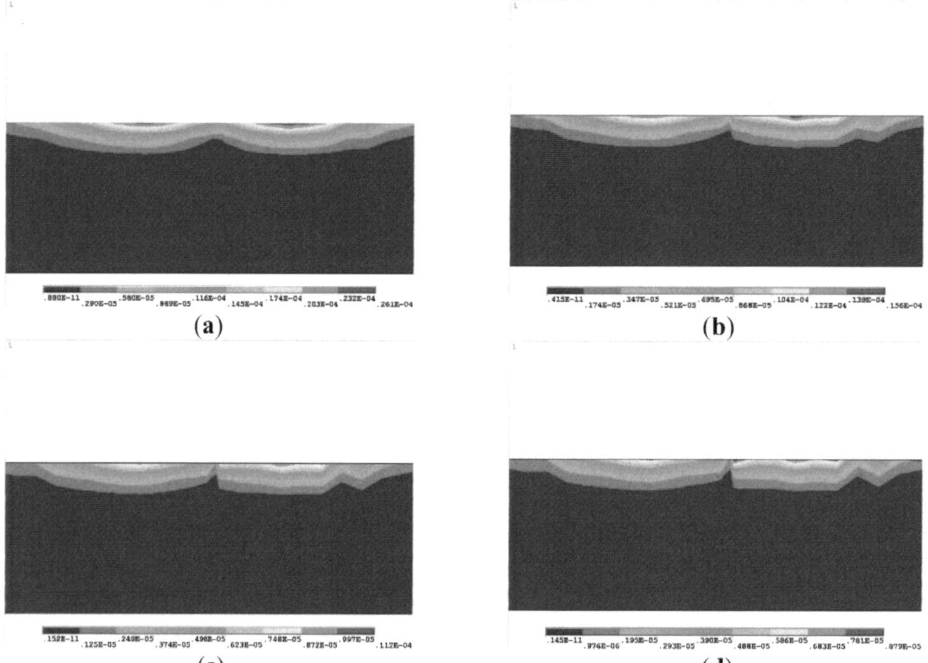

(c) (d)

Figure 10. Eddy current cloud image of steel plate section under different excitation frequencies: (**a**) Cloud diagram of eddy current cross section of steel plate at f=200kHz; (**b**) Cloud diagram of eddy current cross section of steel plate at f=400kHz; (**c**) Cloud diagram of eddy current cross section of steel plate at f=600kHz; (**d**) Cloud diagram of eddy current cross section of steel plate at f=800kHz.

It can be seen from Fig. 10 that the intensity of the magnetic field in the steel plate gradually decays as the penetration depth increases. When the excitation frequency is 800kHz, compared with the excitation frequency of 200kHz, the induced eddy current is mainly concentrated near the metal surface, at this time the magnetic field intensity is the strongest on the surface of the steel plate. Therefore, the detection depth and surface sensitivity of the eddy current sensor are contradictory, and it is difficult to achieve both. The optimal excitation frequency must be selected according to actual needs.

4. Experimental Research

4.1. Test verification

In order to detect defects on the inner wall of the natural gas pipeline, verify the detection effect of the optimized FPCB array eddy current sensor, and make the sensor according to the simulation optimization parameters. The physical map of the single-layer 16-channel FPCB array eddy current sensor is shown in Figure 11. Set up the corresponding eddy current inspection system to inspect the steel pipe. The block diagram of the eddy current inspection system is shown in Figure 12.

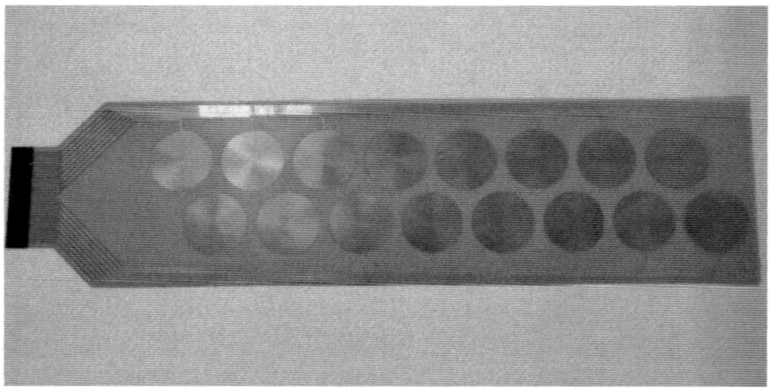

Figure 11. Single layer 16-channel FPCB array eddy current sensor physical map

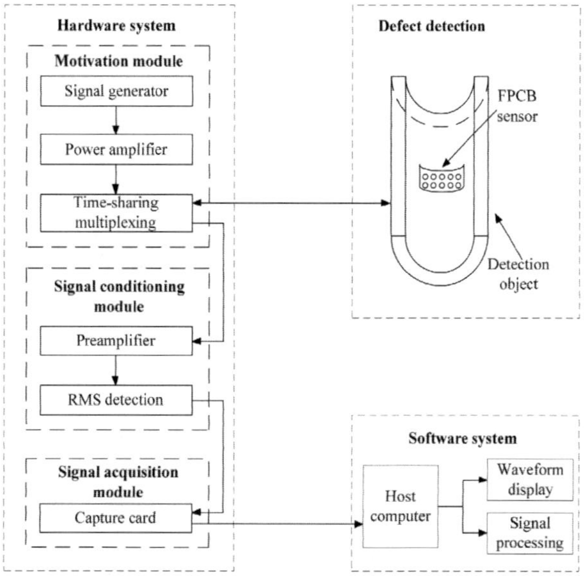

Figure 12. Block diagram of eddy current testing system

The eddy current detection system is mainly composed of signal generator, power amplifier, time division multiplexing module, FPCB array eddy current sensor, signal conditioning module, A/D signal acquisition card and signal analysis software system. After the signal generator is energized, the excitation signal is generated, which is loaded on the coil of the sensor after passing through the power amplifier. The detection coil receives the induced signal, passes through the pre-amplification module and the effective value detection module, and sends it to the A/D data acquisition card, which is processed and analyzed by the computer software.

4.2. Test results and analysis

Use this eddy current inspection system to inspect a 45# steel pipe with a size (length× outer diameter×wall thickness): 400×90×5mm, and process three blind holes with a diameter of 10mm and a depth of 1mm, 2mm, and 3mm are machined on the inner wall of the steel pipe. Set the excitation sinusoidal signal voltage to 5V and frequency to 600kHz, use a single detection channel in the FPCB array eddy current sensor to detect the blind hole, and extract the amplitude of the induced voltage through the effective value detection module. The detection signal change is shown in Figure 13. It can be seen from the figure that when detecting blind holes of different depths, the induced voltages all exhibit double peaks. When the sensor moves to ±3mm from the center of the blind hole, the induced voltage has a maximum value, which is consistent with the simulation result. Comparing with the blind hole depth of 1mm, the maximum induced voltage increases by 132.42mV and 414.43mV when the depth is 2mm and 3mm, indicating that as the depth of the blind hole increases, the detection signal amplitude increases, which verifies the use of the FPCB array The feasibility of eddy current sensor for steel pipe defect detection.

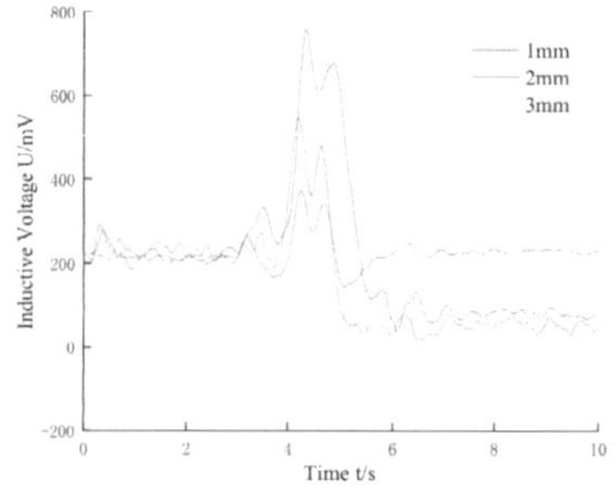

Figure 13. Time-varying characteristic curve of detection signal

5. Results

(1) A new type of array eddy current sensor made by FPCB technology is proposed, and the finite element software is used to analyze the influence of the sensor's size pa-rameters and detection parameters on the detection performance of the sensor. On the premise of keeping the sensor miniaturization, the optimal parameters of wire diame-ter and turn pitch are both 4mil, and the sensitivity of the sensor can be improved by appropriately increasing the coil spacing;

(2) For a single channel, the eccentricity of the upper and lower coils inside the sensor varies within the range of 0.5mm-1.5mm, that is, when the eccentricity is small, it will not affect the detection sensitivity;

(3) The higher the excitation frequency, the higher the sensor sensitivity, and the smaller the depth of eddy current penetration of the steel plate. Therefore, the detec-tion depth and sensitivity of the eddy current sensor are contradictory, and the highest possible excitation frequency must be selected according to actual needs;

(4) In actual testing, the FPCB array eddy current sensor can effectively detect blind hole defects with a diameter of 10mm and a depth of 1mm, 2mm, and 3mm re-spectively. The deeper the defect depth, the greater the detection signal, which verifies the use of the FPCB array eddy current sensor for the feasibility of inner wall defect detection of steel pipes.

Author Contributions:

Q.Z. designed the sensor; G.L. did the simulation experiment; K.H. estab-lished the eddy current detection system; F.C. analyzed the simulation results and reached a con-clusion; K.S. organized the experimental data and test results. All authors have read and agreed to the published version of the manuscript.

Funding:

This research was financially supported by the National Natural Science Foundation of China (Grant No.51865033),the Foundation of Key Laboratory of nondestructive testing technology of Ministry of Education (Grant No.EW201908438).

Institutional Review Board Statement:

Not applicable.

Informed Consent Statement:
Not applicable.

Data Availability Statement:
Not applicable.

Acknowledgments:
The authors would like to thank Professor Song Kai of Nanchang Hangkong University for valuable guidances.

Conflicts of Interest:
The authors declare no conflict of interest.

References

[1] Wang Chongxun.Research and Design of Small Diameter Polyethylene Gas Pipeline Inspection Robot[D]. Beijing Jiaotong University, 2019.
[2] Yang Lijian, Geng Hao, Gao Songwei. Magnetic leakage internal detection technology for long-distance oil and gas pipelines [J].Journal of Instrumentation,2016,37(08):1736-174.
[3] Dengji Zhou,Qinbo Yao,Hang Wu,Shixi Ma,Huisheng Zhang. Fault diagnosis of gas turbine based on partly interpretable convolutional neural networks[J]. Energy,2020.
[4] Wei Tongfeng. Correction coefficient of coupling surface for ultrasonic testing of small-caliber pipeline[J]. Oil and gas storage and transportation,2014,33(06):629-631+652.
[5] Yu Jiangtao. Simulation and experimental verification of small-diameter pipelines based on pulse eddy current technology[D].Beijing Jiaotong University, 2018.
[6] Duan Chenggong, Gao Songwei. Pulsed far-field eddy current detection system for small-caliber pipelines[J].Electronic product reliability and environmental testing,2013,31(05):53-57
[7] Zhang Zhihao, Sun Yinjuan, Yang Tao, Wang Qianzhong, Sun Fangping, Luo Huijuan. Research and Application of Inner Inspection Robot for Small Caliber Pipeline in Changqing Oilfield[J].Petroleum and Natural Gas Chemical Industry, 2020, 49 (01): 93-97.
[8] Xing Lidong, Yu Shenglin, Qu Minxing. Design and experimental study of 3D far-field eddy current sensor[J].Journal of Instrumentation, 2006, 27(11).
[9] Jing Yifei, Wang Xiaofeng, Yang Binfeng, Zhang Hui, Kang Zhibin, Li Shuifang. Design and simulation analysis of pulse far-field eddy current probe for pipeline axial crack detection[J]. Journal of Air Force Engineering University(Natural Science Edition) ,2011,12(06):74-78.
[10] Liu Bo, Luo Feilu, Hou Liangjie. Research on the crack feature extraction method of eddy current array detection[J]. Journal of Instrumentation,2011,32(03):654-659.
[11] Song Kai, Liu Tangxian, Li Laiping, et al. Array Eddy Current Test Simulation of Aeroengine Turbine Blade Cracks[J].Journal of Aeronautics, 2014,35(08):2355-2363.
[12] Li Xuechao, Feng Enxin, Feng Fei. Array-type pulse far-field eddy current pipeline defect detection method [J]. Metrology Technology, 2008 (09): 6-11.
[13] Han Ning, Zhang Zhijie, Yin Wuliang. Design of rotating magnetic field array sensor and simulation study of pipeline defects[J]. Instrument Technology and Sensors, 2019 (12): 17-22.
[14] Yang Binfeng, Luo Feilu, Cao Xiongheng, Xu Xiaojie. Application research of pulsed eddy current corrosion imaging array probe[J]. Journal of Sensing Technology, 2005 (01): 112-115.
[15] Olivera, J.; Aparicio, S.; Hernández, M.G.; Zhukov, A.; Varga, R.; Campusano, M.; Echavarria, E.; Anaya Velayos, J.J. Microwire-Based Sensor Array for Measuring Wheel Loads of Vehicles. Sensors 2019, 19, 4658.
[16] Apicella, V.; Clemente, C.S.; Davino, D.; Leone, D.; Visone, C. Review of Modeling and Control of Magnetostrictive Actuators. Actuators 2019, 8, 45.

Gear tooth breakage analysis based on multi-order modulated sideband RMS trending method

Jian'gang Chen[1,2], Zhi Luo[1,2,*], Ronggen Wu[1,2] and Haiyang Cai[1,2]

[1] Zhejiang Province Key Laboratory of Wind Power Technology, Hangzhou, Zhejiang,310000, China

[2] Zhejiang Windey Co., Ltd, Hangzhou, Zhejiang,310000, China

[*]Corresponding author's e-mail: luoz@chinawindey.net

Abstract. The fault characteristic signal energy for early gear tooth breakage is relatively weak and easily drowned by other signals, which is not conducive to the study of the fault development stage. A multi-order modulated sideband RMS (Root Mean Square) trend analysis method is proposed to analyse the development trend of the broken gear fault characteristics. By using this method to analyse gear breakage faults, the multi-order modulated sideband RMS trend analysis method can effectively determine the fault deterioration and fault stabilisation stages.

1. Introduction

The current use of modulated edge frequency characteristics to diagnose gearbox faults is mainly from the perspective of fault characterization to determine the type and severity of faults [1-4]. The edge frequency analysis method proposed by Chen Yonghui can locate the location and cause of the gear fault diagnosis [5]; Grabill P used the edge frequency parameter SLF to be able to identify the fault and normal state of the planetary wheel of the Black Hawk under the laboratory [6]; Kang Chuanzhang systematically analysed the causes of the modulated edge frequency signal formation from a theoretical perspective [7]. Few, however, have analysed the stages of development of gear breakage faults from the perspective of actual gearbox operating conditions. Combined with the important characterisation of modulated sidebands and RMS trend analysis of gear breakage faults, the "multi-order modulated sideband RMS trend method" is proposed to study the development trend of gear breakage faults, which has certain engineering significance for the fault diagnosis of gear breakage.

2. Gear fault diagnosis

The main types of gear failures are: pitting, shedding, cracking and broken teeth [8]. Broken teeth are late stage failures compared to other types of failures and need to be dealt with immediately. Therefore, the study of the development trend of gear tooth breakage faults is of great importance. After a broken tooth in one of the gear pairs, the time domain diagram exists for the rotational frequency of the broken gear and the meshing frequency shock of the gear pair. The above characteristics of the time domain and spectral diagrams are important fault characteristics of broken gears.

2.1. Full band, meshing frequency and modulation sideband combination RMS trend analysis

The object of the RMS trend analysis is the continuous vibration acceleration signals for the period before, during and after the failure of a pair of broken gear components. In chronological order, the trend of the RMS values of the vibration signals are plotted to analyse the vibration development pattern of the faulty component. The actual complex gearbox condition, the vibration acceleration signal contains all levels of gear meshing frequency, each gear rotation frequency, bearing failure frequency and random noise, etc. The full band RMS trend analysis method cannot effectively display the RMS trend for non-major faults that are easily "flooded" [9]. For example, bearing faults or broken tooth shocks are easily "swamped" by gear meshing and engagement frequency faults, so full band RMS trend analysis is difficult to show the trend of broken teeth.

To address the limitations of the full band RMS trend analysis method, a combined multi-order meshing frequency and modulation sideband RMS trend analysis method is proposed. The method steps are simplified as follows.

1) Selection of vibration acceleration signals in a certain speed range.
2) Calculating the rotational frequency and gear mesh frequency of the faulty gear.
3) To obtain the combined signal of the multi-order gear mesh and its rotational frequency sidebands of the faulty gear pair by band-pass filtering the full frequency band signal.
4) Calculate the RMS value of the combined multi-order gear mesh and sideband signal and plot the timing of the RMS trend.

The full band RMS trend analysis method and the multi-order meshing frequency and modulation sideband combination RMS trend analysis method were used to analyse the gearbox vibration acceleration signal of the 2-stage shaft system pinion broken teeth. 3-stage speed increase gearbox transmission schematic diagram is shown in Figure 1. The rotor Rotor rotates and drives the 1-stage planetary gear system PLS Sun Z3, which in turn drives the 1-stage shaft system to rotate the Z4 gear, the Z5 gear meshes with the Z4 gear and drives the 2-stage shaft system SRS1, the Z6 gear meshes with the Z7 gear and drives the 3-stage shaft system SRS2, and finally drives the generator Generator.

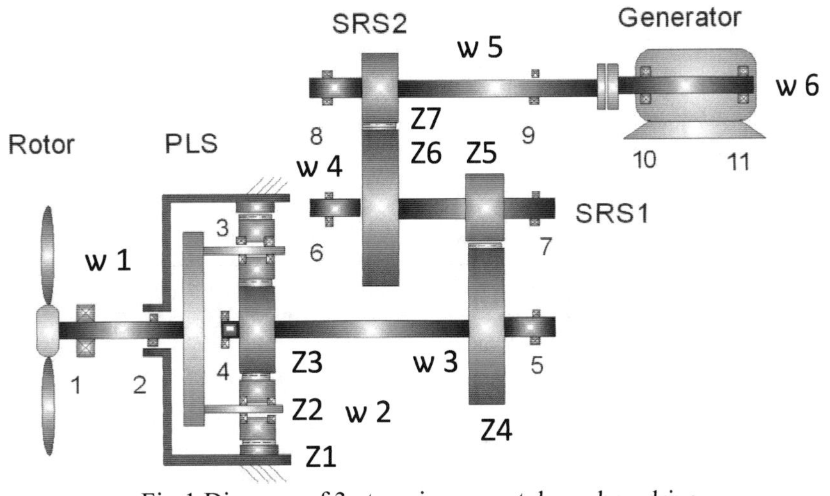

Fig.1 Diagram of 3-stage incremental gearbox drive

Information on key drive chain parameters and vibration acceleration data to be analysed for the 3-stage incremental gearbox are summarised in Table 1, with the broken tooth fault gear being the Z5 gear of the 2nd shaft train.

The acceleration data for the gearbox 3-stage shaft system in the speed range of 1650-1750 rpm for the period from October 1, 2020 to January 15, 2021 was selected for analysis by considering the following five factors.

1) the rated speed of the high-speed shaft of the gearbox is 1800 rpm.
2) the RMS value indicates, to a certain extent, the amplitude of the vibration at the location of the sensor measurement point.
3) the higher the speed the higher the vibration amplitude of the gearbox.

4) the actual speed of the gearbox is combined to ensure adequate analysis data.
5) The fault was found on 20 November 2020 and repaired on 8 December 2020.

Tab.1 Information On Drive Chain Parameters And Data To Be Analysed

Item	Content
Location of sensor measurement points	3-stage shaft system output shaft radial
Date of data collection	2020.10.01-2021.01.15
Sampling frequency (Hz)	25600Hz
Individual data length (points)	102400
Total amount of data (points)	310*102400
R - 3-stage shaft system speed (rpm)	1650-1750
Z7-Number of gear teeth	24
Z6 - number of gear teeth	125
Z5 - Number of teeth (broken gears)	21

The relevant engagement frequencies and frequency of rotation are calculated as shown below.

f_{m1} : The 1st order meshing frequency (Hz) of the Z4 / Z5 gears is given in equation (1).

$$f_{m1} = \left(\frac{Z_7 * R}{60}\right) * \left(\frac{Z_5}{Z_6}\right) \tag{1}$$

f_{band}: Z5 gear rotational frequency sidebands (Hz), calculated as in equation (2).

$$f_{band} = \left(\frac{Z_7 * R}{60}\right) * \left(\frac{1}{Z_6}\right) \tag{2}$$

f_{m2} : Z4 and Z5 gears 2nd order meshing frequency (Hz) is given in equation (3).

f_{m3} : Z4 and Z5 gears 3th order meshing frequency (Hz) is given in equation (3).

$$f_{m1} = \frac{f_{m2}}{2} = \frac{f_{m3}}{3} \tag{3}$$

f_{M1} : The 1st order meshing frequency (Hz) of the Z4 / Z5 gears is given in equation (1).

$$f_{M1} = \frac{Z_7 * R}{60} \tag{4}$$

Taking the 3rd stage shaft system speed R=1700 rpm, f_{m1}=114.20Hz, f_{m2}=228.40Hz, f_{m3}=342.60Hz f_{band}=5.44 and f_{M1}=680.00Hz.

As the broken tooth fault gear is the 2nd axis system Z5 gear, characterised in the spectrum as Z4 and Z5 gear vice multiple order meshing frequency and Z5 gear rotational frequency signal modulation phenomenon, all the RMS trend analysis of the first 3 order meshing frequency of the 2nd level axis system and Z5 gear rotational frequency sideband combined signal. the RMS trend is shown in Figure 2.

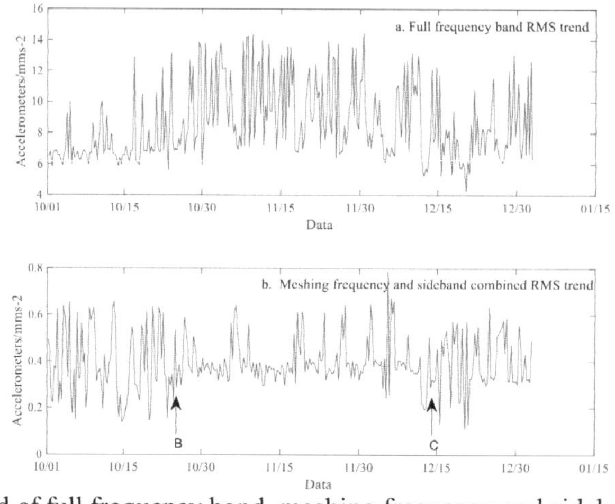

Fig.2 RMS trend of full frequency band, meshing frequency and sideband combination

Figure 2.a shows the RMS trend for the full band signal and Figure 2.b shows the RMS trend for the combination of the meshing frequency and modulation sidebands. Figure 2.a shows a slight overall increase in RMS values from 10/30 onwards, with an unclear stage of gear fault development; the bottom of RMS increases overall in the BC interval in Figure 2.b.

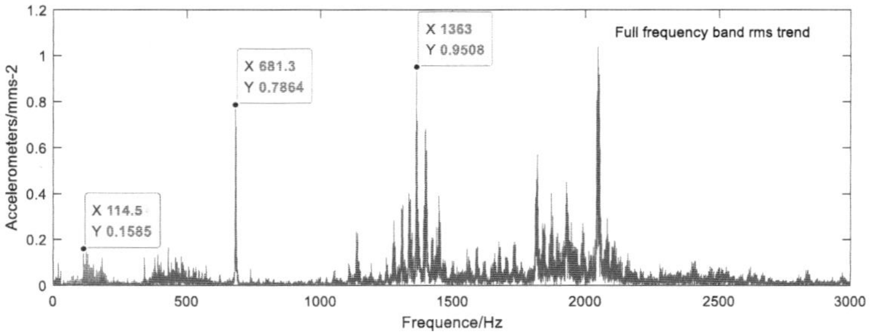

Fig.3 10/30 full band signal spectrogram

For the RMS trend in Figure 2, the 10/30 raw vibration acceleration signal from the BC section was analysed in Spectrum Figure 3, showing that the first 3 steps of the 3-stage shaft system (681Hz) signal is the dominant component in the 0-3000Hz band compared to the first 3 steps of the 2-stage shaft system (114Hz), "drowning out" the The characteristic signal of the Z5 gear of the 2-stage shaft system [9]. The vibration generated by the Z5 gear breakage also contributes to the increase in amplitude of the drive components of the 3-stage shaft system [10], resulting in a slight overall increase in RMS in the BC region in Fig. 2.a, but does not effectively characterise the breakage failure trend. Figure 2.b shows the RMS trend of the combined signal of the first 3rd order engagement frequency engagement frequency and its Z5 rpm sideband for the 2-stage shaft system. After removing the influence of the 3-stage shaft system, the fault occurrence interval of the broken tooth can be found, but the deterioration trend of the fault is not obvious.

2.2. Multi-order modulation sideband RMS trend analysis

In response to the shortcomings of the combined RMS trend analysis method of meshing frequency and modulation sidebands, which can find the fault occurrence interval of broken teeth, but the deterioration trend of the fault is not obvious, the RMS trend analysis method of rotating frequency sidebands is proposed to analyse the trend of the fault. The method steps are simplified as follows.

1) Selection of vibration acceleration signals for a certain speed range.
2) Calculating the rotational frequency and gear mesh frequency of the broken gear.
3) To obtain the combined signal of the multi-stage gear mesh and its rotational frequency sidebands of the faulty gear pair by band-pass filtering the full frequency band signal.
4) Removing the 2-stage shaft system meshing frequency and its harmonic signal to obtain the multi-order stage shaft system faulty gear rotational frequency sideband signal.
5) Calculate the RMS value of the multi-stage 2-stage faulty gears and plot the trend.

The results of the RMS trend analysis of the first 3 steps of the 2-stage shaft system Z5 gear rotation frequency sidebands are shown in Figure 4. The RMS value gradually increased in the AB zone; the RMS value remained stable in the BC zone; the RMS value decreased linearly in the CD zone; the RMS value remained stable after point D and was basically the same in the zone before point A. It is basically consistent with the time point of fault discovery on November 20, 2020 and fault repair on December 8, 2020.

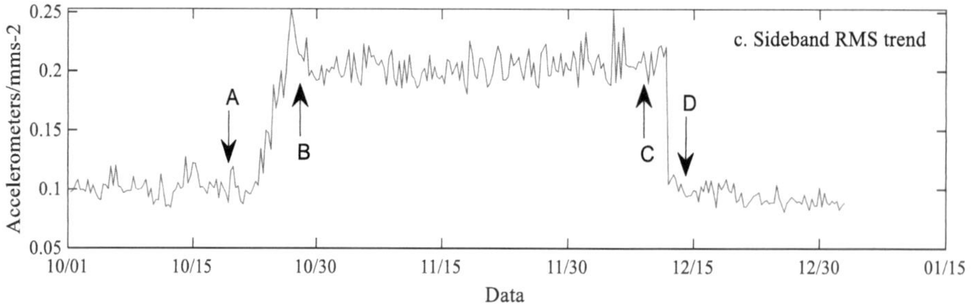

Fig4 Multi-order modulation sideband RMS trend

3. Conclusion
The multi-order modulated sideband RMS trend analysis method is more effective in identifying the deterioration and stabilisation phases of gear breakage faults than the combined multi-order full band and meshing frequency and rotational frequency sideband methods.

References
[1] He Jun. Research on vibration characteristics analysis and intelligent fault diagnosis method of gearbox [D]. Zhejiang University,2018:7-15
[2] Lei Yaguo, He Zhengjia, Lin Jing, Han Dong, Kong Detong. Research progress of fault diagnosis technology for planetary gearboxes[J]. Journal of Mechanical Engineering,2011,47(19):59-67.
[3] Ding Kang. Practical techniques for fault diagnosis of gears and gearboxes [M]. Machinery Industry Press, 2005:12-18
[4] Randall R B.Vibration-based Condition Monitoring: Industrial, Aerospace and Automotive Applications[M]. John Wiley & Sons, 2011:167-169
[5] Chen YH, Jiang X, Li HH, Zhang XL. Gear fault diagnosis based on edge frequency analysis[J]. Mechanical Transmission,2010,34(06):58-61.
[6] GRABILL P. Vibration monitoring of UH-60A main transmission planetary carrierfault[C].59th Annual Forum of American HelicopterSociety.May 6-8.2003.Phoenix.Arizona.USA. 2003:1-11.
[7] Kang Chuanzhang, Chen Bingkui, Zhang Jing, Tan Rulong. Study on the side band of vibration signal of gear system and its application[J]. Me-chanical Transmission,2013,37(04):41-43+76.
[8] Liang X , Zuo M J , Feng Z . Dynamic modeling of gearbox faults: A review[J]. Mechanical Systems and Signal Processing, 2018, 98(jan.1):852-876.
[9] Liu, Tao, Chen, Jin, Dong, Guangming. Research on rolling bearing fault diagnosis based on frequency band entropy[J]. Vibration and shock,2014,33(01):77-80.
[10] Li RF, Wang JJ. Gear system dynamics: vibration, shock, noise [M]. Science Press, 1997. 156-160

Different improvement designs of conventional comparator

Xinmiao Fu[1, a, *, †], **Miao He**[2, b, *, †], **Yuan Zhang**[2, c, *, †]

[1] Electrical and Electronic Engineering School, Tiangong University, Tianjin, China

[2] Information Science and Technology School, Northwest University, Xi'an, China

[*]Corresponding author email: [a]fuxinmiao0812@163.com, [b]2018117355@stumail.nwu.edu.cn, [c]2018117404@stumail.nwu.edu.cn

[†]These authors contributed equally

Abstract. Comparators play a significant role in the semiconductor industry and have become indispensable in the design of ADC. The delay and energy consumption are two important indicators of the comparator. Many designs have been made to reduce the delay and energy consumption, such as separated gata-biasing cross-coupled transistors for a new latching stage, and pMOS is used to replace nMOS in comparators. This paper analyzes the working principle of the proposed comparators designed for different needs reported on different papers. It compares their simulation results about key data such as delay and energy consumption.

1. Introduction

Comparator is at the heart of most integrated circuits. The delay and energy consumption are two important indicators of the comparator. A lot of works have been done to reduce delay and energy consumption, such as the charge-steering system [1], approaches employing body-driven transistors [2], and supply promotion means [3]. Besides, many other methods were proposed to reduce them after that. The representative common-mode insensitive dynamic comparator is described in [4]. Until now, more methods are proposed [5].

This paper has studied five representative papers and summarized the methods or techniques they used to decrease delay and energy consumption. This key article is well established in detail. Section II depicts the system architecture of the Conventional Dynamic Comparator. The methods and effects used in each paper are discussed in Section III, and Section IV offers the experimental findings of each paper. Lastly, Section V shows the reference.

2. Conventional comparator

Fig.1 demonstrates the traditional dynamic comparator, which is comprised of a preamplifier and a latch stage. At the reset phase, resets the 1st and 2nd stages, respectively, to GND and V_{DD}. Then, clk is on and off to start the regeneration phase. The opportunistic capacitors of the preamplifier's output nodes start to be charged differently in the light of the input differential signal. However, with the development of science and technology, the conventional comparator has been unable to meet the needs of the current circuit, so a variety of innovations is carried out based on it to achieve better performance.

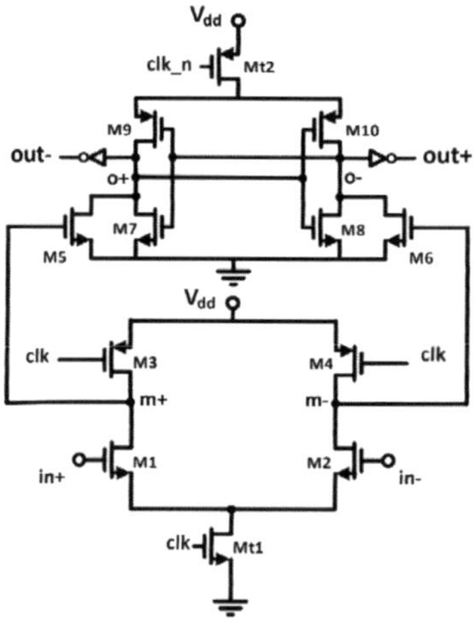

Fig.1 Conventional double tail comparator

3. A novel design for comparator

3.1. Energy-Efficient Comparator with Dynamic FIA

To present an energy-efficient comparator, Xiyuan Tang et al. proposed two methods focusing on preamplifier improvement design: replacing preamplifier with CMOS DB integration [6]. Another is using a floating inverter with a Reservoir capacitor as the preamplifier. By comparing Gain, Noise, and energy efficiency, it is obvious that the proposed circuit is better.

For CMOS DB Integration Pre-Amplifier, a proposed CMOS DB incorporation model is shown in Fig.2.

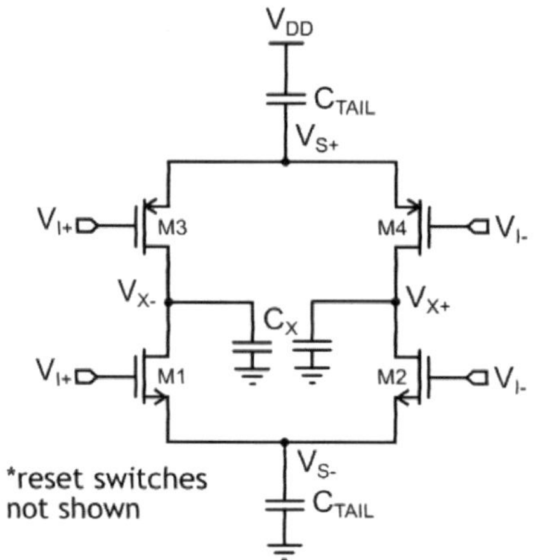

Fig.2 Proposed CMOS DB integration

V_{S-} in the base source node raises during integration phase, while the upper one V_{S+} decreases. It makes a decrease in overdrive voltage of input pairs of nMOS and pMOS. The preamplifier (Gm) transconductance is the sum of the transconduction of the nMOS and the pMOS input couples. As demonstrated in Fig.3, as a result, the integration increases the average G_m/I_D of the preamplifier by more than 2.5 times, and energy efficiency also experiences an increase. And its input common-voltage variations are indicated in Fig.4(a).

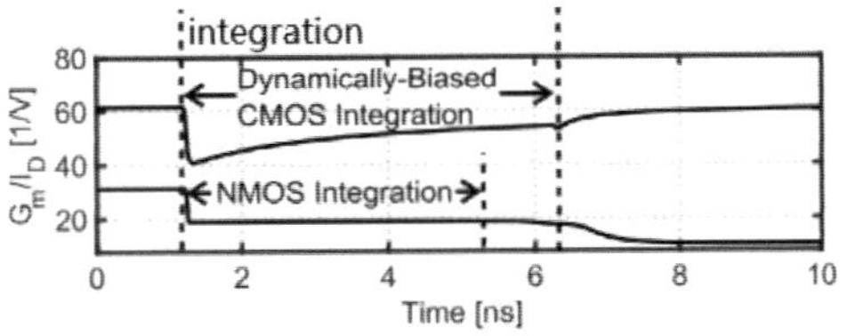

Fig.3 G_m/I_D comparison

For Floating Inverter Preamplifier with Reservoir Capacitor, to decrease the size of capacitance and to provide an isolated voltage domain for preamplifier, this article merges the two tail C_{TAIL} into single floating reservoir capacitor C_{RES} and this operation shown in Fig.5 [7].

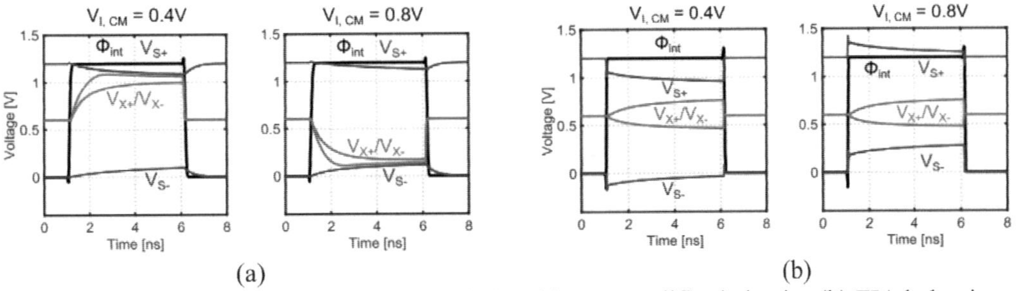

Fig.4 Input common-mode voltage variation (a) Pre-amplifier behavior (b) FIA behavior

This new design guarantees a persistent output common-mode potential. It can make the input-output current of C_{RES} keep until the isolated voltage will be automatically changed with the variation of input common-mode potential. And its input common-voltage fluctuation is displayed in Fig.4(b).

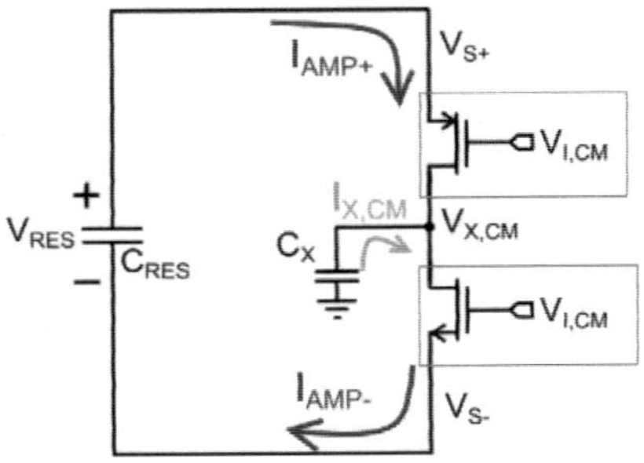

Fig.5 Common-mode comparable of FIA control

Referred to noise analysis in paper [8] is Noise analysis, the input-referred noise of the traditional nMOS combination pre-amplifier can be expressed as

$$\sigma_{in,SA}^2(T_{INT}) = \frac{2nkT}{V_{THN} \cdot C_X} \cdot \frac{I_D}{g_m} \tag{1}$$

Where T_{INT} is the integration phase.

The input-referred noise of DB comparator at the end of T_{INT} is

$$\sigma_{in,DB}^2(T_{INT}) = \frac{2nkT}{C_p \cdot \Delta V_{X,CM}(T_{INT})} \cdot \frac{I_D}{g_m} \tag{2}$$

And it of FIA is calculated as

$$\sigma_{in,FIA}^{2}(T_{INT}) = \frac{2nkT}{C_{RES} \cdot \Delta V_S(T_{INT})} \cdot \frac{I_D}{G_m} \tag{3}$$

A new variate is built to weigh energy consumption and noise, a figure of merit, which is briefly written as FoM. FoM is characterized as the creation of expenditure and the input-referred noise power.

$$\text{FoM} = \text{Energy} \cdot (\text{Noise Power})$$

So, it is obvious that the lower FoM, the higher the comparator efficiency. And energy efficiency of the comparator can be estimated by that of the pre-amplifier [9].

FoM of DB integration is

$$FoM_{DB} = 4nkT \cdot V_{DD} \cdot \frac{I_D}{g_m} \tag{4}$$

FoM of FIA is

$$FoM_{FIA} = 4nkT \cdot V_{DD} \cdot \frac{I_D}{G_m} \tag{5}$$

Comparing the noise, FoM, and inputting common voltage variations in Fig.4(a) and (b), it is no doubt that FIA provides higher integration gain, and it also can release the input-referred noise of latch. As a result, the FIA-based comparator is the best choice.

By observing Fig. 6, it is suitable to find a C_{RES} value that can guarantee a smaller FoM and a smaller noise. So a 2-pF C_{RES} ensures energy efficiency.

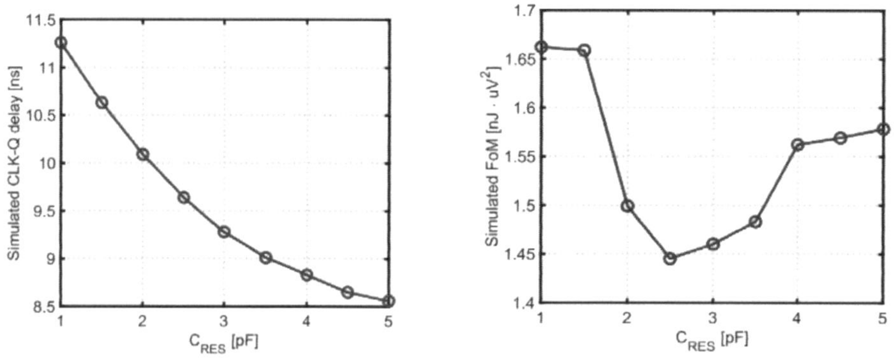

(a) (b)

Fig.6 Simulation of (a) CLK-Q gap with a 1-mV differential input and (b) Energy effectiveness of the suggested comparator vs. CRES value

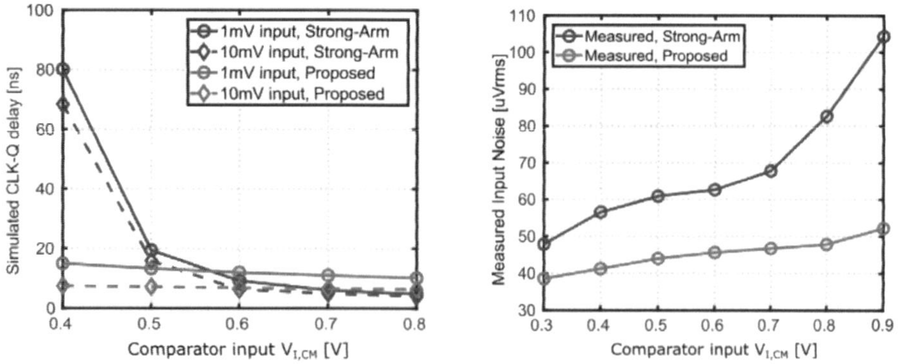

(a) (b)

Fig.7 (a) Modeled CLK-Q interval vs. input common-mode potential. (b) Assessed input-referred noise vs. input common-mode potential.

In Fig.7(a), obviously, with the change of comparator input, the delay of the FIA comparator hardly changes, while the Strong Arm latch changes significantly. It attested its callousness to input common-mode variation.

In Fig.7(b), it is easy to find that for the suggested FIA comparator, the input noise basically keeps in still, while for SA latch, it changes obviously with comparator input.

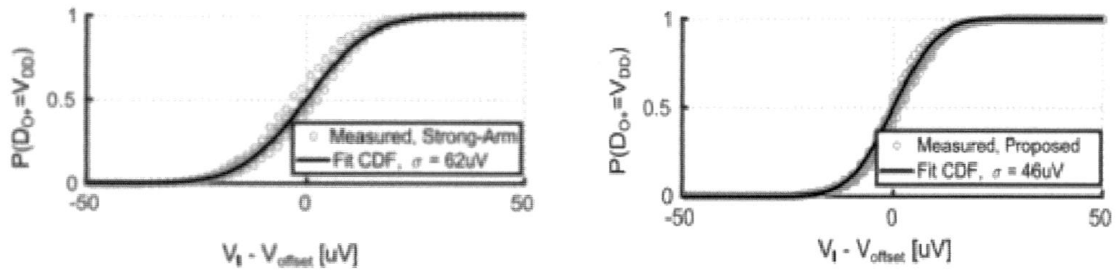

Fig.8 Quantified accumulative likelihood density distribution and obey Gaussian distribution (a) SA latch comparator (b) The anticipated comparator

By comparing the measured rms of the proposed comparator and SA latch comparator, the proposed FIA comparator has smaller rms, so it is more stable (Fig.8).

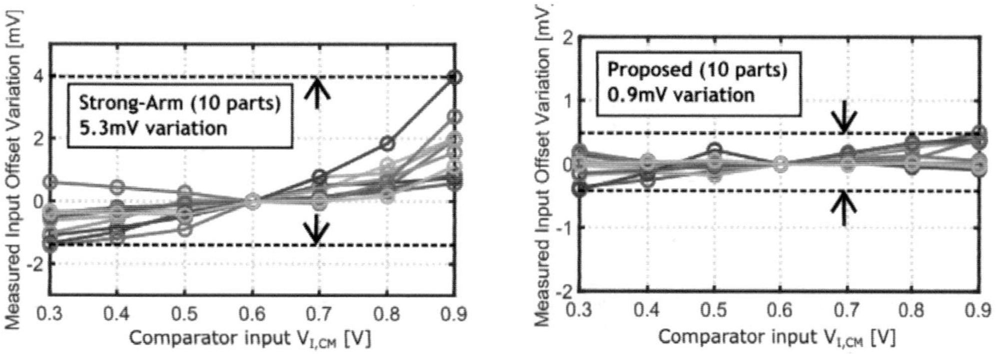

Fig.9 Measured ten-part input-referred offset vs. input CM potential for SA latch and suggested comparator.

Fig.9 shows the relationship between input-referred offset investigation and the input common-mode potential. It's quite clear that the proposed one diminishes the deviation by larger than 5 times.

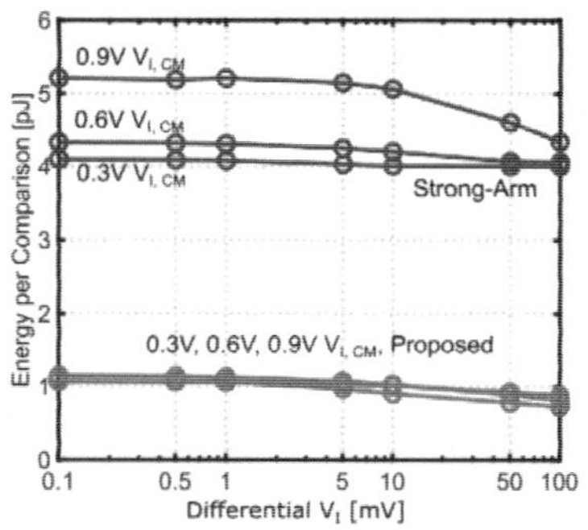

Fig.10 Determined energy expenditure vs. input potential for the SA latch and suggested comparator

Because of the insensitivity of the input common-mode, the proposed comparator has four times less power than the SA latch [10].

3.2. A low-noise PD-based comparator

Chin-Yu Lin et al. proposed a new low-noise phase-detector-base comparator, which enables the reduction of the thermal-induced noise and the possibility of meta stability.

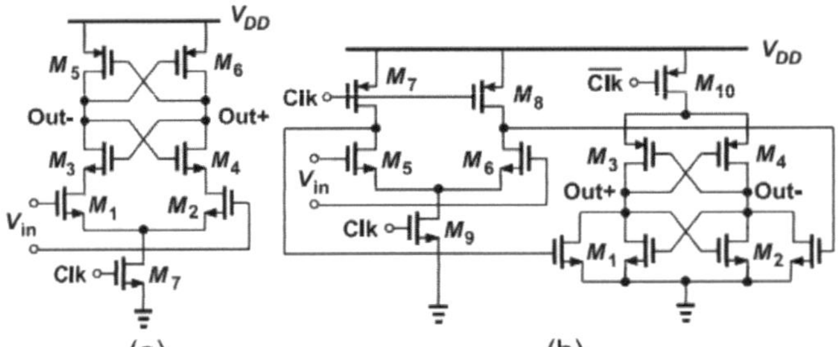

Fig.11 (a) Single-stage comparator (b) Double-tailed comparator

Compared with the traditional single-stage comparator [11] exhibited in Fig.11(a), one double-tailed comparator in Fig.11(b) improves speed, noise, and offset. So, based on these, the PD-based comparator is proposed in Fig.12.

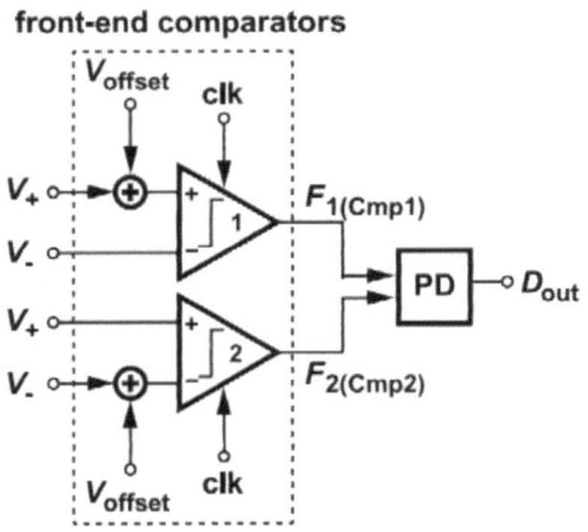

Fig.12 Configuration of comparator in line with PD

This new comparator consists of 2 forward-end dynamical comparators and a back-end phase sensor, and the front stage is made by 2 traditional comparators, demonstrated in Fig.13.

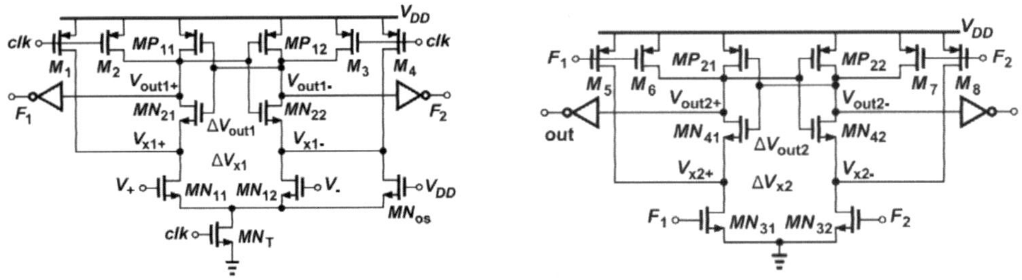

(a) (b)

Fig.13 (a) Illustration of a comparator employed in front-end (b) The dynamic phase detector (PD)

Unlike traditional ones, the proposed comparator can guarantee a shorter delay, which means less occurrence of meta stability (Fig.14).

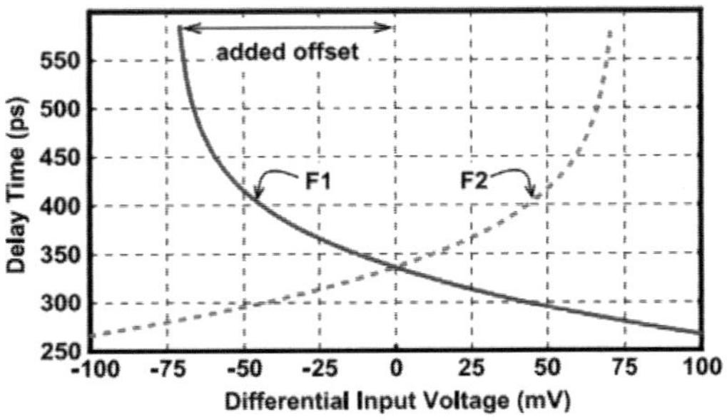

Fig.14 The computer-generated delay time of front-end comparators

3.2.1. Noise and speed analysis

For the two front-end comparators, the output time-domain noise is

$$J^2_{out,1-stg} = \frac{32kT\gamma\tau_{R1}^2\tau_{s1}}{3C_{x1}V_{OS}^2(t_1-t_0)} + \frac{32kT\gamma\tau_{R1}^2\tau_{s1}^2\tau_{s2}}{3C_{x1}V_{OS}^2(t_1-t_0)^3} + \frac{48kT\gamma\tau_{R1}^2\tau_{s1}^2\tau_{s2}^2}{C_{out1}V_{OS}^2(t_1-t_0)^4} + \frac{32kT\gamma\tau_{R1}^2\tau_{s1}^2}{C_{x1}V_{OS}^2(t_1-t_0)^2} \quad (6)$$

For the PD input-referred noise,

$$J^2_{in,2-stg} \cong \frac{16kT\gamma\Delta t_s}{g_{m31}V_{OV}^2} + \frac{8kT\gamma C_{out}}{g_{m31}^2V_{OV}^2} + \frac{16kT\gamma C_{out}}{g_{m31}^2V_{OV}^2} \quad (7)$$

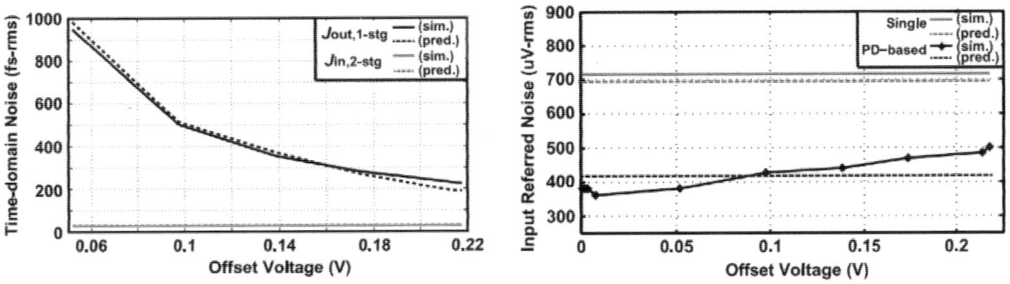

(a) (b)

Fig.15 (a) Replicated and projected time-domain noise at front-end output (b) Replicated and projected input-referred noise for single-stage and PD-based comparators

So, it is easy to say that the noise which is caused by PD is indeed reduced than the front-end circuit, so the PD noise can be ignored while considering the input-referred noise (Fig.15).

As a result, through the above, it is obvious that the anticipated comparator corresponds to a lower noise level than the conventional one.

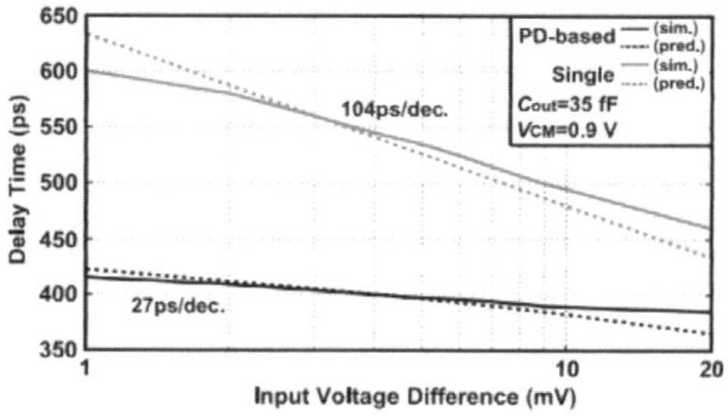

Fig.16 Modeled and projected delays for comparators

To consider the key factor of speed, the most important is the propagation delay. The delay of the front-end comparator is

$$t_{d,1-stg} \cong \frac{2C_{out1}V_{thp}}{I_{MNT}} + \tau_{R1}\ln\frac{V_{mid}\tau_{s1}\tau_{s2}I_{MNT}^2}{2C_{out1}^2V_{thp}^2(\Delta V + V_{OS})} \quad (8)$$

And the delay of PD is

$$t_{d,2-stg} \cong \frac{2C_{out}V_{thp}}{g_{m31}V_{OV}} + \tau_{R2}\ln\frac{2C_{out}V_{mid}}{(g_{m31}V_{OV})(G_{VT}\Delta V)} \quad (9)$$

Based on the equations above and Fig.16, PD can regenerate fast with a small input, and the proposed comparator is improved in the speed within a bounded input range [13].

3.2.2. Measurement result

In Fig.17, the proposed circuit obtains smaller noise.

And from Fig.18(a), the noise perturbation causes little effect on the proposed one.

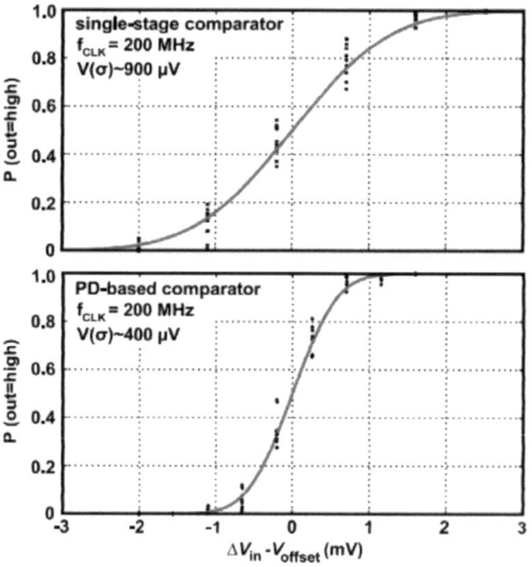

Fig.17 Assessed input-referred noise for comparators

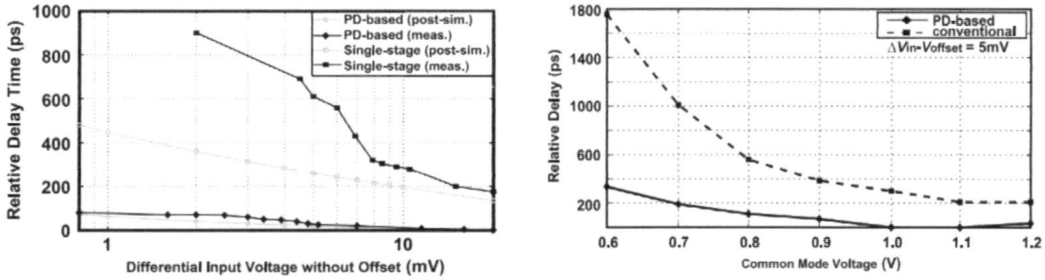

(a) (b)

Fig.18 (a) Determined delay concerning the input potential difference (b) Measured delay in relation to the common-mode voltages

Fig.18(b) shows the PD-based comparator's relative delay in different common-mode-voltage compared with the normal method. This one has a more stable delay.

3.3. Dynamic comparator with a transconductance improved latching stage

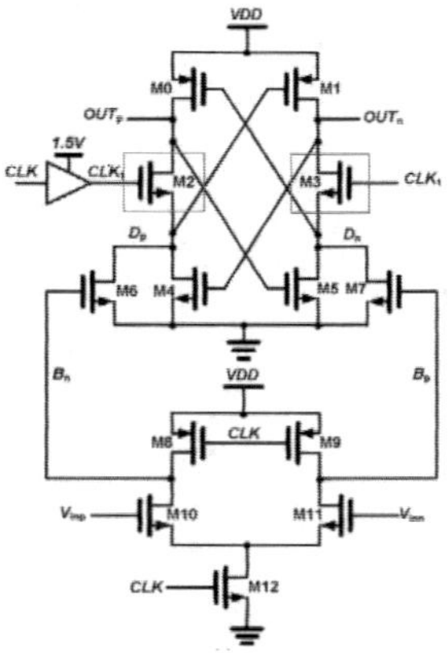

Fig.19 Scheme of the proposed dynamic comparator

The transistors in red in Fig.19 are the optimization part of the proposed design.
The delay of the proposed dynamic comparator is expressed as:

$$t_{delay} = \frac{V_{THN} C_{OUT} \cdot 2\mu_p C_{OX} \frac{W_1}{L_1}}{g_{m,eff}^2} + \frac{C_{OUT}}{g_{m,eff}} \cdot \ln \frac{V_{DD}}{\Delta V_0} \tag{10}$$

The effective total transconductance of the latch is given by

$$g_{m,eff} = g_{mm} + g_{mp} = \mu_n C_{OX} \left(\frac{W}{L}\right) V_{dsn} + \mu_p C_{OX} \left(\frac{W}{L}\right)_P |V_{dsp}| \tag{11}$$

Eq.(10) shows that gm,eff is a highly useful means to decrease the delay. The bigger the gm,eff is, the smaller the delay is. And Eq.(11) shows the width of switches M_2 and M_3 determine the magnitude of gm,eff. According to Fig.20, when the width of M_2 and M_3 is larger than 1 μm, boosting the width cannot help to reduce the delay significantly, unfortunately, with great power expenditure. Therefore, the proposed design sets the width of M_2 and M_3 to 1 μm.

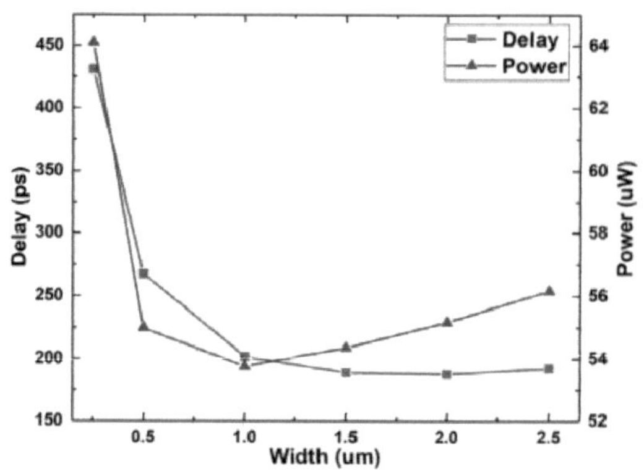

Fig.20 Modeled delay and power expenditure of the suggested comparator vs. the width of M_2 and M_3

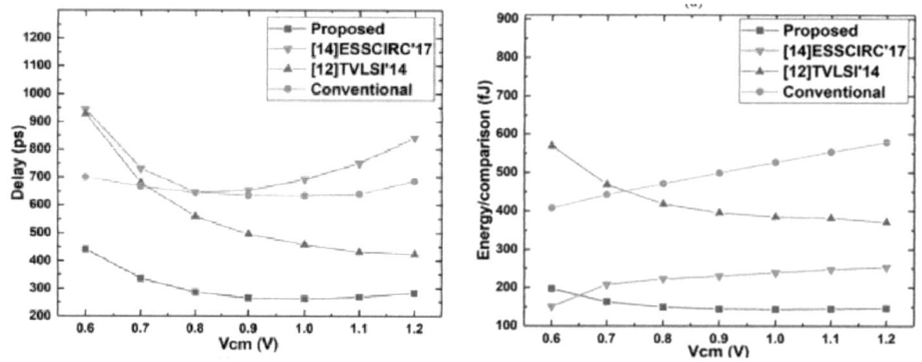

(a) (b)

Fig.21 (a) Post-layout simulated delay and (b) energy consumption versus V_{cm}

Fig.21(a) and Fig.21(b) show that the delay and energy consumption of the suggested design is all the better than the design before. In the application, we may choose V_{cm} between 0.9V and 1.1V. Because in this range, the delay is small, and the power consumption is not too high.

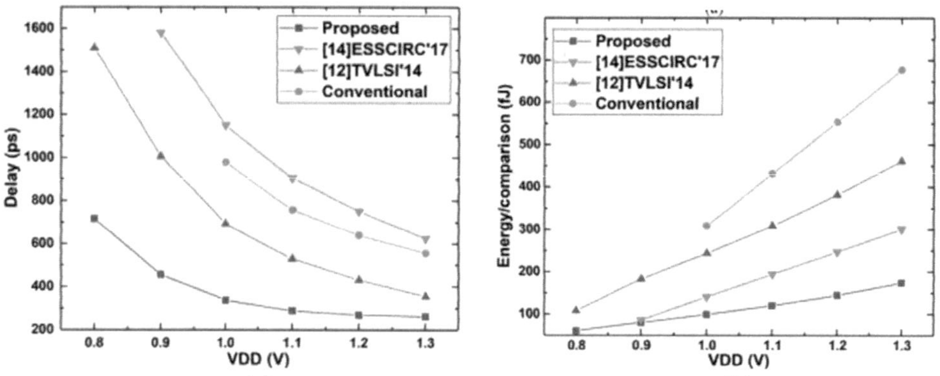

(a) (b)

Fig.22 (a) Post-layout simulated delay and (b) Energy consumption versus V_{DD}

Fig.22(a) and Fig.22(b) show that when V_{id} =50 mV and V_{cm} = V_{DD}–0.1 V, the delay and energy consumption of the suggested design is the least in different V_{DD}. In the application, we choose 1.2V as V_{DD}. Because at this V_{DD}, the delay is small enough, and the energy consumption is not too big.

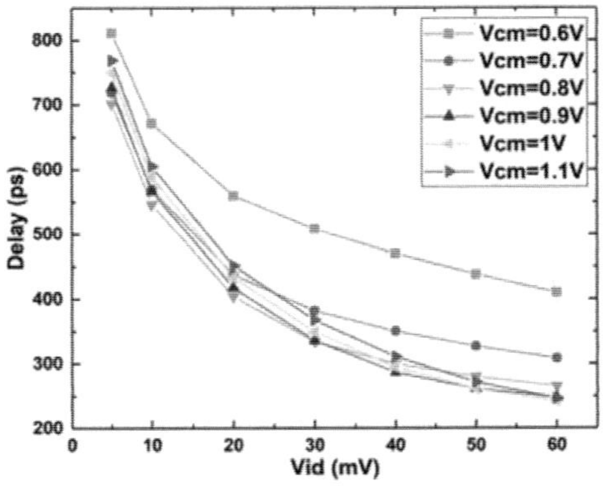

Fig.23 Post-layout simulated delay versus V_{id}

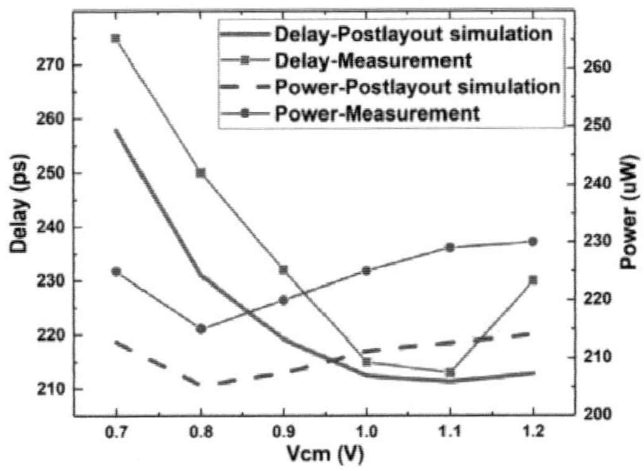

Fig.24 Delay and energy consumption versus V_{cm}

Fig.24 shows the evaluated and modeled delay and energy consumption at 2 GHz and 0.1 V. Their trends are roughly the same, so the test results agree with the simulation results. The numerical difference may be due to measuring instrument accuracy and circuit loss. The measurement energy consumption of the proposed design at 2GHz operating frequency and 1.2V supply voltage is 112.5fJ.

3.4. *Double-Tail comparator*

Samaneh Babayan-Mashadi designs a new dynamic comparator to reduce energy consumption and delay time significantly [14].

When CLK=0, M_{taill1}, and M_{taill2} are off during the reset phase, M_3 and M_4 pull fn and fp nodes to V_{DD}. Consequently, transistors M_{c1} and M_{c2} are cut-off. M_{R1} and M_{R2} reset both latch outputs to the ground.

During the comparison phase, when CLK=V_{DD}, M_{taill1}, and M_{taill2} are on. Transistor M_3 and M_4 turn off. At the beginning of this phase, the control transistor M_{c1} and M_{c2} are still off. fn and fp start to drop with different rates according to input potentials. Supposing V_{INP} >V_{INN}, fn drops faster than fp. The M_{c1} will always pull the fp node back to the V_{DD} and allow fn to be discharged entirely.

Therefore, by time passing △Vfn/fp increases in an exponential. According to Eq.(12) and Eq.(13), V0 and gm are proportional to △Vfn/fp. However, according to Eq.(15), V0 and gm are inversely proportional to delay. In this way, delays will be significantly reduced. Besides, the energy per comparison consumption will be reduced. Because in the proposed design, only node (fn/fp) has to be charged. To overcome stationary energy expenditure, 2 nMOS switches Msw1 and Msw2, are applied below the input transistors, shown in Fig.25(b).

Similar to the △V0 equation of doubled-tail structure, in this comparator, we have:

$$\Delta V_0 = V_{Thn}\frac{\Delta I_{latch}}{I_{B1}} \approx 2V_{Thn}\frac{\Delta I_{latch}}{I_{tail2}} = 2V_{Thn}\frac{g_{mR1,2}}{I_{tail2}}\Delta V_{fn/fp} \tag{12}$$

△Vfn/fp can be calculated by:

$$\Delta V_{fn/fp} = \Delta V_{fn(p)0}\exp\left((A_V - 1)t/\tau\right) \tag{13}$$

△Vfn(p)0 is obtained from:

$$\Delta V_{fn(p)0} = 2\left|V_{Thp}\right|\frac{g_{m1,2}\Delta V_{in}}{I_{tail1}} \tag{14}$$

The overall delay of the proposed comparator is achieved from:

$$t_{delay}=2\frac{V_{Thn}C_{Lout}}{I_{tail2}}+\frac{C_{Lout}}{g_{m,eff}+g_{mr1,2}}\cdot\ln\left(\frac{V_{DD}/2}{\Delta V_0}\right)=2\frac{V_{Thn}C_{Lout}}{I_{tail2}}$$
$$+\frac{C_{Lout}}{g_{m,eff}+g_{mr1,2}}\times\ln\left(\frac{V_{DD}/2}{4V_{Thn}\left|V_{Thp}\right|\frac{g_{mR1,2}}{I_{tail2}}\frac{g_{m1,2}\Delta V_{in}}{I_{tail1}}\exp\left(\frac{G_{m,eff}\cdot t0}{C_{L,fn(p)}}\right)}\right) \quad (15)$$

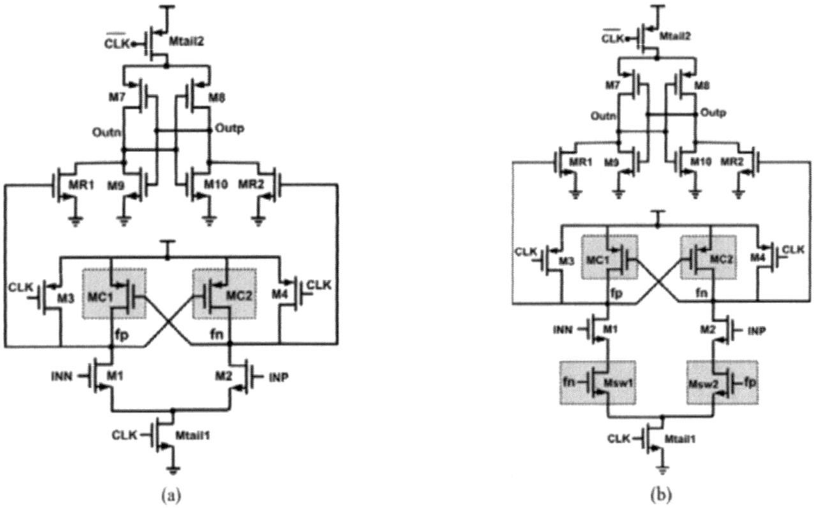

Fig.25 Scheme of the intended dynamic comparator (a) Main idea (b) Final structure

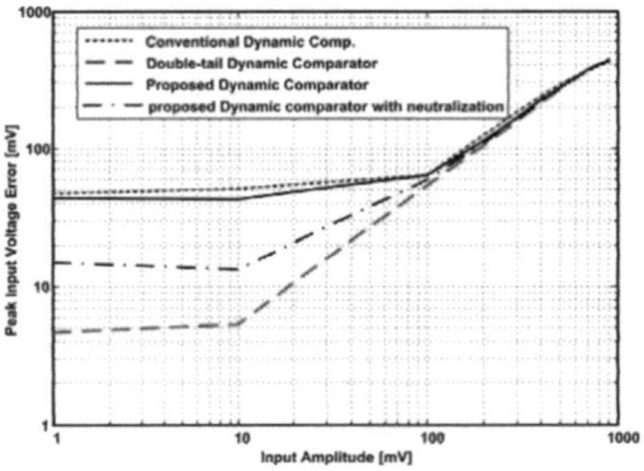

Fig.26 Peak input voltage error due to kickback noise

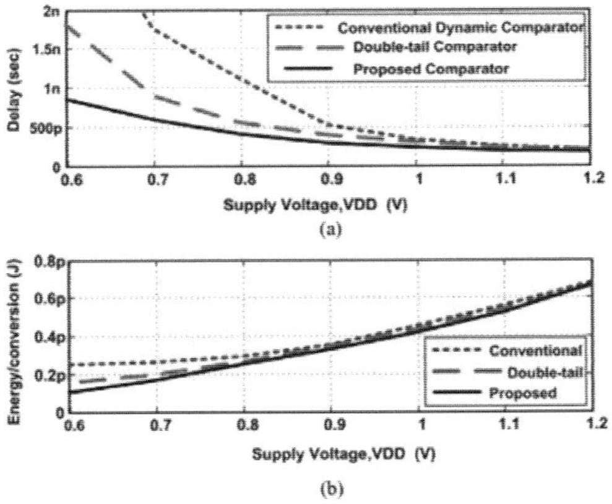

Fig.27 (a) Post-layout modeled delay and (b) Energy per conversion as a function of the supply potential

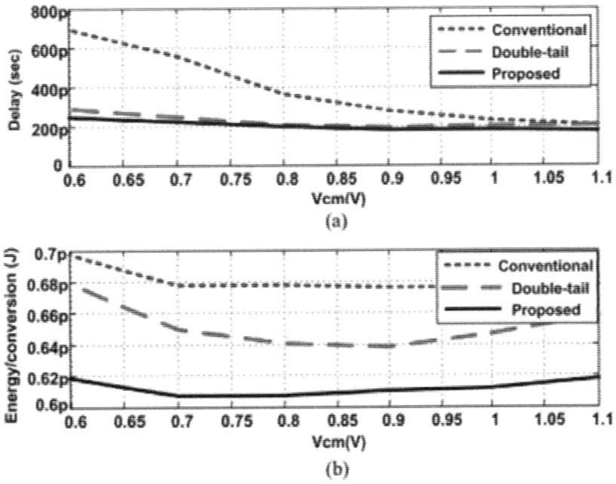

Fig.28 (a) Post layout modeled delay and (b) Energy per conversion as a function of input common-mode potential

Fig.27 reveals the simulated delay and energy consumption of three different comparators in different supply voltages. It is obvious that the proposed comparator has the minimum delay and energy. Besides, when the supply voltage is more than 1V, the delay is nearly stable, but the energy consumption increase quickly. So In practical application, it is reasonable to choose 1V as the supply voltage.

Fig.28 shows the delay and energy expenditure of three unique comparators in different V_{cm}. It is obvious that the suggested comparator has the minimum delay and energy. Besides, when V_{cm} is more than 0.8V, the delay of the suggested comparator is nearly stable, but the energy will increase. So in practical application, it is reasonable to choose 0.8V as the V_{cm}.

This proposed design is simulated in 180 nm CMOS technology. The delay of this comparator is 550 ps, and the energy expenditure of it is 329 μW at 1.2 V and 0.5 GHz.

3.5. A Low-Power High-Speed Comparator

This is a new design that pMOS transistors use at the input of the preamplifier and the latch stage of the comparator proposed by Ata Khorami, which achieves the optimum delay for preamplification and prevent excess energy expenditure [15].

The schematic is displayed in Fig.26(a).

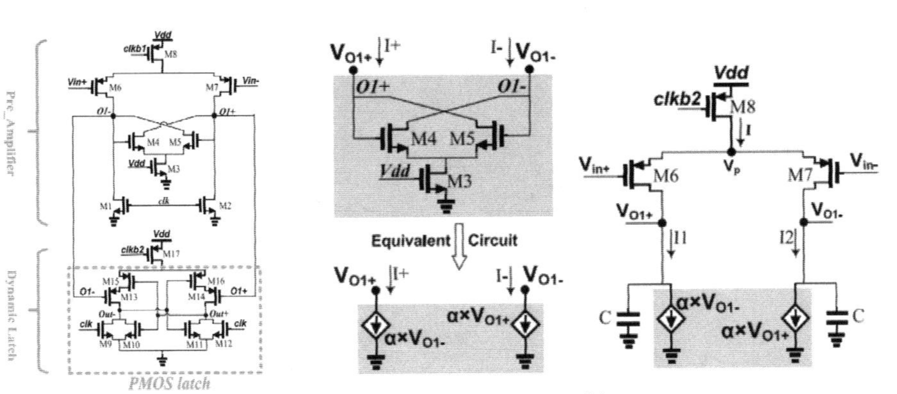

(a) (b)

Fig.29 (a) Circuit of the proposed comparator (b) Equivalent circuit of cross-coupled part and simplified circuit diagram of the preamplifier

 2113 (2021) 012008 doi:10.1088/1742-6596/2113/1/012008

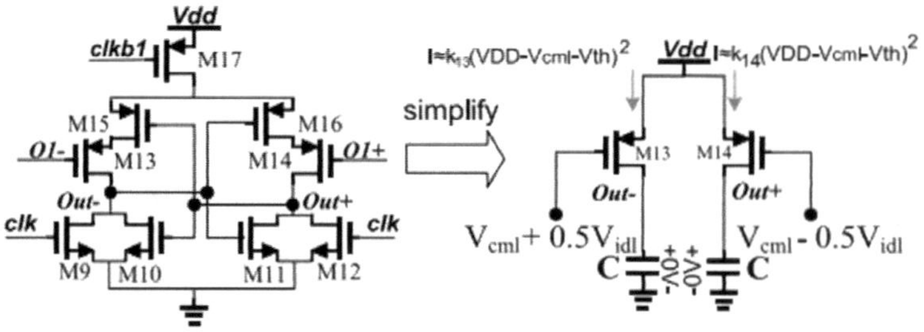

Fig.30 Simplified circuit diagram of the latch right

For the suggested comparator, the latch delay is

$$t_{latch}=\tau_{inv}\times\ln\left(\frac{V_{DD}-GND}{V_{idl}}\right)+\frac{K_{latch}}{(V_{DD}-V_{cml}-V_{thp})^2} \quad (16)$$

V_{cml} is the input common-mode potential, and V_{idl} is the differential voltage of the latch coming from the preamplifier stage[16].

Fig.31 shows the model results and analytical derivations about the delay of the suggested comparator versus input V_{cm} and V_{id}.

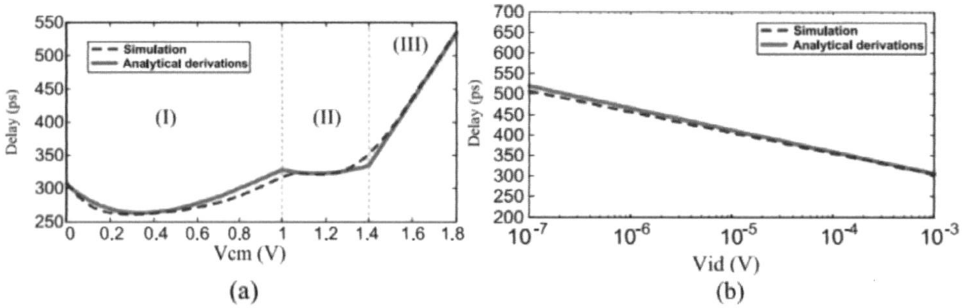

Fig.31 (a) Delay of the suggested comparator vs. input V_{cm} (b) Delay of the suggested comparator vs. input V_{id}

With the increase of V_{cm}, although the variation of delay fluctuated, the overall trend is increasing. And with the increase of Vid, the delay velocity decreases.

Fig.32 shows the simulation results and analytical derivations about the offset voltage of the proposed comparator versus input V_{cm}. With the increase of V_{cm}, the offset voltage is basically stable.

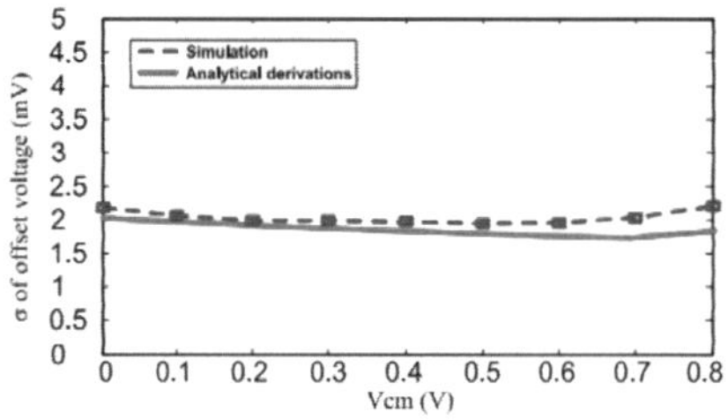

Fig.32 Scheme and analytical model results of offset voltage vs. V_{cm}

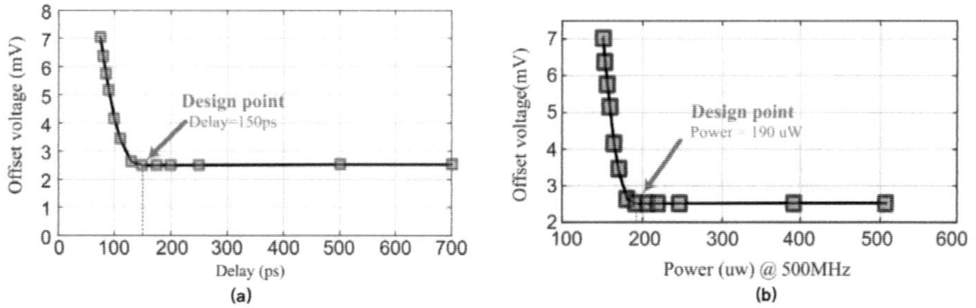

Fig.33 The offset potential of the suggested comparator versus (a) Delay and (b) Power (at 500 MHz) derived from various delay values

Fig.33 shows the offset potential of the suggested comparator vs. tamp and vs. energy expenditure derived from various values of tamp. Evidently, tamp = 150 ps is the optimum design point.

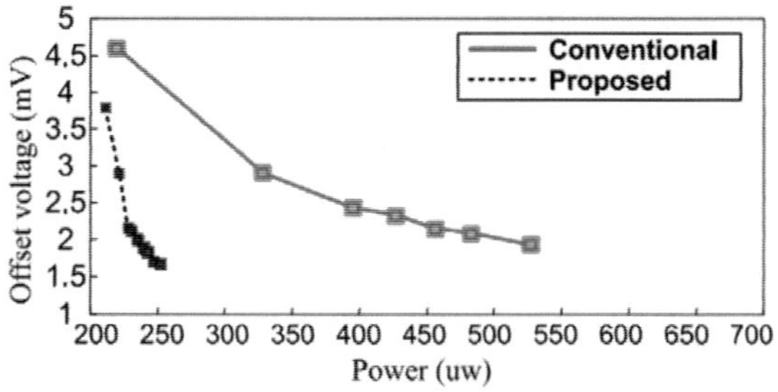

Fig.34 Offset potential vs. energy expenditure for the suggested and the traditional comparators

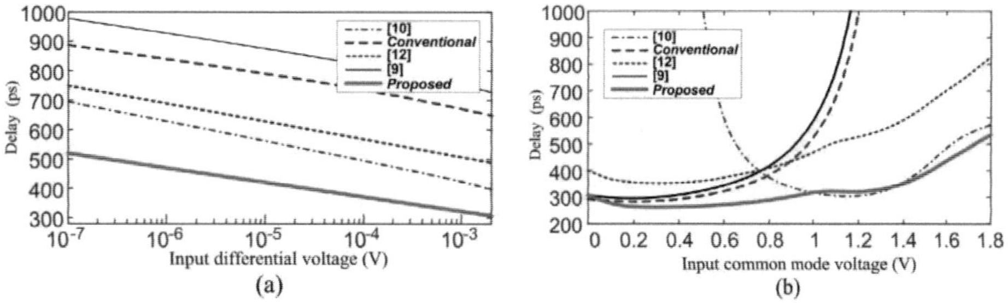

Fig.35 (a) Delay versus V_{id} considering Vcm = 0.7, 1.1 for nMOS input and pMOS input comparators (b) Delay vs. Vcm given V_{id} = 1 mV

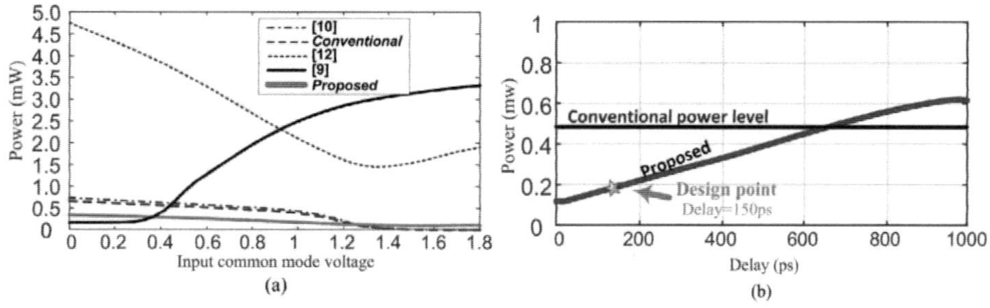

Fig.36 (a) energy expenditure of comparators vs. input V_{cm} (b) energy expenditure of the suggested comparator vs. tamp, V_{cm} = 1.1 V

Fig.36(a) introduces the energy expenditure vs. V_{cm} at the clock frequency of 500 MHz. Neglecting the power consumption of the latch, the total power consumption is calculated as

$$\begin{aligned} Power &= \frac{1}{T}\int_0^{T=T_{clk}/2} v(t)\cdot i(t)\,dt + P_{latch} \\ &= \frac{1}{T}\int_0^{t_{amp}} V_{DD}\cdot(I_1+I_2)\,dt + P_{latch} \cong \frac{t_{amp}}{T} V_{DD}\times I_{MS} \end{aligned} \tag{17}$$

Fig.36(b) presents the power consumption versus delay for V_{cm} = 1.1 V. The conventional power stays pretty much the same, because the preamplification delay is fixed to $T_{clk}/2$. But in the proposed comparator, the power consumption has a linear connection with tamp.

Fig.37(a) shows the power versus input V_{cm}, at F=500 MHz and V_{id}=0.1V, the average power consumption is less than 250 μW for 500-MHz clock frequency. Fig.37(b) shows the power versus clock frequency at V_{cm}=1.1V. With the increase of frequency, the power also increases. And the proposed comparators offer a low power consumption in different frequencies. Fig.37(c) shows the delay versus input V_{cm}, at V_{cm}=0.1V and V_{id}=0.1V.

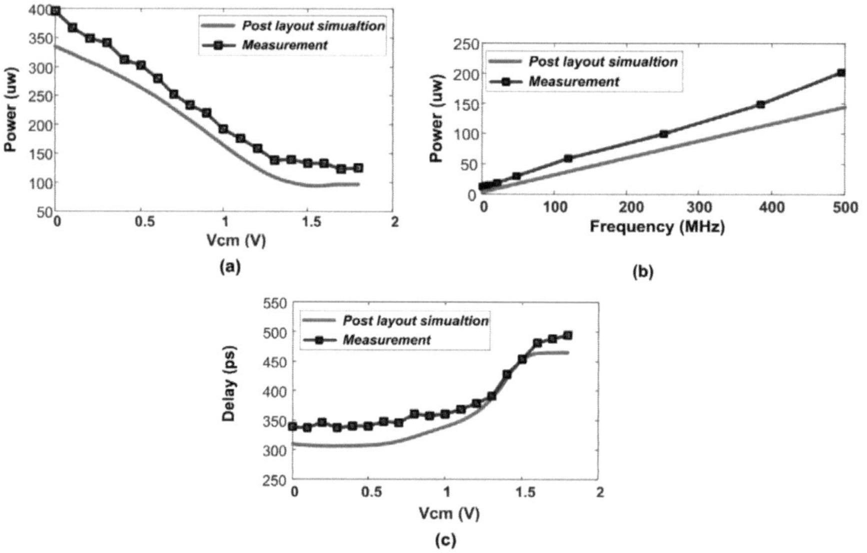

Fig.37 (a) Power versus input V_{cm} (b) Power versus clock frequency (c) Delay versus input V_{cm}

$$\begin{cases} V_{cmI} = \dfrac{V_{O1+}+V_{O1-}}{2} = \dfrac{I\times\frac{t_{amp}}{C}}{I+\alpha\left(\frac{t_{amp}}{C}\right)}, & I = I_1+I_2 \\ V_{idI} = V_{O1+}-V_{O1-} = \dfrac{\Delta I\times\frac{t_{amp}}{C}}{1-\alpha\left(\frac{t_{amp}}{C}\right)}, & \Delta I = I_2 - I_1 \end{cases} \tag{18}$$

TABLE1 shows these improvements clearly. From TABLE 1, it is clear that every comparator has its own improvement,

TABLE 1. PERFORMANCE SUMMARY

parameter	Samaneh'14	Chin-Yu Lin'16	Ata Khorami'18	Yao Wang'19	Xiyuan Tang'20
CMOS technology(nm)	180	180	180	180	180
Noise(μV)		400	28.6		46
supply voltage(V)	1.2	1.8	1.6/1.8/2	1.2	1.2
Energy per conversion(μ W)	329	390	230	72.2	
active area(μm2)		1000	530	252	9800
Maximum operating frequency(Hz)	500M		0.5G	2G	
Offset(mV)	7.8		2	7.3	
Energy efficiency(pJ)	0.66			110-fJ	0.98
Delay(ps)	550	208	263	268.6	

4. Conclusion

This review totally focuses on comparator design. In most of the selected review papers, they have made some changes to the traditional structure and improved the performance. Some replace its pre-amplifier by FIA, which decreases its energy consumption. Some use a transconductance-enhanced latching stage to attain low power high-speed comparator, and some add a special local clock to satisfy low voltage and power.

The challenge nowadays is that how to design a comparator achieve all these improvements at the same time.

References

[1] B. Razavi, ''Charge steering: A low-power design paradigm,'' in Proc.Custom Integr. Circuits Conf. (CICC), Sep. 2013, pp. 1–8.
[2] B. J. Blalock, H. W. Li, P. E. Allen, and S. A. Jackson, ''Body-driving as a low-voltage analog design technique for CMOS technology,'' in Proc. Southwest Symp. Mixed-Signal Design, Feb. 2000, pp. 113–118.
[3] A. Mesgarani, M. N. Alam, F. Z. Nelson, and S. U. Ay, ''Supply boosting technique for designing very low-voltage mixed-signal circuits in standard CMOS,'' in Proc. 53rd IEEE Int. Midwest Symp. Circuits Syst., Aug. 2010, pp. 893–896.
[4] J. Gao, G. Li, and Q. Li, ''High-speed low-power common-mode insensitive dynamic comparator,'' Electron. Lett., vol. 51, no. 2, pp. 134–136, Jan. 2015.
[5] Yao Wang, MengMeng Yao, Benqing Guo, Zhaolei Wu, Wenbing Fan and Juin Jei Liou, A Low-Power High-Speed Dynamic Comparator With a Transconductance-Enhanced Latching Stage, in Digital Object Identifier 10.1109/ACCESS.2019.2927514.
[6] H. S. Bindra, C. E. Lokin, D. Schinkel, A.-J. Annema, and B. Nauta,"A 1.2-V dynamic bias latch-type comparator in 65-nm CMOS with 0.4-mV input noise," IEEE J. Solid-State Circuits, vol. 53, no. 7, pp. 1902–1912, Jul. 2018.
[7] M. S. Akter, K. A. A. Makinwa, and K. Bult, "A capacitively degenerated 100-dB linear 20 -150 MS/s dynamic amplifier," IEEE J. Solid-State Circuits, vol. 53, no. 4, pp. 1115–1126, Apr. 2018.
[8] X. Tang et al., "An Energy-Efficient Comparator With Dynamic Floating Inverter Amplifier," in IEEE Journal of Solid-State Circuits, vol. 55, no. 4, pp. 1011-1022, April 2020, doi: 10.1109/JSSC.2019.2960485.
[9] D. Schinkel, E. Mensink, E. Klumperink, E. Van Tuijl, and B. Nauta, "A double-tail latch-type voltage sense amplifier with 18 ps setup+ hold time," in IEEE Int. Solid-State Circuits Conf. (ISSCC) Dig. Tech. Papers, Feb. 2007, pp. 314–605.
[10] B. Razavi, Design of Analog CMOS Integrated Circuits. Boston, MA,USA: McGraw-Hill, 2001.
[11] J. J. Kang and M. P. Flynn, "A 12 b 11 MS/s successive approximation ADC with two comparators in 0.13 μm CMOS," in Proc. Symp. VLSI Circuits Dig. Tech. Papers, Jun. 2009, pp. 240–241.

[12] D. Schinkel, E. Mensink, E. Klumperink, E. V. Tuijl, and B. Nauta, "A double-tail latch-type voltage sense amplifier with 18 ps setup+hold time," in Proc. IEEE ISSCC Dig. Tech. Papers, Feb. 2007, pp. 314–605.
[13] C. H. Chan et al., "A reconfigurable low-noise dynamic comparator with offset calibration in 90 nm CMOS," in Proc. IEEE Asian Solid-State Circuits Conf., Nov. 2011, pp. 233–236.
[14] Samaneh Babayan-Mashhadi and Reza Lotfi, Analysis and Design of a Low-Voltage Low-Power Double-Tail Comparator, in IEEE TRANSACTIONS ON VERY LARGE SCALE INTEGRATION (VLSI) SYSTEMS, VOL. 22, NO. 2, FEBRUARY 2014.
[15] Ata Khorami and Mohammad Sharifkhani, A Low-Power High-Speed Comparator for Precise Applications, in IEEE TRANSACTIONS ON VERY LARGE SCALE INTEGRATION (VLSI) SYSTEMS, Volume: 26, Issue: 10, Oct. 2018.
[16] A. Nikoozadeh and B. Murmann, "An analysis of latch comparator offset due to load capacitor mismatch," IEEE Trans. Circuits Syst. II, Exp. Briefs, vol. 53, no. 12, pp. 1398–1402, Dec. 2006.

Design and simulation analysis of vacuum fish suction pump

Ziyang Ding[1], Lingfeng Xu[1,2], Ming Gao[1,2*], Guoqing Wang[1]

[1]College of Mechanical and Electronical Engineering, Shandong Agricultural University, Tai'an , China

[2]Shandong Agricultural Equipment Intelligent Engineering Laboratory, Shandong Agriculture University, Tai' an, China

* Corresponding author: gaom@sdau.edu.cn

Abstract: Most of Chinese aquacultural products rely on artificial nets to catch in 2020. Fish suction pump can replace artificial net, is an important mechanical equipment of fish breeding industry. The development of fish suction pump is relatively backward and the level of automation is still to be improved, which is not conducive to the large-scale and rapid development of aquaculture in various countries. This paper designs a single tank vacuum suction fish pump, the suction fish pump mainly uses the tank and the air pressure difference, through the automatic control system on the suction fish pump and each stage of the control, relying on the gravity of the mixture of fish and water to release fish. In this paper, fluent was used to conduct simulation analysis on the flow diagram of fish-water mixture inhaled at work, and the flow diagram and velocity vector diagram of the internal flow field were obtained. The results show that the vacuum fish suction pump can effectively reduce the fish collision damage. The vacuum fish suction pump designed in this paper can effectively complete the fishing work of cultured fish, with the characteristics of high efficiency and low damage, which can provide theoretical reference for the design of vacuum fish suction pump in the future.

1. Introduction

In 2020, China's aquaculture area was 703,611 hectares, and the total output of aquaculture products was 65.4902 million tons, an increase of 1.06 percent over the previous year. And fish suction pump can replace artificial net, is an important mechanical equipment of fish breeding industry. With the continuous development of modern fishery, the market demand for fishing by deep-sea fishing boats and the production and quality of fish raised in cage has been sharply increased. Therefore, it has become an important topic for fishery researchers at home and abroad to realize the mechanization and automation of fish sucking and transportation, reduce the labor intensity of fishermen and improve the survival rate of fish [1]. The present application of net technology and high mechanization of abroad is in stark contrast. Purse Seine, trawl, inefficient and injury mortality rate is high, the comprehensive benefit is relatively low [2]. With the continuous development of manufacturing industry, the structure of domestic fish suction pump has changed a lot, its working principle is more perfect, and its application field is also expanding [3]. Existing equipment is mainly aimed at deep-sea fishing and fish suction and capture in ocean cage farming, while there are relatively few studies on fish suction pump in pond farming [4-5]. Therefore, it is necessary to analyze the foreign fish suction machine and design the fish suction machine suitable for our country.

With the continuous development of China's manufacturing industry, the structure of the fish pump has been improved, the working principle is more perfect, and the field can be used is also expanding.

Such as hyS-60 vacuum fish suction pump, vacuum live fish catching machine, etc. [3]. Hydraulic motor driven submersible centrifugal pump suction fish, every 100 minutes can pump ice 35 t large yellow croaker, also may transport length 10 to 50 cm of live grass carp fish, etc. Foreign fish suction pump mainly depends on the high-speed rotation of the impeller to carry out the fish suction. ETI's TRANSVAC fish pump has a pumping capacity of 300 to 360 tons per hour[6]. Denmark IRAS company's fish pump absorption, by Norway, France, Iceland, Ireland, Japan and other countries[7].

At present, the research of fish suction pump is gradually transferred from centrifugal type to vacuum type. Vacuum type fish suction pump can effectively reduce labor intensity and improve work efficiency. Therefore the design and use of type vacuum pump fish fluent on the analysis of the internal flow field in pump suction fish is very meaningful. This paper carries on the corresponding design to the vacuum type fish suction pump, and carries on the corresponding internal flow field analysis to the vacuum type fish suction pump suction tank fish water mixture in the fish collection tank, so as to obtain some performance parameters of the internal flow field, to provide some reference for the future design of fish suction machine. In order to promote the application and development of fish suction equipment in China and improve the modernization level of fishery machinery.

2. Calculation Model

2.1 Physical Model

Figure 1. Drawing of fish tank parts

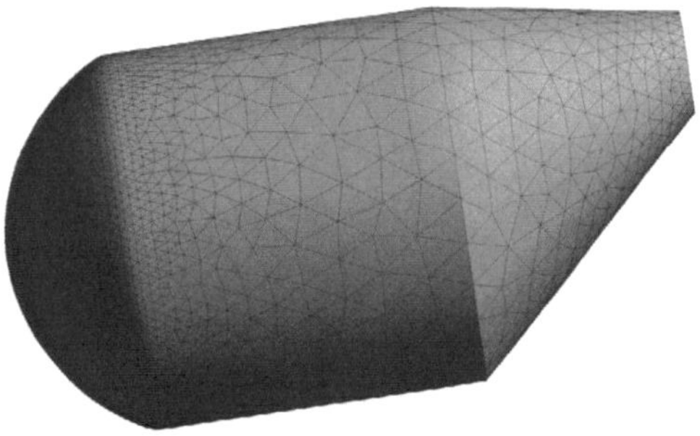

Figure 2. Model grid diagram

According to the current use of vacuum drum fish machine set fish shape are analyzed. The model of the whole container is drawn in the scale of 1:1 using SOLIDWORK software.

The main physical parameters of the model are determined as follows:

(1) The volume of the fish collecting tank is determined according to the suction pipe and suction speed per second. According to the determination of the expected workload $Q=\frac{W+W_{\text{tube}}}{\rho T}$, the corresponding working flow can be obtained as Q=1.65m^3/min. Then, the volume of fish collection tank can be obtained as 0.69m^3.

(2) Considering the overall coordination of the volume of the fish collecting tank, the diameter of the fish collecting tank is 0.7m. On the premise that the diameter has been determined, the length of the fish collecting tank is 1.5m according to the volume formula of the cylinder and the calculation formula of the circular area.

(3) According to the formula for calculating the wall thickness of the cylinder under the external pressure of the vacuum tank: $\delta \geq D^{0.6} \times (\frac{MPL}{2.59E})^{0.4}+C_0$, wall thickness δ=2.64×10^{-2}m can be obtained by corresponding calculation. According to the working environment of the fish pump, the material of the fish tank in the fish pump is 314 stainless steel, and then the corresponding material strength should

be checked. Because the fish suction pump works through the internal and external pressure difference, when the wall thickness of the fish collection tank is insufficient, the phenomenon of deflating the fish collection tank may occur, and the wall thickness is too large will lead to the fish suction pump is too bulky, large consumables. The appropriate wall thickness is particularly important. According to the calculation of the wall thickness also need to carry on the corresponding stress checking, only after checking the right to finish the expected work smoothly. According to the judgment condition of thin-walled cylinder, fish collecting tank can be regarded as thin-walled cylinder, and then the stress analysis of fish collecting tank is carried out. The state of stress on the wall of fish collecting cylinder is in two directions. One direction is atmospheric pressure applied to the outer surface of the vacuum fish collecting cylinder, and the other direction is pressure applied to the inner surface of the fish collecting cylinder after air is pumped from the inside. According to the cylinder section stress calculation formula: $\sigma'=\frac{PD}{4\delta}$ 、$\sigma''=\frac{PD}{2\delta}$. The stress can be obtained $\sigma'=6.9\times10^4$N, $\sigma''=3.45\times10^4$N. According to the fourth strength theory $\sigma_s=7.1\times10^5$N, Refer to the material strength limit table to see that is less than 314 stainless steel stress limit.

(4) Fish suction pump in the fish tank also need to design the corresponding fish inlet and outlet, to complete the complete workflow. Fish inlet and outlet are channels for fish in and out of fish tank, in order to reduce damage to fish and smooth fish in and out of work. The diameter should be as far as possible to ensure that a single successively or accumulative equivalent cross-sectional area of the fish body passes through in an orderly manner, that is, the maximum cross-sectional area of the fish body should be as far as possible to ensure that the maximum cross-sectional area of the fish body takes up half of the suction pipe,

$$\frac{D_{fish}}{D_{tube}}=\sqrt{2}\,.$$

According to the corresponding design, the diameter of the fish collecting tank in the fish suction pump is 0.7m, the length is 1.5m, the wall thickness is 8×10^{-3}m, and the diameter of the fish inlet is 0.2m.

2.2 Flow field model

Sports, as a result of the analysis after the fish into the only consider putting fish movement, so in the ANSYS model is not designed to put the fish in the mouth, as shown in figure 1. Figure 2 is shown in figure 1 is divided by the ANSYS fluent grid model, the mesh with 15943 nodes and 74626 elements, the model after meshing the model of average quality reached 0.82, and can better complete the numerical calculation. Before the fluid analysis of the grid model, it is necessary to check the results of the grid division. Only when there is no unqualified grid, can the corresponding fluid analysis of the grid model be carried out. Because the fish suction machine is the fish from the aquaculture pond through the negative pressure suction into the fish tank, mainly on the fish from the suction mouth into the fish tank, fish in the tank movement track analysis. The standard K-Epsion (2 EQN) model was selected as the calculation model. The turbulence model is accurate and easy to converge, and has been widely used in fluent fluid simulation.

2.3 Boundary constraint

The suction process of fish suction machine is analyzed to determine the corresponding boundary constraints. Through the analysis of the swimming speed of fish in the water, the inlet speed of fish suction machine was selected as 1.2m/s. After the fish entered the fish collecting tank, the fish fell along the Y-axis direction, and the gravity acceleration was set as 9.8m/s^2. Name the working face inside the model in Face and the rest of the non-working face other. After all the parameters are set, the calculation is carried out. After the settlement, the results are saved and analyzed accordingly.

Journal of Physics: Conference Series **2113** (2021) 012009 doi:10.1088/1742-6596/2113/1/012009

3. Results and Analysis

3.1 Streamline trajectory diagram

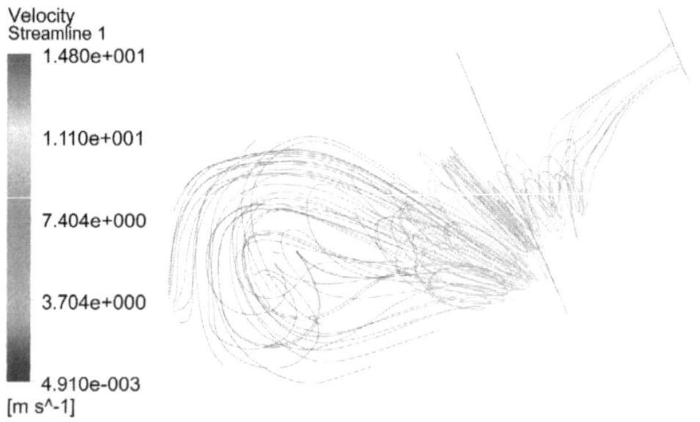

Figure 3. Streamline trajectory diagram

In the actual experiment, the working process of fish suction pump is fast, so it is difficult to measure the fluid trajectory inside the fish collecting tank. After calculation by Fluent, the streamline trajectory inside the fish collecting tank can be observed intuitively and simply. As shown in Figure 3, when the fish-water mixture enters from the entrance of the fish collecting tank, it falls towards the bottom of the fish collecting tank under the action of gravity and initial velocity. After a period of curve movement, it slides to the bottom of the fish collecting tank. It can be seen from the figure that when the fish-water mixture reaches the bottom of the fish collecting tank, the speed is 0.005m/s. The lower speed can reduce the damage caused by the collision between fish and tank wall when they fall into the bottom of the tank, improve the integrity of fish during transportation, reduce the damage caused by the collision between fish and tank wall because of too fast speed, and thus improve the survival rate of fish. In addition, it can be intuitively observed from Figure 3 that the lines of the flow field inside the fish collecting tank are smooth and there is no obvious backflow phenomenon. Therefore, the flow field inside the fish collecting tank is reasonable.

3.2 Velocity vector

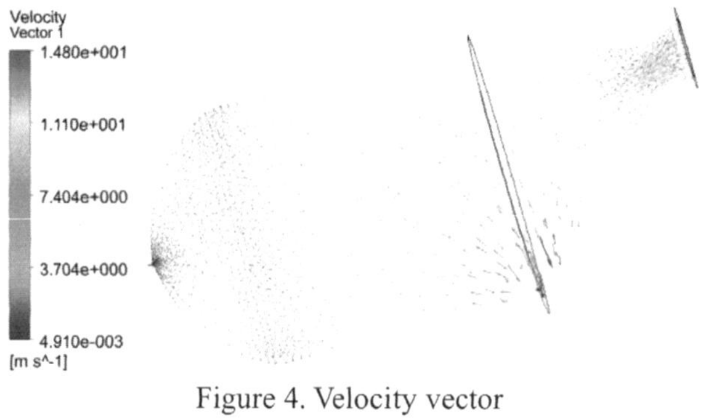

Figure 4. Velocity vector

It can be clearly observed from Figure 4 that when fish and water mix into the fish collecting tank, there is a period of acceleration due to the acceleration of gravity. The velocity of fish-water mixture reaches the maximum value when it first contacts the upper right section of the fish-water tank, and the higher speed is conducive to promoting the fish-water mixture to flow to the arc at the bottom of the bow of the fish-water tank to prevent the blockage in the fish-water tank. Then the velocity gradually decreases to 0.005m/s and becomes stable at the bottom of the fish-water tank. It can effectively avoid the damage caused by collision between fish and bottom of fish collecting tank caused by high speed.

4. Conclusion

This paper designs a vacuum fish suction pump, and uses SOLIDWORK to model the arc-shaped fish collecting tank at the bottom of the bow. Fluent is used to simulate the streamline track and velocity of the liquid inside the fish collecting tank when it works, and the corresponding analysis of the simulation results is made, and the simulation results are as follows:

(1) When the fish-water mixture enters from the entrance of the fish collecting tank, it falls towards the bottom of the fish collecting tank under the action of gravity and initial velocity. After a period of curve movement, it will slide to the bottom of the fish collecting tank. The internal streamline of the fish collecting tank is smooth and stable, and there is no direct collision with the inner wall of the fish collecting tank. The bottom of the arc can better reduce the speed of fish and water mixture and reduce fish damage.

(2) The speed at the inlet of fish collecting tank can absorb fish and water mixture better. Then, after a period of acceleration under the force of gravity, the fish-water mixture flows at a faster rate towards the bottom of the bow of the tank. It can effectively prevent the blockage of fish collecting tank caused by too slow speed, which is conducive to the subsequent entry of fish and water mixture. And when reaching the bottom, the speed is almost zero, which can effectively reduce the damage caused by the collision between fish and fish collecting tank.

In this paper, the vacuum fish suction pump is designed accordingly, and the fluent software is used to analyze the internal flow field of the fish water mixture sucked into the fish collecting tank of the vacuum fish suction pump, so as to obtain some performance parameters of the internal flow field, which provides some reference for the design of the fish suction machine in the future. In order to promote the application and development of fish suction equipment in China and improve the modernization level of fishery machinery.

Acknowledgements
The authors acknowledge the Technical Innovation Guide Plan of Shandong Province: Shandong-Chongqing total cooperation project (Grant No. 2020LYXZ021), the Open Project Program of Tianjin Key Laboratory of Microgravity and Hypogravity Environment Simulation Technology (Grant No. TJWDZL2019KT004).

Reference

[1] Ye X M, Xu J Z. Status of Fish Pump Research at Home and Abroad [J]. Modern Fisheries Information, 2005, 20; 7-8.
[2] Lu J, Qian R C, Huang H L, Li L Z, Feng C L, Liu Z D, Chen S, Wu Y, Xu B. Research Progress of Fish Pump at Home and Abroad [J]. Fishery Information & Strategy, 2013, 40; 57-62.
[3] Lu P. Flow Analysis of Fish Pump Base on ANSYS CFX [J]. Fluid Machinery, 2014, 42; 43-46.
[4] Gehrke P C. Diel Abundance, Migration and Feeding of Fish Larvae (Eleotridae) in a Floodplain Billabong [J]. Journal of Fish Biology, 1992, 40; 695,707.
[5] Mcnabbcd C D, Liston C R, Borthwicksm S M. Passage of Juvenile Chinook Salmon and other Fish Species Through Archimedes Lifts and a Hidrostal Pump at Red Bluff [J]. Trans--actions of the American Fisheries Society, 2003, 132; 326-334.
[6] Davis K B, Newson J, Simco B A. Physiological Stress in Channel Catfish, Ictalurus Punctatus, Harvested by Lift Net, Vacuum Pump, or Turbine Pump [J]. Journal of Aquacult--ure, 1994, 3; 297-310.
[7] Summerfelt S T, Davidson J, Wilson G, Waldrop T. Advances in Fish Harvest Technologies for Circular Tanks [J]. Aquacultural Engineering, 2009, 40; 62-71.

SELNN: Effective Crossover Features by Attention Mechanism and Logarithmic Conversion for CTR

Lu Meng[1], **Li Wang**[1*]

[1] School of Computer Science and Software Engineering, University of Science and Technology Liaoning, Anshan, Liaoning, 114051, China

[*]Corresponding author's e-mail: wangli9966@ustl.edu.cn

Abstract. With the continuous increase of web resources, information overload is becoming more and more serious. Getting the information that meets the needs of a large amount of data has become an urgent problem. One of the popular studies in click-through rate prediction is to construct feature interactions to improve prediction accuracy. Traditional models with low-order interactions cannot adequately represent the intersection information, and deep models with undifferentiated feature intersections lack relevance. In this paper, we propose a model called SELNN, which improves the accuracy of click-through rate prediction through attention mechanism and logarithmic conversion. On the one hand, improving attention mechanisms learns the importance of each feature itself. On the other hand, combining logarithmic conversion structure with feedforward neural networks to learn different orders of feature interactions. The experimental results indicated that the AUC values of SELNN on the Criteo and Movielens-1M datasets were 81.46% and 87.36%, respectively. SELNN effectively improves the prediction accuracy and reduces the number of parameters and computational effort.

1. Introduction

Revenue from advertising has been growing in recent years. Estimating users' clicks on ads and then placing them precisely can greatly increase advertisers' revenue.

The prediction models currently proposed in academia and industry can be divided into two main categories. Traditional models as LR[1], FM[2] are simple and easy to reference. However, it is not possible to perform parallel training, and the cross-correlation between features is neglected. Deep learning-based models as PNN[3], AFM[4], DCN[5], DeepFM[6], and MiFiNN[7]. The Click-Through Rate is improved by strengthening the intensity of feature interactions, combining memory and generalization capabilities, introducing attention mechanisms. However, most models use undifferentiated feature interactions lacking relevance.

The SELNN proposed in this paper improves the accuracy of CTR prediction utilizing attention mechanism and logarithmic conversion. The main contributions of this paper are as follows:

The SELNN cleverly applies the SENet from computer vision research to recommendation algorithms and improvements to achieve feature importance selection.

The SELNN utilizes a logarithmic conversion structure to learn more patterns of feature interactions from valid feature combinations. It enables DNN to expand the input in potential mining information and then guide click-through prediction from feature interactions of different orders.

2. Related work

2.1. Click-through rate prediction

The user and item features are tandem concatenated as an input feature, denoted by $x \in R^n$, where n is the dimensionality of the tandem features. Traditional studies usually input x directly into the model for prediction, such as LR[1], FM[2]. However, because of the sparse and high-dimensional original vector x, the model is prone to overfitting. Therefore, the latest research focuses on extracting key information from feature interactions and making predictions.

2.2. K-order interactive feature

For the input feature $x \in R^n$, the k-order feature interaction is defined as $g\left(x_{i_1}, \dots, x_{i_p}\right)$. Where $g(\cdot)$is a function of the feature interaction, which can be the inner product, outer product, and Hadamard product. Professionals usually perform traditional research for manual feature combinations, such as Wide&Deep[8]. However, it has poor generalization capability, requires more manual operations, and cannot be applied across domains. Therefore, the current popular research is to learn higher-order feature interactions using deep learning methods automatically.

2.3. Attention mechanism

The attention mechanism distributes computational resources to more important tasks by calculating attention weights. It can effectively solve the information overload problem. The features are filtered and judged according to the attention weights to determine the significant impact on the results. The current mainstream approach calculates the attention score based on the relevance of user attributes and item attributes for more targeted recommendations. Such as the application of self-attention[9]. However, its training process is complex with long latency and needs to be optimized.

3. Model description

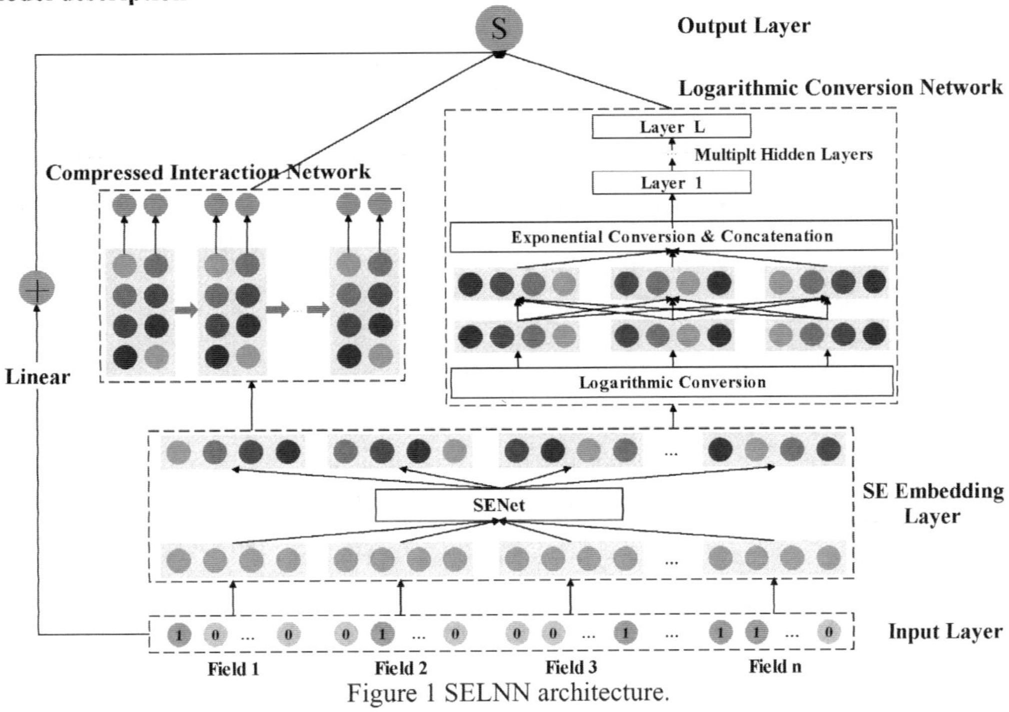

Figure 1 SELNN architecture.

The SELNN architecture is shown in Figure 1. It consists of the components: input layer, SE embedding layer, compressed interaction network, logarithmic conversion layer, multiple hidden layers, and output layer. The data in the input layer has numerical features and categorical features. The SE embedding layer first converts the original sparse features into a vector of dense features, then the weight is learning for each feature by using the attention mechanism. This can suppress the effects of low frequency and invalid features. Next in parallel are the explicit feature interaction component and the implicit feature interaction component. The compressed interaction network implements the explicit feature interaction, combining logarithmic conversion structure with feedforward neural networks. Finally, the output layer predicts the click-through rate by the Sigmoid function. The following section describes the details of SELNN.

3.1. Input layer

The input data consists of sparse numerical and categorical features in the CTR prediction problem for real scenarios. Representing the classification features as one-hot vectors:

$$x = \underbrace{[0,\ 1,\ 0,\ \dots,\ 0]}_{field\ 1},\ \underbrace{[1,\ 0]}_{field\ 2},\ \dots,\ \underbrace{[0,\ 1,\ 0,\ 1,\ \dots,\ 0]}_{field\ m} \tag{1}$$

where m is the total number of features and field i denotes the ith feature.

3.2. SE embedding layer

The feature vectors are dynamically differentiated according to their importance, utilizing an attention mechanism. The learning procedure of SENet[10] is divided into three main steps. First is Squeeze, which compresses the data and summarizes the information for each feature vector. Then comes Excitation, which introduces a narrow two-layer MLP network to learn a feature weight $A = \{a_1, a_2, \dots, a_i, \dots, a_f\}$for each feature, and finally is Re-Weight, which takes the values of f weights and multiplies them back into the corresponding feature vector. This completes the feature importance extraction.

3.3. Compressed interaction network

CIN follows the crossover network in xDeepFM[11]. It accomplishes higher-order explicit feature interactions by recursion. Its core is the summation and pooling of the individual hidden layer feature vectors:

$$p_i^K = \sum_{j=1}^{D} X_{i,j}^k \tag{2}$$

Where$i \in [1, H_k]$, This gives the pooling vector for the k-th layer $p_i^K = [p_1^k, p_2^k, \dots, p_{H_k}^k]$.The pooling vectors of each layer are concatenated as $p^+ = [p^1, p^2, \dots, p^T] \in R^{\sum_{i=1}^{T} H_i}$. From there, the output vector of the CIN component is obtained.

3.4. Logarithmic conversion network

The logarithmic transformation layer learns the power (order) of each feature in the feature interactions, and This layer is composed of multiple logarithmic neurons, each of which is calculated as

$$y = exp(\sum_i w_i \ln x_i) = \prod_i x_i^{w_i} \tag{3}$$

The lack of MLP expressiveness is compensated by converting the input into logarithmic space. It should be noted that the logarithm takes values in the range $(0, +\infty)$. Therefore, a minimal positive number, $1e^{-7}$It is added to the input to ensure that each input is positive.

Next, all the crossover features are connected, and multiple hidden layers are stacked and input into the feedforward neural network.

3.5. Output layer

The SELNN combined Linear, CIN, and LCN, then processed by Sigmoid function to get the final click prediction results. The calculation formula of SELNN is:

$$\hat{y}_{SELNN} = \sigma\left(W_{Linear}^{T}\hat{y}_{Linear} + W_{CIN}^{T}p^{+} + W_{LCN}^{T}\hat{y}_{LCN} + b\right) \quad (4)$$

Where W_{Linear}^{T}, W_{CIN}^{T} and W_{LCN}^{T} denote the weights of Linear, CIN, and LCN, respectively. b is the overall deviation of the model. The output of the Sigmoid function is between 0 and 1, which is convenient for the judgment of click-through rate prediction.

4. Experiments

4.1. Datasets
Criteo dataset contains records of 4500w user clicks on display ads over seven consecutive days, with 13 continuous numerical features and 26 discrete categorical features per sample. Movielens-1M database contains nearly 1 million rating data from 6040 users for 3706 movies.

4.2. Evaluation metrics
CTR prediction is a dichotomous problem. The evaluation metrics adapt AUC and Logloss. The higher the AUC value, the more accurate the prediction. The smaller the Logloss value, the better the performance of the model.

4.3. Hyper-parameter study

4.3.1. Embedding size. It is critical to the model's representation capability and computational cost. Therefore, Embedding sizes are set to 5, 10, 15, and 20; experiments are conducted separately to explore how their sizes affect the prediction performance. The results of the experiment are shown in Figure 2. The AUC and Logloss reach the optimal value on the two datasets when the embedding size is 15. Because it is too small, it will lose the crossover features, and too large will be overfitting.

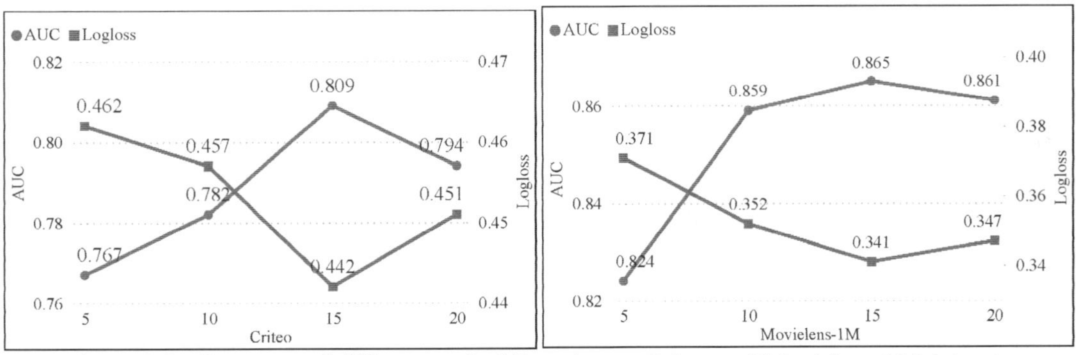

Figure 2 Performance of different embedding size on Criteo and Movielens-1M datasets.

4.3.2. Dropout rate. It can effectively prevent overfitting. Set dropout rate to 0.2-0.6 for each experiment. The results of the experiment are shown in Figure 3. On Criteo, the AUC and Logloss reach the optimal value when the dropout rate is 0.5. On Movielens-1M, the AUC and Logloss reach the optimal value when the dropout rate is 0.3. Because Criteo has a wider variety of features than Movielens-1M, resulting in a different degree of optimization.

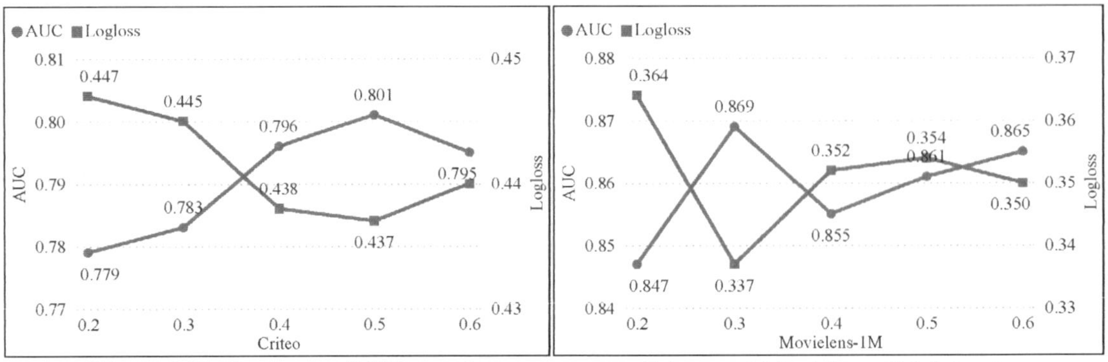

Figure 3 Performance of different dropout rates on Criteo and Movielens-1M datasets.

4.3.3. Activation function. It makes the data linearly separable by adding nonlinear factors. It can solve more complex problems and brings the possibility of the backpropagation of neural networks. Therefore, Relu, Sigmoid, and Tanh were selected to conduct the experiments separately. The results of the experiment are shown in Figure 4. When the activation function is Relu, AUC and Logloss achieve the best performance on the two datasets. Therefore, Relu is the most appropriate activation function in the SELNN.

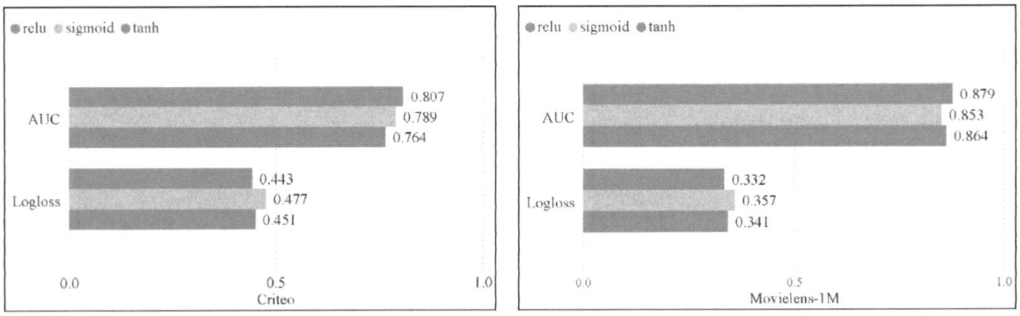

Figure 4 Performance of different activation functions on Criteo and Movielens-1M datasets.

4.4. Model performance comparison

4.4.1. Parameter settings. PyTorch implemented the experiments. In the hyper-parameter setting, the Embedding size is uniformly set as 15, and Batch Size is uniformly set as 256. For the model with DNN, the number of hidden layers is set as 3, the number of neurons in each hidden layer is set as 128, and the activation function is selected as Relu. The dropout rate is set as 0.5. The optimization is performed with Adam optimizer, and the learning rate is set at 10^{-4}.

4.4.2. Experimental results and analysis. To validate the effectiveness of SELNN, several traditional and deep learning models have experimented on two datasets. The results are shown in Table 1.

Table 1 Performance of different CTR models on Criteo and Movielens-1M datasets.

Model	Criteo		Movielens-1M	
	AUC	Logloss	AUC	Logloss
LR	0.7698	0.4974	0.8282	0.3946
FM	0.7833	0.4772	0.8425	0.3725
PNN	0.8009	0.4682	0.8609	0.3519
NFM	0.7903	0.4593	0.8589	0.3485
AFM	0.7962	0.4541	0.8602	0.3442
DeepFM	0.8072	0.4474	0.8627	0.3413
DCN	0.8083	0.4453	0.8679	0.3394
SELNN	**0.8146**	**0.4352**	**0.8736**	**0.3357**

On the Criteo dataset, the SELNN improved the AUC values by 4.48% and 3.13% and reduced the Logloss values by 6.22% and 4.2%, respectively, compared to the traditional models LR and FM. Compared with the models PNN, NFM, AFM, DeepFM, and DCN using deep learning methods, the AUC values improved by 1.37%, 2.43%, 1.84%, 0.74%, and 0.63%, respectively, and the Logloss values decreased by 3.3%, 2.41%, 1.89%, 1.22%, and 1.01%, respectively.

On the Movielens-1M dataset, the SELNN improved the AUC values by 4.54% and 3.11% and reduced the Logloss values by 5.89% and 3.68%, respectively, compared to the traditional models LR and FM. Compared with the models PNN, NFM, AFM, DeepFM, and DCN using deep learning methods, the AUC values improved by 1.27%, 1.47%, 1.34%, 1.09%, and 0.57%, respectively, and the Logloss values decreased by 1.62%, 1.28%, 0.85%, 0.56%, and 0.37%, respectively.

In the field of click-through rate forecasting, an increase of 1‰ in AUC is often considered significant. In the above experiments, SELNN achieved the best performance on the two datasets. On the one hand, SENet is used to measure the importance of sparse features, which helps to improve the discriminability of features. On the other hand, adding a logarithmic conversion structure before the DNN can automatically generate different orders of differentiated intersection features and the corresponding weights.

5. Conclusion

To improve the accuracy of CTR prediction, the SELNN proposed in this paper improves the accuracy of CTR prediction utilizing attention mechanism and logarithmic conversion. The improved SENet in this model dynamically learns the importance of features and suppresses the influence of low-frequency and invalid features. CIN can learn higher-order display feature interactions. Moreover, LCN integration with logarithmic conversion structure, and combined with feedforward neural networks to learn the implicit feature interactions and make predictions.

The main focus of this model is on feature engineering, and it lacks the mining of users' history behavior. More effective ways to capture users' interests will be explored in the future to increase the model's effectiveness.

References

[1] CHAPELLE O, MANAVOGLU E, ROSALES R. Simple and scalable response prediction for display advertising [J]. ACM Transactions on Intelligent Systems and Technology (TIST), 2014, 5(4): 1-34.

[2] RENDLE S. Factorization machines with libfm [J]. ACM Transactions on Intelligent Systems and Technology (TIST), 2012, 3(3): 1-22.
[3] QU Y, CAI H, REN K, et al. Product-based neural networks for user response prediction; proceedings of the 2016 IEEE 16th International Conference on Data Mining (ICDM), F, 2016 [C]. IEEE.
[4] XIAO J, YE H, HE X, et al. Attentional factorization machines: Learning the weight of feature interactions via attention networks [J]. arXiv preprint arXiv:170804617, 2017.
[5] WANG R, FU B, FU G, et al. Deep & cross network for ad click predictions [M]. Proceedings of the ADKDD'17. 2017: 1-7.
[6] GUO H, TANG R, YE Y, et al. DeepFM: a factorization-machine based neural network for CTR prediction [J]. arXiv preprint arXiv:170304247, 2017.
[7] WANG X, DONG H, HAN S. Click-Through Rate Prediction Combining Mutual Information Feature Weighting and Feature Interaction [J]. IEEE Access, 2020, 8: 207216-25.
[8] CHENG H-T, KOC L, HARMSEN J, et al. Wide & deep learning for recommender systems; proceedings of the Proceedings of the 1st workshop on deep learning for recommender systems, F, 2016 [C].
[9] YAN C, CHEN Y, WAN Y, et al. Modeling low-and high-order feature interactions with FM and self-attention network [J]. Applied Intelligence, 2021, 51(6): 3189-201.
[10] HUANG T, ZHANG Z, ZHANG J. FiBiNET: combining feature importance and bilinear feature interaction for click-through rate prediction; proceedings of the Proceedings of the 13th ACM Conference on Recommender Systems, F, 2019 [C].
[11] LIAN J, ZHOU X, ZHANG F, et al. xdeepfm: Combining explicit and implicit feature interactions for recommender systems; proceedings of the Proceedings of the 24th ACM SIGKDD International Conference on Knowledge Discovery & Data Mining, F, 2018 [C].

Estimation of Vehicle Longitudinal Speed Based on Improved Kalman Filter

Qiang Zhang[1], Jun Xiao[1*], Xiuhao Xi[1]

[1]School of Mechanical and Electrical Engineering, Wuhan University of Technology, WuHan, 430000, China

*Corresponding author's e-mail: xjun@whut.edu.cn

Abstract: Estimation of vehicle longitudinal acceleration is very important in vehicle active safety control system. In this paper, two driving conditions of a 4WD off-road vehicle are divided by vehicle signals such as steering angle. Under different working conditions, different estimation algorithms are adopted. In the straight driving condition, the longitudinal speed was estimated by adjusting the variance weight of acceleration Kalman observation noise based on kinematics method. For steering conditions, in order to obtain the longitudinal velocity at the center of mass, by dynamic method, a lateral state estimator was designed and tire sideslip dynamics was modeled. The CarSim-Simulink co-simulation results show that the proposed algorithm has high accuracy and strong practicability.

1. Introduction

Longitudinal speed is an important variable to calculate the wheel slip rate. Accurate reference speed is of key significance to optimize the control effect of ABS, TCS, ASR and other control systems. At present, kinematics and dynamics are the main models for longitudinal speed of vehicle[1]-[3]. The kinematics mainly uses the wheel speed information, inertial measurement unit, etc. To estimate the speed according to the kinematics formula and other operational methods, which requires high anti-interference performance of the sensor signal, and is only applicable to the conventional working conditions. Theoretical method of dynamics relies on the vehicle dynamics model and analyzes the force of the vehicle to realize the estimation of the vehicle longitudinal speed, which greatly increases the difficulty of calculation, and the simplified model will produce estimation errors.

In order to solve the above existing problems, this paper takes a 4WD off-road vehicle as the research object, adopts the hierarchical architecture estimation algorithm. Firstly, a variety of on-board sensors are used to identify vehicle driving conditions. Secondly, the difference of wheel speed was analyzed to obtain the pre-estimated speed, which was used as the measurement value of Kalman filtering algorithm. Finally, in the straight driving condition, the weight value of noise variance was determined by fusing the information of wheel speed and acceleration. In the steering condition, since neither the wheel speed nor the acceleration information can reflect the real longitudinal speed. considering the influence of slip angle of tire on the longitudinal speed, this paper designed a lateral state estimator based on the 3-DOF dynamics model to model the tire sideslip dynamics and obtain the longitudinal velocity at the center of mass. The co-simulation model was established by CarSim-Simulink results show that the estimated vehicle longitudinal speed is consistent with the CarSim output reference vehicle speed, and the algorithm is simple, which is easy to implement in real vehicle ECU.

2. The general structure of the estimation algorithm

The structure of the speed estimation algorithm to be adopted in this paper is shown in figure 1, which consists of three parts: Driving Condition Recognition (DCR), Pre-Estimator of Speed (PES), Longitudinal Speed Calculator (LSC), and the descriptions below are abbreviated.In this paper, with the aid of layered architecture idea, the DCR module makes certain judgments on the current driving conditions according to the longitudinal/lateral acceleration α_x, α_y, front wheel steering angleδ, four-wheel speed ω_{ij}, throttle position a_{thro}, braking signal s_{brake}, ABS working signal s_{ABS} and other on-board signals, and it is divided into two kinds of driving conditions: straight driving and steering. PES module analyzes the wheel speed in different working conditions and puts forward the corresponding algorithm to get the pre-estimated speed v_{gx}, which is close to the real speed.In the straight driving condition, the LSC module obtains the final estimated longitudinal speed through the vehicle dynamics model and the extended Kalman filter algorithm. Under steering conditions, LSC considers the influence of lateral state variables such as sideslip angleβ and yaw rate ω_r on longitudinal speed, and adds slip angle of tire α_f to make a certain degree of correction.

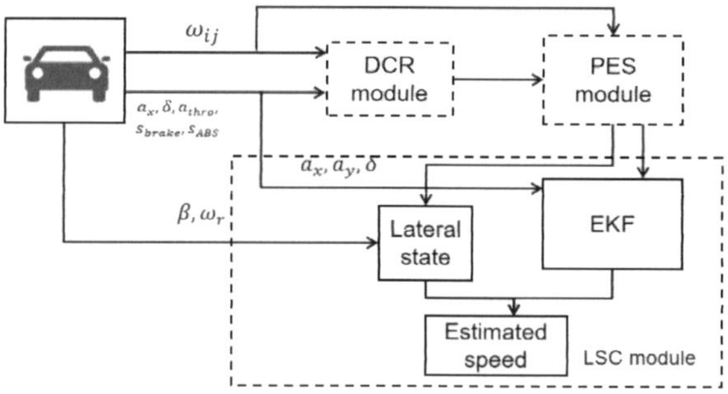

Figure 1. Structure diagram of speed estimation algorithm

3. DCR and PES module

In the process of vehicle driving, the external environment (such as road changes, etc.), and the driver's intention of irregular changes, will lead to a certain difference in four-wheel wheel speed[3].In order to better obtain the pre-estimated speed close to the real speed, it is necessary to fuse the on-board multi-sensor signals to recognize the vehicle driving conditions in real time. In addition, under the steering condition, the vehicle will produce the roll force couple moment, which will change the wheel load and affect the wheel speed. Therefore, the driving conditions are divided into two categories: steering and straight driving. The threshold method is adopted to identify the current driving state, and the judgment logic is shown in formula 1.

$$\begin{cases} \begin{matrix} Steering \\ |\delta| > \varepsilon_1 \end{matrix} \begin{cases} Normal\ steering & |w_{rj}| < \varepsilon_2 \& a_{thro} < \varepsilon_3 \& \overline{v_x} < \varepsilon_4, F = 1 \\ Abnormal\ steering & |w_{rj}| > \varepsilon_2 \& (a_{thro} > \varepsilon_3 || \overline{v_x} > \varepsilon_4), F = 2 \end{cases} \\ \begin{matrix} Straight \\ |\delta| < \varepsilon_1 \end{matrix} \begin{cases} Rapid\ acceleration & a_{thro} > \varepsilon_4 \& \alpha_x > \varepsilon_5, F = 3 \\ Normal\ acceleration/constant\ speed & |w_{rj}| < \varepsilon_2 \& |\alpha_x| < \varepsilon_5, F = 4 \\ Normal\ deceleration & |w_{rj}| < \varepsilon_2 \& |\alpha_x| < \varepsilon_5 \& s_{brake} = 1, F = 5 \\ Sharp\ deceleration & \alpha_x < -\varepsilon_5 \& s_{ABS} = 1 \& s_{brake} = 1, F = 6 \end{cases} \end{cases} \quad (1)$$

Where F is the mark bit defined by the working condition; $\varepsilon_1, \varepsilon_2, \varepsilon_3, \varepsilon_4, \varepsilon_5$ are respectively the front wheel rotation angle, absolute value of wheel speed difference, throttle position, average speed of four wheels and acceleration logic threshold.

According to the different working conditions and the difference of wheel speed, the estimated speed can be obtained through the maximum wheel speed method, minimum wheel speed method, etc.[1], which can be used as the key variable of the estimation algorithm below.

4. LSC module

4.1 Longitudinal Speed estimation of straight driving conditions

4.1.1 Derivation and design of Kalman filter

Under conventional driving conditions, the pre-estimated vehicle speed can be directly used as the estimated value of vehicle speed because the wheel does not slip significantly. However, in unconventional conditions, the deviation of the pre-estimated vehicle speed is large, so it is necessary to introduce acceleration information correction. In this paper, the weight of the estimated vehicle speed and acceleration is changed by adjusting the measurement noise covariance value, so as to adjust the vehicle speed value.

The ideal radius of the wheel without load is R_0. In the case of load, the wheel will deform and the radius will change, denoted as R_a[4]. However, in general, the wheel speed is calculated according to the ideal radius R_0. Here, the offset of wheel radius is defined as ε. That is:

$$\varepsilon = \frac{R_0 - R_a}{R_0} \tag{2}$$

Therefore, the real speed of the vehicle can be expressed as:

$$v_t = v_w(1 + \varepsilon) \tag{3}$$

The pre-estimated speed is composed of the measured noise of the calculated value of the speed, namely:

$$v_m = v_w + \sigma_v \tag{4}$$

$$a_m = a_t + \sigma_a \tag{5}$$

Where v_t is the real speed of the vehicle; v_w is the estimated value of ideal speed; v_m is the speed measurement; a_m , a_t are respectively body measurement and real longitudinal acceleration; σ_v , σ_a is the observation noise of velocity and acceleration.

Due to the influence of vehicle vibration and environment, the measured data of the sensor has a certain error, so σ_v , σ_a is assumed to be Gaussian white noise with an average value of 0.

The state transition equation and observation equation of the system are as follows:

$$x(k+1) = f(x_k, w_k) = A(k)x(k) + B(k)w(k) \tag{6}$$

$$z(k) = h(x_k, n_k) = Hx(k) + n(k) \tag{7}$$

w_1, w_2, w_3 is defined as process noise. The vector and matrix are defined as follows:

$$x(k) = \begin{bmatrix} a_t(k) \\ v_w(k) \\ \varepsilon(k) \end{bmatrix},\ w(k) = \begin{bmatrix} w_1 \\ w_2 \\ w_3 \end{bmatrix},\ H = \begin{bmatrix} 1 & 0 & 0 \\ 0 & 1 & 0 \end{bmatrix}, z(k) = \begin{bmatrix} a_m(k) \\ v_m(k) \end{bmatrix},\ n(k) = \begin{bmatrix} \sigma_a(k) \\ \sigma_v(k) \end{bmatrix} \tag{8}$$

$$A(k) = \begin{bmatrix} 1 & 0 & 0 \\ t & 1 & -ta_m(k) \\ 0 & 0 & 1 \end{bmatrix}, B(k) = \begin{bmatrix} t & 0 & 0 \\ -\frac{1}{2}t^2 & t & \frac{1}{2}t^2 a_m(k) \\ 0 & 0 & t \end{bmatrix} \tag{9}$$

According to Kalman filtering principle[5], it can be obtained:

$$\hat{x}(k|k-1) = A(k-1)x(k-1|k-1) \tag{10}$$

$$P(k|k-1) = A(k-1)P(k-1|k-1)A^T(k-1) + Q(k-1) \tag{11}$$

$$K(k) = P(k|k-1)H^T(k)[H(k)P(k|k-1)H^T(k) + R(k)]^{-1} \tag{12}$$

$$\hat{x}(k|k) = \hat{x}(k|k-1) + K(k)[z(k) - H(k)\hat{x}(k|k-1)] \tag{13}$$

$$P(k|k) = [I - K(k)H(k)]P(k|k-1) \tag{14}$$

4.1.2 Parameter confirmation of Kalman algorithm

Although the fixed observation noise can achieve the filtering effect well, when the vehicle is in the unusual condition and the wheel skids or is locked, the Kalman filter with fixed parameters cannot reflect the real speed[6]. Therefore, according to the different working conditions, this paper changes the measurement noise R in real time, changes the weight of the pre-estimated speed and acceleration, and then revises the estimated speed of the vehicle.

In engineering practice, when the acceleration signal is small and the signal-to-noise ratio of the signal is low, its credibility will be reduced. Therefore, too much trust in the acceleration signal will reduce the authenticity of longitudinal speed. Therefore, the weight of the estimated speed should be appropriately increased, such as in conventional conditions. Similarly, when the acceleration is large and the accuracy of the system dynamics model decreases, the kinematics method should be selected and the weight of the acceleration signal should be appropriately increased, such as the conditions of easy skid such as rapid acceleration and rapid braking. That is:

$$\begin{cases} R_a \uparrow, R_v \downarrow & \text{Unconventional condition} \\ R_a \downarrow, R_v \uparrow & \text{Conventional condition} \end{cases} \quad (15)$$

4.2 Longitudinal speed estimation in steering condition

4.2.1 Vehicle model

Vehicle is a nonlinear system with high complexity, and its related parameters are difficult to calculate.Therefore, this paper grasps the most critical factors, simplifies the model to the greatest extent, and saves calculation time. A 3-DOF vehicle model including longitudinal, lateral and yaw was established, as shown in figure 2.

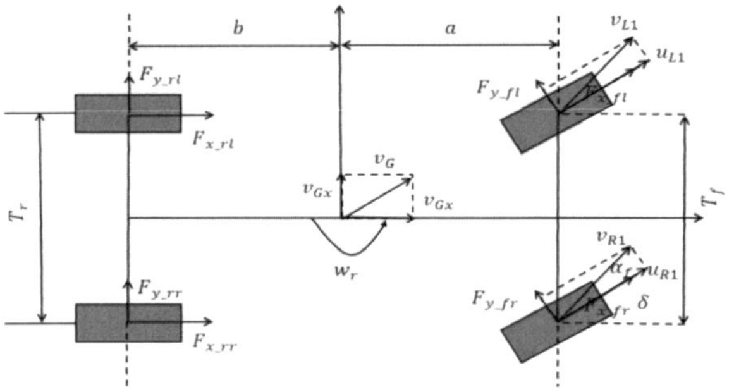

Figure 2. 3-DOF vehicle model

In this model: F_{xyij} is the reaction force of four wheels by the ground; δ is the front wheel rotation angle; w_r is yaw rate; v_{Gx}, v_{Gy} are respectively the longitudinal and lateral velocity at the center of mass; a,b are the distance between the center of mass of the vehicle and the front and rear axles; T_f, T_r is the front and rear wheelbase; α_f, α_r is the slip angle of front and rear tires.

According to the principle of Kalman algorithm, the state transition equation and observation equation of the 3-DOF vehicle model are determined[7].

$$\begin{cases} \dot{w_r} = \dfrac{a^2k_1 + b^2k_2}{I_z v_x} w_r + \dfrac{ak_1 - bk_2}{I_z}\beta - \dfrac{ak_1}{I_z}\delta \\ \dot{\beta} = \left(\dfrac{ak_1 - bk_2}{mv_x} - 1\right) w_r + \dfrac{k_1 + k_2}{mv_x}\beta - \dfrac{k_1}{mv_x}\delta \\ \dot{v_x} = w_r \beta v_x + a_x \end{cases} \qquad \begin{cases} a_y = \dfrac{ak_1 - bk_2}{mv_x} w_r + \dfrac{k_1 + k_2}{m}\beta - \dfrac{k_1}{m}\delta \\ w_r = w_m \end{cases} \quad (16)$$

It is obtained by referring to equations (6) - (14) after the Kalman algorithm derivation process, which will not be repeated here.

4.2.2 Tyre sideslip dynamics

Under the steering condition, the tire sideslip and the body lateral movement will greatly reduce the accuracy of longitudinal velocity estimation, and the information of wheel speed and acceleration can not reflect the change of vehicle speed well[8]. In order to solve this problem, this paper first uses a 3-DOF vehicle model to obtain the yaw rate and sideslip angle of mass center by EKF algorithm, then analyzes the sideslip dynamics of the tire and yaw motion of the vehicle body, and deduces the formula of longitudinal velocity estimation.

The sideslip angle of the two front / rear left and right wheels is regarded as the same, and the wheel speed is the component of the absolute wheel speed, and its relation is as follows:

$$\begin{cases} \alpha_f = \hat{\beta} + \dfrac{a}{v_{gx}} w_r - \delta \\ \alpha_r = \hat{\beta} - \dfrac{a}{v_{gx}} w_r \end{cases} \qquad \begin{cases} v_{L1} = u_{L1}/cos\alpha_f \\ v_{R1} = u_{R1}/cos\alpha_f \\ v_{L2} = u_{L2}/cos\alpha_r \\ v_{R2} = u_{R2}/cos\alpha_r \end{cases} \tag{17}$$

Where $\hat{\beta}$ is the estimated sideslip angle of the center of mass.

Considering the lateral yaw movement of the car body, the absolute speed of the wheel still cannot be taken as the longitudinal speed, which needs to be converted to the center of mass of the vehicle. The conversion formula is as follows:

$$\begin{cases} \hat{v}_{gxL1} = \sqrt{v_{L1}^2 - (\hat{v}_{gy} + w_r a)^2} + w_r \dfrac{T_f}{2} \\ \hat{v}_{gxR1} = \sqrt{v_{R1}^2 - (\hat{v}_{gy} + w_r a)^2} - w_r \dfrac{T_f}{2} \\ \hat{v}_{gxL2} = \sqrt{v_{L2}^2 - (\hat{v}_{gy} + w_r b)^2} + w_r \dfrac{T_r}{2} \\ \hat{v}_{gxR2} = \sqrt{v_{R2}^2 - (\hat{v}_{gy} + w_r b)^2} - w_r \dfrac{T_r}{2} \end{cases} \qquad \hat{v}_{gy} = v_{gx}\hat{\beta} \tag{18}$$

Where $\hat{v}_{gxij}$ is the longitudinal velocity at the center of mass estimated from the four-wheel speed; $\hat{v}_{gy}$ is the lateral velocity estimation at the center of mass.

Since there is no driven wheel in the four-wheel-drive vehicle, the longitudinal velocity of the vehicle is defined as the average estimated speed of four wheels. That is:

$$\hat{v}_{Gx} = \frac{\hat{v}_{gxL1} + \hat{v}_{gxR1} + \hat{v}_{gxL2} + \hat{v}_{gxR2}}{4} \tag{19}$$

5. Simulation verification and analysis

The research object adopted in this paper is a four-wheel-drive off-road vehicle. Firstly, the vehicle model is established by CarSim, and its main parameters are shown in table 1. Then, the vehicle speed estimation system model is established by MATLAB/Simulink, and the CarSim-Simulink co-simulation model is built to prove the effectiveness of the proposed algorithm[9]. The adhesion coefficient of the road is uniformly set at 0.8, which is consistent with the common asphalt pavement.

The actual speed is based on the CarSim output reference speed. The road adhesion coefficient is uniformly set at 0.7, which is consistent with common asphalt pavement.

Table 1 Partial parameters of the vehicle

Parameter	Value of parameter
vehicle mass m/kg	3590
distance from front axis to center of mass a/m	1.18
distance from rear axis to center of mass b/m	1.77
front track T_f/m	1.80
rear track T_r/m	1.80
front wheel side stiffness $k_1/(N/rad)$	-50484
rear wheel side stiffness $k_2/(N/rad)$	-50484
radius of the tire R/m	0.39
the moment of inertia of the body about the Z axis $I_z/kg \cdot m^2$	2687.1

Due to the pre-estimated speed can better simulate the real speed under normal working conditions, the simulation part will not elaborate too much.

5.1 Rapid acceleration/Sharp deceleration condition

Rapid acceleration condition setting: the initial speed of the vehicle is set to 0, and the throttle opening reaches and keeps the maximum value at 0.1s.

Sharp deceleration condition setting: the initial speed of the vehicle is 100km/h, and the braking pressure is linearly increased to the maximum 20MPa between 0-2s, and the maximum is maintained until the vehicle stops.

The simulation results show that, according to figure 3, under the condition of rapid acceleration, all four wheels will slip to a certain extent at the starting stage. In addtion, in the four-wheel drive mode, the wheel speed will alse be slightly greater than the vehicle speed under the condition of vehicle stability, which cannot reflect the real longitudinal speed. By increasing the weight of the added speed of the Kalman filter, the relation between the speed and the wheel speed was reduced. Due to the presence of the integral error, the maximum absolute error of the estimated longitudinal speed was 0.126m/s, which was greatly improved compared with the conventional kinematics estimation method.Similarly,according to figure 4, the maximum absolute error of the estimated longitudinal speed was 0.25m/s.

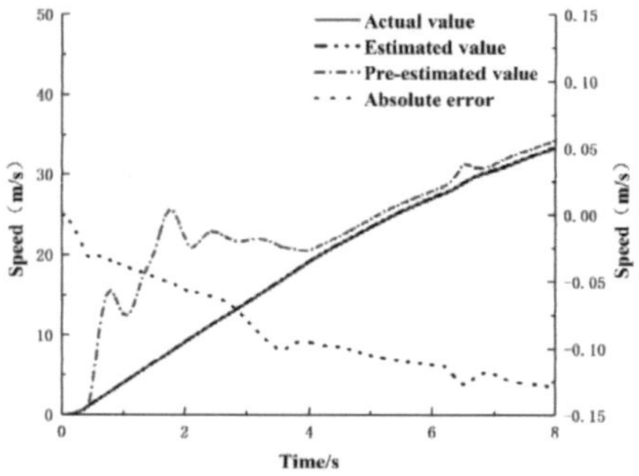

Figure 3.Comparison of longitudinal speed estimation under rapid acceleration conditions

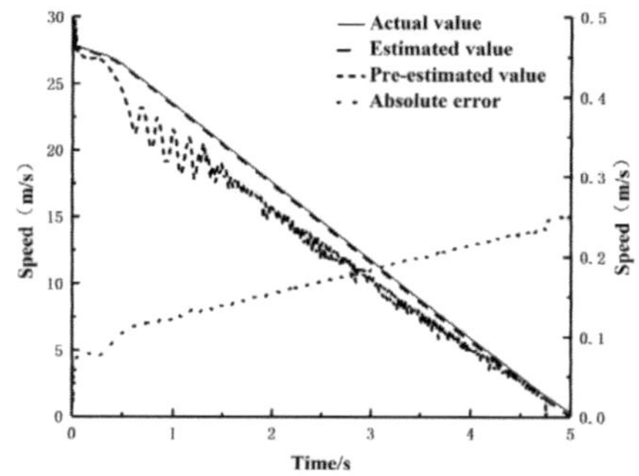

Figure 4.Comparison of longitudinal speed estimation in sharp deceleration conditions

5.2 Abnormal steering condition

Driving condition setting: the vehicle runs at a continuous target speed of 100km/h and the steering wheel turns left at 720°. During the period of 1.5s-3.5s, the steering wheel returns to zero at a uniform speed

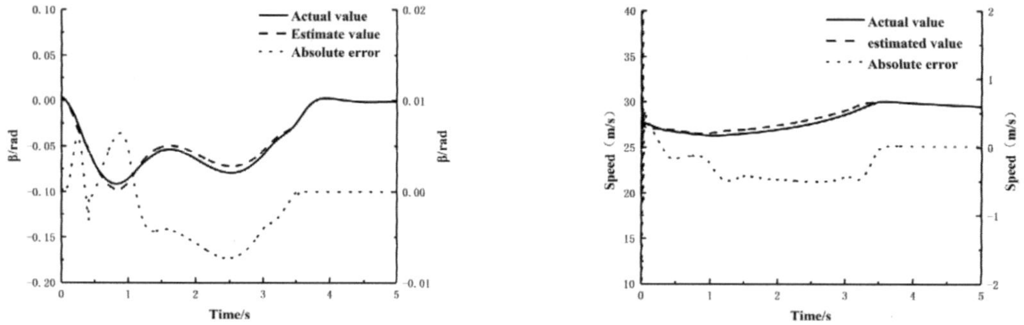

(a)Sideslip angle of mass center (b)The vehicle longitudinal speed

Figure 5.Estimation comparison of abnormal steering conditions

The simulation results show that: according to figure 5, the maximum absolute error of the estimated sideslip angle of mass center is 0.0064. Due to the wheel side angle is large at high speed and large rotation angle,although the accuracy of the established 3-DOF model is reduced, the actual vehicle speed can still be well tracked by adjusting the weights of process noise and observation noise.The absolute error of vehicle speed estimation is 0.46m/s.

6 Conclusion

In this paper, different estimation methods are selected by identifying different working conditions.

The kinematics method was used to adjust the weight value of noise according to the driving state of the vehicle in the straight driving condition, and the Kalman method was used to estimate the transverse state of the vehicle in the dynamic steering condition, so as to obtain the longitudinal velocity at the center of mass. These methods are simple to implement and practical. According to the results of CarSim-Simulink co-simulation test, it can be seen that the estimated longitudinal speed is in good agreement with CarSim output reference speed under various working conditions.

References

[1] Wang Shichao, Qi Zhiquan, Chen Sizhong, Yu Yangmei, Shi Hongsong. Research on Longitudinal Vehicle Speed Estimation Method Based on Vehicle Kinematics [A]. China Society of Automotive Engineers 2019 Annual Conference Proceedings (4) [C].China Society of Automotive Engineers,2019:5.

[2] Wang Zhenpo, Ding Xiaolin, Zhang Lei. Overview on Key Technologies of Acceleration Slip Regulation for Four-wheel-independently-actuated Electric Vehicles [J]. Journal of Mechanical Engineering,2019,55(12):99-120.

[3] Dai Xi. Research on Vehicle Longitudinal Speed Estimation Method Based on Multi-sensor Information Fusion [D]. Yanshan University,2018.

[4] LIU Tong. Research and Hardware Implementation of Absolute Speed Estimation for Vehiclethrough Fuzzy Kalman Filter [D].Jilin University,2011.

[5] Huang Xiaoping.(2015)Principle and Application of Kalman Filter. Publishing House of Electronics Industry.BeiJing.

[6] CHU Wenbo, LI Shen.Vehicle Speed Estimation of All-wheel Independent Electric Drive Vehicle Based on Multi-information Fusion [J]. Automotive Engineering,2011,33(11):962-966.

[7] Zhu Jingna. Research on Yaw Stability Control Strategy of Distributed Drive Electric Vehicle Based on Torque Distribution [D].Beijing Institute of Technology,2017.

[8] Wang G , Li S . A Novel Longitudinal Speed Estimator for Fully Automation Ground Vehicle on Cornering Maneuver[C]// International Conference on Artificial Intelligence & Computational Intelligence. IEEE, 2010.

[9] SHI Yanru. Research on Vehicle Longitudinal and Lateral Velocity Estimation Algorithm Based on UKF Filter [D].Jilin University,2011.

Journal of Physics: Conference Series **2113** (2021) 012012 doi:10.1088/1742-6596/2113/1/012012

QoS-Aware Power Allocation for Multi-UAV Aided Networks

Xiaoya Quan[1]

[1]Guangzhou Foreign Language School, Guangzhou,Guangdong,510000, China

*Corresponding author's e-mail: alinda@wedo365.com

Abstract. UAV base stations (UAVBS's) have been proposed as a revolution for the new architecture of 5G networks. The UAVBS's can be deployed as access points to provide wireless services to users in emergency scenarios. However, it is challenging to solve the highly coupled problem for UAVBS deployment and power allocation. In the meanwhile, the hybrid analog and digital beamforming is leverage to reduce the hardware cost for beamforming in 5G networks. In this work, we first use k-means algorithm to solve the 3D placement of UAVBS's by exploiting the optimal coverage altitude. Next, power allocation problem is resolved using the difference-of-two-convex functions (D.C.) programming algorithm. Furthermore, the quality of service (QoS) for each user is guaranteed by adjusting the transmitted power. Finally, extensive experiments are conducted to demonstrate the feasibility of the proposed algorithm.

1. Introduction

Unmanned aerial vehicle base stations (UAVBS's), as one of the most promising technologies, have been widely used in some specific environments to support the emergency wireless communications. For instance, UAVBS's are deployed in a post-disaster area with unknown user distribution to support the temporary communications by taking the advantage of the high mobility and low cost [1]. Although the proposed algorithm outperforms the baseline spiral path in terms of the total number of users served, it fails to mention the quality of the communication.

In a large area, it usually requires multiple UAVBS's to cover the demanded users. Due to the complexity of the three-dimensional (3D) mobility, it is challenging to optimize the policy for multiple UAVs by considering the co-channel interference [2][3]. In [4], a deep learning method was proposed to construct the optimal UAVBS deployment. However, it definitely requires large-scale training dataset which is not feasible in practical environment. Moreover, neural networks are not capable of reasoning and cannot interpret the results. If special conditions are emerged in an emergency paralyze machine learning, it is severe to handle the corresponding problems. Thus, a low-complexity deployment algorithm is urgently required for UAVBS systems.

Furthermore, the data traffic is extremely high in 5G networks. As a result, the bandwidth is demanded to be improved for the UAVBS to support the large-scale data transmission. As a promising solution, millimeter wave (mmWave) is used in 5G networks since the higher frequency can support a lager bandwidth. Unfortunately, the higher frequency may result in heavy pathloss. Thus, beamforming technique is proposed to overcome the severe pathloss in 5G networks. However, the conventional fully-digital beamforming systems are not intractable due to the high cost and power consumption. More specifically, conventional beamforming requires one RF chain for one antenna, which is unaffordable in practice. To cope with this challenge, hybrid analog and digital beamforming was proposed in [5].

In contrast to the fully-digital beamforming systems, hybrid precoding reduces the number of RF chains by taking the advantage of the multiple phase shifters and mixers. [6] designed the hybrid precoding for UAVBS's using the training data by machine learning-based algorithm. It requires iterative optimization which may cause high latency due to the high complexity.

On the other hand, power allocation is also an important problem for the co-channel interference (CCI) elimination. In [7], an energy-efficient power allocation approach was developed to maximize the energy efficiency (EE). However, the spectral efficiency (SE) is the demanded metric to be maximized. [8] constructed a hybrid precoding design and power allocation using the zero-forcing algorithm while the quality of service (QoS) for each user cannot be guaranteed.

In this work, a hybrid precoding method is proposed for UAVBS's to reconstruct the communications in emergency scenarios. More specifically, we implement a novel deployment method for UAVBS's placement. Next, a hybrid analog and digital precoding method is leveraged to reduce the number of RF chains. After that, a power allocation scheme is developed by using the difference of two convex function (D.C.) programming algorithm.

Finally, the contributions of this work can be summarized as follows:

- The UAVBS's deployment problem is solved using k-means algorithm by considering the practical pathloss channel model. The 3D placement of UAVBS's is separately addressed to optimize the user association by considering the CCI.
- Hybrid precoding and power allocation are jointly optimized using the D.C. programming by considering the large-scale pathloss and beamforming gain.
- Extensive simulations are conducted to demonstrate the effectiveness of the proposed QoS-aware power allocation hybrid precoding algorithms for UAVBS systems.

The remainder content of this paper is organized as follows. Section II presents the system model and formulates the problems for UAVBS systems. After that, the UAVBS deployment is separately optimized in 3D space using k-means algorithm by considering the pathloss channel model in Section III. As the deployment problem is solved, the hybrid precoding and QoS-aware power allocation are jointly optimized using the D.C. programming algorithm in Section IV. Next, the computer simulation results are elaborated in Section V and the conclusion is summarized in the end.

2. System Model and Problem Formulation

In the sequel, the system model as well as channel model for UAVBS are presented. Furthermore, the optimization problem is derived.

2.1 Air-to-Ground (ATG) Channel

As shown in Fig. 1, K UAVBS's are deployed over the users to support the emergency communications for N users. Specifically, N_k users are assigned to the k-th UAVBS. Thus, we have $\sum_{k=1}^{K} N_k = N$.

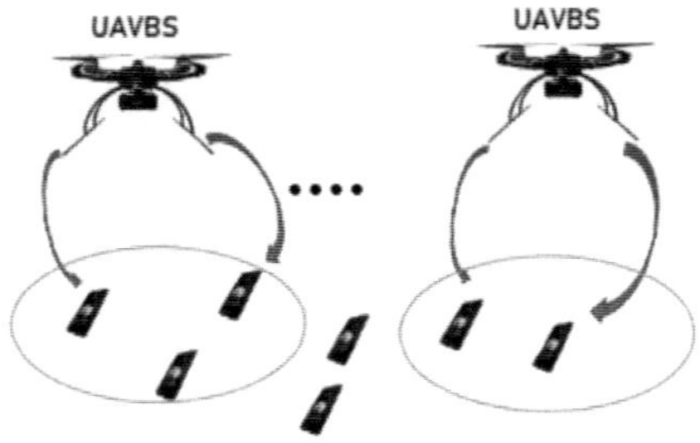

Fig. 1. The schematic diagram of proposed system.

The pathloss between the k-th UAV and the u-th user can be formulated as follows:

$$\mathrm{L}_{k,u}^{dB} = \frac{\eta_{LoS} - \eta_{NLoS}}{1 + ae^{-b(\theta_{k,u} - a)}} + 20log(d_{k,u}) + \tau \tag{1}$$

where

$$\theta_{k,u} = \frac{180}{\pi} \sin^{-1}\left(\frac{h_k^U}{d_{k,u}}\right)$$

$$\tau = 20logf + 20log(4\pi/c) + \eta_{NLoS}$$

Recalling Equation (1), the pathloss can be calculated according to the positions of the users and UAVBS. Denote $\boldsymbol{p}_k^U = [x_k^U, y_k^U, z_k^U]$ and $\boldsymbol{p}_{k,u}^E = [x_k^E, y_k^E, 0]$ by the 3D location of the k-th UAVBS and its u-th user, respectively.

Thus, the horizontal and vertical distance between the k-th UAVBS to its u-th user can be represented by the

$$d_{k,u} = \sqrt{(x_k^U - x_k^E)^2 + (y_k^U - y_k^E)^2},$$

$$h_k^U = z_k^U.$$

2.2 Hybrid Precoding System Design

It is assumed that a base station has N_{BS} antennas and N_{RF} chains when it communicates with N_k users. Each user is equipped with N_{MS} antennas. For the k-th UAVBS, we are concerned with multi-user beamforming case where the UAVBS communicates with each user via one data stream. Thus, the number of minimally required RF chains is equal to the number of users, i.e., N_k.

We assume the digital precoder is $\boldsymbol{F}_k = [\boldsymbol{f}_{k,1}, \cdots, \boldsymbol{f}_{k,N_k}]$ followed by the analog precoder $\boldsymbol{\mathcal{V}}_k = [\boldsymbol{v}_{k,1}, \cdots, \boldsymbol{v}_{k,N_k}]$. The precoded signal transmitted by the UAVBS is

$$\boldsymbol{x}_k = \boldsymbol{\mathcal{V}}_k \boldsymbol{F}_k \boldsymbol{s}_k,$$

with $\boldsymbol{s}_k$ being the data stream $\mathbb{E}[\boldsymbol{s}\boldsymbol{s}^H] = \frac{P}{U} I_{N_k}$.

Then, the decoded signal can be represented as

$$y_{k,u} = \boldsymbol{w}_{k,u}^H \boldsymbol{H}_u \sum_{u=1}^{N_k} \boldsymbol{\mathcal{V}}_k \boldsymbol{f}_{k,u} s_{k,u} + \boldsymbol{w}_{k,u}^H \boldsymbol{n}_{k,u}.$$

The virtual channel model is expressed as

$$\boldsymbol{H}_{k,u} = \sqrt{\frac{N_{BS} N_{MS}}{L_{k,u}}} \sum_{l}^{L_{k,u}} \alpha_{k,u,l} \boldsymbol{a}_{MS}(\theta_{k,u,l}^r) \boldsymbol{a}_{BS}^H(\theta_{k,u,l}^t), \tag{2}$$

where $\alpha_{k,u,l}$ is the complex gain of the l-th path. We assume the UAVBS uses the uniform linear array antennas. The array response vectors in Equation (2) are given by

$$\boldsymbol{a}(\theta) = \left[1, e^{j\frac{2\pi}{\lambda}\sin\theta}, \cdots, e^{j(N_{BS}-1)\frac{2\pi}{\lambda}\sin\theta}\right].$$

2.3 Problem Formulation

The transmitted power is set to be $P_{k,u}^T$. The received signal power is formulated as

$$\beta_{k,u} = P_{k,u}^T - \mathrm{L}_{k,u}$$

Thus, the spectral efficiency of the u-th user under the k-th UAVBS can be presented as follows:

$$R_{k,u} = \log_2\left(1 + \frac{\beta_u |\boldsymbol{w}_u \boldsymbol{H}_u \boldsymbol{F}_{RF} \boldsymbol{f}_u^{BB}|^2}{\sum_{n \neq u}^{U} \beta_n |\boldsymbol{w}_u \boldsymbol{H}_u \boldsymbol{F}_{RF} \boldsymbol{f}_n^{BB}|^2 + \sigma_u^2}\right).$$

The optimization problem in this work is formulated by

$$\begin{aligned} \underset{\boldsymbol{W},\boldsymbol{V},\boldsymbol{F},\boldsymbol{p}}{\text{maximize}} \quad & R_{tot}(\boldsymbol{W},\boldsymbol{V},\boldsymbol{F},\boldsymbol{p}) \\ \text{subject to} \quad & C_1 : |[\boldsymbol{v}_{k,u}]_i|^2 = 1/N_T, i = 1,2,\cdots,N_T; \\ & C_2 : |[\boldsymbol{w}_{k,u}]_j|^2 = 1/N_R, j = 1,2,\cdots,N_R; \\ & C_3 : \|\boldsymbol{\mathcal{V}}_k \boldsymbol{f}_{k,u}\|^2 = 1; \\ & C_4 : \sum_{k=1}^{K}\sum_{u=1}^{M_k} p_{k,u} \leq N_U; \\ & C_5 : R_{k,u} \geq \lambda_{k,u}, \end{aligned} \tag{3}$$

where C_1 and C_2 are the analog constraints for the transmitter and receiver, respectively. C_3 is the power normalization on each RF chain while C_4 constrains the total power consumption. Finally, the QoS of each user is guaranteed with a threshold $\lambda_{k,u}$.

The hybrid beamforming design and power allocation highly depend on the locations of the UAVBS's. Thus, we first need to address the locations of the UAVBS's according to the fixed positions of the users.

3. 3D UAVBS deployment

In this section, the positions of the UAVBS's are separately addressed using the k-means algorithm based on the proposed constraints.

3.1 Horizontal placement by K-Means algorithm

K-means algorithm is the unsupervised method to classify the clusters of objects. Here, a user association method is proposed using k-means by calculating the horizontal distance between the UAVBS and users. First, the altitude of the UAVBS is neglected. We need to deploy the UAVBS's to satisfy the following equation:

$$\min_{\boldsymbol{p}_k^U} \sum_{k=1}^{K}\sum_{u=1}^{N_k} \|\boldsymbol{p}_k^U - \boldsymbol{p}_{k,u}^E\|^2.$$

To solve the positions of the UAVBS, k-means algorithm is implemented to iteratively optimize the center of users. Therefore, the position of the UAVBS can be obtained after several iterations. In the user association problem, the Euclidean distance is employed to find the suitable assignment of each user.

Altitude Optimization

As the horizontal positions of the UAVBS's fixed, we now consider the optimally deployed altitudes. Recalling the Equation (1), the optimal altitude can be derived by

$$\frac{\partial \mathrm{L}_{k,u}^{\max}}{\partial h_k} = 0, \tag{4}$$

where $\mathrm{L}_{k,u}^{\max}$ is the farthest user under the k-th UAVBS. Intuitively, the pathloss of the farthest user should be considered for the altitude adjustment.

In practical, it is particularly challenging to analytically solve the Equation (4). As a result, we can numerically calculate the locally optimum for the altitude derivation.

3.2 Constraints of UAVBS's

Now we consider the constraints for the UAVBS deployment. Since co-channel interference exists among different UAVBS's, the user should be only assigned to one UAVBS at one instant. Furthermore, the maximal coverage radius and altitude of the UAVBS should be also constrained. Thus, the constraints are formulated as:

$$d_{k,u} \leq R_m, \text{for} \quad k = 1, \cdots, K; u = 1, \cdots, N_k,$$
$$z_k^U \leq H_m, \text{for} \quad k = 1, \cdots, K,$$

where R_m and H_m are the given maximal coverage radius and altitude, respectively.

In addition, the safety distance of the UAVBS's should be satisfied. This constraint can be represented as

$$\|\boldsymbol{p}_k^U - \boldsymbol{p}_j^U\|_2 \geq D_m, \text{for} \quad \forall k, j = 1, \cdots, K,$$

where D_m is the given minimal safety distance between two UAVBS's.

4. Hybrid precoding design and QoS-aware power allocation

4.1 Hybrid precoding algorithm for UAVBS

We begin with the analog beamforming design for both transmitter and receiver. It is well known that distinct array response vectors are asymptotically orthogonal as the number of antennas in an antenna array goes to infinity [9], i.e.,

$$\lim_{N \to +\infty} \boldsymbol{a}_T^H(\phi_{k,u}^t, \theta_{k,u}^t) \cdot \boldsymbol{a}_T(\phi_{\ell,v}^t, \theta_{\ell,v}^t) = \delta(k - \ell)\delta(u - v).$$

However, since the antenna number is finite in practice, the residual interference must be considered in the analog precoding design. Recalling the channel model presented in Equation (2), we can asymptotically orthogonalize the transmitted signals by optimizing the design of $\boldsymbol{w}_{k,u}$ and $\boldsymbol{v}_{k,u}$:

$$\begin{aligned}
\{\boldsymbol{w}_{k,u}^*, \boldsymbol{v}_{k,u}^*\} &= \arg\max_{\tilde{\boldsymbol{w}}_{k,u}, \tilde{\boldsymbol{v}}_{k,u}} \sum_{k=1}^{K} \sum_{u=1}^{M_k} \log_2\left(1 + \text{SINR}\left(\tilde{\boldsymbol{w}}_{k,u}, \tilde{\boldsymbol{v}}_{k,u}\right)\right) \\
\text{subject to} \quad & \tilde{\boldsymbol{v}}_{k,u} \in \mathcal{A}_{k,u}^T; \\
& \tilde{\boldsymbol{w}}_{k,u} \in \mathcal{A}_{k,u}^R; \\
& \max\{M_k\}_{k=1}^K < N_{RF},
\end{aligned}$$

where

$$\mathcal{A}_{k,u}^T = \left[\boldsymbol{a}_T(\phi_{k,u,1}^t, \theta_{k,u,1}^t), \cdots, \boldsymbol{a}_T(\phi_{k,u,L_{k,u}}^t, \theta_{k,u,L_{k,u}}^t)\right],$$
$$\mathcal{A}_{k,u}^R = \left[\boldsymbol{a}_R(\phi_{k,u,1}^r, \theta_{k,u,1}^r), \cdots, \boldsymbol{a}_R(\phi_{k,u,L_{k,u}}^r, \theta_{k,u,L_{k,u}}^r)\right].$$

Furthermore, $\text{SINR}_{k,u}$ is given by

$$\text{SINR}\left(\tilde{\boldsymbol{w}}_{k,u}, \tilde{\boldsymbol{v}}_{k,u}\right) = \frac{|\tilde{\boldsymbol{w}}_{k,u}^H \boldsymbol{H}_{k,u} \tilde{\boldsymbol{v}}_{k,u}|^2}{\sum_{j=1, j \neq k}^{N_U} \|\tilde{\boldsymbol{w}}_{k,u}^H \boldsymbol{H}_{k,u} \tilde{\boldsymbol{V}}_j\|^2 + \sum_{t \neq u}^{M_k} |\tilde{\boldsymbol{w}}_{k,u}^H \boldsymbol{H}_{k,u} \tilde{\boldsymbol{v}}_{k,t}|^2 + \frac{1}{\gamma}},$$

With $\gamma = \frac{1}{\sigma_{k,u}^2}$. The optimal analog beamforming precoder can be straightforwardly found by exhaustively searching in the feasible sets of $\mathcal{A}_{k,u}^T$ and $\mathcal{A}_{k,u}^R$.

In contrast to the conventional ZF hybrid beamforming scheme [10] that requires $N_U \leq N_{RF}$, zero-forcing digital precoding scheme is first proposed to transmit data-streams group by group. More specifically, the digital precoder for each block is designed as the inverse of the effective channel of the block:

$$\boldsymbol{\mathcal{F}}_k^{\text{BZF}} = \boldsymbol{\mathcal{G}}_k^H (\boldsymbol{\mathcal{G}}_k \boldsymbol{\mathcal{G}}_k^H)^{-1},$$

With $N_{RF} \geq M_k$, where $\boldsymbol{\mathcal{G}}_k = [\boldsymbol{g}_{k,1}^{(k)}, \boldsymbol{g}_{k,2}^{(k)}, \cdots, \boldsymbol{g}_{k,M_k}^{(k)}]^H$.

To satisfy the constraint C_3 in Equation (3), power normalization is performed on each $\boldsymbol{f}_{k,u}$ derived from $\boldsymbol{\mathcal{F}}_k^{\text{BZF}} = \left[\boldsymbol{f}_{k,1}^{\text{BZF}}, \boldsymbol{f}_{k,2}^{\text{BZF}}, \cdots, \boldsymbol{f}_{k,M_k}^{\text{BZF}}\right]$ as

$$\bar{\boldsymbol{f}}_{k,u}^{\text{BZF}} = \frac{\boldsymbol{f}_{k,u}^{\text{BZF}}}{\|\boldsymbol{\mathcal{V}}_k \cdot \boldsymbol{f}_{k,u}^{\text{BZF}}\|}.$$

Subsequently, this scheme is referred to as the block zero-forcing (BZF) scheme. It is worth noting that BZF degenerates to [10] if $K = 1$, i.e., all users are grouped into one independent group. On the other hand, BZF becomes the analog-only BDMA if $K = N_U$, i.e., each user forms one group and only analog beamforming is performed.

4.2 QoS-Aware Power Allocation Using D.C. Programming

For given analog and digital precoders, we investigate the QoS-aware power allocation $\boldsymbol{p}$ in Equation (3) by using the D.C. programming technique in this section.

We begin with reformulating Equation (3) as

$$\begin{aligned} &\underset{\boldsymbol{p}}{\text{maximize}} \quad \sum_{k=1}^{K}\sum_{u=1}^{M_k} R_{k,u}(\boldsymbol{p}) \\ &\text{subject to} \quad C_1 : \sum_{k=1}^{K}\sum_{u=1}^{M_k} p_{k,u} \le P; \\ &\qquad\qquad\quad C_2 : R_{k,u} \ge \lambda_{k,u}, \end{aligned}$$

Following the procedures in [11], the problem above can be cast as a D.C. programming problem:

$$\underset{\boldsymbol{p}}{\text{maximize}} \quad f(\boldsymbol{p}) - g(\boldsymbol{p}) \tag{5}$$

where

$$f(\boldsymbol{\beta}) = \sum_{k=1}^{K}\sum_{u=1}^{N_k} \log_2\left(\sum_{k=1}^{K}\sum_{u=1}^{N_k} \beta_{k,u}\left|\boldsymbol{g}_{k,u}^{(k)H}\boldsymbol{f}_{k,u}\right|^2 + \sigma_{k,u}^2\right),$$

$$g(\boldsymbol{\beta}) = \sum_{k=1}^{K}\sum_{u=1}^{N_k} \log_2\left(\sum_{\substack{j=1\\ j\neq k}}^{K}\sum_{t=1}^{N_j} \beta_{j,t}\left|\boldsymbol{g}_{k,u}^{(j)H}\boldsymbol{f}_{j,t}\right|^2 + \sigma_{k,u}^2\right).$$

Since the analog and digital precoders are obtained, both $f(\boldsymbol{p})$ and $g(\boldsymbol{p})$ are concave in$\boldsymbol{P}$, i.e., Equation (5) is a D.C. function. Starting from a feasible $\boldsymbol{p}^{(0)}$, the optimal $\boldsymbol{p}^{(n+1)}$ at the n-th iteration is generated as the optimal solution of a convex problem:

$$\max_{\boldsymbol{p}} f(\boldsymbol{p}) - g(\boldsymbol{p}^n) - \langle \nabla g(\boldsymbol{p}^{(n)}), \boldsymbol{p} - \boldsymbol{p}^{(n)} \rangle, \tag{6}$$

which can be efficiently solved by any existing convex programming software, such as CVX [12]. The computational complexity of Equation (6) is $\mathcal{O}(N_{RF}^3)$ in each iteration [11].

As $g(\boldsymbol{p}^n)$ is concave, its gradient $\nabla g(\boldsymbol{p}^{(n)})$ is also super-gradient:

$$\begin{aligned} &f(\boldsymbol{p}^{(n+1)}) - g(\boldsymbol{p}^{(n+1)}) \ge \\ &f(\boldsymbol{p}^{(n+1)}) - \left[g(\boldsymbol{p}^{(n)}) + \langle \nabla g(\boldsymbol{p}^{(n)}), \boldsymbol{p}^{(n+1)} - \boldsymbol{p}^{(n)} \rangle\right]. \end{aligned} \tag{7}$$

The proof is given in Appendix 1.

Finally, since $\boldsymbol{p}^{(n+1)}$ is the solution to Equation (5), it follows that

$$\begin{aligned} & f(\boldsymbol{p}^{(n+1)}) - g(\boldsymbol{p}^{(n)}) - \langle \nabla g(\boldsymbol{p}^{(n)}), \boldsymbol{p}^{(n+1)} - \boldsymbol{p}^{(n)} \rangle, \\ \ge \;& f(\boldsymbol{p}^{(n)}) - g(\boldsymbol{p}^{(n)}) - \langle \nabla g(\boldsymbol{p}^{(n)}), \boldsymbol{p}^{(n)} - \boldsymbol{p}^{(n)} \rangle, \\ = \;& f(\boldsymbol{p}^{(n)}) - g(\boldsymbol{p}^{(n)}). \end{aligned}$$

Therefore, the $(n+1)$-th solution is always better than the previous one. The iterative process terminates after $|f(p^{(n+1)}) - g(p^{(n+1)}) - (f(p^{(n)}) - g(p^{(n)}))| \leq \epsilon$ is achieved with a pre-defined threshold $\epsilon > 0$.

5. Simulation results

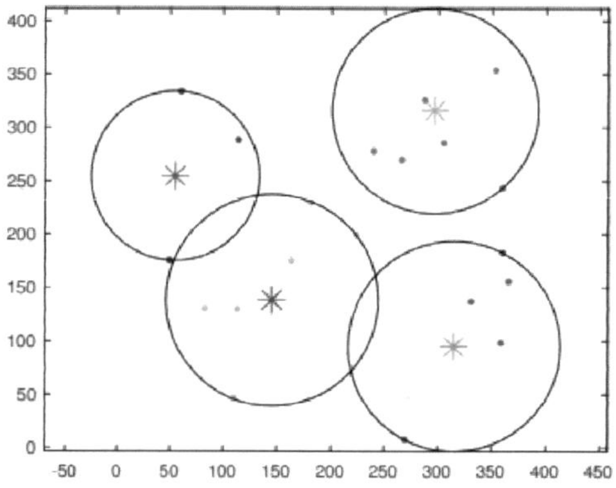

Fig. 2. The deployment of UAVBS's.

In Fig. 2, we can observe that the UAVBS's provide the coverage of the users for emergency communications. We assume 20 users are randomly assigned in a horizontal area with size 400×400 m^2 while four UAVBS's are deployed to support the communications. As proposed in Section III-C, there is no users among the intersection of the UAVBS's coverage. As a result, the interference among different UAVBS's can be minimized by optimizing the deployment.

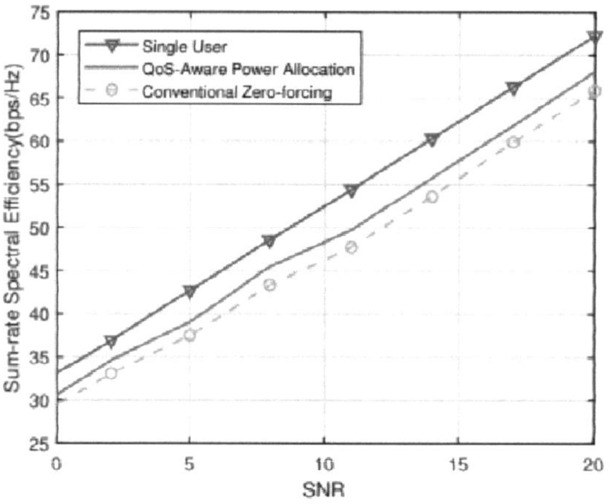

Fig. 3. Sum-rate capacity of users under one UAVBS.

In Fig. 3, it shows the comparison of the sum-rate between the conventional hybrid beamforming system and the proposed scheme with 6 RF chains serving 6 users. We consider the single-user case where only one user is assigned to the UAVBS at each time. Thus, the interference among different UGVs can be totally neglected. The curve labeled as ``Conventional Zero-forcing" refers to the conventional hybrid precoding with uniform power allocation. it is observed that the ``D.C. Power Allocation" outperforms the ``Conventional Zero-forcing" from 0 dB to 20 dB.

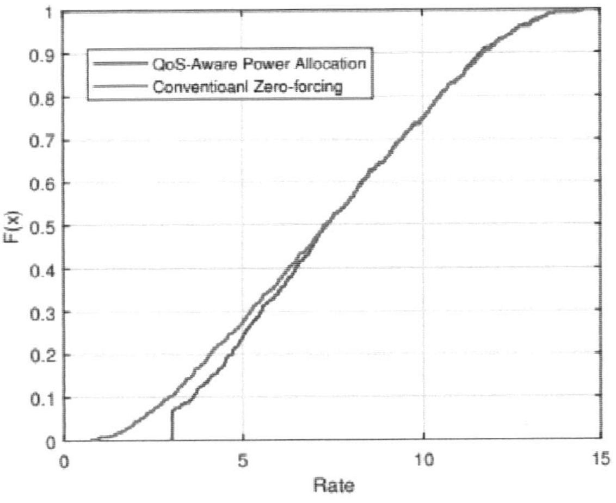

Fig. 4. CDF of users under one UAVBS.

Inspection of Fig. 4 suggests that the proposed D.C. programming algorithm can guarantee the QoS of each user. We set the threshold to be 3 bps/Hz. In contrast, there are approximately 10% of UGVs that cannot be satisfied with required QoS using uniform power allocation algorithm.

6. Conclusion

In this work, a hybrid preocoding scheme is developed to improve the throughput of the UAVBS's. We have proposed a D.C. programming algorithm to ensure the individual QoS. In addition, the deployment problem for the UAVBS's is addressed using k-means with several self-defined constraints. Furthermore, computer simulations are conducted to demonstrate the effectiveness of our proposed algorithms.

Appendix

Proof for Equation (7)

Suppose $f(\boldsymbol{p})$ is a concave function on a convex neighborhood $\mathcal{C}$ and differentiable at $\bm{p}$. Then, for every $\boldsymbol{y} \in \mathcal{C}$, we have the following inequality based on the definition of concavity:

$$\begin{aligned} & f((1-\lambda)\boldsymbol{p}+\lambda\boldsymbol{y}), \\ =& f\left(\boldsymbol{p}+\lambda(\boldsymbol{y}-\boldsymbol{p})\right), \\ \geq& f(\boldsymbol{p})+\lambda\left(f(\boldsymbol{y})-f(\boldsymbol{p})\right), \end{aligned}$$

where $0<\lambda<1$.

Rearranging the terms and dividing both sides by λ, we have

$$\frac{f\left(\boldsymbol{p}+\lambda(\boldsymbol{y}-\boldsymbol{p})\right)-f(\boldsymbol{p})}{\lambda} \geq f(\boldsymbol{y})-f(\boldsymbol{p}).$$

Letting $\lambda \rightarrow 0$, it can be shown that the left hand side of Inequality above converges to $f'(\boldsymbol{p})\cdot(\boldsymbol{y}-\boldsymbol{p})$.

Finally, we have Equation (7) as:

$$f(\boldsymbol{p})+f'(\boldsymbol{p})\cdot(\boldsymbol{y}-\boldsymbol{p}) \geq f(\boldsymbol{y}).$$

References

[1] Y. Lin, T. Wang, and S. Wang, "Uav-assisted emergency communications: An extended multi-armed bandit perspective," IEEE Communications Letters, vol. 23, no. 5, pp. 938–941, 2019.

[2] L. Ruan, J. Wang, J. Chen, Y. Xu, Y. Yang, H. Jiang, Y. Zhang, and Y. Xu, "Energy-efficient multi-uav coverage deployment in uav networks: A game-theoretic framework," China Communications, vol. 15, no. 10, pp. 194–209, 2018.
[3] Z. Wang, L. Duan, and R. Zhang, "Adaptive deployment for uavaided communication networks," IEEE transactions on wireless communications, vol. 18, no. 9, pp. 4531–4543, 2019.
[4] C. H. Liu, Z. Chen, J. Tang, J. Xu, and C. Piao, "Energy-efficient uav control for effective and fair communication coverage: A deep reinforcement learning approach," IEEE Journal on Selected Areas in Communications, vol. 36, no. 9, pp. 2059–2070, 2018.
[5] A. F. Molisch, V. V. Ratnam, S. Han, Z. Li, S. L. H. Nguyen, L. Li, and K. Haneda, "Hybrid beamforming for massive mimo: A survey," IEEE Communications magazine, vol. 55, no. 9, pp. 134–141, 2017.
[6] H. Ren, L. Li, W. Xu, W. Chen, and Z. Han, "Machine learningbased hybrid precoding with robust error for uav mmwave massive mimo," in ICC 2019-2019 IEEE International Conference on Communications (ICC). IEEE, 2019, pp. 1–6.
[7] W. Hao, M. Zeng, Z. Chu, and S. Yang, "Energy-efficient power allocation in millimeter wave massive mimo with non-orthogonal multiple access," IEEE Wireless Communications Letters, vol. 6, no. 6, pp. 782–785, 2017.
[8] X. Yu, F. Xu, K. Yu, and X. Dang, "Power allocation for energy efficiency optimization in multi-user mmwave-noma system with hybrid precoding," IEEE Access, vol. 7, pp. 109 083–109 093, 2019.
[9] C. Sun, X. Gao, S. Jin, M. Matthaiou, Z. Ding, and C. Xiao, "Beam division multiple access transmission for massive mimo communications," IEEE Transactions on Communications, vol. 63, no. 6, pp. 2170–2184, 2015.
[10] O. El Ayach, S. Rajagopal, S. Abu-Surra, Z. Pi, and R. W. Heath, "Spatially sparse precoding in millimeter wave mimo systems," IEEE transactions on wireless communications, vol. 13, no. 3, pp. 1499– 1513, 2014.
[11] H. H. Kha, H. D. Tuan, and H. H. Nguyen, "Fast global optimal power allocation in wireless networks by local dc programming," IEEE Transactions on Wireless Communications, vol. 11, no. 2, pp. 510–515, 2011.
[12] M. Grant and S. Boyd, "Cvx: Matlab software for disciplined convex programming, version 2.1," 2014.

A Novel Post Process Method For Smoothed Particle Hydrodynamics by Interpolation on Specified Slice Plane

Yanli Li, Junhang Guo

Northwest Institute of Mechanical and Electrical Engineering, No.5, Biyuan East Road, Xian Yang, Shaanxi province, 712099, People's Republic of China

nimee@bq202.com

Abstract. The smoothed particle hydrodynamics (SPH) is widely used in numerical simulations as a kind of meshless method. The particles are used to represent object, which is a fully Lagrangian modeling scheme without the need to define a spatial mesh so it can simulate large deformation. The classic post process method cannot visualize the simulation results smoothly, so a novel post processing method is proposed in this paper. By developing a program, the coordinates and field variables of particles were collected to a file. Using the kernel function, the field variable on a specified slice plane was interpolated. By developing a program, the simulation results have been rendered smoothly. This method is illustrated by a penetration example in this paper.

1. Introduction

Smoothed particle hydrodynamics (SPH) uses a collection of points or so called particles to represent a given body[1]. SPH is a typical numerical method that is part of the larger family of meshless (or mesh-free) methods[2]. Smoothed particle hydrodynamics is a fully Lagrangian modeling scheme without the need to define a spatial mesh. For some fluid flow and structural problems involving large deformations and free surfaces, if continuum element is used in simulation, the element may be excessively distortion and causes termination. While smoothed particle hydrodynamics (SPH) as a mesh-free method has been widely used, in some post processors of bushiness software the solving region is rendered as particles. Since nodal connectivity is not fixed, some disadvantages are shown such as field variables are discrete and geometry boundaries are obscure. It is difficult to visualize spatial distribution of variables in internal plane, which is more convenience in finite element method. While these information such as geometry shape, free surface, or spatial distribution are concerned by users or researchers who use SPH method[3]. So it is necessary to develop a post processing method to improve the visual effect for SPH method.

2. SPH discrete method

At its core, the method is not based on discrete particles (spheres) colliding with each other in compression or exhibiting cohesive-like behavior in tension. Rather, it is simply a clever discretization method of continuum partial differential equations. SPH uses an summing interpolation scheme to approximate a field variable at any point in a domain, by summing the weight contributions from neighboring particles, as in Eq.1. The weighting function is called kernel function W[1,2].

$$\langle f(x)\rangle \approx \sum_{j} \frac{m_j}{\rho_j} f_j W\left(\left|x-x_j\right|, h\right) \tag{1}$$

Here, x is the coordinate of any point, j represents neighboring particles, m_j is the mass, ρ_j is the density, f_j is the field variable, x_j is the coordinates, h is the smooth length. An example kernel function is shown in Fig 1. The smoothing length h, determines how many particles influence the interpolation for a particular point.

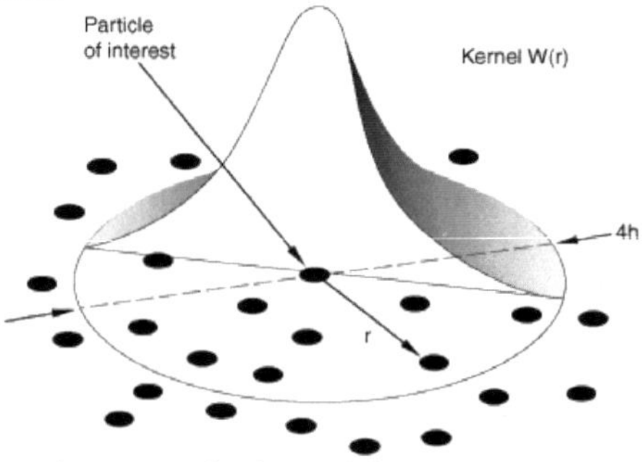

Figure 1. The kernel function in SPH

The common kernel function is shown in Eq. 2 as a cubic spline as the interpolation polynomial

$$W(r,h)=\frac{C}{h^n}\begin{cases}1-\frac{3}{2}d^2+\frac{3}{4}d^3 & \text{when } 0\le d\le 1\\ \frac{1}{4}(2-d)^3 & \text{when } 1\le d\le 2\\ 0 & \text{elsewhere}\end{cases} \tag{2}$$

Where $d=r/h$ and $r=\left|x-x_j\right|$, and C is the normalization factor,

$$C=\begin{cases}2/3 & 1D\\ 10/7\pi & 2D(\text{circular})\\ 1/\pi & 3D(\text{spherical})\end{cases} \tag{3}$$

The shape of kernel function in Eq. 2 can be well illustrated in Fig 2.

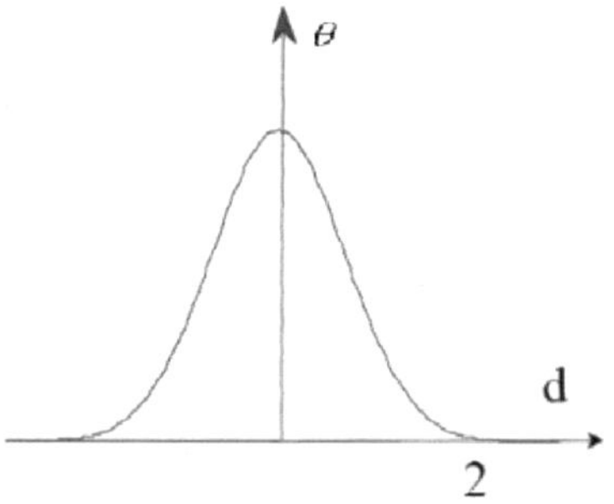

Figure 2. Shape of cubic spline kernel function

The visualization method for SPH can be concluded as following:

1) Gathering particles coordinates and field variables by user interface such as Python[4] script or other programming language
2) Saving the information as text or csv file[5]
3) Defining the slice plane or region which is concerned by the user
4) Finding all the neighboring particles of slice plane or region by smoothing length
5) Defining the destination mesh grid on slice plane
6) Calculation all the value of mesh grid by interpolation of selected particle in step 4

7) For any grid whose distance to selected particles in step 4 larger than given value, maker it as boundary or exclude it

8) Visualizing the results by figure or image.

A GUI based data extractor program developed by Python tkinter[6] is shown in Fig 3. The information such as particles coordinates, strain, stress and et al are saved as a csv file.

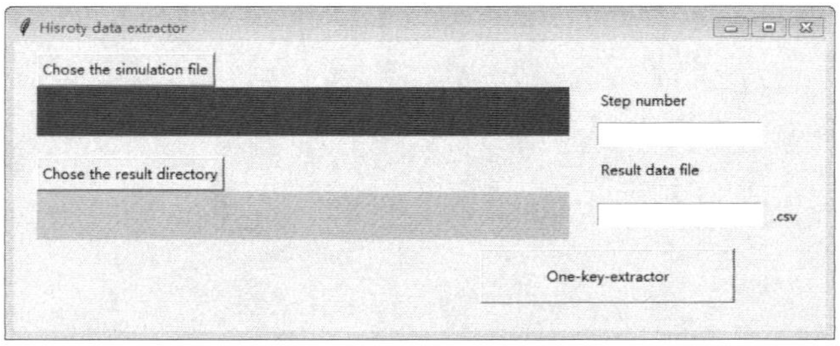

Figure 3. The GUI based data extractor program developed by Python

3. SPH visualization example

Smoothed particle hydrodynamic analyses are effective for applications involving extreme deformation such as ballistics, impacts and et al. The established model for the projectile-target penetration problem is shown in fig 4. The target is split into two parts and bonded in the interface. The outer part may undergo small deformation, so is model by finite element method, while the inner part can be modeled by SPH and they were tied together. The deformed mesh of projectile and target is also shown in fig 5.As the limitation of current post processor function, the result shown in figure 5 is not rendered as a field directly.

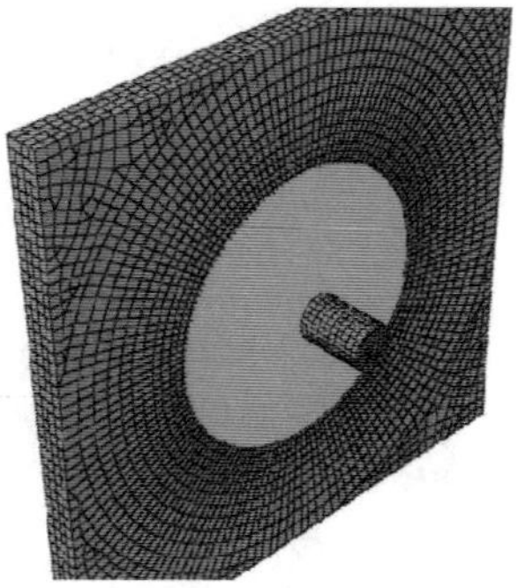

Figure 4. Mesh of rigid projectile penetration into target

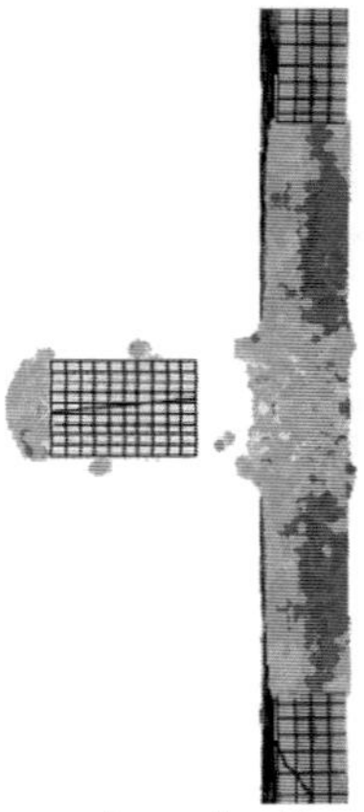

Figure 5. Deformed mesh of projectile and target

By developing a program, fig 6 shows the SPH particle distribution of the target at center cross section region from simulation results, the shape of deformed target can be well rendered.

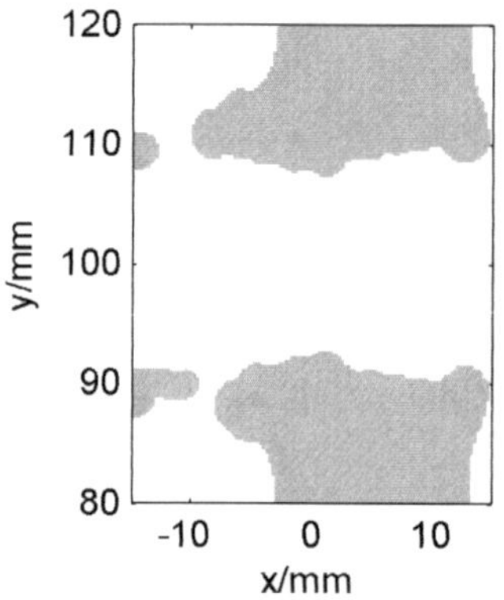

Figure 6. Deformed of the target by developing program

By the above method, the typical variable distributions of equivalent stress and plastic strain are rendered as fields, as shown in fig 7 and fig 8. The shape of target can be well visualized by this method. In this cross section, the maximum equivalent stress is about 1000MPa ,the maximum equivalent strain is about 1.0 locating at shear zone. And the maximum equivalent stress and plastic strain coincides in the same area indicating that material undergoes large plastic deformation and strain concentration.

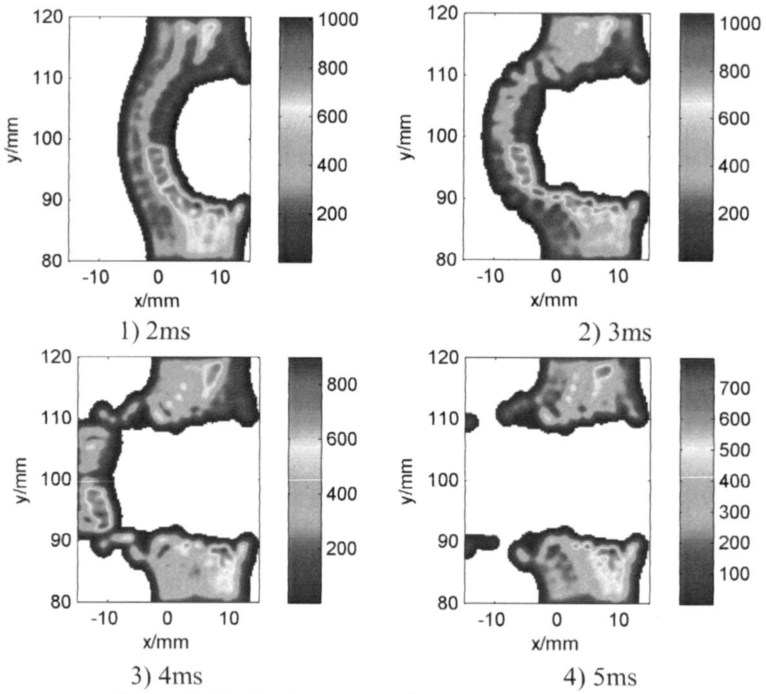

Figure 7. Equivalent stress distribution of the target

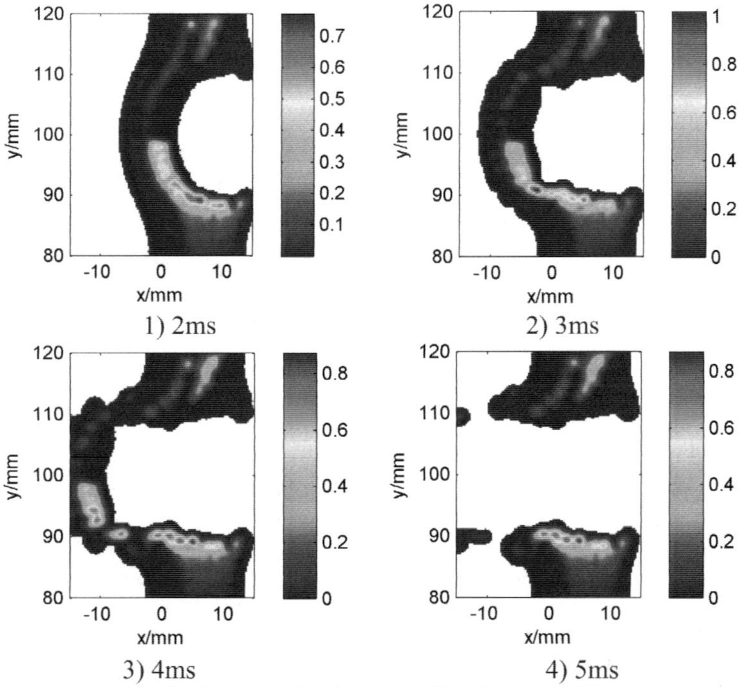

Figure 8. Equivalent plastic strain distribution of the target

4. Conclusions

A novel post process method for smoothed particle hydrodynamics by interpolation on specified slice plane was proposed in this paper. By this method, field variables distributions of SPH region at a specified cross section can be obtained.The projectile penetrates target process with a given velocity was investigated by numerical methods. The deformed shape of SPH region were extracted and rendered by threshold based on smooth length and kernel function type. And the distributions of equivalent stress and plastic strain in a specified cross section was interpolated by kernel function. This method can also be used for other simulations in which smoothed particle hydrodynamics is employed.

References

[1] G.R. Liu, M.B. Liu, Smoothed Particle Hydrodynamics: A Meshfree Particle Method, World Scientific Publishing, 2003

[2] Luca Massidda, Yacine Kadi, SPH simulation of liquid metal target dynamics, Nuclear Engineering and Design 240 (2010) 940–946

[3] J H Guo, X F Yao, J M Qiao et al.An investigation on plastic deformation of rotating band for large caliber gun projectile during engraving process. 2020 J. Phys.: Conf. Ser. 1507 082006

[4] Swaroop, C. H. A Byte of Python. 2003-2005

[5] Timothy J. Barth, Michael Griebel, David E. Keyes et al, A Primer on Scientific Programming with Python. Springer. 2012

[6] Burkhard A. Meier. Python GUI Programming Cookbook. Packt Publishing Ltd. 2015

Feature extraction and recognition of rotating machinery fault noise based on convolutional neural networks

Yongshao Xu[1], Bingzheng Liu[1*], Haotian Shang[2], Yueqi Ge[2]

[1]Shenzhen Branch of China Nuclear Power Maintenance Co., Ltd, Shenzhen, Guangdong, 518000, China

[2]Northeast Electric Power University School of Automation Engineering,

Jilin, Jilin,132013, China

[*]Corresponding author's e-mail: liubingzheng@cnjijx.com

Abstract. Rotating machines are common equipment in industrial production, which may cause failure for a long time. Because of its convenient use and non-destructive to itself, acoustic detection method is suitable for fault diagnosis of rotating machinery. The convolution neural network model is used to identify several typical rotating machine faults. The repeatability experiments and different training sets show that the method has good universality. A visual fault identification system is built, and the effect of the system is verified by experiments.

1. Introduction

With the continuous improvement of science and technology since the industrial revolution, mechanical equipment used in industrial production has become more and more complex. How to detect, diagnose and predict the life of each module of mechanical equipment has become an important research topic over the years[1]. And rotating machinery is basically the core component of all mechanical equipment[2]. It is widely used in many fields (such as aerospace, chemical industry, petroleum, manufacturing, transportation and energy). Once a component fails during operation, it will often involve the entire mechanical equipment and cause unpredictable losses[3]. In severe cases, catastrophic events and casualties occurs multiple times[4-5]. Therefore, fault diagnosis is of great significance to ensure the safe and reliable operation of mechanical equipment[6]. If the equipment operation can be monitored and diagnosed in real time through modern scientific and technological means, the faulty equipment can be alerted immediately or early warning before it occurs. For example, the sound sensor of the array collects the sound of the equipment running state, and diagnoses the location and severity of the fault through the sound[7].

2. Extraction of noise feature of rotating machinery failure

2.1. Sound signal preprocessing

Since the parameters of the equipment vary greatly during operation, even if the same sound sensor and computer recording equipment are used, the sound data collected each time also has a big change, and as the noise continues to be generated during the operation of the equipment, there will be differences between the collected sound samples. So we need to reduce the difference between different samples, so the normalization method is used to make the amplitude of the sound concentrated between 0 and 1, making the data more stable and convenient for calculation and processing.

Commonly used normalization methods include standard normalization and maximum and minimum normalization in others paper[22]. This paper uses maximum and minimum normalization in the experiment process, and the formula is shown in 1:

$$X_m = \frac{(x - x_{min})}{x_{max} - x_{min}} \quad (1)$$

In the formulax_{min} —Minimum signal amplitude

x_{max}—Maximum signal amplitude

The sound signal collected when the equipment is running is an unsteady signal. However, in a very short period of time, the sound has the characteristics of stability, and the sound signal to be processed is required to be smooth and continuous when the short-time Fourier transform is subsequently performed. Therefore, this paper will process the sound signals collected by the capture card into frames.

Because the collected sound signals are affected by the environment, a silent experimental environment cannot be achieved. The abnormal sound reflected on the indoor wall and some other parts are mixed, which has strong noise pollution. Therefore, in order to improve the quality of the sound source, filtering and noise reduction before fault feature extraction becomes very important. The commonly used filtering methods mainly include singular value decomposition filtering[23], wavelet threshold filtering[24] and empirical mode filtering[25].

2.2. Wavelet transform

The wavelet transform turns it into a wavelet base with finite length and attenuation. In this way, the localizing time and frequency can be obtained at the same time. The process of continuous wavelet transform is to obtain a series of wavelet sequence$\varphi_{a,b}(t)$ through the translation and expansion of wavelet basis function $\varphi(t)$, thereby decomposing the original signal into different frequency components, as shown in the formula:

$$\varphi_{a,b}(t) = \frac{1}{\sqrt{a}}\varphi\left(\frac{t-b}{a}\right), a > 0, b \in R \quad (2)$$

In the formula, parameter a represents the scale factor to determine the position of the wavelet time-frequency window in the frequency domain. The parameter b represents the translation factor, which determines the position of the wavelet time-frequency window in the time domain. Therefore, assuming $x(t) \in L^2(R)$, according to the wavelet sequence, the wavelet transform is as shown in the formula:

$$CWT_x(a,b) = \int_{-\infty}^{\infty} x(t)\bar{\varphi}_{a,b}(t)dt = \frac{1}{\sqrt{a}}\int_{-\infty}^{\infty} x(t)\bar{\varphi}\left(\frac{t-b}{a}\right)dt \quad (3)$$

Where$\bar{\varphi}_{a,b}(t)$ is the complex conjugate of $\varphi_{a,b}(t)$, the wavelet transform is reversible, and the inverse of the wavelet transform is:

$$x(t) = \frac{1}{C_\varphi}\int_{-\infty}^{\infty}\int_{-\infty}^{\infty} \varphi_{a,b}(t)CWT_x(a,b)\frac{1}{a^2}dadb \quad (4)$$

In the formula,$C_\varphi = 2\pi\int_{-\infty}^{\infty}\frac{|\hat{\varphi}(\omega)|}{|\omega|}d\omega$.

Common wavelet functions include Haar wavelet, Daubechies (dbN) wavelet, Morlet wavelet, Coiflet wavelet, Meyer wavelet, etc.

2.3. Time-frequency characteristic performance

In order to analyse the time-frequency characteristic method of wavelet, this section collects a section of normal and abnormal sound signals of nuclear power plants for time-frequency analysis. The signal time-domain waveform diagram and frequency-domain waveform diagram of the rotating machinery in normal operation are shown in Figure 1:

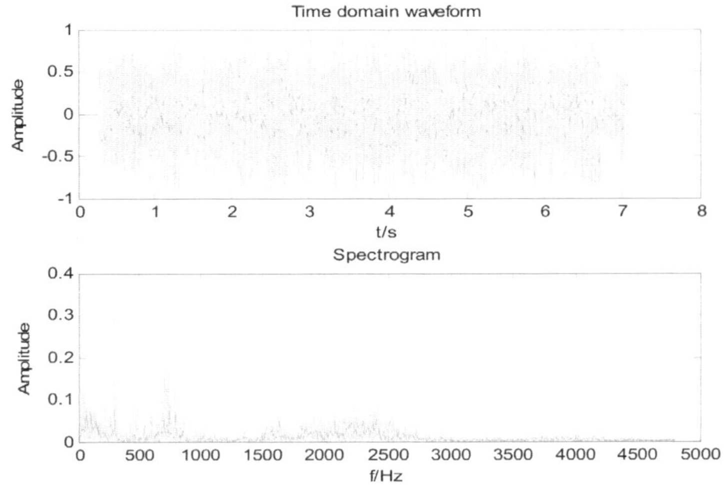

Figure 1. Time domain and frequency domain waveforms of motor fault signal

It can be seen from the figure that if you only look at the relationship between time and amplitude or frequency and amplitude, less information can be obtained. It can only be seen that the signal frequency of the motor rotation is mainly concentrated in the low frequency band. Therefore, we use the wavelet time-frequency analysis method to analyze the signal.

Further analyzing the performance, this paper created a rotating mechanical load fault based on the original rotating mechanical operation, which also contains a normal signal and a fault signal. The time-frequency diagram of the wavelet transform of the fault signal is shown in the figure below:

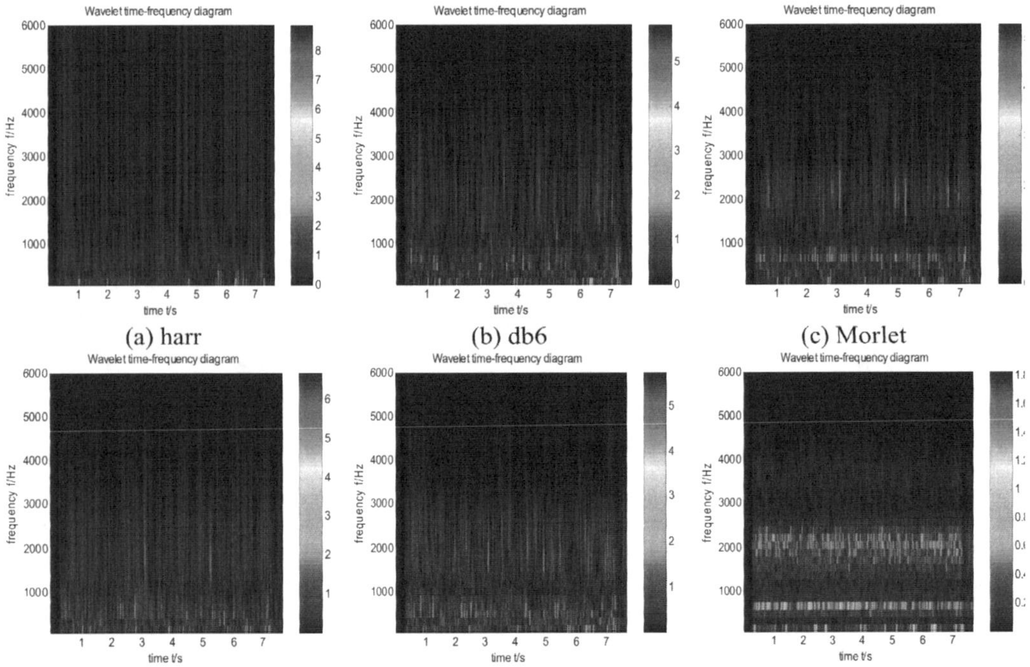

(a) harr (b) db6 (c) Morlet

(d) coif3 (e) meyer (f) cmor3-3

Figure 2. Four time-frequency diagrams of wavelet transform

Through the analysis, it can be seen that the wavelet transform can see the difference between the normal and the fault of the motor, the signal affected by noise is small, it has good time and frequency resolution, and the energy is relatively concentrated, and the frequency fluctuation is small. Through the above analysis, it is shown that the time-frequency diagram can be used as the information of detecting signal failure, but it also has information that cannot be seen by the human eye. We use convolutional neural networks to learn, feature extraction and classification of the collected signal time-frequency maps, and finally achieve the purpose of classifying signal faults.

3. Fault noise classification based on convolutional neural networks

3.1. Neural network

Over the years, convolutional neural networks have been widely promoted in the fields of artificial intelligence, face recognition, voice interaction, and autonomous driving. Convolutional neural network is composed of input layer, convolutional layer, pooling layer and fully connected layer. This section will introduce the basic structure of neural network. Its basic structure is shown in Figure 3.

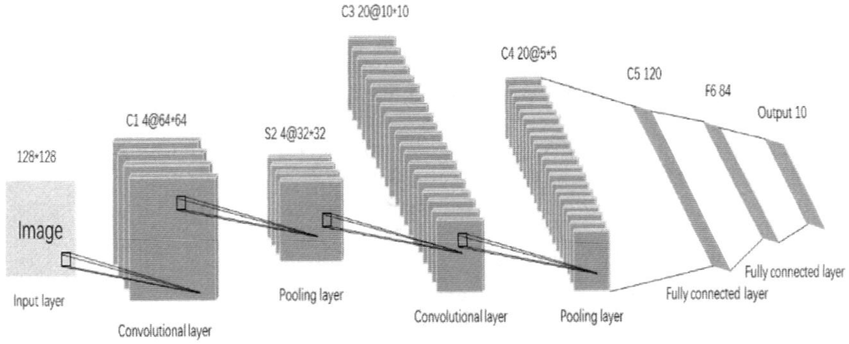

Figure 3. Convolutional neural network structure diagram

3.2. Data set

The experiment collected signals of different time and different conditions for bearing failure, load failure, coupling failure and normal conditions. When the bearing fails, it is considered that the bearing is wounded, and the same four bearings are wounded with depths of 2.0mm, 3.00mm, 4.00mm, and 5.00mm respectively. Load failures are mainly caused by loose fan blades, and fan blades have been wounded with depths of 1.0mm, 2.0mm, and 3.00mm. Coupling failures are mainly caused by wounds on the inner ring, ball and outer ring. The eleven states collected signals at different speeds of the species.

The eleven states are classified into four categories, the collected data is segmented, the sampling frequency is 44100, and the second is divided into one segment, and a time-frequency diagram is made every second. Since the pixel size of the image affects the calculation speed of the convolutional network, we save the time-frequency map as a pixel-by-pixel picture, which collects and analyzes a total of 6000 time-frequency maps of four types of faults and different states, and sets the labels. Among them, the training set is 4800 and the test set is 1200. The recognition rate Acc and the loss function loss are used as the recognition performance standard of the neural network algorithm. The fault types are shown in the following table:

Table 1. Fault type setting

Object	Failure type	Failure size	Number of fault points
		1 mm	1
		2 mm	1
Bearing failure	Bearing wound	3 mm	2
		4 mm	1
		1 mm	1
Load failure	Load fan loose	2 mm	1
	Load fan wound	3 mm	1
			2
	Inner ring wound	1 mm	1
Coupling fault	Outer ring wound	2 mm	1
	Ball wound	3 mm	1
Normal conditions	Normal	None	None

3.3. Test and performance comparison

In order to demonstrate the applicability of the network proposed in this paper, without changing the network structure, this paper collects four states of different motors at different speeds. The same collection equipment as the previous sample has doubled the speed and collected a total of 8280 samples. The specific sample classifications are shown in the following table:

Table 2. Data set collected in the experiment

Failure type	Training data	Verify the data	Test data
Bearing failure	1500	250	258
Load failure	1500	300	300
Coupling fault	1400	300	300
Normal fault	1600	250	250

In the previous experiments, the data set was divided into training set, test set, and verification set in sequence according to the sampling order. This time a crossover experiment is used to randomly select 1000 images from the training set, and then randomly select 200 images from the remaining time-frequency diagrams as the verification set. Repeat the experiment five times. The experimental results are shown in the figure below.

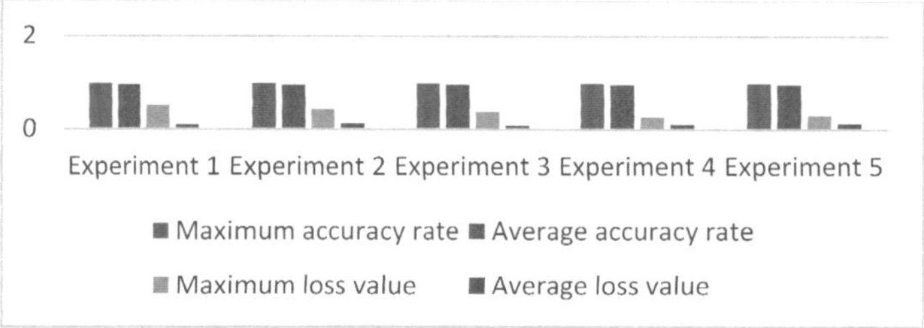

Figure 4. Experimental results of the crossover experiments

It can be seen that the neural network model constructed in this paper has good applicability for the identification of rotating machinery faults. It can also achieve a high recognition rate for different data sets, with high stability, and no messy recognition.

For this research on fault recognition, many scholars have proposed a variety of network model frameworks, such as: Yuan et al. proposed a mechanical fault diagnosis model based on HHT transformation and CNN algorithm. LinKai et al. proposed a fault diagnosis framework based on sparse autoencoders. Zhu et al. proposed a diagnosis method based on support vector machines and improved evidence theory to realize the diagnosis of different faults under mixed conditions. Xie et al. proposed a fault diagnosis method for nuclear detectors based on BP neural network, which effectively improves the fault diagnosis frequency of nuclear detectors. Chen et al. proposed a fault diagnosis model based on the PCA-KNN fusion algorithm. Adjust the parameters of the above algorithm to the optimal, use the data set collected and established in this paper to compare with the neural network model of this paper, the recognition accuracy rate corresponding to the test set is shown in the following table:

Table 3. Experimental results of different algorithms

The Algorithm	Failure type	Accuracy rate:%
SVM	4	93.24
HHT-CNN	4	89.88
BP	4	86.26
PCA-KNN	4	90.12
The framework in this paper	4	96.05

4. Conclusion

By comparing several fault diagnosis algorithms, from the final recognition accuracy respectively, the algorithm framework in this paper has the highest recognition accuracy, which can reach 96.02%. And in terms of convergence speed, the network designed in this paper can reach convergence after more than 100 iterations, which can not only meet the needs of fault identification, but also achieve good results in different rotating mechanical sound signal training sets.

References

[1] Han T, Zhang L, Yin Z, et al. Rolling bearing fault diagnosis with combined convolutional neural networks and support vector machine[J]. Measurement, 2021: 109022.

[2] Xiang Li, Wei Zhang, Qian Ding, Jian-Qiao Sun. Intelligent rotating machinery fault diagnosis based on deep learning using data augmentation[J]. Journal of Intelligent Manufacturing, 2020, 31(2).

[3] Cui L, Huang J, Zhang F. Quantitative and localization diagnosis of a defective ball bearing based on vertical–horizontal synchronization signal analysis[J]. IEEE Transactions on Industrial Electronics, 2017, 64(11): 8695-8706.

[4] Zhou F, Shen J, Yang X, et al. Modified hierarchical multiscale dispersion entropy and its application to fault identification of rotating machinery[J]. IEEE Access, 2020, 8: 161361-161376.

[5] Abid A, Khan M T, de Silva C W. Layered and real-valued negative selection algorithm for fault detection[J]. IEEE Systems Journal, 2017, 12(3): 2960-2969.

[6] Miao Yonghao, Zhang Boyao, Lin Jing, Zhao Ming, Liu Hanyang, Liu Zongyang, Li Hao. A review on the application of blind deconvolution in machinery fault diagnosis[J]. Mechanical Systems and Signal Processing,2022,163:

Virtual DC Motor Stability Control Method Considering New Energy Generation Fluctuation

Yilun Tan*, Yucheng Wang

Xi'an modern Control Technology Research Institute, Xi'an, 710065, China

*Corresponding author's e-mail: Tanyilun@mail.nwpu.edu.cn

Abstract. With the rapid development of new energy generation, the intermittence and randomicity of its power output will have a significant impact on the transmission capacity of DC motor. Therefore, a virtual DC motor stability control method considering the fluctuation of new energy generation is proposed. The natural frequencies and modes of the virtual DC motor shafting rotor are analyzed by means of a steady sinusoidal excitation at zero speed. Considering the transient dynamic response of the shafting rotor of virtual DC motor under the fluctuation of new energy generation, Taylor series and transfer acceleration matrix method are used to calculate the transient dynamic response of shafting rotor under the fluctuation of new energy generation, and the parameters of virtual DC motor are identified and estimated. Based on this, a proportional resonance controller is designed to realize the stability control of virtual DC motor. Experimental results show that the interactive power curve between virtual DC motor and regional distribution network is smoother after optimal control, and this method can effectively improve the power balance ability of virtual DC motor.

1. Introduction

Motor is a kind of electromagnetic equipment which transforms mechanical energy and electric energy according to electromagnetic induction law. In the electricity market environment, the available transmission capacity is an important dispatching basis and market regulation signal for guiding the normal operation of the power system [1]. It can not only indicate the safety and stability margin of power grid operation, but also provide detailed information on power grid usage for system operators and power market participants to guide their market behavior [2]. With the increasingly widespread use of new energy power generation, their permeability in the power system is also increasingly high. The world's new energy power generation is transforming from supplementary energy to alternative energy, and from independent systems to large-scale grid integration. As far as China is concerned, new energy power generation is also developing from decentralized, small-scale development and on-the-spot consumption to large-scale, highly concentrated development and long-distance and high-voltage transmission [3-4]. The large-scale access to new energy power generation and its uncertain characteristics will have a certain impact on all aspects of power system operation. Virtual DC motor stability is one of the important aspects, and is an important research topic for the development of smart grid and power market [5]. With the rapid development of modern motor control theory, the hardware of virtual DC motor is becoming more and more mature, but the load switching control method in the form of software becomes the bottleneck of the application and development of virtual DC motor.

Journal of Physics: Conference Series **2113** (2021) 012015 doi:10.1088/1742-6596/2113/1/012015

2. Virtual DC motor stability control method considering new energy generation fluctuation

2.1. Natural frequency and modal analysis of virtual DC motor shafting rotor

At zero speed, the rotor of high-speed motor shafting is stimulated by steady sinusoidal excitation, as shown in Figure 1.

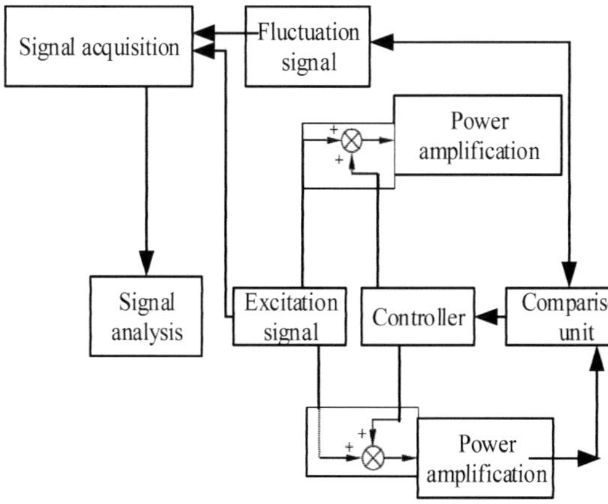

Fig. 1 Specific method of excitation for high-speed motor shafting rotor

The corresponding vibration spectra of 118Hz and 40Hz natural frequencies of rotor of high-speed motor shafting are obtained by excitation. Then the vibration modes and natural frequencies are calculated by finite element method [6], and the magnetic bearings are simulated by spring element, ignoring the influence of small damping.

2.2. Transient dynamicres

The transient dynamic stabiponse analysis of virtual DC motor shafting rotor based on fluctuation of new energy generationlity of shafting rotor under the fluctuation of new energy generation is analyzed by predicting Taylor series and transfer acceleration matrix method, and the instability is avoided by acceleration and force state vector. Firstly, by predicting the Taylor series, the speed and displacement of the rotor of the high-speed motor shafting at the next moment are estimated, and the next moment is expressed by $t+\Delta t$. The specific formula is as follows:

$$\begin{cases} (e)_{n+1} = (e)_n + \Delta t(\dot{e})_n + 0(\Delta t^2) \\ (\dot{e})_{n+1} = (\dot{e})_n + \Delta t(\ddot{e})_n + 0(\Delta t^2) \end{cases} \quad (1)$$

In equation (1), $(e)_{n+1}$ represents the pre-estimate of the speed at the next moment; $(\dot{e})_{n+1}$ represents the pre-estimate of the displacement at the next moment; $(e)_n$ represents the velocity at t moment; Δt represents the increment of the time from t moment to the next moment; $(\dot{e})_n$ represents the displacement at t moment; and $(\ddot{e})_n$ represents the increment of the displacement at t moment to the next moment. The oil film characteristic coefficients of Taylor series are obtained by dynamically analyzing the oil film characteristic coefficients of the rotor, and the nonlinear oil film force is calculated. Thus, the point matrix constructed by the method of transfer acceleration matrix at the neck of the shaft rotor is obtained.

$$\begin{cases} \Delta Q_x = k_{xx}x + c_{xy}\dot{y} \\ \Delta Q_y = k_{yy}y + c_{yy}\dot{y} \end{cases} \quad (2)$$

In formula (2), ΔQ_x and ΔQ_y respectively represent the left and right coordinates of the nonlinear oil film force of the shaft rotor neck; k_{xx}, k_{xy}, k_{yx} and k_{yy} respectively represent the four components of the oil film stiffness matrix K_B, namely, the main stiffness oil film coefficient; c_{xx}, c_{xy}, c_{yx} and c_{yy} respectively represent the four components of the oil film damping matrix C_B, namely, the main damping oil film coefficient; x and y represent the original coordinates of the shaft rotor neck; and $\dot{x}$ and $\dot{y}$ represent the left and right coordinates of the shaft neck after it deviates from its original position due to the disturbance of new energy generation.

2.3. Parameter identification and estimation of virtual DC motor

In order to improve the performance of virtual DC motor speed regulation, the motor parameters need to be accurately identified and estimated, the specific process is as follows: A neural network model is used to identify and estimate the parameters of the virtual DC motor, as shown in Figure 2.

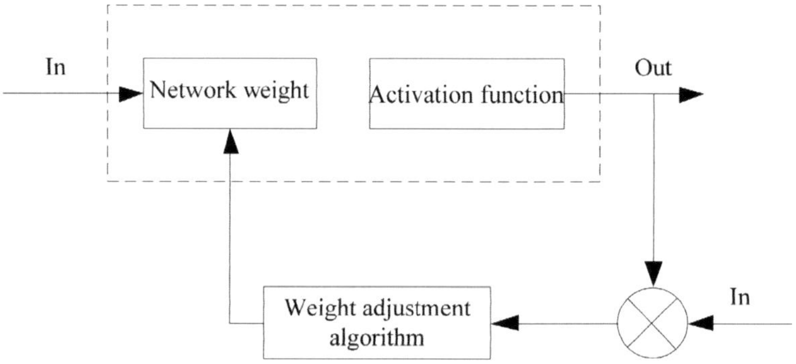

Fig. 2 Schematic diagram of parameter identification and estimation of virtual DC motor

As shown in Figure 2, the specific steps of parameter identification and estimation of virtual DC motor are as follows:

Step 1: distribute all network weights randomly and evenly, and extract the smaller value;

Step 2: collect the measured input and output data, which together constitute the training samples;

Step 3: input the training samples obtained in step 2 into the neural network model, calculate and output;

Step 4: error function calculation, the calculation formula is:

$$e' = (T - Y)^2 = (T - UX)^2 \quad (3)$$

Step 5: modify the network weight according to a variety of weight learning and adjustment algorithms;

Step 6: calculate the network weight error before and after training, and the calculation formula is:

$$e'' = U_{i+1} - U_i \quad (4)$$

Compare the result of formula (4) with the given error, if $e'' \leq e'$, the training of neural network model is stopped. At the same time, comparing the training times, if the training time n is greater than or equal to the given termination time, the training of the neural network model is stopped. This prevents the model from converging.

Step 7: The trained neural network model is used as the model of virtual DC motor parameter identification and estimation.

2.4. Motor digital signal processing

Based on the above identified and estimated virtual DC motor parameters, DSP chip 28335 is used to process the virtual DC motor digital signal for the final virtual DC motor speed control. The 28335 DSP chip has many functions, and it includes many pins.

The 28335 DSP chip adopts the format to calibrate the operation value, and the conversion relationship between the floating-point number and the fixed-point number is as follows:

$$\begin{cases} L' = (\text{int})x \cdot 5^{e} \\ L'' = (\mathit{float})x_q \cdot 5^{-e} \end{cases} \quad (5)$$

In equation (5), x_q represents floating point number; x represents the number of fixed points; (int) and (float) represent the conversion function.

2.5. Realization of low error control

According to the proportional resonance controller designed above, when the DC component in the current passes through the proportional resonance controller, there are:

$$\begin{cases} u_{d_1}^{*} = \left(k_{dd} + \frac{1}{s}\right)\left(i_d^{*} - i_{d_1}\right) \\ u_{q_1}^{*} = \left(k_{qq} + \frac{1}{s}\right)\left(i_q^{*} - i_{q_1}\right) \end{cases} \quad (6)$$

In Formula (6), the upper angle * represents the control variable, the lower angle a represents the output of the DC component, and the k_{qq} and k_{dd} represent the integral coefficient and the scale coefficient in PI respectively. The transfer function of virtual DC motor load switching control is expressed as follows:

$$I_{\mathrm{L}} = \frac{KG(s)}{SZ + KG(s)} r_{\mathrm{ref}} - \frac{1}{SZ + KG(s)} \quad (7)$$

In Formula (7), I_{L} represents the load switching control transfer function of the motor; K represents th control coefficient; $G(s)$ represents the frequency of the proportional resonance controller as the gain; Z represents the discrete factor; and r_{ref} represents the gain approximate equivalent. The difference equation of the proportional resonance controller is obtained by finishi

$$S(\cdot) = b_0 e(h) + b_2 e(h-2) - a_1 y(h-1) - a_2 y(h-2) \quad (8)$$

In conclusion, based on the proportional resonance controller, the steady-state control of load switching of virtual DC motor is realized, and the control program is relatively simple, which provides effective support for the application of virtual DC motor.

3. Design and analysis of simulation experiment

To verify the effectiveness of the proposed model, the following numerical examples are designed. The virtual DC motor used in the experiment is a new energy and cogeneration unit, which can interact with the electric energy in the distribution network. Based on the internal power interaction characteristics of the new energy virtual DC motor, three virtual DC motors are selected as W1, W2 and W3, respectively. Among them, W1 and W2 are the maximum of combined power supply capacity lower than the maximum of load demand, and W3 is the maximum of combined power supply capacity higher than the maximum of load demand. The configuration of the three virtual DC motors is shown in Table 1.

Tab. 1 Source load parameters of each virtual DC motor

Virtual DC motor	Fan installed capacity / MW	PV installed capacity / MW	Power storage capacity / MW · H	Heat storage capacity / MW · H	Electric load max / MW	Heat load max / MW
W1	30	30	62	20	15	28
W2	40	40	45	25	45	80
W3	40	40	95	0	37	60

The new energy output, thermoelectric unit output and combined electric load of three virtual DC motors W1, W2 and W3 in winter are shown in Figure 3.

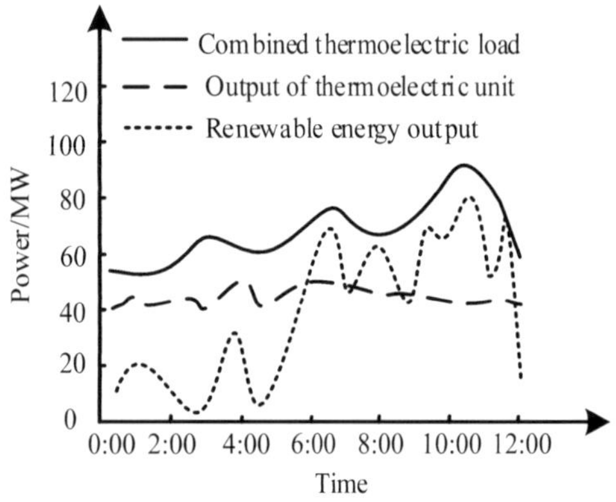

(a) Virtual DC motor W1

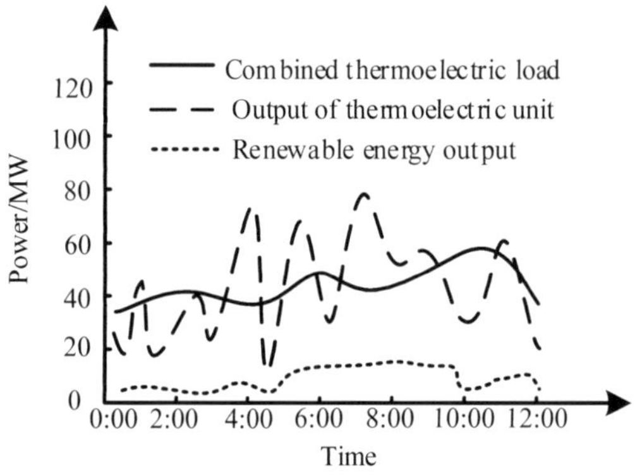

(b) Virtual DC motor W2

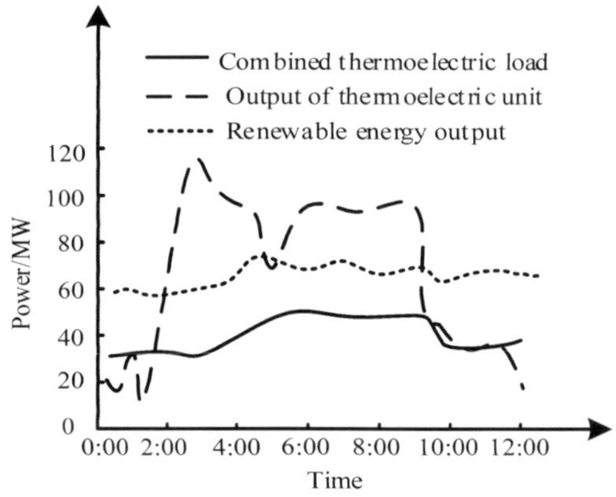

(c) Virtual DC motor W3

Fig. 3 Daily source load curve of different virtual DC motors

The interactive power curve between the new energy virtual DC motor and the regional distribution network after optimizing the operation mode of the new energy virtual DC motor by using the optimization method in this paper is shown in Figure 4.

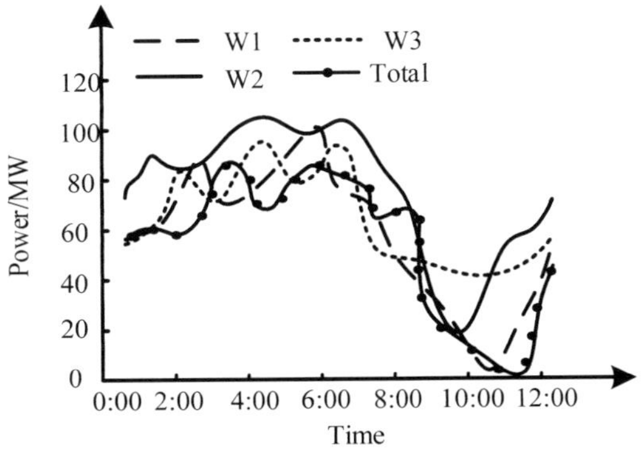

Fig. 4 Interactive power curve

As can be seen from Fig. 4, the interactive power curves of W1, W2 and W3 with the regional distribution network are more smooth after optimization by the method presented in this paper, and the improvement effect is especially obvious in late peak hours. Through the optimization and coordination of the operation mode of the new energy power generation system, the surplus power of R3 is first allocated to R1 and R2, which can effectively improve the power balance ability of the virtual DC motor with new energy.

4. Conclusion

In order to solve the instability of virtual DC motor operation under the fluctuation of energy generation, a method of virtual DC motor stability control is proposed. Considering the transient dynamic response of the rotor of the virtual DC motor, the parameters of the virtual DC motor are identified and estimated. Based on the above analysis results, a proportional resonance controller is designed to achieve the virtual DC motor stability control. The experimental results show that the power balance ability of the virtual DC motor with new energy is improved effectively after the optimal control method is adopted.

References

[1] Rahman A, Zulkifli S A, Pathman E . (2020) A new energy exploration for Malaysia's future electric supply based on waste kinetic energy: ideas, concepts, and possibilities[J]. Electrical Engineering, 102(16).

[2] Dudurych I M . (2021) The Impact of Renewables on Operational Security: Operating Power Systems That Have Extremely High Penetrations of Nonsynchronous Renewable Sources[J]. IEEE Power and Energy Magazine, 19(2):37-45.

[3] Bao J, Yuan T, Song C, et al. (2019) Thermodynamic analysis of a new double-pressure condensation power generation system recovering LNG cold energy for hydrogen production[J]. International Journal of Hydrogen Energy, 44(33):17649-17661.

[4] Kandasamy N K . (2020) Prosumer Site Power Interruption Attacks: Exploiting the Reactive Power Control Feature in Smart Inverters[J]. IET Renewable Power Generation, 14(23):5372.

[5] Gyanwali K, Komiyama R, Fujii Y . (2020) Representing hydropower in the dynamic power sector model and assessing clean energy deployment in the power generation mix of Nepal[J]. Energy, 202(11):117795.

[6] R Vennell, R Major, R Zyngfogel, et al. (2020) Rapid initial assessment of the number of turbines required for largescale power generation by tidal currents[J]. Renewable Energy, 162:1890-1905.

Internal Leakage Predicition of Hydraulic Spool valves Based on Acoustic Emission Technology

Fei Song[1*], Likun Peng[1], Jia Chen[1] and Benmeng Wang[2]

[1]Collage of Power Engineering, Naval University of Engineering, Wuhan, Hubei, 430033, China;

[2]92001 PLA, Qingdao, Shangdong, 266011, China

[*]E-mail: llandsf@nue.edu.cn

Abstract. In order to realize the nondestructive testing (NDT) of the internal leakage fault of hydraulic spool valves, the internal leakage rate must be predicted by AE (acoustic emission) technology. An AE experimental platform of internal leakage of hydraulic spool valves is built to study the characteristics of AE signals of internal leakage and the relationship between AE signals and leakage rates. The research results show the AE signals present a wideband characteristic. The main frequencies are concentrated in 30~50 kHz and the peak frequency is around 40 kHz. When the leakage rate is large, there are significant signal characteristics appearing in the high frequency band of 75~100 kHz. The exponent of the root mean square(RMS) of AE signals is positively correlated with the exponent of the leakage rate only if the leakage rate is greater than 2~3 mL/min. This find could be used to predict the internal leakage rate of hydraulic spool valves.

1. Introduction

The hydraulic spool valve, with its numerous advantages, is widely applied in the hydraulic system. To ensure the good sealing and smooth action of the spool valve, the normal radial clearance between the valve core and the valve sleeve shall range from 3 to 8 μm [1-3]. With the radial clearance, it is inevitable for the internal leakage of the hydraulic spool valve. However, the leakage rate in the spool valve is very small under the normal clearance, and is usually smaller than 0.1% of the rated flow rate [4], which doesn't affect the normal operation of the system. However, when there is any wear in the valve core and the valve sleeve, the leakage rate will be increased drastically due to overlarge radial clearance, which will affect the normal operation of the executing mechanism, resulting in excessive pressure loss and abnormal heating, and even cause the failure of the function of the spool valve in severe cases. Therefore, mastering the leakage in the spool valve will be of great significance to ensure the normal operation of the hydraulic system. However, the leakage fault is concealed, resulting in large diagnosis difficulty. With the traditional means, the valve shall be disassembled for pressurizing detection in the shutdown state in order not to affect the normal operation of the system, but the means requires a lot of labor and time; and in the case of numerous spool valves to be detected, such low-efficiency detection means will not be available.

Many studies indicate that with the AE testing technique, the real-time nondestructive testing of the internal leakage of valves can be achieved, without shutdown and the disassembly of valves. However, the study on the AE testing of the internal leakage of the valve focuses on valves in which gas is used as the flowing medium, and especially ball valves, gate valves, and needle valves [5-7]; and there are a few studies on the AE testing of the internal leakage of the hydraulic spool valve in which hydraulic oil is used as the flowing medium. This article is intended to construct the AE experiment table for the internal leakage of the hydraulic spool valve, by which the internal leakage rate is predicted by analyzing the AE signals of the internal leakage of the spool valve, thereby providing basis for the nondestructive testing of the internal leakage fault of the hydraulic spool valve.

2. Internal leakage simulating experiment table for spool valve

The internal leakage simulating experiment table of the hydraulic spool valve, which is constructed, is shown in Figure 1. The experiment table is composed of two parts: A constant-pressure oil supply system and an internal leakage simulating device for the spool valve.

2.1 Constant-pressure oil supply system

The principle diagram of the constant-pressure oil supply system is shown in Figure 1. The system can provide 0 to 10 MPa of pressure oil for the internal leakage simulating device for the spool valve, and is connected with oil inlet/outlet of the leakage simulating device with the hydraulic hose (length ≥ 5m), which can avoid the interference of the vibration noise of the pump and valves on the collection of the AE signal of leakage. The accumulator arranged on the oil supply line can achieve two effects: One effect is to eliminate pressure pulsation generated during oil supply from the oil source so as to keep stable pressure; and the other effect is to achieve short-time oil supply for the leakage simulating device in extremely quiet environment as the auxiliary oil source upon the oil supply shutoff of the pump, thereby minimizing interference noise during AE testing. Since the internal leakage rate of the spool valve is small, it can be measured in the method of graduated cylinder + high-accuracy flow sensor, and switching between the volumetric cylinder and the high-accuracy flow sensor is achieved via the solenoid directional valve 19.

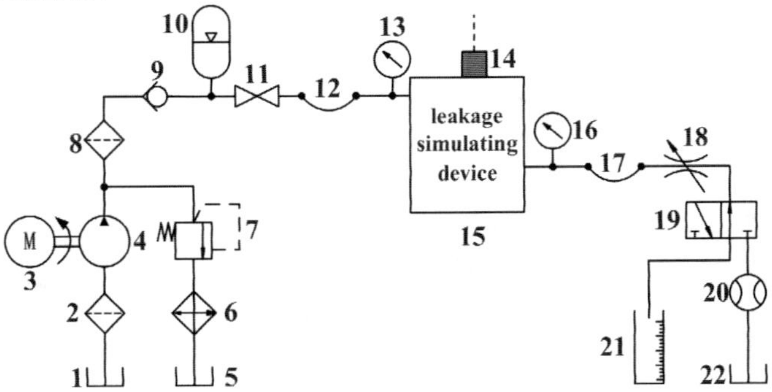

Figure 1. Principle diagram of constant-pressure oil supply system
1, 5, 20-Tank; 2, 8-Filter; 3-Motor; 4-Pump; 6-Cooler; 7-Overflow valve; 9-Check valve; 10-Accumulator; 11-Ball valve; 12, 17-Hydraulic hose; 13, 16-Pressure sensor; 14-AE sensor; 15-Leakage simulating device; 18-Backpressure loading valve; 19-Solenoid directional valve; 20-Flow sensor; 21-Volumetric cylinder

2.2 Internal leakage simulating device for spool valve

The internal leakage simulating device for the spool valve attains the function of simulating the actual condition of the internal leakage of the spool valve under each operating condition to make a study on the influence rule of each factor on the internal leakage rate of the spool valve. The internal structure of the device is shown in Figure 2. The variable parameters of the device include valve core diameter, clearance height, and sealing length; the valve core diameter and the clearance height can be changed by replacing a combination of the valve body 4 and the valve core 3; and the sealing length can be changed by adjusting the spiral micrometer head 1 on the leakage simulating device. The locking mechanism is arranged; and upon adjustment in the sealing length, the valve core can be fastened via the locking nut to prevent the position of the valve core from being changed under the action of hydraulic power.

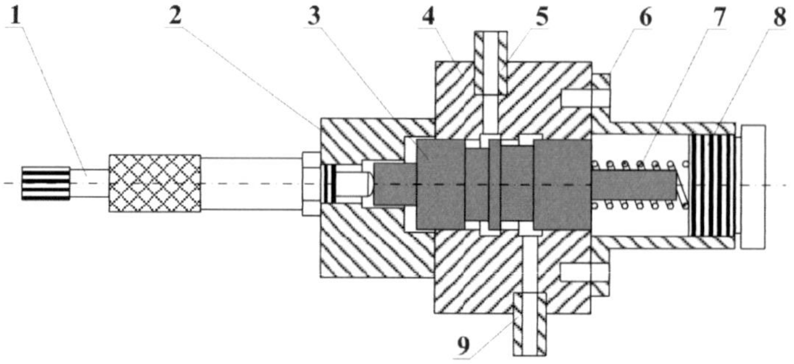

Figure 2. Internal leakage simulating device for spool valve
1-Spiral micrometer head; 2-Check block; 3-Valve core; 4-Valve body; 5-Oil inlet; 6-Sleeve; 7-Return spring; 8. Locking nut; 9-Oil outlet

3. Theoretical basis

3.1 Calculation of leakage rate of spool valve

As hydraulic oil is large in viscosity of hydraulic oil, the leakage gap of the spool valve is micron-sized, and the viscous drag of the wall achieves the principal effect in the oil flowing process, the leakage of the spool valve is, generally, in the laminar flow state. Without the relative movement, the formula of the laminar flow in the concentric annular slot in terms of the internal leakage of the spool valve is [8, 9]:

$$Q_0 = \frac{\pi d h^3}{12 \mu l} \Delta p \tag{1}$$

Where, Q_0 is the leakage rate (L/min) of the concentric annular slot under the laminar flow, and d is valve core diameter (m); h is clearance height (m); Δp is differential pressure (Pa) between upstream and downstream; l is sealing length (m); ρ is oil density (kg/m^3); and μ is dynamic viscosity (Pa·s) of oil.

The valve core of the spool valve is parallel to the valve sleeve. However, in the case of eccentricity, the calculation formula of the leakage rate is [8, 9]:

$$Q_e = \left(1 + 1.5\varepsilon^2\right) Q_0,\ 0 \le \varepsilon \le 1 \tag{2}$$

Where, Q_e is the leakage rate (L/min) in the case of the eccentricity of the valve core, and ε is the eccentricity ratio of the valve core. The internal leakage rate in the case of parallel eccentricity is $(1{\sim}2.5)Q_0$. In addition to eccentricity, there is also an inclination case in the valve core, and the angle of inclination will affect the leakage rate of the spool valve. It can be known from the study of the pertinent literature that the leakage rate is $0.5Q_0$ in the case of complete inclination [9], and thus, with the overall consideration of the influence on the eccentricity and inclination of the spool valve, the leakage rate under the laminar flow shall be:

$$Q = (0.5{\sim}2.5) Q_0 \tag{3}$$

3.2 Relationship between leakage rate and AE signal

The study of the pertinent literature [5-7] indicates the index of the acoustical power of the AE signal of leakage is in positive proportion with the index of the leakage rate, and acoustical power is $P_S \propto AE_{RMS}^2$. Therefore, the index of the RMS of the AE signal is also in positive proportion with the index of the internal leakage.

$$logQ = k_{RMS} logAE_{RMS} + k_c \tag{4}$$

Where, Q is the leakage rate (L/min); AE_{RMS} is RMS (mV) of the leakage AE signal; k_{RMS} and k_c are coefficients.

4. Analysis on test results

In the test, three valve core diameters are 10 mm, 16 mm, and 20 mm, the AE signal of leakage and the internal leakage rate are collected under normal clearance (5 μm) and excessive clearance (40 μm). In the test, variable parameters include sealing length (0.25 mm, 0.5 mm, 1.0 mm, 1.5 mm and 2.0 mm), and differential pressure (2 MPa, 4 MPa, 6 MPa, 8 MPa and 10 MPa). In the test, 46# antiwar hydraulic oil is used, oil density is 850 kg/m^3, and kinematic viscosity is 4.5×10^{-5} m^2/s. To reduce the influence of oil temperature on the test, the test shall be started after the system is started up until it reaches the thermal equilibrium state.

4.1 Frequency spectrum analysis on AE signal of leakage

When the sealing length measured under normal clearance (5 μm) and excessive clearance (40 μm) in the test is 0.25 mm, the frequency spectrum of the AE signal of leakage of each valve under differential pressure of 2 to 10 MPa is shown in Figure 3 and 4.

It can be known from Figures 3 and 4 that in the case of large internal leakage rate, the AE signal of leakage shows the broadband characteristic, which also proves that the AE signal of leakage is caused by turbulent jet flow at the outlet of the spool valve slot. Sealing length and differential pressure hardly affect the frequency spectrum component of the AE signal of leakage, but affect the amplitude of each frequency; and with increase in differential pressure, and reduction in sealing length, the amplitude shows an increase trend. Change in clearance height is significant to the influence on the frequency spectrum, i.e., exerting influence on the frequency spectrum component and the amplitude; and with increase in clearance height and frequency spectrum component, the amplitude is increased significantly.

Regardless of normal clearance or excessive clearance, the primary frequency component of the AE signal of leakage is concentrated within the range of 30 to 50 kHz, with the peak frequency of approx. 40 kHz. Compared with the AE signal of leakage under the normal clearance, the AE signal under the excessive clearance shows the significant signal characteristic at the high-frequency section of 75 to 100 kHz.

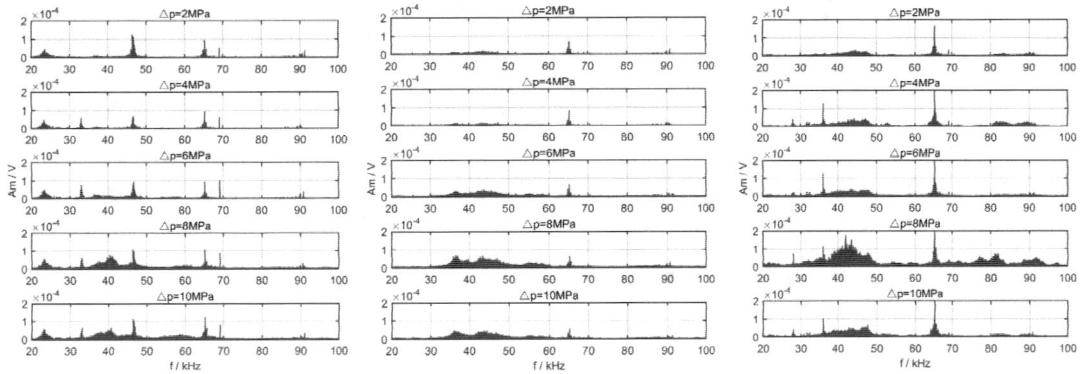

a) Valve core diameter 10 mm b) Valve core diameter 16 mm c) Valve core diameter 20 mm

Figure 3. Spectrogram of AE signal (clearance 5 μm, sealing length 0.25 mm)

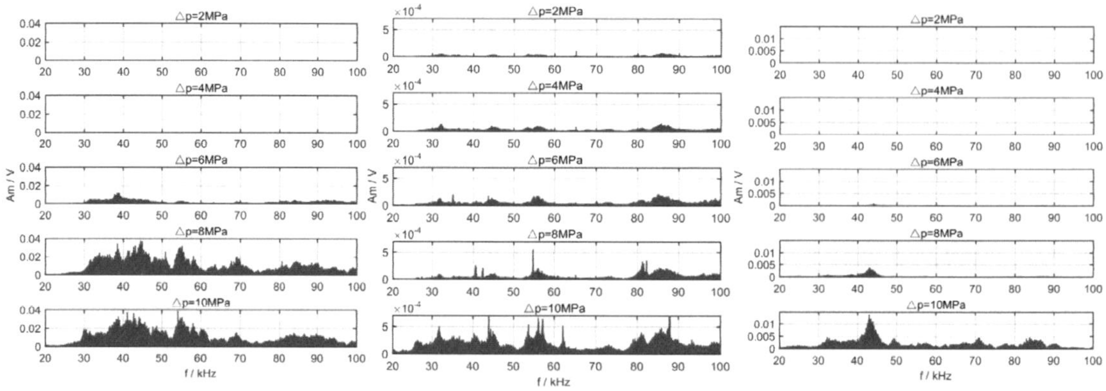

a) Valve core diameter 10 mm b) Valve core diameter 16 mm c) Valve core diameter 20 mm

Figure 4. Spectrogram of AE signal (clearance 40 μm, sealing length 0.25 mm)

4.2 Prediction of leakage rate

It can be known from the formula (4) that the index of the leakage rate is in direct proportion with the index of the RMS of the AE signal, and correlation coefficients between $\log Q$ and $\log AE_{\mathrm{RMS}}$ are calculated by data measured from the test, which are shown in Table 1.

Table 1. Correlation coefficients between $\log Q$ and $\log AE_{\mathrm{RMS}}$

	Clearance height of 5 μm			Clearance height of 40 μm		
	10 mm	16 mm	20 mm	10 mm	16 mm	20 mm
0.25 mm	0.79	0.83	-0.18	-	-	-
0.5 mm	0.99	-0.02	0.97	0.88	0.97	0.96
1.0 mm	-0.39	-0.91	0.99	0.97	0.98	0.88
1.5 mm	-0.19	-0.74	0.94	0.98	0.99	0.91
2.0 mm	-0.75	-0.89	-0.76	0.76	0.93	0.89

It can be known from Table 1 that in the case of small internal leakage, the indexes of the internal leakage rate of the spool valve and the RMS of the AE signal of leakage are poor in correlation, and even, are negatively correlated with each other; and in the case of large leakage rate, the correlation between both is good, and in most cases, the correlation coefficient is larger than 0.9. Therefore, in the case of large leakage rate, it is feasible to predict the internal leakage rate via the RMS of the AE signal. The fitting parameters are obtained by performing fitting on the formula (4) with leakage rate measured in the test and AE data, which are shown in Tables 2 and 3. The actually measured leakage rate and fit internal leakage rate under each pressure in terms of each valve core diameter are shown in Figure 5.

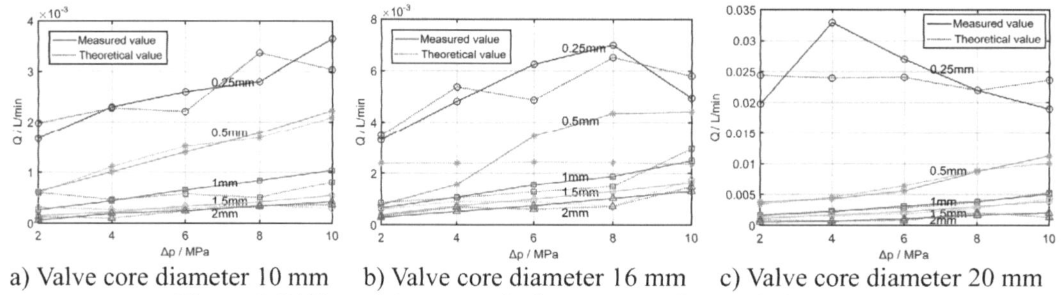

a) Valve core diameter 10 mm b) Valve core diameter 16 mm c) Valve core diameter 20 mm

Figure 5. Differential pressure-leakage curve of normal clearance

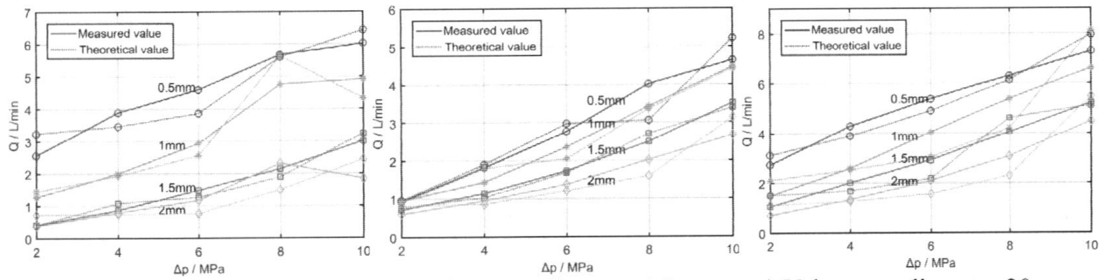

a) Valve core diameter 10 mm b) Valve core diameter 16 mm c) Valve core diameter 20 mm

Figure 6. Differential pressure-leakage curve of excessive clearance

It can be known from Figures 5 and 6 that with rise in pressure, reduction in sealing length, and increase in valve core diameter and clearance height, the internal leakage rate of the spool valve exhibits a rising trend, and the leakage rate is the most sensitive to increase in clearance height; and the internal leakage rates of normal clearance and excessive clearance differ by approx. 3 orders of magnitude. The above-mentioned rule is consistent with the formula (1).

Table 2. Fitting parameters of normal clearance

	10 mm		16 mm		20 mm	
	k_{RMS}	k_c	k_{RMS}	k_c	k_{RMS}	k_c
0.25 mm	0.69	-0.62	0.89	0.22	-0.04	-1.74
0.5 mm	4.67	10.57	-0.1	-2.92	1.01	0.51
1.0 mm	-8.72	-29.34	-16	-51.51	1.33	1.29
1.5 mm	-5.29	-19.55	-9.23	-31.44	2.37	4.51
2.0 mm	-2.85	-12.17	-13.45	-42.54	-2.97	-11.71

Table 3. Fitting parameters of excessive clearance

	10 mm		16 mm		20 mm	
	k_{RMS}	k_c	k_{RMS}	k_c	k_{RMS}	k_c
0.5 mm	0.09	0.73	0.77	2.09	0.27	1.24
1.0 mm	0.34	1.12	0.6	1.71	0.52	1.92
1.5 mm	0.94	2.38	0.63	1.78	1.01	3.16
2.0 mm	0.28	0.73	0.58	1.71	1.22	3.69

It can be known from Tables 2 and 3 that in the case of excessive clearance (40 μm), the fitting parameters k_{RMS} of the diameters of three valve cores under the sealing length ranging from 0.5 to 2.0 mm are larger than zero, indicating that the internal leakage rate of the spool valve is positively correlated with the RMS of the AE signal of leakage. In the case of the normal clearance (5 μm) of the spool valve, the fitting parameters are $k_{RMS} > 0$ under the sealing length ranging from 0.25 to 0.5 mm in terms of the valve core diameter of 10 mm; the fitting parameter is $k_{RMS} > 0$ under the sealing length of 0.25 mm in terms of the valve core diameter of 16mm; the fitting parameters are $k_{RMS} > 0$ under the sealing length ranging from 0.5 to 1.5 mm in terms of the valve core diameter of 20 mm; and the leakage rate curve under the sealing length of 0.25 mm exhibits a declining trend after 4 MPa, resulting in the fitting parameter of $k_{RMS} < 0$, but the value is very small. In conjunction with the formula (3), it is speculated that with rise in pressure, the stress of the valve core is changed, while its eccentricity and inclination degree are also changed, which causes reduction in the leakage rate, resulting in $k_{RMS} < 0$.

The fitting parameters are $k_{\text{RMS}} < 0$ under the sealing length ranging from 1.0 to 2.0 mm in terms of the valve core diameter of 10 mm; the fitting parameters are $k_{\text{RMS}} < 0$ under the sealing length ranging from 0.5 mm to 2.0 mm in terms of the valve core diameter of 16mm; and the fitting parameter is $k_{\text{RMS}} < 0$ under the sealing length of 2.0 mm in terms of the valve core diameter of 20 mm. In the above-mentioned operating conditions, $k_{\text{RMS}} < 0$ indicates that the RMS of the AE signal isn't positively correlated with the internal leakage rate, which means that the AE signal can't effectively predict the internal leakage rate of the spool valve. The leakage rates in such operating conditions are small, and the maximum leakage rate is approx. 2 to 3 mL/min. In view of the above, the minimum internal leakage of the spool valve, which is detected via the RMS of the AE signal, is approx. 2 to 3

5. Conclusions

In the experimental study on the internal leakage of spool valves with three valve core diameters under the normal clearance and excessive clearance through the constructed internal leakage simulating experiment table, the following conclusions can be drawn:

- The AE signal of leakage is caused by turbulent jet flow at the outlet of the spool valve slot, which shows the broadband characteristic.
- The primary frequency component of the AE signal of leakage is concentrated within the range of 30 to 50 kHz, with the peak frequency of approx. 40 kHz.
- Compared with the AE signal of leakage under the normal clearance, the AE signal under the excessive clearance shows the significant signal characteristic at the high-frequency section of 75 to 100 kHz.
- The minimum internal leakage of the spool valve can be detected via the RMS of the AE signal, which is approx. 2 to 3 mL/min.

References

[1] Zhang J J. (2020) Research on the Trajectory of Contaminated Particles in the Chamber of Spool Valve. Lanzhou University of Technology, Lanzhou.
[2] Song F, Chen J, Peng L K, et al. (2021) Experimental Research on the Internal Leakage of Hydraulic Slide Valve. Fluid Machinery, 49(07): 1-6+28.
[3] Chen Y Z. (2021) Research on the Effect of Spool Valve Pairs Fit on Servo Valve Performance. Machine Tool & Hydraulics, 49(05): 107-111.
[4] Chen D W. (2015) Simulation and Eperimental Investigation of Clamping and Leakage in Hydraulic Slide Valves. Shandong University, Jinan.
[5] W. Kaewwaewnoi, A. Prateepasen, P. Kaewtrakulpong. (2010) Investigation of the Relationship between Internal Fluid Leakage through a Valve and the Acoustic Emission Generated from the Leakage. Measurement, 43(2): 274-282.
[6] A. Prateepasen, W. Kaewwaewnoi, P. Kaewtrakulpong, et al. (2011) Smart Portable Noninvasive Instrument for Detection of Internal Air Leakage of a Valve Using Acoustic Emission Signals. Measurement, 44(2), 378-384.
[7] Zhao H, Li Z, Zhu S, et al. (2020) Valve Internal Leakage Rate Quantification Based on Factor Analysis and Wavelet-BP Neural Network Using Acoustic Emission. Applied Sciences, 10(16): 5544.
[8] Zhou Z R. (2010) Research on Water-Powered Rock Drill System and its Key Technology. Central South University, Changsha.
[9] Karl Trutnovsky. (1986) Bruhrungsfreie Dichtungen. China Machine Press, Beijing.

Ship tracking based on the fusion of Kalman filter and particle filter

Wanjin Xu[1*], Jiying Li[1], Junjie Bai[1] and Yingying Zhang[1]

[1] College of Electronic & Information Engineering Lanzhou Jiaotong University, Lanzhou, Gansu,730070,China

[*]Corresponding author's e-mail: 12201670@stu.lzjtu.edu.cn

Abstract. Aiming at the problem of low filtering accuracy and even divergence caused by model mismatch when using extended Kalman filter in ship GPS navigation and positioning state estimation, a positioning ship state estimation algorithm based on the fusion of improved unscented Kalman filter and particle filter is proposed. Compared with the traditional particle filtering algorithm, the algorithm has two improvements: first, the algorithm uses untraced Kalman as the main framework, and uses the optimal estimation of particle updating state by particle algorithm; Secondly, in the resampling process, a resampling algorithm based on weight optimization is proposed to increase the diversity of particles. The simulation results show that not only the particle degradation degree of the particle filter is reduced, but also the particle tracking accuracy is improved.

1. Introduction

Global Position System (GPS) is widely used in military and national economy. Ship GPS navigation and positioning is to install a GPS[1] receiver on the moving target (ship) to carry out navigation and positioning calculation. GPS receiver can receive the signal broadcast by the navigation satellite in orbit in real time, and calculate the position and speed of the receiving carrier (ship). The high frequency oscillatory random jamming signal is artificially added into the ground signal transmitted by GPS navigation satellite in civil neighborhood, which results in the high frequency jitter of all derived ground satellite signals. In order to improve the positioning accuracy, it is necessary to filter the GPS observation signal about the ship's position and velocity. The artificial high frequency random jamming signal in GPS system can be regarded as GPS positioning observation noise. The observation noise intensity (variance) can be obtained from GPS observation signal using system identification method[2].

2. Target motion model

In order to simplify the model, it is assumed that the ship sails in a straight direction when leaving the port. The starting point of the port and wharf is used as the origin of coordinates, and the sampling time is set to T0. S(k) represents the real position of the ship at sampling time kT0, and Y (k) represents the observation value of GPS positioning at sampling time kT0, so there is an observation model.

$$y(k) = s(k) + v(k)$$

Where, v(k) represents GPS positioning error, its zero mean and white noise of σ_v^2 can be assumed, and variance B can be obtained by statistical methods from σ_v^2large number of GPS observation test data. The ship velocity at time kT0 is $\dot{s}(k)$ and acceleration is $a(k)$, which can be obtained from the equation of uniformly accelerated motion

$$s(k+1) = s(k) + \dot{s}(k)T_0 + 0.5T_0^2 a(k)$$

$$\dot{s}(k+1) = \dot{s}(k) + T_0 a(k)$$

Acceleration $a(k)$ is composed of maneuvering acceleration $u(k)$ and random acceleration $w(k)$, namely:

$$a(k) = u(k) + w(k)$$

Where, $u(k)$ is the control signal of the ship power system, which is $u(k)$ known maneuver signal output artificially; $w(k)$ is the random acceleration caused by sea breeze and waves, assuming that it is a white noise independent of $v(k)$ with zero mean and variance of σ_v^2. The state $x(k)$ of the system at sampling time kT_0 is defined as the position and speed of the ship, i.e

$$x(k) = \begin{bmatrix} s(k) \\ \dot{s}(k) \end{bmatrix}$$

The equation of state of ship motion can be obtained as

$$\begin{bmatrix} s(k+1) \\ \dot{s}(k)+1 \end{bmatrix} = \begin{bmatrix} 1 & T_0 \\ 0 & 1 \end{bmatrix} \begin{bmatrix} s(k) \\ \dot{s}(k) \end{bmatrix} + \begin{bmatrix} 0.5T_0^2 \\ T_0 \end{bmatrix} u(k) + \begin{bmatrix} 0.5T_0^2 \\ T_0 \end{bmatrix} w(k)$$

The observed equation is

$$y(k) = \begin{bmatrix} 1 & 0 \end{bmatrix} \begin{bmatrix} s(k) \\ \dot{s}(k) \end{bmatrix} + v(k)$$

Namely, the state space model of the system is

$$x(k+1) = \Phi x(k) + Bu(k) + \Gamma w(k)$$
$$y(k) = Hx(k) + v(k)$$

Type,

$$\Phi = \begin{bmatrix} 1 & T_0 \\ 0 & 1 \end{bmatrix}, B = \Gamma = \begin{bmatrix} 0.5T_0^2 \\ T_0 \end{bmatrix}, H = \begin{bmatrix} 1 & 0 \end{bmatrix}$$

Thus, the Kalman filtering problem of ship GPS navigation and positioning is: based on GPS observation data ($y(1), y(2), \cdots, y(k)$), the optimal estimation $\hat{s}(k \mid k)$ of ship's position $s(k)$ at time K is obtained.

Without considering the dynamic factors of the maneuvering target ($u(k) = 0$), the ship system of uniform speed and straight line is extended to four dimensions, i.e

$$X(k) = \begin{bmatrix} x(k) & \dot{x}(k) & y(k) & \dot{y}(k) \end{bmatrix}$$

The state contains the horizontal position and velocity and the vertical position and velocity, then the system equation can be expressed as follows.

$$\begin{bmatrix} x(k) \\ \dot{x}(k) \\ y(k) \\ \dot{y}(k) \end{bmatrix} = \begin{bmatrix} 1 & T & 0 & 0 \\ 0 & 1 & 0 & 0 \\ 0 & 0 & 1 & T \\ 0 & 0 & 0 & 1 \end{bmatrix} \begin{bmatrix} x(k-1) \\ \dot{x}(k-1) \\ y(k-1) \\ \dot{y}(k-1) \end{bmatrix} + \begin{bmatrix} 0.5T^2 & 0 \\ T & 0 \\ 0 & 0.5T^2 \\ 0 & T \end{bmatrix} w_{2\times 1}(k)$$

$$Z(k) = \begin{bmatrix} 1 & 0 & 0 & 0 \\ 0 & 0 & 1 & 0 \end{bmatrix} \begin{bmatrix} x(k) \\ \dot{x}(k) \\ y(k) \\ \dot{y}(k) \end{bmatrix} + v_{2\times 1}(k)$$

3. Various methods of filtering algorithms

3.1. UKF filter algorithm

The concrete realization process of the algorithm. Consider the nonlinear system model with additive noise as follows[4]:

$$x_k = f\left(x_{k-1}\right) + w_{k-1} \tag{1}$$

$$z_k = h\left(x_k\right) + v_k \tag{2}$$

Where, the state vector of the system at time K is $x_k \in R^n$, and $x_k \in R^n$ is the corresponding observation vector at time K; $f()$ represents the nonlinear state model function, $h()$ represents the nonlinear observation model function; w_{k-1} is the process noise sequence at k-1, and its mean value is 0. Gaussian white noise with variances of Q_{k-1}; v_k is the observation noise sequence at k-1, with a mean of 0 and a variance of R_k, and the two are unrelated[5].

Step 1: Initialize. Given the filter's initial state x_0, initial covariance P_0, process noise variance Q_0, observation noise variance R_0.

Step 2: According to the estimated mean $\hat{x}_{k-1}$ and covariance P_{K-1} at k-1, $\hat{x}_{k-1}$ group of sigma points χ^0_{k-1}, χ^1_{k-1} ... , χ^{2n}_{k-1} and corresponding weights w^0, w^1 ... , w^{2n} (let's think of the sigma set as a vector).

$$\chi^i_{k-1} = \left[\hat{x}^i_{k-1} \quad \hat{x}^i_{k-1} + \sqrt{(n+\lambda)P_{K-1}} \quad \hat{x}^i_{k-1} - \sqrt{(n+\lambda)P_{K-1}} \right] \tag{3}$$

Step 3: Calculate the one-step prediction of the sigma point set.

$$\chi^i_{k|k-1} = f\left(\chi^i_{k-1}\right), \quad i = 0 \ldots, 2n \tag{4}$$

Step 4: Calculate the predicted mean value and covariance matrix of the system state quantity.

$$\hat{x}_{k|k-1} = \sum_{i=0}^{2n} w^i_m \chi^i_{k|k-1} \tag{5}$$

$$P_{K|K-1} = \sum_{i=0}^{2n} w^i_c \left(\chi^i_{k|k-1} - \hat{x}_{k|k-1}\right)\left(\chi^i_{k|k-1} - \hat{x}_{k|k-1}\right)^T + Q_{K-1} \tag{6}$$

Step 5: Calculated predicted mean an $\hat{x}_{k|k-1}$ d covariance matrix $P_{k|k-1}$. Using UT transformation again, new sigma points χ^{0}_{k-1}, χ^{1}_{k-1} ... , χ^{2n}_{k-1} and corresponding weights w^{0}, w^{1} ... , w^{2n}. Note: this step can also be omitted, that is, new sigma points are not generated, but $\chi^{i}_{k|k-1}$ obtained in the prediction stage of Step 3 is directly used, that is, part of the accuracy has been sacrificed to reduce the amount of calculation.

$$\chi^{i}_{k-1} = \left[\hat{x}^{i}_{k-1} \quad \hat{x}^{i}_{k-1} + \sqrt{(n+\lambda) P_{K-1}} \quad \hat{x}^{i}_{k-1} - \sqrt{(n+\lambda) P_{K-1}} \right] \tag{7}$$

Step 6: Perform nonlinear transformation of Sigma points according to the observation model and calculate the observation quantity to predict the sampling points.

$$Z^{i}_{k|k-1} = h\left(\chi^{i}_{k|k-1}\right), \quad i = 0,\dots, 2n \tag{8}$$

Step 7: Calculate the predicted mean value of system observations.

$$\hat{z}_{k|k-1} = \sum_{i=0}^{2n} w^{i}_{m} z^{i}_{k|k-1} \tag{9}$$

Step 8: Calculate the new information covariance matrix, the cross-covariance matrix between the state and the observation.

$$P_{zz,k|k-1} = \sum_{i=0}^{2n} w^{i}_{c} \left(Z^{i}_{k|k-1} - \hat{z}^{i}_{k|k-1} \right)\left(Z^{i}_{k|k-1} - \hat{z}^{i}_{k|k-1} \right)^{T} + R_{k} \tag{10}$$

$$P_{xz,k|k-1} = \sum_{i=0}^{2n} w^{i}_{c} \left(\chi^{i}_{k|k-1} - \hat{x}^{i}_{k|k-1} \right)\left(Z^{i}_{k|k-1} - \hat{z}^{i}_{k|k-1} \right)^{T} \tag{11}$$

Step 9: Calculate the filter gain matrix.

$$K_{k} = P_{xz,k|k-1} P^{-1}_{zz,k|k-1} \tag{12}$$

Step 10: Status update. The mean value and covariance matrix of a posterior state estimate at time K are calculated.

$$\hat{x}_{k} = \hat{x}_{k|k-1} + K_{k}\left(z_{k} - \hat{z}_{k|k-1} \right) \tag{13}$$

$$P_{k} = P_{K|K-1} - K_{k} P_{zz,k|k-1} K^{T}_{k} \tag{14}$$

Step 11: set k= K +1 and repeat step 2 to Step 10 for the next filtering calculation.

Since UKF filter algorithm was put forward, and has extensive application prospect in engineering, but the UKF algorithm also has its disadvantages, first of all, in addressing the problem of high dimension, because at this time no trace filter auxiliary scale parameter in $\kappa < 0$, thus makes some sigma point weights $w < 0$, this leads to the filter covariance appeared in the process of the positive definite, Thus, the filter value is unstable and even diverges. Secondly, the parameter selection problem of UKF algorithm has not been effectively solved, and the filtering effect of the algorithm will also be affected by the initial filtering value.

3.2. Particle filtering algorithm

Importance sampling particle filter[6].

Suppose that the state space model of a continuous nonlinear dynamic system can be expressed as Equation (15) and Equation (16) :

$$x_{k} = f\left(x_{k-1}, u_{k} \right) \tag{15}$$

$$z_k = h(x_k, v_k) \tag{16}$$

Among them, $x_k \in R^n$, $z_k \in R^m$ respectively target state vector and the observation vector, $f(\bullet)$ represents the state of the system of nonlinear transfer function, nonlinear observation $h(\bullet)$ said the state of the system function, u_k and v_k are independent of each other and has nothing to do with the system state, respectively state noise and observation noise, m, n state vector dimension and observation vector dimension respectively.

The standard particle filter (SIR) algorithm is SIS algorithm which introduces resampling and selects the state transition probability distribution with prior properties as the reference distribution. The SIR algorithm process is as follows:

Step 1: Initialize

$\left\{x_0^i\right\}_{i=1}^N$ is sampled from $P(x_0)$, and the initial weight $\left\{w_0^i\right\}_{i=1}^N$ is all 1/N.

Step 2: Importance sampling

(1) Particle set $\left\{x_k^i\right\}_{i=1}^N$ is obtained by sampling from the proposed distribution

(2) Calculate the weight of particles at time K, as shown in Equation (17) :

$$w_k^i = w_{k-1}^i P\left(z_k \middle| x_{k-1}^i\right) \tag{17}$$

(3) Normalized weight, formula (18) is obtained.

$$w_k^i = w_k^i \Big/ \sum_{i=1}^N w_k^i \tag{18}$$

Step 3: Resampling

After resampling, the new particle set $\left\{x_k^i\right\}_{i=1}^N$ is obtained and its weight is normalized to $\left\{w_k^j\right\} = 1_N$

Step 4: State estimation

The estimate is calculated as shown in Equation (19).

$$\hat{x} = \frac{1}{n}\sum_{J=1}^N x_k^j \tag{19}$$

3.3. UKF and PF fusion algorithm

The specific steps of the filtering algorithm are shown in Figure 1[7]:

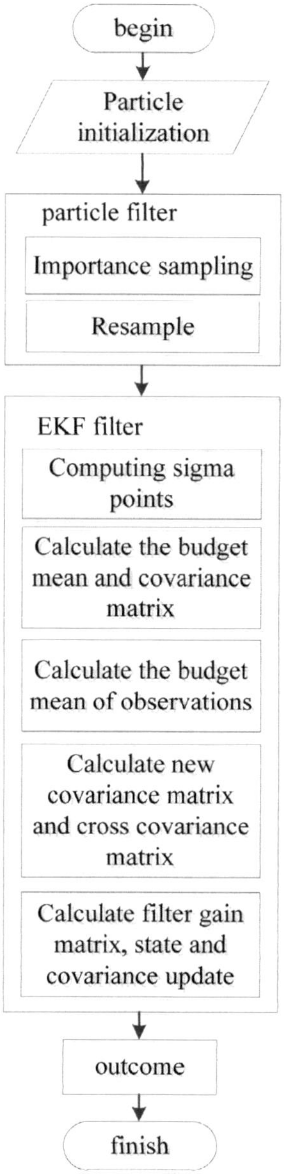

Figure 1: Flow chart of EKF and PF hybrid filtering algorithm

First of all, PF[8] algorithm, according to the known prior distribution of $p\left(x_0\right)$ sample get particle $\left\{x_0^i\right\}_{i=1}^{N}$, then according to the choice of importance density function of particle is updated, the weight of each particle in the further $w_k^i = w_{k-1}^i P\left(z_k \middle| x_{k-1}^i\right)$, then the weight normalized $w_k^i = w_k^i \Big/ \sum_{i=1}^{N} w_k^i$ when effectively estimate the number of training patterns to satisfy certain requirements need to sample, to generate new particles and give weight $\left\{w_k^j\right\} = 1_N$, Then, the variance of state estimation $\hat{X}_k^{PF} = \sum_{i=1}^{N} w_k^i x_k^i$ at time K can be obtained. According to $\hat{X}_{k-1}^{PF}$ at time K-1, 2n+1 sigma points can be obtained, χ_{k-1}^0, $\chi_{k-1}^1 \cdots$, χ_{k-1}^{2n} and corresponding weights w^0, $w^1 \ldots$, w^{2n} (here the set of sigma points is treated as a vector) :

$$\chi_{k-1}^i = \left[\hat{X}_{k-1}^{(PF)i} \quad \hat{X}_{k-1}^{(PF)i} + \sqrt{(n+\lambda) P_{K-1}} \quad \hat{X}_{k-1}^{(PF)i} - \sqrt{(n+\lambda) P_{K-1}} \right] \tag{20}$$

Then calculate the one-step prediction of the sigma point set:

$$\chi_{k|k-1}^i = f\left(\chi_{k-1}^i\right), \quad i = 0 \ldots, 2n \tag{21}$$

Calculate the predicted mean value and covariance matrix of the system state quantity:

$$\hat{X}_{k|k-1} = \sum_{i=0}^{2n} W_m^i \chi_{k|k-1}^i \tag{22}$$

$$P_{K|K-1} = \sum_{i=0}^{2n} W_c^i \left(\chi_{k|k-1}^i - \hat{X}_{k|k-1}\right)\left(\chi_{k|k-1}^i - \hat{X}_{k|k-1}\right)^T + Q_{K-1} \tag{23}$$

The predicted mean $\hat{X}_{k|k-1}$ and covariance matrix $P_{K|K-1}$ are calculated, and UT changes are used again to generate new sigma points $\chi_{k|k-1}^0$, $\chi_{k|k-1}^1 \cdots$, $\chi_{k|k-1}^{2n}$ and corresponding weights w^0, $w^1 \ldots$, w^{2n}. It should be noted that this step can also be omitted, that is, no new sigma points are generated, but $\chi_{k|k-1}^i$ can be directly calculated, that is, the amount of calculation is reduced at the cost of sacrificing part of the accuracy.

$$\chi_{k|k-1}^i = \left[\hat{X}_{k|k-1}^i \quad \hat{X}_{k|k-1}^i + \sqrt{(n+\lambda) P_{k|K-1}} \quad \hat{X}_{k|k-1}^i - \sqrt{(n+\lambda) P_{k|K-1}} \right] \tag{24}$$

Further, the nonlinear transformation of Sigma points was carried out according to the observation model, and the observation quantity was calculated to predict the sampling points:

$$Z_{k|k-1}^i = h\left(\chi_{k|k-1}^i\right), \quad i = 0, \cdots, 2n \tag{25}$$

To calculate the predicted mean value of system observations:

$$\hat{Z}_{k|k-1} = \sum_{i=0}^{2n} W_m^i Z_{k|k-1}^i \tag{26}$$

Calculate the new information covariance matrix and the cross-covariance matrix between the state and the observation:

$$P_{zz,k|k-1} = \sum_{i=0}^{2n} W_c^i \left(Z_{k|k-1}^i - \hat{Z}_{k|k-1}^i\right)\left(Z_{k|k-1}^i - \hat{Z}_{k|k-1}^i\right)^T + R_k \tag{27}$$

$$P_{xz,k|k-1} = \sum_{i=0}^{2n} W_c^i \left(\chi_{k|k-1}^i - \hat{X}_{k|k-1}^i\right)\left(Z_{k|k-1}^i - \hat{Z}_{k|k-1}^i\right)^T \tag{28}$$

Calculation of filter gain matrix:

$$K_k = P_{xz,k|k-1} P_{zz,k|k-1}^{-1} \tag{29}$$

Finally, a status update. The mean value and covariance matrix of a posterior state estimate at time K are calculated[9].

$$\hat{X}_k = \hat{X}_{k|k-1} + K_k\left(Z_k - \hat{Z}_{k|k-1}\right) \quad (30)$$

$$P_k = P_{K|K-1} - K_k P_{zz,k|k-1} K_k^T \quad (31)$$

When the estimated value of the mixed filter A and the covariance matrix B at time K are worked out, the particle at the next moment is updated and the mixed filter calculation at time K +1 continues[10].

4. Analysis of simulation result

It is assumed that the ship is moving on a two-dimensional horizontal plane, the initial position is (-100m,200m), the horizontal movement speed is 2m/s, the vertical movement speed is 20m/s, the scanning period of GPS receiver is T=1s, the mean value of observation noise is 0, and the variance is 100. The closer the target is to uniform linear motion; Otherwise, it is a curved motion.

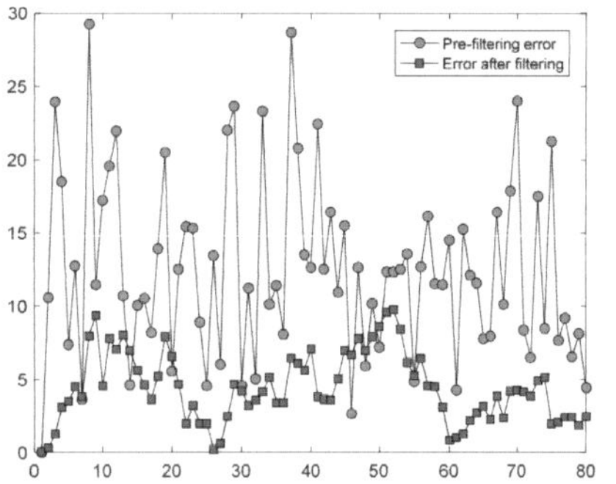

Figure 2. KF filter coordinate position error

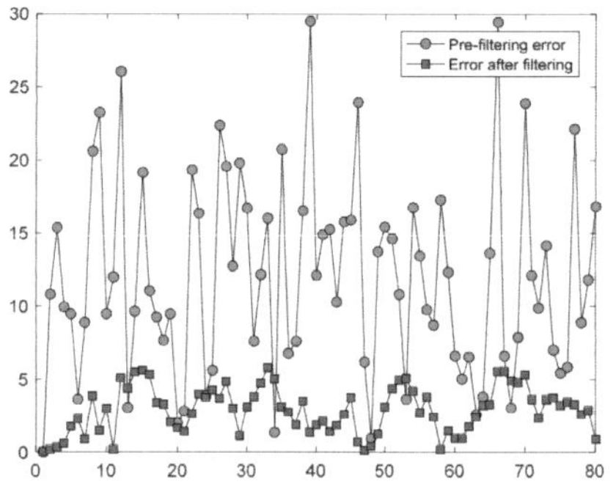

Figure 3. PF-KF filter coordinate position error

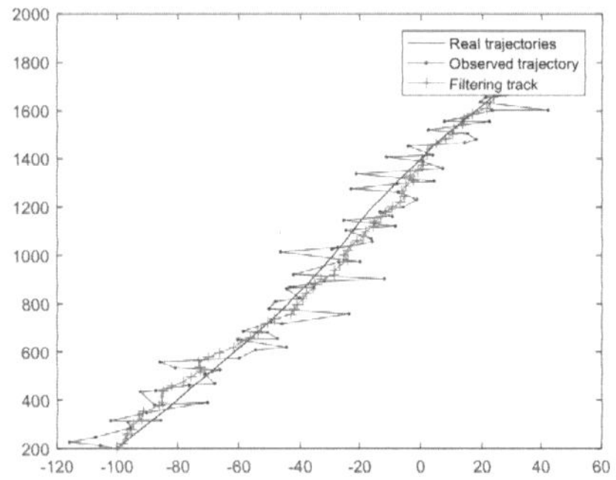

Figure 4. Real target trajectory and KF tracking trajectory

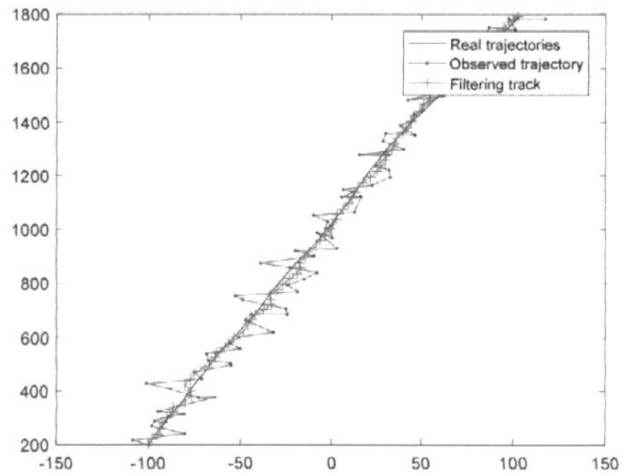

Figure 5. Real target trajectory and PF-KF tracking trajectory

It can be seen that the maximum observed noise of displacement is close to 35m. For the target sports field (about 1800m, about 250m wide), this noise is very large. Of course, it is only for simulation, and the measurement error of the actual sensor cannot be so large. After Kalman filtering, the position deviation is reduced to below 10m. You can see that. Kalman filter can not eliminate noise completely, but it has reduced the influence of noise to the maximum extent.

Therefore, Kalman filter has achieved a good effect. Compared with Figure 2 and Figure 3, it can be concluded that PF-KF filter has a stronger tracking effect, and most of its deviation is almost reduced to less than 5m. It is obvious from FIG. Figure 5 that the filtered tracking track is quite consistent with the real track.

5. Conclusion

This paper mainly studies and analyzes the current main filtering theory and related algorithms. Firstly, linear KF algorithm based on KF framework is introduced. Then, in order to solve the state estimation problem of nonlinear and non-Gaussian systems, PF algorithm based on Monte Carlo method is further introduced. Combining KF algorithm with PF algorithm, a hybrid filtering algorithm combining KF and PF is proposed, which is introduced from ship target tracking related problems. Ship target tracking model based on KF algorithm, KF and PF hybrid filtering algorithm is established respectively. The specific principles and steps of target tracking by using the above two algorithms are described in detail. The simulation experiment is carried out by comparing the trajectory simulation and tracking error. Experimental results show that the estimation result has higher accuracy than that of kalman filter and particle filter alone, and the target tracking algorithm based on KF and PF hybrid filter has the best tracking effect and the smallest error.

Acknowledgments

This paper is supported by the project of Natural Science Foundation of Gansu Province, "Research on the Detection of Beidou Navigation System Interference Sources Based on the Volume Criterion (project number: 20JR5RA407) " .

References

[1] Jwo D J, Yang C F, Chuang C H, Lee T Y 2013 Non linear Dyn. 73377
[2] Zhang Z T, Zhang J S 2010 Chin. Phys. B 19 104601
[3] Arasaratnam I, Haykin S.Cubature Kalman filters[J]. IEEE Transactions on Automatic Control, 2009, 54(6), 1254-1269.
[4] Arasaratnam I, Haykin S, Hurd T R.Cubature Kalman filtering for continuous-discrete systems: theory and simulations[J].IEEE Transactions on Signal Processing,2010,58(10):4977-4993.
[5] Leong P H, Arulampalam S, Lamahewa TA, et al.A Gaussian-sum based cubature Kalman filter for bearingsonly tracking[J].IEEE Transactions on Aerospace & Electronic Systems, 2013,49(2):1161-1176.
[6] Li Q, Sun F.Strong tracking cubature Kalman filter algorithm for GPS/INS integrated navigation system[C] // IEEE International Conference on Mechatronics and Automation, 2013:1113-1117
[7] Bhaumik S.Cubature quadrature Kalman filter[J].International Journal of Control Automation & Systems, 2013, 7(7):533-541.
[8] Hu G G, Gao S S, Zhong Y M, Gao B B 2015 Chin. Phys. B 24 070202
[9] Wang S Y, Feng J C, Tse C K 2014 IEEE Signal Process. Lett. 21 43

Design of the LADRC control system for gravimetric stabilisation platforms

Enfan Lin[1], Jiangning Xu[1], Miao WU[1], Hongyang He[1*]

[1] School of Electrical Engineering, Naval University of Engineering, Wuhan, Hubei, 430033, China

[*]Corresponding author's e-mail: xgdhehongyang@163.com

Abstract. Aiming at the problems of strong non-linearity of gravimeter stabilisation platform system, poor robustness of linear PID control algorithm and non-adaptive control system. This paper designs a LADRC-based gravimetric stabilisation platform control system design and method based on the research of PID controller and ADRC control method, and gives the anti-saturation and anti-noise design applicable to it, and the simulation experiment shows that the method is feasible.

1. Introduction

A high precision marine aviation gravimeter stabilisation platform can provide a high precision horizontal attitude for the gravimeter, ensuring that the gravimeter maintains a high precision vertical pointing under external interference, and is one of the most important instruments for reducing gravity measurement errors. The main factor that really affects the attitude accuracy of the gravimeter stabilisation platform is still the control accuracy of the control system. The appropriate control strategy and reasonable control parameters have a direct impact on the accuracy of the platform.

The gravimeter stabiliser is a very non-linear system due to the uncertainties in the mechanical resonance, electrical parameter fluctuations and torque coupled load variations. How to control the platform to the required accuracy with external disturbances and internal parameter uncertainties is one of the difficulties in the design of a stable platform control system.

Currently, the most common control strategy used by controllers within industrial control is still PID control. However, the PID control algorithm is essentially a linear control method with poor robustness. After a long period of operation of the gravimeter stabilisation platform, the system has time-varying uncertainty due to changes in the working environment and shifts in electrical parameters. When the platform parameters change, the PID requires the need to recalculate the calibrator control parameters and is not adaptive. Therefore, it is essential to find a control algorithm that can effectively suppress the controller's own internal disturbances while maintaining the accuracy of the gravimeter's level.

ADRC is a control method proposed by researcher Han Jingqing based on digital control technology[1], basically absorbing the advantages of PID controller, which has a wide application prospect in the industrial field. In recent years, with the continuous improvement of the theory of ADRC and the development of engineering applications, its influence has become more and more significant.

Despite the good control effect of the ADRC, there are too many adjustable parameters for the ADRC. In the case of a second-order controlled object, there are 11 variable parameters. Although they can be adjusted by means of intelligent optimisation algorithms, in practice the parameters still need to be adjusted according to the actual situation, which is very unfavourable for engineering applications and is one of the reasons why ADRC has not been adopted in large numbers by industry. This is similar to the development of PID controllers, which were likewise not widely promoted until the ZN critical proportionality method was proposed. Professor Gao first proposed the concept of LADRC in 2006, converting ADRC [2-4] from its original non-linear form to a linear form, with a large reduction in controller parameters and only two variable parameters remaining, driving the development of ADRC technology. Currently, most of the engineering applications of ADRC [5-17] are based on the LADRC technique. The biggest difference between LADRC and ADRC is that all the non-linear parts of ADRC are simplified to a linear form. This approach greatly simplifies the structure of the ADRC, and due to the existence of ESO, the estimation of the non-linear part of the system can also be achieved, and in theory, the LADRC can achieve the same control effect of the ADRC. Therefore, this paper attempts to improve the gravimeter stabilised platform control system using the LADRC technique.

2. Design of Cascade ADRC Controller

A gravimeter stabilised platform is, in essence, a type of servo system[18], so the control of the servo system is fully applicable to the stabilised platform, the working principle of which is shown in Figure 1.

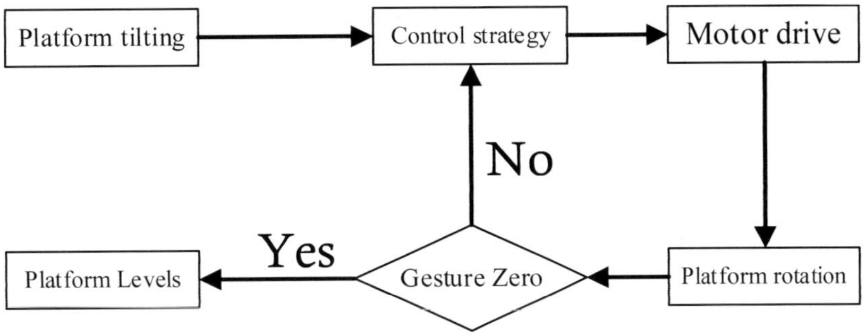

Fig. 1 Principle of control

The structure of a double closed-loop PID control system, using the inner frame as an example, is shown in Figure 2.

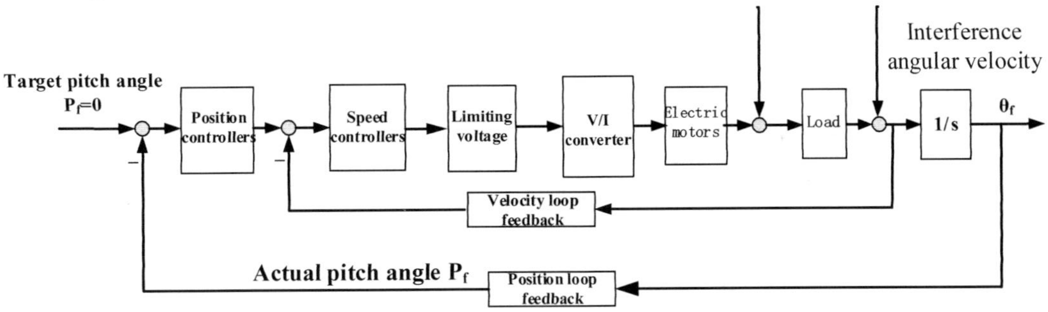

Fig. 2 Structure of control system

Generally speaking, the control of a servo system consists of a current loop, a speed loop and a position loop, respectively, from the inside out. The current loop is built into the motor and is set by the manufacturer without the user's authority to modify it; the speed loop is used to quickly correct the influence of external disturbances on the attitude, speed up the control and reduce overshoot; the

 2113 (2021) 012018 doi:10.1088/1742-6596/2113/1/012018

position loop is used to improve the accuracy of the control, reduce the steady-state error and achieve a high precision control effect.

Gravimeter stabilisation platform model The ideal model of the controlled object is a servo system of second order. Similar to the PID control system, the traditional PID control method is improved using the LADRC technique and the improved model is shown in Figure 3.

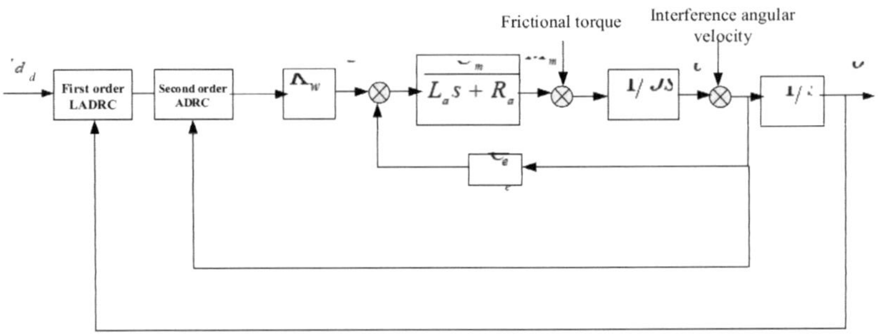

Fig. 3 Stabilized platform LADRC structure diagram

Because the ideal model in the velocity loop is a second-order system, the velocity loop is controlled by second-order LADRC; the velocity loop as a whole is treated as a "black box", which can be approximated as a tracking system where the input equals the output, so the position loop is controlled by first-order LADRC

The speed loop LESO and control rates are as follows:

$$\begin{cases} \dot{z}_1 = z_2 + \beta_1 (y - z_1) \\ \dot{z}_2 = z_3 + \beta_2 (y - z_1) + b_0 u \\ \dot{z}_3 = \beta_3 (y - z_1) \end{cases} \tag{1}$$

$$u_0 = K_p (r - z_1) - K_d z_2 \tag{2}$$

The position ring LESO and control rates are as follows:

$$\begin{cases} \dot{z}_1 = z_2 + \beta_1 (y - z_1) + b_0 u \\ \dot{z}_2 = \beta_2 (y - z_1) \end{cases} \tag{3}$$

$$u_0 = K_p (r - z_1) \tag{4}$$

This results in a first and second order LADRC tandem controller for the gravimeter stabilisation platform.

3. LADRC anti-saturation design

Input saturation must be taken into account in the design of the gravimetric stabiliser control system. As the effective operating voltage of the motor and power amplifier has a certain range, when the gravimeter stabilisation platform is faced with a large shock, the input control voltage may exceed the system's limit, which leads to a longer response time and affects the accuracy of the control system, defining this instability of the control system due to the physical constraints of the controlled system as the saturation of the controller. The most important cause of the saturation phenomenon is the actuator saturation limit, so the so-called anti-saturation design can not fundamentally solve the saturation problem of the control system, but only to alleviate it.

The saturation phenomenon can be seen as a non-linear problem of the system, and the LADRC control technique, despite being composed entirely of linear systems, is able to compensate for this non-linearity by observing it due to the presence of LESO. However, in the original design of the

LADRC control system, the response was extremely poor for the step response of the system, as shown in Figure 4.

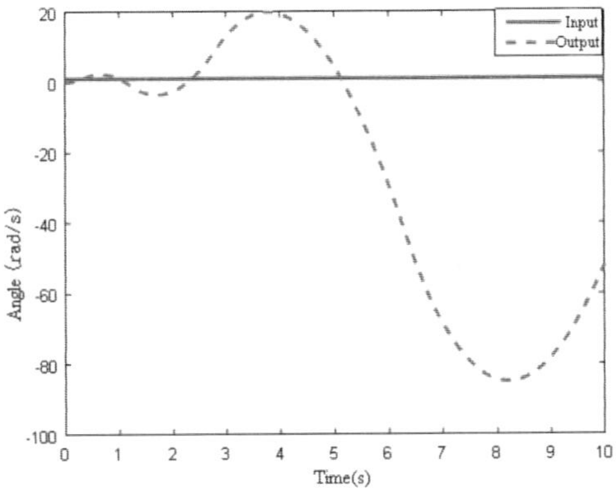

Fig. 4 Principle of control

It can be seen that the control system does not form an effective control before the anti-saturation treatment is applied to the control system, and the control error dissipates more and more as time increases. As shown in Figure 5, analysis of the control system structure shows that it is precisely because the inputs to the LESO are not the real quantities entering the actuator that the control performance deteriorates.

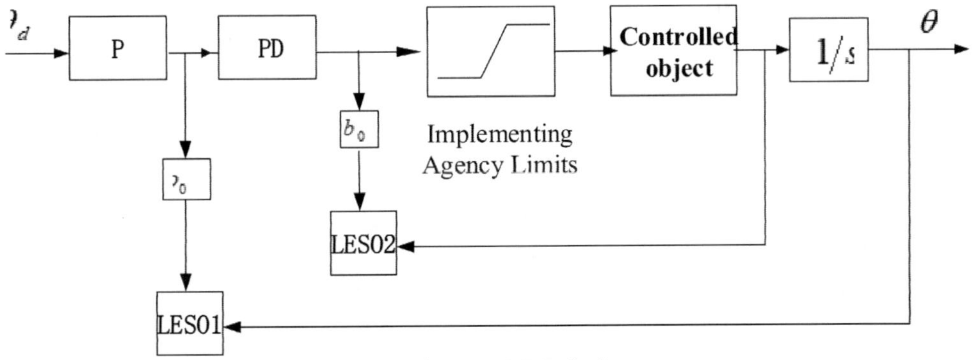

Fig.5 First LADRC design

Therefore, the anti-saturation design of the LADRC is completed by changing the input of the LESO to a real input, and the design results are shown in Figure 6.

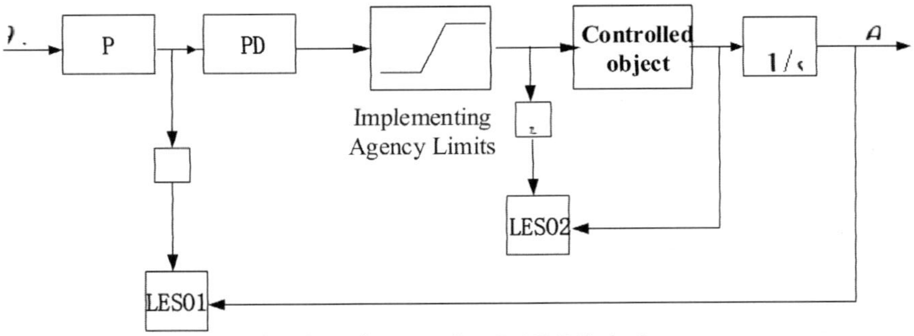

Fig. 6 Anti-saturation LADRC design

The output after the anti-saturation design is shown in Figure 7. As can be seen, with the anti-saturation design, the system no longer oscillates and is able to follow the input signal better.

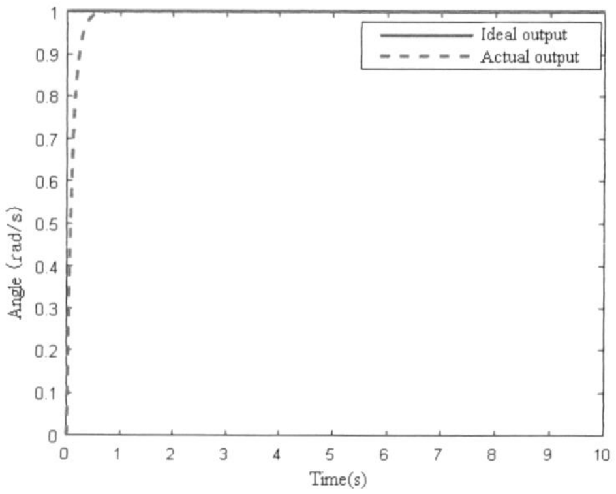

Fig. 7 System step response

4. Noise-resistant design of LADRC based on improved TD

In practical control systems, due to various factors such as mechanical structures and electrical connections, random noise will inevitably exist during the propagation of the signal. Some studies have shown that the presence of system noise will have an impact on LESO and thus reduce the accuracy of LESO. Therefore, if the signal input to LESO can be de-noised, it can improve the control performance of LADRC to a great extent.

Firstly, one of the advantages of the cascaded LADRC design is that it can suppress noise very well, an effect achieved by using the LADRC's own characteristics. Assuming a random noise of 1 rad/s is added to the angular velocity, which, of course, would not actually exist, the step response of the system is shown in Figure 8. As can be seen, despite the addition of such a large amount of random noise to the angular velocity, the final step response of the system is still relatively good and the noise is effectively suppressed. However, the presence of noise reduces the accuracy of the LESO, as shown in Figure 9, where the control accuracy is significantly reduced by adding an external disturbance of 25°/s with a period of 10s to the system. Therefore, noise suppression is necessary.

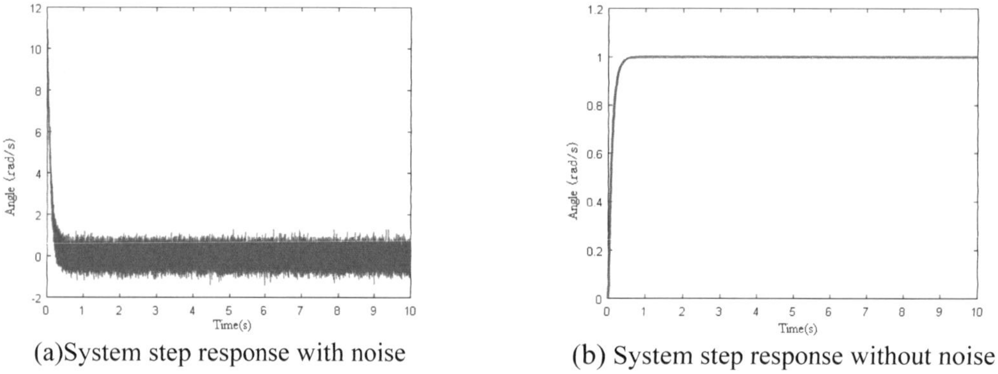

(a)System step response with noise (b) System step response without noise

Fig. 8 System step response

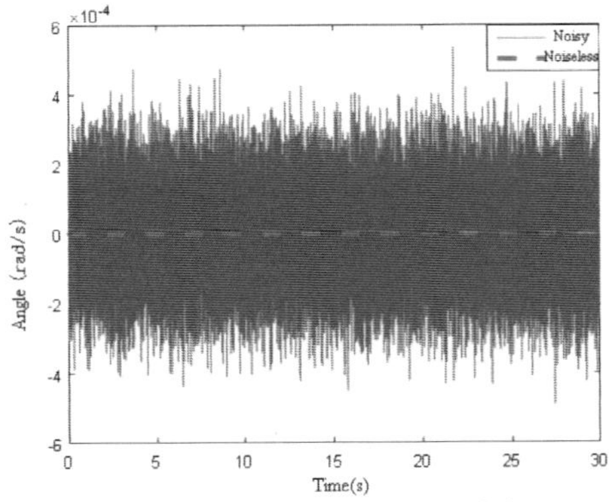

Fig. 9 Control system precision

TD is a good filtering tool that can effectively remove noise from the signal, and the filtering effect and tracking speed can be controlled by adjustable factors. This simplifies the ADRC design while taking advantage of the ADRC to optimise the control performance and accuracy of the LADRC. By adding a TD to the output of the angular velocity of the cascade LADRC, the improved control system structure is shown in Figure 10.

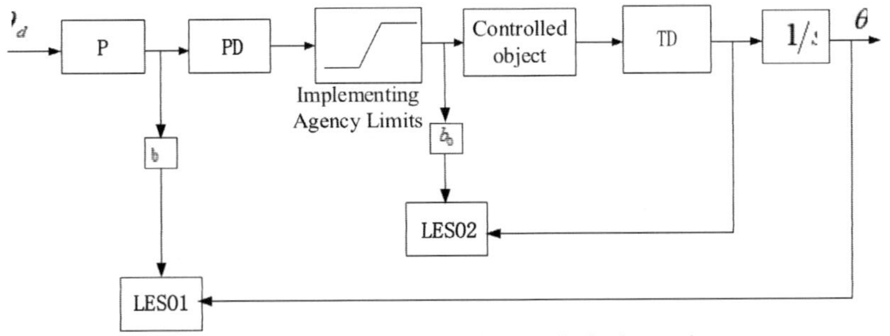

Fig. 10 De-noising LADRC design

While TD achieves the function of filtering, it brings some phase loss, and the better the filtering effect, the greater the phase loss. This phase loss is very detrimental to the control system and leads to a reduction in control accuracy, as shown in Figure 11, which compares the control accuracy with and without TD. As can be seen, the control system with TD not only fails to improve the control accuracy, but even decreases it, the reason for this is the phase loss. If this phase loss can be solved or mitigated on the basis of noise reduction, the control accuracy of the LADRC can be improved by TD.

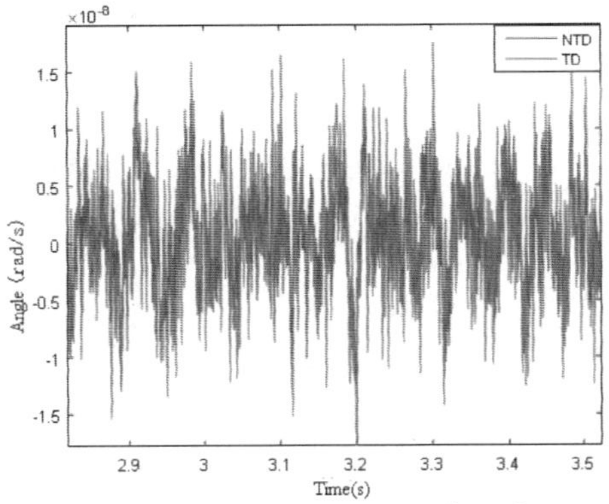

Fig. 11 Comparison of control precision in the disturbance of sine and impulse

Noting that the most primitive purpose of TD is to extract the original and differential signals from the noise-containing signal, and that the differential signal has a certain predictive role, an appropriate predictive correction can be considered using the differential signal extracted by TD to obtain an improved algorithm for the tracking differentiator as follows.

$$\begin{cases} x(k) = x_0(k) + h_1 x_2(k+1) \\ fst = fst(x_1(k) - x(k), x_2(k), r, h_0) \\ x_1(k+1) = x_1(k) + h x_2(k) \\ x_2(k+1) = x_2(k) + h \bullet fst \end{cases} \tag{5}$$

where, h_1 is the prediction time, and extensive simulations have shown that h_1 a filter factor h_0 of 1 to 1.5 times is appropriate, with the maximum not exceeding 2 times. As shown in Fig. 12, the original signal in a sinusoidal signal containing noise is extracted with TD and modified TD, respectively. As can be seen from the partial enlargement on the right, the phase loss is significantly reduced with the improved tracking differentiator, verifying the effectiveness of the algorithm.

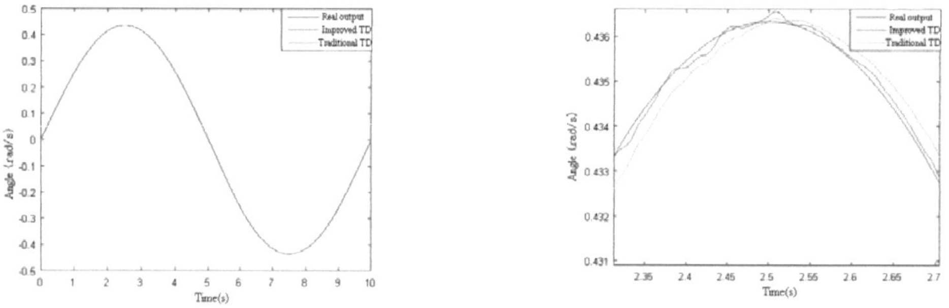

(a) Overall view (b) Partial zoom

Fig. 12 Comparison of control precision in the disturbance of sine and impulse

5. Conclusion

In this paper, the LADRC technique is applied to the control system of gravimeter stabilization platform for the problem of strong non-linearity, poor robustness of linear PID control algorithm, non-adaptive control system and too many ADRC rectification parameters, which is not conducive to engineering applications. The LADRC technology is used to apply to the control system of gravimeter stabilization platform, using LADRC technology to design a series-level controller for gravimeter stabilization platform; based on the actuator (motor and amplifier board), the LADRC anti-saturation function is designed; for the problem of noise in the system, TD is used for filtering to increase the observation accuracy of LESO; as TD filtering will bring phase loss and affect the control accuracy, the method of using TD's differential signal to compensate for the phase is designed.

Acknowledgments

This work was supported by National Natural Science Foundation (42176195) and National Natural Science Foundation (41804076). The authors would like to thank Zhang Cheng for excellent technical support and Professor Jiangning Xu for critically reviewing the manuscript.

References

[1] Han Jingqing. Auto Disturbances Rejection Control Technique [J]. FRONTIER SCIENCE,2007(01):24-31..

[2] Xue W, Huang Y. On performance analysis of ADRC for a class of MIMO lower-triangular nonlinear uncertain systems[J]. Isa Transactions, 2014, 53 (4) : 955-962.

[3] Xue W, Huang Y. Comparison of the DOB based control, a special kind of PID control and ADRC[C]. American Control Conference. IEEE, 2011: 4373-4379.
[4] Pérez-Cano R, Vranckx J J, Lasso J M, et al. Prospective trial of adipose-derived regenerative cell (ADRC) -enriched fat grafting for partial mastectomy defects: the RESTORE-2 trial[J]. European Journal of Surgical Oncology (EJSO), 2012, 38 (5): 382-389.
[5] Li S, Li J, Mo Y. Piezoelectric multimode vibration control for stiffened plate using ADRC-based acceleration compensation[J]. IEEE Transactions on Industrial Electronics, 2014, 61 (12): 6892-6902.
[6] Xue W, Huang Y. On frequency-domain analysis of ADRC for uncertain system[C]. American Control Conference (ACC), 2013. IEEE, 2013: 6637-6642.
[7] Tan W, Zhou H, Fu C. Linear active disturbance rejection control for load frequency control of power systems[J]. Control Theory & Applications, 2013, 30 (12): 1607-1615.
[8] Song K, Xie H, Li L, et al. Disturbance observation and rejection method for gasoline HCCI combustion control[R]. SAE Technical Paper, 2013.
[9] Xie H, Song K, He Y. A hybrid disturbance rejection control solution for variable valve timing system of gasoline engines[J]. ISA Transactions, 2014, 53 (4): 889-898.
[10] Xie H, Song K, Yang S, et al. Model-based decoupling control of VGT and EGR with active disturbance rejection in diesel engines[J]. Advances in Automotive Control, IFAC Proceedings, 2013, 46 (21): 282-288.
[11] Qiu D M, Sun M W, Wang G H, et al. Practical wind disturbance rejection for large deep space observatory antenna[J]. IEEE Transactions on Control System Technology, 2014, 22 (5): 1983-1990.
[12] Zheng Q, Gao Z. On practical applications of active disturbance rejection control[C]. Proceedings the 29th Chinese Control Conference (CCC2010). Piscataway, NJ, USA: IEEE, 2010: 2503-2508.
[13] Zheng Q, Chen Z, Gao Z. A practical approach to disturbance decoupling control[J]. Control Engineering Practice, 2009, 17 (9): 1016-1025.
[14] Hu B, Ying X K, Sun Q L, et al. Linear active disturbance rejection solution for heat treatment resistance furnace[C]. Proceedings the 2014 Chinese Control Conference. Piscataway, NJ, USA: IEEE, 2014.
[15] Sun L, Li D, Hu K, et al. On Tuning and practical implementation of active disturbance rejection controller: A case study from a regenerative heater in a 1000 MW power plant[J]. Industrial & Engineering Chemistry Research, 2016.
[16] Sun B, Gao Z. A DSP-based active disturbance rejection control design for a 1-kW H-bridge DC-DC power converter[J]. IEEE Transactions on Industrial Electronics, 2005, 52 (5): 1271-1277.
[17] Wu D, Chen K, Wang X. Tracking control and active disturbance rejection with application to noncircular machining[J]. International Journal of Machine Tools and Manufacture, 2007, 47 (15): 2207-2217.
[18] LI Ang, LU Xin, LIN Enfan. Model Identification Method of Gravimeter Stable Platform Based on Optimized NARX Neural Network [J]. Ship Electronic Engineering,2021,41(08):151-154.

Journal of Physics: Conference Series **2113** (2021) 012019 doi:10.1088/1742-6596/2113/1/012019

Design of Intelligent Parking Control System

Chunhong He[1], Hongchao Li [1], Bin Ren [2*]

[1]School of Computer and Information, Dongguan City University, Dongguan China 523000

[2]School of Electronic Engineering and Intelligence, Dongguan University of Technology, Dongguan China 523000

*Corresponding author's e-mail: renbin@dgut.edu.cn

Abstract: Along with the historical process of the industrial revolution, the automobile has undoubtedly made life more convenient. But at the same time, various problems also come along: Traffic jams are becoming more and more serious, and problems such as parking are becoming urgent problems. This design uses STC89C52 microcontroller, through the infrared sensor module to detect the information of the garage, then the nRF24L01 wireless module is used to carry out the wireless transmission of master and slave information. Finally, the LCD1602 LCD screen is used to establish the human-computer interaction page, and an intelligent parking space management system is designed. The owners can easily check the parking lot to understand the flow of the parking lot through the system, and judge whether to enter the parking lot, saving time. At the same time, it is easier to integrate with other intelligent systems, which provides a foundation for more advanced management and is the premise for making the parking lot a highly intelligent product in the future.

1. Introduction

Time into the 21st century, as durable cars, undoubtedly greatly improve the quality of life of the people. Especially at present, the automobile industry around the world is in a period of rapid outbreak and promotion, China ' s car ownership is growing rapidly. Nowadays, people ' s habit of using new energy vehicles is becoming more and more common, and the automobile industry ushers in the second spring. At the same time, demand for parking spaces is also rising. Many parking lots still use traditional manual management, which wastes a lot of public and human resources[1-2]. Urban parking difficulties, a difficult, forced parking management system and intelligent technology iterative update. In the future, the user ' s experience should be the top priority of parking management, and intelligent guided parking is one of the main parts.

In this paper, the single chip microcomputer is used to design the intelligent parking control system, which can integrate and configure the resources of the parking system, and effectively improve the parking space problem of the traditional parking lot. Car ownership is growing, and the yard reform is on the string. I am convinced that the intelligent parking lot in each city after the realization of parking problems, to be solved.

2. System principle design

2.1. Functional Structure Design of Intelligent Parking Control System
Nowadays, when the owner enters the car park, due to the lack of effective guidance, parking spaces cannot be easily found in large-scale parking lots[3]. The purpose of this design is to complete the intelligent management of the parking lot, so that users can easily observe the situation in the parking lot. Through the cognition of traffic flow or parking space, users can intuitively understand the actual situation in the parking lot, so as to realize efficient parking and solve the parking dilemma in the city. In this way, the parking resources can be fully utilized, and the social operation is more efficient, so as to improve people ' s happiness. Considering the feasibility, stability, ductility, economy and benefit of the design, Fig. 1 is the main function of the intelligent parking lot control system.

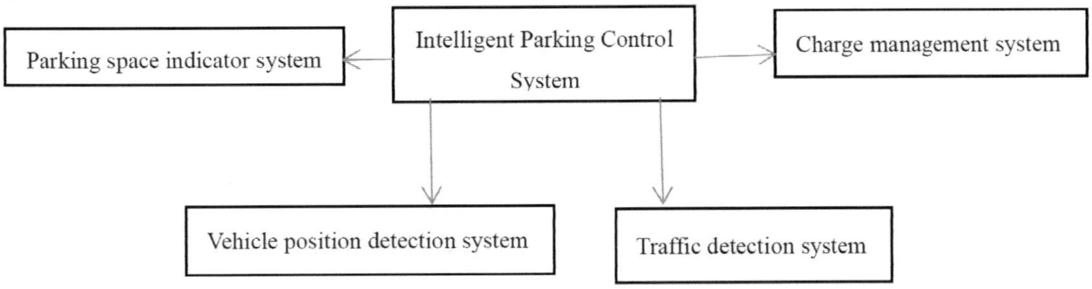

Fig. 1 Function Diagram of Intelligent Parking Control System

2.2. Design and Planning of Intelligent Parking Control System
This design uses STC89C52RC chip, set up a host and two slaves, respectively, to detect the parking space and traffic flow. The sub-modules include multiple system modules such as wireless, sensor, and display modules, which are connected and used to realize the designed functions. At the same time, in order to let users more intuitively feel the specific location of the parking space, in addition to the occupied parking space displayed on the host screen, LED lights are also added to represent the relative position of the parking space. In this design, three LED lights are used to represent three parking spaces for users to view.

When the car enters or leaves the parking lot, the sensor at the door of the parking lot records it, and the data is transmitted to the host computer by wireless, and added to the traffic. When the car enters the parking space in the garage, the corresponding parking space records have been occupied and transmitted to the host computer. At the same time, the LED lights corresponding to the parking space in the host are lit at the same time. Users can watch the specific parking space on the display screen to select whether to enter the parking lot. Fig. 2 is the design of the parking lot.

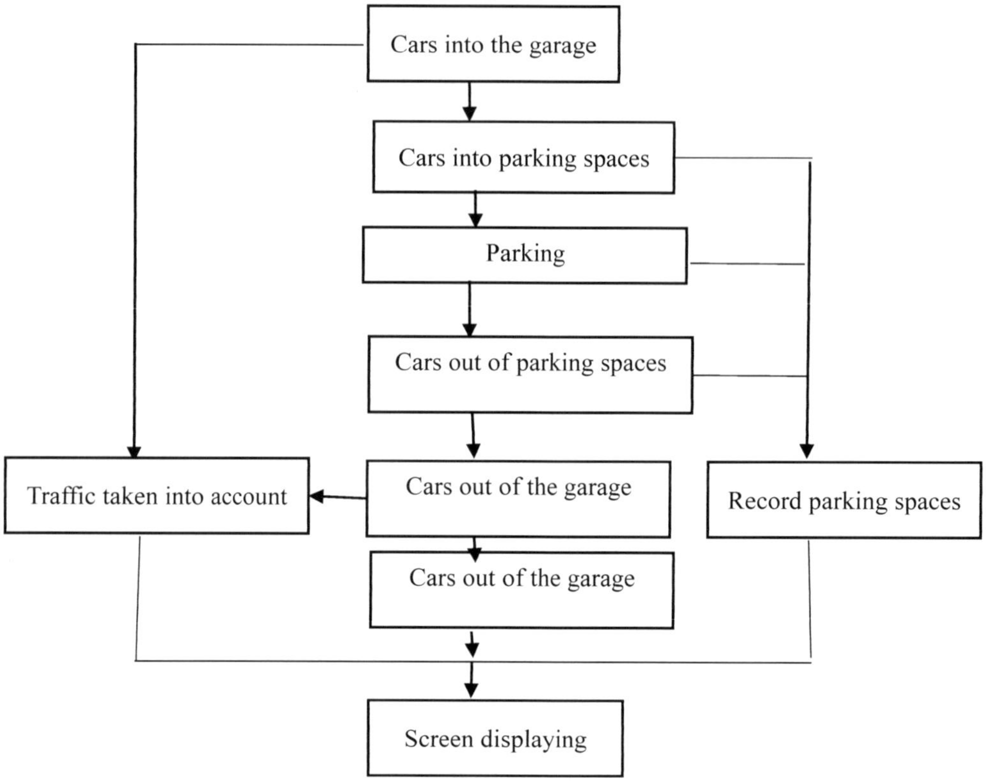

Fig. 2 Design of Intelligent Parking Control System

3. Hardware circuit design

3.1. Hardware Design of Intelligent Parking Control System

The software used in this design schematic is Altium Designer. The software has SOPC development, FPGA development and other functions, which is a development system for electronic product design[4] . In today ' s society, has been widely used in integrated circuit development, chip development and other high-end sophisticated areas. As a hardware designer, mastering Altium Designer software is the basic requirement.

The hardware consists of the following components: STC89C52 microcontroller, LCD1602 LCD screen, NRF24L01 wireless module, TCRT5000 sensor, infrared sensor. Two 89C52 slaves represent parking space detection and traffic flow detection respectively. When the measured signal is effectively received by the sensor, the information from the sensor is received from the microcontroller. After the information is processed, the information is sent to the main controller through the wireless module, and then processed by the main controller to realize the corresponding function. The hardware design is shown in Fig. 3.

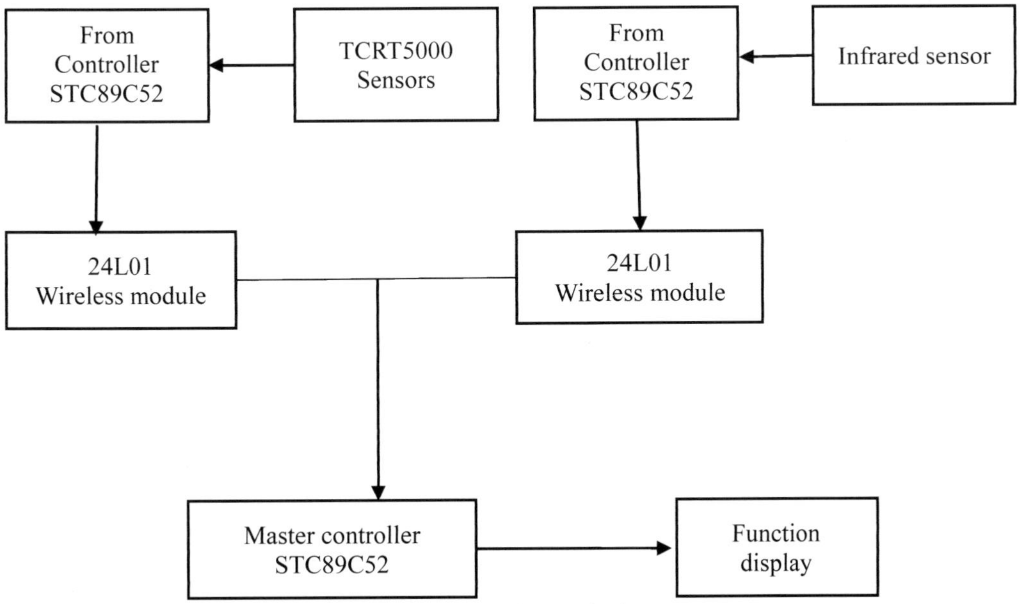

Fig. 3 Hardware design

3.2. Single-chip microcomputer minimum system

The chip processor used in this design is STC89C52, which is widely used in the electronic field. It has 8 - bit central processor (CPU) and 128Byte internal RAM[5]. The power consumption of 52 single chip microcomputer is very low, the price is close to the people, plus high performance processor, for many electronic fields such as embedded design provides a more flexible and economic solution[6].

3.3. Design of Display Module

The LCD1602 LCD screen is used to design the display part, which improves the clarity of the display font. In daily use, most of them are used to display various words and numbers. The screen consists of 16 * 2 display intervals. Users can modify each interval to their own text by programming, which plays a very good role in project research. The working voltage of the chip is 4.5 V ~ 5.5 V, and the working current is 2.0 mA (5.0 V) [7]. There are 16 pins on the screen. Users can transfer data by operating data after enabling the screen. The main data port is D0-D7.

3.4. Wireless module design

Fig. 4 is the schematic diagram of NRF24L01 in this design, the wireless module is 2.4G ISM band. Based on the use of SPI interface, its stability is good, and the price is friendly to the people. It can be used in various scenarios. In this design, 24L01 is driven by a single chip microcomputer to carry out wireless communication between the master and slave computers.

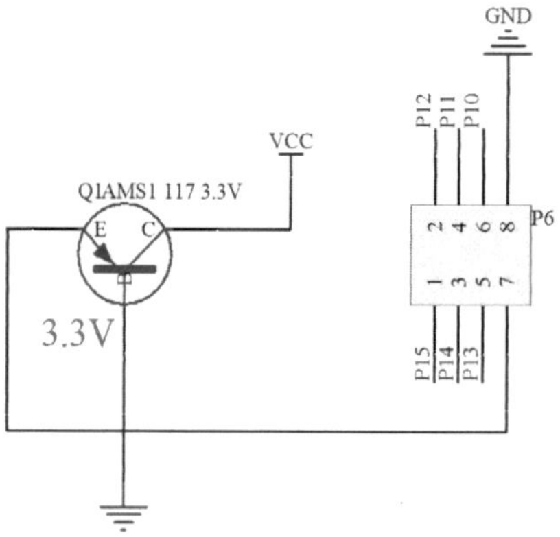

Fig. 4 NRF24L01wiring diagram

3.5. Sensor module design

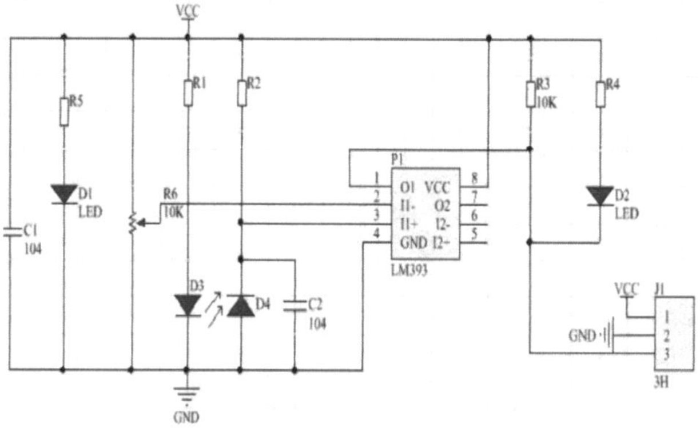

Fig. 5 Internal Diagram of Infrared Sensor

The infrared obstacle avoidance sensor with clean signal, good transmission and strong driving ability is used to detect the parking status and traffic flow, and collect the receiving and transmitting. In today 's development and design, it is commonly used as robot obstacle avoidance, intelligent car and other experimental projects.

4. Software design

The language used in this design is C language, and the software is Keil u Vision5. The software provides configuration optimization of many types and levels[8]. The program design flow chart is shown in Fig. 6.

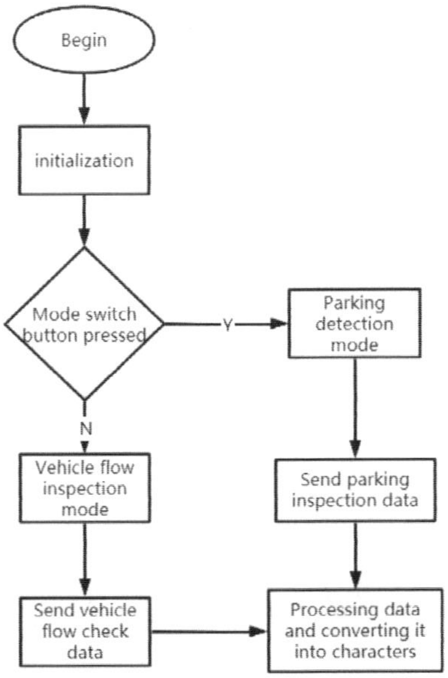

Fig. 6 Programming flowchart

4.1. Design of Sending Mode

When the parking space detection and traffic flow sensor receive the required detection signal, the low level is input to the slave microcontroller. After receiving the low level signal from the single chip computer, the wireless module is controlled to send data to the host. When you need to send data, the data in the array can be sent. In order to realize the wireless communication between master and slave computers, the process is actually controlled by manipulating the register address in NRF24LO1[9]. :

It should be noted that when operating the NRF24L01 register, it needs to be executed in standby or disconnection. Therefore, the CE end should be pulled down at the beginning of the sending function. After the program is completed, the CE end is enabled. Writing the sending mode function, mainly has the following points :

1) Lower CE level. Write receive address enable instruction and write register instruction, and set the address width and receive address. Note whether the sending address and the receiving channel address are the same.

2) Write the data and then enable the receiving channel, where the receiving channel 0 is enabled. In order to ensure that the host receives the sent data completely, the design of automatic retransmission times is ten.

3) Finally, the RF channel, transmitting power, data transmission rate and low noise amplification gain are selected to make the CRC end enable and the CE level can be raised.

4.2. Design of receiving mode

The master-slave communication in this design is realized by operating the register address of the NRF24L01 module, and the wireless module setting of the receiving board is also different. To receive the signal from the host, you need to operate the wireless module register of the receiving board, set to receive mode. It should be noted that the standby mode or disconnection state is also needed in the operation, and the address configuration of the receiving channel should correspond to the sending address, so as to obtain the required data completely.

Similar to the sending mode, the setting steps of acceptance mode are as follows :

1) Draw down the CE level, set the address width and receive address, write the receive address

enable instruction and write the register instruction, so that the address of the receiving channel is consistent with the sending address.

2) Enables the receiving channel, where the receiving channel 0 is enabled.

3)Finally, the RF channel, data transmission rate and low noise amplification gain are selected, and the effective data width of the receiving channel 0 is the same as that of the transmitting channel. Enable the CRC end, and then the CE level is elevated.

5. Analysis of effect

When the design of hardware and software is completed, it is first necessary to ensure that the connection of the hardware part is correct, and whether there is a mistake in the preparation of the software program, and then the program is burned into the microcontroller. Due to there are many master-slave machines, the battery is used to supply power to the master-slave machine. Press the key, master-slave system started, into the initial state. At this point, the traffic flow is 0, three parking spaces are empty, displayed as ' N ', indicating that no car occupies the parking space.

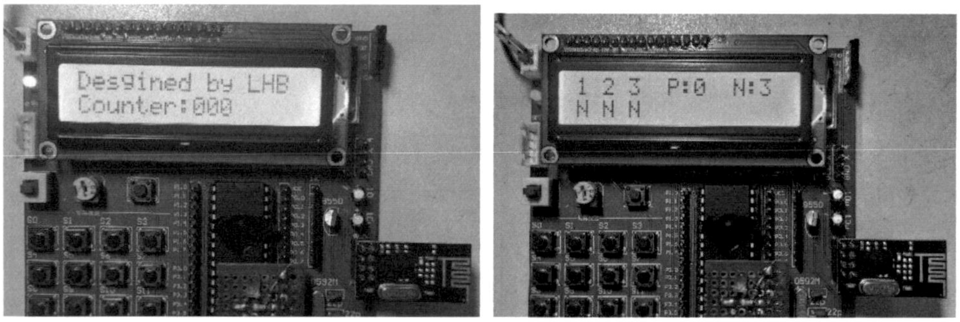

Fig.7 Initial State Diagram

At this time, if a vehicle passes through the gate of the parking lot, the slave machine that detects and measures the traffic flow will sense the vehicle, and the data will be transmitted to the host machine. The user can watch the traffic flow through the host machine to know the specific number of vehicles in the parking lot and determine whether to enter the parking lot. When a vehicle enters or leaves the garage, it is included in the traffic flow, and the traffic flow increases. At the same time, the staff can also reset the button to zero the traffic flow.

Fig.8 Traffic flow detection diagram

When the vehicle enters the parking space in the parking lot, the user wants to see the specific situation of the parking space, you can switch to the parking space detection mode by the key. When the parking space is occupied, if the detection distance is effective, the parking sensor indicates the light, indicating that the vehicle enters the effective detection range, the sensor works effectively, starts working, and transmits the information to the host. After the interactive host receives the information, the specific information of the occupied parking space is displayed on the screen. The parking space information is changed from ' N ' to ' P ', which means that the parking space has been occupied, and the corresponding LED indicator lights are on. When there are multiple occupations, the principle is the same as shown in Fig. 10, screen information and the corresponding indicator lights will also change.

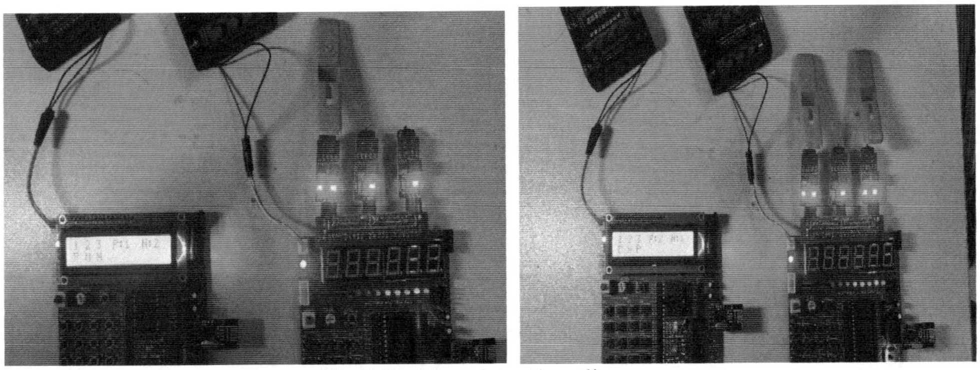

Fig.9 Parking detection diagram

When the car leaves the parking space, the signal detected by the sensor disappears, and the information returned to the host disappears. The parking space in the host shows a recovery from the initial state, from ' P ' to ' N ', and the LED lights corresponding to the parking space are extinguished. Displays the same as the initial state.

Fig. 10 Parking control diagram

6. Conclusion

The concept of this design is born from the dilemma of urban parking difficulties in today ' s society. In order to solve the problem that the majority of owners are hard to find, the backward management system in the parking lot is designed and improved. An intelligent parking control system is designed by combining single chip microcomputer technology, electronic technology, wireless communication technology and LCD display technology.

In order to make the design closer to reality, the master-slave computer abandons the past wired communication mode, the use of wireless module makes the master-slave computer wireless communication, availability has been strengthened. At the same time, the LCD screen is used for human-computer interaction, which greatly improves the quality of display. Users can intuitively understand the required information in the garage on the LCD screen, and the user experience is improved. In order to make the design more humanized and the effect more specific, LED lights are added to assist in the display of specific information. Parking information is no longer a simple character representation, but also becomes the actual specific orientation. Through the design of hardware and software, the detection of traffic flow and parking space is completed, and the use efficiency of the parking lot is improved.

Acknowledgments
The research is supported by the Key Project of Dongguan Science and Technology of Social Development Program under Grant No.20211800901261 and No.20211800902431， Guangdong Provincial Basic and Applied Basic Research Fund Regional Joint Fundt under Grant No. 2020B1515120095, Guangdong Provincial Enterprise Key Laboratory Project under Grant No. 2020B121202001, Guangdong Province Climbing plan project under Grant No.pdjh2020a0786.

References

[1] Yu Ping. The solution of the intelligent parking lot management system in the community [J]. Security Technology, 2003(02): 48-49.
[2] Chen Changzheng, Dou Baoru, Shi Liang. Design and implementation of intelligent parking lot management system based on STM32 single-chip microcomputer[J]. Journal of Xi'an Aeronautical University, 2020, 38(05): 60-64.
[3] Zhao Sheng. Research on the smart parking lot scheme of the super-large comprehensive transportation hub[J]. Railway Communication and Signal Engineering Technology, 2021, 18(04): 47-51.
[4] Wang Qiang. The application of Altium Designer in circuit design [J]. Information Recording Materials, 2020, 21(02): 118-120.
[5] Zhang Xuqiang, Han Jianjie, Sharp Kate Maimat, Maihe Muti Yiming. Design of Intelligent Watering System Based on AT89C51 Single-chip Microcomputer [J]. Industry and Technology Forum, 2019, 18(13): 44-45.
[6] Li Aobo. Design and innovative application based on minimizing single-chip microcomputer system[J]. Science and Technology Innovation, 2019(07): 26-27.
[7] Zhao Huafeng. Research on Chinese Character Display of LCD1602 Module[J]. Modern Information Technology, 2020, 4(17): 35-37.
[8] Yu Qiaoshu. Keil C51 universal accurate time-delay programming [J]. Science and Technology Wind, 2020(10): 24-25.
[9] Chen Cong. Design of embedded wireless download system based on nRF24L01[J]. Mechanical Engineering and Automation, 2021(01): 60-62.

Magnetic Target Position Estimation Method Based on Prior Information

Guangfa Sun*

CSSC Ocean Exploration Technology Institute Co., Ltd., Wuxi, Jiangsu, 214135, China

*Corresponding author's e-mail: csscoeti@csscoeti.com

Abstract. Aiming at the problem of detection and location of magnetic targets in water beach, the acoustic magnetic composite detection method is studied. After the sonar obtains the image of the suspicious object in the target area, the magnetic target recognition and location are realized by using the abnormal magnetic field distribution data near the target area measured by the shipborne magnetic sensor and the multi-sensor information fusion method. A target recognition and location method based on a priori information is proposed to solve the problem that the measurement results of magnetic sensor can not fully reflect the influence of ferromagnetic target on the surrounding magnetic field due to terrain constraints. In order to make up for this lack of information, taking the sonar measurement results as a priori information, the hypothesis test method is adopted to make full use of all the measurement results of different types of sensors to realize the recognition and positioning of magnetic targets.

1. Introduction

The ferromagnetic target in the geomagnetic field will affect the magnetic field distribution in its vicinity. Theoretically, the position of ferromagnetic target can be estimated by measuring the magnetic field distribution. In engineering practice, for the magnetic field measurement of underwater ferromagnetic target, it is necessary to carry magnetic sensor with the help of moving platform, and the movement of the platform will affect the measurement of the sensor; On the other hand, for the target close to the beach, the underwater platform, such as various measuring ships, cannot measure vertically above the target, and can only measure part of the magnetic field generated by the magnetic target [1-3].

Figure 1. Schematic diagram of motion path of magnetic sensor measurement carrier platform.

As shown in figure 1, the ship cannot measure along the path shown in AA' in the figure, so the "over the top" measurement data about the target cannot be obtained, It can only be measured along BB' path and its parallel path, so the measurement results must not reflect the full impact of ferromagnetic target on the surrounding magnetic field. In order to make up for this lack of information, it is necessary to take the sonar measurement results as a priori information and adopt the information fusion method in order to make full use of all the measurement results and realize the classification and positioning of targets.

2. Magnetic field generated by ferromagnetic target

In a small range, it can be considered that the geomagnetic field is an "invariant field". The existence of ferromagnetic targets will "disturb" the geomagnetic field. The magnetic induction vector generated by a single magnetic dipole in the surrounding space is [4]:

$$\vec{B} = \frac{\mu}{4\pi}\left(\frac{3(\vec{m}\cdot\vec{r})\vec{r}}{r^5} - \frac{\vec{m}}{r^3}\right), \tag{1}$$

where, $\vec{m}$ represents the magnetic moment of the magnetic dipole, $\vec{r}$ represents the vector diameter of the source point to the field point, and the direction of $\vec{r}$ points from the center of the magnetic dipole to the calculated field point.

According to the above formula, the magnetic field distribution results shown in figure 2 can be obtained by simulating four independent ferromagnetic targets. In the area of x-axis [- 20,20] m and y-axis [- 20,20] m, there are four ferromagnetic targets, whose positions are: (5,15) m, (4,13) m, (3,12) m and (2,10) m. figure 2 (a) ~ (d) respectively show the magnetic field "disturbance" generated by a single target, and figure 2 (e) shows the results of the combined magnetic field "disturbance" of these targets.

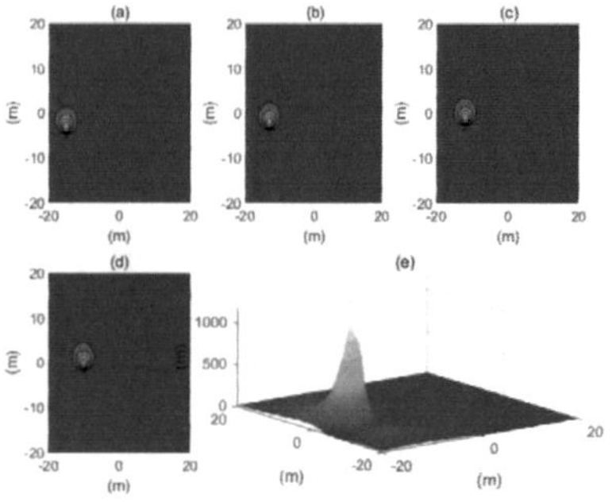

Figure 2. Schematic diagram of independent magnetic field and combined magnetic field generated by ferromagnetic target.

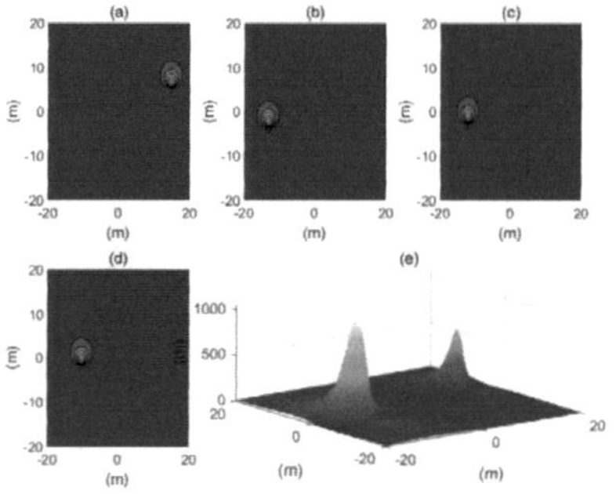

Figure 3. Schematic diagram of independent magnetic field and combined magnetic field generated by ferromagnetic target (after target position change).

If the distribution positions of ferromagnetic targets change, namely: (-5, -15) m, (4,13) m, (3,12) m, (2,10) m, the magnetic field distribution is shown in figure 3.

Comparing figure. 2 and figure. 3, it can be intuitively concluded that the distribution information of the joint magnetic field contains the information of the distribution of independent ferromagnetic targets. It is a kind of inversion problem to obtain the target distribution by measuring the joint magnetic field. Its analytical solution is affected by many factors, and it is difficult to draw a deterministic conclusion. At the same time, the actual constraints shown in figure 1 also pose a challenge to the measurement of the complete information of the joint magnetic field. On the premise that only part of the information of the joint magnetic field can be measured, how to estimate the position of the ferromagnetic target is the following problem to be discussed.

3. Influence of ship motion on measurement results of magnetic sensor

The constant magnetic field generated by the hard magnetic material of the carrier platform, the induced magnetic field generated by the soft magnetic material under the action of the geomagnetic field, and the eddy current field generated by cutting the geomagnetic field at the conductive part will interfere with the magnetic field measurement of the magnetometer, resulting in the vector superposition of the target signal magnetic field, geomagnetic field, platform related magnetic field and environmental magnetic field.

It is assumed that the ship's heading angle is within 0°~ 360°, and roll angle and trim angle have peak values of 5°, the error caused by the attitude angle error to the measured value of the magnetic sensor can be calculated, as shown in figure. 4.

4. Target classification and location fusion algorithm based on partial magnetic field measurement results and sonar sensor prior information

It can be seen from the above two sections that the magnetic sensor mounted on the moving platform is difficult to directly measure and obtain the target magnetic field distribution information, which is due to the interference of motion and limited to the special position of the target. Under such conditions, the information obtained by directly using the measurement results of the magnetic sensor is very limited. It is necessary to combine the measurement results of sonar sensor and adopt the method of signal decomposition to estimate some information related to the target from the complex magnetic field information obtained by magnetic sensor, so as to realize information fusion and target classification and positioning. The following describes the execution steps of this fusion process using simulation data.

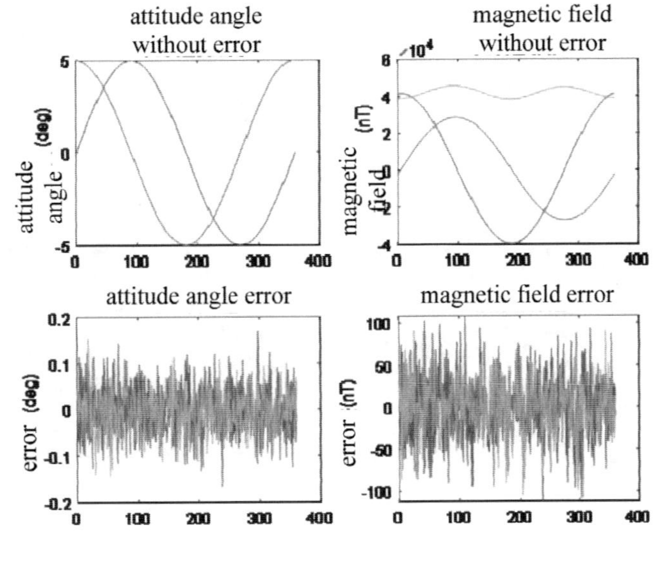

Figure 4. Influence of attitude angle error on magnetic field strength measurement error.

The simulation scene is set as follows: in the measured environment, there are four objects suspected to be targets detected by sonar sensor. According to the principle of sonar detection, the positions of these four objects can be given. On this premise, it is hoped to further determine which of the above four objects are objects with ferromagnetic properties through the measurement data of the magnetic sensor, so as to classify the sonar measurement results.

In this process, the prior information used includes the following categories: first, the sonar detection results, including the size and position information of the object; The second is the prior knowledge of target characteristics, including the approximate range of target ferromagnetic characteristic parameters; The third is the calculation assumptions of magnetic field distribution, such as geomagnetic characteristics. The implicit information is that the targets detected by sonar include two categories: targets with ferromagnetic characteristics and targets without ferromagnetic characteristics. In the context of this study, it can be understood that the targets detected by sonar may be man-made obstacles (with ferromagnetic characteristics) or natural objects such as reefs. With the support of this prior information, the hypothesis test [5] method can be used to estimate which of the above targets have ferromagnetic properties.

If the four targets are all ferromagnetic targets, after their respective positions are known (measured by sonar), their respective magnetic field distribution can be calculated, as shown in figure 5; Their joint magnetic field distribution is shown in figure 6, and the measurement results of the joint magnetic field by the magnetic sensor are also shown in figure 6. It can be seen that it is difficult to directly distinguish the actual distribution of the combined magnetic field [6] from the measurement results of the magnetic sensor.

If target 3 does not have ferromagnetic characteristics, that is, only three of the four targets measured by sonar have ferromagnetic characteristics, the joint magnetic field distribution is shown in figure 7. Now, the problem to be solved can be described as "how to judge the number and position of ferromagnetic targets after obtaining the measured values of magnetic sensors and sonar measurement results".

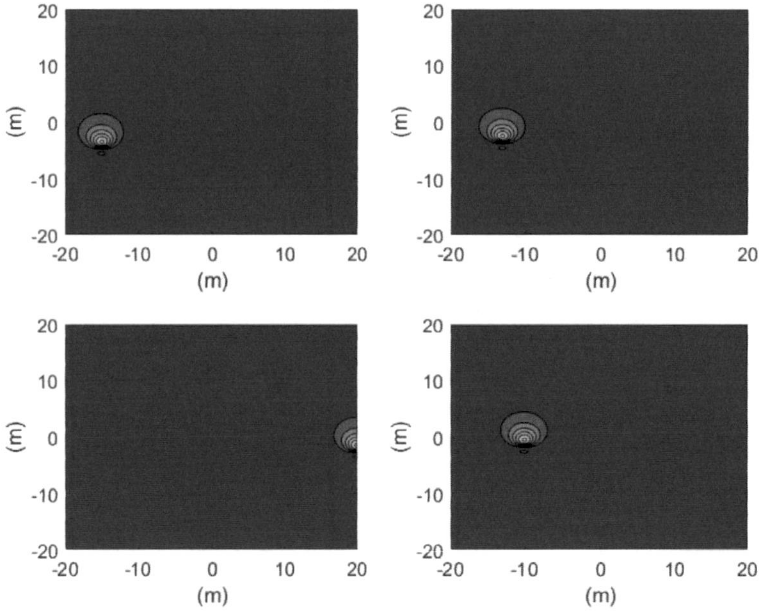

Figure 5. Schematic diagram of target magnetic field distribution.

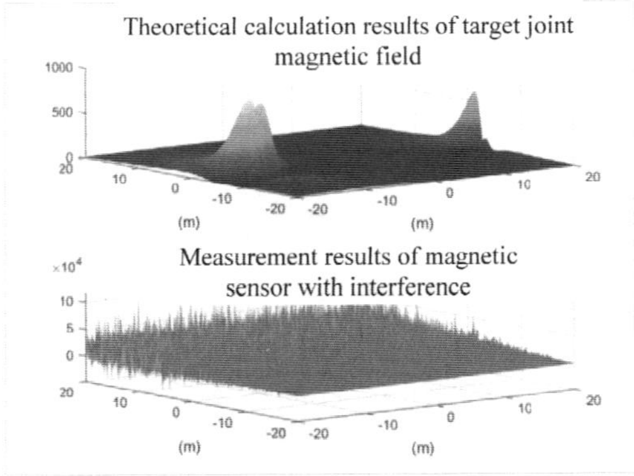

Figure 6. Schematic diagram of theoretical calculation results and measured value simulation results of combined magnetic field.

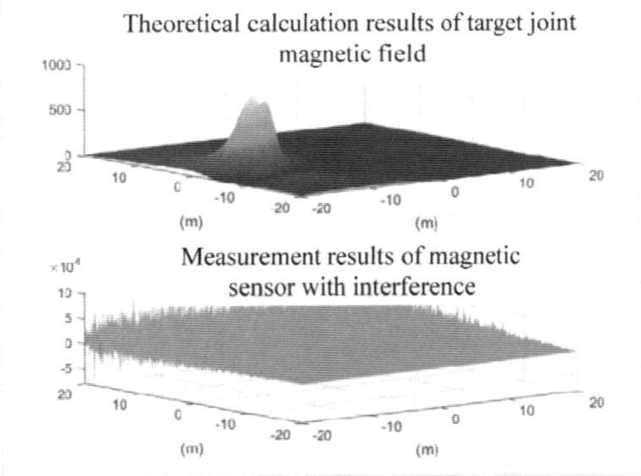

Figure 7. Schematic diagram of joint magnetic field distribution under the assumption that target 3 does not have ferromagnetic characteristics.

To solve the above problems, we first need to decompose the measured value obtained by the magnetic sensor, in order to decompose the signal with the lowest frequency from other high-frequency interference. The reason for this is that the target we want to judge is a stationary target. The magnetic field can be considered as a fixed and low-frequency signal during observation task span. Compared with the target magnetic field, the periodic variation magnetic field of the measured value caused by the platform motion is a high-frequency signal. The calculation results are shown in figure 8. (a) shows low frequency magnetic field distribution corresponding to measured values of three targets, (b) shows low frequency magnetic field distribution corresponding to three target theoretical values, (c) and (d) represents the corresponding situation of the four objectives.

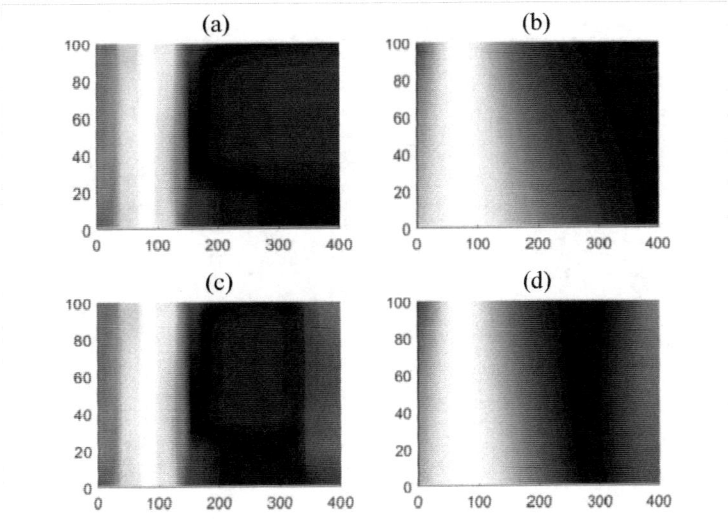

Figure 8. Schematic diagram of low frequency magnetic field distribution.

As can be seen from figure 8, the measured values and their corresponding theoretical values have strong similarity. If the measured results are as shown in the first figure in figure 5, and the two theoretical values shown in Figure 5 can be calculated through the prior information, then by comparing the first figure and the two theoretical values, we can intuitively conclude that the measured results tend to be consistent with the theoretical values in the upper right corner. In this way, we can compare the measured data with the prior information. The ferromagnetic target and its distribution are judged.

5. Conclusion

Aiming at the situation that the measuring ship cannot approach the target in the process of underwater magnetic target detection, this paper proposes target recognition and positioning method based on acoustic/magnetic fusion. The target in the area is measured by sonar sensor and the suspicious targets are marked. On this basis, some magnetic field information measured by magnetic sensor is used. The magnetic field theoretical calculation results closest to the measured values are sought by hypothesis test to find the distinguishing magnetic targets from multiple targets measured by sonar. The limitation of this method is that it is difficult to eliminate the inherent magnetic substances in the environment through the algorithm, which affects the accuracy of target recognition.

Acknowledgments

The project was supported by Stable Foundation Project of the Science and Technology on Near-Surface Detection Laboratory with funding number 6142414190101.

References

[1] F. Maussong, J. Chanussot, A. Hétet. (2003) Automated Segmentation of SAS Images using the Mean-Standard Deviation Plane for the Detection of Underwater Mines. MTS/IEEE OCEANS, 16: 2155–2160.
[2] S. Kumar. (2005) Real-Time Tracking Gradiometer for use in an Autonomous Underwater Vehicle for Buried Minehunting. MTS/IEEE OCEANS, 28: 228–235.
[3] S. Hettich, S. D. Bay. (1999) The UCI KDD Archive. http://kdd.ics.uci.edu.
[4] Masek J G , Vermote E F , Saleous N E , et al. (2006) A Landsat surface reflectance dataset for North America. IEEE Geoscience & Remote Sensing Letters, 3(1):68–72.
[5] Payne R J . (2012) Volcanic impacts on peatland microbial communities: A tephro palaeoeco logical hypothesis-test. Quaternary International, 268(1): 98–110.
[6] Budgifvars G , Bergstrom I. (1998) Combined magnetic field detector and activity detector employing a capacitive sensor, for a medical implant: US, US5709225 A.

Study on dynamic characteristics of rotor-bearing system under pitching condition

Pu Songcheng[1],He Yumin[1*],Yang Qijiang[1],Cheng Ming[1],Wang Yupeng[1],Wang Zheng[1],Lin Yaodi[1]

([1.] School of Marine Engineering, Guangzhou Maritime University, Guangzhou, Guangdong 510725, China)

Corresponding author's e-mail:504897094@qq.com

Abstract: In many cases, ships will encounter special conditions, such as bad sea conditions, and the effect of various disturbance factors, such as wind and waves, will cause low-frequency longitudinal and transverse rocking motion of ships. This would affect the rotor-bearing system of the power plant, which is directly attached to the hull of the ship, and would roll along with the hull in the same longitudinal and transverse motion. This will have a significant impact on the stability of the rotor bearing system of the ship power plant and the safety of the ship. In this paper, a rotor-bearing test rig is mounted on a multi-degree-of-freedom swing test rig simulating sea conditions. The multi-degree-of-freedom rocking test bed simulates sea conditions. The rotor-bearing test bed is equipped with an eddy current sensor and an acceleration sensor to obtain experimental data. The programmable loader is used to control the rotor speed so as to carry out the experiment process at different speeds. Finally, the dynamic characteristics of rotor-bearing system under swinging condition are obtained.

1. Introduction

When a ship sails at sea, it will always encounter some special conditions, such as bad sea conditions. Under the action of wind, wave, current and other factors, the ship will experience low-frequency longitudinal and transverse rocking motion. The foundation of the rotor-bearing system of the ship's power device is fixed on the ship's hull. The longitudinal and transverse rolling motion of the hull caused by wind waves and other factors will also produce low-frequency and large-displacement vibration, which will be transmitted to the sliding bearings through the foundation, and then to the rotor of the power device. Compared with the high-speed rotation of the rotor of the ship power plant. The vibration frequency of the longitudinal and transverse rolling motion of the ship hull is far lower than the rotating speed of the rotor of the ship power plant, but these vibrations will still affect the whole rotor-bearing system[1].

They will first pass through the foundation of the system to the sliding bearing, have an impact on the dynamic characteristics of the sliding bearing, and then through the sliding bearing oil film force transfer to the rotor, and finally have a significant impact on the vibration characteristics of the entire rotor-bearing system.

These influences have significant influence on the stability of the rotor-bearing system and the safety of the ship. Therefore, for the rotor-bearing system of the ship power plant under the action of longitudinal and transverse rolling, it is essential to consider the influence of the ship hull's own longitudinal and transverse rolling.

2. Ship longitudinal rocking dynamic model

Usually when ships sailing the sea encounter severe sea condition and swaying motion occurs, sailing ships sailing speed, change direction, swaying motion of ship hull is a complex movement in 3 d space, because at this time, not only around the axis of rotation movement, hull also coordinates with the change of position, velocity and acceleration[2]. In this paper, we mainly discuss the ship's pitching, that is, the rocking motion in which the ship rotates about x, as shown in Figure 1.

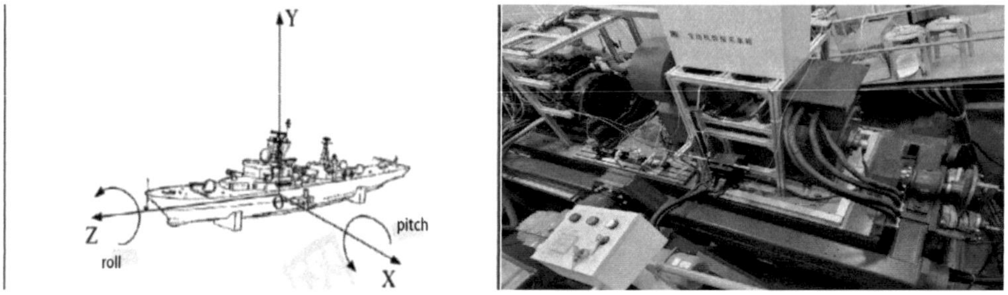

Fig 1. Ship model and sway experiment

In this paper, a series of effects of ship pitching motion on ship rotor-bearing system are analyzed. By comparing with or without pitching motion, the influence of rotor-bearing system[3] under different rotational speeds and different swing periods is studied.

In this paper, Frontback diagram and FFT diagram in analysis software of LMS modal tester are used to analyze the changes of rotor-to-bearing system with or without pitching motion under pitching motion. The influence of pitching motion on rotor motion attitude under multiple speeds is studied by using near-acceleration analysis, time-domain response and spectral response. Finally, the influence of amplitude and frequency ratio of pitching motion on the dynamic characteristics of rotor system under certain system parameters is discussed.Analysis of dynamic characteristics of ship pitching bearing - rotor system[4] .

3. Analysis of dynamic characteristics of ship pitching bearing - rotor system

3.1 Influence of different rotor speeds on rotor-bearing system under the same longitudinal yaw condition

When the ship pitches, the swing amplitude of pitching movement is 18°, the swing period is 8s, and the rotor speeds are 2000rpm, 2400rpm and 2800rpm respectively. Using the LMS modal tester[5] analysis software, the vertical and horizontal spectrum of the eddy current sensor can be displayed, as shown in Figure 2.

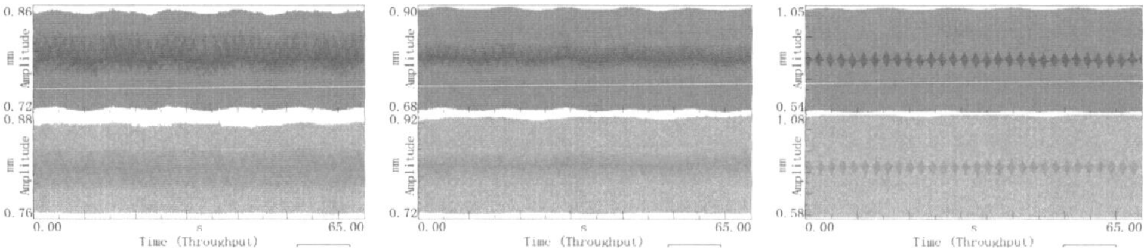

Cycle 8s, speed 2000rpm Cycle 8s, speed 2400rpm Cycle 8s, speed 2800rpm

Fig2. Spectrum of vertical and horizontal eddy currents at different rotor speeds

According to the above figure, we can see from the output analysis diagram of the eddy current sensor in the longitudinal yaw condition.

When the rotor speed is at 2000rpm, the eddy current data fluctuates significantly under the condition of 8s periodic pitching. This reflects that the distance between the sensor and the rotor changes significantly, and the rotor with a speed of 2000rpm is greatly affected by the working condition.

When the rotating speed is 2400rpm, the eddy current data fluctuates significantly. This reflects that the distance between the sensor and the rotor changes greatly, and the rotor with a speed of 2400rpm is greatly affected by the working condition.

When the rotating speed is 2800rpm, the eddy current data will fluctuate slightly. This reflects that the distance between the sensor and the rotor changes little, and the rotor with a speed of 2800rpm is less affected by the working condition.

Therefore, the influence of changing the rotor speed on the rotor-bearing system is relatively large. It can also be seen from the figure that under the longitudinal yaw condition, the influence on the rotor in the vertical direction is more obvious than that in the horizontal direction. The smaller the rotor speed is, the more obvious the influence of the rotor system is under the longitudinal sway condition.

3.2 Influence of different pitching conditions on near acceleration of rotor-bearing system

A near acceleration sensor is arranged in the part close to the motor on the rotor-bearing test bench. Select the vertical sway Angle of 18°. First, Frontback diagrams with rotor speeds of 2000rpm, 2400rpm and 2800rpm when the pitch is 18° are analyzed. Then, Frontback diagrams with rotor speeds of 2000rpm, 2400rpm and 2800rpm are selected for periods of 8s and 12s. Using the LMS modal tester analysis software, the data analysis of the near acceleration sensor can be displayed, and the corresponding Frontback diagram generated is shown in Figure 3.

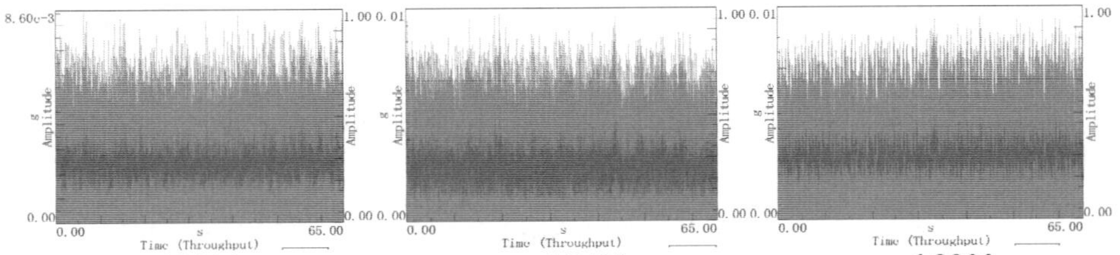

speed 2000rpm speed 2400rpm speed 2800rpm

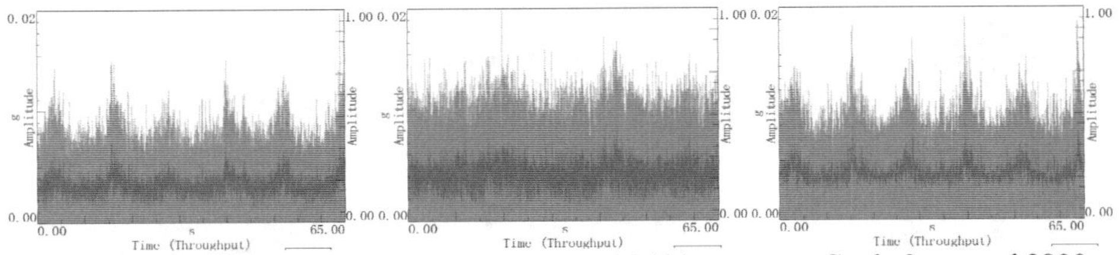

Cycle 8s, speed 2000rpm Cycle 8s, speed 2400rpm Cycle 8s, speed 2800rpm

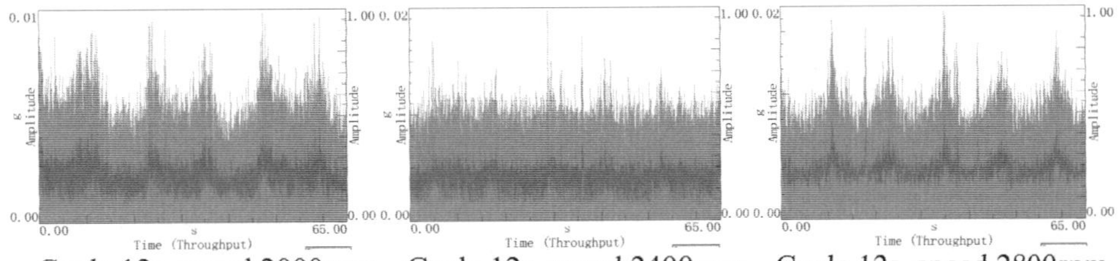

Cycle 12s, speed 2000rpm Cycle 12s, speed 2400rpm Cycle 12s, speed 2800rpm

Fig3. Frontback diagram of near acceleration under different working conditions

According to the analysis in the above figure, when the rotor-bearing system is in the condition of 18° pitch, because the near acceleration sensor is relatively close to the motor and is affected by the motor, the change of near acceleration is irregular, and the near acceleration reflects the vibration of the rotor. With the increase of rotor speed, the near acceleration also increases.

Under the condition of pitching 18°, the near acceleration will change periodically, and the period of change is about 12s. The periodic variation is obvious at the rotor speeds of 2000rpm and 2800rpm, but not obvious at the rotor speeds of 2400rpm.

4. Conclusion

The dynamic characteristics of rotor-bearing system were analyzed by simulating the real longitudinal sway condition with a swing test bed, and the following conclusions were drawn:

(1) The control swing amplitude is 18°. Under the condition that the swing period is not changed, the rotor-bearing system is less affected by the working condition as the rotor speed increases.

(2) When the control swing range is 18°, the vibration of the rotor-bearing system will form periodic changes with the longitudinal sway condition, and the period of such periodic changes is about 12s.

(3) When the control swing amplitude is 18°, the periodic variation of the vibration of the rotor-bearing system will be affected by the irregular rotor speed.

Acknowledgments

Authors wishing to acknowledge assistance or encouragement from colleagues, In addition, the author would like to thank the Guangdong University Students Climbing Program (C21070002) for supporting this paper.

References:

[1] Wei Zuolin, Chen Guanyu, Wang Lijuan.Simulation and Experimental Study on Multi-system Parallel Parallel of Swing Test Bench [J]. Jiangsu Ships,2020,37(05):30-33.

[2] Zhang Lei, Pei Shiyuan, Xu Hua.Research on Safety and Stability of Two Ship Rotor Bearing Systems under Swing Condition [J]. Science China: Technical Science,2018,48(04):369-381.

[3] Zhang Zhiming, Zhang Yanyang, Xie Youbai et al.Hydrodynamic Lubrication Theory of Plain Bearing [M]. Beijing: Higher Education Press,1986.275-305.

[4] Li Chuanjiang, Lan Ge, Wang Xiaoliang, Cui He, Zhu Chenzhe. Research on Dynamic Behavior of Rotor System Supported by Sliding Bearings [J]. China Equipment Engineering,2021(09):139-141.

[5] Zhang Gangben, Chen Gang, Peng Xuexuan, et al.Test and Analysis of Compressor Pipeline Vibration Based on LMS Test. Lab [J].Compressor Technology,2006(05);5-7.

Robust optimal control of a nonlinear impulsive time-delay system for 1,3-PD fed-batch culture

Chao Sun[1*]

[1] School of Computer Science and Technology, Shandong Technology and Business University, Yantai, Shandong, 264000, China

2019410041@sdtbu.edu.cn

Abstract. In this paper, taking the feeding process as a form of impulsive and considering the time-delay in fermentation process. A robust model with the time-delay system as the control variable and the time-delay system as the constraint is established. In order to solve this optimal control problem, we have propose an particle swarm optimization method to solve problem. Numerical results show that 1,3-PD yield at the terminal time increases compared with the experimental result.

1. Introduction

The main raw of a new polyester material is 1,3-Propanediol (1,3-PD) is the production, antifreeze and protective agent. Traditional chemical synthesis of 1,3-PD requires expensive catalysts and complex environment. In recent years, microbial fermentation technology has received attention from experts and scholars at home and abroad.

There are three types of microbial fermentation for the production of 1,3-PD. Fed-batch fermentation includes both batch and continuous fermentation modes. Compared to other fermentation methods, fed-batch fermentation has the ability to overcome the inhibition of catabolic metabolites produced during the chemical production process. Therefore, the industrial production of 1,3-PD has been widely used in the replenishment fermentation of wholesale fermentation[1]. During fed-batch fermentation materials are added to the fermenter in batches to maintain a suitable fermentation environment and improve the yield of 1,3-PD. Therefore, the optimal control of glycerol and alkali addition in fed-batch fermentation has received much attention. The literature[2] propose the optimal switching control problem for the fermentation processes of the replenishment batch is studied. However, the above optimal control studies ignore the time-delay phenomenon in the reaction process and the robustness of the intracellular material.

In this article, a impulsive time-delay differential equation with control is proposed to describe the time-delay replenishment batch, using the addition of glycerol and alkali as control vectors. In order to maximize production, we use the end-moment 1,3-PD production as a performance indicator, we promose a optimal control model with time-delay.

2. Nonlinear impulsive time-delay system

In the production of replenishment sub-wholesale fermentation, alkali and glycerol are added to the fermenter in batches to provide sufficient nutrients. Based on the fermentation process, this paper assumes that :

(H1) : The concentration of substance in the container is uniform, ignoring the inhomogeneity of spatial distribution.

(H2): In the batch flow process, only alkali and glycerol are added to the fermenter.

(H3): The concentrations of extracellular material at moment t- τ .

Under assumptions (H1)-(H3), biomass, substance and reaction in an intermittent process can expressed bythe differential equation

$$\dot{x}_1(t) = \mu(x(t))x_1(t-\tau), \tag{1}$$

$$\dot{x}_2(t) = q_2(x(t))x_1(t-\tau), \tag{2}$$

$$\dot{x}_3(t) = q_3(x(t))x_1(t-\tau), \tag{3}$$

$$\dot{x}_4(t) = q_4(x(t))x_1(t-\tau), \tag{4}$$

$$\dot{x}_5(t) = q_5(x(t))x_1(t-\tau), \tag{5}$$

$$\dot{x}_6(t) = \frac{1}{V_s}[J_{\max}\frac{x_2(t)}{x_2(t)+K_m} + \frac{1}{A_s}(x_2(t)-x_6(t)) - q_2(x(t))] - \mu(x(t))x_6(t), \tag{6}$$

$$\dot{x}_7(t) = \beta C_{protein}[K_{GDHt}U_{GDHt}(x(t))\frac{x_6(t)}{k_m^{GDHT}(1+\frac{x_6(t)}{K^{GDHt}})+x_6(t)} - k_{PDOR}U_{PDOR}(x(t))$$

$$\frac{x_7(t)}{k_m^{PDOR}+x_7(t)(1+\frac{x_7(t)}{K_{PDOR}})}] - \mu(x(t))x_7(t), \tag{7}$$

$$\dot{x}_8(t) = \beta C_{protein}k_{PDOR}U_{PDOR}(x(t))\frac{x_7(t)}{k_m^{PDOR}+(1+\frac{x_7(t)}{K^{PDOR}})x_7(t)} - K_{PD}(x_8(t)-x_3(t))$$

$$-\mu(x(t))x_8(t) \tag{8}$$

where τ is the delay arguments; $J_{\max}$ is a maximal of glycerol; K_m represents mMichaelisMenten constant of permease; k_{GDHt} and k_{PDOR} are the ration of activities of 1,3-PD oxydoreductase (PDOR) and glycerol dehydratase (GDHt) in vitro, respectively; $C_{protein}$ is the average in vitro protein concentration; β is a the enzyme for activity coefficient; K_m^{GDHt} and K_m^{PDOR} are GDHt and PDOR for respectively Michaelis-Menten constants; K^{GDHt} and K^{PDOR} represent inhibitor constants of 3-hydroxypropionaldehyde, respectively; and $U_{GDHt}(x(t))$ and $U_{PDOR}(x(t))$ are the specific activities , respectively. Based on the work[2], functions $\mu(\cdot)$, $U_{GDHt}(\cdot)$ and $U_{PDOR}(\cdot)$ are expressed as

$$\begin{cases} \mu(x(t)) = u_m\dfrac{x_2(t)}{x_2(t)+K_s}\prod_{j=2}^{5}(1-\dfrac{x_j}{x_j^*}), \\ U_{PDOR}(x(t)) = U_{PDOR}^0 - \mu(x(t))U_{PDOR}^1 - \Delta U_{PDOR}^m\dfrac{x_2(t)}{x_2(t)-K_{PDOR}^s}, \end{cases} \tag{9}$$

the above parameters values are given in the literature[3].

Now, let $f_i(x(t),x(t-\tau),u(t)), i=1,2,\ldots,8$, be the right-hand sides of delay differential equation. Moreover, define

$$f(x(t),x(t-\tau),u(t)) := (f_1(x(t),x(t-\tau),u(t)),\ldots,f_8(x(t),x(t-\tau),u(t)))^T. \tag{10}$$

Then, the whole fed-batch process can be written as

$$\begin{cases} \dot{x}(t) = f(x(t), x(t-\tau), u(t)), & t \neq l_k, \\ \Delta x(t) = I(x(t)), & t = l_k,\ k = 1,2,\ldots,N-1, \\ u(t) \in U(t), \\ x(t) = \phi(t), & t \leq 0, \end{cases} \tag{11}$$

where $u(t) = (u_1(t), u_2(t))^T \in R^2$ is control function N is the number of jumps; $l_k, k = 1,2,\ldots,N-1$, are given impulsive moments; $T = l_N$ is a given terminal time; and $\phi : R \to R^8$ is a given continuous function. Additionally, to meet the actual conditions of cell growth, the substance concentrations in the biological reaction process must be strictly limited in a certain range. Hence, define

$$W_{ad} := \prod_{i=1}^{8} [x_{*i}, x_i^*], \quad \forall t \in [-\tau, \mathrm{T}].$$

3. Optimal control model

3.1. Optimal control model

During batch fermentation process, the producer expects to maximize the yield at end moment. Therefore, with the addition of glycerol and alkali as the control function and the yield of 1,3-PD at the terminal moment as the performance index, the following optimal control model is proposed in this paper:

$$\begin{aligned} (OCP)\quad & \max J(u,T) = x_3(T \mid u, T) \\ & s.t.\ x^T(t \mid u, T) \in W_{ad}, \\ & u(t) \in U(t), \\ & T \in [T_{\min}, T_{\max}]. \end{aligned} \tag{12}$$

3.2. Numerical solution method

In this article, in order to overcome the localization of the particle swarm algorithm, a new velocity and position strategy is proposed. In addition, a new strategy of dealing with position outside parameter bounds is also introduced.

(1) (Updating strategy) In the previous iterations, balance development and exploitation by modifying the update speed and location as follows.

$$\begin{cases} v(m) = (-1 + c_1 \times r_1) \times v(m-1) + c_2 \times (pbest(m-1) - \sigma(m-1)) + r_2 \times c_2 \times (gbest - \sigma(m-1)), \\ \sigma(m) = \sigma(m-1) + w(m) \times v(m), \end{cases}$$

$w(k) := w_2 - \dfrac{(w_1 - w_2) I_{ter}}{M_{Iter}}$; w_1 and w_2 are the inertia weights.

(2) (Processing parameter bounds) Assuming that the position component of the particle violates the parameter bound constraint, the particle update position component reflected back by parameters.

$$\sigma(m) = \begin{cases} 2\sigma^* - \sigma(m), & if\ \sigma > \sigma^*, \\ 2\sigma(m) - \sigma_*, & if\ \sigma < \sigma^*. \end{cases}$$

4. Numerical results

In the numerical experiments, we use the same settings initial condition for the experimental to optimize the feeding amount. The initial concentration, initial volume of fermenting material, and end time are $x_0, V_0 = 5L$, and $T = 24.16h$, respectively. Based on experiment, the fermentation process is divided into eight stages. Material are added at 100s in each stage. Table 1 and Table 2 given the upper and lower bounds of the amounts of glycerol and alkali added at each stage. Applying the PSO algorithm to solve OCP, the obtained optimal feeding strategy of material are listed in Table 3. From Figure 1, we can obtain that product is better than experiment data.

Table 1. The bounds of feeding amount of glycerol.

Stage	I	II	IIII	IV	V	VI	VII	VIII
Upper	0.0017000	0.0024000	0.0025000	0.0020000	0.0014000	0.0012000	0.0008000	0.0004000
Lower	0.0004206	0.0005888	0.0006374	0.0005074	0.0003542	0.0002924	0.0002308	0.0001019

Table 2. The bounds of feeding amount of alkali.

Stage	I	II	IIII	IV	V	VI	VII	VIII
Upper	0.0012750	0.0018000	0.00187500	0.0015000	0.0010500	0.0009000	0.0006000	0.0003000
Lower	0.0003154	0.0004416	0.00047805	0.0003805	0.0001593	0.0002193	0.0001731	0.00007642

Table 3. Optimal feeding strategy of glycerol and alkali.

Stage	I	II	IIII	IV	V	VI	VII	VIII
glycerol	0.00167 50	0.0017140	0.0024750	0.000650	0.0010500	0.0009000	0.0002160	0.0003947
alkali	0.00031 54	0.0004416	0.00047805	0.0003805	0.0001593	0.0002193	0.0001731	0.00007642

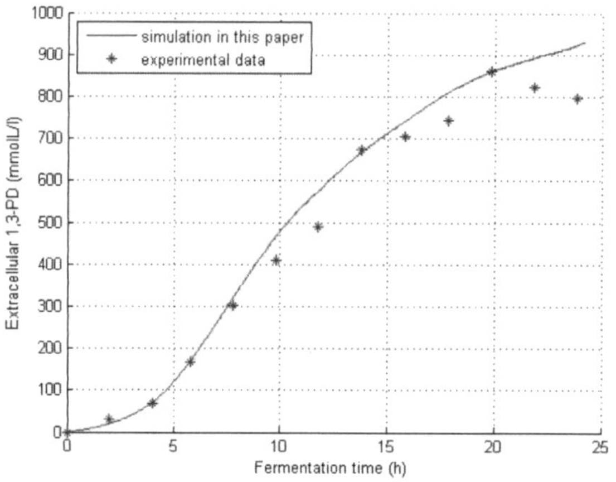

Figure 1. The changes of extracellular 1,3-PD concentration with time.

5. Conclusions

In this article, an impulsive time-delay system with multiple control variables to descr-ible the 1, 3-PD fed-batch process. Taking the 1, 3-PD at terminal time as performance index, the optimal control model of impulsive time-delay is established. The improved PSO is applied to sol-ve the optimal control problem. Numerical shows that the production of 1,3-PD is significantly hi-gher than experimental data at terminal time.

References

[1] Zeng, A.P. (1994) Multiple product inhibition and growth modeling of Clostridium butyricum and Klebsiella pneumoniae in glycerol fermentation. Biotechol. Bioeng., 44: 902–911.

[2] Liu, C.Y. (2017) Robust multi-objective optimal switching control arising in 1,3-propanediol microbial fed-batch process. Noblinear Anal-Hybri., 25: 1-20.

[3] Kennedy, J., Eberhart, R.C. (1985) Particle swarm optimization. Proceedings of the IEEE International Conference on Neural Networks, Perth, Australia, 1942-1948.

Data detection technology of wireless sensor network based on non-stationary filtering

Xinghua Lu[1, 2 *], Jiahao Huang[1] and Guohua Luo[1]

[1] Guangzhou Huali College, Guangzhou, Guangdong, 511325, China

[2]St. paul University Philippines, Tuguegarao , 3500, Philippines

*Corresponding author's e-mail: xhlu@gdut. edu. cn

Abstract. By optimizing the data detection performance of distributed wireless sensor networks, the data sensing and collecting ability of wireless sensor networks can be improved. Traditional methods adopt statistical characteristic parameter detection algorithm for distributed wireless sensor network data detection. Distributed wireless sensor network data has strong time-frequency coupling, so it is difficult to realize frequency domain spatial parameter clustering in frequency domain, and the detection performance is not good. A distributed wireless sensor network data detection algorithm based on non-stationary filtering and high-order statistical feature peak retrieval is proposed. The data model of distributed wireless sensor network is constructed under the interference of color noise. The weak vibration signal is subjected to time-frequency analysis and noise separation by non-stationary filtering, and the spectral peak of distributed wireless sensor network data is searched by the fourth-order cumulant slice post-operator to realize the optimal detection of signals. The simulation results show that the algorithm has a high probability of accurate detection, and has a good ability of suppressing noise and noise sidelobe information interference, which improves the probability of accurate detection of distributed wireless sensor network data under low signal-to-noise ratio.

1. Introduction

With the development of sensor technology, using sensors to collect signal data, so as to guide the sensor system to realize intelligent control, will have a good application prospect in navigation control, guided weapon design, flight control and other fields. Sensor gyro is a kind of high-precision sensor, and distributed wireless sensor is a kind of sensor sensitive to angular velocity and angular deviation. The output data of distributed wireless sensor has linear feature fitting. By constructing an optimized collection model of distributed wireless sensor, the signal can be collected and analyzed. Distributed wireless sensors are widely used in aircraft inertial navigation design, missile attitude sensitivity control and torpedo inertial navigation control. Distributed wireless sensors are used in complex inertial navigation control environment. By collecting data information of distributed wireless sensors as control information input, the weak detection signals in distributed wireless sensors can be optimally detected, and the ability of collecting and analyzing signals of distributed wireless sensors can be improved. People pay great attention to the research of related data detection algorithms [1-3].

At present, data detection algorithms for wireless sensor networks mainly include data detection algorithms based on fractional Fourier transform, data detection algorithms based on high-order cumulant feature extraction and data detection algorithms based on time-frequency analysis. The principle of the algorithm is to construct the data model of wireless sensor network and adopt the corresponding signal processing algorithm to realize the data detection of wireless sensor network, and some research results have been obtained. Literature [7] proposes a data detection algorithm of wireless sensor network based on IMF component decomposition of multiple narrowband signals, which adopts instantaneous frequency filtering detection to realize signal purification, thus improving the detection probability of wireless sensor network data. However, the algorithm has high computational cost and poor real-time detection effect on wireless sensor network data. Literature [8] proposes a data detection algorithm for wireless sensor networks based on Gaussian cascade filtering and time-frequency analysis, which filters the interference generated in low frequency domain, and realizes adaptive beam focusing of wireless sensor network data combined with time-frequency analysis, and the detection effect is good. However, this method needs to accurately estimate the prior time, frequency and other parameter information of wireless sensor network data, and the detection performance is not good when the knowledge of parameter information collection is insufficient. Literature [9] proposes to use statistical feature parameter detection algorithm. Distributed wireless sensor network data has strong time-frequency coupling, so it is difficult to realize frequency domain spatial parameter clustering in frequency domain, and the detection performance is not good. To solve the above problems, this paper proposes a distributed data detection algorithm for wireless sensor networks based on fractal Fourier transform (FRFT) and fourth-order cumulant post-peak search. The data model of distributed wireless sensor network is constructed under the interference of color noise. The weak detection signal is analyzed in time-frequency and noise separation by non-stationary filtering, and the peak of distributed wireless sensor network data is searched by high-order statistical feature post-operator, so as to realize the optimal detection of signals. Finally, the performance is verified by simulation experiment, which shows the superior performance of the algorithm in weak detection data detection, and draws a conclusion of effectiveness.

2. Data Model and Noise Separation Pretreatment of Distributed Wireless Sensor Network

2.1 Data Model of Distributed Wireless Sensor Network under Color Noise Interference

Distributed wireless sensor adopts multi-turn optical fiber loop to enhance effect, and adopts strong coherent light source to enhance resonance effect of resonant cavity for signal acquisition. Micro-detection signals collected by distributed wireless sensor are interfered by strong color noise. Under the interference of color noise, the data of distributed wireless sensor network is a group of wide and stationary Gaussian narrow-band signals with time-frequency coupling performance, and the acquisition nodes of detection signals of distributed wireless sensor are distributed in the array space in the form of near-field source uniform linear array distribution[10]. The near-field source uniform linear array distribution model of wireless sensor network data is shown in Figure 1.

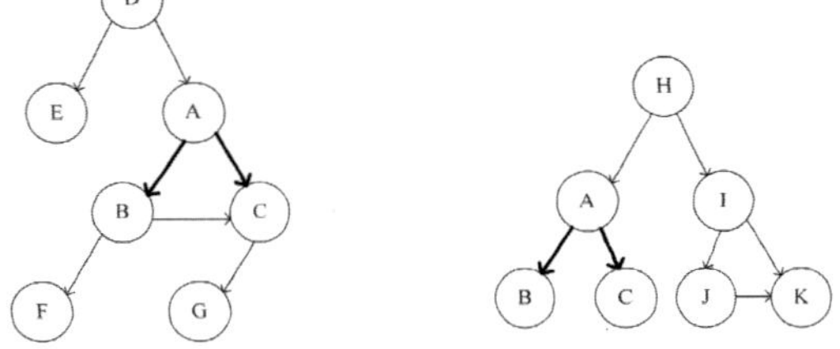

Figure 1 Distribution model of near-field source uniform linear array for wireless sensor network data

In the data distribution structure of wireless sensor network shown in fig. 1, the channel vectors between the distributed sources and array elements of wireless sensor network data have strong time-frequency coupling, the sampling time series $s_1(t), s_2(t), \cdots s_L(t)$ of wireless sensor network data is a zero-mean, non-Gaussian, statistically independent narrow-band stationary process, the cross terms between wireless sensor network data have non-zero peaks, and the spatial distribution distance of detection sensors in the center of the array meets the following requirements:

$$\begin{aligned} T &= F\boldsymbol{w}\left[v(1),v(2),\ldots,v(L)\right] \\ &= \begin{cases} \left[1,e^{j2\pi d\sin(\theta(i))/\lambda},e^{j2\pi 2d\sin(\theta(i))/\lambda},\ldots,e^{j2\pi(M-1)d\sin(\theta(i))/\lambda}\right]^{\mathrm{T}} & d_i \le 0 \\ 2[\boldsymbol{T}^{\mathrm{H}}(\boldsymbol{n})\boldsymbol{x}(\boldsymbol{t})+\Delta y(x_{i,r}+1+\frac{1}{2}-\Delta xB)] & d_i > 0 \end{cases} \end{aligned} \tag{1}$$

In which, D represents the radiation radius of the continuous signal emitter of wireless sensor network data and the wavelength of the information source. Assuming that the detection sensor of distributed wireless sensor is a near-field source distribution model composed of array elements, and the array element coordinate at the center of the reference array element is 0, the wireless sensor network data time series model obtained by sensing the first array element is as follows:

$$G_t = AF_{t-1} + \boldsymbol{T}^{\mathrm{H}}(n)\boldsymbol{RT}(n) \tag{2}$$

Wherein, $s_i(t)$ is the signal complex envelope on the time-frequency coupling plane, $n_m(t)$ is the radial phase of the wireless sensor network data, and g represents interference color noise. Through the above analysis, the analytical signal model of wireless sensor network data constructed by discrete Fourier transform is as follows:

$$A = \begin{bmatrix} \boldsymbol{BP_tB}^{\mathrm{H}} & \sigma_n^{\ 2}\boldsymbol{I}_M \\ \boldsymbol{U}_{sB}\boldsymbol{\Lambda}_{sB}\boldsymbol{U}_{sB}^{\ \mathrm{H}} & \cos\theta \end{bmatrix}, t = \begin{bmatrix} \boldsymbol{U}_{nB}\boldsymbol{\Lambda}_{nB}\boldsymbol{U}_{nB}^{\ \mathrm{H}} \\ \boldsymbol{U}_{sB}\boldsymbol{\Lambda}_{sB} \end{bmatrix} \tag{3}$$

Assuming that the vertical directivity of the signal transceiver does not change much in the distributed wireless sensor, the envelope and phase of the wireless sensor network data under the background of color noise are as follows:

$$\begin{aligned} N(x,y) &= -Cx(k) + DKx(k-\tau_k)\sigma\nabla U \\ &= -\sigma hT + \int h(t)R(x,y;t) \\ &= -\sigma[G_x(x,y;t)i + G_y(x,y;t)j] \end{aligned} \tag{4}$$

Because the spectrum of sensor output resolution signal can effectively reflect the structural characteristics of wireless sensor network data, the amplitude of distributed wireless sensor network data under color noise interference is related to target characteristics, channel loss and array characteristics. By the above signal model construction, the signal source foundation is provided for realizing wireless sensor network data detection[11].

2.2 Signal noise separation preprocessing based on non-stationary filtering

On the basis of the data model construction of distributed wireless sensor networks, in order to achieve accurate detection of signals, noise separation is needed in the background interference environment of color noise with low signal-to-noise ratio. In this paper, non-stationary filtering method is used to realize signal noise separation by time-frequency analysis of signals. The order non-stationary filtering of distributed wireless sensor network data defined in the domain can be expressed as $X_p(u)$ or $F^p x(u)$. In the time period of continuous sampling of signals, there are:

$$k = \sum_{i=1}^{q} s_i(t)\boldsymbol{b}_i(\boldsymbol{\theta}_i) \tag{5}$$

The frequency spectrum characteristics of the signal in frequency domain and time domain are constructed, and the calculation formula of $K_p(t,u)$ is obtained as follows:

$$K_p(t,u)=\begin{cases}\sqrt{\dfrac{1-j\cot\alpha}{2\pi}}\exp\left(j\dfrac{t^2+u^2}{2}\cot\alpha-jtu\csc\alpha\right) \\ \qquad\qquad \alpha\neq n\pi \\ \delta(t-u),\ \alpha=2n\pi \\ \delta(t+u),\ \alpha=(2n\pm1)\pi\end{cases} \tag{6}$$

Where n is an integer, i. e $n\in Z$. For any non-zero distributed wireless sensor network data, in any two non-orthogonal fractional Fourier domains, the time-frequency focusing decomposition of the signal in time domain and frequency domain is adopted, and the signal order non-stationarity filtering after noise separation can be expressed as:

$$t(x)=\begin{bmatrix}\boldsymbol{R}_x(1,1), & \boldsymbol{R}_x(1,2), & \dots & \boldsymbol{R}_x(1,M) \\ \boldsymbol{R}_x(2,1), & \boldsymbol{R}_x(2,2), & \dots & \boldsymbol{R}_x(2,M) \\ \cdots & \cdots & \cdots & \cdots \\ \boldsymbol{R}_x(M,1), & \boldsymbol{R}_x(M,2), & \dots & \boldsymbol{R}_x(M,M)\end{bmatrix} \tag{7}$$

To sum up, it can be seen that the color noise in distributed wireless sensor network data can be effectively separated by using the rotational additivity and reversibility of non-stationary filtering, and the energy distribution of weak detection signals among different LFM cores has significant characteristics. Based on this, the data detection algorithm is designed[12].

3. Improve implementation of data detection algorithm

3.1 Propose the algorithm of improving the thought description spectrum peak search

On the basis of preprocessing the distributed wireless sensor network data model and noise separation, the data detection model is optimized and modeled, and the data detection algorithm is improved. By analyzing the above non-stationary filtering method, it can be seen that when the distributed wireless sensor detection signal is interfered with by strong color noise, the distributed wireless sensor network data shows strong time-frequency coupling in the time-frequency domain. The time-frequency coupling makes it difficult to realize frequency-domain spatial parameter clustering in the frequency domain and leads to poor detection performance. In order to overcome the disadvantages of traditional methods, this paper uses the insensitivity and natural blindness of high-order cumulant diagonal slices to Gaussian noise, and purifies a distributed wireless sensor network data detection algorithm based on non-stationary filtering and high-order statistical feature peak retrieval. On the basis of time-frequency analysis and noise separation of weak detection signals by non-stationary filtering, the fourth-order cumulant operator of distributed wireless sensor network data is constructed as follows:

$$U(x)=\begin{bmatrix}\bar{r}(1), & \bar{r}^*(2), \cdots, \bar{r}^*(M) \\ \bar{r}(2), & \bar{r}(1), \cdots, \bar{r}^*(M-1) \\ \cdots, & \cdots, \cdots, \cdots \\ \bar{r}(M), \bar{r}(M-1), & \cdots, \bar{r}(1)\end{bmatrix} \tag{8}$$

In the formula, x (.) is the input distributed wireless sensor network data, and $\hat{c}_{2x}(\tau)=\dfrac{1}{N}\sum_{i=1}^{N}x(i)x(i+\tau)$ represents the second-order cumulant of the signal. After the non-stationary filtering of the distributed wireless sensor network data, the spectrum peak of the signal is generated in the time-frequency domain space, and the focus analysis with obvious kurtosis value is carried out in the frequency domain.

$$I_l(n_1,n_2)=\frac{1}{4}\sum_{i_1=0}^{1}\sum_{i_2=0}^{1}I_{l-1}(2n_1+i_1,2n_2+i_2) \tag{9}$$

By using the insensitivity of high-order cumulant to Gaussian noise, the output weak detection signal after noise separation is matched with the kernel function of high-order cumulant at a certain angle, which can suppress the interference of Gaussian white noise and Gaussian color noise, and realize the post-focusing of signal spectrum and the suppression of noise sidelobe information. According to this principle, in the fractional Fourier domain, the signal energy after mixed slice frequency modulation by fourth-order cumulant is:

$$\hat{c}_{4x}(n,\tau)=\hat{c}_{4s}(n,\tau)+\hat{c}_{4w}(n,\tau)=\hat{c}_{4s}(n,\tau) \tag{10}$$

Wherein, $\hat{c}_{4s}(n,\tau)$ represents the energy of pure detection signal in distributed wireless sensor network data, $\hat{c}_{4w}(n,\tau)$ represents the energy of noise part in distributed wireless sensor network data, and if the interference noise is Gaussian color noise, the fourth-order mixed cumulant slice of distributed wireless sensor network data is obtained as follows:

$$I_{(k)}(i,j)=\begin{cases}0, & otherwise\\ 1, & [I(i,j)/2^{k-1}],\bmod 2=1\end{cases} \tag{11}$$

On the basis of Doppler modulation processing, the signal is output by discrete non-stationary filtering, and the fourth-order mixed cumulant slice is used for post-focusing processing, then the output detection signal is:

$$I(i,j)=\sum_{k=1}^{P} I_{(k)}(i,j)\times H(v)R^{k-1} \tag{12}$$

In which γ represents the energy kurtosis of distributed wireless sensor network data. The kurtosis of a weak detection signal interfered by Gaussian color noise by linear integral projection. Adopting signal-to-noise kurtosis ratio (SNKR) can effectively describe the performance of data detection. To sum up the design principle, the distributed wireless sensor network data detection is carried out through the fourth-order cumulant post-peak search. The flow chart of the improved algorithm is shown in Figure 2.

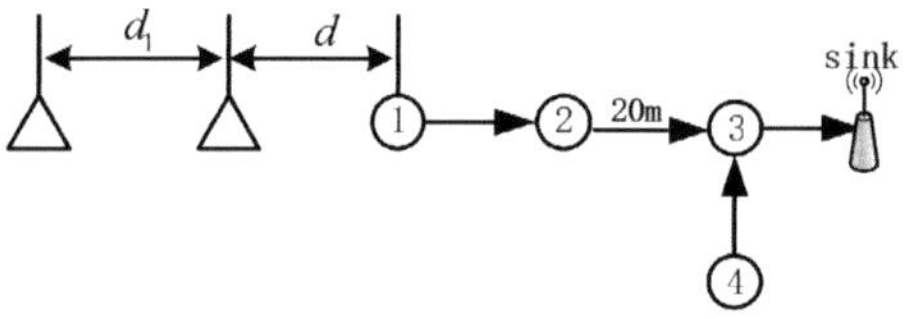

Figure 2 Flow chart of improved implementation of data detection algorithm

3.2 Implementation of distributed wireless sensor network data detection

Specifically, the implementation process of the distributed wireless sensor network data detection modeling is described as follows: let the received signal $y(t)$ of the distributed wireless sensor be:

$$r(n_1,n_2)=\frac{1}{N_1N_2}\sum_{k_1k_1}R(k_1,k_2)e^{j\frac{2\pi k_1n_1}{N_2}}e^{j\frac{2\pi k_2n_2}{N_2}} \tag{13}$$

The nonstationary filtering of the received distributed wireless sensor network data is as follows:

$$\begin{aligned}M&=\begin{bmatrix}u_{11} & u_{12}\\ u_{21} & u_{22}\end{bmatrix}=u(x,y,\sigma_I,\sigma_D)\\ &=\sigma_D^2G(\sigma_I)*\begin{bmatrix}L_x^2(x,y,\sigma_D) & L_xL_y(x,y,\sigma_D)\\ L_xL_y(x,y,\sigma_D) & L_y^2(x,y,\sigma_D)\end{bmatrix}\end{aligned} \tag{14}$$

The non-stationary filtered distributed wireless sensor network data is subjected to post-peak search through a fourth-order cumulant post-operator, which is described as:

$$V(u)=G(x,y;t)+\mathbf{w}=[1\ \ 0]^{\mathrm{T}} \tag{15}$$

Based on the above algorithm design, the appropriate rotation angle is selected to match the fourth-order cumulant with the time-frequency characteristics of the processed distributed wireless sensor network data, and the time-frequency coupling is removed and the noise is effectively suppressed, thus realizing the effective signal detection.

4. Simulation experiment and detection performance analysis

In order to test the performance of this algorithm in realizing effective data detection in wireless sensor networks, a simulation experiment is carried out. Matlab.7 programming software is used to design the algorithm of wireless sensor network data detection model. firstly, the original data of wireless sensor network data is sampled, where the sampling frequency is 23KHz, t=12ms, the number of sampling points, the input signal-to-noise ratio is 21dB, and the center frequency of color noise is 12. 35KHz the time domain waveform of wireless sensor network data under color noise interference is obtained as shown in figure 3.

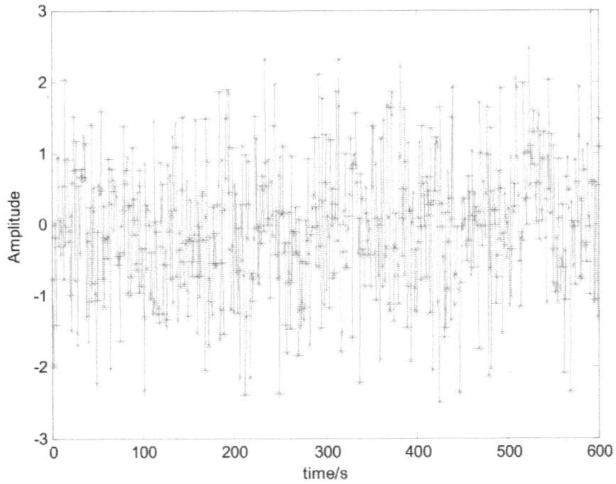

Figure. 3 Time domain waveform of wireless sensor network data

It can be seen from the figure that the original wireless sensor network data is disturbed by large noise, which makes it difficult to realize effective detection. The method in this paper is used for non-stationary filtering noise separation, and the high-order statistical feature post-operator is used to search the spectrum peak of distributed wireless sensor network data, so as to realize the optimal detection of signals. In order to compare the algorithm performance, the spectral peak modulus of wireless sensor network data detection is obtained by using this method and the traditional method, as shown in Figure 4.

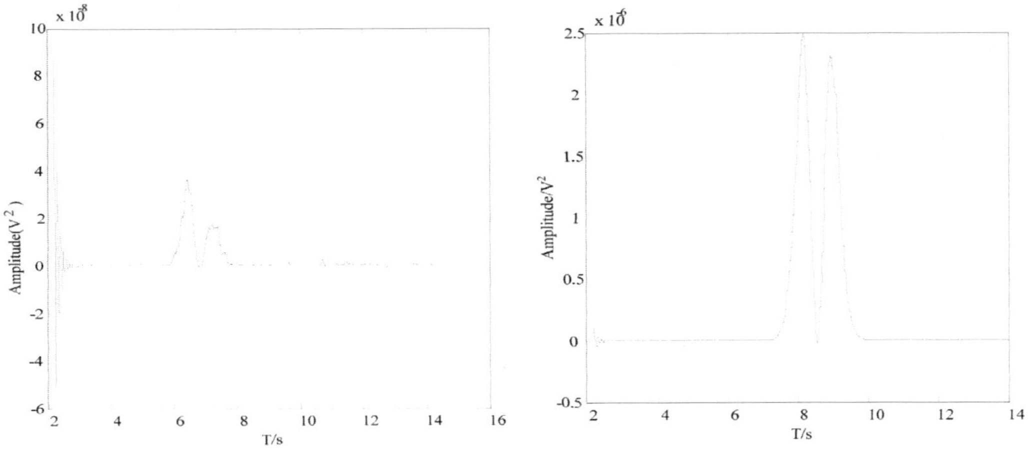

(a) Traditional detection algorithm (b) The algorithm in this paper

Fig. 4 Data detection results of wireless sensor network

It can be seen from the figure that the spectral peak focusing performance of wireless sensor network data detection using this method is better, and the sidelobe suppression ability is higher. However, when the traditional method is used for data detection, the waveform is completely submerged in noise, and the noise is not suppressed. In order to quantitatively analyze the performance of the algorithm, 1000 Monte Carlo tests are used to obtain the detection performance comparison results of wireless sensor network data with false alarm probability less than 0. 001, and the related data are sorted and analyzed, and the detection performance curve comparison results are shown in Figure 5. From the results in Table 1, it can be seen that the accurate detection probability of this method for wireless sensor network data detection is higher than the traditional method, which shows the superiority of this algorithm in improving the detection performance.

Table 1 Detection Probability of Various Methods under Different SNR

SNR/dB	SVM	FrFT-	This method
-30	0	0	0
-25	0	0	0. 276
-20	0. 045	0. 065	0. 287
-15	0. 067	0. 167	0. 644
-10	0. 745	0. 867	0. 979
-5	0. 967	0. 956	1
0	1	1	1

5. Conclusions

In this paper, the data detection performance of distributed wireless sensor networks is optimized to improve the data sensing and collecting ability of wireless sensor networks, this paper proposes a data detection algorithm of distributed wireless sensor networks based on non-stationary filtering and high-order statistical feature peak retrieval. The data model of distributed wireless sensor network under color noise interference is constructed, the signal noise is separated, and the high-order statistical feature post-operator is used to search the spectrum peak of distributed wireless sensor network data, so as to realize the optimal signal detection. The research shows that the accurate detection probability of distributed wireless sensor network data detection with this algorithm is higher, and the ability of suppressing noise and noise sidelobe information interference is better, and the accurate detection probability is improved at low signal-to-noise ratio.

Acknowledgement

This paper was supported Guangdong Province Science and Technology Innovation Strategy Special Fund Project, The project number: pdjh2020b0778, Researching on Data Reconstruction Model of Wireless Sensor Network Based on Dynamic Data Compression. Thank you Guangdong University of Technology Huali College Student Innovation and Entrepreneurship Training Program,(project number: 202113656085)for supporting this paper.And Thank you for the above project's financial support for this study, and thank you for the review of the organizing committee.

References

[1] LIU Wanghua, LIU Guangshuai, CHEN Xiaowen, LI Xurui. Robust texture representation by combining differential feature and Haar wavelet decomposition. Journal of Computer Applications, 2020, 40(9): 2728-2736.

[2] GAN J Y, ZHAI Y K, XIANG L, et al. Spatial-temporal texture cascaded feature method for face liveness detection[J]. Pattern Recognition and Artificial Intelligence, 2019, 32(2):117-123.

[3] LIU L, LONG Y, FIEGUTH P W, et al. BRINT:binary rotation invariant and noise tolerant texture classification[J]. IEEE Transactions on Image Processing, 2014, 23(7):3071-3084.
[4] GUO Z, WANG X, ZHOU J, et al. Robust texture image representation by scale selective local binary patterns[J]. IEEE Transactions on Image Processing, 2016, 25(2):687-699.
[5] EBTEHAJ I, BONAKDARI H, ES-HAGHI M S. Design of a hybrid ANFIS-PSO model to estimate sediment transport in open channels[J]. Iranian Journal of Science and Technology-Transactions of Civil Engineering, 2019, 43(4):851-857.
[6] GHASEMI M, AKBARI E, RAHIMNEJAD A, et al. Phasor particle swarm optimization:a simple and efficient variant of PSO[J]. Soft Computing, 2019, 23(19):9701-9718.
[7] SHU W, QIAN W, XIE Y, et al. An efficient uncertainty measure-based attribute reduction approach for interval-valued data with missing values[J]. International Journal of Uncertainty, Fuzziness and Knowledge-Based Systems, 2019, 27(6):931-947.
[8] YAN Haisheng, MA Xinqiang. Feature construction algorithm for multi-target regression via radial basis function. Journal of Computer Applications, 2021, 41(8): 2219-2224.
[9] ZHEN X T, YU M Y, HE X F, et al. Multi-target regression via robust low-rank learning[J]. IEEE Transactions on Pattern Analysis and Machine Intelligence, 2018, 40(2):497-504.
[10] DING Yin, SANG Nan, LI Xiaoyu, WU Feizhou. Prediction method of capacity data in telecom industry based on recurrent neural network. Journal of Computer Applications, 2021, 41(8): 2373-2378.
[11] BELOGLAZOV A, BUYYA R. OpenStack Neat:a framework for dynamic and energy-efficient consolidation of virtual machines in OpenStack clouds[J]. Concurrency and Computation:Practice and Experience, 2015, 27(5):1310-1333.
[12] TRAN N, NGUYEN T, NGUYEN B M, et al. A multivariate fuzzy time series resource forecast model for clouds using LSTM and data correlation analysis[J]. Procedia Computer Science, 2018, 126:636-645.

A High Speed Redundant IO Bus for Energy Power Controller System

Qinghui Lou[1]*, Liguo Sun[1] , Haisong Lu[1],Weifeng Xu[1] , Zhebei Wang[1], Dan Cai[1],Xiangjian Shi[1]

[1] NR Electric Co., Ltd, Nanjing, 211102, China

* Corresponding author's e-mail: louqinghui@nrec.com

ABSTRACT. This paper designs and implements a High Speed Redundant IO Bus for Energy Power Controller System. The physical layer adopts multi-point low-voltage differential signal standard. This bus has the characteristics of high real-time, high throughput and easy expansion. The controller communicates with IO module by A/B bus alternately, monitors link status in real time and collects IO module data. Non real time slots can be used to control non real time messages for IO modules such as time synchronizing and memory monitoring. The controller ARM core runs QNX real-time operating system, and transmits the message needed to communicate with IO modules to the FPGA through DMA. After receiving the message, the FPGA parses the message and automatically fills in the CRC check code and frame end flag at the end of the message. When the FPGA receives the data feedback from the IO module, it performs CRC verification. If the verification passes, it fills the corresponding module receiving buffer. Otherwise, it fills the CRC verification error flag in the register of the corresponding IO module to reduce the load of the arm core.

1. Introduction

Control system, including distributed control system (DCS), programmable logic controller (PLC), etc., is a combination of computer technology and automation control technology, in power, petrochemical, metallurgical, chemical, etc. It has a wide range of applications in large industries [1]. Compared with the booming fieldbus (such as Profibus, FF, CAN, LonWorks, etc. [2] [3]), the development of the local bus of the control system appears to be relatively slow. The reason is mainly that the control system and the local input/output (IO) module communication are relatively closed, involving the architecture of the control system and all the IO modules, so each manufacturer keeps the protocol confidential, making other manufacturers unable to get their communication information. The paper [4] proposed a communication protocol between controller and IO based on general object oriented substation event (GOOSE), which limits its scope of use due to the use of network communication. The paper [5] proposed a communication bus based on PCI Express, but because of the compatibility, reliability and scalability of computer serial buses (such as PCI Express, Universal Serial Bus (USB), IEEE1394), The requirements of the control system application are not well met [6]-[9].

Aiming at the problems of the above bus, this paper proposes a high-speed and reliable redundant IO communication bus based on field programmable gate array (FPGA). The physical layer uses multi-point low-voltage differential signals (M-LVDS [10]) standard, realizes the real-time data transmission of the data link layer and the monitoring of the channel link status, ensuring the reliability and real-time performance of data transmission. The IO module uses direct memory access (DMA) to analyze the communication packets of the universal asynchronous receiver transmitter (UART). The detailed design is cited from paper [11].

2. Overall System Plan

The current mainstream control system architecture can be generally divided into four levels: the field instrument layer, the control device unit layer, the factory (workshop) layer, and the enterprise management layer [12], where the control unit layer includes the controller and the IO module. The interconnection between the controller and the IO module is the hub of data transmission of the whole system. It has very high requirements in real-time, reliability and scalability. The efficiency of the bus directly affects the performance of the entire control system [1]. The controller designed in this paper adopts the scheme of active/standby redundancy, and the IO unit adopts the A/B redundant IO bus scheme. The active and standby controllers communicate with the IO module through the A/B redundant bus, but only the active controller sends the query data and control messages to the IO module alternately through the A/B redundant bus, and the standby controller passes the A/B redundancy. The bus monitors the response data of the IO module to determine the continuity of the link. After the standby controller is upgraded to the active controller, the original active controller no longer sends query data and control packets, and the newly-based controller sends query data and control packets. The system structure is shown as in Fig. 1, wherein the IO module can be analog input (AI), analog output (Analog Output, AO), digital input (DI), digital output (Digital Output, DO) or some other special IO module. A pair of controllers can carry 6 redundant IO buses, and each IO redundant bus can mount 10 IO modules. Therefore, the controller designed in this paper can mount 60 local IO modules.

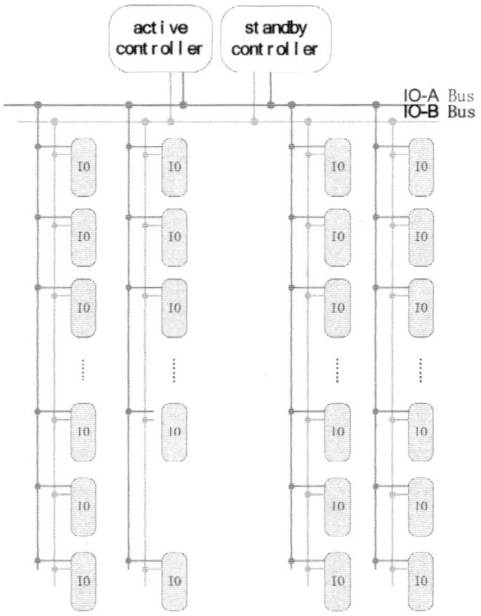

Fig.1 Main structure of the system

3. Protocol Design And Implementation

The real-time and reliability of communication between the controller and the IO module mainly depends on the communication rate of the physical layer and the scheduling strategy of the data link layer. The communication protocol designed in this paper specifies the information exchange mode between the controller and the IO module. This protocol complies with the open system interconnection reference model (OSI/RM) and uses the application layer, data link layer and physical layer. The application layer implements the real-time data or non-real-time data analysis of the IO module, the data link layer implements a cyclic redundancy check (CRC), and the physical layer implements the codec function of the serial port data.

Each IO module designed in this paper has the same communication interface and frame structure. The IO module includes a micro control unit (MCU), which realizes data acquisition and analysis through DMA, and detailed reference [11]. The controller uses Xilinx's Zynq chip, which has two (advanced risc machine, ARM) chips and one FPGA. The ARM chip runs the QNX real-time operating system and extends the 6-way redundant IO bus through the FPGA. Each bus realizes 5Mbps high-speed serial communication, including M-LVDS transmission/reception module, 10bit codec module and CRC check module. The QNX real-time operating system interacts with the FPGA through configuration registers and DMA.

3.1 Physical Layer Protocol
The physical layer specifies the mechanical, electrical, functional, and procedural characteristics of the communication between the controller and the IO module. This communication protocol adopts the multi-point low-voltage differential signaling (M-LVDS) standard to implement a single-master multi-slave communication mode. M-LVDS has the characteristics of high-speed transmission, low voltage swing, anti-interference, low power consumption, multi-point communication, and fault safe isolation [1]. Because M-LVDS has the above characteristics, the IO modules can be online hot-plugged to establish or disconnect real-time communication with the controller, and the communication functions of other IO modules on the IO bus will not be affected. The hot-plugged function of the bus improves the reliability and scalability of the whole system.

3.2 Data Link Layer Protocol
The data link layer is responsible for performing link layer protocol data unit (PDU) interaction between the controller and the IO module on the link. The data link layer defines the protocol data units in details as shown in Fig2.

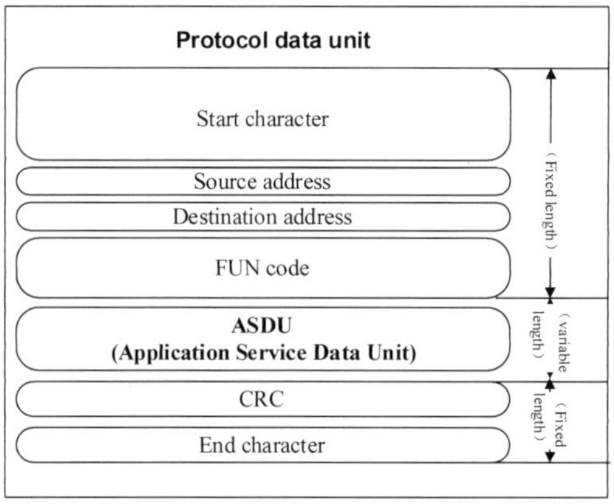

Fig.2 Protocol data unit

3.3 Application Layer Protocol
The application layer protocol mainly includes 3 parts:

1) The controller groups the query or control message sent to the IO module according to the configuration file of the IO module;

2) Parses the real-time data sent by the IO module, and sends the data to fill in the IO receiving buffer;

3) Group or parse non-real-time packets.

Since the QNX real-time operating system transmits the message delivered to the IO module to the FPGA through DMA, it is required to design the issued message format to ensure the correctness of the data received by the FPGA. The data format defined by ARM to FPGA is shown in Fig 3. The "info" uses 6 bits as the subnet mask. Each bit represents each branch of the 6 buses. For example, 0x1 means sending on the first bus, 0x10 means sending on the fifth bus, and 0x3f means that all the 6 branches will send. The "len_byte" represents total length of the data packet in this frame. The "content" is the data link layer protocol data unit (PDU), but there is no CRC checksum and frame tail flag. These two items will be automatically added by FPGA to reduce the load on the ARM chip.

Because sending data by DMA will lead to DMA interrupts, after the FPGA receives the data, it does not transmit the data to the QNX real-time operating system through DMA, but stores the data in the global data area by means of the circular buffer. This design has two advantages: 1) Reducing the number of DMA interrupts, the QNX real-time operating system does not need to generate an interrupt to process data after the FPGA receives each IO module data; 2) IO modules under 6 buses can be processed uniformly, and the data of the IO module can form a cross-section, which is beneficial to the algorithm page processing.

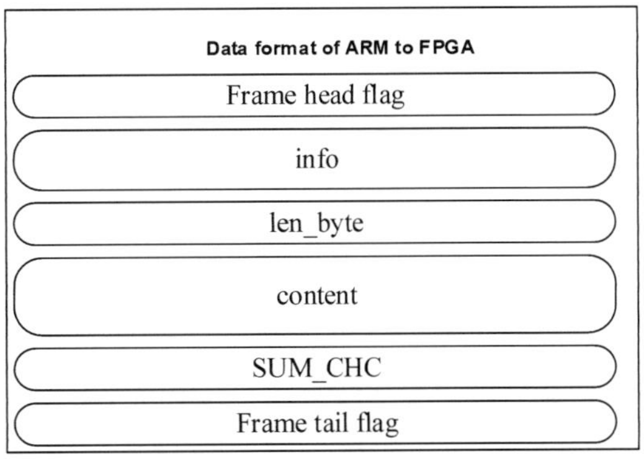

Fig.3 Data format of ARM to FPGA

4. Software Design

The software program mainly includes:

1) IO module configuration file analysis;
2) Serial port register configuration;
3) Serial port driver configuration;
4) FPGA register configuration;
5) Real-time message transmission;
6) Non-real-time message transmission.

After the controller is powered on, the parameter configuration will be set first. After the setting is completed, the FPGA generates a 5ms pulse signal at a fixed period. After the QNX real-time operating system captures this pulse signal, it performs packetization and analysis of real-time and non-real-time messages.

4.1 Real-time Message Transmission

After the QNX real-time operating system captures the timing interrupt generated by the FPGA, it first analyzes the message that the FPGA transmits to the application layer through the circular buffer, and fills the real-time acquisition data of the IO module into the memory area of the IO module according to the analysis result. Then the controller groups control or query messages sent to the IO module according to the IO module configuration information. Finally, after scanning 60 IO modules, the formed sent message will be transmitted to the FPGA through DMA. After the FPGA receives the message, it parses it in the format listed in Table 2, and fills in the CRC checksum and frame tail flag at the end of the link protocol data unit (PDU). Please refer to Fig. 4 for the specific flow.

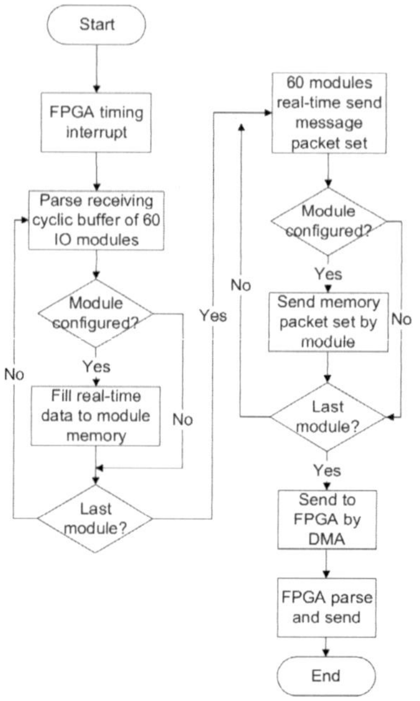

Fig.4 Flowchart of real-time transmission

4.2 Non-Real Time Message Transmission

Because the FPGA generates 5ms pulse signal at a fixed period to receive and send real-time messages, it is required to consider adding non-real-time messages to these real-time messages, e.g.: time synchronization, IO module memory query, specific IO module upgrade program, etc. These non-real-time messages will not affect transmission of real-time messages. The design scheme of this paper is: If the branch has non-real-time data, a frame of non-real-time message will be added at the end of the real-time message transmission period of this branch. If the branch has no non-real-time data, time slot of this branch will be ignored. With this scheme, conflicts between real-time message transmission and non-real-time message transmission can be effectively resolved.

5. Performance Analysis

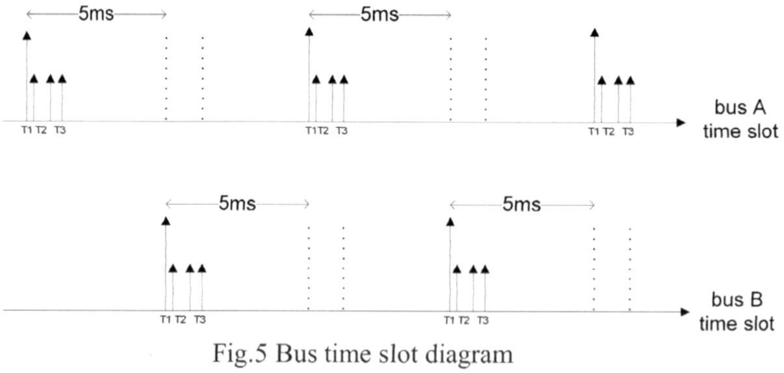

Fig.5 Bus time slot diagram

Through the bus time slot diagram in Fig. 5, the communication between the controller and the IO module through the A / B redundant bus can be introduced in details.

In this figure: T1 represents the time when the QNX real-time operating system parses the data sent by the IO module, T2 represents the packetization time when the QNX real-time operating system issues the query or control packet according to the IO module configuration file, and T3 represents the time that DMA sent data from the ARM chip to the FPGA. After actual measurement, the processing time of T1, T2 and T3 all reaches microsecond level.

When T3 is reached, the FPGA analyzes the delivered data message according to the format in Tab.2, and automatically adds a CRC checksum and a frame tail flag to issue a query or control message. In theory, the communication time can reach a cycle of 5ms, because T1, T2, and T3 do not occupy the bus bandwidth time. If this cycle is a message sent by the A bus, then the next cycle is switched to the B bus to send a message. In this way, switching and sending can not only reduce the communication load of the IO module, but also monitor the status of the A / B bus communication link. However, the disadvantage of this operation is to process the data fed back by the IO module. It needs to wait for 2 query cycles (i.e.:, 10ms). If the A-bus or B-bus communication link is disconnected, it will cause the entire branch query period to double. However, the controller and the IO module are grouped in a panel cabinet, and the controller will alarm when the communication link is disconnected, so this situation can be avoided well.

6. Conclusion

This paper presents the design and implementation of a high-speed redundant IO bus based on FPGA. The system combines the advantages of the embedded ARM chip and FPGA perfectly, and realizes the development of embedded platform combining FPGA and ARM chip, which improves the real-time performance, reliability and stability of the system. This design can guarantee the real-time concurrent processing of the 6-channel IO module bus, and it does not occupy the processing time of the ARM chip, so it has excellent value and promotion significance.

References

[1]MI Xiao-ling, HUANG Wen-jun, JIN Jian-xing, SHI Yi-ming. Design and implementation of FPGA based high-speed bus for control system [J]. Journal of Zhejiang University, 2011, 45(11):2043-2049.

[2]CHEN Ji-ming, WANG Zhi, SONG Ye-qiong. Study on schedule problem of aperiodic message traffic in foundation fieldbus [J]. Journal of Zhejiang University, 2003, 37(03):273-277.

[3]YANG Xianhui Fieldbus technology and its application [M]. Beijing: Tsinghua University Press, 2008.

[4]ZANG Feng, XU Weifeng, WU Bo, NIU Honghai. Development of Industrial Process Controller Based on GOOSE Protocol [J]. Process Automation Instrumentation, 2016, 37(6):51-54.

[5]JIANG Qunxing, HU Lisheng. Research on High-speed Communication Bus Technology of Host Controller in Nuclear-level Control System [J]. MICROCOMPUTER APPLICATIONS, 2010, 26(4):6-8.

[6]DHAWAN S K. Introduction to PCI Express a new high speed serial data bus[C] //Nuclear Science Symposium Conference Record. Fajardo: IEEE, 2005, (2):687-691.

[7]DONG Xinwei, WU Guozhong, ZHAO Rongxiang. Analysis and design of power electronic building blocks' communication interface [J]. Journal of Zhejiang University, 2007, 41(11):1857-1861.

[8]DEPARI A,FLAMMINI A,MARIOLI D, et al. USB sensor network for industrial applications[J]. Instrumentation and Measurement, 2008, 57(7):1344-1349.

[9]SCHOLLES M,FROMMHAGEN K, KLEINMANN L.IEEE1394 "FireWire" system design for industrial and factory automation applications[C] //IEEE Proceedings of the 8th IEEE International Conference on Emerging Technologies and Factory Automation. AntibesJuan les Pins: IEEE, 2001, (2): 627-630.

[10]ANSI/TIA/EIA-899. Electrical characteristics of multipoint-low-voltage differential signaling (M-LVDS) interface circuits for multipoint data interchange[S].USA TIA/EIA， 2000.

[11]NIU Honghai, ZANG Feng, ZHOU Xugui. Design and Implementation of High Speed UART Based on DMA [J]. Process Automation Instrumentation, 2018, 39(9):45-48.

[12]WANG Changli. Current Status and Development Trends of Distributed Control System [J]. ELECTRIC AGE, 2004, (01):82-86.

Sum Rate Analysis for Massive MIMO Downlink With MRC Beamforming and User Selection

Yiyang Wu[1,a*], Chang Chang[2,b], Fei Xie[3], Dacheng Ju[4], Yilun Pan[5]

[1]University of Electronic Science and Technology of China, Chengdu, Sichuan, 611731, China

[2]School of Electronic Information Wuhan University, Wuhan, Hubei, 430070, China

[3]School of Engineering and Applie Science University at Buffalo, SUNY Buffalo, NY, 14228, U.S.A

[4]Hefei NO.1 High School Hefei, Anhui, 230000, China

[5]Nanning NO.3 High School Nanning, Guangxi, 530000, China

email: 2018190504019@std.uestc.edu.cn

Abstract. Average allocation of data rate to each user is inefficient since the resource a base station can allocate is limited. Thus, user selection and user scheduling need to be applied into multi-user massive multiple-input multiple-output (MIMO) downlink system. In this paper, we mainly focus on the methods of user selection. First, we establish a downlink system model including transmission model and channel model. Then, two user-rate based user selection algorithms via the signal-to-interference-plus-noise-ratio (SINR) are proposed, where the SINR is generated by MRC beamforming. Finally, simulation results are provided to compare the performance of two proposed algorithms and their fairness towards selected users. In the simulation results, location-based selection algorithm and random selection algorithm are jointly compared. The second proposed algorithm possesses the highest total sum-rate and is the optimal algorithms among the four algorithms.

1. Introduction

Nowadays, 5G has become a popular technique for its high bit rate and higher system capacity in order to accommodate the widespread use of smart devices and the qualitative change in association with the Internet of Things (IoT). However, because of the great number of users waiting for signal communication with a Base Station (BS), the resource a BS can provide to each user will be extremely limited in a multi-users Massive MIMO system. Therefore, it is important to distribute data rate effectively.

In that case, we need to find a solution to select and schedule users to receive the signal from the BS. Those users who have better communication channels or higher data rate should be given more power. To achieve this purpose, sum-rate based user selection are considered. This is related to the beamforming technique in Massive MIMO system [1-3].

When a base station has a large number of antennas, the phase of each received signals by the base station can be adjusted to form a superimposed signal, which usually means each transmitted antennas in this massive MIMO system need lower power to transmit signals. Beamforming is an important technique, by performing different beams points to large number of users, the interference between users and beams can be effectively reduced.

However, as everyone knows, there are no systems with no resources limitation exists in the real world, the massive MIMO system have very limited resources compare to the huge amount of users it need to support, it just do not have enough antennas to handle tens of thousands communication requirements between users, to make the overall performance of this massive MIMO system better, we have to make a choice, we select those users have better communication with the base station(BS) and give them more power [4].

This paper is more focused on the downlink user selection. Which means we are thinking a way to let the BS to think which user it needs to provide more power, pays less attention to those users have worse signal or have noisier channel, by doing the power allocation control, this system can run more efficient and have much better performance.

By saying the word better performance, we need a standard to see what we are talking about exactly, we are doing the sum rate analysis of the massive MIMO system, we conduct quantitative studies on the sum-rate performance of the massive MIMO system has limited antennas.

To achieve higher efficiency, we choose to optimize the plan of user selection, so we focus on downlink situation. First we need to consider about different users in one same channel. Several users exist in one channel will cause interference and higher path loss during signals transition. A method of user selection based on zero-forcing beamforming is provided in [4]. Their method is useful to be an alternative choice of Dirty Paper(DP)-based schemes. For less numeration, we choose this method as part of our project. It proposes a user selection algorithm considering about signal-to-leakage-and-noise-ratio(SLNR) and derived sum-rate expression with contrast low-complexity. Also, simulation results show the advantage of this special method compared with existed algorithm. A two-step method of user selection has been proposed in [5]. First, cut the high-frequent users and low-quality users. Second, use channel estimation to make an order of the other users. A special algorithm used in Massive MIMO and OFDMA system is mentioned in [6], which will maximum the total sum-rate of all sub carriers. In practical situations, there are more users than antenna arrays. Thus, dealing with the numeration of choosing users online becomes a big problem for base station.

Solving the problem of user selection, our method's purpose is to provide a better power allocation. Assumed the system is working in time division duplex(TDD) mode with beamforming training(BT) method to transmit downlink pilots, it is possible to achieve higher power rate with different power allocation on different users [7]. Then, a deep understanding of this kind of user selection is provided in [8]. It uses machine learning to learn the connection between the positions of user equipments (UEs) and better power allocation. The special thing of this method is that it does not need compute any statistical average, which means higher accuracy because of the base station is able to collect accurate data of UEs. We learn the method of how to compute the distance between base station and UEs from [9]. This method helps our method to reduce a large amount of work need to measure the distance and angles between base station and UEs. With all the algorithms mentioned before, thus, our method could help the process of forming a better power allocation at base station with all numerations finished online. The result of code simulation proves the feasibility of our method.

Based on the result appears after our analysis, we have a more straight forward looking on the system design and performance. We are using two different user selections algorithm compared to the random selection to see which one is better, as the selected number of users increased, the performances of three algorithms are changing, one could be better when there are less users, but as the number of users increasing, the performance is surpassed by another algorithm.

The rest of this paper is organized as followed. In section II, the system model including a transmission model and a channel model will be presented. Section III proposes two user selection algorithms. And the following section IV and V are the results analysis and conclusion of this paper.

2. System Model

This section introduced the transmission model in massive MIMO downlink and the channel model based on location.

2.1. Transmission model

In this paper, we consider a single cell multi-user downlink system. The single cell system model is a Massive MIMO system with beamforming technology. Consider our main application area is IoT (internet of things), generally a great number of different devices and users should be taken into account. However, the resources that a BS (Base Station) can provide to users are limited. The BS need to make a selection and schedule (not presented in this paper) to serve the users.

A downlink user selection and a scheduling are executed in the BS by choosing the users possessing the highest data rate. And the BS will send signal to the selected users in sequence. Then, the overall UE (User Experience) can be better.

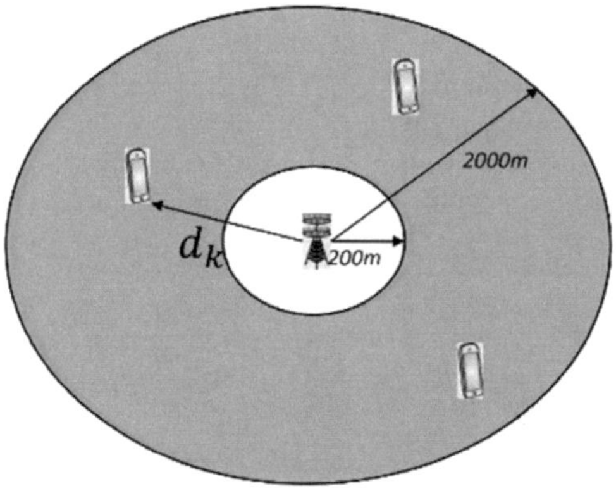

Fig 1. Single cell transmission model

We assume the range of BS coverage to be 2000 meters, users are randomly distributed in the circle like Fig.1.

The pass loss of the channel depends on the distance between users and the BS. In this paper, the pass loss coefficient is set to be 3.4. The wavelength is set to be 108 mm as is like in 5G communication network.

The received signal of each user at BS is described as followed:

$$y_{D,k}(n) = \mathrm{h}_k^*\left(\sum_{l=1}^{K} v_l\sqrt{P_{D,l}}x_{D,l}(n) + w_k(n)\right). \quad (1)$$

In this formula, $x_{D,l}(n)$ is the unit power data symbol transmitted by UE-l in downlink, $P_{D,l}$ is the allocated transmit power by BS for UE-l.

2.2. Notation

Table 1. Symbols explanation

h_k^*	Complex conjugate transpose of h_k
M	Total number of antennas of BS
U	Total user set
K	Number of all users
rho	path loss coefficient
σ_w^2	Noise factor
N	Number of selected users

Table 1 shows the explanation of the notations and symbols used in this paper; these notations will appear in the equations in the following sections.

2.3. Channel model

The channel model is built based on the description in [10].

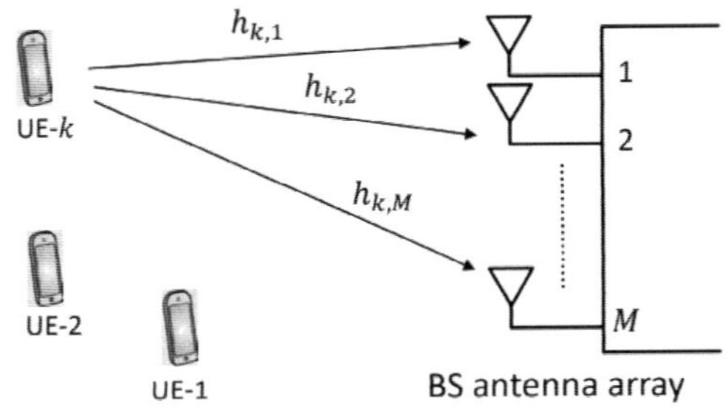

Fig 2. A typical relationship between UE-k and BS antenna array

An example of channel is depicted in Fig.2. $\boldsymbol{h_k}$ is the channel vector between UE-k and BS antenna array. It is given as

$$\begin{aligned}\boldsymbol{h_k} &= \left[h_{k,1}, h_{k,2}, \dots, h_{k,M}\right]^T \\ &= \beta_k[1, e^{j\pi \sin(\theta_k)}, \dots, e^{j\pi(M-1)\sin(\theta_k)}]^T,\end{aligned} \qquad (2)$$

where θ_k is the angle from the UE-k to the BS. It is a random value from 0 to 2π. And β_k is the pass loss between UE-k and BS. It is given as

$$\beta_k = \sqrt{\frac{1}{2} {d_k}^{-rho}}, \qquad (3)$$

where d_k is the distance between the UE-k and BS. It is a random value from 200 to 2000. And rho is the pass-loss exponent.

A MIMO channel matrix H with size M×K between K UEs and BS antenna is given as

$$\boldsymbol{H} = [\boldsymbol{h_1}, \boldsymbol{h_2}, \dots, \boldsymbol{h_K}]^T. \qquad (4)$$

Therefore, we can determine the location of the user based on the angle and distance information, and then indicate its channel.

3. User Rate Calculation for MRC Beamforming Scheme and User Selection Schemes

In this section, an analytical user-rate expression of MRC Beamforming is derived. This expression enables user selection with the direct user-rate criterion. Then, two user-rate-based user selection schemes is proposed for the massive MIMO downlink.

3.1. MRC Beamforming Scheme

The pre-coding schemes for downlink transmission is MRC (Maximal Ratio Combining) beamforming scheme which aims at maximizing the gain towards the selected user.

According to [10], the beamformer v_k satisfies the following conditions

$$Maximum \quad |h_k^* v_k|^2 \tag{5}$$

$$\|v_k\| = 1 \tag{6}$$

Thus, the beamforming vector can be expressed as:

$$v_k = \frac{h_k}{\|h_k\|} \tag{7}$$

This method is to maximize the gain of the wanted signal without taking care of the inter-user interference. The inter-user interference can be found in the equation of SINR:

$$\gamma_{D,k}^{MRC} = \frac{P_{D,k}\|h_k\|^2}{h_k^*(\sum_{l=1,l\neq k}^{K} P_{D,l} v_l v_l^*) h_k + \sigma_w^2}, \tag{8}$$

where σ_w^2 is the noise power assumed to be fixed and $P_{D,k}$ is the transmit power for User k with the total power constraint $\sum_{i=1}^{K} P_{D,k} \leq P_D$. In this paper, we assume the power allocation among the users will be equal. That is, $P_{D,k} = P_D/K$.

The ergodic rate of User k is given by [11]

$$R_k = E\left\{\log_2\left(1 + \gamma_{D,k}^{MRC}\right)\right\}, \tag{9}$$

and the ergodic sum-rate is

$$R_{sum} = \sum_1^K R_k\ . \tag{10}$$

3.2. User Selection Schemes

In this subsection, two user selection algorithms are proposed. The first one is User-Rate Based User Selection Addition Algorithm. The second one is User-Rate Based User Selection Subtraction Algorithm. Both algorithms select users with higher data rate.

In Algorithm 1, the users with higher data rate will be selected one by one to form the user set. And a new beamforming will be executed after the whole selection to derive the real user-rate set which eliminate the interference of unselected users.

In Algorithm 2, the users with lower data rate will be selected and move out the user set. Beamforming will be executed each time a user is selected to eliminate the interference of unselected users.

The difference between these two algorithms is whether to do beamforming while selection or to do beamforming after the selection. The selected user-rate results of these two algorithms are what we are truly interested in so as to enable the users to derive the highest user-rate.

Algorithm 1 User Rate Based User Selection Addition Algorithm

1) Initialization:

- Set n = 0 and $S_0 = \emptyset$.

2) While n < N:

- Increase n by 1.
- Find the user with index s_n, such that

$$s_n = \arg \max_{u \in U \setminus S_{n-1}} R_{MRC}(S_{n-1} \cup \{u\}) \tag{11}$$

- Set

$$S_n = S_{n-1} \cup \{s_n\} \tag{12}$$

3) Re-beamforming: $v_k \rightarrow v_n$

4) New user-rate set: $\{R_{MRC}\}$ generation based on v_n

5) Sum-rate:

$$R_{sum} = \sum_{1}^{N} R_n \tag{13}$$

Algorithm 2 User Rate Based User Selection Subtraction Algorithm

1) Initialization:

- Set n=0 and $S_0 = \emptyset$.

2) **While n** <K-N:

- Increase n by 1.
- Find a user s_n, such that

$$s_n = \arg \min_{u \in U \setminus S_{n-1}} R_{MRC}(S_{n-1} \cup \{u\}) \tag{14}$$

- Set

$$S_n = \complement_U\{s_n\} \tag{15}$$

- Re-beamforming: $v_k \rightarrow v_n$
- Derive new user-rate set $\{R_{MRC}\}$ based on v_n.

3) Sum-rate:

$$R_{sum} = \sum_{1}^{N} R_n \tag{16}$$

4. Result

In this paper, we got two pictures through MATLAB. In each picture, we evaluate the two algorithms mentioned in section III. The first one is the user-rate based user selection addition algorithm, which will run beamforming after the whole selection. The second one is the user-rate based user selection subtraction algorithm, which will run beamforming while the selection. Fig. 3 visually shows the average sum of the data rates of 10 users from 50 users selected by these two algorithms, and in order to show their fairness, we introduce Fig. 4 to show the average data rate of individual users selected out.

In order to make a comparison with our algorithm and show its efficiency, now there are two new user selection methods.

The first method is to randomly select users. For example, we have 50 users. This algorithm does not consider the relationship between users and BS antennas, and selects 10 users completely at random. The purpose of this group is mainly to make a basic comparison.

According to [12], the second method is to select users based on their channel. Unlike the previous screening method based on user data rate, this method is based on the user's channel to filter, and our channel is defined based on location information. Therefore, this algorithm also has a lot to do with location.

We will screen out users with higher channel quality. For example, the algorithm found the kth user, s_k.

$$s_k = argmax\ \boldsymbol{h_k}\boldsymbol{h_k}^* \tag{17}$$

Then select users based on higher s_k, so that we can pick the highest channel quality ten users. After beamforming, we will show the rates of these users.

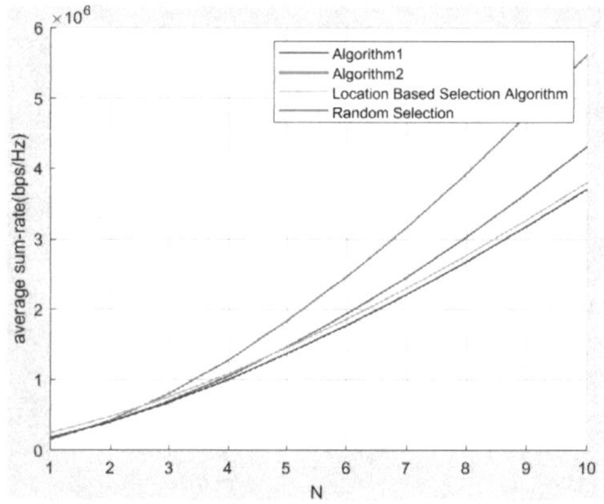

Fig 3. Average sum-rate of users selected by four algorithms

Fig.3 compares the average sum-rates of 10 users selected by four different algorithms. On this figure, we can clearly draw the conclusions that the both algorithm1 and algorithm2 can select users with higher user rates than the random selection.

At the same time, although the data rates of users selected by channel are higher than those of randomly selected users, they are much lower than those of users selected by algorithm1 and algorithm2.

More importantly, User-Rate Based User Selection Subtraction Algorithm can select users with higher user rates than the User-Rate Based User Selection Addition Algorithm. This is a good demonstration of the efficiency of our algorithm.

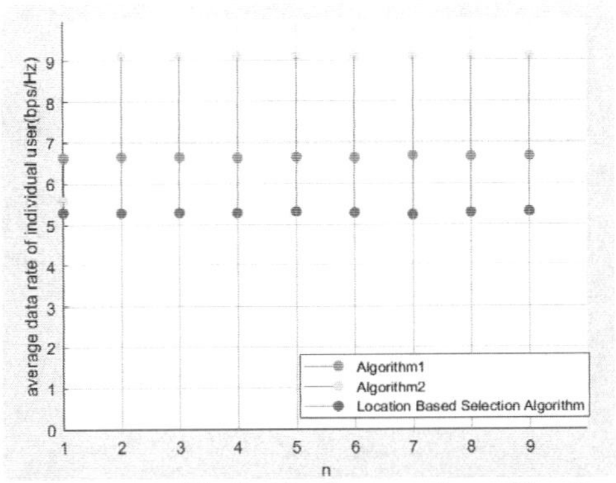

Fig 4. Average data rate of individual user selected by three algorithms

Fig.4 illustrates the general fairness of three algorithms. We averaged the data rate of users selected by each algorithm. On this figure, it can be clearly shown that the data rate of each user is not much different. It can be seen from the first point of Algorithm 2 that although the user data rate selected by Algorithm 2 is higher, it is not as stable as Algorithm 1. This result proves that the choice of Algorithm 1 is fair.

5. Conclusion

In this paper, the computer simulation proposes a new channel model and user selection scheme for Massive MIMO Downlink with MRC Beamforming and User Selection. The algorithm is aimed at maximizing the sum rate of served users, which improves user experience in getting better Internet quality. This paper analyzes and examines the performance of all proposed solutions by comparing the data rate of selected users.

The results show that the user selection algorithm based on user rate is better than the user selection algorithm based on channel and the random selection algorithm. Moreover, User-Rate Based User Selection Subtraction Algorithm is better than the User-Rate Based User Selection Addition Algorithm. Finally, through averaging the data rate of individual user selected by three algorithms, the fairness of Algorithm 1 has been proved. In the future work, user selection in complicate environment such as multi-cell systems and user fairness should be further studied.

References

[1] Rusek, F., Persson, D., Buon Kiong Lau, Larsson, E.G., Marzetta, T.L., Edfors, O., Tufvesson, F. (2013) Scaling Up MIMO: Opportunities and Challenges with Very Large Arrays. IEEE Signal Processing Magazine, 30: 40-60.

[2] Suyama, S., Jiyun Shen, Suzuki, H., Fukawa, K., Okumura, Y. (2014) Evaluation of 30 Gbps super high bit rate mobile communications using channel data in 11 GHz band 24×24 MIMO experiment. In: 2014 IEEE International Conference on Communications (ICC). New York. pp. 5203-5208.

[3] Osseiran, A., Monserrat, J. F., Marschh, P. (2016) 5G Mobile and wireless Communication Technology. Cambridge University Press, Cambridge.

[4] Dimic, G., Sidiropoulos, N.D. (2005) On downlink beamforming with greedy user selection: performance analysis and a simple new algorithm. IEEE Transactions on Signal Processing, 53: 3857-3868.

[5] Nonaka, Nobuhide., Muraoka, Kazushi., Suyama, Satoshi., Okumura, Yukihiko. (2019) Two-Step User Selection Algorithm for Multi-User Massive MIMO with Hybrid Beamforming.

In: 2019 22nd International Symposium on Wireless Personal Multimedia Communications (WPMC). Lisbon. pp. 1-5
[6] Bogale, T.E., Le, L.B., Haghighat, A. (2015) User Scheduling for Massive MIMO OFDMA Systems with Hybrid Analog-Digital Beamforming. In: IEEE International Conference on Communications. London. pp. 1757-1762.
[7] Izadi, A., Razavizadeh, S. M., Saatlou, O.(2020) Power Allocation for Downlink Training in Cell-Free Massive MIMO Networks. In: 2020 10th International Symposium on Telecommunications. London. pp. 111-115
[8] Sanguinetti, L., Zappone, A., Debbah, M. (2019) Deep Learning Power Allocation in Massive MIMO. Dipatimento di Ingegneria dell/Informazione, Pisa.
[9] Muppirisetty, L. S., Wymeersch, H., Karout, J., Fodor, G. (2015) Location-Aided Pilot Contamination Elimination for Massive MIMO Systems. In: 2015 IEEE Global Communications Conference. Gothenburg. pp. 1-5.
[10] Kang, Y., Lu, T. (2019) Channel estimation and data transmission in massive MIMO TDD systems. International Journal of Communications, Network and System Scienes, 12: 170-187.
[11] Zhang, C., Huang, Y., Jin, S., Yang, L., Jing, Y. (2017) Sum-rate analysis for massive MIMO downlink with joint statistical beamforming and user scheduling. IEEE Trans. Wireless Communications, 16: 2181-2194.
[12] Dimic, G.; Sidiropoulos, N.D. (2005) On Downlink Beamforming With Greedy User Selection: Performance Analysis and a Simple New Algorithm. IEEE Transactions on Signal Processing, 53: 3857-3868.

Application in photovoltaic MPPT based on improved hysteresis loop comparison method

Liqun Shang[1], Fan Li[1*]

[1] School of Electrical and Control Engineering, Xi'an University of Science and Technology, Xi'an, Shaanxi, 710054, China

*Corresponding author's e-mail: dkyjs@xust.edu.cn

Abstract. In order to overcome the misjudgement and oscillation of the traditional perturbation observation method when tracking the maximum power point of the PV system, this paper proposes a fixed-zone variable-step hysteresis comparison algorithm. The bi-directional perturbation of the hysteresis loop can effectively avoid the occurrence of misjudgement; after setting the partition voltage threshold, different step lengths can be used to track specific regions, which can ensure the tracking speed and maintain the system stability. Finally, the proposed algorithm is validated in Matlab/simulink and the results show that the method is effective in tracking the maximum power point of the PV system in all aspects compared to conventional algorithms.

1. Introduction

With the gradual maturity of science and technology, people's demand for energy is gradually transforming from traditional fossil fuels to new sources of clean, safe, efficient and renewable energy. With the background of the times and government support, solar energy has become mainstream in the power generation industry. However, solar power plants require high upfront investment and the efficiency of solar power generation is easily affected by the external environment. In order to keep the photovoltaic system working efficiently at all times, researchers have proposed the maximum power tracking technique (MPPT) for photovoltaic systems[1]. MPPT algorithms with different tracking accuracies and development costs are used in different applications, the most commonly used being Perturbation and Observation (P&O) and Incremental Conductivity (INC). Both of these algorithms track the maximum power point by adding a small amount of perturbation to the system and are simple to implement. However, the need to continuously apply perturbations results in a system that is constantly in a state of dynamic change, and the oscillations caused by this dynamic change are very detrimental to the stability of the system. Therefore, the choice of the size of the added perturbation becomes particularly important[2]. In addition, when the operating environment of a PV system changes, the tracking of traditional algorithms can lead to misjudgements, causing the tracking point to deviate infinitely from the actual maximum power point, thus reducing the energy conversion efficiency[3]. In this paper, we analyse the shortcomings of the traditional P&O algorithm and propose a fixed-zone variable-step hysteresis comparison algorithm. The unidirectional perturbation of the traditional P&O algorithm is replaced by a bidirectional one to avoid the problem of misjudgement due to environmental changes; then different perturbation steps are selected according to the partitioning of the voltage power curve, so that the system can maintain maximum system stability while ensuring rapidity.

2. Photovoltaic cell model and output characterisation

Photovoltaic cells are able to generate electricity due to the photovoltaic effect[4]. When light is shone on the surface of a photovoltaic panel, a potential difference is created by the directional movement of electron and hole pairs within the semiconductor material, which can generate an electric current if an external circuit is connected. The individual photovoltaic cell model is usually represented in Figure 1.

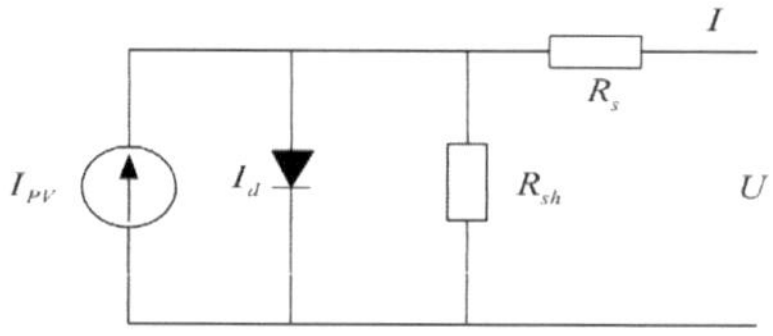

Figure 1. Photovoltaic cell equivalent model circuit diagram.

Applying Kirchhoff's law in Figure 1 yields the PV cell output current-voltage relationship:

$$I = I_{ph} - I_d \left[\exp\left(\frac{q(U + IR_S)}{KAT} \right) - 1 \right] - \frac{U + IR_s}{R_{sh}} \tag{1}$$

Where I_{ph} is the current flowing from the photogenerated current source and its value magnitude is related to the intensity of the light; I_d is the current flowing on the diode; A is the diode quality factor; T is the absolute temperature; and K is the Boltzmann constant.

To obtain the output characteristic curve of the PV cell, the PV cell is modelled in the simulation conditions of Matlab/simulink. The main parameters of the PV panel are: I_{SC}=6.83A, U_{OC}=43.92V, I_m=5.69A, U_m=35.13V. Figure 1(a) and Figure 1(b) show the output characteristics of the PV cell with constant temperature and variable light at 25 ℃. Figure 1(c) and Figure 1(d) show the output characteristics of the PV cell when the light intensity is kept constant at 1000W/m^2 and the temperature is changed.

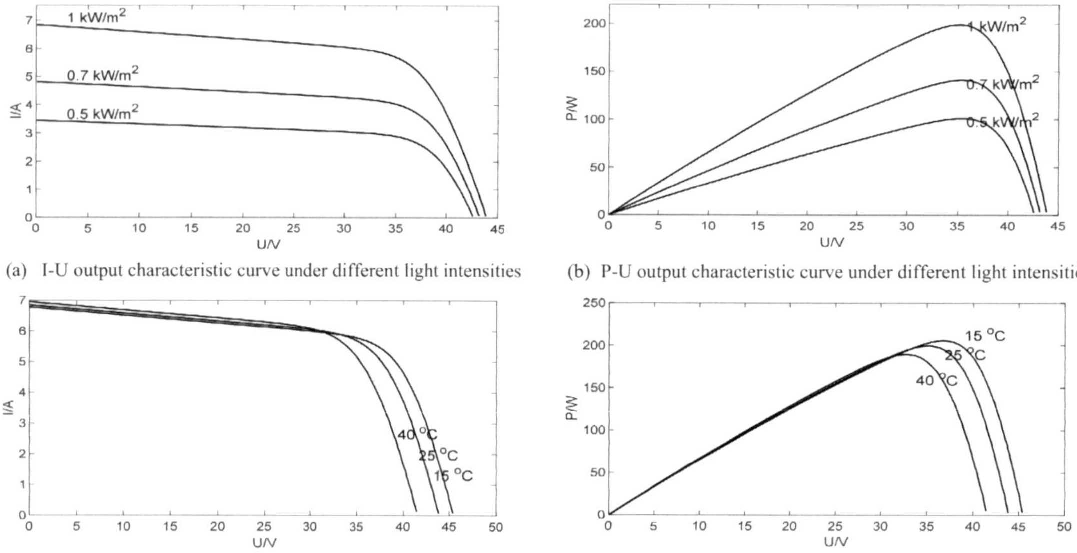

(a) I-U output characteristic curve under different light intensities (b) P-U output characteristic curve under different light intensities

(c) I-U characteristic output curve at different temperatures (d) P-U output characteristic curve at different temperatures

Figure 2. Photovoltaic cell output characteristic curve.

Analysis of the output of the PV cell during the four environmental changes in Figure 2 shows that: when the ambient temperature is a fixed value, the maximum power point of the PV system will increase with the increase in light intensity; when the light is fixed, the position of the maximum power will gradually decrease with the increase in temperature. The output characteristics of a PV cell will change to varying degrees following changes in the external environment, resulting in a maximum power point that cannot be fixed at a standard value. However, as can be seen in the P-U output characteristic curve, the maximum value is uniquely present on a separate curve. Therefore, the use of maximum power tracking technology can maintain the output power of the PV system near the maximum value at all times in a dynamically changing environment, avoiding the consequences of low efficient output with high economic input.

3. Traditional P&O algorithm and their problems

The P&O algorithm is one of the most frequently used algorithms in MPPT for photovoltaic power generation[5]. It has few objects to control and a simple tracking process. It does not require too many calculations, but simply adds a certain step of perturbation to the detected PV cell output voltage periodically. The power difference is then calculated before and after the cycle to determine the direction of the next perturbation until the power output of the system reaches its maximum. The voltage change pattern during the perturbation process is shown in (2).

$$V_{k+1} = V_k + \mathrm{sgn}\left(\frac{dP}{dV}\right)\Delta V \tag{2}$$

Where dP is the amount of change in power, dV is the amount of change in voltage and ΔV is the perturbation voltage step added. It can be seen that a forward perturbation needs to be added when the amount of change in power and voltage are of the same sign, and when the sign of change in both is different, the perturbation is made to proceed in the opposite direction.

When using a fixed step size for power tracking of PV systems, the P&O algorithm is unable to meet both high accuracy and fast speed due to the choice of step size. If the step size is chosen to be large, the system can quickly reach near the maximum power point, but the system will operate with more serious power oscillations; if a small voltage is used for the step size, the oscillation problem is suppressed, but the tracking speed will be much slower. At the same time, the photovoltaic system is always in a changing environment, and according to the output characteristics curve of the photovoltaic cell, the maximum power point will change as the output curve moves. If the tracking rule of the traditional disturbance observation method is still used to judge the direction of the disturbance addition when the curve is shifted, the system is likely to gradually move away from the real maximum power point due to misjudgement[6], thus causing a large amount of energy loss.

4. The fixed-zone variable-step hysteresis comparison algorithm

In order to solve the problem of possible misjudgement of the traditional P&O algorithm when the environment changes, a hysteresis loop comparison is introduced on the basis of the basic disturbance. The hysteresis loop comparison method[7] adds a two-way power difference comparison to the control principle so for the system to make a judgement on whether the external environment is changing, the basic control logic is:

Let the starting point of the perturbation be point A. Take a point B and a point C in equal steps before and after this point, and record the power values P_A, P_B and P_C at the three points. compare the power values at points B and C with those at point A. When P_A>P_C, the symbol is "+", when P_B>P_A, the symbol is "+ ", in the rest of the cases the symbol is recorded as "-". A common situation is shown in Figure 3.

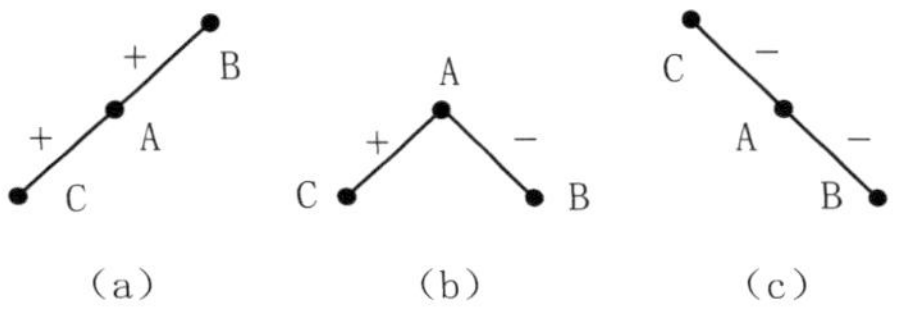

Figure 3. Three-point power comparison diagram.

After two perturbations the power comparison results are obtained: if the situation in Figure 3 (a) is positive for both perturbations, the voltage is on the left side of the maximum power point and needs to be perturbed further; if the situation in Figure (c) is positive, the voltage is already on the right side of the maximum power point and the perturbation direction needs to be changed; if the situation in Figure (b) is positive, the voltage is already running at the maximum power point If the situation shown in (b) means that the voltage is already running near the maximum power point or that the external environment is changing, it is sufficient to keep the voltage perturbation situation unchanged.

The hysteresis loop comparison method can be used to determine whether to continue adding perturbations to the system by means of symbols, for the purpose of avoiding misjudgements. However, the hysteresis loop comparison method has the same problem of step size selection as the P&O algorithm. In order to accelerate convergence while providing system operation accuracy, the concept of fixed partitioning with variable step size is proposed. Figure 4 shows the partitioning of the power-voltage output curve of the PV cell.

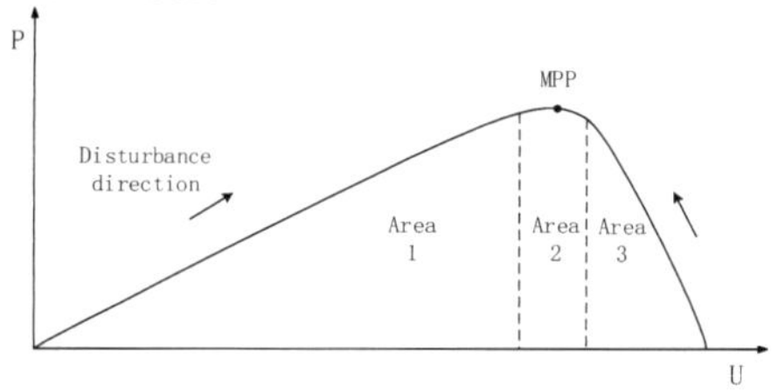

Figure 4. Partition diagram of photovoltaic cell output curve.

In Figure 4 the power-voltage curve is divided into three zones. zones 1 and 3 are farther away from the MPP and the step size should be chosen to match the need for accelerated convergence, and the slope of the curve in zone 3 is The slope of the curve in zone 3 is greater, as illustrated in [8], and the slope of the curve in zone 3 is 4 times that of zone 1, so zone 3 should be higher than zone 1 in the choice of step size. In zone 2, the power has gradually converged around the MPP and the step size should be reduced to reduce the power oscillation back and forth during operation and improve accuracy. In order to minimise the oscillations around the MPP, set Minimum power difference. When the power difference obtained from the first two perturbations meets the specified value, the system will judge that the power is already at its maximum value, thus outputting the current voltage, the basic flow chart of the algorithm is shown in Figure 5.

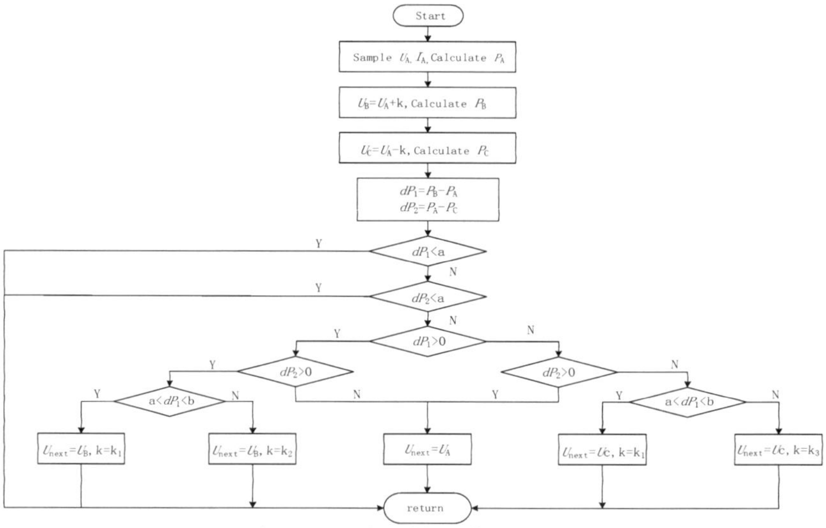

Figure 5. Algorithm flow chart.

5. Analysis of simulation results

In order to verify the effectiveness of the fixed-zone variable-step hysteresis comparison algorithm for tracking in PV MPPT, the conventional P&O algorithm was compared with it by simulation. The simulation model was built in Matlab/simulink based on the system diagram shown in Figure 6, where the model parameters in the PV panel are shown in Table 1. The Boost converter is used to adjust the load resistance to match the internal resistance of the PV panel, so as to output the maximum power. The parameters of the Boost circuit are: C_1=0.1mF, C_2=0.1mF, L=0.01mH and R=10Ω.

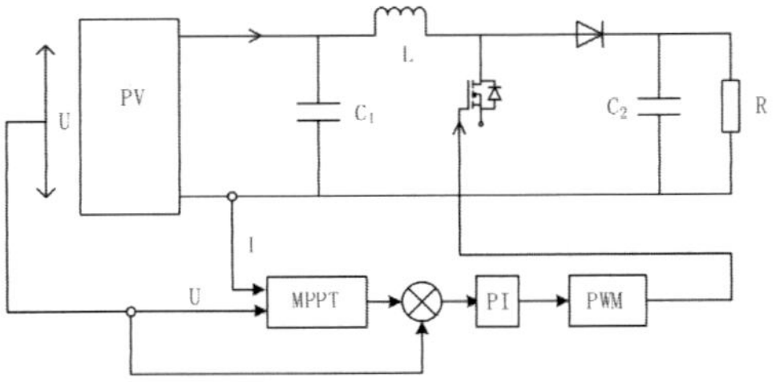

Figure 6. Photovoltaic power generation system diagram.

The simulation time was set to 0.2s, the initial environment was 25℃, and the light intensity was 700W/m^2. The step size of the perturbation observation method was chosen to be 0.5V after combining rapidity and stability, and the light intensity suddenly changed to 1000W/m^2 at 0.1s. The simulation output waveforms of the PV cell for the perturbation observation method and the hysteresis comparison method with constant partitioning and variable step size are shown in Fig. 7(a) and Fig. 7(b) respectively.

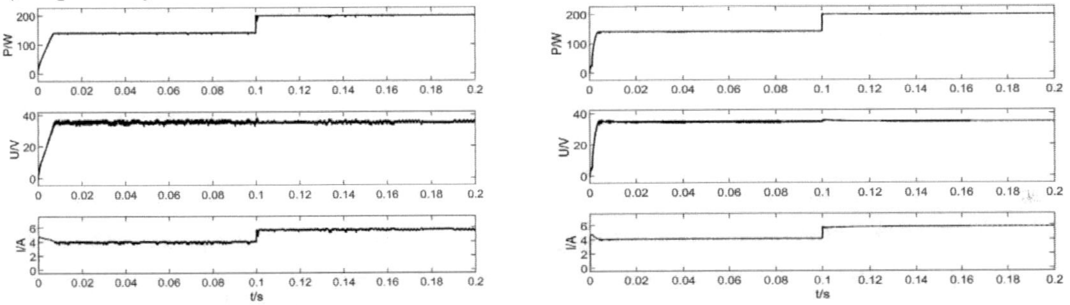

(a) the disturbance observation method (b) the fixed-zone variable-step hysteresis comparison algorithm

Figure 7. Photovoltaic cell simulation output waveform.

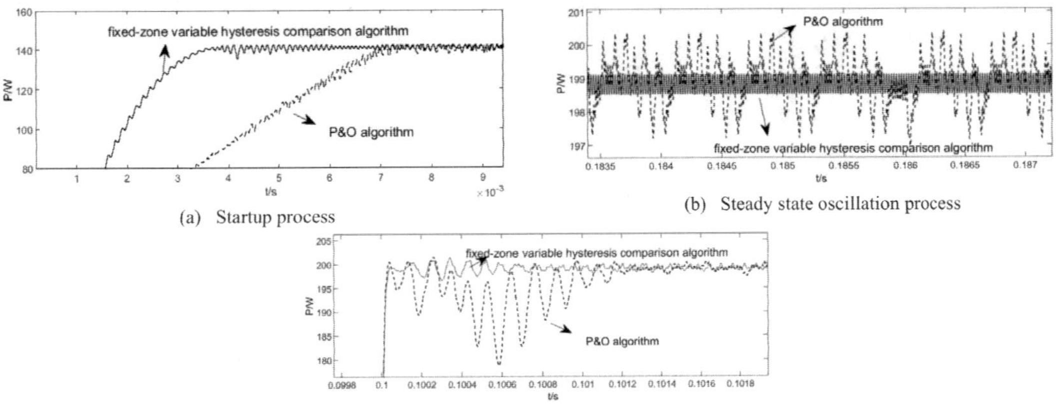

(a) Startup process

(b) Steady state oscillation process

(c) Light intensity change process

Figure 8. Simulation local details.

From Figure 7 (a) simulation results can be seen, the traditional P&O algorithm because of the fixed value of the step, it is difficult to meet the growth rate of the voltage in different stages, in the tracking system maximum power point when the output power of the volt battery produced a significant oscillation. The simulation waveforms of the fixed-zone variable-step hysteresis comparison algorithm in Figure 7(b) can all see that the voltage grows at a rate up to the maximum power point in the pre-start period, and the power output is smoother, and the voltage and current waveforms of the PV cell output have been improved. Compared to the P&O algorithm, the improved algorithm not only has improved dynamic response capability, but also has very good stability performance. Figure 8 shows the comparative local details of the two algorithms. In Figure 8(a) it can be seen that the fixed-zone variable-step hysteresis comparison algorithm uses a larger step size in the first stage of tracking, so the tracking speed is improved by a factor of 1 compared to the P&O algorithm. Figure 8(b) shows a graph of the oscillation waveform of the algorithm at the maximum power point attachment. The improved algorithm has reduced the oscillation rate to a very low level and the oscillation step is present to continue tracking the maximum power point that may change under sudden environmental changes. As can be seen in Figure 8(c), the power tracked by the P&O method is shifted more when the light amplitude is changed and the energy loss is more severe. The improved algorithm stabilises the voltage within the correct perturbation range because of the hysteresis loop, ensuring that the system can track quickly and accurately to the new maximum tracking point. Experiments have demonstrated that the method is effective in overcoming misjudgements when tracking the maximum power point of the PV system, with fast and accurate tracking and also stability, resulting in a steady increase in the energy utilisation of PV power.

6. Conclusions

In this paper, after analysing the problems of the traditional perturbation observation method, we propose a fixed-zone variable-step hysteresis comparison algorithm. The power-voltage curve of the PV cell is partitioned so that a large step size can be used to speed up the tracking speed at a distance from the maximum power point, while a small step size can be used nearer to the maximum power point to make the system output more stable. Simulation results demonstrate the effectiveness of the method in PV maximum power point tracking.

References

[1] Fu Z, Zhang Z. (2019) A review of maximun power point tracking techniques for photovoltaic system. Chinese Journal of Power Sources, 43(12): 2067-2069+2073.
[2] Xu J, Shen J, Wang H, Li K, Chen K. (2018) Maximun power point tracking strategy based on novel variable step-size incremental conductance algorithm. Renewable Energy Resources, 36(9): 1305-1313.
[3] Zhao H, Zhang S, Zhang G. (2020) Segmented adaptive variable step size maximum power tracking method. Chinese Journal of Power Sources, 44(09): 1326-1330.
[4] Bellia H, Youcef R, Fatima M. (2014) A detailed modeling of photovoltaic module using matlab. Nriag Journal of Astronomy and Geophysics, 3(1): 53-61.
[5] Wang H, Xiao Y, Xiao R, Chen X, Ding H. (2019) Photovoltaic MPPT based on improved voltage perturbation method. Chinese Journal of Power Sources, 43(11): 1847-1851.
[6] Zhu W, Shang L, Li P, Guo H. (2018) Modified hill climbing MPPT algorithm with reduced steady-state oscillation and improved tracking efficiency, 2018(17): 1878-1883.
[7] Liu S, Wang Y, Yang L. (2020) Research on application of switching boost converter in photovoltaic power generation system. Acta Energiae Solaris Sinica, 40(6): 272-277.
[8] Tu Y, Yu A, Xue H, Wang Y. (2018) Maximum power point tracking for photovoltaic systems based on exponential variable step size incremental conductance algorithm. Electric Power, 51(11): 132-138.

Solving Multiple Triggering Problem of Ultrasonic Distance Measurement by Debouncing

Zhenhua Xiang

Three Gorges University, Xiantao City, Hubei Province, Postcode: 443000, China

*Corresponding author's e-mail: Billy_xzh@163.com

Abstract. In recent years, multiple measurement systems fusion technologies have developed rapidly. Ultrasonic ranging is paid much attention because it has the advantages of directly measuring short-distance targets, high longitudinal resolution, and a wide application range. However, when the ultrasonic distance measurement is carried out by assembling the circuit board, it is found that the switch is triggered multiple times, and the measurement data is inaccurate. This paper proposes a new method that adding debounce to improve the accuracy of measurement. Besides, Some suggestions are provided to decrease the operation difficulty in the actual measurement process. Based on the experimental result, the accuracy and efficiency of ultrasonic distance measurement are greatly improved via this method proposed in this paper. By the way, the single-chip computer simulation technology simulates the ranging phenomenon in the experiment.

1. Introduction

With the rapid development of science and technology, ultrasound is widely used in rangefinders. However, as far as the current level of technology is concerned, the available ranging technology is still limited. The ultrasonic rangefinder will have a lot of room for development in all aspects. It will develop in the direction of higher positioning and high precision to meet the growing social needs. However, the original circuit drawings have multiple triggering problems [1][2].

This paper proposes to add a debounce circuit to solve this problem. To test the accuracy of this new method, we simulated the total circuit diagram in Proteus. Specifically, we divided the circuit diagram into the circuit diagram with debounce and the non-joined circuit diagram to run the program. The 74LS279 was selected to achieve the purpose of debouncing. Finally, we found that adding a de-jitter ranging system can effectively measure the distance accurately in a specific range.

In short, the efficiency of the system after adding de-jitter is dramatically improved. Besides, this system has satisfying software and hardware design, strong anti-interference ability, and good real-time performance.

2. Ultrasonic ranging

2.1. Ultrasonic ranging principle

The most commonly used ultrasonic ranging method is the echo detection method [3][4]. Its working principle is to make the transducer emit acoustic pulse to the medium. The reflected acoustic wave (echo) must act on the transducer when the acoustic wave meets the measured object (target). Suppose the acoustic velocity of the medium is known to be v and the time difference between the arrival time of the first echo and the time of the emission pulse is t. In that case, the distance between the transducer and the target can be calculated by the formula d = vt/2 [5].

2.2. Design block diagram of an ultrasonic ranging circuit

The block diagram of an ultrasonic ranging circuit is given in Fig.1. It comprises an oscillating circuit, a gate circuit, a HC-SR04, a debouncing switch, a reset switch, and three decoders [6][7]. Since the sound velocity is 340 m per second, the module measures the return and return distance. An oscillator circuit that can provide a stable 17 kHz is required. Firstly, the trig port of HC-SR04 is connected in parallel with a 1K resistor on the switch. The ultrasonic wave starts from the transmitter and returns after reaching the measured object, which is received by echo (receiver) and generates a square wave about distance. The ultrasonic signal from the module and the clock pulse signal provided by the oscillation circuit is received by the gate (74LS00), resulting in a new pulse signal received by the counter (74LS90). We added a switch in it to clear zero. The counting method is decimal; 74LS90 gets the signal and converts it into the distance. The counter further sends out the signal to be received by the decoder. The number is directly converted into the display number of the digital tube through decoder decoding. So you can see that the number on display is the measured distance. In centimeters, the module theoretical measurement distance is 4 meters to 4.5 meters. Since there are too many external interference factors in ultrasonic ranging, when the ultrasonic wave is emitted, the receiving sensor will be subject to strong interference signals, resulting in unstable experimental data measured each time. After research, the problem is that the switch is triggered repeatedly. To prevent the system's error, we add a debouncing circuit before the trigger switch to further improve the experiment to make the measurement data more accurate.

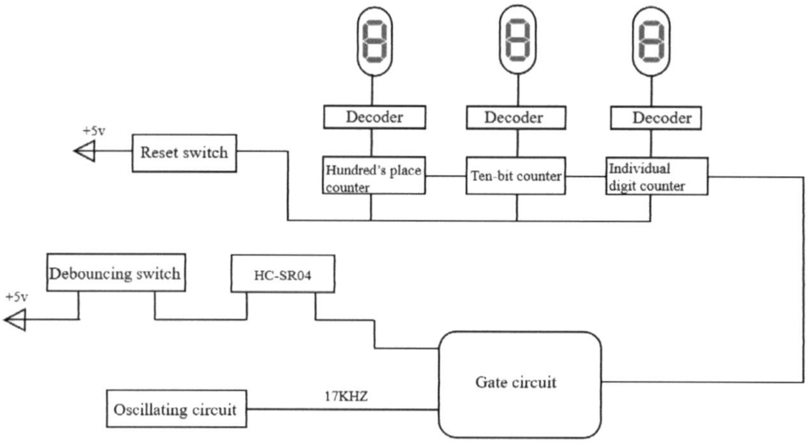

Fig. 1: block diagram of an ultrasonic ranging circuit

2.3. Simulation of the debouncer circuit

Proteus conducted the debouncer circuit simulations. The overall simulation model of the debouncer circuit can be seen in Fig.2.

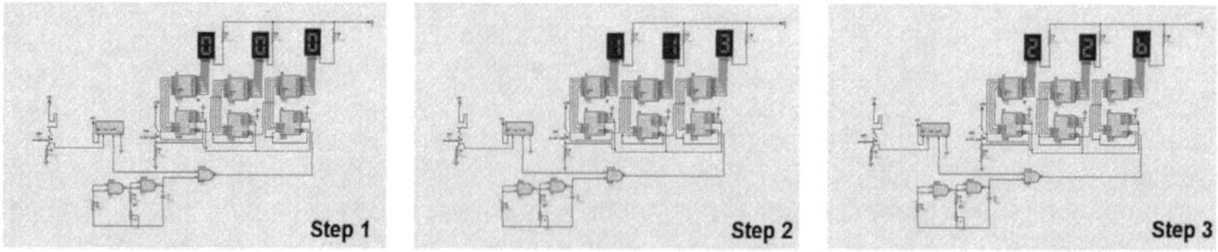

Fig. 2: The overall design drawing of the first generation circuit simulation

The simulation steps are provided in the following :

Step 1: build the whole circuit in Proteus.

Step 2: start the switch. Press the switch to perform the first circuit simulation experiment, and the digital tube shows the number is normal.

Step 3: measurement. Press the switch again, the experimental distance remains the same, but it is found that the number displayed by the digital tube is different from the first experimental data and has a multiple relationships

The simulation result shows that multiple runs in the case of a specific current and the given distance. The experimental data are inconsistent, and each measurement found that the experimental data are multiples of a certain distance. Then, it is speculated that there are many trigger switch phenomena because the switch is too sensitive, and we cannot improve it under artificial conditions. Therefore, we add a debouncing circuit to prevent this phenomenon from happening again.

2.4. Experimental test

The overall test method is as following statements: First, turn on the power supply, slide the knob, let CH2 emit 5.0 V voltage, access the breadboard, and press the output button to provide 5.0 V voltage to the breadboard. Then, the monitor is cleared by pressing the clearing switch, and the ultrasonic emission switch is pressed to detect whether the number on the monitor is consistent with the actual distance, and multiple experiments are carried out to eliminate the error. The detailed test steps are explained in the following. Fig.3 also gives the diagrams.

Step 1: turn on the power and connect to the breadboard;

Step 2: clear to zero;

Step 3: press the ultrasonic test switch, measure the selected position distance;

Step 4: measure the actual distance with a tape measure;

Step 5: error correction again, using the same method to measure a close distance.

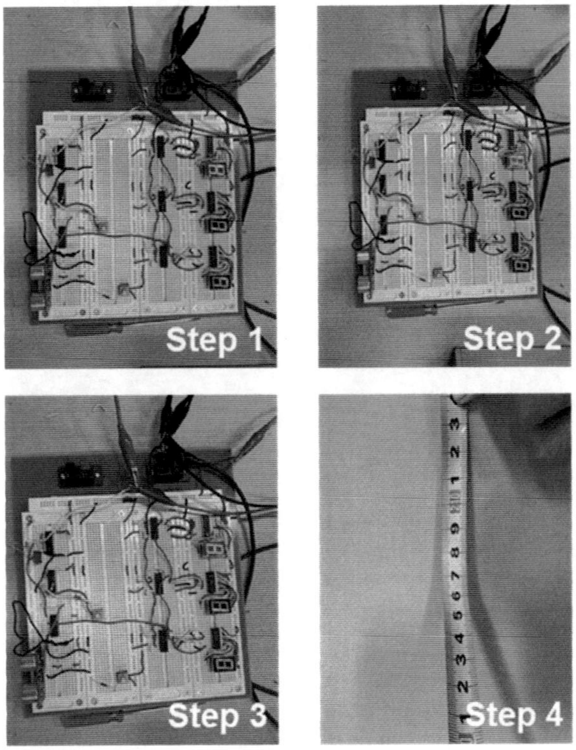

Fig. 3: The detailed test steps

The experimental result shows that there is not much error between actual distance and measured distance, no matter the exact distance is long or short.

3. Conclusion

In this ultrasonic-ranging experiment, it was found that the experimental data appeared many times. After many experiments, it was found that there was a problem of multiple triggering of the switch. Therefore, it is recommended to add a debounced switch on the original basis. To reduce the number of experimental failures, the experiment adopted a single-chip simulation. A de-jitter circuit is added on the premise of successful simulation, and the experiment can complete the goal. It shows that debouncing can effectively realize the problem of multiple trigger switches in the circuit, solve the interference of unnecessary factors, and provide reliability and accuracy for many aspects. Adding debounce is significant because the ultrasonic distance measuring device can effectively avoid physical problems such as unstable current, susceptible switches, and excessive temperature drop due to external factors. In this way, the setting can perform distance measurement well and provide convenience for people in different environments. In addition, according to the problems in the experiment, we summarize the following points.

1. HC-SR04 should be placed horizontally. Otherwise, the measurement distance will be too large.
2. Because the air's humidity, density, and sound velocity are not necessarily 340m / s, the frequency is not necessarily 17kHz; it needs to be adjusted.
3. Display wiring due to equipment defects, resulting in a poor display of part of the display strokes.
4. Be careful not to trigger many times when you press the startup switch.

References

[1] Gao Feiyan (2005) Design of Ultrasonic Ranging System Based on Single-chip Microcomputer Information Technology 128-129.
[2] Kang Huaguang （2014） Electronic Technology Fundamental Digital Part, Higher Education Press, Beijing.
[3] Qin Zenghuang (1990) Electrical Engineering (Fourth Edition) Volume 2 Higher Education Press, Beijing.
[4] Chen Daqin (1900)Electronic Technology Experimental Foundation-Electronic Circuit Experiment, Design, Simulation-(Second Edition) Department of Electronics, Huazhong University of Science and Technology (Editor) Higher Education Press, Beijing.
[5] Xie Zimei (2000) Electronic Circuit Design·Experiment·Test (Second Edition) Huazhong University of Science and Technology Press, Wuhan.
[6] Ji Peirong, She Xiaoli (Editor) (2016) Principles of Circuits China Electric Power Press, Beijing.
[7] Su Wei, Gong Bijian, Pan Xiao (2004) Error Analysis of Ultrasonic Ranging Sensor Technology Volume 23 Issue 6, 8-11.

Design of High Distance Diameter Ratio Wireless Transmission System Based On LCC-S Compensation Network

Aiwei Yang[1,a], Wei Xie[1], Tao Liu[1], Xiaobin He[1]

[1] State key laboratory of space power-sources technology, Shanghai Institute of Space Power-Sources, Shang Hai, China

[a]email: aiwei_yang@126.com

Abstract. A wireless power transmission system based on LCC-S compensation network is designed. The transmission system is composed of power frequency power supply, LCC-S compensation resonant converter and DC / DC converter, and the output power supply for the battery system. Through the design of transmission coil and compensation network parameters, zero current input is realized. A wireless transmission coil with outer diameter of 47cm and spacing of 30cm is designed. A 2kW wireless transmission system prototype is built. Under 311V input, the actual transmission efficiency of high frequency inverter / wireless transmission / high frequency rectifier reaches 88.5%, and the wireless energy transmission with high distance to diameter ratio is realized.

1. Introduction

Inductive wireless power transmission system is widely used in electric vehicle charging, mobile phone charging and commercial UAV charging because of its electrical isolation, convenience and reliability.

In order to improve the efficiency of wireless energy transmission, the transmitter and receiver of wireless power transmission system will use compensation circuit to achieve higher power and maximum efficiency of energy transmission. The four basic compensation networks of wireless power transmission system are S-S, S-P, P-P, P-S topology. In order to improve the transmission efficiency and input-output characteristics of the system, LCC and LCL compensation networks are proposed. Through the analysis of wireless power transmission system. Wireless charging coil has been a lot of research[1][2]. Through a lot of research on coil configuration, core shape and number, there are various coil designs to choose from in different application scenarios[3].

In this paper, in order to meet the demand of non-contact charging, a wireless power transmission system based on LCC-S compensation network is designed. This system is used for non-contact wireless charging at a long distance, using smaller volume and lighter weight coils to achieve high efficiency charging of high voltage lithium batteries. The parameters of LCC-S compensation network and a wireless charging coil with high distance to diameter ratio are designed and analyzed. A 2kW prototype is made, from 311V DC input to 280V DC output. The full load wireless transmission efficiency is 87.5%.

2. Design of wireless power transmission system

This system uses 220 V / 50 Hz AC power supply to 100 V Li-ion battery pack. The full bridge inverter with constant frequency and duty cycle is used in the primary side, and the uncontrolled rectifier circuit is used in the secondary side. In order to achieve faster and more stable voltage and current regulation,

DC / DC converter is connected in the back stage to realize constant voltage and current charging of battery.

For different compensation networks, the output of wireless power transmission system can be divided into constant voltage output and constant current output. Considering the demand, LCC-S compensation network with constant voltage output is adopted. The designed wireless power transmission system is shown in Figure 1. Among them, the AC rectification circuit is omitted, and Vin is the DC voltage after rectification.

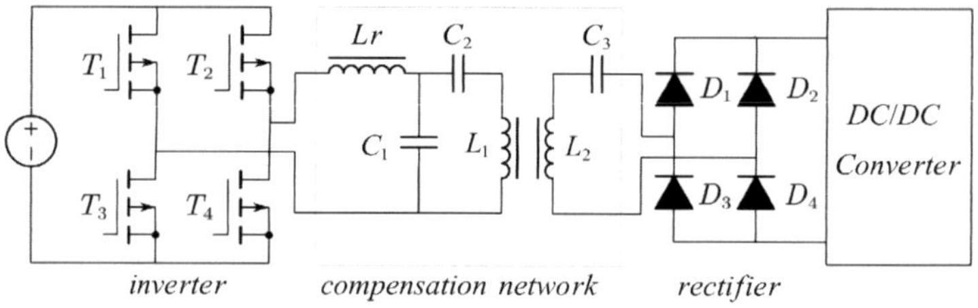

Figure 1 wireless power transmission system

2.1. Parameter design of LCC-S compensation network

The structure of LCC-S compensation network is shown in Figure 1. In order to achieve the highest efficiency of wireless energy transmission, the three resonant loops of this topology are Lr and C1; C1, C2 and L1, L2 and C3. The resonant frequencies of the three resonant circuits are the same[4]. Therefore, there are:

$$\begin{cases} j\omega L_r - j\dfrac{1}{\omega C_1} = 0 \\ j\omega L_1 - j\dfrac{1}{\omega C_1} - j\dfrac{1}{\omega C_2} = 0 \\ j\omega L_2 - j\dfrac{1}{\omega C_3} = 0 \end{cases} \tag{1}$$

The circuit topology is analyzed without considering the parasitic parameters of inductance and capacitance. The topological loop equation matrix is obtained as follows:

$$\begin{pmatrix} U_S \\ 0 \\ 0 \end{pmatrix} = \begin{pmatrix} j\omega L_r + \dfrac{1}{j\omega C_1} & -\dfrac{1}{j\omega C_1} & 0 \\ -\dfrac{1}{j\omega C_1} & j\omega L_1 + \dfrac{1}{j\omega C_1} + \dfrac{1}{j\omega C_2} & -j\omega M \\ 0 & -j\omega M & j\omega L_3 + \dfrac{1}{j\omega C_3} + R_L \end{pmatrix} \begin{pmatrix} I_S \\ I_1 \\ I_2 \end{pmatrix} \tag{2}$$

The output voltage can be obtained as follows:

$$U_{out} = \frac{M}{L_r} U_S \tag{3}$$

Obviously, the LCC-S compensation network is a constant voltage output topology, and the output voltage is related to the input voltage, inductance Lr and coil mutual inductance M. Similarly, the efficiency and output power of LCC-S topology system can be deduced as follows[5]:

$$\begin{cases} P = \dfrac{\left(\omega_0^2 L_1 M\right)^2 U_S^{\,2} R_L}{\left[Z_{in} Z_s \left(R_1 + Z_r\right)\right]} \\ \eta = \dfrac{\left(\omega_0^2 L_1 M\right)^2 R_L}{Z_{in} Z_s^{\,2} \left(R_p + Z_r\right)^2} \end{cases} \tag{4}$$

Here, we choose the parameters of compensation network as follows: resonant frequency fc = 120kHz, secondary side capacitance C2 = 10nF, secondary side inductance L2 = 178.9uH, primary side capacitance C1 = 10nF, C1 = 100nF, we can get primary side inductance L1 = 196.8uH, resonant inductance Lr = 17.89uH. At this time, the three resonant circuits of the compensation network work at the resonant frequency, and the transmission efficiency can reach the maximum. According to equation (4), the curve of the output power and transmission efficiency of wireless transmission network can be obtained with the load.

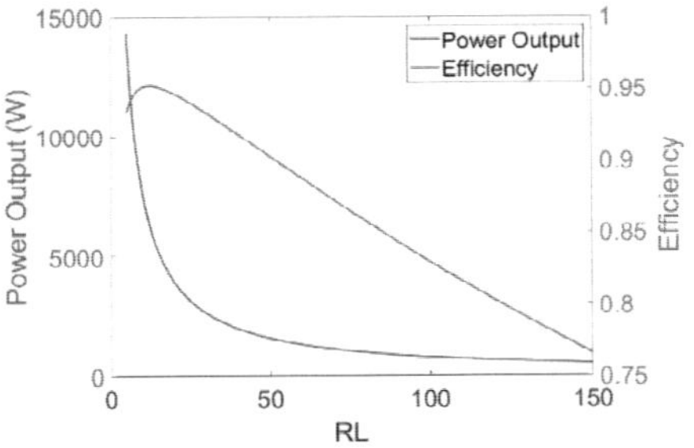

Fig. 2 power and efficiency curve with load

Fig. 3 coil 3D model

2.2. Design of wireless energy transmission coil

In order to realize wireless power transmission with high aspect ratio, the coil needs to be optimized. In practical applications, the transmission distance is given, and the wireless energy transmission coil has weight and size requirements. Therefore, it is necessary to select the optimal coil design scheme under the condition of meeting the size and weight.

For a circular coil, the self inductance, mutual inductance and equivalent series resistance of the resonant coil can be expressed as:

$$\begin{cases} L_1 = N_1^2 L_{10} \\ L_2 = N_2^2 L_{20} \\ M = N_1 N_2 M_0 \\ R_1 = N_1 R_{10} \\ R_2 = N_2 R_{20} \end{cases} \tag{5}$$

Among them, N1 and N2 are the turns of transmitting coil and receiving coil, L10, L20 and M0 are the equivalent self inductance and mutual inductance of single turn transmitting coil and single turn receiving coil respectively, R1 and R2 are the resistance of transmitting coil and receiving coil, R10 and R20 are the equivalent resistance of single turn transmitting coil and single turn receiving coil.

The equivalent mass of primary and secondary coils is as follows:

$$m=\rho\pi r^2 Nl \tag{6}$$

Where m is the mass of coil, ρ is the equivalent density of coil conductor, r is the equivalent radius of conductor, N is the number of coil turns, and l is the equivalent length of single coil turn. In the iterative design of the coil, it is necessary to consider the weight limitation of the coil.

By changing the coil turns and coil diameter, the magnetic field between the two coils can be enhanced as much as possible while keeping the self inductance of the coil unchanged. By increasing the mutual inductance of the coil appropriately by increasing the magnetic field, the output voltage can be increased, and the output current can be reduced under the same power, so as to improve the transmission efficiency.

Usually, the magnetic field strength of the coil is increased by adding ferrite core or making the coil configuration more complex. But adding ferrite core or changing coil configuration will greatly increase coil weight. Under the same conditions, the self inductance of square coil is larger than that of circular coil, but the coupling coefficient is the same. For the core design of the coil, in the same case, with the increase of the coil distance, the coupling coefficient of the coil with magnetic core decreases rapidly, and is close to the coupling coefficient of the coil without magnetic core. When the coil distance is longer, the coil mass will be significantly increased. Meanwhile, the coupling coefficient will be slightly improved. Therefore, a square coil without magnetic core is designed.

Through several iterations, the coil as shown in Figure 3 can be obtained. The coil spacing is 30 cm, the outer diameter is 47cm, the primary side is 14 turns, and the secondary side is 13 turns. The primary inductance of actual coil is L1 = 199uH, the secondary inductance is L2 = 180uH. And the coupling coefficient of primary and secondary sides is k = 0.087, the mutual inductance is M = 16.32uH.

3. Simulation and experiment

The simulation circuit of wireless power transmission system is built by PLECS to verify the design of LCC-S compensation network.

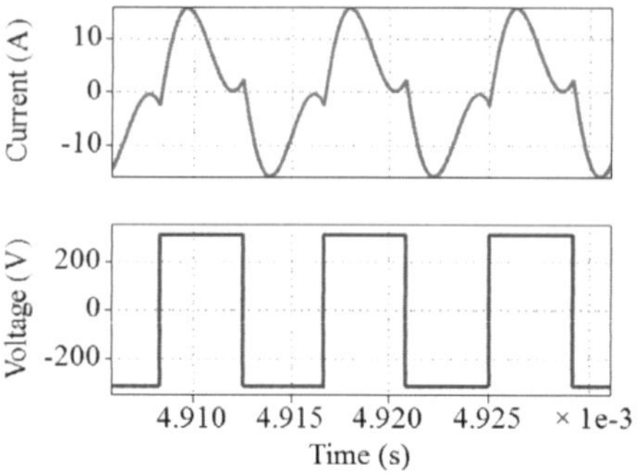

Fig. 4 LCC-S input voltage and current waveform

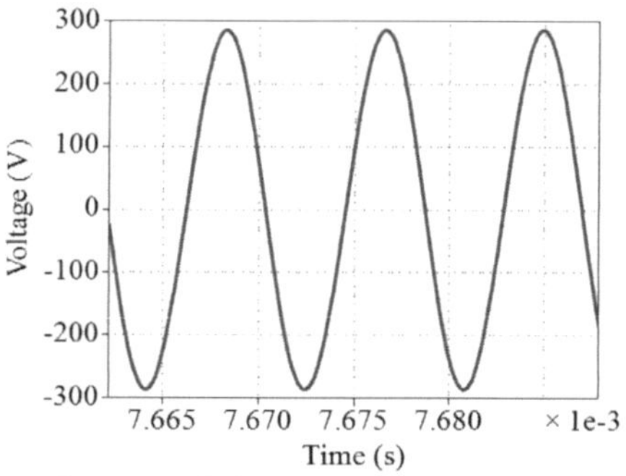

Fig. 5 LCC-S output voltage waveform

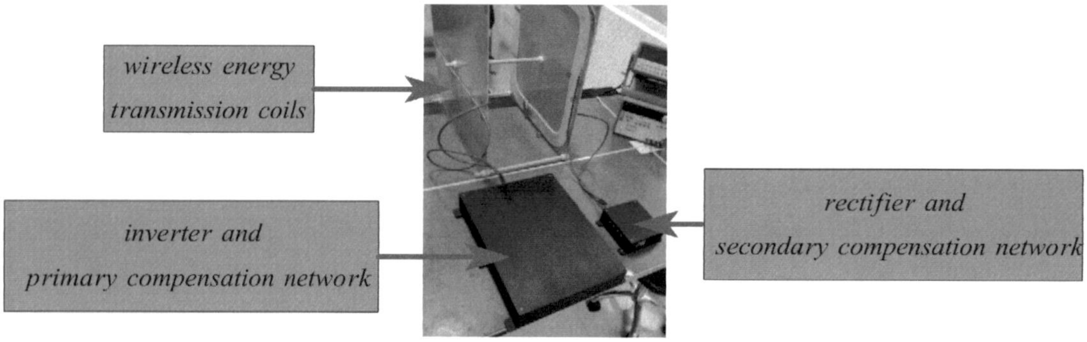

Fig. 6 physical system diagram

Figure 4 shows the input voltage and current waveforms of LCC-S compensation network. It can be seen that due to a small deviation of inductance and capacitance parameters during design, the input is weak inductive at 120kHz, and the circuit realizes zero current turn-on, reducing the switching loss of the circuit.

According to equation 3, the theoretical output value of DC voltage is 283V, which is consistent with the peak value of output voltage in Figure 5.

By building the actual circuit and winding the coil, the physical diagram of the system as shown in Figure 6 is obtained. By testing the circuit of the system, we can get the input voltage and current waveform shown in Figure 7, which is basically consistent with the simulation waveform in Figure 4.

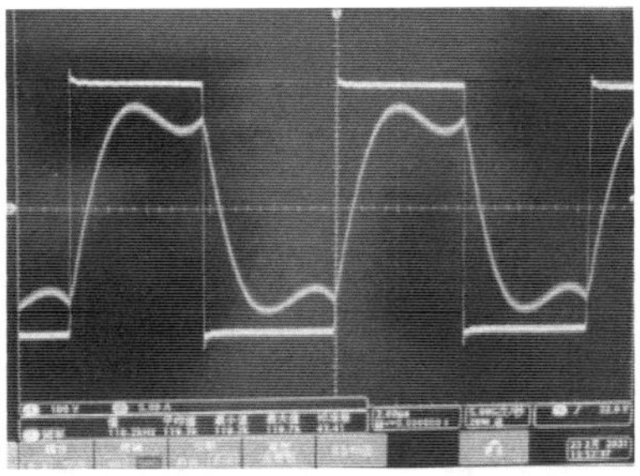

Fig. 7 waveform of input voltage and current

When the system is full load output, the output power is 2003W, the input power is 2264W, and the wireless transmission efficiency is 88.5%.

4. Conclusion

In this paper, a coil with a diameter of 47 cm is used to achieve high efficiency wireless power transmission at a distance of 30 cm. By designing the coil reasonably, considering the coil size, weight and coupling coefficient, a coil with high aspect ratio can be obtained. The design of compensation network can make the high frequency circuit work at zero current and reduce the loss of the system. After the comprehensive optimization design of the system, high transmission efficiency can be obtained at the receiving end even if the uncontrolled rectifier is used.

In order to further improve the transmission efficiency, we can consider adding magnetic core on the coil, or using synchronous rectifier circuit to further improve the wireless transmission system from the circuit design and coil design. By increasing the magnetic core, the coupling strength of the transmitting coil and the receiving coil can enhanced, so as to improve the energy transmission efficiency. Using synchronous rectifier circuit can reduce the energy loss of the receiving circuit, so as to improve the energy transfer efficiency of the system. Therefore, there is still a lot of room to improve the performance of this system.

References

[1] Y. Chen, H. Zhang, C. Shin, K. Seo, S. Park and D. Kim. (2019) A Comparative Study of S-S and LCC-S Compensation Topology of Inductive Power Transfer Systems for EV Chargers. In: IEEE 10th International Symposium on Power Electronics for Distributed Generation Systems (PEDG). Nanchang. pp. 99-104.

[2] J. Lu, G. Zhu, Y. Shou and F. Liu. (2018) Coupling- and Load-Independents Output Voltage and ZPA Operation in LCC-Series Compensated IPT System. In: IEEE Energy Conversion Congress and Exposition (ECCE). Portland. pp. 1172-1176.

[3] D. Yinliang, S. Yuanmao and G. Yougang. (2011) Design of coil structure achieving uniform magnetic field distribution for wireless charging platform. In: 4th International Conference on Power Electronics Systems and Applications. HongKong. pp. 1-5.

[4] R. Tavakoli and Z. Pantic. (2018) DC Modeling of an LCC Resonant Compensation Network in Wireless Power Transfer Systems. In: IEEE Energy Conversion Congress and Exposition (ECCE). Portland. pp. 6187-6132.

[5] Y. Wang, H. Wang, T. Liang, X. Zhang, D. Xu and L. Cai. (2017) Analysis and design of an LCC/S compensated resonant converter for inductively coupled power transfer. In: IEEE Transportation Electrification Conference and Expo, Asia-Pacific (ITEC Asia-Pacific). Harbin. pp. 1-5.

Research on Frequency Characteristics of VSG Virtual Parameter Adaptive Control Strategy Based on Fuzzy Control Theory

Jie Jin[1], Lan Li[2*], Haiyang Yu[3],and Shengzhou Feng[4]

[1]Taiyuan University of Technology Taiyuan, China

[2] Lan Li Taiyuan University of TechnologyTaiyuan, China

[3]Shengzhou Feng Taiyuan University of Technology Taiyuan, China

[4]Haiyang Yu *Taiyuan University of Technology*Taiyuan, China

*Corresponding author's e-mail: lilan@tyut.edu.cn

Abstract—Traditional virtual synchronous generators (VSG) control inverters. Inverter output frequency characteristic of the virtual inertia (J) and virtual damping (D) coefficient , and the virtual parameters need to be modified and adjusted according to the purpose. To solve this problem, this paper proposes a virtual parameter adaptive control strategy based on fuzzy control theory to adjust the frequency characteristics of VSG. MATLAB /Simulink is used to build a simulation model to verify the correctness of the proposed fuzzy control theory's adaptive virtual parameter theory.

1. Introduction

With the introduction of distributed energy, the inverter transforms dc input into three-phase AC input to large power grid, optimizing the operation of power system[1-2]. However, traditional inverters lack the same rotational inertia[3] and damping [4]characteristics as synchronous generators, when the power grid fails, the inverters do not provide sufficient inertial support. With the introduction of VSG control strategy, the inverter can simulate the external characteristics of the synchronous generator, and can show certain inertia and damping characteristics when the voltage and frequency fluctuate[5-6].

The document [7-8] proposes a method to determine the virtual rotation inertia by damping ratio, but the method has limited ability to regulate the frequency of the system. By analyzing the frequency fluctuation of multiple VSG parallel operations, document [9] points out that primary frequency modulation is insufficient and the secondary frequency modulation control strategy is put forward, but the VSG parameters are fixed and the frequency regulation is not flexible enough. Dcument [10] is in silo mode, the adaptation of J control, but did not consider the effect of D on frequency. In document [11], J and D coordinated adaptive control strategy applied to VSG is proposed, and the over-tuning of frequency and the rate of change are realized under the condition of grid-connected, but the effect of adaptive change of VSG parameters on frequency is not analyzed in silo mode.

This paper proposes a VSG virtual parameter adaptive control based on fuzzy control theory[13-14]. Firstly, the mathematical model of the inverter VSG control strategy is established, the function frequency controller and excitation controller of VSG are designed, then the fuzzy control theory is introduced to realize adaptive control of virtual parameters, the frequency characteristics of the inverter are adjusted, and the simulation model is built to verify the correctness of the control strategy.

1.1 VSG Structure diagram

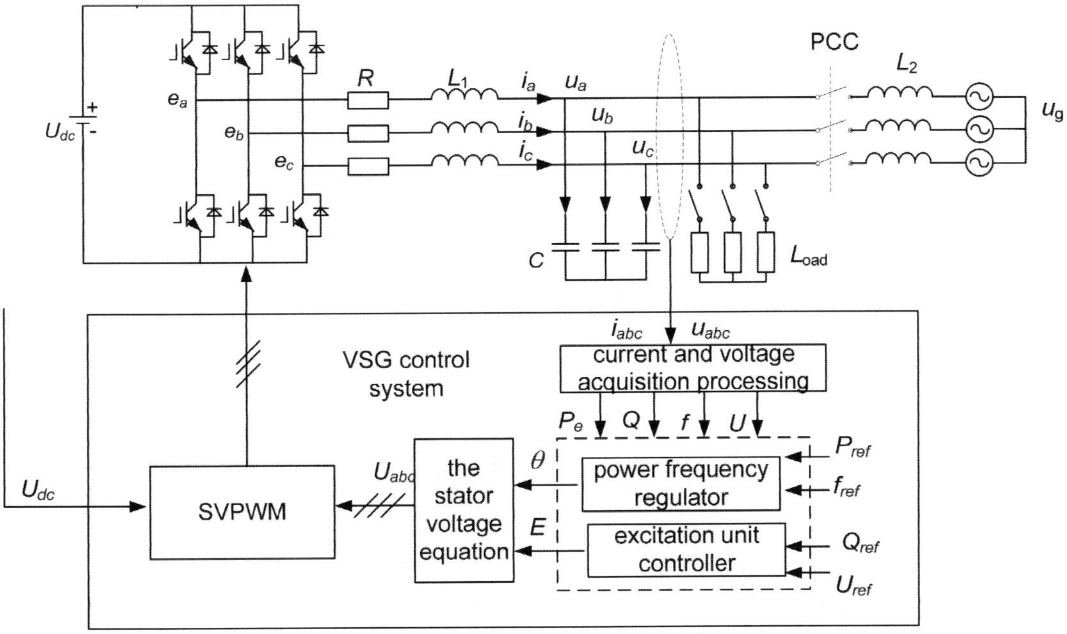

Fig.1 VSG structure block diagram

Figure 1 is a VSG structural block diagram, consisting of the main circuit and the control circuit. U_{dc} is the DC input voltage, e_a , e_b , e_c is the equivalent inductive electromotive force of the INverter AC side output, R is the filter inductor internal resistance, L_1 s the filter inductor, C is the filter capacitor, u_a , u_b , u_c is the equivalent end voltage of the synchronous generator, L_2 is the line inductor, u_g is the grid side voltage, L_{oad} is the load of the inverter. The introduction of rotating inertia and damping coefficients to simulate synchronous generator characteristics in the control circuit, combined with SVPWM modulators to achieve control of the inverter.

1.2 Power frequency controller

The synchronous generator rotor motion equation is:

$$\begin{cases} J\dfrac{d\omega}{dt} \approx \dfrac{P_m}{\omega_N} - \dfrac{P}{\omega_N} - D\Delta\omega \\ \dfrac{d\delta}{dt} = \omega \end{cases} \tag{1}$$

Where, the δ is the working angle of the synchronous generator, J is the rotational inertia, D is the damping coefficient, $\Delta\omega=\omega-\omega_N$, ω is the actual angular velocity, the ω_N is the rated angular speed, P_m is mechanical power, and P is electromagnetic power.

The function frequency controller of the virtual synchronous generator simulates the motion equation of the synchronous motor rotor, and simulates the mechanical power p_m input of the synchronous generator's prime mover through the phase angle of the output excitation motor force correspondingly controlled by the rotor motion equation. p_m consists of the active power reference value P_{ref} and the frequency variation difference difference ΔP

$$P_m = P_{ref} + \Delta P = P_{ref} + D_P\left(f_{ref} - f\right) \tag{2}$$

Where, D_p is the frequency modulation coefficient. f_{ref} is the frequency reference value.

The power frequency controllers of VSG are available by formulas (1) and (2) as shown in Figure 2.

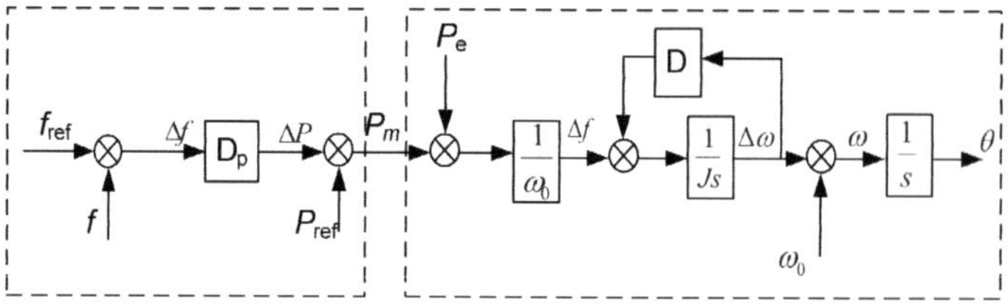

Fig. 2 Block diagram of VSG power frequency controller

1.3 Excitation controller

The active power and end voltage regulated by the excitation system in the synchronous generator. Therefore, the excitation controller needs to control both voltage difference and reactive power, in order to simplify the control process, the first-order inertia link is used to model. So the excitation formula for VSG is derived

$$K\frac{dE}{dt} = Q_{ref} + D_q(U_N - U) - Q \qquad (3)$$

Where, K is the inertial coefficient, E is the magnitude of the modulated voltage, Dq is the sagging factor, U_N is the rated voltage of the output, U is the actual voltage of the output, Q_{ref} is the reactive power given value, and Q is the reactive power detection value.

$$Q = [(u_a - u_b)i_c + (u_b - u_c)i_a + (u_c - u_a)i_b]/\sqrt{3} \qquad (4)$$

Where, u_a , u_b , u_c is the three-phase voltage output of the inverter, and i_a,i_b,i_c are the three-phase current output of the inverter.

The excitation controllers of VSG are available by formulas (3) and (4) as shown in Figure 3.

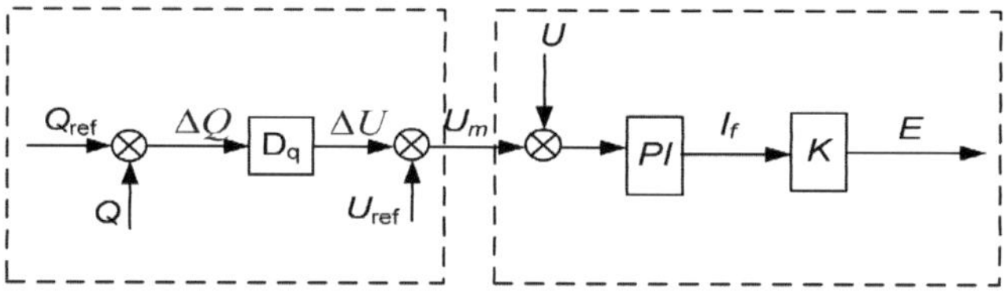

Fig. 3 Block diagram of VSG excitation unit controller

In summary, after obtaining the output voltage phase angle and modulation voltage, the command voltage of VSG is obtained

$$\dot{E} = \begin{bmatrix} \sqrt{2}E\sin\delta \\ \sqrt{2}E\sin(\delta - 2\pi/3) \\ \sqrt{2}E\sin(\delta + 2\pi/3) \end{bmatrix} \qquad (5)$$

2.The adaptive parameter control of fuzzy control theory

Power angle and angular frequency oscillation curve of synchronous generator

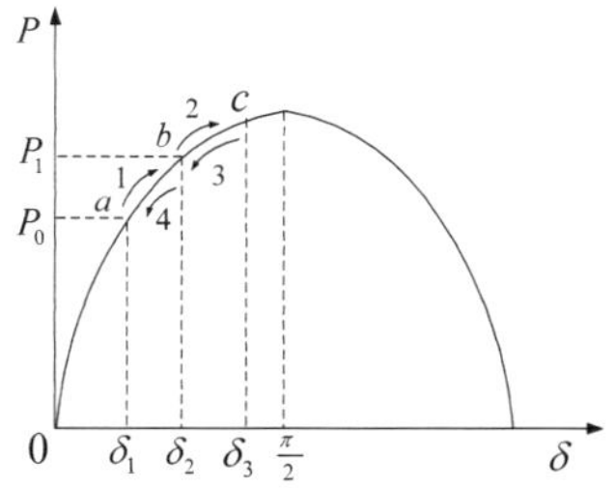

(a) Power angle

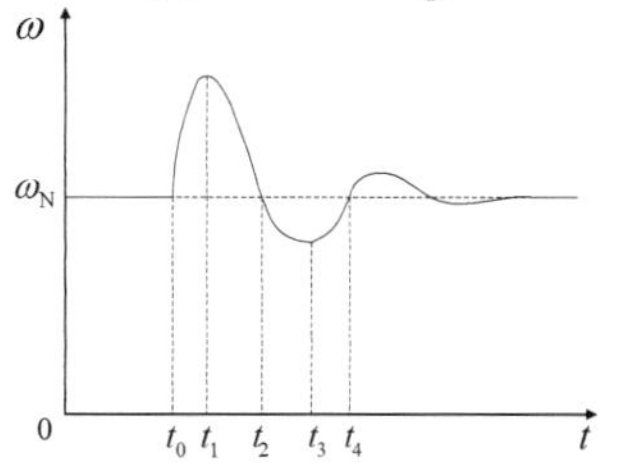

(b) Angular frequency oscillation curve

Fig.4 Power angle and angular frequency oscillation curve of synchronous generator

In the t0-t1 stage, the inverter output angle frequency is higher than the rating due to the increase in output power, and the angular frequency is increasing. At this stage, the stability of the output angular frequency can be maintained by reducing the rate of change of the angular frequency. In the t1-t2 phase, the inverter output angle frequency value is continuously reduced from the maximum value, and eventually reaches the rated angular frequency of ω_N, which requires increasing the rate of change of the angular frequency and reducing the adjustment time of dynamic response, and the t2-t3 and t3-t4 stages are similar to the analysis process of the t0-t1 and t1-t2 stages, respectively.

When the angular frequency change curve is in the t0-t1 stage, the VSG angle frequency is increasing and the angular frequency change rate ($d\omega/dt$) is greater than zero, at which stage, the formula (1) can be obtained, if the rotational inertia J is increased, The rate of change of angular frequency can be reduced, thus reducing the increase of angular frequency, and at this time the amount of angular frequency change (ω-ω_N) is greater than zero, and if the damping coefficient D increases, the rate of change of angular frequency can also be reduced, thereby reducing the increase rate of angular frequency. In the t1-t2 stage, The angular frequency change rate or VSG is less than zero, in this stage, if the rotational inertia J is reduced, it will be reduced, but its absolute value increase, can speed up the process of angular frequency reduction; The t2-t3 stage is similar to the t0-t1 stage in that the angular frequency changes in the direction of departure from the rated angle frequency, so it is also necessary to increase the rotational inertia and damping coefficient.

In order to take into account the influence of angular frequency change rate and angular frequency change on virtual parameter selection, this paper adopts a method of controlling virtual parameters by fuzzy controller, so that virtual parameters can adjust the output frequency of the system according to the dynamic change of angular frequency rate and angular frequency change. Therefore, the angular frequency change rate and the amount of angular frequency change are used as the input of the fuzzy controller, and the output is set to the virtual parameters J and D.

Set the angle frequency change rate and the amount of angular frequency change input fuzzy set selection { NB,NS,Z,PS,PB } , output virtual parameters J and D fuzzy set pick {NB,NM,NS,Z,PS,PM,PB} , membership function selection isosceles triangle, de-fuzzy selection center of gravity method[15].

The control idea of fuzzy controller is to increase the rotational inertia and damping coefficient when the angular frequency changes in the direction of divergence from the rated angle frequency, and change in the direction of the angle frequency towards the recovery rating, reduce the rotational inertia and increase the damping coefficient. So the fuzzy control rules tables for virtual parameters J and D are shown in Tables 1 and 2.

Table 1 J Fuzzy control rule table

J		$\Delta\omega$				
		NB	NS	Z	PS	PB
$d\omega/dt$	NB	PB	PM	PS	NM	NB
	NS	PM	PS	PS	NS	NM
	Z	PS	Z	Z	Z	NS
	PS	Z	NS	Z	PS	Z
	PB	NB	NM	NS	PM	PB

Table 2 D Fuzzy control rule table

D		$\Delta\omega$				
		NB	NS	Z	PS	PB
$d\omega/dt$	NB	PB	PM	PS	PM	PB
	NS	PM	PS	Z	PS	PM
	Z	PS	Z	Z	Z	PS
	PS	PM	PS	Z	PS	PM
	PB	PB	PM	PS	PM	PB

The actual angular frequency change rate and angular frequency change amount are blurred as the input of the fuzzy controller, and the actual output value is defunctioned by the reasoning of the fuzzy rules to the output values J and D. Wherein the range of values of $d\omega/dt$ is [-0.6, 0.6], the range of value of $\Delta\omega$ is [-0.5, 0.5], the range of value of J is [0.2, 0.4], and the range of value of D is .[2,4].

3. Simulation verification

In order to verify the adaptive selection of virtual parameters proposed above, an inverter system is set up on matLAB/Simulink platform, and its control circuit is using virtual synchronous generator, and the simulation is carried out in silo mode. The system parameters are shown in Table 3.

Table 3 The simulation parameters

Parameter	***Parameter value***
DC voltage u_{dc}/V	330
Inverter side filter inductor L_2/mH	3
Line resistance R/Ω	0.01
Filter capacitor C/μF	70
VSG virtual inertia J/kg·m^2	0.3
VSG damping coefficient	3
Frequency modulation coefficient D_p	400
Reactive inertia coefficient K_q	20
Rated frequency f_N/Hz	50

In silo mode, the inverter output is connected to a active load of 10000W and a reactive load of 5000Var in the initial state. At 1.5s, the inverter adds an active load of 5000W to the given power. At 2s, cut off the increased 5000W active load. Observe the frequency waveforms output of the system in different J and D situations.

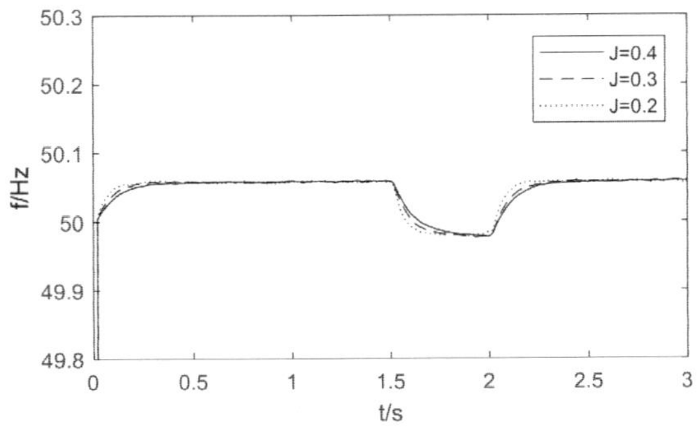

(a) Change the inertia.

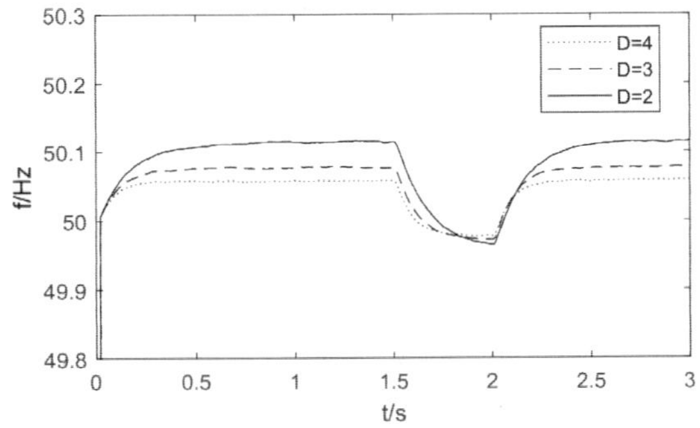

(b)Change damping coefficient.

Fig.5 Frequency wave form of different J and D

As can be seen from Figure 5, if the rotation inertia increases, the initial frequency change rate decreases, but the adjustment time is longer, and the damping coefficient increases, the frequency deviation can be reduced. Therefore, selecting the appropriate rotational inertia and damping coefficient can improve the dynamic change process of frequency.

In order to verify the effect of the proposed fuzzy control virtual parameter adaptive regulation on frequency change. The remaining simulation conditions remain unchanged, the rotational inertia and damping coefficient are added to the fuzzy control, and then the two parameters are used at the same time fuzzy control, the output frequency simulation results of the system as shown in the figure 6.

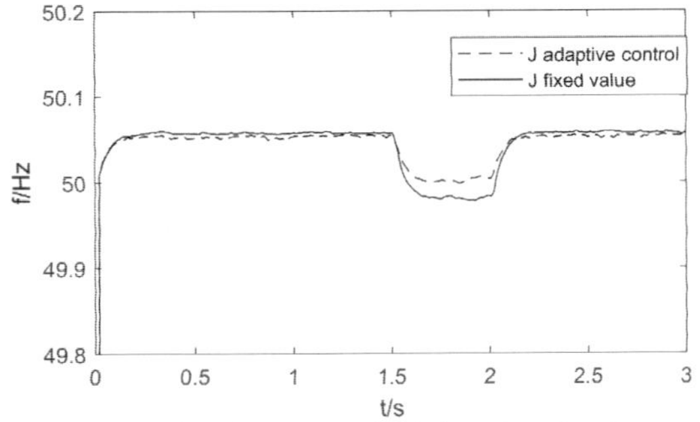

(a) Adaptive control of inertia adaptive.

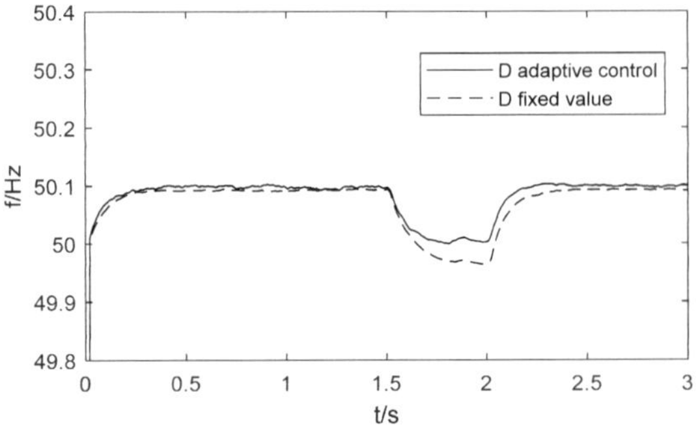

(b) Adaptive control of damping coefficients.

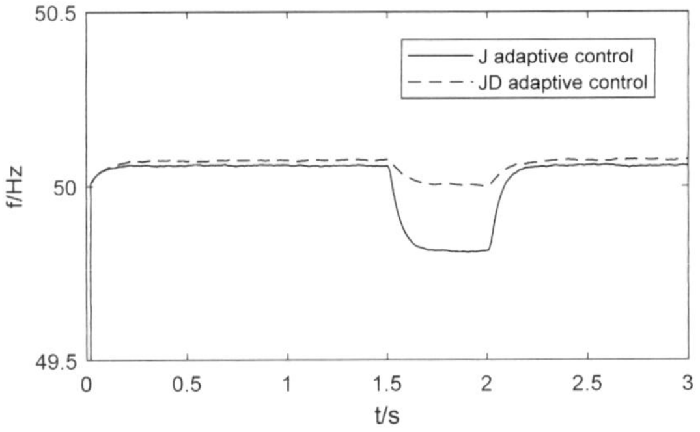

(c) Adaptive control of virtual parameters.

Fig.6 Frequency wave form with adaptive change of J and D,

As can be seen from Figure 6, when the system cuts into a active load at 1.5s, the frequency fluctuates. As can be seen from the frequency change waveform of Figure 6 (a), when the damping coefficient is fixed and the rotation inertia adaptive changes, it can not only reduce the frequency change rate but also shorten the adjustment time of the transient process, so that the frequency can reach the stable value more quickly. Figure 6 (b) compares the frequency change waveform when the rotation inertia is fixed and the damping coefficient adaptive changes, and Figure 6 (c) compares the frequency change waveforms of J, D simultaneous adaptive changes and J and adaptive changes, on the basis of J changes, D adaptive changes can effectively reduce the frequency variation, thereby improving the stability of frequency changes.

4. 。Concliion

In view of the problem that the frequency characteristics of the inverter output are affected by the different frequency characteristics of the virtual rotation inertia and virtual damping coefficient in the control of the traditional VSG control strategy, and the virtual parameters need to be constantly modified and adjusted, this paper proposes a virtual parameter adaptive adjustment method based on fuzzy control, and adaptively adjusts the frequency response of the output of the virtual rotation inertia J and damping coefficient D regulation system. The simulation proves that adaptively adjusting the virtual rotation inertia J and damping coefficient D can suppress the fluctuation of the system frequency. It can reduce both the frequency change rate and the frequency change. The dynamic performance of this control policy frequency recovery is further optimized.

References

[1] Zhang, T.W.,Chen,L.J.,Chen,T.Y.,et al.(2015)Technology and prospect of virtual synchronous generator.Power System Automation, 39(21):165-175.
[2] Lv,Z.P.,Sheng,W.X.,Liu,H.T.,et al.(2017)Application and challenge of virtual synchronous machine technology in power system.Proceedings of the CSEE,37(02):349-360.
[3] Kenichi,S.,Yushi,M.,Toshifumi,I., (2014) Stabilization of a Power System Including Inverter-Type Distributed Generators by a Virtual Synchronous Generator. Electr Eng Jpn ,4:258-263
[4] Hassan,B.,Toshifumi,I.,Yushi,M.Virtual Synchronous Generators: A Survey and New Perspectives. International Journal of Electrical Power and Energy Systems ,2013.
[5] Lu,X.J.,Duan,J.T. (2021)A Hybrid Microgrid Control Strategy Based on VSG Technology. Journal of Solar Energy,42(03):243-250.
[6] Tang,N.,Sheng,C.,Chen,M.,et al.(2016)Summary of virtual synchronous generator technology, Southern Power Grid Technology,10(11):30-38.
[7] Zhang,F.D.,Pu,Z.G.,Guo,Y.Q.,et al. (2020)Research on adaptive control strategy of VSG moment of inertia.Journal of Solar Energy,41(10):93-100.
[8] Gao,C.W.,Liu,X.M.,Fan,A.L., (2020)Research on Control Strategy of Improved Virtual Synchronous Generator in Distributed Inverter.Journal of Solar Energy,41(12):33-39.
[9] Li,J.Chu,B.Q.Yu,J.M., (2019)Secondary frequency modulation control strategy based on VSG in micro-grid. Information Technology,43(11):62-66.
[10] Zhang,W.M.Zhao,Q.Liu,Y.M.et al. (2019)VSG inverter power source control strategy with adaptive rotary inertia in island mode.Proceedings of the CSUEPSA31(02):95-100.
[11] Yang,Y.Mei,F.Zhang,C.Y.et al. (2019)Coordinated adaptive control strategy of rotational inertia and damping coefficient for virtual synchronous generator,Electric Power Automation Equipment, 39(3):125-131.
[12] Yang,Q,Li,L.,Zhao,R.L.,et al. (2020)Low voltage through-through excitation control of doubly-fed wind turbines based on fuzzy control virtual resistance.Power capacitors and reactive power compensation,41(02):158-163.
[13] Chai,L. (2019)Micronet inverter control based on virtual synchronous generator.Taiyuan University of Technology,
[14] Wang,Y.N,Sun,W. (2008)Intelligent control theory. Machinery Industry Press, Beijing.

The application of Deep Reinforcement Learning in Coordinated Control of Nuclear Reactors

Jing Li[1], Yanyang Liu[1], Xianguo Qing[1], Kai Xiao[1], Ying Zhang[1], Pengcheng Yang[1], Yue (Marco) Yang[2]

[1] Science and Technology on Reactor System Design Technology Laboratory, Nuclear Power Institute of China, Chengdu, 610213, China

[2] Department of Engineering, The University of Sydney, Sydney, 2006, Australia

*WinkOvO2021@yeah.net

Abstract. The nuclear reactor control system plays a crucial role in the operation of nuclear power plants. The coordinated control of power control and steam generator level control has become one of the most important control problems in these systems. In this paper, we propose a mathematical model of the coordinated control system, and then transform it into a reinforcement learning model and develop a deep reinforcement learning control algorithm so-called DDPG algorithm to solve the problem. Through simulation experiments, our proposed algorithm has shown an extremely remarkable control performance.

1. Introduction

The nuclear power plant control system automatically maintains the relevant parameters of the nuclear power plant within the limits specified in the operating conditions and can automatically adapt the operating state of the nuclear power plant, so as to match the energy generated by the nuclear reactor with the load demand [1]. The design of nuclear power control system can be complicated. The characteristics of the controlled object in the nuclear power plant are multivariable, nonlinear, and strong coupling. As a consequence, it is hard to formulate the thermal process and involved characteristics by an accurate and concise mathematical model. The traditional model-based control method is not perfectly suitable for the above problem.

In recent research of the nuclear power plant, the widely used method is single-input, single-output, closed-loop, and cascade loop PID control [2-3]. Although the PID-based method can capture the operation characteristics in nuclear power plants, the improvement is still required considering its response speed, overshoot and other indicators. On the other hand, the traditional PID-based method is often designed for a single subsystem with a single controlled variable, the coupling and cooperation among multiple systems cannot be tackled by this method. Therefore, the PID-based method cannot be applied to the coordinated control of nuclear power plant systems.

To address the above challenges, the dynamic scheduling problem of the nuclear reactor coordinated control system can be modelled as a stochastic sequential decision problem, which can be solved by reinforcement learning. Reinforcement learning is a classical and critical method in control theory and machine learning, that focuses on how agents take actions in the environment to obtain the maximum cumulative return [4]. Unlike traditional methods modelling the control system in advance, Model-free reinforcement learning does not depend on any prior knowledge. Since most of the control parameters in the control system studied in this paper are continuous variables. there will be a dimension disaster problem by the traditional reinforcement learning method (e.g., Q-learning).To this end, the deep reinforcement learning algorithm is further introduced to model and solve the coordinated control problem of the nuclear reactor.

In this paper, we aim to develop an intelligent control method for the small nuclear reactor. According to the deep reinforcement learning algorithm, the collaborative control of reactor power regulation system and steam generator liquid level regulation system is studied in this work. These two systems can be modelled as the environment in reinforcement learning and Deep Deterministic Policy Gradient (DDPG) is further introduced to solve the problem. Through extensive simulation experiments, the proposed method demonstrates a remarkable performance compared to several benchmarks.

2. Control system modelling

2.1. Description of nuclear reaction coordinated control system

The complete reactor control system mainly includes power control system, steam generator liquid level control system, main pump control system, pressurizer pressure control system, valve control system, etc. This paper makes further design and research on the reactor power control system and steam generator liquid level control system.

In the reactor power control system, the signals to be considered include nuclear power N , demand power N_0 , measured value of average temperature of primary coolant $T_{avg}(t)$, set value of average temperature of primary coolant T_{ref} , feed-water flow of steam generator F_W and control rod drive signal Rv .The control rod drive signal is used to drive the action of the control rod. The action of the control rod should not only adjust the reactor power, but also adjust the average coolant temperature. Therefore, the purpose of the reactor power control system is to make the reactor power deviation ΔN and the average coolant temperature deviation ΔT close to zero. The control principle of reactor power control system based on classical cybernetics is shown in Figure 1:

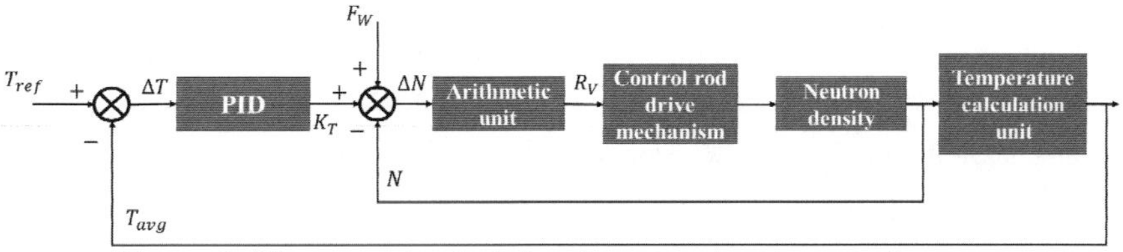

Figure 1. The control principle of reactor power control system.

In the steam generator liquid level control system, the signals to be considered include the measured value of steam generator liquid level L liquid level setting value L_S steam generator feed water flow F_W , steam flow F_S and feed-water regulating valve opening signal V .The opening signal of feed water regulating valve is used to control the action of feed water regulating valve. The purpose of controlling the action of feed water regulating valve is to make the feed water flow deviation of steam generator ΔF_W and the water level deviation of steam generator ΔL tend to zero. The control principle of steam generator liquid level control system based on classical cybernetics is shown in Figure 2:

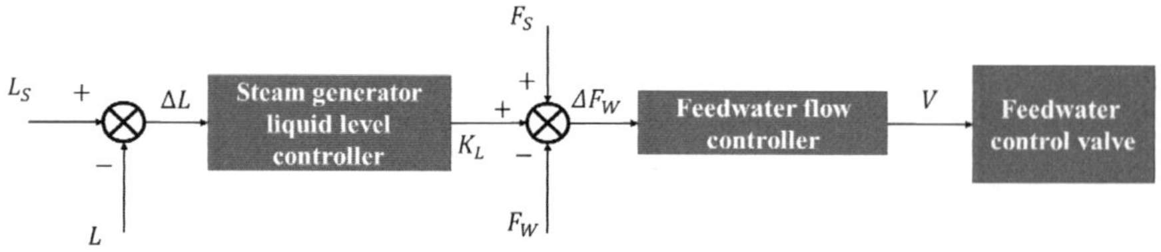

Figure 2.The control principle of liquid level control in steam generator system.

2.2. Mathematical description of control problems

The optimization objective of this paper is to minimize the deviation signals of reactor power control system and steam generator level control system. Each deviation signal can be expressed as:

$$\Delta T = T_{ref} - T_{avg} \tag{1}$$

$$\Delta N = K_T + F_W - N \tag{2}$$

$$\Delta L = L_S - L \tag{3}$$

$$\Delta F_W = K_L + F_S - F_W \tag{4}$$

Among them, ΔT, ΔN, ΔL and ΔF_W are the average temperature deviation, power deviation, steam generator water level deviation and feed-water flow deviation respectively. The reactor coordinated control system changes the reactivity in the reactor through the control rod drive signal, and changes the feed-water flow and steam generator liquid level through the feed-water regulating valve opening signal, so as to eliminate the deviation between the measured feedback signal in the system and the secondary circuit load reference signal, so as to restore the primary and secondary circuits to the state of energy balance. The objective function to be selected in this paper is:

$$F = \min f(\Delta T, \Delta N, \Delta L, \Delta F_W) \tag{5}$$

In this paper, we focus on the control optimization problem under discrete-time steps. A complete time control interval can be discretized into $\{1,2,...,M\}$. In each time step of control decision, we can write the objective function in (5) above as:

$$f_t(\Delta T, \Delta N, \Delta L, \Delta F_W) = e_T(t) + e_N(t) + e_L(t) + e_{FW}(t) \tag{6}$$

Among them, ΔT, ΔN , ΔL and ΔF_W are the average temperature deviation, power deviation, steam generator water level deviation and feedwater flow deviation at timerespectively.

Therefore, the optimization objective of this problem is to optimize the system deviation under the whole control time window:

$$F = \min \sum_t f_t(\Delta T, \Delta N, \Delta L, \Delta F_W) \tag{7}$$

3. Reinforcement learning modelling

3.1. Reinforcement learning modelling

Using the advantage that reinforcement learning is very suitable for solving the optimization decision problem with uncertain factors, this paper models and solves the control decision of reactor control system in discrete time window. Here, we first need to transform the mathematical description constructed in the previous section into a reinforcement learning framework.

The basic components of reinforcement learning include state set S representing environment, action set A representing agent action and reward r for agent. Based on the current state $s, \forall s \in S$, the agent selects an action $a, \forall a \in A$ to act on the environment. After the environment accepts the action, it changes and generates a reward signal r to feed back to the agent. The agent then continuously selects the next action according to the signal and the current state of the environment. In this paper, the reinforcement learning model is described as follows:

(1) State s: The observed states of nuclear reactor coordinated control system include nuclear power $N(t)$, average temperature of primary coolant $T_{avg}(t)$, feed-water flow $F_W(t)$, steam flow $F_S(t)$, steam generator liquid level $L(t)$, etc

$$s(t)=\{N(t),T_{avg}(t),F_W(t),F_S(t),L(t)\} \tag{8}$$

(2) Action a: In decision time step t, the action of the coordinated control system can be represented by control rod drive signal $Rv(t)$ and feed-water regulating valve opening signal $V(t)$,

$$a(t)=\{Rv(t),V(t)\} \tag{9}$$

(3) Reward r: In this paper, the problem of minimizing the total deviation of the system is transformed into the classical reward maximization form of reinforcement learning. Therefore, the reward obtained by the agent in period 1 is expressed as (where 1 / 1000 is the scaling of the objective function):

$$r_t(s(t),a(t))=-\frac{1}{1000}\cdot\left[e_T(t)+e_N(t)+e_L(t)+e_{FW}(t)\right] \tag{10}$$

(4) Action value function $Q(s,a)$: When a certain state $s(t)$ in the reactor coordinated control system is determined, the advantages and disadvantages of system action $a(t)$ can be evaluated using action value function $Q(s,a)$:

$$Q(s(t),a(t))=E\left(\sum_{k=0}^{M}\gamma^k\cdot r(s(t+k),a(t+k))\right) \tag{11}$$

Among them, $E(\cdot)$ is the expectation function, $\gamma\in[0,1]$, γ is the discount factor, which represents the impact proportion of the reward in the cumulative reward at a certain time in the future.

(5) Optimal strategy π^*: The objective of the reactor coordinated control system is to find the optimal strategy to maximize the action value function:

$$\pi^* \leftarrow \arg\min_a Q(s,a) \tag{12}$$

3.2. Solution algorithm based on deep reinforcement learning

The traditional reinforcement learning method performs well in the problems of small-scale discrete actions and state space. However, when the number of states and actions increases exponentially, the dimensional disaster occurs [5]. The traditional reinforcement learning method cannot effectively deal with this situation. In the reactor coordinated control system studied in this paper, the state space and action space are continuous variables. Therefore, this paper uses deep neural network (DNN) [6] to approximate the function of reinforcement learning. The specific selection of the algorithm is DDPG algorithm based on actor critical, which estimates the optimal strategy function through deep neural network, It cannot only avoid dimension disaster, but also save the information of the whole action domain.

The principle of reactor coordinated control scheme based on DDPG algorithm is shown in Figure 3. For DDPG algorithm, the input of policy network is state vector $s(t)$ as shown in formula (7), and the output is action vector $a(t)$ as shown in formula (8); The input of the value network is state vector $s(t)$ and action vector $a(t)$, and the output is action value function $Q(s(t),a(t))$. DDPG algorithm uses the empirical playback mechanism in deep Q network. It enters the storage of experience playback by storing the agent's experience $e_t=\{s(t),a(t),r(t),s(t+1)\}$ in each period. During training, min-batch (size M) experience samples will be extracted from the experience pool each time, and the network parameters will be updated based on the gradient rules. Then, DDPG algorithm trains the value network and policy network respectively to obtain the optimal parameters in the network.

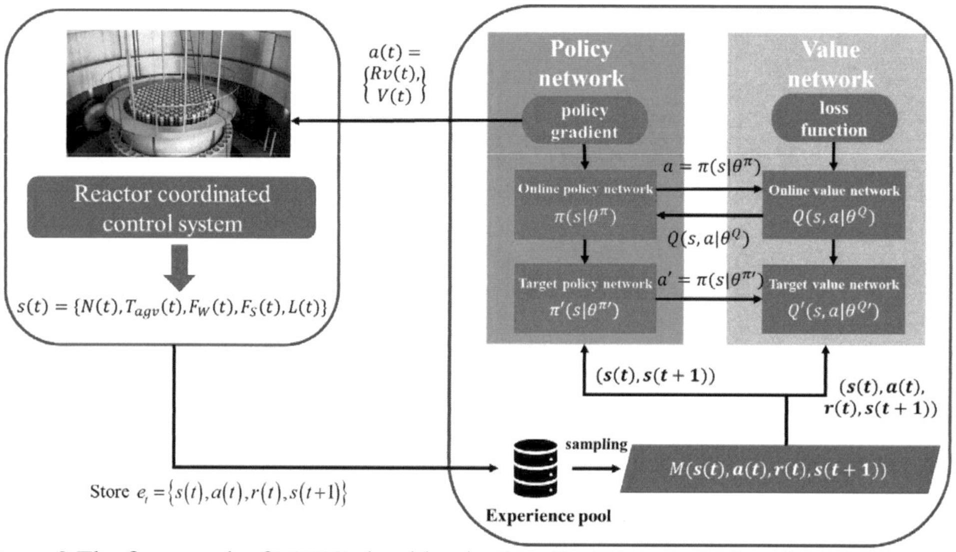

Figure 3.The framework of DDPG algorithm in Coordinated control of nuclear reactors.

For the value network, we optimize the parameters by minimizing the loss function $L(\theta^Q)$:

$$L(\theta^Q)=E\left(y_t-Q\left(s(t),a(t)\mid\theta^Q\right)\right)^2 \tag{13}$$

Among them, y_t is the target Q value

$$y_t=r(t)+\gamma Q'\left(\left(s(t+1),\pi'\left(s(t+1)\mid\theta^{\pi'}\right)\mid\theta^Q\right)\right) \tag{14}$$

We update the value network parameters according to the gradient rules:

$$\theta^Q\leftarrow\theta^Q-\mu_Q\nabla_{\theta^Q}L(\theta^Q) \tag{15}$$

For the policy network, the gradient information $\nabla_a Q\left(s(t),a(t)\mid\theta^Q\right)$ provided by it is used as the direction of action improvement. According to the deterministic policy gradient, the parameters of the policy network are updated as follows:

$$\theta^{\pi} \leftarrow \theta^{\pi} - \mu_{\pi}\nabla_{\theta^{\pi}}L\left(\theta^{\pi}\right) \tag{16}$$

Parameters $\theta^{Q'}$ and $\theta^{\pi'}$ of the target network use soft update technology to improve the stability of learning:

$$\theta^{Q'} \leftarrow \tau\theta^{Q} + (1-\tau)\theta^{Q'} \tag{17}$$

$$\theta^{\pi'} \leftarrow \tau\theta^{\pi} + (1-\tau)\theta^{\pi'} \tag{18}$$

Among them, τ is the soft update parameter, $\tau \ll 1$.

4. Numerical experiment

In order to evaluate the effectiveness of the proposed reactor coordinated control system based on deep reinforcement learning DDPG algorithm, this paper takes the small reactor as the control object, including point reactor neutron dynamics model, reactor and primary loop main equipment, steam generator and its secondary side boundary, and uses the same transient simulation model as the engineering design to build the basic simulation environment.

The control parameters of the reactor coordinated control system are shown in Table 1. We mainly conduct the control simulation test under the working condition of 100% - 85% FP linear load reduction, in which the change rate of linear load reduction is selected as 1.33% FP / s.

Table 1. Ranges of control parameters.

Deviation signal	Fluctuation range of steady state
Power deviation (%)	≤±0.6
Temperature deviation (℃)	≤±0.5
Liquid level deviation (%)	≤±0.3
Feed water flow deviation (%)	≤±0.1

4.1. The training of reinforcement learning

Before applying the deep reinforcement learning network to the reactor coordinated control system, the parameters of deep reinforcement learning are trained through historical data to obtain the deep reinforcement learning network. In order to better evaluate the performance of the DDPG algorithm, we introduce deep Q network learning (DQN) as a reference algorithm for further comparison.

In the value network of DDPG, we set up two hidden layers, each layer has 200 neurons, and the activation function of the hidden layer is ReLU. In DDPG's policy network, we also set two hidden layers, and the activation function is Tanh. Other training parameters of DDPG are shown in Table 2. For the DQN network, the input is a 5-Dimensional state vector $s(t)$ and the output is the Q value of the state-action pair. In this paper, $Rv(t)$ and $V(t)$ are discretized according to the step size of 0.1. thus the discretized action space is obtained. DQN network has two hidden layers, each layer has 200 neurons, and the activation function of the hidden layer is ReLU. In order to show the convergence performance of the proposed method, Figure 4 shows the average reward value curve every 60 minutes during agent training. The algorithm converges after about 40 minutes of training, and the optimal control strategy is obtained. It can be observed that because the agent is not familiar with the environment at first, the reward value obtained by the agent after executing the scheduling decision is small. With the continuation of the training process, agents constantly interact with the environment and gain experience. Therefore, the overall trend of reward value is to gradually increase and finally converge. This shows that the agent has learned the control strategy to minimize the deviation signal cost.

Table 2. Parameters of DDPG method.

Parameter	Value
Experience playback scale	5000
Experience playback batch	32
Discount factor γ	0.9
Network learning rate	0.01
Soft update parameter τ	0.05

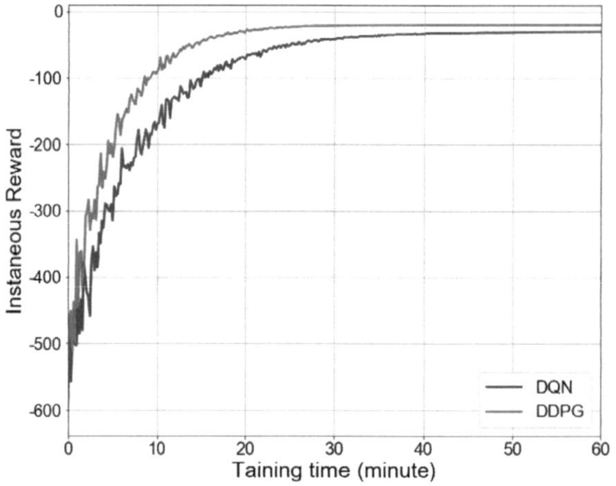

Figure 4.The reward function in the training of DDPG and DQN.

4.2. Control effect

In order to verify the effectiveness of the reactor coordinated control system based on DDPG algorithm proposed in this paper, we compare the scheduling method based on DDPG algorithm with the scheduling method based on DQN [7] and the scheduling method based on random strategy control. In the random control strategy, the controlled parameters $Rv(t)$ and $V(t)$ are randomly taken within their value ranges. Table 3 shows the simulation data of the three methods. Among them, the system deviation based on random strategy is 174.3, the system deviation based on DQN strategy is 43.1, and the system deviation of our proposed DDPG strategy is 17.3. Further, we draw the system deviation of the reactor coordinated control system under the DDPG strategy in Figure 5. We can observe that the DDPG algorithm can basically control the four system deviations within the range of nearly 0 after the system runs for about 40 minutes, showing a superior control effect. Therefore, the scheduling method based on DDPG algorithm proposed in this paper is more suitable to solve the control problem of the reactor coordinated control system.

Table 3. Evaluation of different methods

Strategy	System deviation F
Random	174.3
DQN	43.1
DDPG	17.3

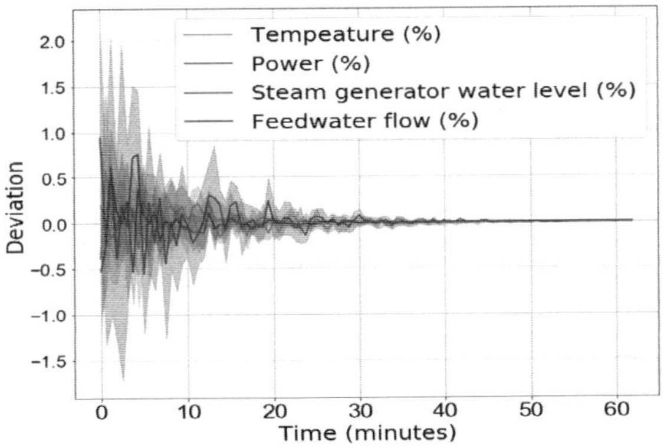

Figure 5.The systemic deviation controlled by DDPG.

5. Conclusion

Based on the control problem of reactor coordinated control system, we propose a mathematical model to minimize the system deviation. Considering the uncertainty of the system and the requirements of dynamic real-time optimization, the mathematical model is further transformed into a reinforcement learning control model. Then we use DDPG deep neural network reinforcement learning algorithm to realize interactive learning and problem solving of reactor coordinated control system. In the experimental part, we fully compare the control effects of different deep reinforcement learning algorithms and control schemes formed under random strategies and verify the superior control effect and performance of DDPG algorithm.

References

[1]Zhang, Y., Jiang, L., Cai, Q. (2019) Research on sensitivity analysis of passive system for nuclear power plants based on HFR. Nuclear Power Engineering, 40: 150-154.

[2] Jiang, Q., Zeng, W.(2020) Application of T-S fuzzy switching controller in core power control. Nuclear Power Engineering, 41: 126-130.

[3] Teng, L., Wang, S., Wang, X. (2020) Design of high pressure water jet decontamination device with pressure and flow synchronous control function. Nuclear Power Engineering, 41: 153-157.

[4] Sutton, R.S, Barto, A.G. (1998) Introduction to Reinforcement Learning.MIT Press, Cambridge.

[5] Yang, T., Zhao, L., Li, W., et al. (2020) Reinforcement learning in sustainable energy and electric systems: A survey. Annual Reviews in Control, 49: 145-163.

[6] LeCun, Y., Bengio, Y., Hinton, G. (2015) Deep learning. Nature, 521: 436-444.

[7] Mnih, V., Kavukcuoglu, K., Silver, D., et al. (2015) Human-level control through deep reinforcement learning. Nature,518: 529-533.

Layout Guide for Design of a Novel Wheelchairs with Circumferential Linkage using Microsoft Word

Zhilin Zhao

NUAA, Nanjing, Jiangsu, ZIP code, China

*Corresponding author's e-mail: 412864306@nuaa.edu.cn

Abstract. In recent decades, the aging of the world's population has intensified, and the number of people with disabilities caused by various disasters and diseases has gradually increased. Most of the elderly, frail and disabled people will choose wheelchairs as their means of transportation. However, ordinary wheelchairs are unable to climb stairs, especially in cities, which seriously limits the range of activities of wheelchair users and affects their daily life. Hence, it is of great significance and value to design an intelligent building climbing wheelchair with appropriate price, stability and safety. This paper designs a multi-functional planetary height adjustable building climbing wheelchair with simple structure and low price. Firstly, a multi-functional building climbing wheelchair based on planetary gear train is designed by using modular design idea, including ground walking mechanism, in place steering mechanism, center of gravity adjustment mechanism and Seat leveling mechanism. The working principle and design characteristics of each mechanism are analyzed in detail. Then, through the analysis of the mathematical model of the climbing mechanism, the dimensional parameters of the three-star wheel set are determined. According to the power demand of wheelchair, the power system is selected and analyzed. It provided a new possibility for the wheelchair application and also for those in need.

1. Introduction

After entering the new century, aging problem is increasingly serious in the world. According to the forecast data of the United Nations, from 2015 to 2055, the proportion of China's population aged 60 and over will increase at an average annual rate of 2.35%.[1]

At the same time, due to heredity, accidents, natural disasters etc., the number of disabled people is also increasing year by year. According to the data analysis of 2018 statistical yearbook of China's undertakings for the disabled, the total number of disabled persons in China was 85.02 million at the end of 2017, including 24.72 million with physical disabilities.[2]

With more intense aging and more people with physical disabilities, the demand for auxiliary medical devices has increased significantly. Nowadays, wheelchair is still the most important auxiliary travel equipment with a variety of functions. The problems of traditional wheelchairs are obvious: the scope of activities is small, and the work efficiency is relatively low. In particular, it is difficult to meet the needs of the elderly and the disabled to go up and down the stairs freely. Hence, the design and research of intelligent building climbing wheelchair can help users travel more conveniently, enjoy life better, and reduce the burden of family and society, which has positive social and practical significance.

Researchers have developed different wheelchairs according to the application fields, including crawler type, walking type, planetary wheel type, and the compound mechanism which is the combination of the two or three of the above ones. The research of crawler climbing mechanism is relatively mature. The advantages are the small subsidence, large contact area with the ground and strong adaptability to the terrain. However, the speed of the crawler climbing mechanism is slow when driving on flat ground and it is difficult to turn around, limiting the use in narrow corridors. [3]The walking building climbing wheelchair is mainly designed based on the principle of bionics and the movement form of animal legs. But the bearing capacity is poor, and the structure is complex, which is cumbersome to control. [4]The planetary wheel stair climbing wheelchair uses the alternating turnover of the planetary wheel set to realize the floor climbing action. The planetary wheel climbing wheelchair has simple structure and flexible movement, but due to the influence of climbing mode, the center of gravity will fluctuate in the process of climbing.[5] In addition, the adaptability of planetary wheel climbing wheelchairs to step height and size is poor, and most domestic planetary wheel wheelchairs do not have anti-skid devices. The compound mechanism integrates the advantages of the corresponding building climbing mechanism, makes up for the shortcomings of a single building climbing mechanism, and has good obstacle crossing function, but it brings the problem of coordinated control among various mechanisms. [6]

According to the analysis of the existing building climbing wheelchairs at home and abroad, the design principle of planetary wheel building climbing wheelchairs is the simplest, with low cost and easy operation, but it also has an obvious disadvantage as poor adaptability. With the rapid development of the construction industry and the great difference in the height of stairs in different buildings, it is difficult for the users to adapt to all kinds of scenarios. Based on the existing planetary climbing mechanism, this paper proposed a novel design of wheelchair with circumferential linkage, which can adjust the climbing height of planetary gears. Considering the safety and stability of wheelchairs, the difficulty of controlling and the comfort of passengers, a multi-functional building climbing wheelchair based on planetary gear train is designed by using modular design idea, including ground walking mechanism, in place steering mechanism， center of gravity adjustment mechanism and Seat leveling mechanism. Then the dimensional parameters of the three-star wheel group and the height of the stairs are determined. Finally, the motor power required by wheelchair climbing stairs is calculated.

2. The holistic model of the intelligence wheelchair

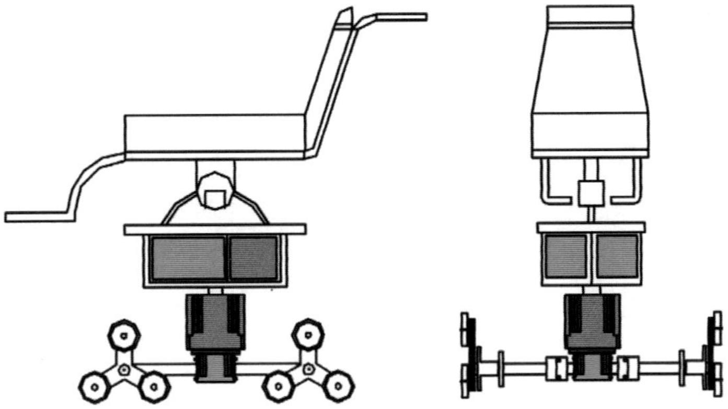

Figure 1. Holistic model of the wheelchair

As shown in Figure 1, the wheelchair is mainly composed of a stair climbing mechanism and a seat leveling mechanism. The star wheel mechanism includes of planetary gears and planetary gear carrier.

Each planetary gear can not only revolve around the central axis of the gear carrier, but also rotate around its own axis. When driving on the flat ground, the planet carrier is fixed relative to the main shaft, and the grounding planet wheels rotate around their respective axes; When walk on stairs, each star wheel is locked, and the star wheel in the wheel set rotates around the main shaft with the planet carrier. According to the number of planetary wheel groups, planetary wheel climbing wheelchairs can be divided into single wheel group and double wheel group. Single wheel set type means that only one pair of planetary wheel sets is used for the front and rear axles of the wheelchair, and the other pair uses ordinary wheels. Double wheel set type means that a pair of planetary wheel sets are used on the front and rear axles of the wheelchair. In this paper, we adopt the double wheel style.

When the wheelchair climbs the stairs, if the seat has no adjustment ability, it will cause discomfort to the passengers. In serious cases, it may fall due to the change of center of gravity. Therefore, the wheelchair shall have seat leveling function to ensure the comfort and safety of passengers.

Different from the previous research for wheelchair with planetary gears, the adjustable stair climbing mechanism is the focus of this paper. The planetary wheel structure of the building climbing wheelchair is based on the ordinary planetary wheel and added with a circumferential connecting rod mechanism, so that the height of the stairs it can climb can be adjusted according to the actual situation. It enhances the adaptability of planetary wheel climbing wheelchair.

3. The mechanism design of the wheelchair with adjustable planetary gears

As the core part of the building climbing wheelchair, the three-star wheel group is mainly composed of star wheel as simplified in Figure 2, circumferential linkage and gearbox. The dimensional parameters of the star wheel and gearbox will determine the stair climbing trajectory of the wheelchair.

When the wheelchair climbs the stairs, all the star-wheels are self-locking and cannot rotate freely. The synchronous turnover of the front and rear three-star wheel groups drives the whole body to complete the climbing action, ignoring the relative sliding between the star wheels and the step plane. Because the three-star wheel group of wheelchair has the same overall dimensions and climbs the stairs at the same turnover speed under the action of chain transmission system, it is only necessary to study the dimensions of one group of three-star wheels. The schematic of the overturning process of the three-star wheelchair group is shown in the Figure 2.

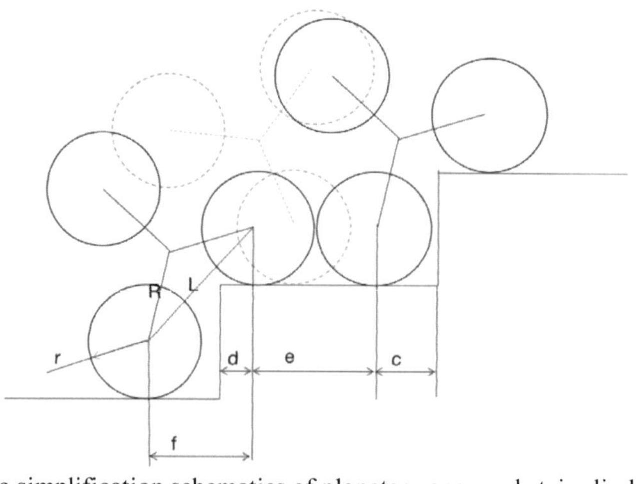

Figure 2. The simplification schematics of planetary gear and stair climbing process

The solid line in the figure describes the initial state and end state when the three-star wheel group is overturned, and the dotted line describes the state during the overturning of the three-star wheel group. Assuming that the stair climbing wheelchair can keep the front and rear position relative to each step unchanged after one cycle of turnover when the three-star wheel group turns synchronously and the landing star wheel does not slip, this is the ideal climbing state of the wheelchair. In this state, it can not only better control the whole process of climbing stairs, but also reduce the friction and collision between the star wheel and the steps, avoid the sliding of the wheelchair, and greatly improve the stability of the wheelchair.

In a climbing cycle of the ideal climbing state, the star wheel landing point of the step above the three-star wheel group is the turning point for turning. In the process of turning, the other two-star wheels have left the ground, and only the landing star wheel rolls purely on the steps. After completing a turnover cycle, the landing star wheel rolls a total of 120 degrees, which can be obtained according to the geometric relationship,[7]

$$R=\{\sqrt{3}*[(a-2\pi r/3)^2+b^2]\}/3 \quad (1)$$

Where R is the radius of the star wheel and set as constant for the following analysis.

It can be seen from the above that there is a nonlinear inverse relationship between the star wheel radius R and the planet carrier length r when the stair size is determined. When the star wheel radius R is determined, the size of the whole three-star wheel set is determined. However, in reality, the size of three-star wheel group of wheelchairs and the size of stair steps often cannot meet the relationship in equation (1). When the climbing height remains unchanged and the width changes, the climbing state of the three-star wheel set needs to change accordingly. The specific changes are as follows:[7]

1.When the step width changes little, the end state of each cycle of the three-star wheel set will deviate from the front and rear position of the step compared with its initial state. During a period of building climbing, the deviation of the wheelchair at each step accumulates. After the building climbing process, the distance between the three-star wheel group and the side of the stairs is compared with that at the initial turnover of the three-star wheel group. If the accumulated deviation is not large and the wheelchair will not slip during the whole climbing process, this climbing state can also be regarded as an ideal climbing state.

2.When the step width is less than the original design size, the three-star wheel group will move forward a certain distance than the ideal climbing position after completing a cycle of turnover. When the wheelchair turns over to the nth step in the process of climbing the building, and the cumulative absolute deviation is within the acceptable range, the previous n-1 turns of the three-star wheel group can be regarded as an ideal climbing state. From the nth turn of the Samsung wheel group, before another star wheel contacts the step, the star wheel with pure rolling has touched the side of the step. When the Samsung wheel group continues to turn, the star wheel will rub against the side of the step until the end of the turning process. Although the climbing process cannot reach the ideal climbing state after that, when the star wheel is rubbed, the wheelchair is in a free-falling state, there is no sliding phenomenon in the whole climbing process, and the safety of the wheelchair can be guaranteed. Therefore, this climbing state is acceptable.

3.When the step width is larger than the original design size, the three-star wheel group will move backward for a certain distance than the ideal climbing position after completing a cycle of turnover. During the climbing process of Samsung wheel group, the previous M-1 turnover can also be regarded as an ideal climbing state. When the three-star wheel set is overturned for the m time, the star wheel that should have been built on the step cannot be built on the step due to the cumulative absolute deviation in the process of wheelchair climbing the stairs. At this time, the three-star wheel set collided with the edge and corner of the step, and the wheelchair slipped to the previous step. This climbing process failed. The collision and sliding in the whole building climbing process are uncertain, resulting in the unpredictable overturning process of the M + 1 of the three-star wheel set, and the wheelchair cannot climb the stairs according to a certain preset track. The stability and mechanical strength of wheelchair will be affected by sliding collision. The unpredictable climbing process cannot ensure the comfort and safety of passengers. Therefore, when the wheelchair climbs the stairs with excessive step width, it is necessary to adjust the dimensions of the three-star wheel set to meets the really conditions. As shown in Figure 3, the importing of the circumferential linkage can resolve this problem by changing the radius of the star wheel, which greatly increases the applicational field of the wheelchair for stairs with different height.

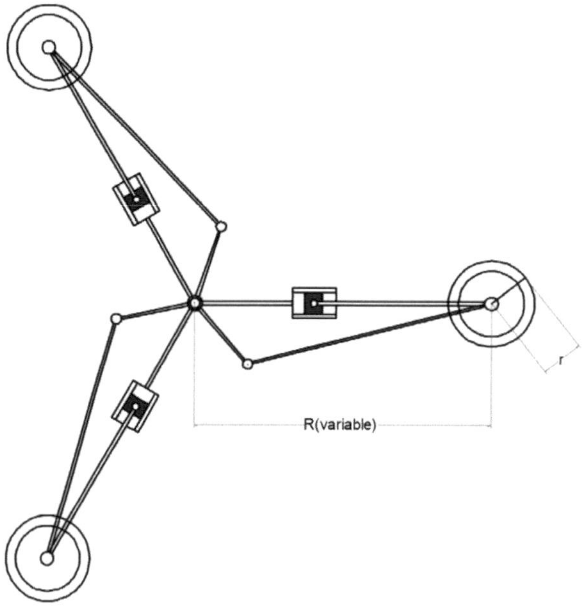

Figure 3. The schematic of the circumferential linkage

4. Structural parameter design and theoretical analysis

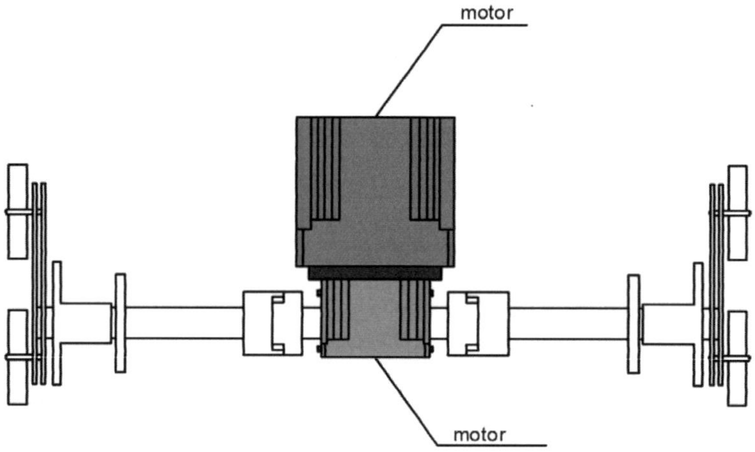

Figure 4. The driving part of the intelligence wheelchair

Figure 4 presents the driving part of the wheelchair, where the motor in the middle provides the power for motion (moving and climbing etc.). The stair climbing process of wheelchair is converting electric energy into gravitational potential energy. During the turning process of the three-star wheel set on the stairs, the wheels in contact with the ground only roll, ignoring the small rolling friction couple, then the average output power of the motor required by the wheelchair to climb the stairs is the absolute value of the output power of gravity to the wheelchair. Therefore, the average output power P3 required by the wheelchair when climbing stairs is

$$P3=Gv\sin\alpha \quad (2)$$

Where V is the average speed of the wheelchair rising along the slope of the stairs, and the average speed in the vertical direction when the wheelchair climbs the stairs is the ratio of the height H2 of the wheelchair climbing the stairs per minute to the time t,

$$V\sin\alpha=h2/t \quad (3)$$

According to the designed wheelchair climbing parameters, the wheelchair climbs 9 steps per minute and each step is 150mm. Therefore, when h2 = 1350mm and T = 60s are brought into equations (2) and (3), the average power required for the wheelchair to climb the building is:

$$P3=32W \quad (4)$$

When the wheelchair climbs the stairs, the overall center of gravity of the wheelchair and the passenger fluctuates up and down with the turnover of the three-star wheel group, and the torque effect of the total gravity of the wheelchair and the passenger on the front and rear three-star wheel groups also changes. Therefore, when the wheelchair climbs the stairs, the output torque of the motor is also changing, and it is difficult to determine the size of the motor output torque at each time, but the maximum value of the motor output torque can be solved. Assuming that the seat is always horizontal during the turnover of the three-star wheel set, ignoring the initial rotation acceleration of climbing the building, the maximum torque to be output by the motor when the wheelchair climbs the stairs is the product of the maximum force borne by the central axis of the three-star wheel set and the maximum action distance. Obviously, the maximum force borne by the three-star wheel set is the total gravity g of the passenger and the wheelchair, and the maximum action distance is the length r of the planetary carrier of the three-star wheel set. Therefore, the maximum instantaneous power P4 output by the motor when the wheelchair climbs the stairs is:

$$P4=53W \quad (5)$$

According to the power requirements and many factors such as price and assembly, yb3-280s-4 motor is selected, and the maximum output power of each motor is 75W.

5.Conclusion

In view of the increasing number of the elderly and the disabled, this paper designs a multifunctional planetary wheel stair climbing wheelchair, which has simple structure and diverse functions. It can change the climbing height according to the actual situation of the stairs, which has good practical value and research significance. Firstly, according to the motion characteristics of the stair climbing mechanism, the ideal climbing state of the wheelchair was put forward. Then, the relationship between the dimensional parameters of the three-star wheel group and the height of the stairs is determined based on the dimensional constraints of the wheelchair in the motion state. Finally, the motor power required by wheelchair climbing stairs is obtained as 53W and the certain type was determined. Although the multifunctional stair climbing wheelchair designed this time can realize the basic functions, it still has a long way to go from market promotion. We believe that it provides a new possibility for the elder and disabled people who need the assistance of intelligence wheelchair.

References

[1] CAI Fang, Supply-and Demand-side Impacts of Aging on China's Economic Growth. Economic Perspectives,2021(1)
[2] China Disabled Persons' Federation. China Disabled Persons' cause statistical yearbook. 2018 [M]. China Disabled Persons' cause statistical yearbook. China Disabled Persons' Federation Information Center, 2017
[3] Pan Sheng,Research on wheel shoe switching automatic ladder wheelchair [D] .Zhejiang agriculture and Forestry University, 2018

[4] Shi Kang,Design and analysis of a new type of manned climbing wheelchair[D].Beijing Jiaotong University,2019
[5] Zhang Fuqiang, Zhang Yu. Design and research status and development trend of building climbing wheelchair in China[J]. Modern industrial economy and informatization, 2019, 9 (8): 38-39
[6] Cai Jingyu, Feng Liyan, Li Desheng. Design and analysis of a wheel shoe composite stair climbing wheelchair [J]. Mechanical engineering and automation, 2021 (01): 106-107 + 110
[7] Dai Bensheng. Design and research of multifunctional planetary wheel stair climbing wheelchair[D]. Xuzhou: China University of mining and technology, 2017

Design and Implementation of an Augmentative and Alternative Communication System Based on MATLAB GUI

Zihe Wang[1], Boyu Si[1*] and Jingwen Wang[1]

[1] School of Medical Instrument, Shanghai University of Medicine & Health Sciences, Pudong District, Shanghai, 201318, China

[*]Corresponding author's e-mail: siby@sumhs.edu.cn

Abstract. In clinical practice, augmentative and alternative communication (AAC) is an effective training tool to improve the communication skills of individuals with language impairment, hearing impairment, and autism. However, there are few studies on the development of augmentative communication systems based on Mandarin Chinese in China. This study presents an augmentative communication system developed using the MATLAB GUI platform. The system consists of two parts: a learning part and a communication part, where the learning part provides the basis of the vocabulary content for the system, covering 5558 words, and its content is mainly a word list designed based on the content of special school language textbooks. The communication part provides the system with a complete discourse training content, mainly by arranging and combining the vocabulary content of the learning part to form a complete discourse. This aided communication system can expand the vocabulary of people with language expression disorders and improve their social communication skills. In addition, the effectiveness of the system was tested in a single-subject experiment, and the results showed that the augmentative communication system is effective in learning pronunciation, expanding vocabulary, and improving the social communication.

1. Introduction

Among many AAC systems there are few that are currently developed based on Mandarin and can be widely used in mainstream operating systems to achieve a system that allows patients with expressive language disorders to install and use the system for learning correction with a simple operation on an electronic device[1-3]. As the society has been developing fast, people with speech disorder need to improve their life quality by receiving treatment and training [4-5]. Therefore, this study proposes an AAC system based on the MATLAB GUI platform. The system is easy to operate for the Chinese Mandarin speakers. It facilitates the use of patients with language expression disorders in therapeutic training in China.

2. System Framework

The system consists of four major components: communication symbols, communication aids, communication techniques and communication strategies, which are complementary to each other. Based on the MATLAB GUI technology, the system is designed with an interface, pictures and audio that match the word list content.

Communication symbols: A symbol that collects the required information through tactile, visual and other sensory symbols, and then transmits it to others through body language or other devices, such as gestures, pictures, words, etc.

Communication aids: Patients who are unable to express their ideas to others through the use of language to achieve the purpose of communication need to use other assistive devices to complete communication.

Communication technique: The way and method for patients to use communication aids. It can be divided into direct selection and indirect selection. Direct selection means that the patient selects options such as buttons on the augmentative communication device through his or her body parts (e.g., fingers, mouth, etc.), while indirect selection means that the patient operates the augmentative communication system through other objects such as pens and chopsticks instead of his or her body parts.

The communication strategy in this system consists of communication symbols, communication aids, and communication techniques. The communication strategy mainly enables the rehabilitation physician to provide a specific augmentative rehabilitation plan based on the individual patient's real situation, thus improving the patient's use and rehabilitation process.

3. MATLAB GUI Programming

First open MATLAB software, enter lowercase GUIDE in Command Window command box and enter.

In the New GUIDE file interface, choose to create a blank file 'Blank GUI', select the file storage location, store it in a folder, click OK to generate the GUI file and has been saved.

To display the background photo, add the following corresponding code to function figure1_CreateFcn(hObject, eventdata, handles).

```
ha=axes('units','normalized','pos',[0 0 1 1]);
uistack(ha,'down');
ii=imread('background.jpg');
image(ii);
colormp gray
set(ha,'handlevisibility','off','visible','on');
```

Click on the image to make it sound, insert the button, double-click on the button logo, enter the property editor, delete "button" from the property string, change the property Units to pixels, and add the following two procedures to the OpeningFcn in the editor.

```
icon_name=importdata('filename.jpg');
set(handles.pushbutton1,'CDAT'',icon_name);
```

After the image is added, the sound insertion, right click 'call back' to enter the program writing interface, the software directly to the corresponding program framework is built and displayed, add the corresponding sound insertion procedures below it, the procedure is as follows.

```
[y,Fs]=audioread('filename.mp3');
Soud(y,Fs);
```

To switch between the two interfaces, you need to insert a button, right click 'call back' to enter the program writing interface and insert the corresponding program.

```
set (gcf, 'visible', 'off');
UI_name ('visible','on');
```

The above is the completion of the basic writing of the program, on the basis of this pilot test, you can again write the secondary enrichment content as needed.

4. Construction of AAC system

The interface of the system is shown in Figure 1. To attract users' interest, a simple and clear layout is used and the color scheme of the interface is slightly warm. The two functional modules of the system are linked in the interface. In this interface, users can click on either the Learning section or the Communication section for different purposes of learning and training.

Augmentative communication systems can be targeted to help use patients to achieve better rehabilitation training purposes, can stimulate their willingness to want to communicate, and reduce the sense of failure and frustration they suffer when they cannot successfully communicate with others. It improves their communication skills and enables them to have a healthy psychological development. This auxiliary system consists of two parts, the learning chapter and the communication chapter, which form the main body, and generates a multi-level and multi-content learning and communication interface through the function of interlinking between interfaces and interfaces, while keeping the interface simple and not easily cumbersome for users.

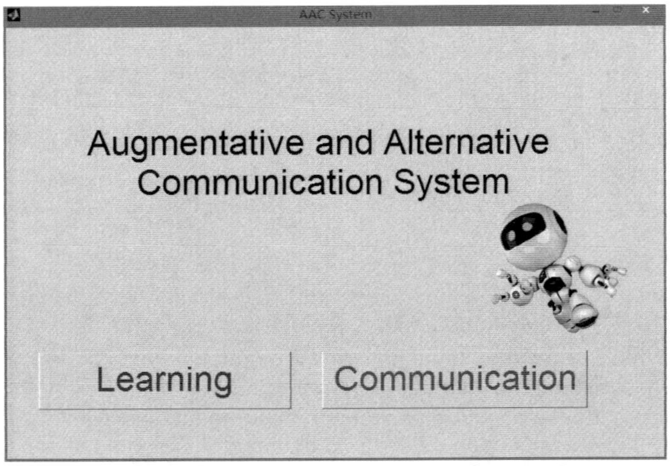

Figure 1. User Interface of Components of an AAC system.

The content of the learning chapter is based on the content of the special school language materials and the corresponding design of the learning word list, which covers almost all the high-frequency basic words involved in daily life. For example (Table 1): numbers, people, colors, sports, weather, common plant and animal names, etc. After the user clicks the learning chapter on the initial interface to enter the learning, the vocabulary content will be presented to the user in a combination of voice and picture, and the user can learn and train the pronunciation by clicking the picture or text. The user can click on the same pronunciation repeatedly during the training process to make the training effect more obvious.

Table 1. Contents of the learning chapter.

Number	People	Animal	Item	Color
Sports	Body	Direction	Food	Weather

The content in the communication chapter was mainly derived from reviewing related literature and research. In the research, the types of communication were identified by observing the common words used by preschool and school-age children. In the communication chapter, patients are taught how to link, use, and combine vocabulary, and learn the high-frequency vocabulary in the chapter to make a complete communication statement. For example: statements that express the patient's wishes: I want to (can be combined with statements such as: eat apples, eat bananas, buy cars, wear clothes, etc.) / I don't want to (can be combined with statements such as: take a plane, make bread, eat rice, drink water). Commonly used statements in life; (dad, mom, teacher) hello, thank you (classmate, uncle, aunt), (I'm) full, I'm hungry, and a full range of example sentences to meet the different needs of users. The background of the communication module is different from that of the learning module, so that it can be better differentiated. The communication module uses warm and beautiful photos, which can make patients take the initiative to learn and experience different feelings from the learning module. The function of the communication module also adds a picture function that matches the text content, and adds a voice function, which makes it easier for patients to learn to master the pronunciation of statements and enhance their ability to understand the meaning of statements when using the communication module, so as to train them to express complete statements.

5. Experiment

A single-subject experimental design method was used. The single-subject experimental design is mainly aimed at quantitative assessment of single or few subjects and can compensate for the inaccuracy of statistical results due to insufficient sample size. The improvement of users' (i) correct pronunciation vocabulary and (ii) correct communication rate over time was directly observed.
The experimental design was specified as follows.

5.1. Subjects' basic information.

Two children with hearing impairment: A and B, male, both 5 years old, received cochlear implant surgery, had basically recovered hearing function, did not attend kindergarten, had received speech rehabilitation training for 1 month, had established initial awareness of speech, and had some learning ability.

5.2. Experimental methods.

(1)*Measurements of correct pronunciation vocabulary*: The experiment used an A-B single baseline experimental design, with A as the baseline period and B as the treatment period.

First, the baseline experiment was conducted. The vocabulary list was obtained from the learning module of the communication aid system designed in this paper, and a total of 545 words were screened, and each word was worth 1 point.

Second, a treatment period experiment was conducted. Each subject was asked to use the communication aid system to learn the pronunciation of the vocabulary. A learning period of 7 days was used, and at the end of each period, the correct pronunciation vocabulary score was performed. The entire intervention period was 21 days, and the correct pronunciation vocabulary scores for the intervention period were obtained in 3 sets of scores in chronological order.

Finally, the A-B single-baseline line graphs were drawn based on the "correct pronunciation vocabulary" scores obtained in the baseline and treatment periods, and significance tests were performed on the data from both periods.

(2)*Communication correctness*: The experiment still adopts the A-B single baseline experimental design, where A is the baseline period and B is the treatment period. The "communication correctness" measure covered 10 communication scenarios, each of which consisted of 2 sets of communication statements. These scenarios were taken from the communication module of the communication aid system, and subjects were scored 1 point for each scenario in which they correctly used AAC to complete 1 set of communication. Communication correct rate = correct score / total score of communication scenarios × 100%.

First, a baseline period experiment was conducted. The "communication correctness" of the two subjects was counted three times in each period, and the scores were recorded once a day. The results of communication correctness were calculated and recorded.

Second, a treatment period experiment was conducted. Each participant was asked to use the communication module of the communication aid system to learn the vocabulary of communication terms. A learning period of 5 days was used, and at the end of each period, the communication correctness score was performed. The entire intervention period was 15 days, and the communication correctness scores for the intervention period were obtained in 3 sets of scores in chronological order.

Finally, a single baseline A-B line graph was drawn based on the "communication correctness" scores obtained for the baseline and treatment periods, and the data from the two periods were subjected to significance.

5.3. Significance tests of indicator data in both periods were performed.

Significance tests of indicator data for the baseline and treatment periods were analyzed by t-test with SPSS 19.0, and the results of significant differences are shown in Table 2.

As can be seen from Table 2, for both subjects A and B, their two indicators (i) the amount of correct pronunciation vocabulary and (ii) the rate of correct communication were significantly different between the baseline and treatment periods, especially the rate of correct communication for subject B was extremely significantly different between the baseline and treatment periods. The results indicate that the communication aid system can play a more significant role in pronunciation learning and improving the communication success rate for patients with language expression disorders.

Table 2. Data significance tests for baseline and treatment periods.

	Subject 1 (Sig.)	Subject 2 (Sig.)
Correct pronunciation vocabulary	0.02*	0.030*
Communication correctness	0.018*	0.009*

6. Conclusion

This study is based on the MATALB platform, which is easily changeable, and can be upgraded at a later stage, such as adding to the augmentative communication system, types of pictures, categories, styles, etc., as well as improving the functions of the system, adding recording functions, game animation functions, search functions, and other related functions such as the patient's own combination of statements, adding personalization for different kinds of users with different interests The system is designed for different kinds of users with different interests to choose the content they want to learn, etc. On the basis of this system, we can improve the system individually for different individuals to better mobilize patients' interest in learning, so that they can learn in entertainment and achieve better training effect. It is also possible to find professional announcers to modify and enrich the system's audio resources and add corresponding pronunciation, word meaning and sentence meaning to provide users with correct and guaranteed quality audio resources. On the basis of this system, it is also possible to add a networking function to connect the system with mainstream electronic mobile devices in the market, so as to assist users in learning and training through mobile terminals or to install the system on mobile devices for easier and faster use. In the connection with mobile devices, it can also be combined with artificial assistants, VR, AR and other functions, through a variety of ways to make users more willing to use the aided communication system, improve communication and learning ability, and achieve good training results, so as to better help users achieve the purpose of rehabilitation.

References

[1] Fu, Z. (2020) Discussion on the application of augmentative communication system in mainland China. Journal of Xiangyang Vocational and Technical College,19:42-44,52.

[2] Wang, X. (2020) Research on the application of augmentative communication system in special education. Science and Technology Wind, 9:211-213.

[3] Ye, L., Sun X.,Yang X. (2021) Effect of simulated school daily life scenario combined with augmentative communication system on improving communication skills of school-aged children with epilepsy. Nursing Practice and Research,18:567-569.

[4] She, L., Zou Y., Hu, Y. (2021) An intervention study of augmentative communication system on functional vocabulary learning of preschool children with developmental delays. Modern Special Education, 2:28-33.

[5] Wang, F.(2019) A preliminary investigation on the strategy of the 'prelude' stage of communication and expression development for children with severe autism--based on the activity-based augmentative communication system. Modern Special Education, 13:61-64.

Automatic Assembly of Bolts and Nuts Based on Machine Vision Recognition

Haocheng He

College of Mechatronics and Control Engineering, Shenzhen University, Shenzhen, 518060, China

Author's E-mail: hehaochengliuxue@126.com

Abstract: In this study, based on the computer vision technology, we developed a recognition system for nuts positioning to complete the automatic bolt assembly part of the automatic production line. The actual image of the nut was captured by an industrial camera, which would be processed by the following edge detection and Hough circle transformation. After that, the coordinates of the nuts were obtained in the pixel scale. Finally, the real position of nuts would be fed back to the robot arm, according to which the automatic assembly of the bolt would be completed. This computer vision based recognition system is an indispensable part for the efficiency and accuracy improvement of automatic production line.

1. Introduction

Assembly is one of the important parts of machinery manufacturing, which occupies a considerable proportion in the whole manufacturing process [1-3]. The assembly process is usually divided into component assembly, component assembly, assembly stages, and so on. The threaded connection is a disassembled connection, which has the advantages of easy disassembly, good reliability and reusability, etc. Bolted connection is almost inevitable in all stages of assembly [4-6]. At present, in the traditional assembly line, the bolting and adjustment are still done manually. However, there are some problems such as high error rate, high labor intensity and low efficiency during manual work. With the continued requirements of high efficiency and low cost labor for the production line, the shortcomings of manual tightening operations become increasingly apparent.

In recent years, with the increasing demand for automation in industrial production, machine vision inspection technology has been used in a large number of industrial, medical, telemetry detection and intelligent transportation fields. Among them, online inspection and positioning of products is one of the most widespread applications of machine vision inspection technology. The use of machine vision-based positioning methods could significantly improve production efficiency and positioning accuracy, making the automation and intelligence of the production line greatly improved. For example, Z.A. Jaffery and his coworkers [7] proposed a method of emulating the visual ability of human operator for the automation of the process based on a machine vision approach, which has high reliability and robustness with less computational complexity. A.H.A. Mohammed and M.A. Alshameri [8] developed a novel automated engagement intelligent spanner which was capable of autonomously changing its jaws' size according to the diameters of the bolt's/nut's heads. Q. Geng et al [9] presented an automated visual inspection system based on space-scale normalization to inspect the status of fastening bolts on the sides of China Railway High-speed, which performs excellently in monitoring the missing fastening bolts.

In this study, we developed a recognition system for nuts positioning to complete the automatic bolt assembly part of the automatic production line based on the computer vision technology. The actual image of the nut was obtained by a high speed industrial camera, which would be processed by the following edge detection and Hough circle transformation. After that, the coordinates of the nuts were obtained in the pixel scale. Finally, the real position of nuts would be fed back to the robot arm, according to which the automatic assembly of the bolt would be completed. This recognition system could effectively improve the accuracy and efficiency of automatic assembly of bolts and nuts.

2. Experimental Setup

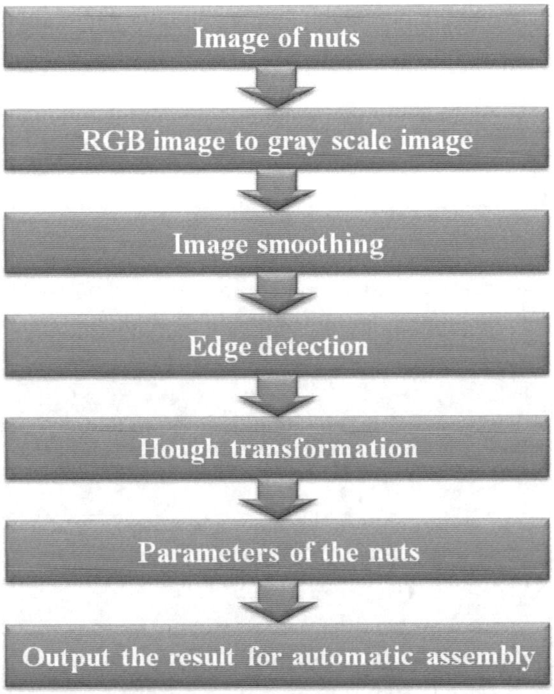

Figure 1. The identification process of nuts for automatic assembly.

Fig. 1 shows the identification process of nuts for assembly. First, an industrial high-speed camera was used to capture the actual image of the nuts. Then, the RGB model images were converted into the grayscale images for subsequent processing. In order to highlight the features of the image and reduce the negative impact of redundant pixel points on the detected edges, the image filtering and smoothing was further carried out by using Gaussian filtering and morphological gradients. Finally, the contour of the nuts was separated out by Hough transformation [10]. The parameters of the nuts circle coordinates and diameters were obtained with the relationship between the number of pixel points and the actual size. Based on the detection results, the location information of nuts was identified, according to which the robot arm would automatically complete the assembly procedure of bolts and nuts.

3. Results and discussion

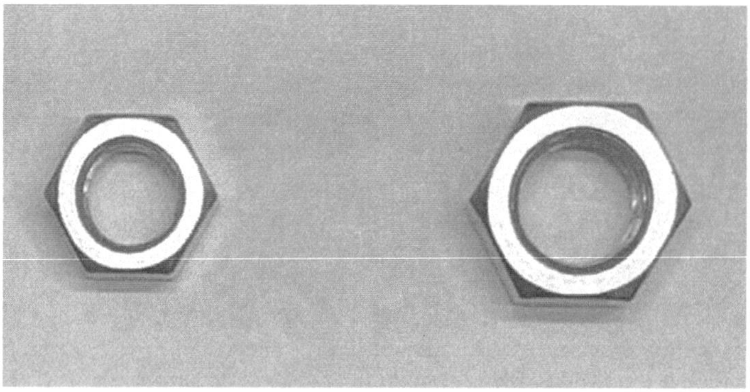

Figure 2. The image of two nuts with different size.

Fig. 2 shows the image of two nuts with different size taken by a high-speed camera on the actual production line. This kind of image was very common on the assembly line, which could be used as a typical example for analysis. It is found that two nuts with different size lying in the production line. The purpose of the inspection was to determine the data including the location of the center and diameter of the nuts from the image for automatic assembly.

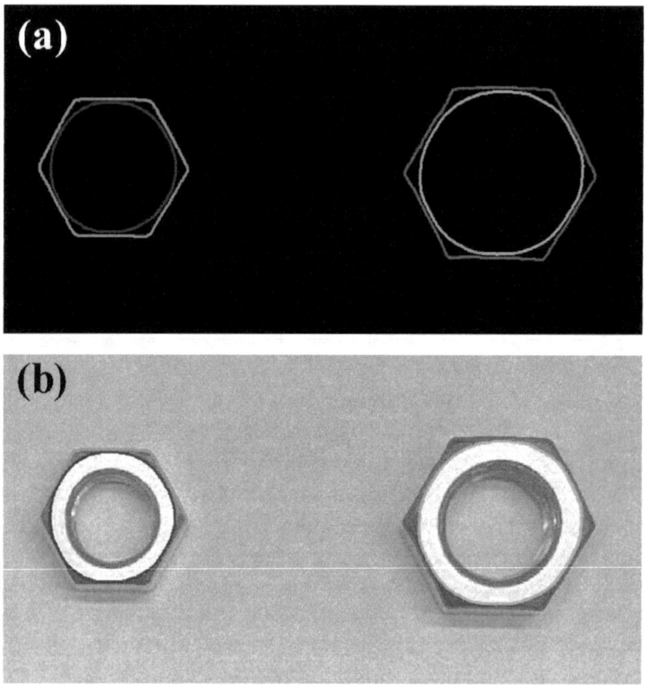

Figure 3. The contour of nuts after image processing.

Fig. 3 shows the processing results of nuts by the machine vision recognition algorithm. According to Fig. 3, the RGB format image was first converted into a grayscale image. Then, the image was smoothed by Gaussian filtering to reduce the effect of noise on image recognition. After that, the edge detection was performed on the above processing image by the Canny edge detection. It is found that the hexagonal edges of the outer circle of the nuts and the circular edges of the inner circle were accurately detected in Fig 3(a). Fig. 3(b) presents the fusion of edge detection and original image.

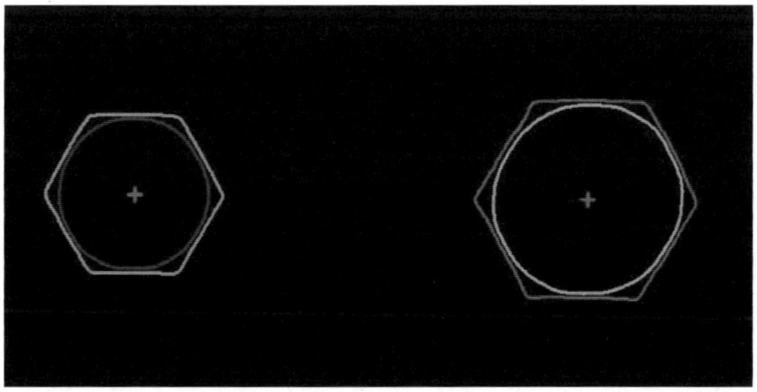

Figure 4. The detection result of nuts after Hough circle detection.

After edge detection, the Hough circle detection was carried out on the image, as exhibited in Fig.4. According to Fig. 4, it is observed that all the nuts were accurately detected. In order to detect two centers of the circle, the centers were marked by red cross, as revealed in Fig. 4. At the same time, the center coordinates of these circles were (843.028, 1231.625) and (3725.336, 1265.075), respectively. In order to achieve high accuracy, the pixel coordinates kept three decimal places for the circle centers. According to the tested diameters of nuts, the length value corresponding to the coordinate value would be obtained, which was used for the calculation of the real position. Based on the detection results, the location information of nuts was identified, according to which the robot arm would automatically complete the assembly procedure of bolts and nuts. Therefore, it is demonstrated that the designed recognition system based on computer vision could effectively improve the accuracy and efficiency of automatic assembly of bolts and nuts.

4. Conclusion

In summary, this study designed a recognition system based on computer vision for automatic assembly to complete the precise positioning of nuts. The high-speed camera could be perfectly adapted to the real assembly line and clearly capture the image of each nut. Then, the impact of noise was reduced and the image features were highlighted through image morphology processing and smoothing. After that, the extraction of edges was completed by edge detection. Finally, the parameters of nuts were accurately obtained by the Hough circle transformation. With the assistance of the automatic inspection system, the production efficiency and accuracy could be significantly improved.

Acknowledgement

This study was supported by the Scientific Research and Training Program.

References

[1] Wang L., Mills J.K., Cleghorn W.L. (2008) Automatic microassembly using visual servo control. IEEE T. Electron. Pack., 31(4), 316–325.

[2] Das A.N., Murthy R., Popa D.O., Stephanou H.E. (2012) A multiscale assembly and packaging system for manufacturing of complex micro-nano devices. IEEE T. Autom. Sci. Eng., 9(1), 160–170.

[3] Liu S., Xu D., Zhang D., Zhang Z. (2016) High precision automatic assembly based on microscopic vision and force information. IEEE T. Autom. Sci. Eng., 13(1), 382–393.

[4] Wang T., Song G., Liu S., Li Y., Xiao H. (2013) Review of bolted connection monitoring. Int. J. Distrib. Sens. N., 9(12), 871213.
[5] Lacey A.W., Chen W., Hao H., Bi, K. (2019) Review of bolted inter-module connections in modular steel buildings. J. Build. Eng., 23, 207–219.
[6] Galińska A. (2020) Mechanical joining of fibre reinforced polymer composites to metals-A Review. Part I: Bolted joining. Polymers, 12(10), 2252.
[7] Jaffery Z.A., Sharma D., Ahmad N. (2017) Detection of missing nuts & bolts on rail fishplate. 2017 International Conference on Multimedia, Signal Processing and Communication Technologies (IMPACT), 36-40.
[8] Ali M.A.H., Alshameri, M.A. (2019) An intelligent adjustable spanner for automated engagement with multi-diameter bolts/nuts during tightening/loosening process using vision system and fuzzy logic. Int. J. Adv. Manuf. Tech., 101(9), 2795–2813.
[9] Geng Q., Liu W., Dai Y. (2020) Automated visual inspection of multiple missing fastening bolts on the sides of high-speed train body in railway maintenance. IEEE Access, 8, 187060–187071.
[10] Mukhopadhyay P., Chaudhuri B.B. (2015) A survey of Hough Transform. Pattern Recogn., 48(3), 993-1010.

Which one is better: Assessment and Locomotion imitation strategies in Changing Environments

Cunying Chen[1*] and Hua Zhang[2]

[1] Department of IT Development & Management, Hua Qiao University, Quanzhou, Fujian Province, 362021, China

[2] Business School, Hua Qiao University, Quanzhou, Fujian Province, 362021, China

[*]Corresponding author's e-mail: chencunying@hqu.edu.cn

Abstract. Imitation is ubiquitous, yet what self-regulation orientations' role played in imitation strategies is poorly understood, which is particularly challenging in dynamic and uncertain environments. According to regulatory mode theory, we model two imitation strategies: assessment and locomotion. Assessment pays more attention on comparation among different alternatives, they repeatedly measure, evaluate, and compare desired means and try to find out the 'best' one. Contrariwise locomotion refers to 'keep moving', once choosing one alternative, they change some choices and learn from the resulting performance feedback. Using a computational model, we explore the performance implications of dynamic environments for these two imitation strategies. Consequently, when environment is stable, assessment is more effective in maintaining the lead, whereas locomotion prevails as environmental changes become more frequent and substantial. We contribute to the literatures on strategy, imitation, and NK studies.

1. Introduction

The strategy and organization literature has paid substantial attention to imitation not only because imitation is ubiquitous, but also because imitation is a frequently-used approach to learning and central way to get competitive advantage [1]. However, what has yet to be considered is the possibility that self-regulation orientations play a role in how their imitation. Yet the efficiency of imitation strategies is poorly understood, which is particularly challenging in dynamic and uncertain environments.

Regulatory mode theory posits that people have two key self-regulation orientations towards movement: assessment and locomotion [2]. When people adopt assessment imitation strategy, they make comparisons among alternatives. While when people use locomotion imitation strategy, they prefer to change some choices and learn from the resulting performance feedback once they pick up an alternative. Moreover, although several studies have used NK model to examine the imitation outcomes, they have paid less attention to dynamic and uncertain environments [3]. Such environments are often characterized by uncertainty arising from sudden environmental shifts, that is, landscape changes every f period with w magnitude. In this paper, we build on this burgeoning stream of research to examine the efficacy of assessment and locomotion imitation strategies in changing environments.

The main contributions of this paper are summarized as follows:

- To the best of our knowledge, ours is the first study to simulate locomotion and assessment orientations in a computational model to explore the non-linear relationship between imitation strategy and dynamic environment, extending the traditional NK model.
- We extend the current research on regulatory mode theory in the context of dynamic and uncertain environment. Most studies on regulatory mode are rooted in psychology. We introduce this theory to Artificial Intelligence research and highlight its value in predicting the performance implication of agents' problems solving strategy.

This paper is organized as follows. First, we review the literature and introduce the related work which provide the foundation for our work. Second, we develop a computational model of imitation through which we examine the performance implication for the two imitation strategies. Next, we implement the extended NK model and analyze the final results. At last, we conclude by briefly highlighting key theoretical insights resulting from our analysis and discussing limitations of our research.

2. Related work

2.1. NK Performance Landscape and environmental changes

While in the traditional NK model the environment is typically assumed to be stable, recently some scholars considered the environmental change which constitutes a break with the past. Csaszar and Siggelkow [4] altered the landscape at regular intervals by changing the payoffs of each solution to introduce the changing performance feedback. Since Csazar's work, scholars construct the changing environment model with modifying the payoffs. Sharapov [5] created another landscape using the same N and K, and introduced the weight w representing the magnitude of environmental changes, and the frequency f representing the length of changing intervals. With the changing environment model the author investigated how different leader-rival imitation strategies in a changing environment affect the leader's performance over time. Similarly Dirk M [6] introduced a coefficient which can take any value between -1 and +1 to reflect the correlation between the payoffs pre- and post-environmental change. Particularly if the correlation is one, the environment is stable, and if the correlation is positive, payoffs which are bad in the past become good through the change (and vice versa). Changing environment model has both important practice and theoretical meanings, by exploring the role of environmental changing for different searching strategies, we relax the assumption of static environments in prior research on exploring the performance implication of alternative strategies.

2.2. Regulatory Mode Theory

Regulatory mode theory suggests that locomotion and assessment orientations are universal tendencies. Assessment pays more attention on comparation among different alternatives, they repeatedly measure, evaluate, and compare desired means and try to find out the 'best' one. Contrariwise locomotion refers to 'keep moving', they pursue changes from one state to a desired other state even final state leads to a worse situation[7]. Prior studies also indicated that either can be used in research and popular in practice. In fact, locomotion and assessment are quite different in the nature and consequences and they are independent [8]. That is, there are individuals who are high in locomotion and low in assessment, vice versa, while there are also individuals who are both high in locomotion and assessment or both low. We should note that locomotion and assessment would compete for an individual's resources. Although there may be many possible combinations of locomotion and assessment, we shall consider each individually. Even an individual who is both high in locomotion and assessment, however, in a given situation, he or she will emphasize one or the other mode. Specific activities would be token by any individual at any time point. This paper is concerned with the performance implication when individuals emphasize locomotion or assessment in their goal pursuits.

2.3. Imitation Strategy and Regulatory Mode

To solve a problem, regardless of its complexity and uncertainty, a searching strategy must be identified. Two alternative strategies are typically modeled in prior literature: experimentation strategy and imitation strategy [9]. Local search can be seen as a typical experiential strategy that involves changing only one choice at a time and decide whether to accept the new solution or deny it from the performance feedback. Rather than exploring the landscape itself, imitation strategy relies on imitating the other agents' solutions. Imitation strategy is a substitute for experimentation strategy in the sense that it allows for simultaneously changing several choices. Gavetti and Levinthal [10] introduced cognitive representations in search strategy research. Cognitive representations are modeled as a part of the real landscape. Although it is rough and partial it can improve the effectiveness of search. In this case, experimentation strategy can be seen as a looking-forward strategy depending on the cognitive map while imitation strategy belongs to looking-backward strategy depending on the performance feedback.

This study benefits from the above research. Self-regulation orientations capture the human nature and there is little we know about the performance implication of imitation strategies based on the two orientations in a changing environment. Aiming at exploring which imitation strategies have a better performance in a changing environment, the locomotion and assessment are introduced and the related strategies are defined in detail. In this paper, the two parameters: the frequency and magnitude are defined to measure the changing environment. Our computational model extends the original NK model and provides a more comprehensive and in-depth understanding of problem-solving process.

3. Model

To investigate how different imitation strategies in a changing environment affect the performance over time, we use an adapted version of an NK simulation model. In our model, agents use locomotion and assessment imitation heuristics to travers a fitness landscape. According to NK model, N represents the number of binary choices, each of which is interdependent with K other choices (drawn randomly in our model).

We use a moderately rugged landscape with $N = 10$ and $K = 4$ to reflect important interdependencies between choices, which are present in many previous literature settings. To introduce environmental changes into our simulation, we alter two parameters w and f. the weight w reflects the magnitude of environmental changes, while the frequency f of changes corresponds to the length of intervals between periods in which the landscape is altered.

Agents conduct two different imitation strategies: locomotion and assessment. According to regulatory mode theory, assessment takes the superior configuration as a reference point in a distant search process while locomotion tries to change some choices at a time and learning from the resulting performance feedback.

4. Simulation Results

4.1. Implementation

We consider two agents randomly seeded on the landscape at the start of each run of the simulation. The performance is updated in every simulation period. Our simulation last for 100 periods which make sure the two imitation strategies get final performance level, and we perform 1000 runs of each simulation. Alongside the two imitation strategies, we vary the frequency ($f = 10$ to 100, corresponding to the landscape changing every f periods) and magnitude ($w = 0.1$ to 0.9, corresponding to the weight given to the newly-generated performance landscape, with 1-w being the weight assigned to the previous one) of environmental changes. Note that high levels of environmental variations are associated with low values of f and high values of w. the detail parameters are following.

Table 1. Summary of Model Parameters.

	Parameter Definition
Environment	N: the number of choices
	K: the number of other choices which influence each N
	f: the length of intervals between periods in which the landscape is altered
	w: the magnitude of environmental changes
Locomotion	Keeping compare all the superior alternatives
Assessment	Once choose one alternative try to change some choice to learn the feedback

4.2. Analysis
We use this analytical structure of NK model and assessment and locomotion imitation strategies to engage in two sets of analyses. The first examines the differences in performance of the assessment and locomotion imitation strategies under stable and dynamic environment, while the second set extends this comparison to explore the final performance level considering both magnitude and frequency of environmental changes.

4.2.1. The differences in performance of the assessment and locomotion imitation strategies under stable and dynamic environment. Figure 1 shows, in stable environments, assessment outperforms locomotion all the time. By contrast, in dynamic environments, assessment imitation strategy can not regain the lead all the time. Indeed, as figure 2 shows, assessment outperforms locomotion in the short run while locomotion get a much higher performance level holding the frequency and magnitude of changes fixed (f=10, w=0.1).

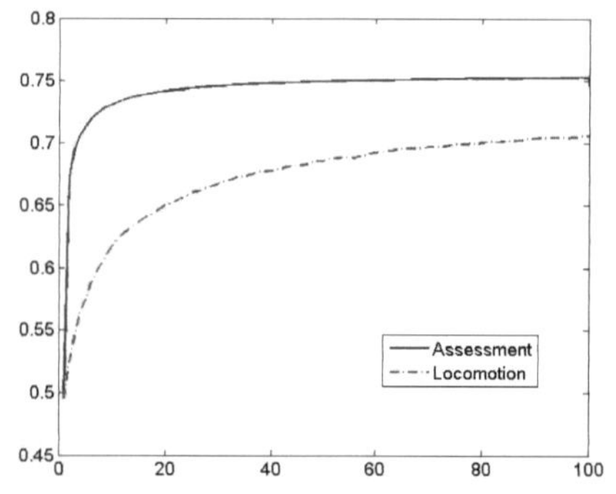

Figure 1. Performance of assessment and locomotion imitation strategies in a stable environment.

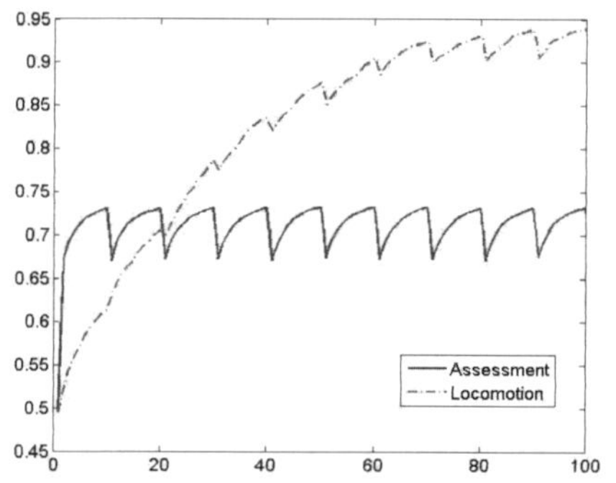

Figure 2. Performance of assessment and locomotion imitation strategies in a dynamic environment.

4.2.2 Final performance gaps under different magnitude and frequency of environmental changes. We take the performance gaps between the locomotion and assessment in final time, this measure of performance captures the fluctuations of these relative performances and is a function of environmental changes. Figure 3 shows, an increase in the magnitude of environmental changes appears to increase the performance level of locomotion imitation strategy relative to assessment imitation, with this effect being most pronounced when magnitude increases from moderate to high (f = 50 to 90). The results are more complexity under different frequency of changes. As figure 4 shows, low and moderate frequency leads to a large performance gap, when magnitude increases from low to moderate (w=0.1 to 0.5) low frequency (f=10) get a better performance gap, while when magnitude increases from moderate to high (w=0.5 to 0.9) moderate frequency (f=50) outperforms.

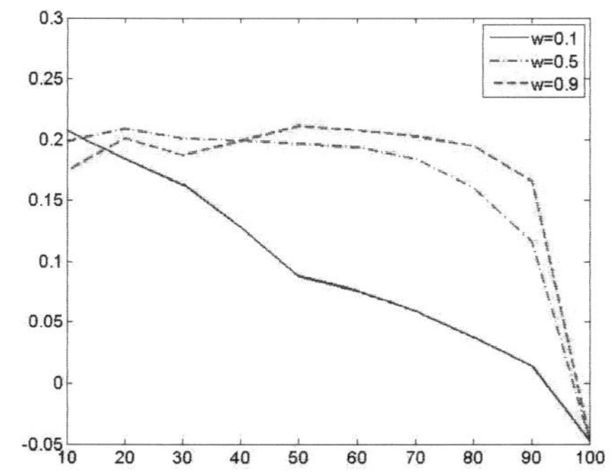

Figure 3. Final performance gap between locomotion and assessment when w=0.1, 0.5 0.9.

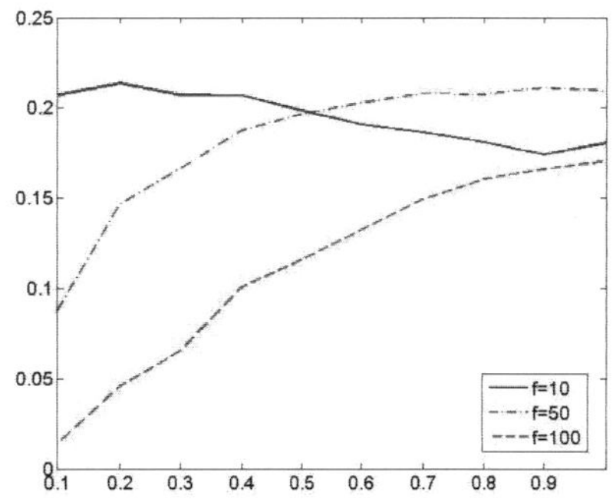

Figure 4. Final performance gap between locomotion and assessment when f=10, 50, 100.

The above findings can be explained by considering the trade-off between exploration and exploitation. The most pronounced difference between locomotion and assessment is that locomotion can be seen as a series of experiments due to their strong willing to keep moving. when environment is stable rugged-landscape models offer the results that locomotion is local and fails to consider the superior alternatives, it will likely lead to a low local peak. That is the reason why assessment outperforms under stable environments. When environments change, assessment is no more a better substitute for experimentation. In this case it is not a good ideal for decision making without considering the experimental learning process. Dynamic and complex changing environment make it difficult to only rely on imitation. Assessment imitating too much precludes the potential for subsequent improvements through local search. Our simulation results echo March's (1991) [11] argument for sustaining diversity in learning systems.

5. Conclusion

Making use of a simulation model, we examine how the locomotion and assessment imitation strategies perform in changing environments. We find that assession outperforms locomotion in a stable environment, however, increases in the magnitude and frequency of environmental changes increase the attractiveness of locomotion imitation strategy over assessment. When environmental changes are frequent and moderate, locomotion imitation is more effective in maintaining the lead. We reveal that the relative effectiveness of locomotion is driven by the trade-offs between exploration and exploitation. Complex problems make it difficult to rely only on assessment. This is because when benchmarking is changing, even small fluctuations in environment may lead to significant performance penalties. What is more, assessment precludes the potential for subsequent improvements through local search. Especially in a frequent and moderate dynamic environment, locomotion who keeps changing choices and learning from the resulting performance feedback can get a better final performance level. Our insights extend research on the imitation-performance relationship, which has largely focused on legitimacy- and learning-based arguments [12], and address calls to incorporate the notion of environmental uncertainty more completely into management theory [13].

A limitation of this study is that we design locomotion and assessment separately. According to self-regulation research locomotion and assessment can work together in an interdependent manner. High achievement needs both a correct direction to target (high assessment) and a commitment consistently for a movement (high locomotion). So there is an interesting opportunity for future research could be to design searching strategy combined both locomotion and assessment. Furthermore, we didn't consider the co-evolution of self-regulation and environmental change which can take many forms. This dynamic process would end up with finding a perfect independence and it is worthy of further development.

Acknowledgments

This study is supported by grants from the Ministry of Education in China Project of Humanities and Social Sciences (Project No.17YJA630132), the 68th General Funding from China Post Doctoral Science Foundation (2020M682068), National Natural Science Foundation (No. 71802137), Project of The Young Researcher Developing Foundation in Shanghai (SISUQJZZJH2016006), and Project of SISU (20161140001; 2018JG020). The authors are extremely grateful for the valuable comments and suggestions from two anonymous reviewers and the editor of the journal, as these significantly helped improve the quality of this paper.

References

[1] Posen, HE, Martignoni, D. (2018) Revisiting the imitation assumption: Why imitation may increase, rather than decrease, performance heterogeneity. Strat Mgmt J. 39: 1350– 1369.
[2] Li, G., Liu, H. and Luo, Y. (2018), Directive versus participative leadership: Dispositional antecedents and team consequences. J Occup Organ Psychol, 91: 645-664
[3] Baumann, O., Schmidt, J., & Stieglitz, N. (2019). Effective Search in Rugged Performance Landscapes: A Review and Outlook. Journal of Management, 45(1), 285-318.
[4] Csaszar, F. A., & Siggelkow, N. (2010) How much to copy? Determinants of effective imitation breadth. Organization Science, 21: 661-676.
[5] Sharapov, D, Ross, J-M. (2019) Whom should a leader imitate? Using rivalry-based imitation to manage strategic risk in changing environments. Strat. Mgmt. J. 1– 32.
[6] Martignoni, D, Keil, T. (2021) It did not work? Unlearn and try again—Unlearning success and failure beliefs in changing environments. Strat Mgmt J, 42: 1057– 1082.
[7] Mannetti, L., Leder, S., Insalata, L., Pierro, A., Higgins, T., & Kruglanski, A. (2009). Priming the ant or the grasshopper in people's mind: How regulatory mode affects inter-temporal choices. European Journal of Social Psychology, 39(6), 1120–1125.
[8] Higgins, E. T. (2005) Value from regulatory fit. Current Directions in Psychological Science, 14, 209–213.
[9] Huber, G. P. (1991) Organizational learning: The contributing processes and the literatures. Organization Science, 2: 88-115.
[10] Gavetti, G., & Menon, A. (2016) Evolution cum agency: Toward a model of strategic foresight. Strategy Science, 1: 207-233.
[11] March, J. (1991). Exploration and Exploitation in Organizational Learning. Organization Science, 2(1): 71-87.
[12] Coad, A. (2021). Econometrics and the Growth of Firms: Perspectives from Evolutionary Economics. Strategy Science.
[13] Posen, HE, Yi, S, Lee, J. (2020) A contingency perspective on imitation strategies: When is "benchmarking" ineffective? Strat. Mgmt. J. 41: 198– 221.

Development of a 125% Load Brake Test Instrument for Elevator

Pengzhao Zhang[1]*, Shuangchang Feng[2], Liang Fang[2], Shengjie Liu[2]

[1] Quality management department, Shanghai Institute of Special Equipment Inspection and Technical Research,Shanghai,Shanghai,200062,China

[2] Elevator inspection department, Shanghai Institute of Special Equipment Inspection and Technical Research,Shanghai,Shanghai,200062,China

*Corresponding author's e-mail: zhangpz@ssei.cn

Abstract. In order to solve the problem of elevator braking test inspection, this paper combs the requirements of relevant national standards and inspection rules on braking performance, and develops an elevator braking test inspection instrument. With the help of the instrument, the elevator braking effect can be expressed quantitatively, and the inspectors can conveniently judge whether the braking test results meet the requirements of elevator inspection rules remotely. A way to standardize and rationalize the test is given.In order to solve the problem of elevator brake test inspection, this paper sorted out the national standards and inspection rules on braking performance, and developed an instrument of elevator brake test inspection. With the help of the instrument, we could quantitatively give the expression on the elevator braking effect, and the inspectors can conveniently judge whether the braking test results meet the requirements of the elevator inspection rules in the remote.

1. Introduction

According to the Circular of the State Administration for Market Regulation on the Safety situation of Special Equipment in China in 2020, there were 7,865,500 elevators registered, and 107 special equipment accidents occurred, of which 25 were elevators, accounting for 23.36% of the total accidents. According to relevant statistical analysis in recent years, 85% of passenger death in vertical elevator accidents is caused by unexpected movement of elevator car, which lead to shear, extrusion and falling, and the main reason for the above accidents is brake failure or insufficient braking force.[1]

In research of the elevator accident case analysis and risk, as well as the elevator on technical development, the former General Administration of Quality Inspection and Quarantine issued implementation made by General Office of aqSIQ < Regulations of Elevator Inspection and Regular Inspection>, including 6 safety technical specification noticed on no. 2 modified issues (referred to as the "inspection regulations"), it is required to add brake test items in elevator inspection. In view of the inspection requirements, the key parameters of the car deceleration is difficult to measure, at the same time, the project inspection workload is heavy, so the inspection personnel are generally confused by the lack of inspection equipment. Therefore, it is very urgent to polish the test method and corresponding equipment.

By analyzing the mechanism of elevator brake test, this paper summarizes the internal characterization index of brake test, and develops a set of brake test inspection instrument, we hope that this instrument can quantitatively analyze whether the elevator brake test meets the requirements of "Inspection regulations", and reduce the workload of inspectors.

2. Elevator brake test requirements

2.1. Requirements of elevator brake test in Inspection Regulations

For the traction passenger elevator, first of all, the area of the car is strictly limited, which controls the number of passengers from the spatial position. Secondly, the elevator is required to set up overload protection device, which prevents the normal start when it reaches to 110% of the rated load. The inspection requirements of the brake test are as follows: when the car goes down at the normal running speeds without the motor and brake, it can carry 125% of the rated load capacity maximally, and the brake should be able to stop the driving host, and the car shall have no obvious deformation and damage after the test. In this way, enough allowance is given for the design of the working brake, and the safety of passengers can be guaranteed theoretically under the worst working conditions.[2]

2.2. Technical requirements for elevator braking in standard

"Safety Code for Elevator Manufacturing and Installation" GB7588-2003 (referred to as the "Standard")[3] has the following requirements for elevator braking performance:

2.2.1. Running downward with rated speed

Operating the brakes shall stop the tractor when the car is carrying 125% of the rated load and moving downward at rated speed. In such cases, the car should not decelerate beyond the deceleration caused by the safety clamp action or striking the buffer.

2.2.2. In the case of free fall

In the case of free falling, the average deceleration rate of the car with rated load is $0.2g_n$ to $1.0\ g_n$ when the progressive safety clamp is moving.

2.2.3. Energy dissipation buffer

The buffer should conform to the principle that when a car with rated load falls freely and hits the car buffer at 115% of the rated speed, the average deceleration during the buffer operation should not be greater than $1g_n$.

The deceleration time should be no more than 0.04s when costing $2.5g_n$ or more.

2.3. The analysis

Comprehensive comparison of the requirements of "Inspection Regulations" and "Standards" for brake test can be concluded that the inspection item is in essence to verify the elevator braking ability and effect. In the current inspection, it is more judged by the inspectors' experience for the braking effect, we use marking method, which is a simple way to measure the stop distance to find whether it meets $0.2g_n$~$1.0g_n$ after comparing with the standard requirements of the value, However, it is impossible to determine the requirement that the deceleration time above $2.5g_n$ should not be greater than 0.04s. And the way of marking, in the elevator with no room or the elevator with high speed, the judgment of the brake point completely depends on the reaction of the inspector, and it would be cost large error relatively.

In order to meet the requirements of the standard brake performance finally, so as to achieve improving the accuracy and effectiveness of the test data and reducing the error caused by subjective reasons at the same time, this paper research and develop brake test testing instrument from the combination with average deceleration, maximum deceleration, modern network and computer technology.

Journal of Physics: Conference Series **2113** (2021) 012035 doi:10.1088/1742-6596/2113/1/012035

3. Development of testing instruments

The system integrates the perception layer, network layer and application layer of the Internet of Things structure, and the basic architecture is shown in Figure 1. Among them, it involves hardware composition, software development, data transmission&storage, analysis and application.

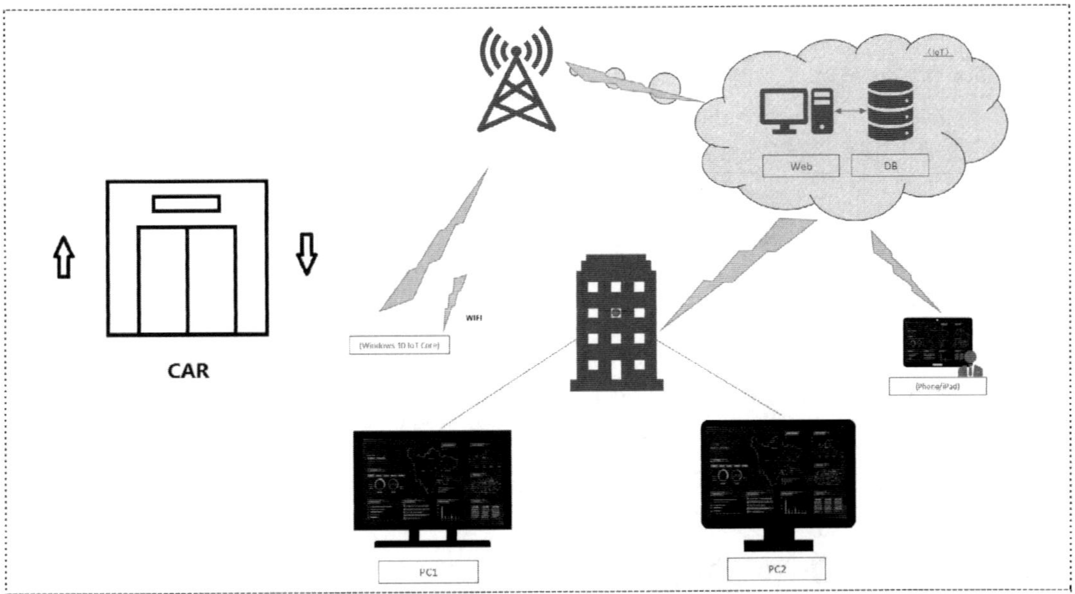

Figure1. Instrument architecture diagram

3.1. Hardware components

The hardware of the instrument is composed of the camera installed in the car, the machine room, and the inspection instrument terminal.

Based on the main data collection is made by acceleration. Now there are mature acceleration sensors on the market. We will integrate the acceleration sensors into the inspection instrument terminal.

3.2. Software selection and development

The selection of software technology should be considered comprehensively in many ways, including the combination of hardware equipment and operating system; the combination of software and operating system; the combination of software and Internet of Things (IoT), and software engineering. The operating system of the inspection instrument terminal is Windows 10 IoT core edition, and the software is based on. NET/C# custom development.

We introduce the Azure IoT Center to the long-term solution for IoT services, it is a comprehensive and mature IoT Center. The ecological side is equipped with common cloud services, Iot centers, Iot device development languages&communities, database systems, and more. Based on the completeness of the scheme, the development cycle is 20% shorter than other schemes, benefiting from more direct callability of common modules, visualization of configuration management and debugging of common problems, and visualization of performance monitoring and debugging. In addition, there is integration with the third party (sensors, video equipment SDK), Microsoft has the correlation and inertia of all technologies with upgrade background services and monitoring analysis software, which to achieve the Internet of things more widely and deeply enabling in the inspection process of elevator equipment.

3.3. Software selection and development

In the field test, the installation and maintenance personnel can adjust the instrument with the support foot, and place it at the center of the car to the ground. We installed a camera in the face of traction machine in the computer room, and use it to record test of traction machine braking condition; we installed a camera in the the capsule to confirm the weight and when the test capsules can change, this camera can synchronizes the video data to the local storage of the inspection instrument terminal through connecting with the inspection instrument terminal. Equipment installation is simple without debug, which can be delivered to the site installation and maintenance personnel to complete instead of intervention of inspection personnel.

At the beginning of the test, it just need to click the start button manually on the screen of the instrument, and the operator exits the car to ensure safety, then the operator controls the elevator to run down from the top terminal station, and then makes an emergency stop. The software will automatically make the judgment on running status. At the beginning, the real-time acceleration data of the elevator during the braking process was record by time unit of 0.01s, and the elevator can stop collecting data after the software automatically stop, then the instrument will collect data through analysis and standard comparison, as shown in figure 2, the data includes in stopping distance, average deceleration and breaking time, and finally to find the test results whether it is qualified or not. If the test is not qualified, the field operator can query the brake data, and adjust the brake to prepare for the next test. At the same time, the instrument will automatically upload the test data and video as shown in FIG. 3 to the cloud server, and inspectors can query the data and video of the test results from the server without going to the scene, and make a judgment on the project. At the same time, the data stored in the server can provide support for the later analysis of brake usage.

Figure2. Partial elevator braking test system

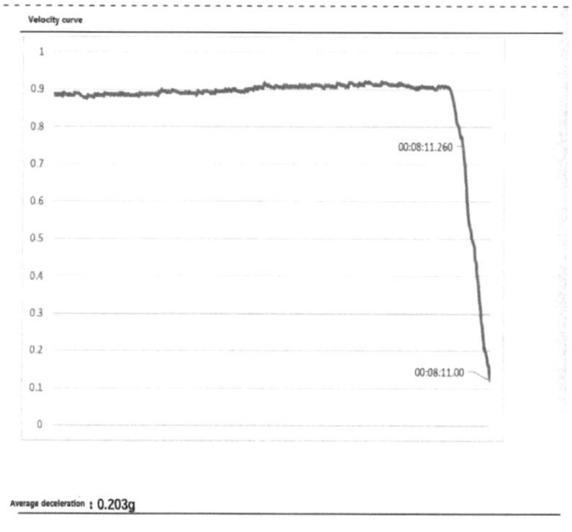

Figure3 Braking test curve

3.4. Verification of test data

In order to test the reliability and accuracy of the instrument, we selected 5 elevators for braking test, and compared the instrument measurement data with the traditional manual inspection data, than get the results as shown in Table 1.

Table 1 Comparison of measurement data

	Stopping distance (mm)			Average deceleration (g_n)			Maximum deceleration (g_n)		
	Instrument to measure	Artificial measurement	difference	Instrument to measure	Artificial measurement	difference	Instrument to measure	Artificial measurement	differe nce
Elevator#1	203	190	-13	0.203	0.217	+0.014	0.3347778		/
Elevator#2	247	262	+15	0.529	0.498	-0.031	0.7603149		/
Elevator#3	241	228	-13	0.541	0.573	+0.032	0.81561279	You can't measure	/
Elevator#4	226	245	+19	0.578	0.533	-0.045	0.7611694		/
Elevator#5	183	212	+29	0.714	0.616	-0.098	0.7974243		/

As shown in Table 1, elevator no. 5 has the biggest stopping distance deviation of 29mm, and Elevator No. 1 and Elevator No. 3 have the smallest deviation of 13mm. The average deceleration deviation of elevator No. 5 is 0.098g_n, and that of Elevator No. 1 is 0.014g_n. The maximum deceleration obtained in the instrument test is 0.8156279 g_n of no. 3 elevator, and the rest elevators are also meet the standard requirements of more than 2.5 g_n and less than 0.04s. The test results obtained by the two test methods are all qualified, and the error is within the acceptable range. In addition, the maximum deceleration cannot be measured by manual measurement, therefore, the conclusion of instrument measurement is more scientific and accurate.[4]

4. Conclusion

Based on the requirements of relevant national standards and inspection rules for braking performance, an elevator brake test instrument is developed. The instrument is convenient to be installed and can be used by non-professionals, which reduces the workload of inspectors.

The instrument can quantify the performance of the elevator brake test, and the output data is intuitive and easy to be read, which can make the test results more accurate and reliable. The instrument usage can avoid the inspection risk brought by the subjective judgment of inspectors, and give a way for the test of standardization and rationalization.

- The data measured by the instrument is rich, including a variety of real-time elevator operation information, such as stop distance, deceleration, average deceleration, elevator speed and so on. Through further data analysis, we can predict the braking force decline or failure of the brake. Thus, it can provide the test method and basis for the scrapping judgment of brake in the national standard "Technical Conditions for scrapping of Main Parts of Elevators" (GB/T 31821-2015).[5]

Reference

[1] Ma Jinghao , Yuliang Li.(2021)Talking about the sliding distance of elevator car in 125% brake test. Special Equipment Safety Technology, 023:34-36,51.
[2] TSG T7001-2009 Rules for Supervision, Inspection and Periodic Inspection of Elevators -- Traction and Forced Drive Elevators (including Amendment No. 2).
[3] GB 7588-2003 "Standard for Elevator Manufacturing and Installation Safety" (including Amendment No. 1). [S].
[4] Xin Hongbin, Xu Xiaoxiao, Niu Ben, Wang Jiawei, Xu Zhi. (2021)Test Method for Brake Test of Traction Drive Elevator. Quality Safety Inspection and Testing, 31:106-109.
[5] GB/T 31821-2015"Technical Conditions for scrapping of Main Parts of Elevators". [S].

Optimization of Inspection Plan for Transformer Verification Line Under Uninterrupted State

Shunsheng Guo[1,2], Bitao Yin[1,2*],Xiang Sun[1,2],Zhao Peng[1,2],Xiaobin Tu[3]

[1] Department of Mechanical Engineering, Wuhan University of Technology, Wuhan, Hubei, 430070, China

[2] Hubei Digital Manufacturing Key Laboratory, Wuhan University of Technology, Wuhan, Hubei, 430070, China

[3] Hubei Provincial Bureau of Market Supervision, Wuhan, Hubei, 430299, China

[*]Corresponding author's e-mail: 1066347580@whut.edu.cn

Abstract. At present, transformer verification line of metering centre adopts fixed cycle inspection method manually. This method requires downtime for detection, which costs a lot of time and cost. Moreover, the inspection cycle is determined based on experience and lacks rigorous basis. To solve this problem, a hybrid delivery of inspection devices is proposed to realize non-stop detection and reduce the cost of inspection time. Considering impact of cost and false detection risk on inspection cycle, a multi-objective optimization model of inspection cycle based on inspection and false detection cost is proposed. Based on NSGA-II algorithm, perturbation population is introduced to enhance the global search ability, which aims to minimize the cost of inspection and false detection. Taking the verification line's inspection plan of the metering centre as an example. It is solved by ENSGA-II algorithm, and feasibility of hybrid delivery mode is verified, which reduced downtime by 14.58%. A more reasonable inspection cycle is obtained, inspection cost is reduced by 29.57%, and false detection cost is reduced by 6.34%. It provides a reference for the formulation of inspection plan in the actual production process.

1. Introduction

Since 2010, the State Grid Corporation of China has vigorously promoted the construction of low-voltage current transformer verification line. At present, 27 provinces and cities have built low-voltage current transformer verification line [1,2]. The health status monitoring of verification line mainly depends on inspection. The inspection is regular or quantitative [3]. However, with the rapid growth of power demand in recent years, the long-term full load operation of the verification line makes the health status of equipment change rapidly [4], improving monitoring quality and adopting a more flexible inspection method has gradually attracted attention.

Reasonable formulation of the inspection plan can reduce inspection cost and false detection cost. Aiming at the problem of inspection plan, Jingshu Liu proposed variable control inspection method and applied to the inspection of electric energy measurement standard equipment. The inspection cycle and frequency are considered in the proposed inspection method, but these are still based on the status of verified equipments and the experience of metering personnel [5]. Bo Gao studied the monitoring method of the three-phase watt hour meter verification line, and proposed a monitoring method combining inspection and verification. Where inspection cycle adjust appropriately according to the number of tasks and the stability of equipment [6]. Libin Wang integrated the reliability and importance of equipment, proposed the concept of maintenance priority coefficient. And an optimal maintenance frequency model of verification line equipment based on the maintenance priority coefficient was established [7].

On the inspection issue of verification line, fewer personnel in the power measurement industry have studied the inspection cycle in the process of inspection. However, it has not been found to study the inspection problem from the delivery mode, delivery quantity and experimental times of the inspection device. Based on the analysis of the verification line's workflow, a hybrid delivery strategy of putting the inspection device and transformer into the line together is proposed. The model with optimization objectives of inspection cost and false detection cost is established. And solved by NSGA-II algorithm (ENSGA-II) with disturbance population.

2. Problem description and related strategy

2.1. Problem description

The body of low-voltage current transformer verification line is composed of a series of standard devices. Under long-term full load operation, the standard devices will have different degrees of faults [8]. Failure of standard device will lead to false detection. In order to monitor the health status of the line, the method of inspection is generally adopted. The inspection task brings costs. In the process of inspection, if it is found that the health status of the equipment does not meet the requirements, it is necessary to recover all the inspected transformers between this inspection and the last inspection, and re-verify them after the problems are eliminated. This process will produce recovery, re-inspection costs and maintenance costs. The composition of cost is shown in Figure 1.

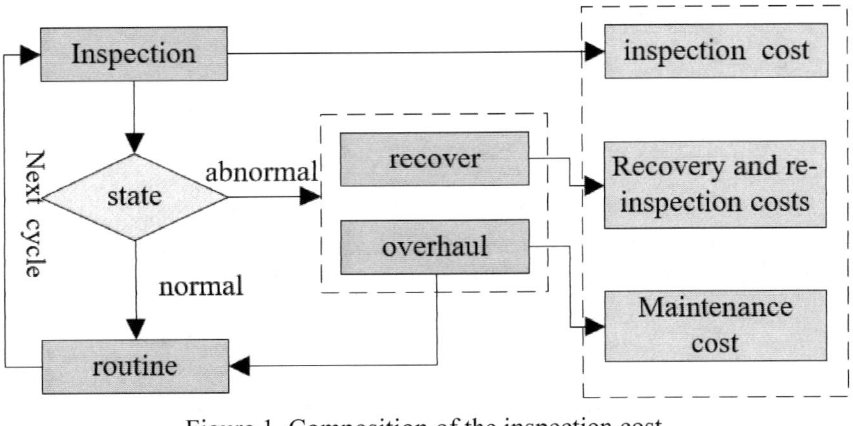

Figure 1. Composition of the inspection cost

Cost in the optimization problem of inspection plan is affected by cycle, quantity and experiment times of inspection devices. Short inspection cycle can reduce the probability of false detection, but the inspection cost will increase. Therefore, the inspection plan optimization problem aims to minimize various costs and study the impact of the above factors.

2.2. Hybrid delivery strategy

Traditional inspection method manually put the inspection device into each warehouse for experiment after shutdown. In this case, when a multi-functional verification warehouse performs inspection tasks, the other multi-functional verification warehouses are affected and generally do not work. In order to reduce the impact of daily tasks, an inspection device with higher accuracy is developed according to relevant verification regulations, which accuracy level is 0.05s. The hybrid delivery strategy as shown in Figure 2 is proposed.

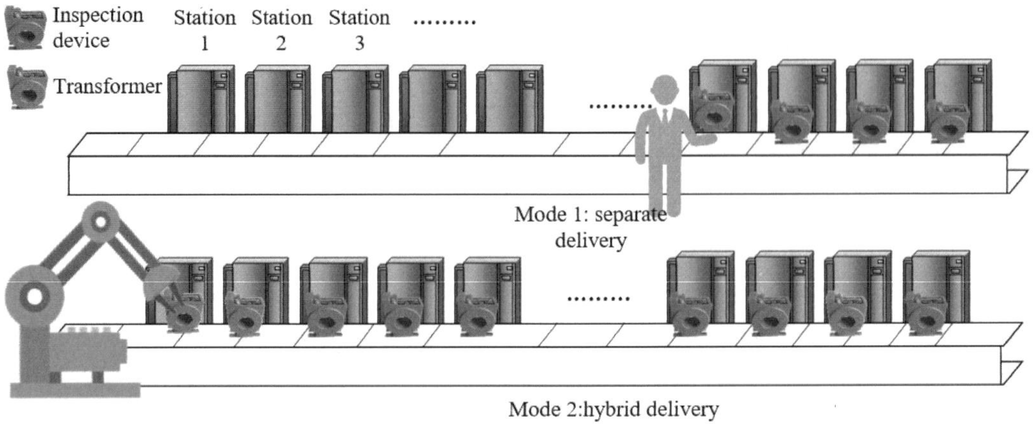

Figure 2. Two delivery methods

In the hybrid delivery strategy, after inspection device information is entered into the system, loading is completed by the manipulator together with the transformer. On the verification line, the inspection device enters each multi-functional verification warehouse in turn, and the other processes are the same as transformer. In the hybrid delivery scheme, it is not necessary to stop the normal task. It is only necessary to obtain the experimental data of inspection deviceand draw the quality control chart to meet the needs of line health status evaluation.

3. Optimization model of inspection plan

The inspection plan optimization model improves the inspection scheme at the level of production plan formulation, so as to reduce the impact on normal tasks and various costs without reducing the monitoring quality.

3.1. Assumptions

Based on above analysis, the basic assumptions of the model are as follows.

- verification/inspection time of transformers with different ratios is the same
- at the same time, a multi-functional verification warehouse can only verify the same batch of transformers
- in the process of verification/inspection, the time for the transformer/inspection device to enter and leave the station shall be considered within the transportation time
- inspection cannot be interrupted, and the inspection device can enter any station in any multi-functional verification warehouse
- when inspection device and transformer are in the same multi-functional verification warehouse, experiment of inspection device shall be completed first. After inspection device experiment is completed, inspection device and transformer shall enter next link at the same time.
- control center can control different stations of the multi-functional verification warehouse, but the tested equipment can be out of the warehouse together only after the experiments on all stations in the warehouse are completed.

3.2. Objective function and constraint

In this section, the mathematical formula of the problem will be described. Table 1 illustrates the symbols used in the optimization model of inspection plan.

Table 1. Notations used in models.

notations	significance
Parameters	
I	total working days of the year designed
J	number of inspections in a year
T	inspection cycle
H	number of inspection devices placed each time
X	whether there is inspection on that day, 0/1 variable
S	number of stations in the multi-functional verification warehouse of the verification line
M	indicates the delivery strategy of the inspection device, 0/1 variable
Cap	characterize production related capabilities
t	time spent on different experiments
c	cost of different experiments
Subscript	
ins	inspection related
ver	verification related
mt	maintenance related
l	task batch

(1) Inspection cost: cost of performing experiments by the inspection device during the inspection processand cost caused by the loss of capacity. The objective is to minimize the cost of inspection.

$$F_1 = \min(\sum_{i=1}^{I}\sum_{j=1}^{J} H_j * X_i * (n * c_{ins} + (S - H_j) * \frac{t_{ins}}{t_{ver}} * c_{ver} * M_m)) \tag{1}$$

(2) False detection cost: failure of the standard device in the verification line but not found in time, after it is found in the next inspection process, all the last batch of transformers verified by the station shall be recovered and the equipment shall be repaired. The objective function is to minimize the false detection cost.

$$F_2 = \min(\sum_{s=1}^{S} f_b(t)_s * f(H_{j,s}) * (\frac{Cap_a}{T} * f_{rel}(n) * c_2 + \frac{t_{mt}}{t_{ver}} * c_3) \tag{2}$$

Constraints

$$J = floor(\frac{I}{T}) \tag{3}$$

$$f_b(t) = \frac{m}{\eta}(\frac{t}{\eta})^{m-1} \tag{4}$$

$$Cap_a = Cap_d - \sum_{i=1}^{I}\sum_{j=1}^{J} H_j * X_i * n * \frac{t_{cer}}{t_{ver}} \tag{5}$$

$$f_{rel}(n) = k_{rel}\ln(pn + q) + o \tag{6}$$

$$\max((e_{l,s} + t_{l,s}) * \sigma_{l,s}) \le b_{l+1} \tag{7}$$

$$e_l = b_l + \max(t_{cer}, t_{ver}) \tag{8}$$

Formula (3) represents the determination of annual inspection times; Formula (4) is the equipment failure probability based on Weibull distribution on station s; Formula (5) describes the relationship between Cap_a(actual production capacity) and Cap_d(design capacity) of verification line; Formula (6) is to verify the relationship between the times of experiment execution and the reliability of experimental results, which is fitted based on historical empirical data; Formulas (7) and (8) indicate that the next batch of tasks can be carried out only after the prerequisite inspection is completed.

4. ENSGA-II algorithm

The inspection optimization problem is essentially an assembly line production scheduling problem, which belongs to NP-hard problem. It has the characteristics of computational complexity, multi-objective, multi-constraint and discreteness [9]. To solve the problem, the fast non dominated sorting genetic algorithm based on elite strategy (NSGA-II) is effective [10]. In the inspection plan optimization model, solution space of the model is discontinuous, resulting in the use of NSGA-II algorithm, which is easy to produce many identical individuals. This is not conducive to the gene exchange and evolution of the population, also easy to fall into local optimization. Therefore, the disturbed population is introduced to improve the diversity of the population and avoid the algorithm falling into local optimization [11], Schematic diagram of the algorithm is shown in Figure 3.

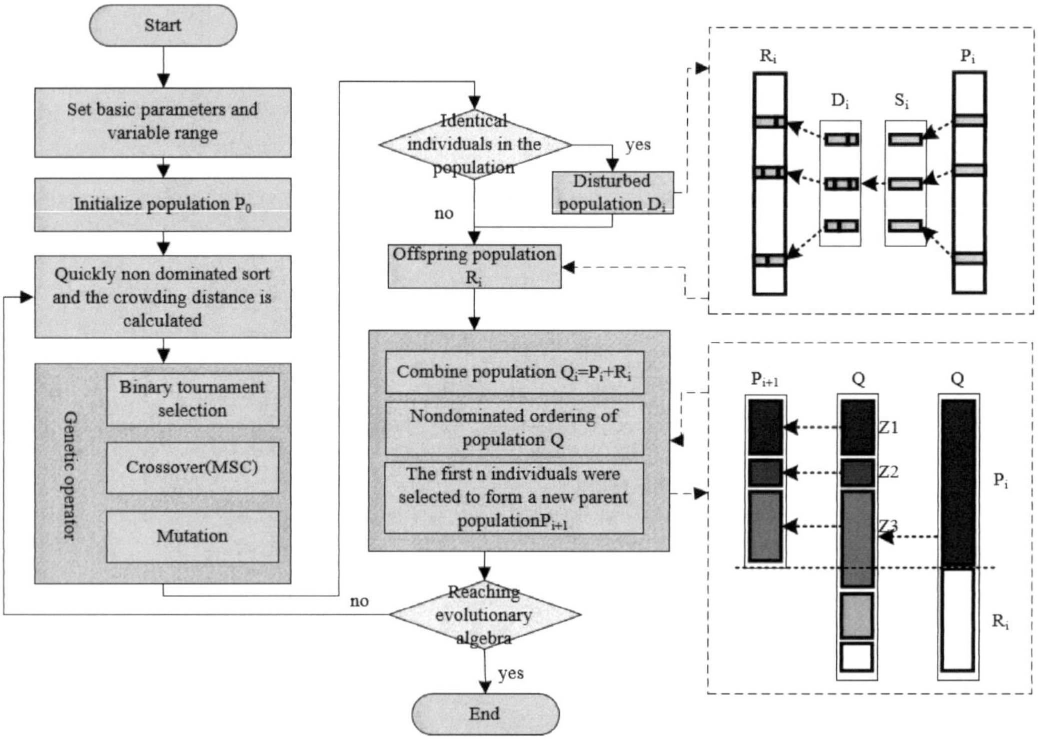

Figure 3. Flowchart of NSGA-II that coupled with disturbed population

When genetic composition of the population is relatively simple, the same individuals are easy to appear in the process of evolution, which leads to the decline of population quality and premature entry into the state of local convergence. In order to avoid the adverse effects of the same individual on population diversity, the concept of disturbed population is introduced to avoid decline of search ability. Introduction and operation process of disturbed population is shown as Figure 3 and steps 1 to 4.

(1) In the population P_i after completing the genetic operation, the individual S_i with exactly the same gene is selected and the same individual is deleted;

(2) A certain number of individuals are randomly selected from the disturbed population D_i and supplemented to the population P_i excluding the same individuals, so as to restore the population to its original size and obtain the offspring population R_i;

(3) Whether there are still the same individuals in the population R_i. If so, repeat steps 1 to 3 until there are no same individuals in the population.

(4) Each figure should have a brief caption describing it and, if necessary, a key to interpret the various lines and symbols on the figure.

5. Case Study

The case is selected from the laboratory of a metering center. This laboratory has built a set of low-voltage current transformer verification line compatible with through core and multiple turn current transformers. The system is equipped with one appearance inspection bin, two insulation and withstand voltage bins and three multi-functional verification warehouse, with a total of 36 multi-functional verification stations and a design annual verification capacity of 200000 low-voltage current transformers.

5.1. Result and analysis

Based on the proposed optimization model and solution method, taking the verification line of low-voltage current transformer in a metering center as the object. NSGA-II and ENSGA-II are used respectively, combined with the separate delivery and hybrid delivery strategy of inspection device. The pareto front solutions obtained experimentally are shown in Fig. 4 and Fig.5 respectively.

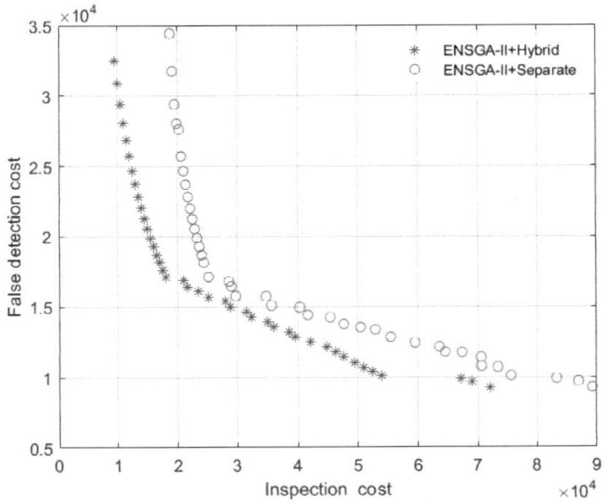

Fig4. ENSGA-II solution set

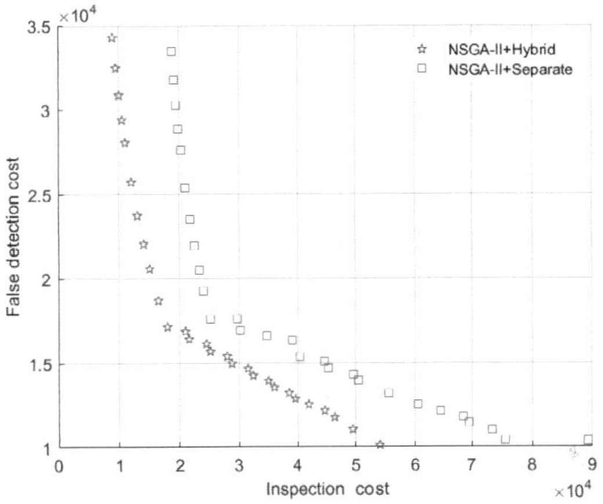

Fig5. NSGA-II solution set

In the Pareto frontier distribution of two algorithms, scheme using hybrid delivery method obtains better solutions on the two objectives. Which shows that compared with the separate delivery strategy of the inspection device, the hybrid delivery strategy can better reduce the cost in all aspects. In order to verify the effect of introducing disturbed population, the proposed ENSGA-II is compared with the original NSGA-II. In the evaluation indicators, IGD, Spacing, Pareto solution mean PC_ave and program execution time mean t_ave are selected. Results are shown in Table 2.

Table 2. Evaluation indicators of algorithms

	IGD	Spacing	PC_ave	t_ave/s
ENSGA-II	50.55	94.76	37.46	15.12
NSGA-II	506.89	113.29	26.81	11.12

It can be seen from table 2 that ENSGA-II is better than NSGA-II in the convergence and diversity index IGD of pareto solution, indicating that the convergence and diversity of the algorithm have been greatly improved after the introduction of disturbed population. In the solution set uniformity index Spacing and number mean index PC_ave, ENSGA-II can not only obtain more leading-edge solutions, but also have a more uniform distribution of solutions. However, after the introduction of disturbed population, the steps of finding, deleting duplicate individuals and adding new individuals are added on the basis of the process of NSGA-II, which increases the time complexity of algorithm execution.

Table 3. Costs of different methods

Algorithm	Delivery mode	Cycle/d	Inspection device quantity	Inspection time/h	Inspection cost	False detection cost
ENSGA-II	Separate	67	[8,12,10,11]	192	24900	17130
ENSGA-II	Hybrid	70	[10,12,9]	93	25100	15680
NSGA-II	Separate	54	[11,12,9,12]	192	25200	17630
NSGA-II	Hybrid	66	[11,10,11,9]	164	24500	16130
	Manual	60	[12,12,12,12]	216	35640	16742

The selected solutions obtained by different algorithms are compared with traditional method. The cost and relevant data are shown in Table 3. Compared with the traditional method, system under the hybrid strategy does not need to shut down and inspection time is reduced by 28 hours, a 14.58% drop. On the other hand that using algorithm with different delivery methods can achieve a large degree of inspection cost reduction on the basis of small fluctuation of false detection cost. A more reasonable inspection cycle is obtained, which inspection cost is reduced from 35640 to 25100, about 29.57%, and false detection cost is reduced from 16742 to 15680, about 6.34%.

In order to analyze the impact inspection cycle on the cost of inspection plan, the pareto solution obtained by using ENSGA-II algorithm and hybrid delivery scheme is shown in Figure 6.

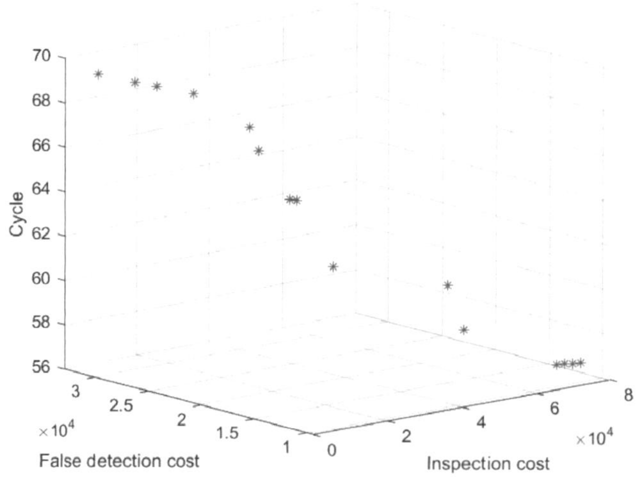

Figure 6. Relationship between cycle and cost

As Figure 6 shows, with the increase of cycle, cost of inspection will continue to decrease, but the cost of false detection will increase. The main reason is that in case of false detection, a large number of transformers verified in the previous batch need to be recovered, resulting in a large increase in cost. However, when the system is in better working condition, the probability of false detection of the system is lower. By appropriately increasing the inspection time interval, a better inspection cost reduction effect can be obtained.

6. Conclusion

Firstly, a hybrid delivery strategy of inspection device and transformer is proposed in this paper to atoptimize the inspection plan. Automation of inspection process is improved through the application of high-precision inspection device. On this basis, inspection cycle is taken as the main influencing factor. In order to minimize the cost of inspection and false detection, an optimization model of verification plan is established.

Secondly, the ENSGA-II is used to solve the problem and verified it can effectively solve the inspection optimization problem. Experimental results show that the hybrid delivery strategy has a better cost reduction effect and realized inspection under uninterrupted state.

Finally, inspection cycle plays a decisive role in the process of optimize inspection plan. Under different working states of the system, selecting an appropriate inspection cycle can effectively reduce costs.

The optimization model and algorithm of inspection plan proposed in this paper can provide theoretical basis and reference for the establishment and implementation of inspection plan. In the follow-up research, preventive maintenance based on equipment health status will be introduced into the influencing factors of inspection plan optimization to enter a deeper discussion.

References

[1] Lijun D, Enguo Z, Qian Z, et al.(2015) Design and application of site calibration system for automatic verification pipeline. 52(17): 118-123.
[2] Guishan L, Lei L, Ying L, et al.(2013) Research on Auto-Verification and Intelligent Storage System for Smart Watt-Hour Meter. 50(05): 95-100.
[3] (2014) Q/GDW11278-2014 Calibration specification for 0.4kV metering current transformer automatic testing system. Place of publication unknown.
[4] Haibin C, Shunan Y, Liwen C, et al.(2021) Research on current transformer error testing technology based on field signal simulation technology. 58(02):133-138.
[5] Jingshu L, Fei L, Lei Z.(2014) A Period Verifi cation Method for Standard Electrical Energy Measuring Equipment. (06): 33-34.
[6] Bo G, Zhenming Z, Hongwu Z, et al.(2019) Application Research on Quality Monitoring of Automatic Verfication Production Line of Three-phase Intelligent Watt-hour Meter.47(04): 37-40.
[7] Libin W, Hongying W, Chao Z.(2017) Research on the optimal maintenance frequency of automatic verification pipeline equipment in electric energy meter.54(08): 89-92.
[8] Lihua Z, Zhaoxin D.(2015) Verification quality monitoring of metering low-voltage current transformer automatic verification line. 52(19):43-48.
[9] Youhua J, Jinwan Y, Le Z, et al.(2021) Optimal Scheduling Strategy for Limited Electricity in Multiple Parks in the Absence of Power Supply from Grid .42(01):105-116.
[10] DEB K, PRATAP A, AGARWAL S, et al.(2002) A fast and elitist multiobjective genetic algorithm: NSGA-II.6(2): 182-97.
[11] Zhenzhen Z, Xingshi H, Qinglin Y, et al.(2021) Computer Engineering and Applications: 1-13[2021-07-09]. http://kns.cnki.net/kcms/detail/11.2127.TP.20210428.1801.021.html.

A spatial-temporal correlation based method for advanced persistent threat detection

Luoli Wang[1]

[1] Pre-Science, University Of Washington, Seattle, Washington, 98105, United States

*Corresponding author's e-mail: luoliwang0612@outlook.com

Abstract: Advanced Persistent Threats (APT) have caused severe damage to the core information infrastructure of many governments and organizations. APT attacks usually remain low and slow which makes them difficult to be detected. In this case, the way of correlatively analyzing massive logs generated by various security devices for effectively detecting the new type of cyber threat turns out to be more and more significant. In this paper, on the basis of analyzing the principles and characteristics of APT, we propose an intelligent threat detection method based on the expanded Cyber Attack Chain (CAC) model and the long short-term memory network (LSTM) autoencoder to extensively correlate malicious behaviors from spatial and temporal dimensions, which provides a brain new idea for the application and practice of complex network attack detection.

1. Introduction

Recently, Advanced Persistent Threats, one that is launched by professional and resourceful organization and that targets on specific and valuable objects, have gradually caught people's attention. To be more specific, compared to traditional models of attacks, APTs possess more threats to us with the following characteristics: its targets are specific and clear; its ways of attacking are complex and diverse; its action of attacking stays low and slow; its purposes are political and financial; and it remains destructive and contagious[1].

Speaking of conventional detection of APT attacks, signature-based detection (uses security devices, such as Intrusion Detection System (IDS) in which behaviors that match the rules are detected as attacks, to analyze events in network)[2-4] and anomaly-based detection (applies statistical machine learning methods on training models by collecting normal data, in which the one that deviates from threshold of the model will be considered as an attack)[5-7] are less effective on dealing with APTs attacks. The reasons are followings. First of all, it's difficult for signature-based detection to detect an attack that is unknow or has no record in the database. To be more specific, IDS pays more attention on one single stage of the attack instead of focusing on the combination of network and system information from different attack stages or links to establish an attack model. In this case, some behaviors that seem to be normal might be a bridge step in the APTs attacks, which might also cause us to neglect some harmful but seemingly low risk warning messages[8]. On top of that, speaking of anomaly-based detection, since the action of APT attacks possess the feature of low and slow, in which it's hard for the model of traditional machine learning to effectively correlate this type of covert, time-lapse attack[9]. Moreover, most of the data sets for anomaly-based detection are private and proprietary. Given that many malicious sample sources are commercial in nature and the issue of intellectual property set a high threshold for this field, researchers can only train their model with limited amount of data[10].

To deal with this complex attack, we have dig into the principles of attacks of APTs and corresponding ways of defending and we have proposed a design for APTs detection approach based on expanded attack chain and LSTM-autoencoder, which improves our ability to recognize APTs attacks by corelating network behaviors both in time and space. The advantages of this model are followings. To begin with, by inputting the network event sequence to the LSTM and mapping the event to the expanded attack chain model, we can amplify the attack behaviors that remains low. On top of that, by using LSTM-autoencoder, even if the behavior of the frequency of APTs attacks remains slow, we can detect attacks that are abnormal, and even in an unsupervised way. Last but not least, by training offline and detecting online, we can even capture APTs attacks in the production environment, which shows this detection approach is pragmatic.

The rest of this paper is organized as follows. Preliminaries are given in section 2. Details of the proposed method is given in section 3. A case study is presented in section 4. Section 6 concludes the paper and gives the future work.

2. Preliminaries

In advanced persistent threat attacks, the attack behavior is not only a direct destructive behavior but also a continuous penetrating process with multiple stages, which allows attacks that stay secretly and persistently in the target system. On top of that, unlike regular attacks APTs attacks remain low and slow. Owing to these features, characterizing the behavior of each APT attack individually is difficult. Thus, it's necessary to introduce two key models, cyber attack chain and LSTM, which allows us to correlate them in time and space respectively to capture attacks.

2.1 Cyber attack chain

To begin with, attack chain, one that contains a collection of paths and means of attacks taken by the attacker and that is carried out based on different steps of the attack in target system to divide and associate each stage to form a complete attack[11], makes the detection of APT possible. For instance, a classic chain model of cyber-attack, is proposed by Lockheed Martin in 2011 from the perspective of the attacker, assisting security in identifying suspicious activities at each stage of a potential attack. The advantage of CAC model is that if security systems fail to find signs of an attack in a certain step, they can still find clues in the subsequent stages, which maximizes the chance of defending and allows security system to detect attacks in advance. Consequently, analyzing attack chains to depict and capture the logic behind malicious attacks provides conditions for network security event tracking and tracing, relationship extraction and attack prediction.

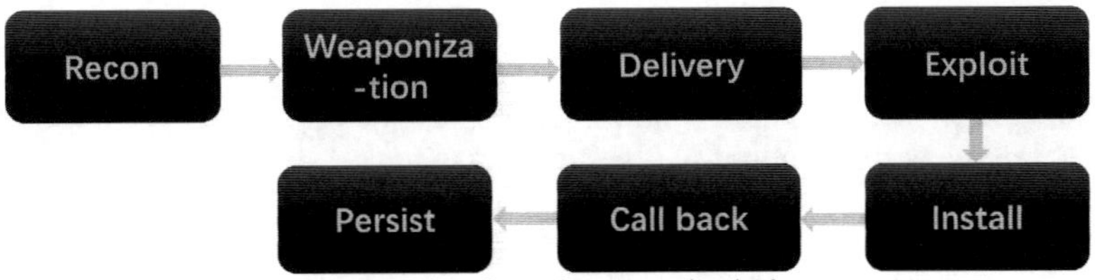

Fig 1. Classic cyber attack chain

As shown in figure 1, in reconnaissance stage, intruder screens the target and looks for vulnerabilities; in weaponization stage, intruder develops malware designed to exploit the vulnerability; in delivery stage, intruder transmits the malware via a phishing email or another medium; in exploit stage, the malware begins executing on the target system; in the install stage, the malware installs a backdoor or other ingress accessible to the attacker; in the call back stage, the intruder gains persistent access to the victim's network; in the persist stage, intruder initiates end goal actions, such as data theft, data corruption, or data destruction.

2.2 Long short-term memory network

Long Short-Term Memory (LSTM) network, a refined version of the Recurrent Neural Network (RNN), has been explored, successfully handling the trouble that potential exploding gradient and vanishing gradient problem[12]. Speaking of LSTM, even though all recurrent neural networks have the form of a chain of neural network repeating modules, the repeating module of LSTM has a different structure. In other words, there are three gates, namely input gate, forget gate and output gate, as well as memory cells with identical shapes in LSTM, which allows it to record additional information. The unit of LSTM is shown in figure 2.

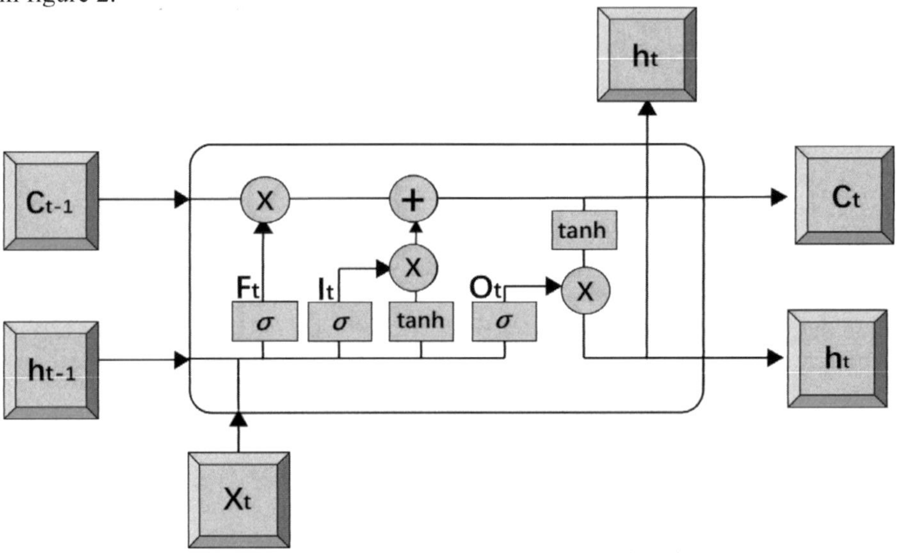

Fig 2. Long short-term memory network unit

Specifically, let the number of hidden units be h, the given time step is t, the input is X_t, the hidden state of the previous time step is h_{t-1}, and the input gate I_t, forgetting gate F_t and output gate O_t of time step t are calculated separately as formula 1:

$$I_t = \delta(X_t W_{xi} + H_{t-1} W_{hi} + b_i)$$
$$F_t = \delta(X_t W_{xf} + H_{t-1} W_{hf} + b_f)$$
$$O_t = \delta(X_t W_{xo} + H_{t-1} W_{ho} + b_o)$$
$$C_t = tanh(X_t W_{xc} + H_{t-1} W_{hc} + b_c)$$
$$H_t = O_t \cdot tanh(C_t)$$

Among these, W_{xi}, W_{xf}, W_{xo}, W_{xc} and W_{hi}, W_{hf}, W_{ho}, W_{hc} are the corresponding weight parameters, and b_i, b_f, b_o and b_c are the corresponding bias parameters.

3. Our method

3.1 Expanded attack chain

The classic attack chain model described above is only suitable for modeling external attacks, but not for internal threats[13]. In APT attacks, a large number of attack behaviors are carried out after obtaining internal system privileges. In this case, the classic attack chain model needs to be expanded. The expanded attack chain model is shown in Figure 3.

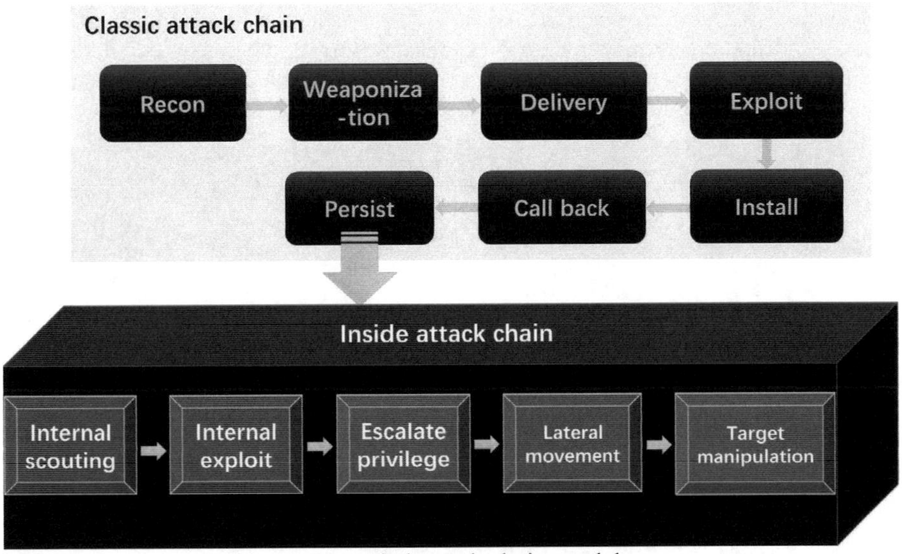

Fig 3. Expanded attack chain model

Compared with the classic attack chain, the expanded attack chain can be subdivided into the internal behavior and added the internal attack chain during the process of getting to the goal. In other words, the internal attack chain is to gain access to the target system, including internal scouting, internal exploit, escalate privilege, lateral movement and internal target manipulation.

3.2 LSTM autoencoder

The encoder-decoder model is often applied to the sequence modeling problem. It is an unsupervised learning model. The encoder converts the input sequence into a fixed-length vector representation, inversely the decoder, convert the previously generated representation into an output sequence. The performance of the model is evaluated based on the model's ability to recreate the input sequence. The essence of the autoencoder model is to realize the mapping between intuitive representations (such as network event sequences) and vector representations.

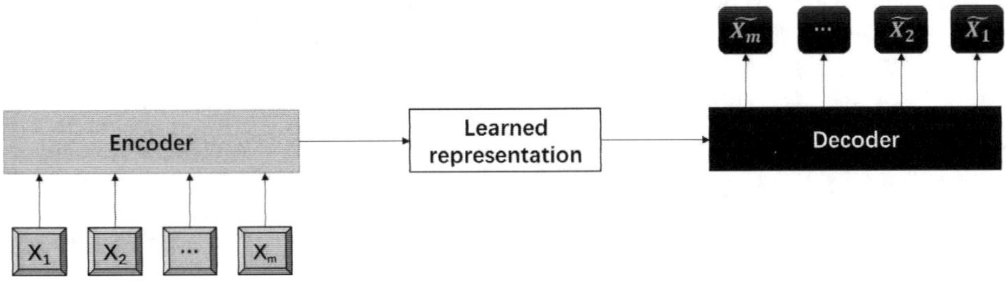

Fig 4. LSTM autoencoder model

As illustrated in figure 4, there are two recurrent neural networks in the model, the encoder LSTM and the decoder LSTM. The input of the model is the vector sequence (network events). When the last input is read in, the internal state and output state of the encoder will be directly fed into the decoder. The decoder reconstructs a sequence the same as the target sequence, except that the order is reversed. Reversing the order can make the optimization easier because the output of LSTM is also the reverse.

3.3 APT detection based on expanded CAC and LSTM-autoencoder

In our design, we propose a framework for APT detection using the expanded attack chain model and the LSTM autoencoder model. Figure 5 shows the overall framework of the method, which can be divided into the following four steps.

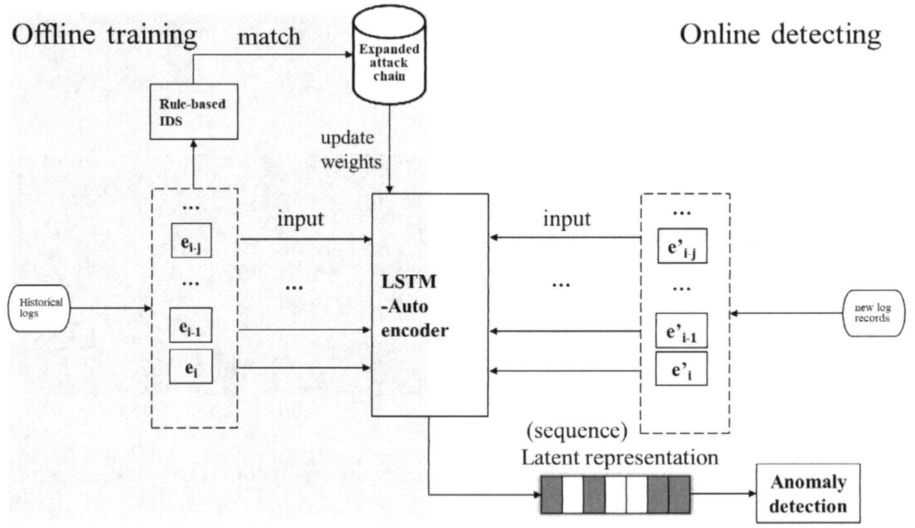

Fig.5 APT detection framework based on attack chain and LSTM autoencoder

Step1 Expanded attack chain mapping:

Input the network event sequence $e_{i-j},...e_{i-1},e_i$ in a window with length j to LSTM-autoencoder, and at the same time, map the event detected by a rule-based IDS to expanded attack chain model;

Step 2 Event weight update:

Obtain the weight w_i of each event e_i according to the result of attack chain mapping, and update it to the event sequence that has been input to the LSTM-Autoencoder model in step 1;

Step 3 Offline latent representation learning:

Train the LSTM autoencoder model with updated weight sequence generated by the step 2 to combine the information of events at different times to extract the changes of characteristics of the event, and perform the learning of the representation vector, and represent the entire sequence of events in the window as a fix-length vector (Entire training process conducted offline);

Step 4 Online anomaly event detection：。

Use the mature anomaly detection algorithm RRCF[14] on the data stream to give anomaly scores. In order to determine the sequence of anomalies, a threshold can be set and it is considered that the sequence whose score exceeds the threshold is anomaly. (Entire detecting process conducted online).

4. Case study

We deployed the above-mentioned prototype system on the server of the information technology center of Campus Network. The center provides fundamental network services with the purpose of scientific researching, is responsible for maintaining and managing more than 10,000 pieces of information, 500 network equipment, and 40,000 campus online accounts. Programs running on the host server on Campus Network carry the main function of many service systems and development systems. Once a cyber invasion occurs, it's inevitable that the operation of the Campus Network system will be affect in a large extent. Consequently, the protection of the host server and ability to detect cyber-attacks are particularly significant. Following is the study we've carried out based on the scenario mentioned above.

The monitoring objects of the detection application cover Linux hosts in a laboratory of the campus network. The subject of the detection is the operation of Command line (CMD request) after logging in

Shell remotely and the interfaces of the remote terminal to the host Call (URL request). For example, a CMD operation sequence is [*cd, netstat, ls, ll, chage, ll, ifconfig, ll, chpasswd, chage, chage, more, exit, ll, chpasswd, exit, ll, ll, more, chage, grep, grep*]; and an interface call request (url) is "*ip:port /login/login_frame.php?site=default*".

Speaking of the case of operation of CMD, the detection found a large number of SCP commands input behavior. The abnormal behavior occurred on a host from 16:00 to 17:00 on April 8, 2021. Afterwards, the time of occurrence of the abnormal behavior of the host and the sequence of operations in the past six months were counted separately. The distribution showed that within the six months, there were only 28 SCP operations, however, all of which happened within this one hour. This case is very likely to be abnormal in human opinion, but unfortunately, the existing detection system did not notice this anomaly.

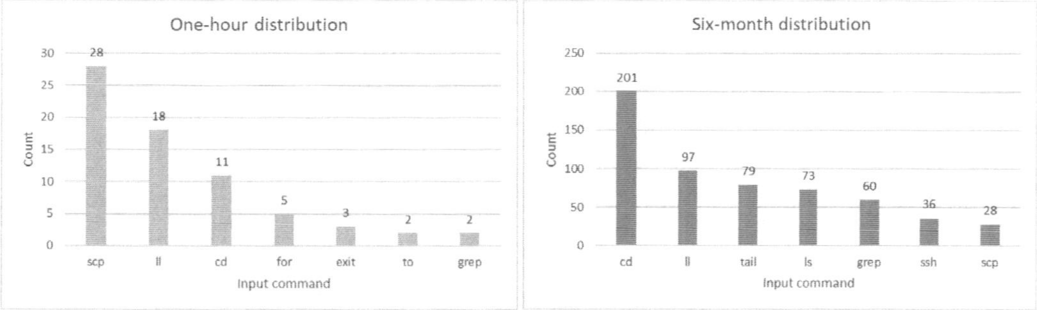

Fig 6. Top7 distribution of shell command (One-hour and Six-month)

Speaking of the interface call request, the anomaly detection engine found that a certain host interface was accessed abnormally 13 times with the abnormal URL that was "*ip:port/_vti_bin/shtml.exe/_vti_rpc*" on May 20, 24, and 27, 2021. It has been verified that the call is a buffer overflow vulnerability in Microsoft FrontPage, which can cause the host-side server to crash and have the possibility of executing arbitrary code.

Consequently, both cases of the CMD operation and URL interface did not generate corresponding alarm in existing security system, which causes the internal threat behavior successfully to evade the monitoring system. Under this circumstance, further attacks on system will occur.

5. Conclusion and future work

Given that existing defensive approaches of network attack mainly focus on the detection of regular network attack behaviors and the detection of APT attacks is ineffective, we have proposed a detection model combined with expanded attack chain and LSTM-autoencoder which makes broadly correlating network events possible.

By correlating network events in APTs attacks with respect to time and space in our approach, there is no doubt that the ability to recognize APTs attacks has been improved. In fact, the fine granularity of the expanded attack chain can still be refined. On the other hand, while the length of the network events keeps increasing, the accuracy of the detection will drop. Thus, we will keep digging into the detection of APTs attacks and dealing with it as effective as possible.

Reference

[1] M. Li, W. Huang, Y. Wang, W. Fan, and J. Li, "The study of APT attack stage model," in *2016 IEEE/ACIS 15th International Conference on Computer and Information Science (ICIS)*, Jun. 2016, pp. 1–5. doi: 10.1109/ICIS.2016.7550947.

[2] I. Karim and Q.-T. Vien, *Snort-based intrusion detection system for practical computer networks: implementation and comparative study*. LAP LAMBERT Academic Publishing, 2017. Accessed: Sep. 26, 2021. [Online]. Available:

https://www.morebooks.de/store/gb/book/snort-based-intrusion-detection-system-for-practical-computer-networks/isbn/978-3-659-69329-8

[3] S. More, M. Matthews, A. Joshi, and T. Finin, "A Knowledge-Based Approach to Intrusion Detection Modeling," in *2012 IEEE Symposium on Security and Privacy Workshops*, May 2012, pp. 75–81. doi: 10.1109/SPW.2012.26.

[4] R. Rajendran, S. V. N. Santhosh Kumar, Y. Palanichamy, and K. Arputharaj, "Detection of DoS attacks in cloud networks using intelligent rule based classification system," *Cluster Comput*, vol. 22, no. 1, pp. 423–434, Jan. 2019, doi: 10.1007/s10586-018-2181-4.

[5] J. Hong, C.-C. Liu, and M. Govindarasu, "Integrated Anomaly Detection for Cyber Security of the Substations," *IEEE Transactions on Smart Grid*, vol. 5, no. 4, pp. 1643–1653, Jul. 2014, doi: 10.1109/TSG.2013.2294473.

[6] D. S. Terzi, R. Terzi, and S. Sagiroglu, "Big data analytics for network anomaly detection from netflow data," in *2017 International Conference on Computer Science and Engineering (UBMK)*, Oct. 2017, pp. 592–597. doi: 10.1109/UBMK.2017.8093473.

[7] A. L. Buczak and E. Guven, "A Survey of Data Mining and Machine Learning Methods for Cyber Security Intrusion Detection," *IEEE Communications Surveys Tutorials*, vol. 18, no. 2, pp. 1153–1176, 2016, doi: 10.1109/COMST.2015.2494502.

[8] P. Gao *et al.*, "An Intelligent Threat-Detection Method for Power Monitoring System Based on Attack Chain Knowledge," in *2021 IEEE 4th International Electrical and Energy Conference (CIEEC)*, May 2021, pp. 1–6. doi: 10.1109/CIEEC50170.2021.9510852.

[9] "Advance Persistent Threat Detection Using Long Short Term Memory (LSTM) Neural Networks | SpringerLink." https://link.springer.com/chapter/10.1007/978-981-13-8300-7_5 (accessed Sep. 26, 2021).

[10] R. Harang and E. M. Rudd, "SOREL-20M: A Large Scale Benchmark Dataset for Malicious PE Detection," *arXiv:2012.07634 [cs]*, Dec. 2020, Accessed: Sep. 26, 2021. [Online]. Available: http://arxiv.org/abs/2012.07634

[11] "Intelligence-Driven Computer Network Defense Informed by Analysis of Adversary Campaigns and Intrusion Kill Chains : National CyberWatch Center." https://www.nationalcyberwatch.org/resource/intelligence-driven-computer-network-defense-informed-by-analysis-of-adversary-campaigns-and-intrusion-kill-chains-2/ (accessed Sep. 26, 2021).

[12] S. Hochreiter and J. Schmidhuber, "Long Short-Term Memory," *Neural Computation*, vol. 9, no. 8, pp. 1735–1780, Nov. 1997, doi: 10.1162/neco.1997.9.8.1735.

[13] T. Dargahi, A. Dehghantanha, P. N. Bahrami, M. Conti, G. Bianchi, and L. Benedetto, "A Cyber-Kill-Chain based taxonomy of crypto-ransomware features," *J Comput Virol Hack Tech*, vol. 15, no. 4, pp. 277–305, Dec. 2019, doi: 10.1007/s11416-019-00338-7.

[14] S. Guha, N. Mishra, G. Roy, and O. Schrijvers, "Robust Random Cut Forest Based Anomaly Detection on Streams," in *Proceedings of The 33rd International Conference on Machine Learning*, Jun. 2016, pp. 2712–2721. Accessed: Sep. 26, 2021. [Online]. Available: https://proceedings.mlr.press/v48/guha16.html

An Absolute-value Detector with Threshold Comparing for Spike Detection in Brain-machine Interface

Mingzheng Yuan*

Miami College, Henan University, Kaifeng 475000, China

*Corresponding author e-mail: 1838020028@vip.henu.edu.cn

Abstract. This research designs an absolute-value detector with the function of threshold comparing. Specifically, it is an essential device in the spike detection of the brain-machine interface. The optimized design in the research can accomplish the main functions in spike detection and has good performance in both delay and energy consumption. It comes up with two types of design at the beginning. To make the design reliable and comprehensive, it decides to discuss both methods in this paper. The first design is using a full adder, multiplexer and comparator. The concept of its logic circuit is adding the logic one to the input when the given input data is negative, keeping the original information as the given input data is positive. To achieve the function of adding, this study chooses the full adders. The primary purpose of using multiplexers is to select from the processed input and original input, and the choice depends on the most significant bit (MSB) of the input data. To compare the absolute value of the input data with a given threshold, this research used a multi-bit comparator. The second design is based on the fundamental algorithms of calculating total numbers. It indicates that this study can operate it with the threshold value through a subtractor when the input is negative. On the contrary, an adder can be used when the information is positive. Based on the concept of logic optimization, this study chooses to use the only subtractors, and it just needs to focus on the borrow bit, which can indicate the more significant number. By connecting the MSB of the input with the subtractors through XOR gates, the selection can be achieved without using any multiplexer. In the process of removing and replacing the devices, it reached the optimization of the design. Then, this paper compared the minimum delay by calculating each stage's size and finding that the second design is better. Finally, based on the dual design, this essay computed the energy consumption in the circuit and implement V_{DD} optimization to obtain the minimum energy.

1. Introduction

The human brain is one of the most significant components in the human body, and it is responsible for most human activities. The human brain consists of billions of neurons and they are communicating with each other through electrochemical processes. Based on the electro-chemical properties, this research can obtain practical information by processing and analyzing the neural signals. The neurons in the human brain can excite the action potentials, and this mechanism allows neurons to reliably transmit information over long distances without transmission attenuation. Additionally, the action potentials can be detected as spikes when exceeding a specific threshold value. However, it is inevitable that the extracted signals may have multiple components, including background noise signals and specific neural signals. To obtain accurate signals and describe correct neural activities, it is essential to introduce spike sorting, which separates the electrical activity of individual neurons from noisy recorded signals [1].

The Spike-sorting algorithms have gained much interest in the research community with the recent advancements in neural signal acquisition systems. Considering the existence of the background noise signals and chaotic spike activities, it is essential for researchers to detect the accurate and effective neural signals. However, farraginous methods are applied to accomplish spike detection in current studies. This research is to design one of the commonly used spike-detection algorithms, a 4-bit threshold-value comparator with absolute-value detection.

The logic function of this design mainly includes two parts, one is to detect and process the absolute value of the signal, and the other is to compare the processed signal with the specific threshold value. To achieve these logic functions, it comes up with two types of designs. The first design uses full adders to add logic when the value of neural signal is negative, which is actually the function of absolute-value detection. In addition, multiplexers are used to choose between the processed neural signal and the original neural signal. The second design abandons full adders and multiplexers for better performance of the circuit as adders and multiplexers can lead to more significant energy consumption and longer delay. However, to make this research effective and comprehensive, it analyzes both of the designs. This study finds the critical paths of these two circuits and calculated the corresponding delay of the two critical paths. After calculating the minimum delay of each critical path, it can be illustrated from the results that the second design has better performance. Based on the second design, this research implements further analysis, including computing the minimum energy consumption and discussing the relationship between delay and energy consumption.

2. Spike Detection

Brain-machine interface (BMI) is a direct connection path established between human or animal brain (or brain cell cultures) and external devices. In a one-way brain-machine interface, the computer either accepts commands from the brain or sends signals to the brain. In the early stages of research in the field, recognizing user's intentions through a BMI was exploited mainly in the communications domain [2]. However, more recently, researchers have opened up new directions for BMI, especially in the field of brain-controlled robotic devices [3]. This technique also has many vital applications in biological, medical and other academic fields. Considering the practical value of BMI technique in multidisciplinary areas, it has become one of the most valuable interdisciplinary techniques in the current research.

Figure 1 demonstrates the working principle of a classic BMI through a block diagram. Firstly, the activities of neurons are detected by microelectrode arrays which can accurately separate the electrical signals of a single cell, and record the signs of almost all neurons in the sample. Using the current technique of the CMOS-based high-density microelectrode arrays (HD-MEAs), the neurites of brain cells can be drawn, and electrical signals of substructures such as cell bodies and axons can be detected as well [4]. Then after amplifying and filtering, the neural signals are still combined with background noise signals. On account of this, the spike sorting is implemented to eliminate the surrounding noise, and it can also regulate and cluster the processed signals. Finally, the decoded signals can issue specific instructions and commands to external devices to achieve corresponding control.

However, this research focuses on spike detection, which is mainly responsible for eliminating ambient noise. Spike detection is commonly used in two stages, pre-emphasis and thresholding. The neural signals are processed for the pre-emphasis by detecting the absolute value at first [5]. As the neural signals are emitted and received through a discrete waveform, this study has to implement the non-linear energy operator (NEO) and discrete wavelet transform (DWT) as the processes of neural signals [6,7]. The next step is thresholding. The main idea is to compare the neural signal, or the pre-emphasized signal, to a threshold value. A spike is detected when the neural signal crosses an assumed threshold [8]. Figure 2 is the overall procedure for constructing the design that achieves the functionality needed. It can be seen from this figure that this design is mainly divided into two stages, detecting and thresholding. However, this procedure has formed the overall idea and framework of the design. At the next stage of this research, it focuses on two main steps, including the absolute-value detection and thresholding comparison.

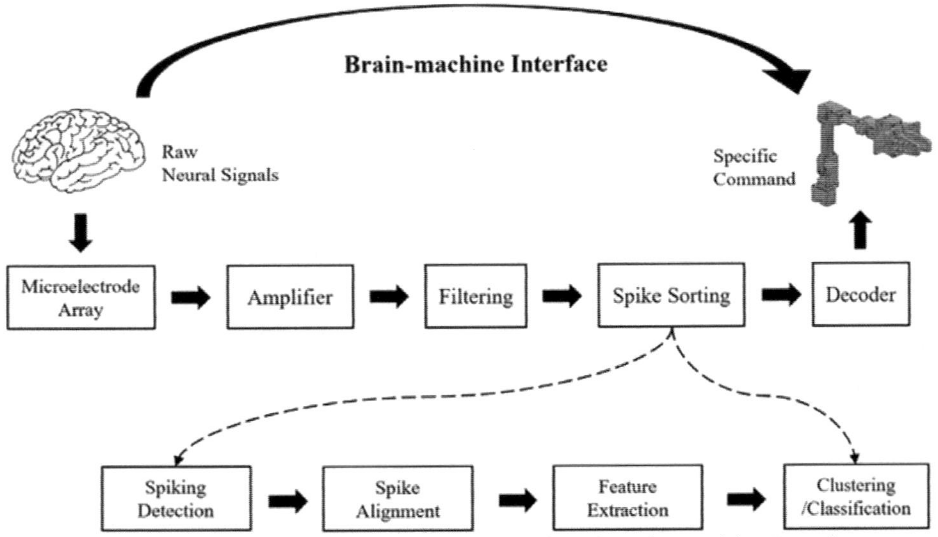

Figure 1: The Working Procedure of a Conventional Brain-machine Interface

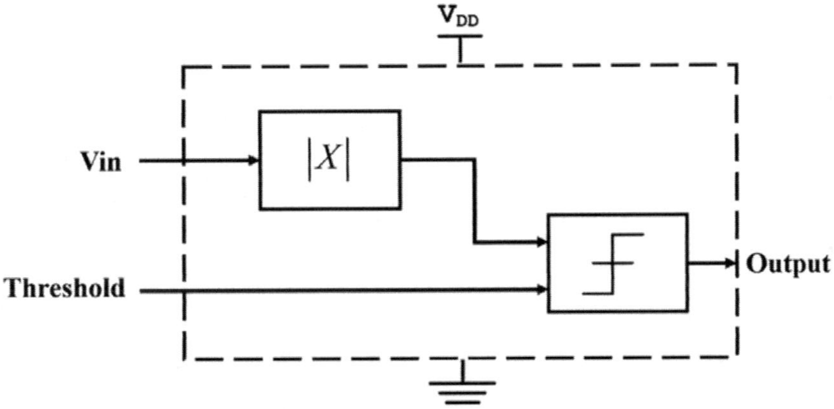

Figure 2: Overall Procedure in Construction the Design

3. Circuits and Topology

This section concludes the crucial algorithms for constructing an absolute-value detector. Then two types of designs are introduced to fulfill the implementation of the absolution value detection and comparison. After that, it discusses both methods by calculating the delay and energy consumption of the critical path. Based on the calculation results, this part is able to choose the optimal one and do further research and analysis.

3.1. Algorithms of Absolute-value Detector

For constructing a design that can detect the absolute values of neural signals and compare them with specific thresholds, it is essential to sort its key algorithms and main ideas [9,10]. For the absolute-value detector, the most significant algorithm is adding a logic "1" to the inverted input number when the input is negative. To achieve the function of inverting and adding, this study has to introduce the fundamental algorithm which is the magnitude of 2's complement number. In the 2's complement representation of binary numbers, the most significant bit (MSB) is represented as the flag bit to determine positive and negative input numbers. Then it can determine the magnitude of the 2's complement number by checking the best bits of the input numbers. On account of this, it can decide on the optimal design, and further analysis can be implemented based on it.

3.2. *First Design*

To accomplish the functions of absolute-value detection and thresholding comparison, this paper arises two types of designs at the early stage of constructing circuits and topology. First design chooses to use the devices including full adder, invertor, multiplexer and comparator. As shown in Figure 3, this research separates the input neural signals into two parts, MSB and magnitude. What is mentioned above is that the volume has to be inverted, and added "1" for the input is negative, and it remains still for the positive input. Thus, this study uses inverters and full adders when processing the input data. Then the processed data and original data are presented as the input data of the multiplexer, and the MSB is acting as the select signal, which can determine whether it is a positive signal or a negative signal. Finally, the output of the multiplexer is sent to a comparator and compared with a specific threshold.

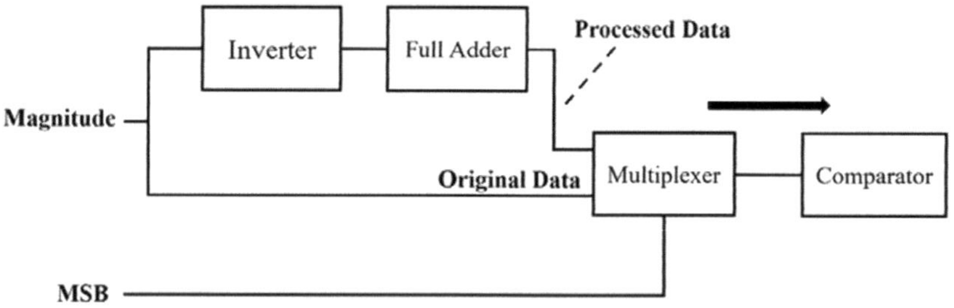

Figure 3: Basic Flow Chart of the First Design

This research assumes that the input neural signal is a 4-bit binary number and is given in 2's complement format. In this case, it needs a 3-bit full adder (Figure 4) to add the magnitude part of the input data. To do the addition operation of the 3-bit number bit by bit, the carry bits also have to be considered. The 3-bit full adder consists of six input signals which include three bits of the input neural signal (A0, A1, A2) and three bits of the adding number "1" (B0, B1, B2). In addition to the sum of two numbers in the home position, this research has to consider adding the rounding from the lower part, which are the carry bits (C0, C1, C2, C3). Lastly, the S0, S1, S2 are represented as the output data of this 4-bit adder [11].

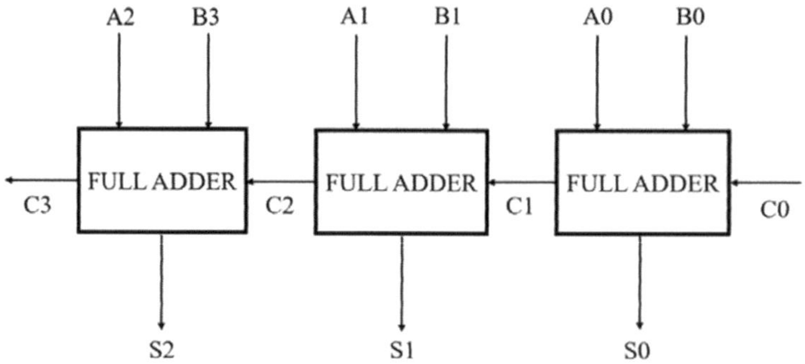

Figure 4: A 3-Bit Full Adder

Multiplexers are used to choose between the processed input signal and the original signal, which is called absolution-value detection. To obtain the correct output, choices made by the multiplexers depend on the Most Significant Bit (MSB) of the input data. When the MSB of the input data is "1" which indicates the input data to be harmful, the processed input signal is selected by the corresponding multiplexer. Otherwise, the original signal is determined by the multiplexer as the MSB of the input data is "0". Thus, this research builds a 2-to-1 multiplexer, and its logic circuits schematic is shown in Figure 5.

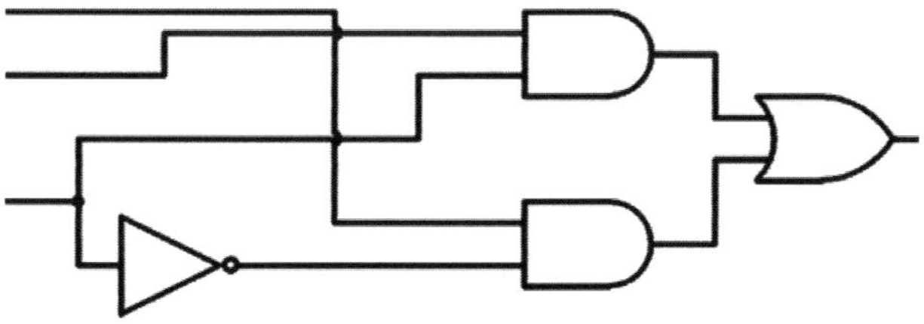

Figure 5: A 2-to-1 Multiplexer

After the multiplexer processes the input data, the output data is directly compared with the given threshold in the comparator which is described in Figure 6. This figure demonstrates the logic circuit schematic of a 3-bit comparator, and the data processed previously become the input of this comparator. A single-port output of the comparator presents the result. If the magnitude of the input signal is greater than the threshold, its output should a "1" (high logic value), otherwise the production should be a "0" (low logic value) [12].

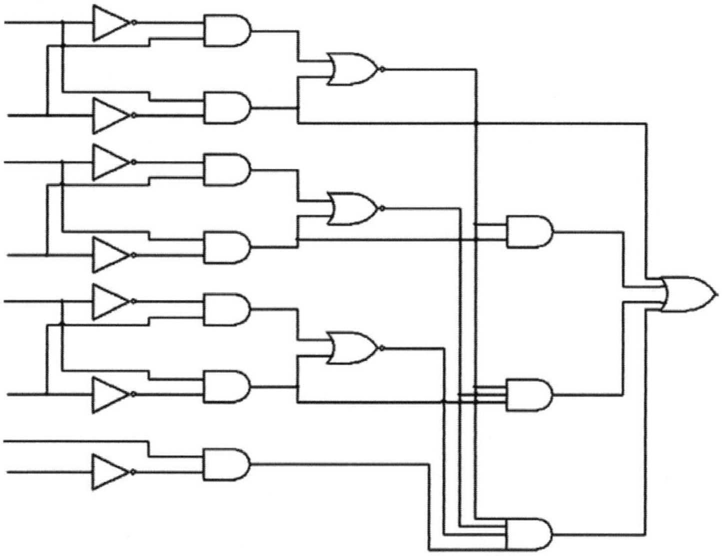

Figure 6: A 3-Bit Magnitude Comparator

Based on the devices above, this paper designs an absolute-value detector with the function of threshold comparison. Figure 7 is the complete logic circuit schematic of the first design, and it is apparent that it is divided into three main parts: full adder, multiplexer, and comparator. When the input neural signals are applied, the multiplexers read the MSB of inputs and select corresponding branches, which allow the absolute-value data to transmit. Meanwhile, the absolute-value data are processed in the 3-bit full adder before being determined by the multiplexers. Finally, the outputs of multiplexers are compared with the given thresholds in a 3-bit magnitude comparator.

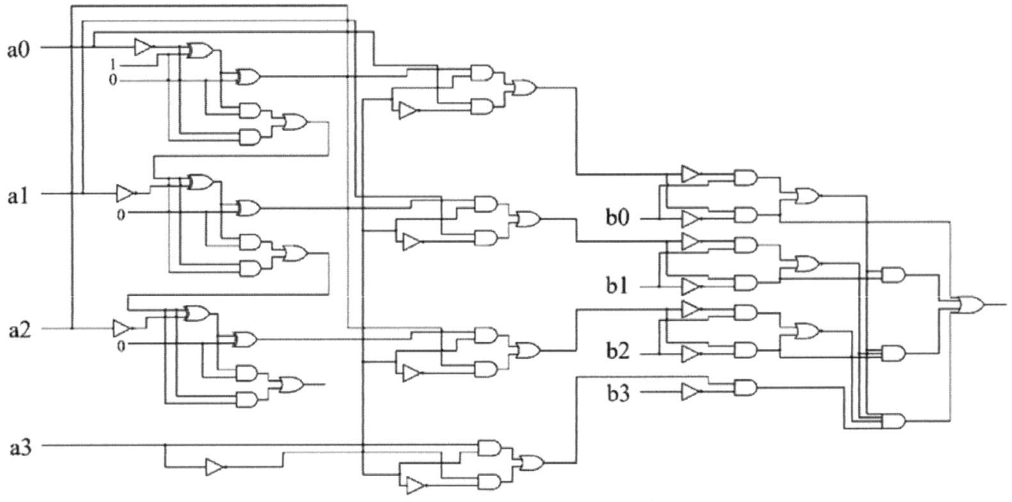

Figure 7: The Logic Circuit Schematic of the First Design

3.3. Second Design

Previous part has accomplished the logic circuit schematic of the first design and described the main components. However, as it reviewed the schematic of the first design, a large number of gates and logic devices are used. Consequently, the performance of this design may be affected negatively as gates increase delay and energy consumption. On account of this, it comes up with another type of design based on logic optimization.

Firstly, this essay reviews the basic idea of the design. In calculating the absolute-value numbers in the fundamental algorithm, it denotes the threshold and the input by T and A. Thus, differencing these two numbers (information is in the form of absolute value) it has the equation expressed in (1).

$$T-|A|=\begin{cases} T-(-A)=T+A \\ T-A \end{cases} \tag{1}$$

Therefore, this study compares the input with the given threshold by determining the difference between them. It can be seen in Figure 8 that this paper uses subtractor and adder to do the calculations above. Similar to the first design, it utilizes a multiplexer to implementing the selection between positive input and negative input. Finally, it can accomplish the comparison by checking out the difference between input and threshold.

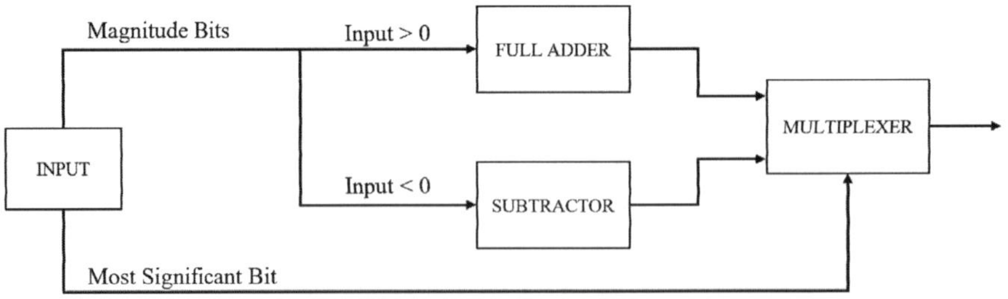

Figure 8: The Procedure to Implement the Minus

Based on the concept of logic optimization [13], this research puts more emphasis in optimizing the circuit in the second design. By comparing the adder and subtract, this study finds the only difference is that one is counting a borrow bit and another is counting a carry bit. For threshold is removed by input data, the borrow bit of the highest bit exists when the input data is greater than the threshold. On account of this, this study chooses to use only subtractor and it just needs to focus on the borrow bit which can indicate the greater number. Moreover, by connecting the input with the subtractors through XOR gates, it can achieve the judging without using any multiplexer. In the process of moving and replacing devices, this paper performs the optimization of the design.

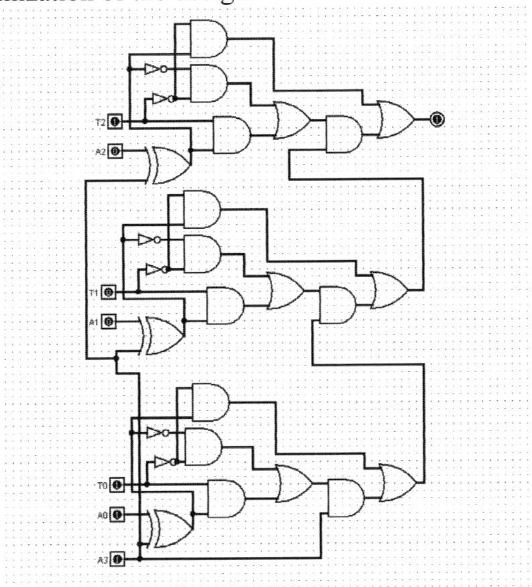

Figure 9: The Logic Circuit Schematic of the Second Design

4. Critical Path Analysis

In the last stages, it has come up with two types of design and described them in detail. To discuss and compare both designs, this paper analyzes their critical paths respectively which means the single path with the worst slack [14]. The critical approach of a circuit reflects the overall performance in delay and energy consumption. Therefore, it performs the critical path analysis for both of the designs.

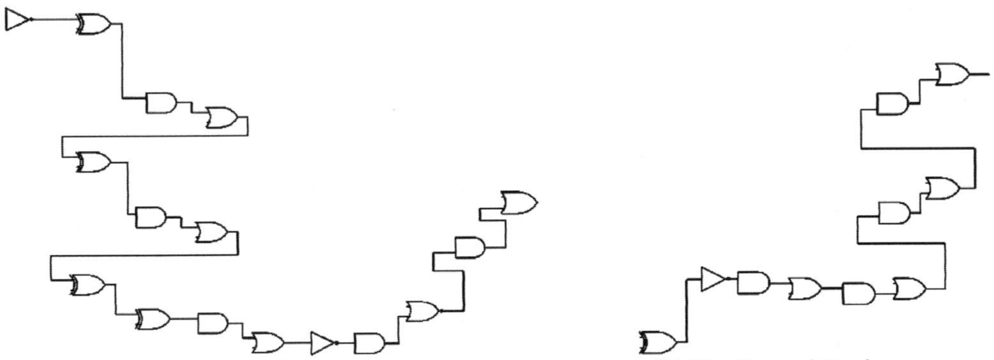

(a) The First Design (b) The Second Design

Figure 10: The Critical Path of the First design and the Second Design

To calculate and analyse the critical path delay, this research performed derivation of relevant algorithms at first.

$$f^* = \sqrt[N]{PathEffort} \tag{2}$$

$$f^N = \prod g \cdot h = PathEffort \tag{3}$$

Therefore, the study starts with the last stage of the schematic and derives the sizing of each stage from back to front.

$$C_{in} = g\frac{C_{out}}{f^*} \tag{4}$$

Based on the parameters above, this paper can have the derivation of the minimum path delay which is the critical parameter it is looking for. And the minimum path delay is expressed as Equation (5).

$$D_{\min} = \sum (g_i \cdot h_i + p_i) = N \cdot f^* + P \tag{5}$$

However, it still needs to calculate and analyse the energy consumption of the critical path circuit based on the delay time. For the calculations above, "delay" refers to the worst-case propagation delay. The energy this research is about to compute refers to the total energy drawn from V_{DD} for a given input probability distribution. To find the optimal V_{DD}, this study has to derive from the relation between minimum delay and energy. It can be assumed that the delay is proportional to V_{DD} as Equation (6) to find optimal V_{DD}. V_{DD}^{Opt} refers to the V_{DD} at the minimum energy consumption, while V_{DD}^{Mdelay} is the V_{DD} at minimum delay. Finally, the energy consumption can be solved by the Equation (7).

$$\frac{\dfrac{V_{DD}^{\ Opt}}{(V_{DD}^{\ Opt} - V_T)^2}}{\dfrac{V_{DD}^{\ Mdelay}}{(V_{DD}^{\ Mdelay} - V_T)^2}} \tag{6}$$

$$P = C_L \times V_{DD}^{\ 2} \times (W_P + W_N) \tag{7}$$

5. Results

Based on the derivations in the previous analysis, this study can compute the specific numbers and elicit the result of this research. To specify the calculation and rationalize the research findings, this research declares the conditions and constraints of this design. It is assumed that the $C_{parasitic}/C_{gate}$ equals to 1 and V_T equals to 0.2V. Moreover, the input signal is a 4-bit number represented in 2's complement format. The threshold value (3- bit binary) is fixed and you can assume a mid-range value for this Threshold. Assume that each bit of the input has an equal probability of being 0 or 1 and the bits are mutually independent. The input capacitance of all inputs (A0-A3) is less than or equal to 2-unit sized inverters (see below for the definition of the unit-sized inverter). For design purposes, the inputs to adder are driven by a unit sized buffer (chain of two-unit sized inverters). The delay is measured as the delay after the input driver (2 inverters) to before the load (32 times unit sized inverters). The output bit is loaded with CL = 32 unit-sized inverters. V_{DD} is 1V when the minimum delay is applied. Lastly, it is assumed that the unit-sized inverter is W_P= 650nm, W_N= 430nm, Lp = Ln = 100nm.

For the minimum delay of both designs, this paper has the result as follows. Sizes of each stage is shown in Table 1. The minimum delays are presented in Table 2, and it finds that the second design has better performance in delay. Thus, the further computations and analysis are based on the dual design. The V_{DD} at the minimum energy consumption and the minimum energy consumption are presented in Table 3. Each logic gate in Table 3 is labelled in order from the end of the critical path to the beginning.

Table 1: Size of Each Stage

Gates in the 1st Design	Size	Gates in the 2nd Design	Size
C_{in15}	41.92	C_{in10}	27.28
C_{in14}	30.51	C_{in9}	18.60
C_{in13}	22.20	C_{in8}	15.86
C_{in12}	12.90	C_{in7}	10.82
C_{in11}	5.63	C_{in6}	9.22
C_{in10}	4.10	C_{in5}	6.29
C_{in9}	2.39	C_{in4}	5.36
C_{in8}	4.17	C_{in3}	3.66
C_{in7}	7.30	C_{in2}	1.87
C_{in6}	5.30	C_{in1}	3.83
C_{in5}	3.08		
C_{in4}	2.24		
C_{in3}	1.30		
C_{in2}	2.31		
C_{in1}	1.01		

Table 2: Minimum Delay for Both Designs

Minimum Delay for the 1st Design	Minimum Delay for the 2nd Design
81.35	48.5

Table 3: Minimum Energy Consumption for the 2nd Design

V_{DD} at the Minimum Energy Consumption	Minimum Energy Consumption
0.7754	2076

6. Conclusion

This research comes up with two types of design at the beginning. To make the design reliable and comprehensive, it decided to discuss both designs in this paper. The first design is using full adders, multiplexers and a comparator. The concept of this circuit is adding the logic one to the input when the given input data is negative, and keeping the original input as the given input data is positive. To achieve the function of adding, this study chose the full adders. The main purpose of using multiplexers is to choose from the processed input and original input, and the choice depends on the most significant bit (MSB) of the input data. To compare the absolute value of the input data with a given threshold, it used a multi-bit comparator. The second design concept is based on the fundamental algorithms of the calculation of the absolute numbers. It means when the input is negative, this study can operate it with the threshold value through a subtractor. On the contrary, it can use an adder when the input is positive. By comparing the adder and subtract, this research found the only difference is that one is counting a borrow bit and another is counting a carry bit. On account of this, it decided to use the only a subtractor, so that can delete the borrow bits when it does not need to consider the borrow bits. By connecting the input with the subtractors through XOR gates, this paper can achieve the judging without using any multiplexer. In the process of moving and replacing devices, it reached the optimization of this design. Finally, this essay compared the minimum delay by calculating each stage's sizing and finding that the second design is better. Based on the second design, it calculated the energy consumption in the circuit and do V_DD optimization to get the minimum energy consumption.

References

[1] Lewicki M S. A review of methods for spike sorting: the detection and classification of neural action potentials[J]. Network, 1998.
[2] Birbaumer N, Ghanayim N, Hinterberger T, et al. A spelling device for the paralysed. [J]. Nature, 1999, 398(6725):297-298.
[3] Noninvasive Brain–Machine Interfaces for Robotic Devices[J]. Annual Review of Control Robotics and Autonomous Systems, 2021, 4(1).
[4] Single-Cell Electrical Stimulation with CMOS-based High-Density Microelectrode Arrays[J]. Frontiers in Cellular Neuroscience, 2018, 12: -.
[5] Obeid I, Wolf P D. Evaluation of spike-detection algorithms for a brain-machine interface application. [J]. Biomedical Engineering IEEE Transactions on, 2004, 51(6):905-911.
[6] Kaiser J F. On a simple algorithm to calculate the 'energy' of a signal[C]. International Conference on Acoustics. IEEE, 2002.
[7] Kim K H, Kim S J. A wavelet-based method for action potential detection from extracellular neural signal recording with low signal-to-noise ratio. [J]. IEEE transactions on bio-medical engineering, 2003, 50(8):999-1011.
[8] D Valencia, Alimohammad A. Neural Spike Sorting Using Binarized Neural Networks[J]. IEEE Transactions on Neural Systems and Rehabilitation Engineering, 2020, PP(99):1-1.
[9] Sikandar Kumar,Subha P. Eswaran. Improved Spectrum Sensing Schemes Using Prewhitening and Weights Under Spatially Correlated Noise[J]. Wireless Personal Communications,2020,115(prepublish).
[10] M. Hirata,S. Nagashima,T. Cho,J. Kohagura,M. Yoshida,H. Ito,S. Tokioka,T. Numakura,R. Minami,Y. Nakashima,T. Kondoh,K. Yatsu,S. Miyoshi. A Novel Electrostatic Ion-Current Absolute-Value Detector Under Circumstances with Simultaneous Electron Incidence by The Use of a Proposed "Self-Collection" Method for Secondary Electron Emission[J]. Fusion Science and Technology,2003,43(1T).
[11] S. Prema. Comparative Analysis of Different Full Adder Circuits[J]. Programmable Device Circuits and Systems,2013,5(2).
[12] Efstathiou Constantinos,Kitsos Paris. Efficient majority logic magnitude comparator design[J]. Microprocessors and Microsystems,2021,82.
[13] Lee Seong Bong, Chong Jong Wha. A Rule-Based System for VLSI Gate-Level Logic Optimization[J]. Journal of the Korean Institute of Telematics and Electronics,1989,26(1).
[14] Chang-Lin Li,Hyun Joong Kim,Seo Weon Heo,Tae Hee Han. Voltage and Frequency Tuning Methodology for Near- Threshold Manycore Computing using Critical Path Delay Variation[J]. Journal of Semiconductor Technology and Science,2015,15(6).

Realization and optimization of a new spike detection algorithm logic circuit

Yuefeng Gao[1, †], Qingxing Liu[2, †], Hengji Ye[3, *, †]

[1] College of Information Sciences and Technology, Beijing University of Chemical Technology, Beijing 100000, China.

[2] Changkong college, Nanjing University of Aeronautics and Astronautics, Nanjing 320100, China.

[3] School of Optical and Electronic Information, Huazhong University of Science and Technology, Wuhan 430074, China

[*]Corresponding author email: 2019040306@mail.buct.edu.cn

[†]These authors contributed equally.

Abstract. At present, artificial intelligence has increasingly become the most promising and essential major. The interdisciplinary cooperation between the integrated circuit and artificial intelligence brings infinite possibilities. In the convolutional neural network area, in order to achieve the valuable output, the comparison between values is often encountered. The output is obtained by comparison between the result after the convolution calculation with the designed threshold. Therefore, our team not only design a 4-bit binary comparator hardware logic circuit to complete the task but also discuss and verify the feasibility and performance characteristics of the program from the perspective of energy and time delay. As for the overall framework, we design a convenient circuit that converts the complement code into the original code and chooses a CLA adder to accomplish this part. Using this kind of adder can ideally help us reduce the time delay at the expense of a complicated circuit schematic. In the comparator part, we design a high-quality circuit framework. The strategy of our circuit is to compare the relationship between the four-bit binary code and the threshold bit by bit from MSB to LSB, which performs better than the original 4-bit comparator, and we design two outputs that can legibly illustrate the relationship between two values. We use logic effort to discuss the normalized delay in our project. Besides, we find the connection between the energy and the delay by calculation. Finally, we design a trade-off function to make the optimization of energy and delay together with respect to voltage.

1. Introduction

Nowadays, the rapid development of electronic technology has brought tremendous changes to human life. This is an emerging technology that began to develop in the late 19th century and early 20th century [1]. In the twentieth century, electronic technology developed the fastest and most widely used, becoming an important symbol of the development of modern science and technology [2].

The first generation of electronic products is centered on electron tubes. At the end of the 1940s, the world's first semiconductor transistor was born [3]. It was quickly used in various countries for its compactness, lightweight, power saving, and long life, replacing the electronic tube in a wide range. In the late 1950s, the world's first integrated circuit appeared [4]. It integrated many electronic components such as transistors on a silicon chip, making electronic products more compact. Integrated circuits have rapidly developed from small-scale integrated circuits to large-scale integrated circuits and ultra-large-scale integrated circuits so that electronic products are moving towards high-efficiency, low-consumption, high-precision, high-stability, and intelligence.

Among the integrated circuits, the most outstanding is the metal-oxide-semiconductor (MOS) integrated circuit. This is an integrated circuit composed of MOS field-effect transistors as the main components, referred to as MOSIC [5]. In 1964, the insulated gate field effect transistor was developed. It was not until 1968 that the stability of MOS devices was solved.

MOS integrated circuit is a commonly used integrated circuit. The smallest unit is an inverter, which consists of two metal-oxide-semiconductor field-effect transistors. In addition, the MOS tube is particularly easy to produce AND gates and NOT gates in digital logic circuits and has special features such as high speed and simple structure [5]. This kind of integration process is mainly used for the manufacture of digital integrated circuits, and the degree of circuit integration can be very high.

The electronic adder is a digital circuit component used to perform additional operations and is the basis of the arithmetic logic unit in the core microprocessor of an electronic computer. In these electronic systems, the adder is mainly responsible for calculating the address, index, and other data. In addition, the adder is also an important part of some other hardware, such as the binary number multiplier. Personal digital assistants and notebook computers require ultra-large-scale integrated circuits, as well as ultra-large-scale integrated designs with improved power delay characteristics [6]. The adder is one of the most basic cornerstones in all the above-mentioned circuit applications. Over the years, its optimization has been particularly important, with the main goal of reducing energy consumption and time delay and reducing structural complexity. Among them, there are generally two optimization schemes: 1. Logic layer optimization, rearranging a Boolean equation to obtain a circuit with a faster speed or a smaller area. 2. Circuit layer optimization, changing the size of the transistor or the topology of the circuit [7].

The more traditional CMOS full adders now have better indicators for voltage scaling and can provide full swing output. But the disadvantage is that this method will have a larger area and slower speed due to the larger input capacitance [8]. In order to better understand the advantages and disadvantages of different types of adders, this article compares the power and delay between different types of full adder circuits so that we can have a more comprehensive understanding of some problems of adders [9].

At present, a more popular optimization method is to carry out the comprehensive and detailed design for each subsystem, which helps to improve the overall performance of complex circuits not only directly but also easily realizes the arithmetic operations in the ultra-large integrated-scale integrated circuit system [10]. In addition, size optimization is also the focus of optimization of the MOS adder. This paper proposes a new way to reduce power consumption and demonstrates that reducing area is an effective way to reduce power consumption [11].

Integrated circuits also play an important role in artificial intelligence, and one of the more commonly used applications is to design a circuit to detect spikes from the nervous system. The algorithm is a commonly used spike detection algorithm, also known as absolute value detection. Its basic diagram is shown in Fig.1. Here x[n], Thr, and Output respectively represent the detected value, the preset value (threshold), and whether it is recognized as valid (1 means valid, 0 means invalid).

The simple idea is that after the signal is converted into a digital signal, it is first corrected by absolute value and compared with a preset threshold. If it is greater than the threshold, it means that a spike signal is detected. And in order to enable the device to work for a long time to detect the nerve impulse of patients who have a mental illness such as violent fever, we need to reduce its power consumption as much as possible; to make the device sensitive to nerve impulse, we also hope to reduce its detection time (also known as delay time). We plan to use an adder and a comparator to accomplish this goal, reduce energy consumption, and delay time through a series of measures.

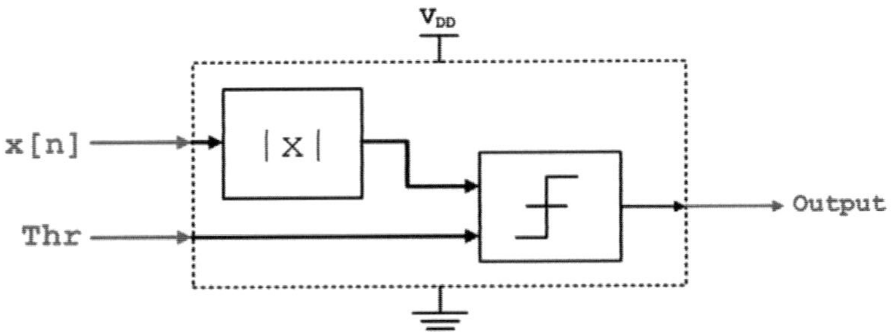

Fig.1 The logic diagram of spike detection algorithm (absolute value detection)

2. Mechanism

2.1 Optimization of 2's Complementary Calculation

Because the input is a 4-bit binary number in complement form, in order to make the comparison between our input and the threshold number, we need to convert the complement code into the original form.

There are two cases of the input. For convenience, we useA_3 A_2 A_1 A_0 (4-bit, from MSB to LSB) to describe our information:

IfA_3=0, which means that the input itself is positive), then the original code form of the input remains the same.

If A_3=1, which means the input itself is negative), then the original form of the input is to make the bitwise negation and plus 1.

To this end, our team designs a practical strategy that can effectively avoid the discussion about whether A3 is positive or negative. Our team decided to make XOR between A_3 and A_3, A_2, A_1, A_0 in turn, and then add $000A_3$.. We can use the following formula to express 1's complement:

$$A_i' = A_3 \oplus A_i \quad (i = 0,1,2,3) \tag{1}$$

Tab.1 An Example of 1's Complement Calculation

Cases	$A_3A_2A_1A_0$	$A_3'A_2'A_1'A_0'$
$A_3 = 1$	1001	0110
$A_3 = 0$	0011	0011

As we can see, if $A_3 = 1$, it means that the 4-bit binary number is less than zero. We assume that the input's complement is 1001, and the number obtained after the complement conversion operation is 0110. According to our strategy, XOR A_3 and $A_3A_2A_1A_0$ respectively, the number obtained is 0110, plus $000A_3$, which is 0001, and the final result is also 0110. If $A_3 = 0$, that is to say, this 4-bit binary number is greater than zero. After the operation of converting the complement to the original code, its complement form is the same as itself. As shown in the figure, we take 0011 as an example, and the number obtained after the operation is still 0011. Using our strategy, we perform exclusive OR operations separately, plus $000A_3$, which is 0000, and the final result are also 0011. Our strategy avoids the discussion of A_3 value and saves hardware resources. The 2'complement can be expressed as follows:

$$A_3''A_2''A_1''A_0'' = A_3'A_2'A_1'A_0' + 000A_3 \tag{2}$$

The final result is shown in Tab.2 below:

Tab.2 The Final Result of 2's Complement Calculation

Cases	$A_3A_2A_1A_0$	$A_3''A_2''A_1''A_0''$
$A_3 = 1$	1001	0111
$A_3 = 0$	0011	0011

2.2 Optimization of carrying Lookahead Adder

In order to achieve great performance in the time delay calculation, we choose the Carry Lookahead Adder(CLA) rather than the Ripple Carry Adder(RCA). We inspect these two adders from various of perspectives.

We made the comparison between the CLA adder and the RCA adder in Tab.3.

Tab.3 Comparison Between CLA and RCA

Advantages and Disadvantage	RCA	CLA
Structural Features	C_{out} of the low full adder is connected to the high full adder C_{in}	The Cin of each full adder comes from the logic of the advanced bit
Advantage	Simple circuit layout and convenient design	The delay is fixed as a three-level gate delay, has nothing to do with the number of bits
Disadvantage	The high-order must wait for the low-order operation to complete, and the delay time is long	If the number of bits of the adder is further widened, the circuit becomes complicated

And this is the schematic of our logic circuit, and we add two intermediate variables to accelerate the calculation. Finally, we design this adder based on the Boolean expression optimization of the complement calculation and the concept of carrying lookahead adder as shown in Fig.2. Equation (3) is the Boolean expression of the adder.

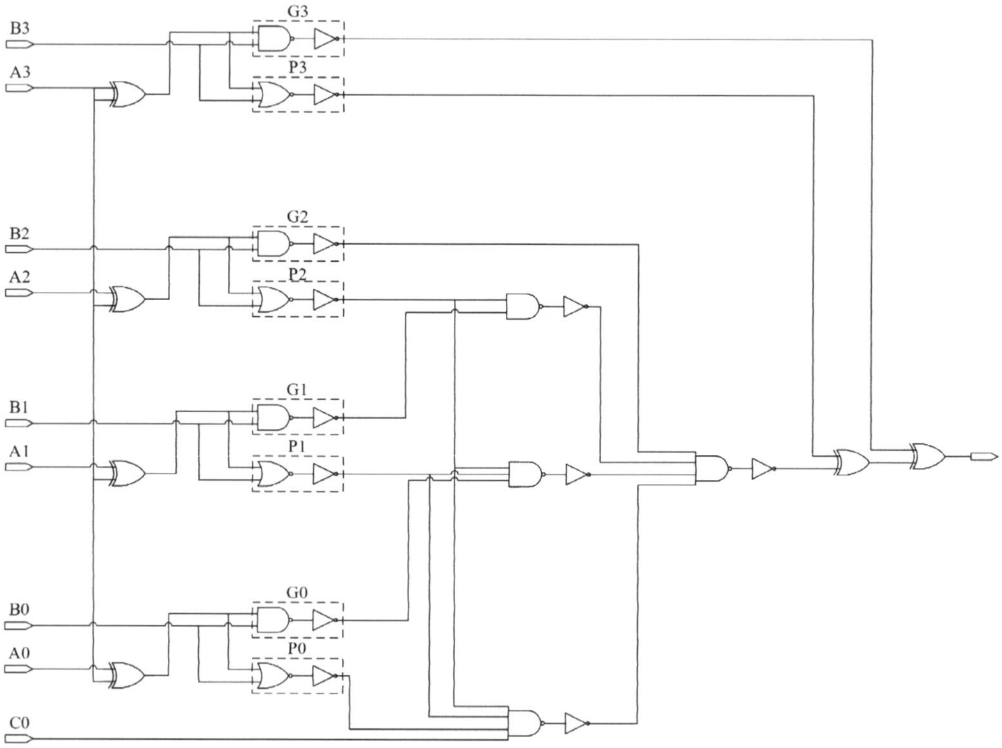

Fig.2 Schematic of Adder Logic Circuit

$$\begin{aligned}
C_1 &= G_0 + P_0 \cdot C_0 \\
C_2 &= G_1 + P_1 \cdot C_1 \\
&= G_1 + P_1 \cdot (G_0 + P_0 \cdot C_0) \\
&= G_1 + P_1 \cdot G_0 + P_1 \cdot P_0 \cdot C_0 \\
C_3 &= G_2 + P_2 \cdot C_2 \\
&= G_2 + P_2 \cdot (G_1 + P_1 \cdot G_0 + P_1 \cdot P_0 \cdot C_0) \\
&= G_2 + P_2 \cdot G_1 + P_2 \cdot P_1 \cdot G_0 + P_2 \cdot P_1 \cdot P_0 \cdot C_0 \\
C_4 &= G_3 + P_3 \cdot C_3 \\
&= G_3 + P_3 \cdot (G_2 + P_2 \cdot G_1 + P_2 \cdot P_1 \cdot G_0 + P_2 \cdot P_1 \cdot P_0 \cdot C_0) \\
&= G_3 + P_3 \cdot G_2 + P_3 \cdot P_2 \cdot G_1 + P_3 \cdot P_2 \cdot P_1 \cdot G_0 + P_3 \cdot P_2 \cdot P_1 \cdot P_0 \cdot C_0
\end{aligned} \tag{3}$$

In these formulas, $P_i = A_i'' + B_i$, $G = A_i'' \cdot B_i$.

2.3 Optimization of Comparator

The traditional 4bit comparator is obtained by simplifying the design based on the truth table. The strategy of the 4bit comparator designed by our team is to compare the MSB to the LSB. If the high-value bit can be judged, the subsequent low Numerical bit comparison does not need to be carried out, which maximizes the speed of operation and reduces the time delay. According to our ideas, we can get the following logic circuit diagram as shown in Fig.3.

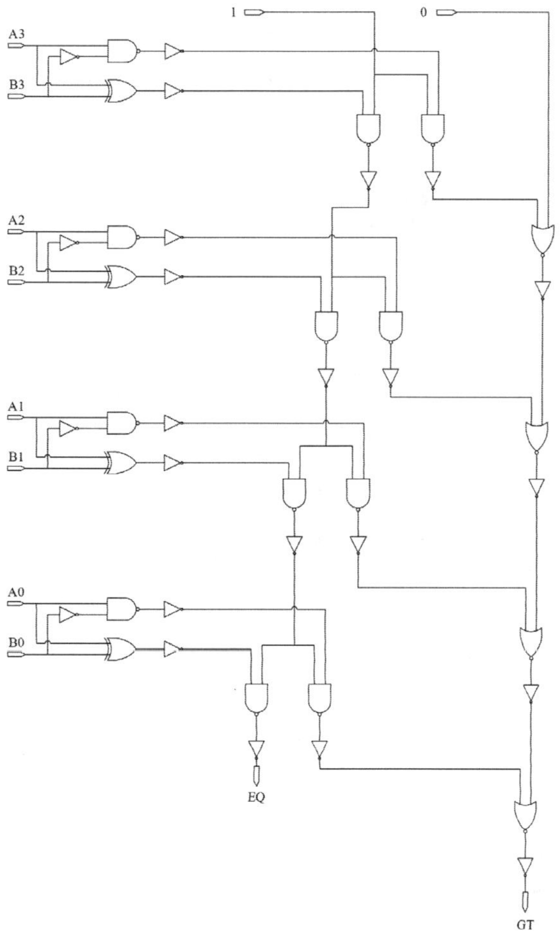

Fig.3 Schematic of Comparator Logic Circuit

We can see that there are two outputs in the entire circuit diagram. We call them EQ (equal) and GR (greater than). When EQ=1, the size of the 4-bit binary input representing the entire circuit is equal to the threshold voltage. At this time. In this case, GR must be 0. When EQ=0, which means that the 4-bit input signal and the threshold voltage of the entire circuit are not equal. At this time, when GR=1, it means that the 4-bit input signal is greater than the threshold voltage. When GR=0, it means that the 4-bit input signal is less than the threshold voltage. It can be seen from the circuit schematic that the entire system is similar in form from top to bottom so that we can observe from A_3B_3 to examine the feasibility and logic of the entire circuit. The figure below shows the different results represented by different values of A_i and B_i. For the GT signal, when $A_3B_3 = 10$, through multiple logic gates, the GT signal can be directly determined to be 1, and it has nothing to do with A_2B_2, A_1B_1, and A_0B_0. When $A_3B_3 = 10$ or 10, the judgment of this layer of the circuit is equivalent to invalid, continue to look at the relationship between A_2 and B_2 in the next layer. When $A_3B_3 = 01$, the GT signal can be determined to be 0 after passing through the logic gate, and the EQ signal is also 0. The whole circuit system is compared from the high position, which saves the circuit resources and reduces the circuit delay. In the end, we can get four situations, as shown in Tab.4.

Tab.4 Four cases of comparator

A_iB_i	Case
00	Equal
01	Less than
10	Greater than
11	Equal

2.4 Logic effort

We use the concept of logic effort to get the normalized delay of our Absolute-Value Detector. Compared with the traditional way of delay calculating, the logic effort can normalize the delay to a fictitious time constant, which is more practical and convenient for examining the performance of the circuit.

We usually use formula (4) to calculate delay.

$$t_p = 0.69R_gC_{in,\,gate}\left(\frac{C_{par,gate}}{C_{in,gate}} + \frac{C_{out}}{C_{in,\,gate}}\right) \tag{4}$$

We make the following definition

$$\tau_{gate} = 0.69R_gC_{in,\,gate} \tag{5}$$

$$\gamma_{gate} = \frac{C_{par,gate}}{C_{in,gate}} \tag{6}$$

So that we can have formula (7):

$$\frac{Delay}{\tau_{INV}} = \frac{\tau_{gate}}{\tau_{INV}}\left(\gamma_{gate} + \frac{C_{out}}{C_{in,\,gate}}\right) \tag{7}$$

To simplify this formula, we have

$$g = \frac{R_{gate} \cdot C_{in,gate}}{R_{INV} \cdot C_{in,INV}} \tag{8}$$

$$h = \frac{C_{out}}{C_{in,\,gate}} \tag{9}$$

$$p = \frac{C_{par,gate}}{C_{par\,,INV}} \cdot \gamma_{INV} \tag{10}$$

Where g means logic effort, h means electrical fanout, p means parasitic delay, d means normalized delay

Then we have the normalized delay in formula (11):

$$d = gh + p\gamma_{INV} \tag{11}$$

Here, we assume that γ_{INV} is equal to 1, then we get:

$$d = gh + p \tag{12}$$

To calculate the logic effort of each gate, first, we need to choose an input of the gate, then find the total device width driven by that input, Find W_P, the pull-up device width of a single device that has equivalent drive strength as a gate's pull-up of that input. For a reference inverter with Equal Rise/Fall, we determine the total gate widths of the inverter devices with W_P, finally divide the total device width by total gate widths to get *g*.

When we applied logic effort to inverter chain as shown in Fig.4:

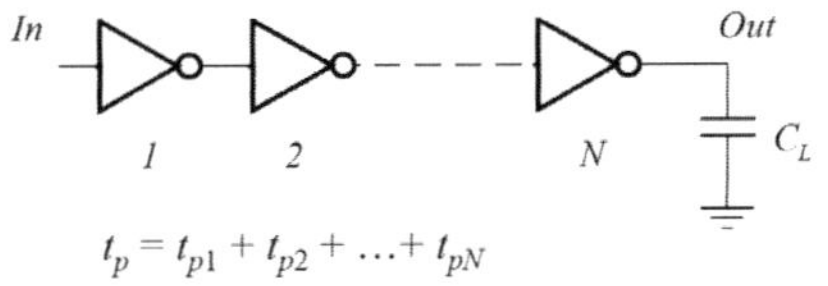

Fig.4: Inverter Chain

$$t_{pj} = R_{gate,unit}\, C_{in\,gate,unit} \left(\gamma + \frac{C_{g\,in,j+1}}{C_{g\,in,j}}\right) \tag{13}$$

$$t_p = \sum_{j=1}^{N} t_{p,j} = \tau \sum_{i=1}^{N} \left(\gamma + \frac{C_{gin,j+1}}{C_{gin,j}}\right), C_{gin,N+1} = C_L \tag{14}$$

$$C_{gin,N+1} = C_L \tag{15}$$

To find minimize delay, we find N-1 partial derivatives, we have

$$\frac{C_{gin,j+1}}{C_{gin,j}} = \frac{C_{gin,j}}{C_{gin,j-1}} \tag{16}$$

When each stage is sized by f and has same fanout *f*,

$$f^N = F = \frac{C_L}{C_{gin,1}} \tag{17}$$

Effective fanout of each stage:

$$f = \sqrt[N]{F} \tag{18}$$

And the minimum path delay is:

$$t_p = N\tau\left(\gamma + \sqrt[N]{F}\right) \tag{19}$$

So in the generalization for multistage networks, we have minimum path delay:

$$D_{min} = \sum(g_i \cdot h_i + p_i) = N \cdot f + P \tag{20}$$

Path Effort Delay:

$$D_F = \sum_i f_i \tag{21}$$

Path Parasitic Delay:

$$P = \sum_i p_i \tag{22}$$

Total Path Delay:

$$D = \sum_i d_i = D_F + P \tag{23}$$

If the circuit has a branch, we need to take branching effort into consideration:

$$b = \frac{C_{\text{on-path}} + C_{\text{off-path}}}{C_{\text{on-path}}} \tag{24}$$

And we have path branching effort:

$$B = \prod b_{\mathrm{i}} \tag{25}$$

We can change formula (18) to:

$$\text{Path effort}: F_{\text{path}} = G_{\text{path}} \cdot H_{\text{path}} \cdot B_{\text{path}} \tag{26}$$

2.5 Trade-off between Energy Consumption and Delay time

According to the previous analysis, we know that when V_{DD} increases, the energy consumption will increase, and the delay time will decrease simultaneously. In order to explore the optimal trade-off relationship between delay and energy, I intend to construct an optimization function (also known as a trade-off function) from the perspective of mathematical analysis to find the optimal voltage.

The steps are as follows:

- Construct an appropriate trade-off function;
- Normalize the variables that need to be measured;
- Find the domain of the independent variables (possibly multiple) of the trade-off function;
- Using the optimization method to solve the maximum value of the trade-off function (the function of one variable is to find the extreme point).

We use the average trade-off function:

$$y = \sqrt{(E' - \bar{E})^2 + (D' - \bar{D})^2} \tag{27}$$

Where E' and D' are respectively the energy and delay after normalization

The normalization formula is as follows:

$$x' = \frac{x}{max(x) - \min(x)} \tag{28}$$

The energy expression is:

$$E = C{V_{DD}}^2 \tag{29}$$

The delay expression is:

$$D = \frac{C}{I_D} = \frac{CV_{DD}}{I_D} = \frac{CV_{DD}}{k'\frac{W}{L}(V_{DD} - V_T)^2} = \frac{kV_{DD}}{(V_{DD} - V_T)^2} \tag{30}$$

When $V_{DD} = 0.78V, min(E) = 0.6084C$; When $V_{DD} = 1V, max(E) = C$;

The normalized energy is:

$$E' = \frac{E}{max(E) - min(E)} = \frac{C{V_{DD}}^2}{C - 0.6084C} \approx 2.5{V_{DD}}^2 \tag{31}$$

The delay is a single valued function of the voltage, so we can get the derivative as follows:

$$\frac{dD}{dV_{DD}} = -k\frac{V_{DD} + V_T}{(V_{DD} - V_T)^3} < 0 \tag{32}$$

When $\mathrm{V_{DD}} = 1\mathrm{V}, \max(\mathrm{D}) = 1.5625\mathrm{k}$; When $\mathrm{V_{DD}} = 0.78\mathrm{V}, \min(\mathrm{E}) = 3\mathrm{k}$;

The normalized delay is:

$$D' = \frac{D}{max(D) - \min(D)} = \frac{\frac{kV_{DD}}{(V_{DD}-V_T)^2}}{3k - 1.5625k} \approx \frac{0.7V_{DD}}{(V_{DD} - V_T)^2} \tag{33}$$

The average values of energy and delay are as follows:

$$\bar{E} = \frac{\int_{0.78}^{1} {V_{DD}}^2 dV_{DD}}{0.22} \approx \frac{0.175}{0.22} = 0.795 \tag{34}$$

$$\bar{D} = \int_{0.78}^{1} \frac{0.7V_{DD}}{(V_{DD} - V_T)^2} dV_{DD} = \frac{0.2914}{0.22} \approx \frac{0.29}{0.22} = 1.32 \tag{35}$$

The final trade-off function is as follows:

$$y = \sqrt{\left(2.5{V_{DD}}^2 - 0.795\right)^2 + \left(\frac{0.7V_{DD}}{(V_{DD} - V_T)^2} - 1.32\right)^2} \tag{36}$$

3. Result

3.1 Energy and Delay

We have the following results through the optimized calculation of logical efforts, as shown in Tab.5.

Tab.5 The result of four variables

Variable	Result
Final Delay [FO4(1V)]	$7031.25C_L$
Final Energy [E_unit(1V)]	$19.47C$ (inverter)
Critical Path Delay at 1V	93.5
The energy at 1V and Min Delay	32C(inverter) $7031.25C_L$

3.2 Calculation of V_{DD}

We use the delay of the basic RC network to characterize the time delay model, and the equivalent resistance R_{eq} can be calculated like this.

$$R_{eq} = \frac{1}{\frac{V_{DD}}{2}} \int_{\frac{V_{DD}}{2}}^{V_{DD}} \frac{V}{I(1+\gamma V)} dV \approx \frac{3}{4}\frac{V_{DD}}{I} \tag{37}$$

In this formula, we assume $\gamma = 0$.

And in the saturation region, we can calculate the current in the formula,

$$I = \frac{1}{2}\mu_n C_{ox} \frac{W}{L}(V_{DD} - V_T)^2 \tag{38}$$

So, we bring this equation in the last equation, and we can achieve the time delay(t_p)

$$t_p = 0.69R_{eq}C_L = 0.69 \times \frac{3}{4} \frac{C_L V_{DD}}{\frac{1}{2}\mu_n C_{ox} \frac{W}{L}(V_{DD} - V_T)^2} \sim \frac{kV_{DD}}{(V_{DD} - V_T)^2} \tag{39}$$

3.3 Calculation of Sizing

According to the calculation method of intrinsic delay, we can get the following formulas:

$$t_{p0} = 0.69R_{eq}C_{int} \tag{40}$$

$$t_p = 0.69R_{eq}(C_{int} + C_{ext}) = t_{p0}\left(1 + \frac{C_{ext}}{C_{int}}\right) \tag{41}$$

We define sizing parameter S, which is a constant proportional to the size of the MOS tube. Generally, we think that the equivalent capacitance of a MOS tube is proportional to its size area, and the equivalent resistance is inversely proportional to its area, so we can get equations as follows:

$$C_{int} = SC_{iref} \tag{42}$$

$$R_{eq} = \frac{R_{eq}}{S} \tag{43}$$

$$\begin{aligned} t_p &= 0.69(R_{eq}/S)(SC_{iref})\left(1 + \frac{C_{ext}}{SC_{iref}}\right) \\ &= 0.69R_{eq}C_{iref}\left(1 + \frac{C_{ext}}{SC_{iref}}\right) \\ &= t_{p0}\left(1 + \frac{C_{ext}}{SC_{iref}}\right) \end{aligned} \tag{44}$$

As a result, when S increases, delay decreases. When V_{DD} decreases, delay increases. In addition, the intrinsic delay is:

$$t_{p0} = \frac{Q}{I} = \frac{C{\cdot}V_{DD}}{I} = \frac{C{\cdot}V_{DD}}{K\frac{W}{L}(V_{DD} - VT)^2} \sim \frac{kV_{DD}}{(V_{DD} - VT)^2} \tag{45}$$

Next, we get the following formulas in the calculation process:

$$k\frac{1}{(1-0.2)^2} = k\frac{25}{16} \tag{46}$$

$$\frac{delay'}{delay} = \frac{3}{2} = \frac{\frac{V'}{(V'-0.2)^2}}{\frac{25}{16}} \tag{47}$$

After calculation, because the equation is a quadratic equation with one variable, we get two results: $V' = 0.78\text{V}$, $V'' = 0.05\text{V}$. We need to make sure the MOS is working in the appropriate region, so $V' = 0.78\text{V}$.

Our final energy is:

$$E = C_L V^2 \tag{48}$$

3.4 *Logic Effort*

We use static complementary CMOS to structure different gates. The inputs to our adder are driven by a unit-sized buffer (chain of two-unit sized inverters), and the output bit is loaded with $C_L = 32$ unit-sized inverters. W_p/W_N of the unit-sized inverter is 2:1.

As we all know, it's more convenient to structure the gates that have NOT logic with static complementary CMOS, such as NAND gate and NOR gate. In order to simplify the inner COMS construction of each gate and the calculation of Normalized delay, in our design, we use a NAND gate plus a NOT gate to replace a AND gate, and we use a NOR gate and a NOT gate to replace an OR gate.

In order to size the gate for the minimum delay, we need to Identify the input vectors that will exercise critical path, the critical path is the path that goes through most gates and has the most branches, and here we've labeled the critical path whose color is red of the adders and comparators and marked the g, h, b corresponding to each logic gate in Fig.5.

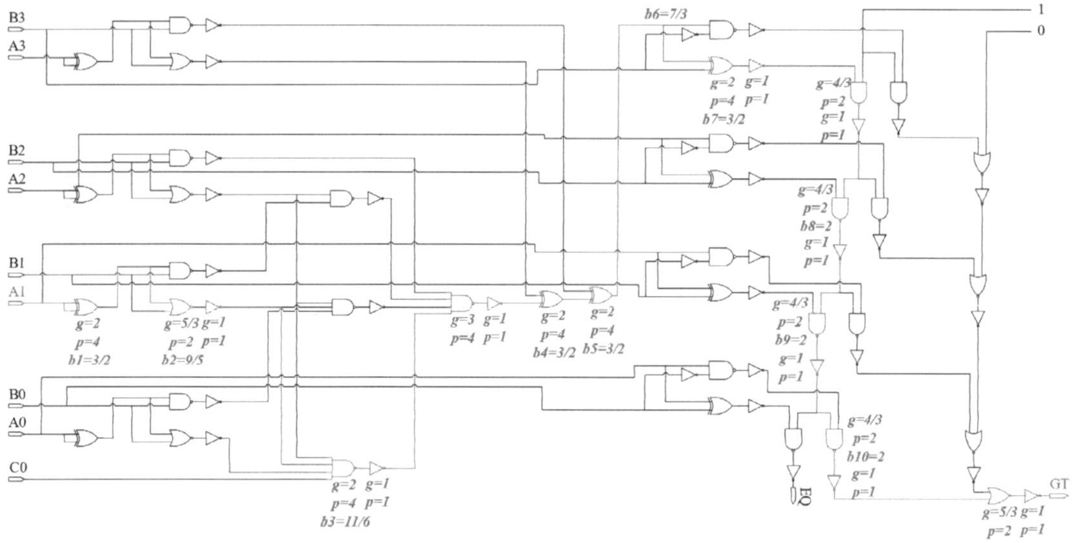

Fig.5 The overall logic circuit of the spike detection algorithm

To obtain the total normalized delay of the critical path, we need to calculate the logic effort(g) and Parasitic delay(p) of each gate. Through the method we mentioned in the Mechanism part, we obtain the *g* and *p* of each gate as shown in Tab.6:

Tab.6 *g* and *p* of each gate

Gate	*g*	*p*
XOR gate	2	4
NAND gate	4/3	2
NOR gate	5/3	2
Four-input NAND gate	2	4
Four-input NOR gate	3	4

The branch effort also needs to be considered when calculating logic effort, To construct an XOR gate by static COMS construction, we have to get a negative logic of the input, that lead to a branch effort in the inner construction of the XOR gate, and it can be calculated to be 1.5. To sum up, there are ten branches in the critical path as shown in Tab.7.

Tab.7 Ten branches in the critical path

b_1	b_2	b_3	b_4	b_5	b_6	b_7	b_8	b_9	b_{10}
1.5	9/5	11/6	1.5	1.5	7/3	1.5	2	2	2

and)

Finally, we have:

Path effort:

$$F = \prod g \cdot h \cdot b = 1868989.63 \quad (49)$$

Fanout:

$$f = g \cdot h \cdot b = \sqrt[25]{F} = 1.78 \quad (50)$$

Normalized delay:

$$D = 25 \cdot f + \Sigma P = 93.5 \quad (51)$$

4. Conclusion

Based on the results and discussions presented above, the conclusions are obtained as below:

According to formula (36), when VDD is equal to 0.81, the y has a minimum value, which means that the optimal trade-off between the two is reached at this time. The combination of Carry Lookahead Adder and comparator has a good performance in both delay and energy consumption.

Acknowledgments

This work was financially supported by the Applied Research Project of Public Welfare Technology of Zhejiang Province (2017C33013).

References

[1] Wang Jianping. The development history of electronic technology application system[J]. Electronic Technology and Software Engineering, 2014(05): 125.
[2] Zhou Zixin. Briefly describe the development process of electronic technology[J]. Science Education Journal: Electronic Edition, 2016(16):141-141.
[3] Ye Wenhao. Deep love and rapid development of China's lamp triode semiconductor industry [J]. China Lighting, 2005, 000(006): 72-75.
[4] Chen Biao. The development of integrated circuit technology[J]. Microprocessor, 2011, 32(3):1
[5] Pan Guizhong. MOS integrated circuit structure and manufacturing technology [M]. Shanghai Science and Technology Press, 2010.

[6] C.-K. Tung, Y.-C. Hung, S.-H. Shieh, and G.-S. Huang, "A low-powerhigh-speed hybrid CMOS full adder for embedded system," in Proc.IEEE Conf. Design Diagnostics Electron. Circuits Syst., vol. 13.Apr. 2007, pp. 1–4.
[7] Zhou Dapeng, Long Anwen, Xu Guorong, et al. Structural parameter constraints for the optimal design of the advance position adder[J]. Journal of South-Central University for Nationalities (Natural Science Edition), 2006, 25(001): 58-61.
[8] Waste, N. H. E., & Harris, D. (2010). CMOS VLSI design: A circuits and systems perspective. (4th ed.). Boston: Addison Wesley Publishing Company.
[9] Panda S , Maji A , Mukhopadhyay A . Power and Delay Comparison in between Different types of Full Adder Circuits[J]. International Journal of Advanced Research in Electrical Electronics & Instrumentation Engineering, 2013, 1(3).
[10] Goal, S., Kumar, A., & Bayoumi, M. A. (2006). Design of robust, energy efficient full adders for deep-submicrometer design using hybrid-CMOS logic style. IEEE Transactions on Very Large Scale Integration (VLSI) Systems, 14(12), 1309–1321
[11] Rabaey, J. M., Chandrakasan, A., & Nikolic, B. (2003). Digital integrated circuits: A design perspective. (2nd ed.). New Delhi: Pearson Education.

Design and Application of Automatic Calibration Device for Multi-Channel Resistance Strain Gauge Indicator

Xiaoyin Hu*, **Ye Li, Haoyu Zhang, Yueling Yu, Zhangyi Kang**

Shanghai Institute of Measurement And Testing Technology, Shanghai 201203, China

*Corresponding author's e-mail: huxy@simt.com.cn

Abstract. In this paper, an automatic calibration device for multi-channel resistance strain gauge indicator is designed and its applicability and measurement accuracy are verified at laboratory. The calibration done by original resistance bridge calibrator is time-consuming for its manual operation and complex calibration process. With the intent to increase calibration efficiency, an automatic channel switch device was developed, and the resistance bridge calibrator was automated. The designed calibration device is completely computer controlled enabling a sequence of unmanned measurements. The calibration device was verified at laboratory that the maximum of error is 0.072%. It was applied to calibrate a 60-channel resistance strain gage indicator to approve its practical applicability. The result shows that the designed calibration device can realize automatic calibration and the efficiency is increased by 40%.

1. Introduction

Strain gage is one of the most reliable strain measurement devices in laboratory and industrial field [1-3]. Dozens of and sometimes thousands of strain gages are used in material and structural stress analysis [4-5]. The strain gage indicator is the measuring device of the strain gage by means of Wheatstone bridge (shown in Figure 1). Calibration of multi-channel strain gage indicator done with the resistance bridge calibrator (shown in Figure 2) is time consuming, because an operator is needed to manually manipulate the switches on the calibrator and switch channel [6]. The traditional calibrator is designed for strain gage indicator with just several channels. It's a tedious work to do the calibration of an indicator with dozens of channels by the traditional calibrator. Kreuzer designed a new circuit to realize automatic calibration. However, the problem of channel switch is not taken into consideration and the calibrator is not applicable for half bridge and full bridge [7]. Xu designed an automatic calibration. The device is developed for a specific strain gage indicator and not suitable for others [8]. For this reason, an automatic calibration device for the multi-channel strain gage indicator is designed, and its applicability and performance are validated at laboratory.

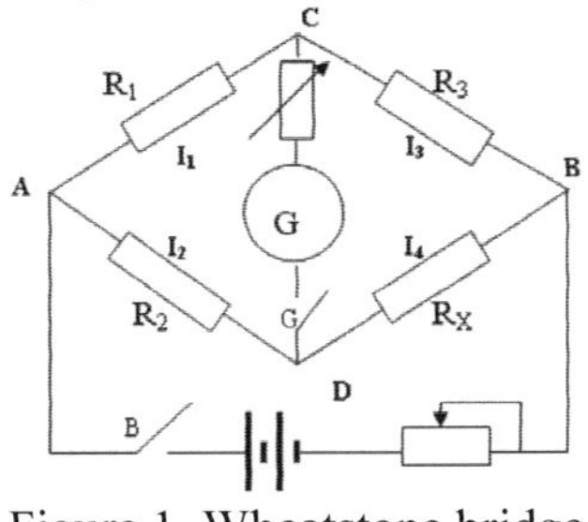

Figure 1. Wheatstone bridge

Figure 2. Traditional resistance bridge calibrator

Journal of Physics: Conference Series **2113** (2021) 012040 doi:10.1088/1742-6596/2113/1/012040

2. Design of the calibration device

The designed calibration device presented in Figure 3 is made up of three parts: calibrator automation system, channel switch system and computer software.

The calibrator automation system automates the traditional resistance bridge calibrator. Six motors are installed on the calibrator through mechanical connection. Instructions from the computer software actuate the motors to rotate. Each rotary switch on the calibrator can generate ten fixed-strain loading values by fixed-angle rotation. The angles corresponding to strain loading values are measured to ensure motors can reach correct position. The calibration process is controlled by the computer software. The strain loading values of six motors in software is defined. The motors rotate according to the edited calibration documents to generate strain loading. The designed calibrator automation system can be applied to any resistance bridge calibrator that has similar mechanical structure. The main differences are the number of motors and setup of angels.

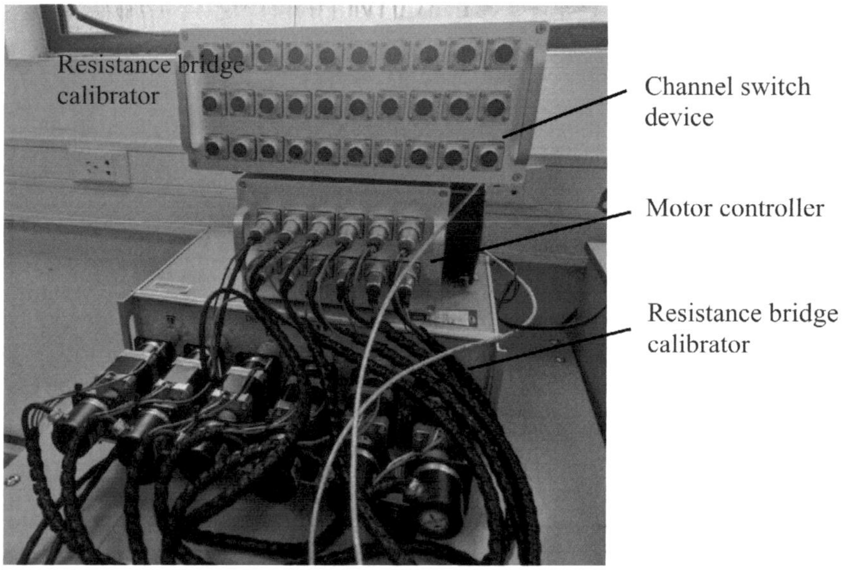

Figure 3. Designed calibration device

The channel switch system has 60 channels and is applicable for the majority of the strain gage indicator. The circuit is shown in Figure 4. The open and close of the channel is controlled by a relay. The relay will close when receiving the current from the photoelectric coupler. The power supply is controlled by computer program to determine the channel number to switch.

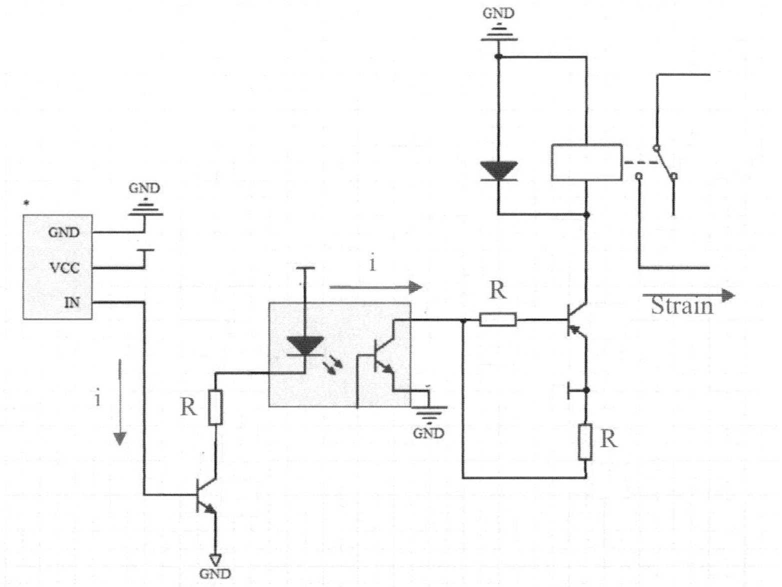

Figure 4. Channel switch circuit

3. Verification and application of the calibration device

The error of the designed calibration device is greater than the original calibrator for the electron component applied in the channel switch system. To verify the performance of the designed calibration device, especially the error introduced by channel switch system, a high-class resistance strain gage indicator the relative uncertainty of which was 0.005% (k=2) was employed. The parameters of the resistance strain gage indicator are listed in Table 1. Measurement results are shown in Table 2 and the max error is 0.072%.

To prove the applicability of the designed calibration device, a 60-channel resistance strain gage indicator (model: EDX-5000A-60) was calibrated by the device shown in Figure 5. All channels were connected to the channel switch system before the calibration and a calibration file is edited according to requirements. The setup parameters include channel number, strain loading value and silent period time. In this calibration, strain loading values of each channel were 200με, 500με, 1000με, 2000με and 5000με, and motors kept still for 20 seconds after reaching the position. Then, the calibration started. The designed calibration device completed the calibration process according to the excel file in 105 minutes with no manipulation from the operator and data was saved in computer. The source record was produced according to the calibration excel file and data file. It is known from practice that 3 hour is the minimum time to do the calibration of an indicator with 60 channels by traditional bridge calibrator.

Table 1. Parameters of Resistance Strain Gage Indicator

Model	Measurement Range	Relative Uncertainty (k=2)
DMP40	$(0.01\sim10^4)$με	0.005%

Table 2. Measurement Results

	Actual value/με				
	× 10^3με panel	× 10^2με panel	× 10^1με panel	× 10^0με panel	× 10^{-1}με panel
1	999.8	100.0	10.0	1.0	0.1
2	1999.5	199.9	20.0	2.0	0.2
3	2998.2	299.9	30.0	3.0	0.3
4	3997.8	399.8	40.0	4.0	0.4
5	4996.4	499.7	50.0	5.0	0.5
6	5996.2	599.7	60.0	6.0	0.6
7	6995.7	699.6	70.0	7.0	0.7
8	7995.2	799.6	80.0	8.0	0.8
9	8995.2	899.5	90.0	9.0	0.9
10	9995.5	999.5	100.0	10.0	1.0

Figure 5. 60-channel strain gage indicator

4. Conclusions

In this paper, an automatic calibration device for multi-channel resistance strain gauge indicator is designed and verified. The traditional manual-operation resistance bridge calibrator was automated, and an automatic channel switch device was developed. The result of verification implemented at laboratory shows that the maximum error of calibration device is 0.072% and the calibration efficiency can increase by 40%.

References

[1] MA L, YANG B, TIAN Y Z, XIAO J T, HUA X Q, SUN L N. (2017) Design of 3-DOF planar nano-positioning platform with 3-PRR structure, Optics and Precision Engineering, 25: 1866-1873

[2] CHEN L W, ZHOU X W, PENG W P, LI B, TONG J. (2018) A method of measuring the strain of geotextile, Journal of Yangtze River Scientific Research Institute, 35:70-74

[3] Xiao H E, Xiang X U, Sen Y. (2018) Design of control system for dynamic calibration device of eddy current sensor based on FPGA. DEStech Transactions on Computer Science and Engineering

[4] Ding Y, Du L L, Li H T. (2017) Challenges for multi-Channel strain measurement system, Metrology and Measurement Technology, 37:190-192

[5] Huang X L. (2015) On-site Calibration of large-scale strain measurement system, Metrology and Measurement Technology, 35:70-72

[6] Podgornik T, Bojkovski J, Batagelj V. (2009) Investigation of resistance-ratio bridges using the automated resistance bridge calibrator. Measurement, 42:1065-1072.

[7]Kreuzer M(Germany). (2017) Strain gauge ¼ bridge calibration with shunt signals. https://www.hbm.com/en/3197/tips-and-tricks-strain-gage-bridge-calibration-with-shunt-signals/

[8] Xu L J. (2016) Auto-calibration system design for strain gauge, Shanghai Jiao Tong University

Design of high precision mechatronic-hydraulic leveling system for vehicle-mounted radar

Jun Xiao*, **YanChao Wang, XiangYu Li, Jie Zheng**

School of Mechanical and Electrical Engineering, Wuhan University of Technology, Wuhan, 430000, China

*Corresponding author's e-mail: xjun@whut.edu.cn

Abstract: In order to improve the control performance of vehicle-mounted radar, a high precision leveling control system is designed through mechatronic-hydraulic integration technology. A leveling method combining angle error leveling method and height error leveling method is designed, and inclinometers are set in the front and rear of the vehicle body platform to effectively eliminate the virtual outrigger phenomenon in the leveling process. The mathematical model of hydraulic system is developed, the fuzzy PID control method is introduced, and the four-point support hydraulic leveling control system is developed by using STM32 microcontroller as the control core. The field test proves that the control system has good control performance, high leveling accuracy, short time, and meets the control requirements of the radar vehicle.

1. Introduction

In modern information warfare, radar plays an important role as a clairvoyant on the battlefield. The work of vehicle-mounted radar requires a high level of car body, so a high precision car body leveling control scheme needs to be designed. Liao Yide et al from Huazhong University of Science and Technology designed a set of fast leveling device controlled by high-speed on-off valve [1]. Xia Xin et al. from Southwest Jiaotong University designed a set of 4-point electro-hydraulic leveling system with PLVC as the control core, with a precision of 2.5 '[2]. In this context, to improve the leveling control performance of a radar vehicle, the original control system is improved, a set of high-precision automatic leveling control system is designed, and virtual outrigger phenomenon is effectively dealt with. In this control system with ARM STM32F407 as control core, fuzzy PID control method is adopted to develop a four-point support hydraulic leveling control system, and the system is tested in the field. Under the premise that the initial angle is close to 3°, the final leveling accuracy reaches 0.015°. The adjustment time is less than 400s (including the time of outrigger synchronous pre-support), which proves the control performance of the system is better than that of the same type of vehicle.

2. Leveling system modeling

2.1. Introduction to leveling System

The 3D model of radar vehicle platform is shown in figure 1. It is necessary to control the movement of four hydraulic outriggers to adjust the platform to the level, and the inclination angle of the platform is measured by the inclinometer installed at the bottom of the platform. Among them, the hydraulic leveling outrigger is drived by the proportional directional control valve, and the inclinometer signal is processed by STM32 microcontroller, and the control signal in the form of current is input to the proportional directional control valve to control the opening of the proportional directional control valve and then control the movement speed of the outrigger. The block diagram of the radar vehicle leveling control system is shown in figure 2.

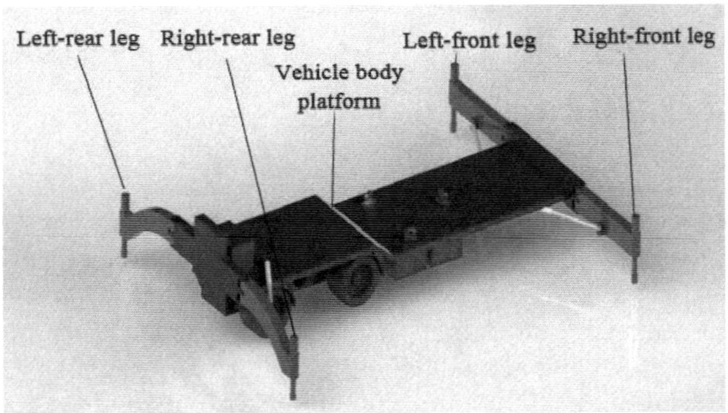

Figure 1. 3D model of radar vehicle platform

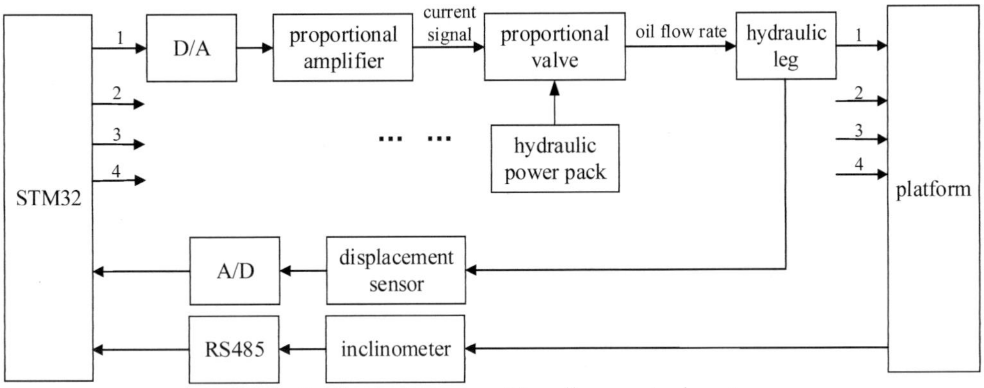

Figure 2. Block diagram of leveling control system

2.2. Mathematical modeling of hydraulic outrigger

In this system, the outrigger movement speed v_l is directly controlled by the oil flow Q_v, and the proportional valve spool displacement X_v and the external load size P_l are the direct factors affecting the oil flow Q_v, in which the control plate controls the proportional valve spool displacement X_v through the current signal I. The system transfer function and its derivation are shown below.

$$v_l = \frac{4Q_v}{\pi d^2} \tag{1}$$

$$Q_v = K_q X_v - K_p P_l \tag{2}$$

$$X_v = \frac{IK_q}{\frac{s^2}{\omega_v^2} + \frac{2\zeta_v s}{\omega_v} + 1} \tag{3}$$

By combining equations (1), (2) and (3), we can obtain:

$$v_l(s) = \frac{4K_q^2 I(s)}{\left(\frac{s^2}{\omega_v^2} + \frac{2\zeta_v s}{\omega_v} + 1\right)\pi d^2} - \frac{K_p P_l(s)}{\pi d^2} \tag{4}$$

Where, $\boldsymbol{K_q}$ is the flow gain of the proportional valve near the steady-state operating point, $\boldsymbol{m^2/s}$; $\boldsymbol{K_p}$ is the flow-pressure coefficient of the proportional valve near the steady-state operating point, $\boldsymbol{m^3 \cdot MPa/s}$; $\boldsymbol{\omega_v}$ is the equivalent undamped natural frequency of the proportional valve, $\boldsymbol{rad/s}$; $\boldsymbol{\zeta_v}$ is the equivalent damping coefficient of proportional valve, dimensionless; $\boldsymbol{d}$ is the rodless cavity diameter of hydraulic outrigger, $\boldsymbol{m}$.

3. Control strategy design and simulation

3.1. Design of leveling method

Automatic leveling system is a negative feedback control system. According to the error type, leveling methods can be divided into angle error control method and height error control method[3]. Angle error control method, according to the platform dip angle controls outriggers to move. The control method is relatively simple, and can control the movement of two outriggers at the same time, fast but with low precision. Height error control method, according to the platform dip angle calculates the needed stroke of the outriggers, and then control the corresponding outriggers to extend the corresponding stroke, can control the three outriggers at the same time with high precision[4].

Leveling method of design need to consider to the processing of virtual outrigger phenomenon. Virtual outrigger phenomenon is that one outrigger does not force or stress is small in the process of leveling. Because of the large platform span, when a virtual outrigger, the four outrigger uneven bearing will inevitably lead to deformation of the platform, at this point does not obtain in angle accurate leveling needed for each outrigger movement, thus height error control method can't process the virtual outrigger phenomenon. The angle error leveling method can directly adjust the inclination angle of the platform. When inclinometers are set on the front and rear of the vehicle platform, it can control the angle difference of vehicle platform in a small range and the virtual outrigger can be accurately excluded. The sensor layout is shown in figure 3.

After comparing the two leveling methods, a leveling method combining angle error control and height error control is designed. Specifically, the angle error control method is used to conduct rough adjustment, adjust the platform inclination to a certain angle threshold value, and then adjust the angle to the final demand angle by position error control method.

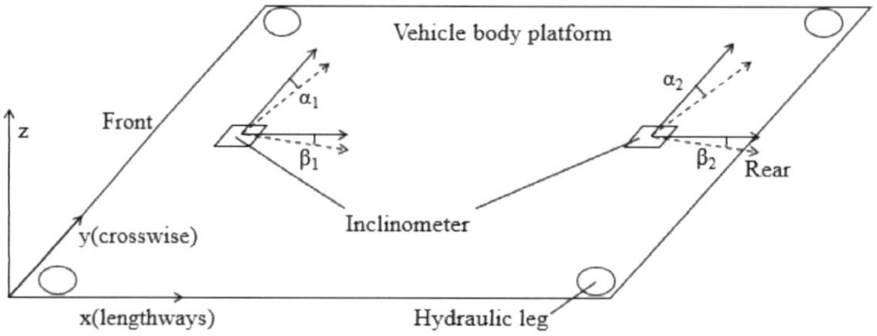

Figure 3. Schematic diagram of inclinometer layout

3.2 Fuzzy PID control principle

Fuzzy control is a modern intelligent control algorithm based on fuzzy mathematics, which can be used to simplify the solution of complex systems. As an intelligent PID, fuzzy PID control can solve the problems of poor parameter setting and poor performance while retaining the strong robustness of conventional PID[5], its control principle diagram is shown in figure 4.

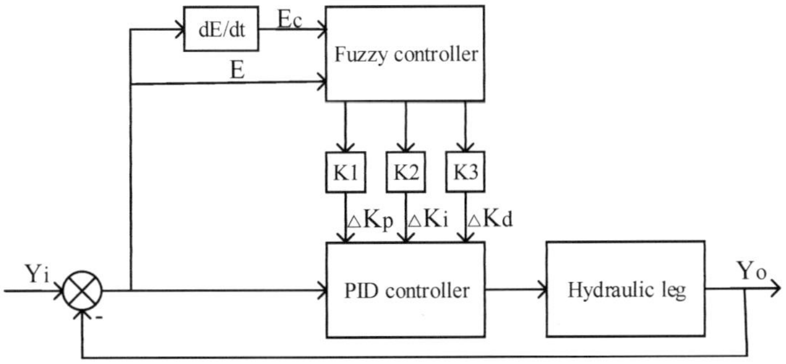

Figure 4. Fuzzy PID control schematic diagram

The fuzzy PID control process is as follows[6]:

- Input the deviation and deviation change rate into the fuzzy controller after fuzzy processing.
- The fuzzy controller queries the fuzzy control rule table according to the membership function of the input parameter and obtains the corresponding fuzzy value output.
- Fuzzy processing of output parameters of fuzzy controller.
- Input the value of the fuzzy solution to the PID controller for system control.

3.3 Fuzzy PID controller design

Let's take the deviation E and the rate of change E_c is the input, and variables, and PID parameter change $\Delta K_p/\Delta K_i/\Delta K_d$ are the output. The continuous variables were discretized into NB (negative large), NM (negative medium), NS (negative small), ZO (zero), PS (positive small), PM (positive middle), And PB (positive large).Among them, PB is the s-type membership function, NB is the Z-type membership function, and the others are triangular membership functions. The domain of E and E_c is [-3,3], the domain of variable $\Delta K_p/\Delta K_i$ is [-0.3,0.3], and the domain of variable ΔK_d is [-0.06,0.06]. The formulation of fuzzy control rules is shown in the table 1.

Table 1. Table of fuzzy control rules

△Kp/△Ki/△Kd		Ec						
		NB	NM	NS	ZO	PS	PM	PB
	NB	PB/NB/PS	PB/NB/NS	PM/NM/NB	PM/NM/NB	PS/NS/NB	ZO/ZO/NM	ZO/ZO/PS
	NM	PB/NB/PS	PB/NB/NS	PM/NS/NB	PS/NM/NM	PS/NS/NM	ZO/ZO/NS	NS/ZO/ZO
	NS	PM/NM/ZO	PM/NM/NS	PM/NS/NM	PS/NS/NM	ZO/ZO/NS	NS/PS/NS	NS/PS/ZO
E	ZO	PM/NM/ZO	PM/NM/NS	PS/NS/NS	ZO/ZO/NS	NS/PS/NS	NS/PS/NS	NM/PM/ZO
	PS	PS/NM/ZO	PS/NS/ZO	ZO/ZO/ZO	NS/PS/ZO	NM/PS/ZO	NM/PM/ZO	NB/PB/ZO
	PM	PS/ZO/PB	ZO/ZO/NS	NS/PS/PS	NM/PM/PS	NM/PM/PS	NM/PB/PS	NB/PB/PB
	PB	ZO/ZO/PB	ZO/ZO/PM	NM/PS/PM	NM/PM/PM	NM/PM/PS	NB/PB/PS	NB/PB/PB

The output of fuzzy controller needs to be defuzzified to get the accurate output value, so as to correct PID parameters. The centre of gravity method is commonly used to defuzzify[7], as shown in the equation (5).

$$y_i = \frac{\sum_{i=1}^{j} x_i \mu_i(x_i)}{\sum_{i=1}^{j} \mu_i(x_i)} \tag{5}$$

Where, j is the number of fuzzy set elements; x_i is the element value in the fuzzy set; μ_i is the weighting coefficient of membership degree.

Input the value of the fuzzy solution to the PID controller for parameter $K_p/K_i/K_d$ in order to meet the requirements of different error and error rate on the control system.

4. Control system design

4.1 Hardware Structure

The control principle of leveling system is shown in figure 5.The main control chip is STM32F407, which receives input signals from all sides and generates corresponding output signals to control the action of the hydraulic system. The inclinometers through the RS485 communication mode input platform dip angle into levelling control circuit board which communicates other system through the RS485 communication mode and display real-time control status on the OLED display platform. Operators can control body automatic leveling or leg separate action, etc. via buttons on the panel.

4.2 Software Design

The software of the control system is developed on the Keil5 platform, and the control program is written in C language. It is mainly composed of solenoid valve control module, button input module, digital sensor input module, analog sensor input module, RS485 communication module, OLED display module, timer module and logic control module. The functions of the system include automatic movement of the extension arm and outrigger, detection of outrigger landing, automatic leveling of the body, synchronous recovery of outrigger and display of the control status. The automatic leveling program is mainly divided into two parts, coarse and fine, respectively using different leveling strategies, angle error leveling method and position error leveling method[8], the leveling flow chart is shown in figure 6.

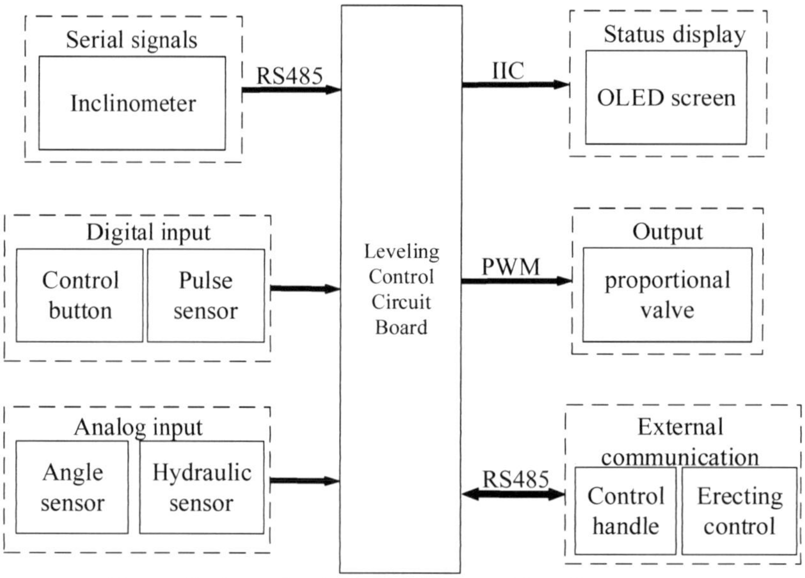

Figure 5. Control principle diagram of leveling system

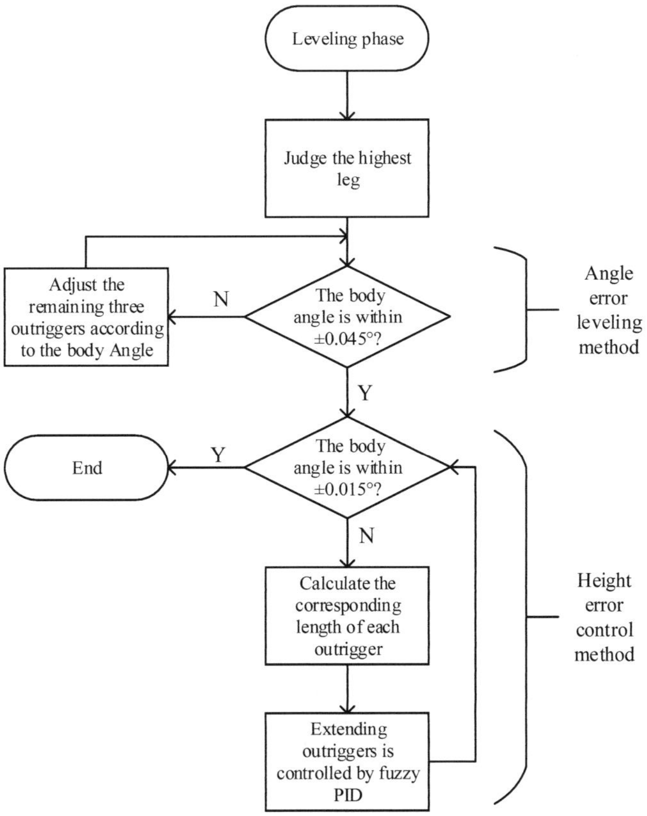

Figure 6. Control diagram of leveling process

5. System test verification

5.1. Test system

The design scheme and theoretical research of the automatic leveling control system have been successfully applied to a radar vehicle, and the system has been verified by field test. In this radar vehicle leveling system, the software is developed on Keil5 platform, and the control program is established by C language. LIQZO-L-503 /L4 proportional valve of ATOS company is used for outrigger control, and HCA546T of RION Company is used for inclinometer, which has high stability, high precision and strong anti-interference. As mentioned above, in order to exclude virtual outriggers, inclinometers are set in front and back of the vehicle body respectively.

5.2. Experimental Results

After outrigger to support of synchronous control system according to the control logic and algorithm for automatic leveling, in order to verify the validity of the leveling strategy and control algorithm, has carried out many experiments, the car body initial angle adjustment for different values, the following will be data graph from two trials, as shown in figure 7.

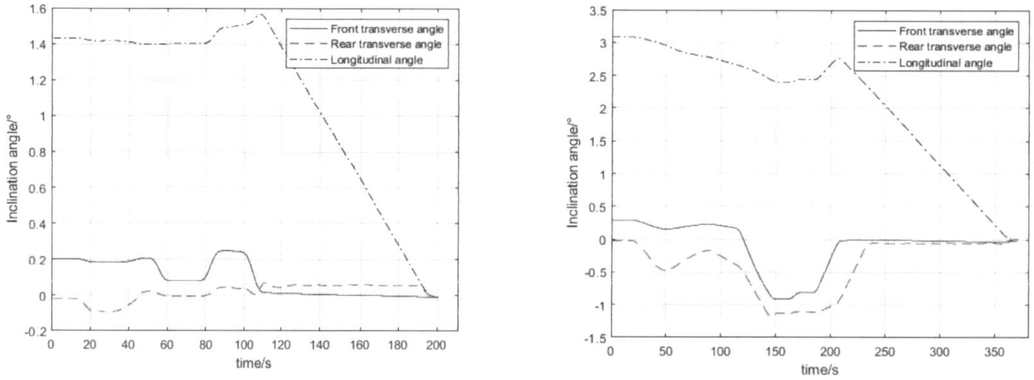

(a) First leveling (b) Second leveling

Figure 7. Tilt angle change during leveling process

In Figure 7(a), the initial longitudinal inclination is about 1.43° and the transverse inclination is about 0.1° on average. Combined with the actual observation, the body outriggers all landed in about 105 seconds, the complete leveling process was less than 200 seconds, and the final leveling accuracy reached 0.015°.In Figure 7(b), the initial longitudinal inclination is about 3.09°, and the transverse inclination is about 0.15° on average. Combined with the actual observation, the body outriggers all landed in about 180 seconds, the complete leveling process was less than 370 seconds, and the final leveling accuracy reached 0.015°.In the whole leveling process, there is only little difference between the front transverse angle and rear transverse angle, which indicates that this leveling strategy can effectively eliminate the virtual outrigger. When the initial angle reaches 3°, the leveling time is within 400s, and the leveling accuracy reaches 0.015°, which can fully meet the requirements of accuracy and time. The test results verify that the leveling control system has excellent performance, its control method has good repeatability, and the inclination angle change process is almost the same in the two leveling tests.

6. Conclusion

- Through the test, the hydraulic leveling control system applying the leveling method combining angle error control and height error control, and fuzzy PID control method, has good control performance, whose leveling precision can reach 0.015°, and even though on the premise of the initial angle 3°, can complete the leveling process (including preliminary support phase) in the 400 s, which fully meets the radar vehicle leveling requirements.
- In this leveling system, inclinometers are arranged on the front and rear horizontal sides, which can effectively avoid the occurrence of virtual outriggers in the leveling process. And the system has the characteristics of high leveling precision, short time, easy operation and universality, and combined with the actual engineering improvement, can be extended to other vehicle-mounted platform leveling occasions.

References

[1] Liao Yide, Yang Wenrui.Horizontal Adjustment System of Radar Antenna Platform Based on High-speed On-off Valve [J]. Machinery & Electronics,2003(03):53-54.

[2] Xia Xin, Wang Hai-bo, Li Xue-feng, He Jian.Research on High Precision Automatic Leveling Control System of a Military Radar Vehicle [J]. Machinery Design & Manufacture,2019(09):158-160+164.

[3] Xu Song. Research on Modeling and Control Algorithm of Six-point Leveling System of Launch Vehicle [D]. Harbin Institute of Technology,2017.

[4] Li Bin, LI Hong, Yu De-ping, Yao Jin.Research on Accurate Leveling Method of Heavy Duty Platform [J]. Machinery Design & Manufacture,2008(09):99-101.
[5] Li Xiang-yu, XIAO Jun, PAN Yun-ping, LIU Shi-yong, DENG Xing.Research on Fuzzy PID Leveling Control System of Vehicle Platform Based on PSO [J]. Modern Manufacturing Engineering,2021(02):58-65.
[6] Luo Yanlei, Tu Songting, Shi Liming.Research on fuzzy PID control of large damping system based on hydraulic leveling [J]. Machine tool & hydraulics,2020,48(15):118-121.
[7] Tang Pingjian, Sun Zelin, Song Peng.Research on Hydraulic Automatic Leveling and Lifting Control System Based on Fuzzy PID [J]. Journal of Ordnance Equipment Engineering, 201,42(02):189-193.
[8] He Binghuai. Research on Multi-point Leveling System [D]. University of Electronic Science and Technology of China,2014.

Rotating equipment fault noise location based on adaptive beamforming algorithm

Yongshao Xu[1], Bingzheng Liu[1*], Haotian Shang[2], Mingduo Wang[2]

[1]Shenzhen Branch of China Nuclear Power Maintenance Co., Ltd, Shenzhen, Guangdong, China

[2]Northeast Electric Power University School of Automation Engineering, Jilin, Jilin, China

[*]Corresponding author's e-mail: liubingzheng@cnjijx.com

Abstract. Rotating machinery often produces continuous impact during operation due to the change of load and speed, which shows the characteristics of unsteady state and time-varying. Its working state can not be comprehensively judged by a single vibration state parameter. Therefore, this paper proposes to use acoustic sensors to collect the fault noise signal of rotating machinery, and use the whole column of sensors to detect the fault noise signal. Based on the microphone array, this paper studies the adaptive beamforming algorithm (MVDR) to locate the fault source of rotating machinery in space. The effect of fault source location is verified by simulation and equipment measurement experiments. The acoustic sensor does not in contact with the equipment, which will not damage the generator set, but also provide more effective information for fault source location and fault diagnosis and analysis.

1. Introduction

Array signal processing technology was applied in the 1930s[1]. It was first applied in the field of radar detection and achieved remarkable results. After decades of development, array signal processing technology has formed a relatively perfect system. Array signal processing technology has become the basis of modern signal processing technology[2-3]. Therefore, it is of great significance to locate mechanical equipment faults through array signals[4-5]. For example, Hou et al, used holographic array to measure azimuth acoustic signal, directly obtained the amplitude distribution of two-dimensional sound field sound pressure at the holographic measurement position, and reconstructed the two-dimensional sound field information of azimuth sound source position by using NAH technology[6-7]. To sum up, with the development of positioning technology, the purpose and significance of this paper is to use microphone array to study and improve the sound source enhancement algorithm based on adaptive beamforming, so that the improved algorithm has stronger noise suppression ability[8-9].

2. Sound source localization based on microphone array

2.1. Sound source and microphone array model

The sound source localization should consider the far-field and near-field model according to the nature of the sound wave. The distance between the sensor and the sound source and the wavelength of the main sound source affect the selection of the model. The far-field model and the near-field model represent two different geometric calculation models.

When the sound source signal is a near-field model, the sound propagates outward in the form of spherical waves. If the sound source is a far-field source model, then the sound source can be regarded as a plane wave.

Compared with the near-field model, the far-field model can be regarded as a kind of near-field model, which is less difficult to calculate. The difference caused by the amplitude attenuation in the sound propagation process can be ignored. In general, when the sound source is close to the microphone array, the incident angle of the signal received by each microphone array element is different, and a more accurate sound position is required. At this time, the near-field model is used. In most cases, the far-field model is used.

The choice of microphone array is also an important factor that affects the localizing accuracy. The topological structure of the array is one-dimensional, two-dimensional and three-dimensional, with common wire arrays, circular arrays, and spherical arrays. Compared with the linear array and the ball array, the circular array has a higher localizing accuracy than the linear array and less calculation than the ball array. In this paper, the circular array is selected as the microphone array for sound source localization.

2.2. Sound source localization algorithm

Sound source localization algorithm is an important part of real-time fault localization. This paper uses beamforming algorithm and MUSIC localization algorithm.

The basic principle of the beamforming algorithm after beamforming the signal received by the microphone, the output data is represented by $y(t)$:

$$y(t) = \boldsymbol{w}^H \boldsymbol{x}(t) = \boldsymbol{w}^H \alpha(\theta, \varphi) s(t) + \boldsymbol{w}^H \boldsymbol{u}(t) \tag{1}$$

The MUSIC localization algorithm is a high-resolution azimuth spectrum estimation algorithm, which overcomes the defects of beamforming and is superior to beamforming algorithm in localization accuracy. But it cannot overcome two or more sound sources that are closer together in the same direction. In recent years, the various high-resolution azimuth estimation methods that have been continuously improved theoretically overcome the limitations of the Rayleigh criterion.

2.3. Simulation experiment results and analysis

Comparing the performance of the adaptive beamforming algorithm and the MUSIC algorithm, this section compares and analyzes the differences between the two methods through simulation. First of all, this paper simulates a linear array and simulates the incident angle of a spatial sound source. Set the incident angle of the sound source to 10°, 30°, 60°, and the microphone array to be a linear array with 10 elements.The simulated signal-to-noise ratio of the array is 10, and the ratio between the radius and the wavelength is 0.5. The obtained angle and amplitude comparison chart is shown in the figure 1 below:

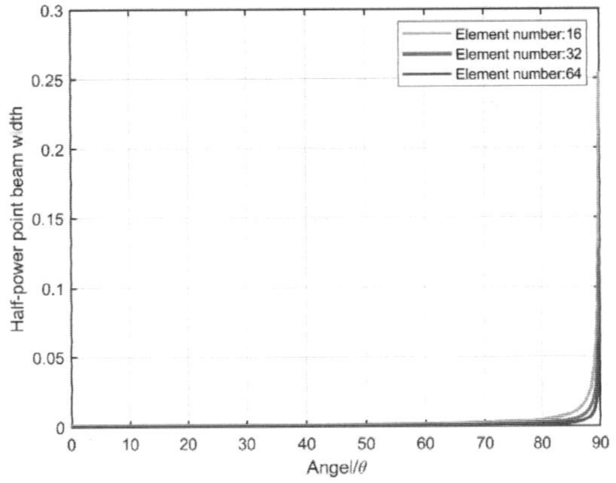

Fig 1. Localization algorithm performance comparison chart

It can be seen from the figure that both algorithms can locate the angle of incidence, but compared to the beamforming algorithm, the MUSIC algorithm has a higher resolution. But its array error has a great influence on its localizing accuracy, and the sound source and noise are required to be statistically stable. The ability to deal with interference sounds such as reflections and noises in a complex environment is weak, and high sound source signal energy is required for long-distance sound source localizing. The beamforming accuracy is not as high as that of the MUSIC localization algorithm, but the direction of arrival can be enhanced to suppress noise in other directions.

In addition, the localizing effect is also affected by the number of array microphones and the size of the array aperture. This paper uses a circular array and uses an adaptive beamforming localization algorithm. As the number of element microphones of the circular array increases, the effective time delay of the entire system will increase, and the localizing accuracy will also increase, but the amount of calculation will also increase. The number of microphones in the array can be reasonably and effectively selected within the range allowed by the hardware. When the sound source wavelength is set to 0.17m, the localizing effects of the three arrays with the number of array elements of 16, 32, and 64 are shown in Figure 2 below:

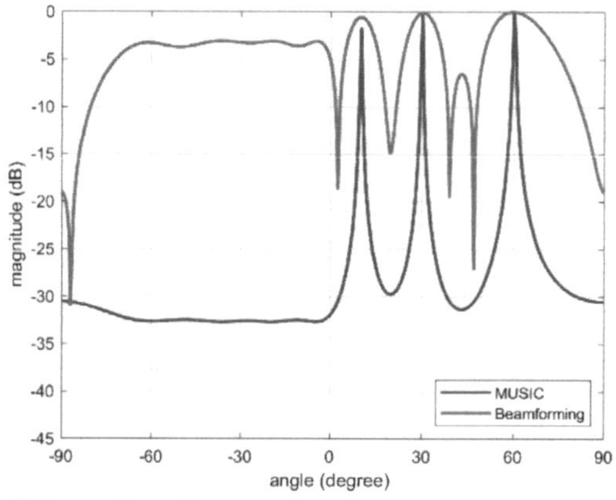

Fig 2. The localizing effect diagram of the different number of array elements

In addition, the parameter of the array aperture that affects the localizing accuracy is also the parameter. We simulated the actual situation and set the number of array microphones to 16 to explore the influence of the array aperture on the localizing accuracy of the azimuth angle.In this paper, we have selected three different array apertures of 1.36, 2.72, and 5.44 to observe the overall changes. The effect diagram is shown in Figure 3 below:

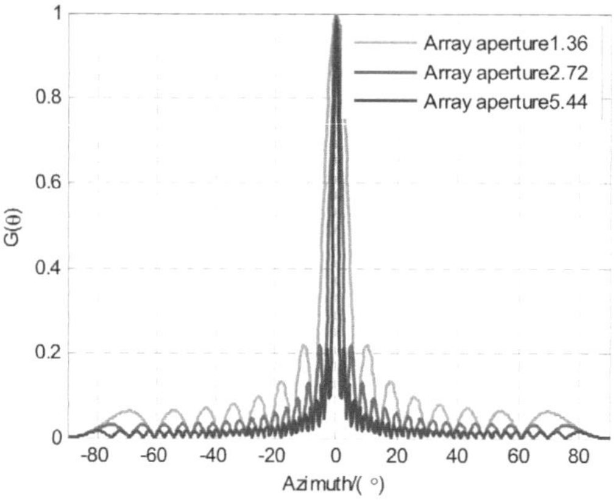

Figure 3. Simulation renderings of array aperture and localizing accuracy

Since white noise is added to the simulated sound source, it can be clearly seen that there are side lobes in the localizing simulation, but it does not affect the localizing accuracy. The three-dimensional simulation is shown in Figure 4.

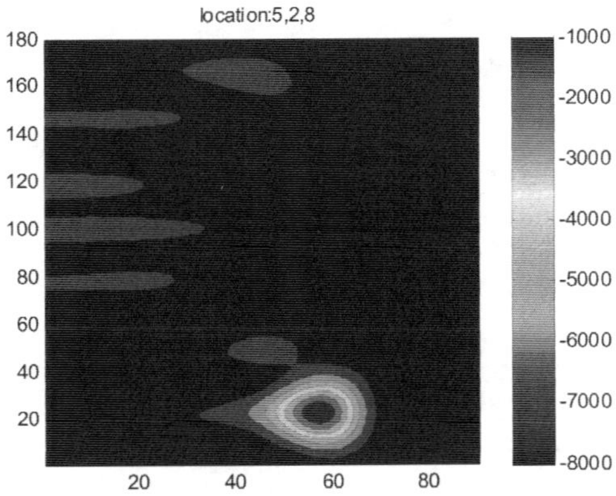

Fig 4. Capon localizing simulation result graph

It can be seen from the three-dimensional localizing simulation diagram that the peak of the sound source direction is significantly higher than other directions, and the localizing effect is better. In order to compare and select the appropriate algorithm, we compared the localizing effect of the MUSIC localization algorithm and the beamforming algorithm in the case of different positions of the spatial sound source in the simulation situation. The comparison error table is shown in Table 1:

Table 1. Two algorithm simulation comparison table

Calibrated incident angle	MUSIC algorithm measurement results	MUSIC algorithm measurement error	Beamforming measurement results	Beamforming measurement result measurement error
(20.21,13.65)	(22,15)	(1.79,1.35)	(21,13)	(0.21,0.65)
(35.5,22.8)	(37,20)	(1.5,2.8)	(35,24)	(0.5,1.2)
(48.2.36.2)	(48,37)	(0.2,0.8)	(49,35)	(0.8,1.2)
(61.8,50.6)	(60,52)	(1.8,1.4)	(61,50)	(0.8,0.6)
(76.2.60.4)	(75,61)	(1.2,0.6)	(76,60)	(0.2,0.4)
(82.56,76.58)	(82,76)	(0.56,0.58)	(83,76)	(0.44,0.58)

3. Conclusion

In this paper, according to analysed and compared the difference of near-field sound and far-field sound, narrow-band and broadband signals. Comparing to the advantages and disadvantages of MUSIC positioning algorithm and beam formation algorithm, selecting adaptive beam formation algorithm as the positioning algorithm, and proposing improvement Comparing the positioning results of the improved algorithm and the traditional beam formation algorithm through simulation experiments, it is proved that the improved algorithm is better in accuracy.

References

[1] Chen X, Feng Z. Order spectrum analysis enhanced by surrogate test and Vold-Kalman filtering for rotating machinery fault diagnosis under time-varying speed conditions[J]. Mechanical Systems and Signal Processing, 2021, 154: 107585.

[2] Jean A, Francois C, Jean P G. KSE: A Real-Time Expert System to Diagnose Nuclear Power Plant

Failures [J]. Transactions on Nuclear Science, 1991, 38(4): 70-76P.
[3] Huhn Kim, Wan Chul Yoon, Sangsup Choi. Aiding fault diagnosis under symptom masking in dynamic systems[J]. Ergonomics, 1999, 42(11): 1472-1481P.
[4] Joon O Y, Soon H C. A Diagnostic Expert System for the Nuclear Power Plant Based on the Hybrid Knowledge Approach [J]. Transactions on Nuclear Science, 1989, 36(6): 2450-2458P.
[5] Reifman J, Wei T Y C. Prodiag: a process-independent transient diagnostic system—i: theoretical concepts[J]. Nuclear science and engineering, 1999, 131(3): 329-347.
[6] Jaques Reifman, Gleun E. GRAHAM. Flexible Human Machine Interface for Proeess Diagnostie [C]. Proeeedings NPIC&HMIT, 1996:1437-1444P.
[7] Dieter Wach, Werner Bastl. On-line condition monitoring of large rotating machinery in NPPs [C]. Proceedings NPIC&HMIT, 1996: 1313-1320P.
[8] Ghaderi H, Kabiri P. Automobile engine condition monitoring using sound emission[J]. Turkish Journal of Electrical Engineering & Computer Sciences, 2017, 25(3): 1807-1826.
[9] Pan N, Yi Z. Research of brake sound acoustic features extraction based on frequency-domain blind deconvolution[C]//2015 IEEE International Conference on Information and Automation. IEEE, 2015: 2892-2895.

Improved Design of a 4-bit Absolute-Value Detector Using Simplified Chain Carry Adder

Xinhang Dong[1, a, *, †], **Boyuan Jing**[2, b, *, †], **Xiang Yang**[3, c, *, †]

[1] School of Information Science and Technology, Shanghai Tech University, Shanghai, Shanghai 201210, China

[2] School of Electronic and Control Engineering, North China Institute of Aerospace Engineering, Langfang, Hebei 065000, China

[3] Henry Samueli School of Engineering, University of California, Irvine, California 92614, United States

[*]Corresponding author's e-mail: [a]dongxh@shanghaitech.edu.cn, [b]jingby@outlook.com, [c]yxiang10@uci.edu

[†]These authors contributed equally

Abstract. 4-bit absolute-value detector (AVD), as one of the basic implementations of bit arithmetic with logic circuits, can help grab a better understanding about digital integrated circuits. Conventional 4-bit AVDs scheme in a multi-comparator and multiplexers, or need to consider multiple situations of overflow and carry-in, both of which could make the final circuit to be complex, labyrinthine and inefficient in the meantime. In this paper, a new design of 4-bit AVD is proposed, the topology of which includes a 2's complement calculator and a specially designed logic circuit known as chain carry adder (CCA). The whole circuit is concise and the critical path is rigorously considered to make it as short as possible. The delay is set to 1.5 times its minimum, which is positively corresponding to the length of the critical path, the energy accordingly reaches its lower limit. Gate sizing and Device Voltage (VDD) optimization are proceeded for the exact purpose of proving that the circuit energy is minimized.

1. Introduction

This paper is in purpose of showing a new design of 4-bit absolute-value detector (AVD), which has a more concise layout and minimum energy consumption.

1.1. Basic concept

Spike-sorting algorithms, as a vital progress that recently made in the neural signal acquisition systems, have gained a lot of interest in the research community. Among all, we choose one of the commonly used spike-detection algorithms known as absolute-value detection, to study and make further improvement to its design.

Due to the many uncertainties of the inputs in the actual situation, the AVDs mentioned in the following context are limited to 4-bit AVDs. In our opinion, 4-bit AVDs are more representative, which takes into account the interaction and interconnection between all inputs and presents the typical characteristics of AVDs in a relatively crisp way. In a more general case, 4-bit AVDs can be easily extended to higher-bits AVDs with more inputs [1].

1.2. Conventional designs
There are already some conventional designs of AVD [2]. In most circumstances, there exist two main types of AVD, also called magnitude comparator with signed inputs, which are known as signed magnitude comparator and bit operation based multiplexer.

1.2.1 Signed magnitude comparator
One of them is the improvement of the unsigned comparator, shown in Fig.1, see p462-463 of [3]. Basically speaking, the 4-bit unsigned magnitude comparator shown as follows is built from a carry-ripple adder and a 2'complementer. The input is A and B while the output is the sign of B-A. Since A and B are unsigned, the sign of B-A can easily show that whether A is smaller than B or not.

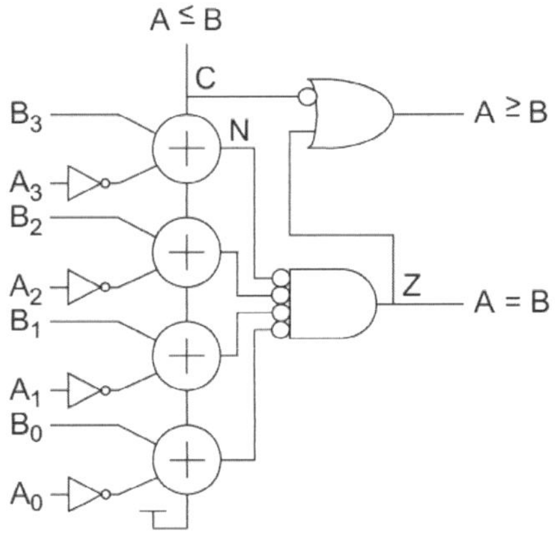

Fig.1 Unsigned magnitude comparator.

The biggest difference that the signed magnitude comparator holds from the circuit of the unsigned comparator is that the signed magnitude comparator introduces a one-bit signal, V, to highlight the overflow situation when subtracting the two inputs, which means the B-A in this case [4]. V is true when the inputs A and B have different signs and the output has a different sign with B. The final output is the sign of the subtraction of the two inputs, with only its most significant bit (MSB) being specially considered. The MSB of the subtraction which is known as S in this circumstance, is obtained by the following formula:

$$S = N \oplus V. \tag{1}$$

Here N is the inverse of the result's original MSB. When V is true, we find that the overflow flips the sign of subtraction.

Same as unsigned magnitude comparator, the sign of the subtraction, also known as S, shows that whether A is smaller or not than B.

1.2.2 Bit operation based multiplexer
In another case, we put the inputs of the AVD to be a subject and a threshold. In this traditional design, the circuit includes an isolated bit operation on the MSB of the subject. MSB should be designed to be 1 or 0, so that the sign of the subject is decided. If MSB is 1, then the subject is negative; if it is 0 then the subject would be positive. In this circumstance, a multiplexer is a traditional choice. Then the absolute value of the input should be extracted, which would include at least 3 inverters and a 1-bit full adder in the process.

The second procedure is to construct a 4-bit comparator. Fig.2 shows a classic design of 1-bit comparator, using complementary pass transistor logic. The problem of this circuit, other than its complexity, is that the output may not be distinct enough to decide whether the output is 0 or 1 without pulling up or down. The output could be at a potential just between the high level and the low level. To avoid this flaw, a large amount of redundancy would have to be added to the circuit [5].

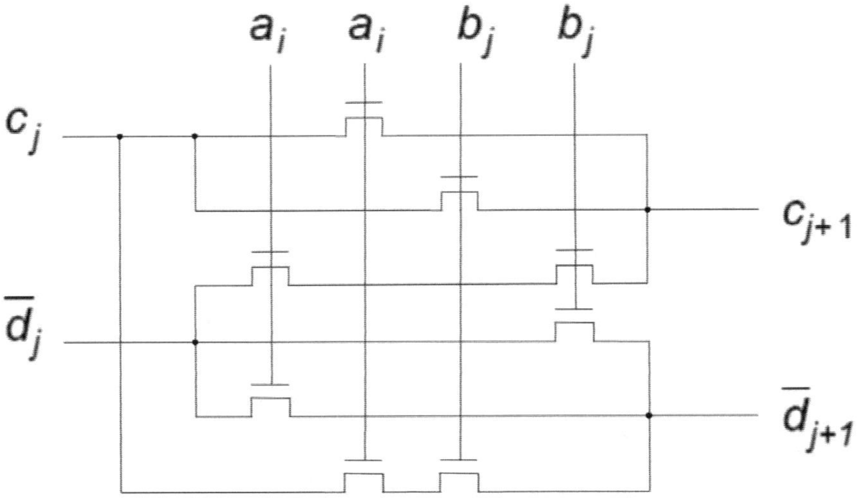

Fig.2 Construction of 1-bit comparator

In this case, a new design of 4-bit AVD is made to avoid complexity and increase efficiency.

1.3. Design description

The goal of our design is to find a new layout of the 4-bit AVD with minimum energy and without any registers like pipelining. The "energy" refers to total energy that is drawn from the upper high level, which means Device Voltage (VDD) here.

Basically speaking, absolute-value detector as a special design of magnitude comparator, takes signed value as inputs and outputs 0 or 1 to illustrate whether the absolute value of one input is larger than the other. Fig.3 shows the basic diagram of a 4-bit AVD. The inputs are shown in blue as follows. The circuit of AVD, which is shown in black, including an absolute value calculator and a comparator, is to be designed. The output of the circuit is shown in red to give that whether input $A_{[n]}$ is larger or not than the threshold value, known as Thr. If $A_{[n]}$ is larger than Thr, the output should be high, and vice versa.

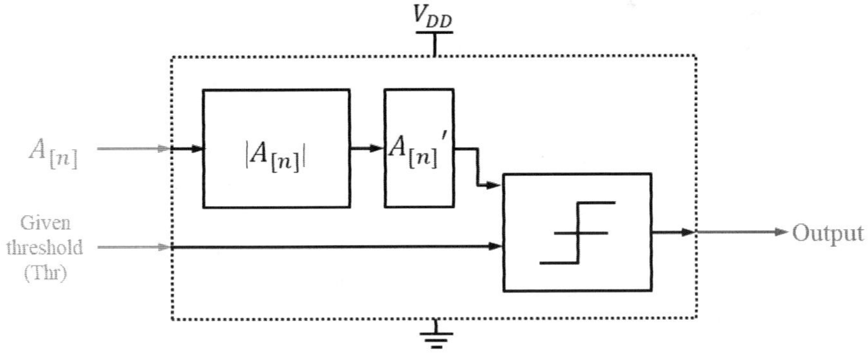

Fig.3 Basic diagram for an AVD

Our work of design experienced two main phases. In phase 1, we found an architecture that best optimizes the speed-energy goal. We tried several feasible options and figured out the best architecture and circuit style to meet the goal, even mixed the circuit styles. A sketch of the optimized design is made.

In phase 2, the block-level schematic of the designed AVD is obtained and an evaluation of the critical path is made. The gates along the path and the input operands that cause worst-case delay between input and output bias are determined in the process. We also consider the VDD scaling to ensure the least-energy-case is acquired.

2. Absolute-value Detector

2.1. Algorithm of Absolute-value Detector

We compare the input signals with the given threshold. When the input signal threshold is larger than or equal to the given signal threshold, the out result of absolute-value detector is 1. When the input signal threshold is less than the given signal threshold, the output result is 0.

In the process of neural signal transmission, we consider that the direction of input signal and the direction of neural signal transmission are different. Our input digital signal has 4 bits, which is the sign bit $A_3$and the input signal bits $A_2, A_1, A_0$from high to low. When the sign bit A_3 is 0, it means that the delivery direction is positive (we can understand it as a positive input signal); when it's 1, it means the direction is opposite (we can understand it as a negative input signal). Therefore, the representation range of our input digital signal is divided into positive and negative parts, from -7 to +7. We need to calculate the absolute value of the input signals. As the absolute value result is used for making comparison in the comparison parts. In response to this problem, we propose two corresponding solutions.

(1) When the input signals are positive, we don't process the input signals. When the input signals are negative, we convert them to the corresponding positive value by using the 2's complement method. Then the absolute-value calculation parts output the processed signals ($A'_{[n]}$)) and compare them with the given thresholds.

(2) When the input signal is negative, we don't calculate the absolute value of this input signal. When the input signal is positive, we use the 2's complement methods. Then it's compared with a given threshold. The comparison results can be obtained according to the sign of the calculation results.

With the comparative analysis, we choose the second option. Both of these schemes also use the 2's complement operation method and the second scheme is more advantageous. The reason is that the second scheme uses logic operations to replace the function of the comparator. The second scheme fully improves the utilization of circuit devices through logic operations and reduces the circuit degree of complexity. At the same time, it's easy to implement in engineering [6].

The algorithm principle of the absolute value detector is as follows.

(1) Determine whether the input signal is positive or negative. We collectively refer to the 4-bit input signal as A. A' is the output result of the 2's complement operation. We judge the positive or negative of the input signal according to the sign bitA_3. For the negative value input signal, output doesn't make calculation. For the positive value input signal, the 2's complement calculation is performed to obtain the corresponding negative value.

$$A' = - |A|. \quad (2)$$

(2) Detect and compare A' with the given thresholds. We collectively refer to the output result as B. When the input signal is a negative value, due to the direct output, a full adder is used to achieve the addition operation with the given value. When the input signal is positive, the full adder implements the subtraction operation by using the 2's complement method. Here we simplify the circuit through logic operations. Formula (2) shows the calculations.

$$B = T + A' = T - |A|. \quad (3)$$

Journal of Physics: Conference Series **2113** (2021) 012043 doi:10.1088/1742-6596/2113/1/012043

2.2. Circuit design of taking 2's complement

Our input signal is A. From high to low, the sign bit and the input signal bit areA_3, A_2, A_1, A_0. The given thresholds are collectively referred to as T. The given threshold bits areT_3, T_2, $T_1$and $T_0$from high to low. $A'$is the result of 2's complement operation of the A. When $A_3$is equal to 0, then A is a positive value. We take 2's complement operation on it. The logical operation formulas are as follows.

$$A'_0 = A_0. \tag{4}$$

$$A'_1 = A_1 \oplus A_0. \tag{5}$$

$$A'_2 = A_2 \oplus (A_1 + A_0). \tag{6}$$

$$A'_3 = A_3 \oplus (A_2 + A_1 + A_0). \tag{7}$$

Some examples of calculation results are as follows:

Table 1: Take 2's complement logic operation example

A_3 (Sign bit)	$A_3A_2A_1A_0$	$A'_3A'_2A'_1A'_0$
1 (Negative number)	1011 (-7)	1011
0 (Positive number)	0010 (+2)	1110

When $\boldsymbol{A_3}$=1, the input signal A is a negative value. We don't take processing with output.

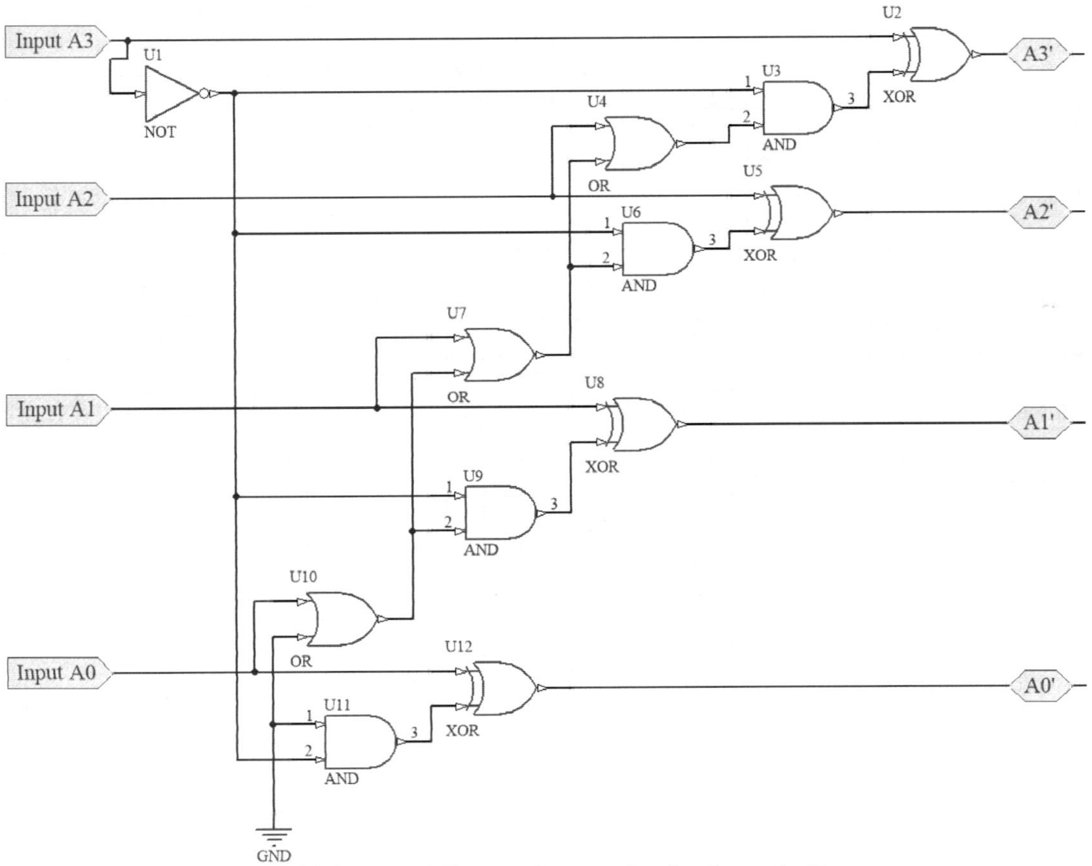

Fig. 4 4-bit input and 2's complement circuit schematic diagram

2.3. Circuit design of chain carry adder

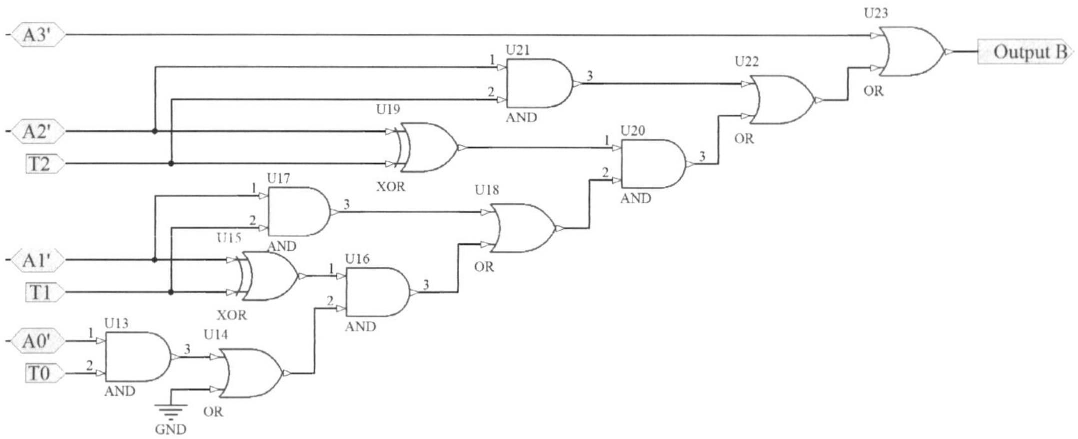

Fig.5 Schematic diagram of chain adder circuit

We design a chain adder to implement subtraction by using the 2's complement method. For design parameters, the chain adder input has a buffer of one unit, driven by two inverter chains of unit size. The chain adder consists of three full adders, which are connected in series with carry output. For a single full adder, the input signal A, the logical operation result B, C_{in} and the output is C_{out}. The formula shows the results of a single full adder.

$$C_{out} = AB+(A \oplus B)\, C_{in}. \tag{8}$$

Since we have connected the carry output in series, the output of the chain carry adder is B_3. As shown in the formula.

$$B_3 = T_2A_2' + (T_2 \oplus A_2')\,(T_1A_1' + (T_1 \oplus A_1')\,(T_0A_0')). \tag{9}$$

The final result output is $A_3'+B_3$ (in this case, $A_3'=1$)

2.4. Schematic diagram of Absolute-value Detector

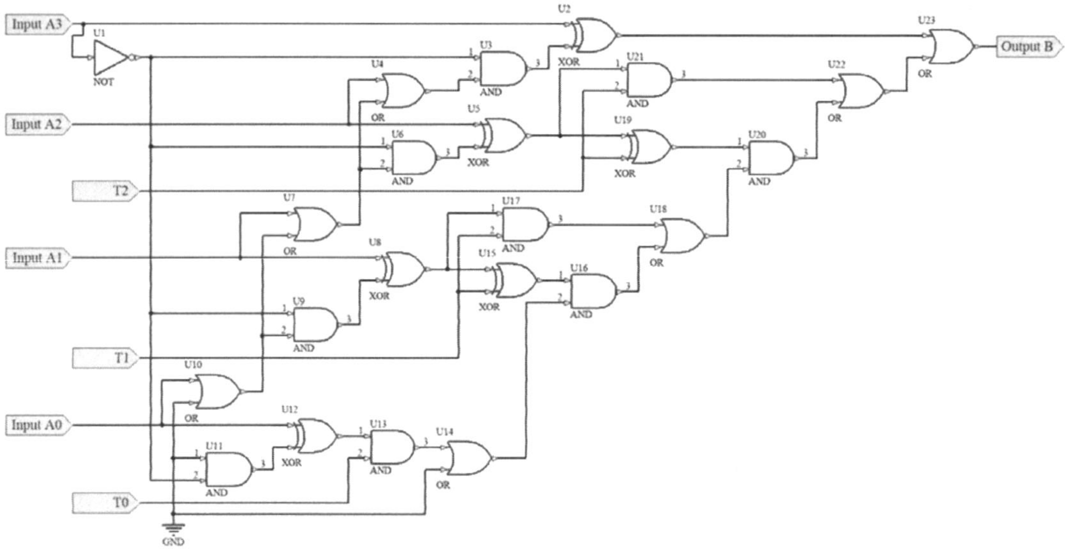

Fig.6 Absolute value detector circuit schematic diagram

3. Key parameters of Absolute-value Detector Circuit

3.1. Critical path delay

The critical path is the logical path with the longest delay from input to output in the circuit diagram. In the logic circuit we designed, the critical path is marked by the red path. As shown in fig.7, the red path is the critical path.

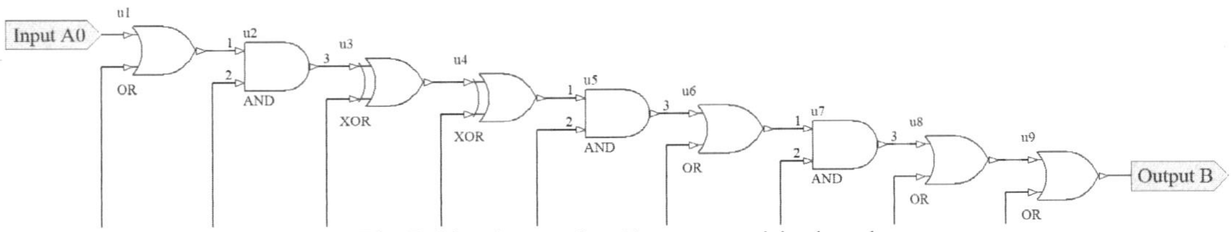

Fig.7 Absolute-value Detector critical path

As shown in table 3-1, several known parameters will be used in the calculations of the formulas [7].

Table 2: Logical effort of common gates, from p 156 of [3]

Gate Type	Number of Inputs				
	1	2	3	4	n
Inverter	1				
NAND		4/3	5/3	6/3	(n+2)/3
NOR		5/3	7/3	9/3	(2n+1)/3
Tristate, Multiplexer	2	2	2	2	2
XOR,XNOR		4,4	6,12,6	8,16,16,8	

3.1.1 V_{DD}=1v, Critical path delay

According to the path logic effort, we calculate the logic effort in the multi-level path [8]. As the following formula shown.

$$G = \prod g_i = \frac{4^6 \times 5^3}{3^7}. \tag{10}$$

Path electrical effort is calculated:

$$H = \frac{C_{out(path)}}{C_{in(path)}} = 32. \tag{11}$$

Path branch effort is calculated:

$$B = 1. \tag{12}$$

Path effort is calculated:

$$F = GBH. \tag{13}$$

$$\hat{f} = \sqrt[N]{F} = 2.69. \tag{14}$$

Path parasitic delay is calculated:

$$P = 22 \text{ PS}. \tag{15}$$

Total path delay is calculated:

$$D = N\hat{f} + P = 46.25 \text{ PS}. \tag{16}$$

3.1.2 Critical path gate sizing approach

The C_l of the output bit of the critical path is a 32-bit inverter. We calculate the size of the critical path. According to the formula calculation, the corresponding result can be obtained.

We know the below formula about the path effort F how to calculate.

$$F = \prod(g \times h) \tag{17}$$

The stage effort is:

$$f^* = g \times h \tag{18}$$

We can now size the gates:

$$C_{in} = g \times \frac{C_{out}}{f^*} \tag{19}$$

We get the sizing of the critical path gates from left to right (labelled u1-u9):

$$C_{in\,u1} = 1.27;\ C_{in\,u2} = 2.05;\ C_{in\,u3} = 4.13;$$
$$C_{in\,u4} = 2.78\ ;\ C_{in\,u5} = 1.87;\ C_{in\,u6} = 3.02;$$
$$C_{in\,u7} = 6.09;\ C_{in\,u8} = 9.83;\ C_{in\,u9} = 19.83.$$

3.1.3 V_{DD}^{opt} Calculation

The V_{DD}^{opt} is optimal voltage for the V_{DD} in the range of 0v -1v. This voltage also satisfies the requirements that the maximum delay is 1.5 minimum delays.

As the following formula shown, we calculate the V_{DD}^{opt} by using t_p:

$$t_p = \frac{CV_{DD}}{K'(V_{DD} - V_T)^2},\ K' = \frac{W}{L}K. \tag{20}$$

According to the situation that the maximum delay is 1.5 minimum delays, we preset V_T= 0.2v:

$$D \propto \frac{V_{DD}}{(V_{DD} - V_T)^2}. \tag{21}$$

$$V_{DD}^{opt} = 0.7758\text{ V}. \tag{22}$$

When V_{DD}^{opt} = 0.7758V, the critical path delay is calculating:

$$D_1 = 1.5\text{ D} = 69.38\text{ PS}. \tag{23}$$

3.2. The overall energy consumption of the circuit

3.2.1 Energy consumption when V_{DD}=1v

$$E = C_l {V_{DD}}^2 = 32.00. \tag{24}$$

3.2.2 The overall energy consumption of the circuit

$$E = C_l V_{DD}^{opt^2} = 19.26. \tag{25}$$

3.3. Summary of key circuit parameters

When the V_{DD} changes from 0V to 1V, we consider reducing energy consumption. We plan to recalculate the critical delay and energy consumption. After optimization calculations, we find that the energy consumption when V_{DD} is 1V is 39.81% lower than the energy consumption when V_{DD}^{opt} is 0.7758V. At the same time, the delay is 1.5 times the previous V_{DD} = 1V delay. According to the analyses of experimental results, the energy consumption of V_{DD}^{opt} at this time is the smallest energy consumption in the range of 0V to 1V.

In the calculation above we can also see that we can have a better energy performance while its delay is not being increased dramatically [9] as an individual static logic circuit [10], and this is an essential feature considering its usage in reversible neural signal processing the Arithmetic and Logic Units (ALUs) [11].

In addition to our conclusion above, since the change of dynamic energy in Complementary Metal Oxide Semiconductor (CMOS) is in proportion to its operation temperature and in inverse proportion to VDD, we conclude that for our AVD being set to a lower VDD, its energy changes more significantly with its change in operating temperature when being part of a dynamic circuit [12].

4. Conclusion

In this paper, a static-logic 4-bit AVD has been designed, implemented, and optimized with respect to the worst-case propagation delay and the total energy drawn from VDD. This new AVD is designed with the minimum energy for which the delay is 1.5 times the minimum delay, and both its delay and energy outperform traditional comparator based AVDs due to its relatively shorter critical path and replacement of comparator by the four carry bits of adders. We have only used 22 logic gates and an inverter in total, avoiding those high-fan-in gates like four-input nor gate.

Aside from the improvements we've made, there are still some flaws existing in our design. The XOR gate we used a lot in our design consumed the most energy and contributed a large part of the delay. What's more, the performance could be improved by mirror adder if we use it at some point on the critical path.

References

[1] S. Iranmanesh, G. Raikos, Z. Jiang and E. Rodriguez-Villegas, "CMOS implementation of a low power absolute value comparator circuit," 2016 14th IEEE International New Circuits and Systems Conference (NEWCAS), 2016, pp. 1-4.

[2] S. Lubaba, K. M. Faisal, M. S. Islam and M. Hasan, "Design of a Two-Bit Magnitude Comparator Based on Pass Transistor, Transmission Gate and Conventional Static CMOS Logic," 2020 11th International Conference on Computing, Communication and Networking Technologies (ICCCNT), Kharagpur, India, 2020, pp. 1-5.

[3] Neil H.E. Weste, David Money Harris. CMOS VLSI Design A Circuits and Systems Perspective Fourth ed.PEARSON.pp.156

[4] H. Fukuda, "Signed digit CMOS (SD-CMOS) logic circuits with static operation," Proceedings. 34th International Symposium on Multiple-Valued Logic, 2004, pp. 128-134.

[5] J. M. Rabeay, Digital Integrated Circuits: A Design Perspective, scoend ed., Upper Saddle River N. J. : Prentice Hall, c 1996.

[6] X. Lin, Y. Wang, S. Nazarian and M. Pedram, "An improved logical effort model and framework applied to optimal sizing of circuits operating in multiple supply voltage regimes," Fifteenth International Symposium on Quality Electronic Design, 2014, pp. 249-256.

[7] R. M. Anacan and J. L. Bagay, "Logical Effort Analysis of various VLSI design algorithms," 2015 IEEE International Conference on Control System, Computing and Engineering (ICCSCE), 2015, pp. 19-23.

[8] Jingyue Xue, D. Al-Khalili and C. N. Rozon, "A normalized intrinsic delay model of static CMOS complex gates for deep submicron technologies," The 2nd Annual IEEE Northeast Workshop on Circuits and Systems, 2004. NEWCAS 2004., 2004, pp. 17-20.

[9] H. Q. Dao, B. R. Zeydel and V. G. Oklobdzija, "Energy minimization method for optimal energy-delay extraction," ESSCIRC 2004 - 29th European Solid-State Circuits Conference (IEEE Cat. No.03EX705), 2003, pp. 177-180.

[10] K. Subramanyam, S. Shaik and R. Vaddi, "Tunnel FET based low voltage static vs dynamic logic families for energy efficiency," 18th International Symposium on VLSI Design and Test, 2014, pp. 1-2.

[11] J. B. Chacko and P. Whig, "Low Delay Based Full Adder/Subtractor by MIG and COG Reversible

Logic Gate," 2016 8th International Conference on Computational Intelligence and Communication Networks (CICN), 2016, pp. 585-589.

[12] P. Kocanda and A. Kos, "Static and dynamic energy losses vs. temperature in different CMOS technologies," 2015 22nd International Conference Mixed Design of Integrated Circuits & Systems (MIXDES), 2015, pp. 446-449.

A New Type of Sliding Mode Observer for Permanent Magnet Synchronous Motor Control

Sheng Cheng [a], **Yu-Fa Xu** [b*]

School of Electrical Engineering, Shanghai Dianji University, Shanghai, China

[a]email:cs1942218015@163.com

[b*]email: xuyf@sdju.edu.cn

Abstract: Aiming at the high-frequency oscillation and estimation accuracy problems of traditional sliding mode observers in the control process of permanent magnet synchronous motors, a sensorless control method for permanent magnet synchronous motors based on a new approaching law sliding mode observer is proposed. Based on the construction of a permanent magnet synchronous motor two-phase static coordinate system model, a sliding mode observer is used to estimate the back electromotive force, and then the rotor speed and position information are obtained. Finally, a simulation experiment is carried out. The results show that the new sliding mode observer based on the new approaching law effectively reduces the high frequency chattering of the system, improves the estimation accuracy of the system, and has better control performance.

1. Introduction

Permanent Magnet Synchronous Motor (PMSM) has attracted more and more attention because of its combination of power density and high torque density, and superior speed regulation performance. Especially in the current new energy field. Because the traditional mechanical sensor will not only increase the cost and volume, it is easy to be damaged in some special environments, resulting in failure to work normally. Therefore, the control method of sensorless control technology of permanent magnet synchronous motor has become the current research focus.

At present, PMSM sensorless control methods are roughly divided into two categories: one is the signal injection method is mainly suitable for zero-speed and extremely low-speed conditions, including rotating high-frequency voltage signal injection method and pulse high-frequency voltage signal injection method [1]. The other type is mainly suitable for medium and high speed conditions. Commonly used are model reference adaptive method [2], extended Kalman filter method [3] and sliding mode observer method [4-10]. Literature [11] uses a piecewise function to replace the traditional sign function to reduce the chattering of the system to a certain extent, and introduces a division link in the traditional phase-locked loop to improve the accuracy of the observer. Literature [12] adjusts the sliding mode gain through fuzzy control, and designs an adaptive low-pass filter, which reduces the chattering and phase lag problems of the system. Literature [13] selects a continuous switching function at the zero point, combined with an exponential approaching law, to improve the chattering of the system. Literature [14] suppressed the chattering of the system by designing a new exponential approaching law and adaptively changing sliding mode gain, and reduced the error of the rotor position estimation when the motor was running stably.

This paper proposes a sliding mode observer based on a new approaching law. The sliding mode observer uses a special power function $fal(x, \mu, \delta)$ [15-16] to design a new approaching law, and uses the Lyapunov function to prove its stability. . Finally, through simulation comparison and verification, the results show that the method improves the observation accuracy, significantly reduces the high-frequency chattering of the system, and has good control performance.

2. Mathematical model of permanent magnet synchronous motor

Assuming that the stator windings of the permanent magnet synchronous motor are three-phase symmetrical, the magnetic circuit is linear, there is no damping winding on the rotor, the core loss and eddy current loss are ignored, and the core saturation is ignored, then the surface mount permanent magnet synchronous motor is in a two-phase coordinate system. The mathematical model is:

$$\begin{cases} \dfrac{di_\alpha}{dt} = \dfrac{1}{L_s}(-R_s i_\alpha + u_\alpha - E_\alpha) \\ \dfrac{di_\beta}{dt} = \dfrac{1}{L_s}(-R_s i_\alpha + u_\beta - E_\beta) \end{cases} \tag{1}$$

$$\begin{cases} E_\alpha = -\psi_f \sin\theta \\ E_\beta = \psi_f w_e \cos\theta \end{cases} \tag{2}$$

Where, i_α、i_β is the component of the stator current on the α、β axis, R_s is the stator resistance, L_s is the stator inductance, u_α、u_β is the voltage component of the stator voltage on the α、β axis, E_α、E_β is the induced electromotive force of the α、β axis, ψ_f is the permanent magnet flux linkage, w_e is the electrical angular velocity , θ is the rotor position angle.

It can be seen from equation (1) that the back EMF contains the speed information and position information of the rotor. Therefore, the estimated values of the induced electromotive force of the $\alpha-\beta$ axis can be obtained to obtain the speed and rotor position.

3. Traditional sliding mode observer

In order to obtain the value of induced electromotive force, the traditional sliding mode observer is designed as:

$$\begin{cases} \dfrac{d\hat{i}_\alpha}{dt} = \dfrac{1}{L_s}(-R_s \hat{i}_\alpha + u_\alpha - V_\alpha) \\ \dfrac{d\hat{i}_\beta}{dt} = \dfrac{1}{L_s}(-R_s \hat{i}_\beta + u_\beta - V_\beta) \end{cases} \tag{3}$$

$$\begin{cases} V_\alpha = \lambda\,\mathrm{sgn}(\hat{i}_d - i_d) \\ V_\beta = \lambda\,\mathrm{sgn}(\hat{i}_q - i_q) \end{cases} \tag{4}$$

Among them, $\hat{i}_\alpha$、$\hat{i}_\beta$ are the current observation values of the stator α axis and β axis respectively; λ is the sliding mode gain.

From equation (1) and equation (3), the state equation of the current error system can be obtained as:

$$\begin{cases} \dfrac{d\tilde{i}_\alpha}{dt} = \dfrac{1}{L_s}(-R_s \tilde{i}_\alpha - V_\alpha + E_\alpha) \\ \dfrac{d\tilde{i}_\beta}{dt} = \dfrac{1}{L_s}(-R_s \tilde{i}_\beta - V_\beta + E_\beta) \end{cases} \tag{5}$$

Among them, $\tilde{i}_\alpha = \hat{i}_\alpha - i_\alpha$, $\tilde{i}_\beta = \hat{i}_\beta - i_\beta$ are current observation errors. The sliding mode surface function is defined as $\tilde{\boldsymbol{i}} = [\tilde{i}_\alpha\ \tilde{i}_\beta]^T$. When the observer enters the sliding mode, $\tilde{\boldsymbol{i}} = \boldsymbol{0}$. According to the equivalent principle of sliding mode control, get:

$$\begin{cases} E_\alpha = V_\alpha = \lambda\,\mathrm{sgn}(\hat{i}_\alpha - i_\alpha) \\ E_\beta = V_\beta = \lambda\,\mathrm{sgn}(\hat{i}_\beta - i_\beta) \end{cases} \tag{6}$$

In order to obtain a continuous back-EMF, it is necessary to perform low-pass filtering on equation (6), and the filtered induced electromotive force is:

$$\begin{cases} \hat{E}_\alpha = \dfrac{w_c}{s + w_c} V_\alpha \\ \hat{E}_\beta = \dfrac{w_c}{s + w_c} V_\beta \end{cases} \tag{7}$$

In the formula, w_c is the cut-off frequency of the low-pass filter, and the estimated value of the back electromotive force can be obtained. From the formula (2), the observed values of the rotor position and speed can be obtained:

$$\begin{cases} \hat{\theta} = -\arctan(\dfrac{\hat{E}_\alpha}{\hat{E}_\beta}) \\ w_e = \dfrac{\sqrt{\hat{E}_\alpha^{\,2} + \hat{E}_\beta^{\,2}}}{\psi_f} \end{cases} \tag{8}$$

Since the use of a low-pass filter will cause phase lag and amplitude conversion, it is necessary to compensate the rotor position and angle to obtain more accurate rotor position information, namely:

$$\hat{\theta} = -\arctan(\frac{\hat{E}_\alpha}{\hat{E}_\beta}) + \arctan(\frac{\hat{w}_e}{w_c}) \tag{9}$$

4. New sliding mode observer

Traditional sliding mode observers mainly use symbolic functions, and the discontinuity of symbolic functions at the origin leads to system chattering. This paper uses a special power function $fal(x, \mu, \delta)$ to design a new type of approaching law. The nonlinear function $fal(x, \mu, \delta)$ is:

$$fal(x, \mu, \delta) = \begin{cases} |x|^\alpha\,\mathrm{sgn}(x), & |x| > \delta \\ \dfrac{x}{\delta^{1-\alpha}}, & |x| \le \delta \end{cases} \tag{10}$$

The image is shown in Figure 1:

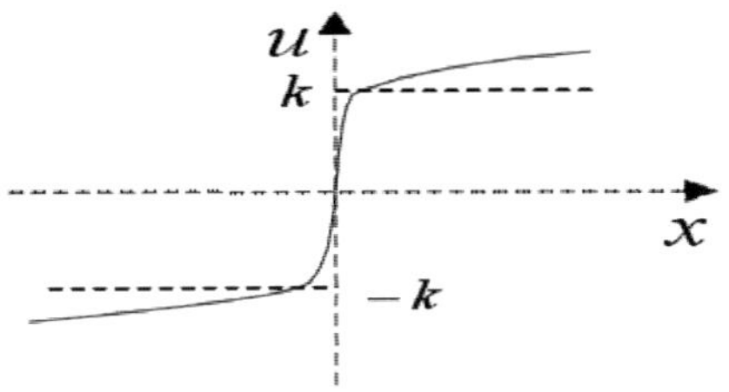

Figure 1 Image of function $fal(x, \mu, \delta)$

Among them, $0 < \delta < 1$, $\alpha > 0$, δ is the switching factor. It can be seen from equation (10) that when the value of the independent variable x is greater than δ , the gain x of the independent variable $\delta^{1-\alpha}$ is larger; otherwise, the gain is smaller.

Take the law of approaching:

$$\dot{s} = -\varepsilon \cdot fal(s, \ \mu, \ \delta) - karsh(s) \tag{11}$$

Among them, $s = \tilde{i} = \hat{i} - i$ is the sliding mode function, $\tilde{i}$ is the observation error of the stator current, $\hat{i}$ is the observation value of the stator current, i is the actual value of the stator current, and $\varepsilon > 0$, $k > 0$, $\mathrm{arsh}(s)$ is the inverse hyperbolic sine function.

In formula (11), when $|s| \leq \delta$, the $\varepsilon \cdot fal(s, \ \mu, \ \delta)$ term accelerates the approach speed, ensuring that the system can quickly approach the sliding mode surface, and the $fal(s, \ \mu, \ \delta)$ function is a continuous function near the origin, which can effectively reduce the high-frequency oscillation problem of the system. At the same time, the hyperbolic tangent is a smooth continuous function. When $|s| > \delta$, the $k \cdot arsh(s)$ term can not only provide a faster approach to the sliding surface, but also have a smoothing and limiting effect, so that the value of e will not be too large. The sliding mode controls the amplitude of the input signal. The design of the new approaching law sliding mode observer is as follows:

$$\begin{cases} \dfrac{d\hat{i}_\alpha}{dt} = -\dfrac{R_s}{L_s}\hat{i}_\alpha + \dfrac{u_\alpha}{L_s} - \dfrac{\varepsilon \cdot fal(\hat{i}_\alpha - i_\alpha, \ \mu, \ \delta) + k \cdot arsh(\hat{i}_\alpha - i_\alpha)}{L_s} \\ \dfrac{d\hat{i}_\beta}{dt} = -\dfrac{R_s}{L_s}\hat{i}_\beta + \dfrac{u_\beta}{L_s} - \dfrac{\varepsilon \cdot fal(\hat{i}_\beta - i_\beta, \ \mu, \ \delta) + k \cdot arsh(\hat{i}_\beta - i_\beta)}{L_s} \end{cases} \tag{12}$$

The error equation of sliding mode observer is obtained by subtracting equation (10) and equation (1):

$$\begin{cases} \dfrac{d\tilde{i}_\alpha}{dt} = -\dfrac{R_s}{L_s}\tilde{i}_\alpha + \dfrac{1}{L_s}(E_\alpha - (\varepsilon \cdot fal(\tilde{i}_\alpha, \ \mu, \ \delta) + k \cdot arsh(\tilde{i}_\alpha))) \\ \dfrac{d\tilde{i}_\beta}{dt} = -\dfrac{R_s}{L_s}\tilde{i}_\beta - \dfrac{1}{L_s}(E_\beta - (\varepsilon \cdot fal(\tilde{i}_\beta, \ \mu, \ \delta) + k \cdot arsh(\tilde{i}_\beta))) \end{cases} \tag{13}$$

To verify its stability, define the Lyapulov function as:

$$V = \frac{1}{2}s^2 \tag{14}$$

According to the stability condition, it is necessary to satisfy $\dot{V} < 0$, derivate the formula (14), and then bring the formula (11) into the derivation formula to obtain:

$$\dot{V} = s\dot{s} = -\varepsilon \cdot s \cdot fal(s, \ \mu, \ \delta) - k \cdot s \cdot arsh(s) \tag{15}$$

From the characteristics of functions $fal(s, \ \mu, \ \delta)$ and $arsh(s)$, we can see that the independent variable has the same sign as the function value, and $\varepsilon > 0, k > 0$, so $\dot{V} < 0$ is satisfied, that is, the system is stable.

5. Simulation results and analysis

The sensorless control block diagram of the permanent magnet synchronous motor in the synchronous rotating coordinate system based on the new sliding mode observer is shown in Figure 2. Use Matlab/Simulink to build a simulation model, and select $i_d = 0$ control strategy. The corresponding motor parameters are shown in Table 1.

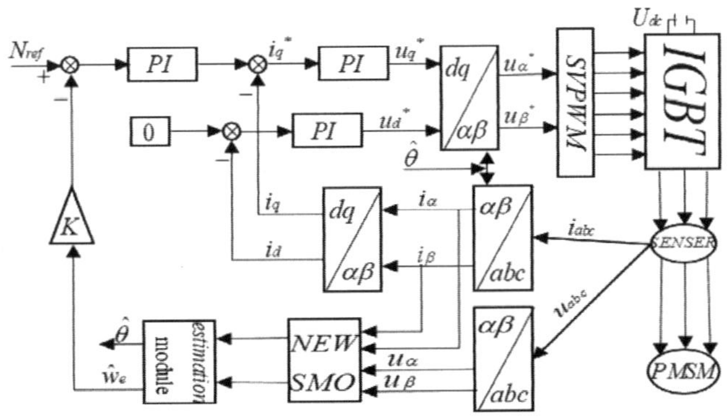

Figure 2 Block diagram of sensorless control

Table 1 Motor parameters

R/Ω	2.875
L_s/mH	0.0085
ψ_f/wb	0.175
$J/(kg\cdot m^2)$	0.001
$B/(N\cdot m)$	0
P_n	4
$N_{ref}/(r\cdot \min)$	1000

Figure 3 shows the actual speed, estimated speed and speed error waveforms using a traditional sliding mode observer. Figure 4 shows the actual speed, estimated speed and speed error waveform using the new sliding mode observer. It can be clearly seen from the figure that the chattering phenomenon of the traditional sliding mode observer system is obvious, and the speed error is about $\pm 10\ r/\min$ after stabilization. Based on the new sliding mode observer, the chattering phenomenon of the system is effectively reduced, and the actual speed can be better followed, the response speed is faster, and the speed error after stability is about $\pm 2\ r/\min$, which improves the observation accuracy of the system.

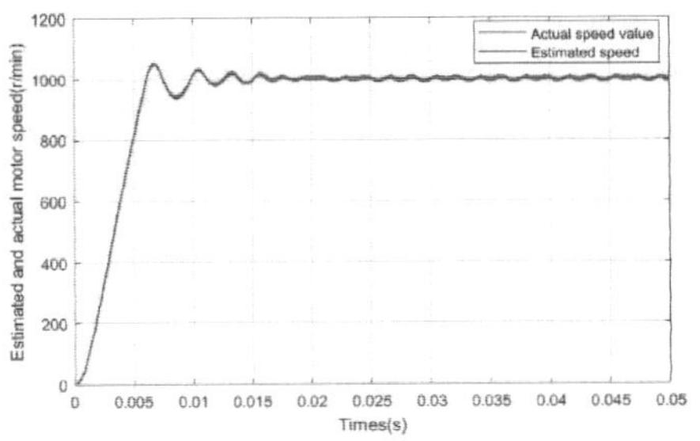

(a) Actual and estimated speed

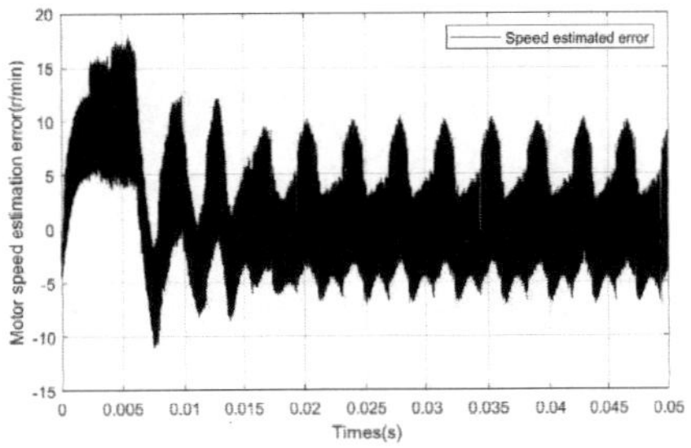

(b) Speed error

Figure 3 Rotational speed information waveform of traditional sliding mode observer

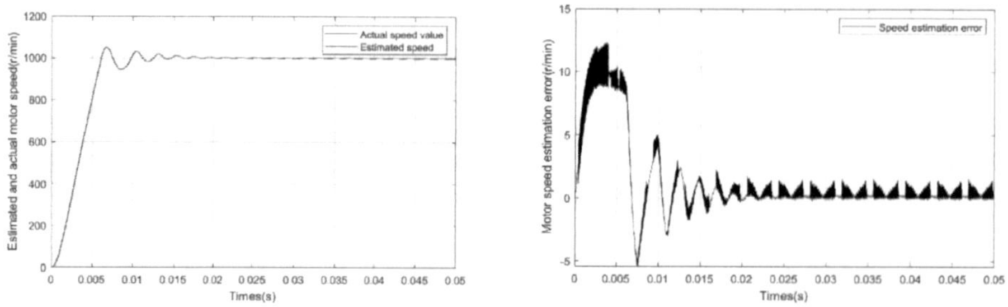

(a) Actual speed and estimated value (b) Speed error

Figure 4 Rotational speed information waveform of the new sliding mode observer

Figure 5 shows the estimated value, actual value and estimated error waveform of the rotor position using a traditional sliding mode observer, and Figure 6 shows the estimated value, actual value and estimated error waveform of the rotor position using a new sliding mode observer. It can be seen from the figure that the chattering phenomenon of the system when the new sliding mode observer is adopted is obviously lower than that of the traditional sliding mode observer, and the observation accuracy is higher.

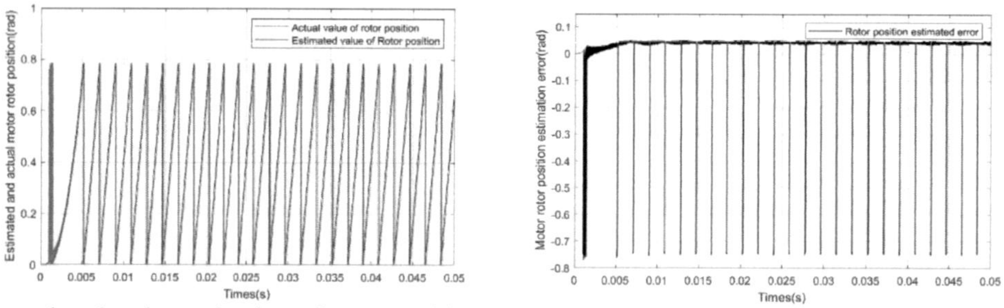

(a) Actual and estimated value of rotor position (b) Error of rotor position

Figure 5 The rotor position information waveform of the traditional sliding mode observer

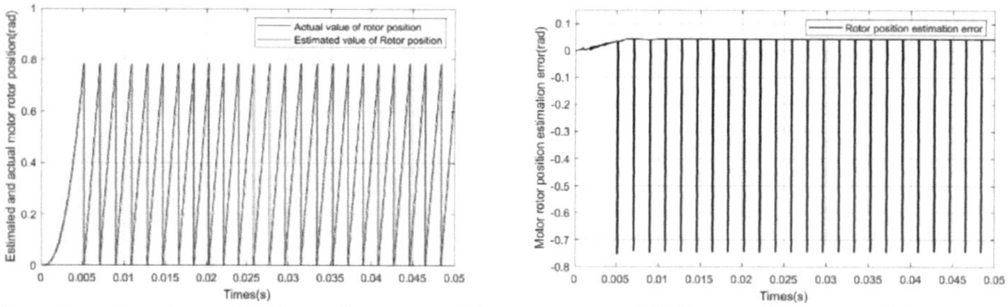

(a) Actual and estimated value of rotor position (b) Error of rotor position

Figure 6 Rotor position information waveform of the new sliding mode observer

6. Conclusion

Aiming at the high-frequency chattering and observation error problems of traditional sliding mode observers, this paper uses a special power function $fal(x, \mu, \delta)$ as the approaching law to construct a new sliding mode observer. The advantage of this law of approach is that when $|s| \le \delta$, $\varepsilon \cdot fal(s, \mu, \delta)$ can increase the approach speed, thereby ensuring that the system state can be arbitrarily limited Time to reach the sliding mode surface $s = 0$. And, near the origin, $k \cdot arsh(s)$ is a smooth continuous function, which can effectively reduce the high-frequency tremor of the control input.

At the same time, the inverse hyperbolic sine function is smooth and continuous Function, when $|s| > \delta$, the $k \cdot arsh(s)$ term can ensure that the system state approaches the sliding mode at a greater speed. In addition, the inverse hyperbolic sine function $arsh(s)$ plays a smoothing and limiting role. The value of the $k \cdot arsh(s)$ term in the reaching law can not be too large, so as to reduce the amplitude of the sliding mode control input signal. The simulation results show that this new sliding mode observer is effectively weakened. The system's high-frequency chattering problem can estimate the speed and rotor position more accurately than traditional sliding mode observers, and has good control performance.

References

[1] Liao, J. Research on PMSM speed sensorless control strategy based on high frequency signal injection[D]. Hefei University of Technology, 2010.
[2] Qi, F., Deng, Z.Q., Qiu, Z.J., Wang, X.L. Speed sensorless permanent magnet synchronous motor based on MRAS[J]. Transactions of the China Electrotechnical Society, 2007(04): 53-58.
[3] Xue, S.G., Wei, L.S., Ling, Y.Z. Sensorless vector control of permanent magnet synchronous motor based on extended Kalman filter[J]. Motors and Control Applications, 2011, 38(08): 15-18.
[4]Wang, H.X., Xiao, F., Ma, W.M., Chen, M.L., Liu, Y. PMSG SENSORLESS CONTROL BASED ON SLIDING MODE OBSERVER AND SPLL[J]. Journal of Electrical Machines and Control, 2011, 15(01): 49-54.
[5] Zhang, L.W., Li, X., Song, P.P., Zhang, P., Yun, L.S. Permanent magnet synchronous motor sensorless vector control system based on a new sliding mode observer[J]. Transactions of the Chinese Society of Electrical Engineering, 2019, 34(S1): 70- 78.
[6]Wu, H. Research on Sensorless Control Strategy of Surface Mount Permanent Magnet Synchronous Motor Based on Sliding Mode Observer[D]. Zhejiang University, 2021.
[7] Lin, J.C., Liu, S.M., Lin, M.Y. Research on MTPA control of permanent magnet synchronous motor with an improved sliding mode controller [J]. Electrical Engineering, 2021(02): 12-16.
[8] Xu, Y.Q. Research on PMSM position sensorless vector control based on sliding mode observer [D]. Tianjin University of Technology, 2020.
[9]Sun, N. Research on PMSM sensorless position based on sliding mode observer[D]. Northeast Petroleum University, 2020.
[10]Tang, C.M. Research on Permanent Magnet Synchronous Motor Vector Control System Based on Sliding Mode Observer Position Estimation[D]. Chongqing University of Posts and Telecommunications, 2018.
[11] Bai, T.Y., Liu, J., Xu, Z.M., Tang, W.Q. Permanent magnet synchronous motor control based on power approachaching law sliding mode observer [J]. Modular Machine Tool and Automatic Manufacturing Technology, 2021(01): 122-125.
[12] Zhu, X.Y., Zeng, G.H., Huang, B., Liu, J., Wei, Y. Speed sensorless vector control of permanent magnet synchronous motors[J]. Control Engineering, 2021, 28(05): 898-904.
[13] Lu, D.G., Li, Z.H. Improved sliding mode observer control for surface-mount permanent magnet synchronous motors [J/OL]. Journal of Electrical Machines and Control: 1-10 [2021-08-05]. http://kns.cnki. net/kcms/detail/23.1408.TM.20210518.1346.008.html.
[14] Wang, G.P., Zhu, L.G. Design of a new sliding mode observer for PMSM sensorless control[J]. Power Electronics, 2021, 55(04): 12-15.
[15] Pu, M., Liu, P., Xiong, A. Improvement of Fal function and three new nonlinear extended state observers [J]. Control and Decision, 2021, 36(07): 1655-1662.
[16] Zhou, T. Adaptive sliding mode control based on a new type of approaching law [J]. Control and Decision, 2016, 31(07): 1335-1338.

Research on Robustness of Related Images for Power Equipment Inspection

Chunlei Zhou [1,a], Xiangzhou Chen [1,b], Wenli Liu [1,c], Tianyu Dong [2,d], Huang Yun[2,e]

[1]Big data center of State Grid Corporation of China, Financial Street China Re Center, No.18, Luomashi Street, Hufangqiao, Xicheng District, Beijing, China

[2]State Grid information & Telecommunication Group Anhui Jiyuan Inspection and Testing Technology Co., HeFei, AnHui, China

[a]email: clzhou@sgcc.com.cn ,[b]email: xiangzhou-chen@sgcc.com.cn ,[c]email: wenli-liu@sgcc.com.cn ,[d]email: dongtianyu@sgitg.sgcc.com.cn ,

[e]email: dongtianyu@sgitg.sgcc.com.cn ,

Corresponding author's e-mail: [a]email: clzhou@sgcc.com.cn

Abstract. With the increase in the number of traction substations year by year, manual inspections are gradually being replaced by unattended inspections. Target detection algorithms based on deep learning are more widely used in intelligent inspections of power equipment. However, in practical applications, it is found that due to the small target to be detected, the accuracy of the deep learning model will decrease when the shooting angle is inclined and the light conditions are poor. This is because the algorithm's robustness is low, and the detection ability of the model will be seriously affected when the angle or illumination difference with the sample is large. Based on this, the feature fusion part of the YOLOv3 algorithm and the selection of the loss function and the size of the anchor frame are improved, and the improved ASFF fusion method is used to classify various images in the power equipment. Actual measurement and repeated experiments show that the proposed method can be effectively applied to image recognition of various power equipment, optimize robustness, and greatly improve the image recognition efficiency of power equipment.

1. Introduction

In recent years, the number of substations has increased year by year, and the number of equipment in substations has also gradually increased. Due to the large number of power equipment and the remote location, it requires more manpower, financial and material resources to maintain these equipment, so it is urgent to realize the intelligent inspection of power equipment. Intelligent inspection uses digital image processing technology, shallow machine learning or deep learning methods to classify objects on power equipment, and improves the efficiency of inspection fault discovery, which has important engineering practical value.

At present, computer vision methods based on deep learning theory have become a research hotspot in image target detection, which can be divided into two types: one-step method and two-step method.The two-step method includes Area Convolutional Neural Network R-CNN and its optimization algorithms Fast R-CNN[1] and Faster R-CNN[2],The difference between the two-step method and the one-step method is that the two-step method first generates a number of target candidate domains through a region suggestion strategy, performs feature extraction, uses a support vector machine to identify the target category, and finally outputs the target type in the image and marks its location.

Aiming at the adaptability, generalization and accuracy of digital image processing, shallow machine learning, one-step and two-step methods. At the same time, considering the high requirements for real-time performance in power equipment detection, this paper proposes an improved YOLOv3 algorithm to optimize robustness and improve the ability of power equipment inspection to image detection and positioning.

2. YOLOv3 algorithm

YOLOv3 is a target detection algorithm improved by Redmon on the basis of YOLOv2. In terms of network structure, YOLOv3 uses a new network structure Darknet53 as shown in Figure 1, instead of Darknet19 in YOLOv2. Since there are 53 convolutional layers, it is named Darknet53.

	Type	Filters	Size		Output	
	Convolutional	32	3	3	256	256
	Convolutional	64	3	3 / 2	128	128
1	Convolutional	32	1	1		
	Convolutional	64	3	3		
	Residual				128	128
	Convolutional	128	3	3 / 2	64	64
2	Convolutional	64	1	1		
	Convolutional	128	3	3		
	Residual				64	64
	Convolutional	256	3	3 / 2	32	32
8	Convolutional	128	1	1		
	Convolutional	256	3	3		
	Residual				32	32
	Convolutional	512	3	3 / 2	16	16
8	Convolutional	256	1	1		
	Convolutional	512	3	3		
	Residual				16	16
	Convolutional	1024	3	3 / 2	8	8
4	Convolutional	512	1	1		
	Convolutional	1024	3	3		
	Residual				8	8
	Avgpool		Global			
	Connected		1000			
	Softmax					

Figure 1. Darknet53 structure diagram

In addition, drawing on the idea of residuals [3] uses a large number of residual blocks in the network, thus avoiding the phenomenon of gradient explosion in the training process of too deep networks;In terms of loss function, YOLOv3 uses multiple independent logistic regression classifiers. Because when objects appearing in a target box belong to multiple categories at the same time, the softmax loss function in YOLOv2 can only consider them as one category.Therefore, multiple independent logistic regression classifiers are used in YOLOv3 to achieve multi-label classification;In terms of feature fusion, YOLOv3 draws on the idea of FPN[4] and predicts on 3 different sizes, each size predicts 3 target borders.For a picture, if it is initially divided into a $K \times K$ grid, and C categories need to be predicted, then the final tensor obtained for each scale is $K \times K \times [3 \times (4+1+C)]$,It includes the 4 offset coordinates of the target frame and the confidence score.Because of the fusion of the feature maps of the first two layers, the model can obtain more low-level and high-level image semantic information.

3. Related work

3.1. Data set

In the detection process of this article, the accuracy of detection also depends on the quality of the data set, and the number of data sets also affects the detection effect.At present, the open source data set does not contain images in the power equipment scene, and only similar instrument images can be filtered from the COCO data set [5], as shown in Figure 2, and a total of 2076 images have been screened out.In addition, in addition to the open source data set, I collected, screened and marked the targets to be detected in the power equipment scene, and the collected images included images obtained from multiple angles. Cut out 1000 images from the video collected on site.In addition, in order to avoid the over-fitting problem caused by insufficient data of the network, the sample was augmented as shown in Figure 3, including scaling, mirroring, cropping and Gaussian noise. Through augmentation, the number of images was increased to 5000. Finally, this experiment uses 3 data sets with a total of 15,380 images for target detection model training.

Figure 2. Power data equipment

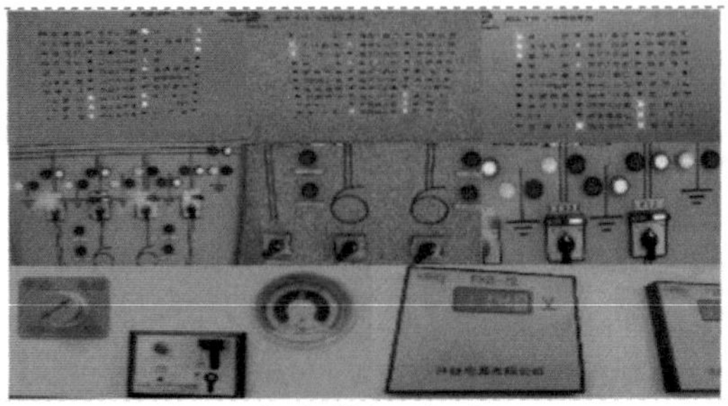

Figure 3. Self-made data set

3.2. K-means combined with a priori to set the anchor box

K-means sets the anchor box in combination with a priori. The Anchor box is a number of boxes of different sizes obtained by statistics or clustering from the ground truth of the training set, constraining the range of the predicted object, and adding a priori experience of size, So as to achieve the purpose of multi-scale learning, and avoid blindly finding the model during training, which helps the model to converge quickly. The k-means clustering method is used in the original YOLOv3 to obtain an anchor box that matches the size of the object in the original data set. After the size of the anchor frame is determined, the number of anchor frames needs to be determined, but the specific number is obtained after the original author's experiment, and the author uses various numbers of anchor frames in the model. Finally, by comparing the recall rate between different numbers, it is found that the recall rate is the highest when 9 anchor boxes are selected, and the final 9 anchor boxes are the best as shown in Figure 4.They are [10, 13], [16, 30], [33, 23], [30, 61], [62, 45], [59, 119], [116, 90], [156, 198], [373, 326].

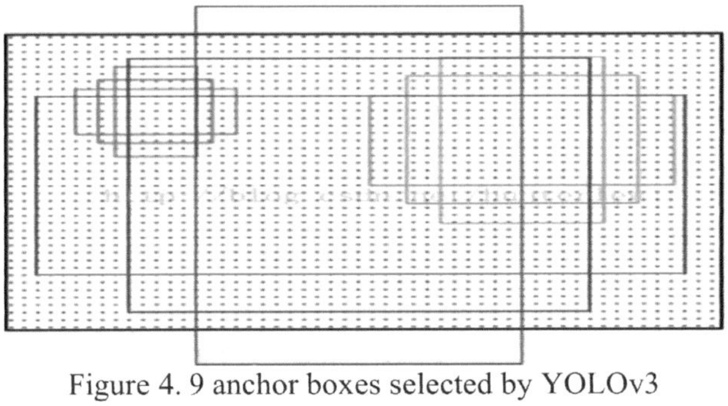

Figure 4. 9 anchor boxes selected by YOLOv3

For objects on power equipment, the Anchor of the YOLOv3 original data set has a large gap, including a large anchor box, which is not suitable for the detection of small targets on power equipment. In order to converge better and faster, and improve the detection accuracy, a set of Anchors specifically suitable for the data set used in this paper is designed. The specific parameters are [10, 17], [16, 27], [17, 31] , [17, 75], [18, 31], [27, 34], [31, 53], [40, 65], [71, 75].

3.3. Adaptive Feature Fusion (ASFF)

Previously, because the number of layers of the target detection network was relatively shallow, the features after feature extraction were not obvious, but the specific location of the object was more accurate. With the development of deep learning, the number of network layers has gradually increased, and the development of network fusion has also improved, among which multi-scale fusion has developed rapidly. The FPN feature fusion method used in YOLOv3 [4] is shown in Figure 5. The difference from other feature fusion is that its prediction is performed independently at different feature layers.However, this fusion method mostly uses concatenation direct connection, which does not make full use of the characteristics of different scales.Therefore, this paper uses adaptive feature fusion fusion [6] to replace concatenation. Adaptive Spatial Feature Fusion (ASFF) , most of the previous fusions rely on element feature fusion methods and different levels of feature fusion methods that rely on cascading each other. It is a new pyramidal feature fusion that relies on the relationship between different data.The idea is to adapt and learn the map fusion of different scales of spatial weights. It is composed of two steps: the same rescaling and adaptive fusion, and function adjustment. Because the resolution and the number of channels between the three convolutional layers that need to be fused in YOLOv3 are different, the up-sampling and down-sampling methods of different layers should be corrected.For layers that need to be upsampled, for example, if you want to get P3, you need to adjust C5 to the same size as C3. The method used is to first adjust the number of channels to be consistent with level3 through 1×1 convolution, and then use interpolation to resize to the same size ; For the layers that need to be downsampled, for example, if you want to get P5, you only need to use a 3×3 convolution with stride=2 for C4 to C5.If it is C3 to C5, you need to add a maximum pooling of stride=2 on the basis of 3×3 convolution, so that the sizes of C3 and C5 can be adjusted to be the same.Through the adaptive spatial feature fusion method, we can see that for small objects, we need more fine particles in the underlying features.

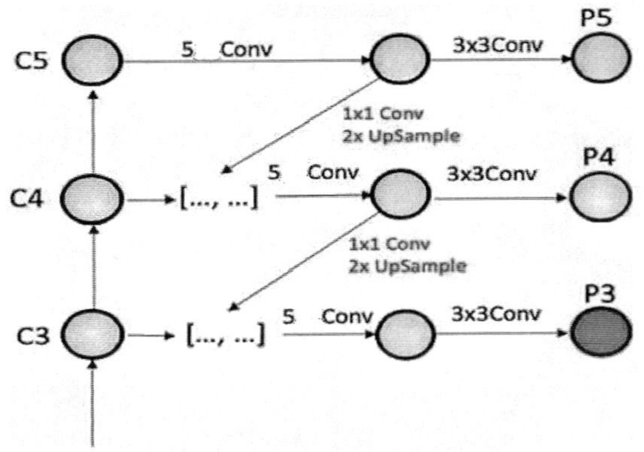

Figure 5. FPN feature fusion

3.4. TSE loss function

As a basis for the deep neural network to punish misdetected samples, the loss function can greatly affect the effect of model convergence. In the final output of the YOLOv3 network, the (x, y) object confidence and category confidence are all activated by the Sigmoid function, and then SSE is used to calculate the final loss. In the network needed in this article, due to the relatively high similarity of object features in power equipment, there is a situation where the positive and negative samples are very unbalanced. It is easy to appear that the output of the network is 0 at the beginning of training. At this time, the negative sample means that the real value is 0 and the error is close to 0, and the real value is 1 has a large error. For predx and predy, the difference is (-1,1) instead of (0,1), Directly use the formula grad=tx−predx to calculate the gradient, which is exactly the derivative form of the squared loss function. Following the same idea, this article adopts the calculation formula of the derivative form of TSE (Tan-Squared Error) [3]. The derivative form is as Formula 1:

$$\textbf{Loss Grad}=\tan(t-\sigma(z))/\tan(1) \qquad (1)$$

In YOLOv3, $t-\sigma(z) \in (-1,1)$, since t is also an indeterminate term, this difference has nothing to do with the value of $\sigma(z)$ itself, Let $x=t-\sigma(z)$ and make the function graph as shown in Figure 6.

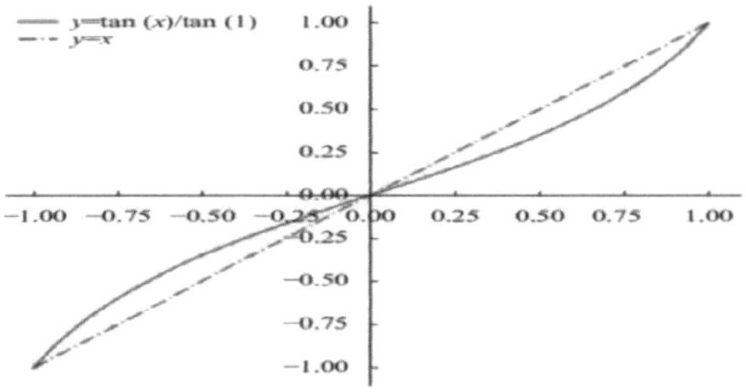

Figure 6. TSE and SSE derivative image

From the graph above, it can be found that the value range of the derivatives of the two functions are both (-1,1), and the absolute value of the derivative of the sum squared loss function must be greater than the absolute value of the derivative of the Tan square loss function.When there is a big difference between the actual value and the predicted value, the error can be reduced. When the result passes the sigmoid function, the partial derivative of w and the partial derivative of b can be gradually increased. This not only increases the speed of convergence at the beginning of the experiment,At the same time, the impact of the disappearance of the gradient is minimized. Because the value of the sum square loss function is slightly larger than the value of the Tan square loss function, for the case where the error between the true value and the predicted value is close to 0, in order to get better convergence of the model, the adjustment range of the weight of the output layer can be made smaller.This can better solve the situation that the training model does not converge and the results diverge during the training process, and the adjustment range can be controlled within a small range. The derivative images of the two functions are shown in Figure 7

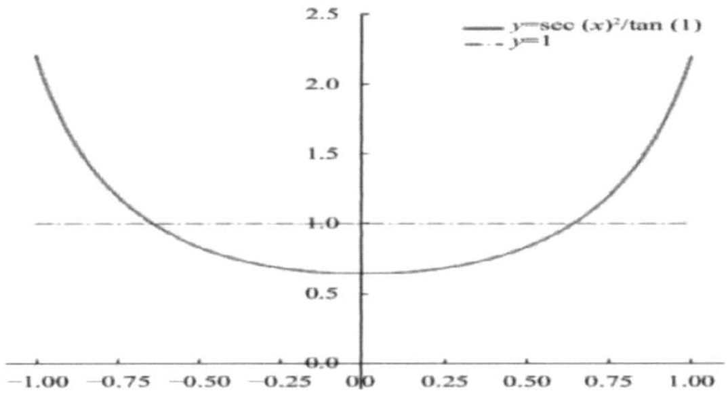

Figure 7. The gradient of the derivative of TSE and SSE

4. Experimental results

The experiment was conducted on a server with Ubuntu 16.04 operating system, and the graphics card used was NVIDIA GTX 1080Ti. Using 2076 pictures filtered from the COCO data set and a self-made data set, a total of 15380 pictures are formed. In order to compare whether to use clustering to select the initial candidate frame on the level of model detection performance, first train according to the original anchor box parameters of YOLOv3, and then train according to the anchor box obtained by clustering. The training sets use the same data set. The performance of the final model on the test set is shown in Table 1.

Table 1 The impact of the initial anchor box on model performance

Initial candidate frame	**mAP**	**FPS**
YOLOv3	72.77	33
Improved candidate border	82.30	39

The results in Table 1 show that using the original YOLOv3 anchor box parameters compared to the model trained using the clustered anchor box parameters, the latter has a significant improvement in detection accuracy or detection speed. Compared with the original YOLOv3, its mAP has increased by 9.53 percentage points, and it can detect 6 more pictures per second. This experiment uses two different loss functions, SSE and TSE, and tests the same samples in different environments (including normal, inclined, reflective, and dim). It is concluded that using the TSE loss function can get better accuracy in different environments.

This article modified the original YOLOv3 anchor box, FPN feature fusion and SSE loss function, and obtained a higher accuracy rate through comparative experiments. The experimental results are shown in Table 2.

Table 2 Comparative experiment

	Yolov3	**Yolov3 or anchor**	**Yolov3+ anchor+ASSF**	**Yolov3+anchor+ ASSF+TSE**
mAP	71.9	74.1	81.7	82.2
Pointer table	84.5	86.0	88.1	89.3
Counter	69.3	80.6	81.4	83.2
Digital meter	71.2	70.1	78.8	80.1
Round table	75.2	57.3	80.8	84.3
Indicator light	76.6	82.9	84.7	86.7
Square lamp	65.3	73.2	77.6	79.3
Rotary switch	75.7	76.1	80.1	81.3
Air switch	78.2	80.4	84.7	84.9
Fixed switch 1	82.0	82.0	84.1	85.4
Fixed switch 2	40.7	52.2	76.7	78.2

5. Conclusion

Based on the improvement of the YOLOv3 network, this paper aims at the problem of low detection accuracy and low robustness caused by the small target to be detected in the intelligent inspection of power equipment, when the shooting angle is inclined and the light conditions are poor. When the K-means algorithm is used to cluster the initial target frame, the TSE loss function and adaptive feature fusion method improve the network detection accuracy, and solve the current common problem of low network detection accuracy. On a data set combining the same self-made data set and open source data set, the modified network in this article and the original network are trained and compared. The experimental results show that:

The improved method proposed in this paper has achieved a greater degree of improvement in image accuracy, optimized robustness, and verified the effectiveness of the network framework.

References

[1] Girshick R. Fast r-cnn[C]//Proceedings of the IEEE International Conference on Computer Vision. 2015: 1440-1448.

[2] Ren SQ, He KM, Girshick R, et al. Faster R-CNN: Towards real-time object detection with region proposal networks[C]. Proceedings of the 28th International Conference on Neural Information Processing Systems. Montréal, Canada. 2015:91–99.

[3] He KM, Zhang XY, Ren SQ, et al. Deep residual learning for image recognition[C]. Proceedings of 2016 IEEE Conference on Computer Vision and Pattern Recognition. Las Vegas, NV,USA. 2016:770–778.

[4] Lin TY, Dollar P, Girshick R, et al. Feature pyramid networks for object detection[C]//IEEE Conference on Computer Vision and Pattern Recognition, 2017:936-944.

[5] Chen XL, Fang H, Lin TY, et al. Microsoft COCO captions:Data collection and evaluation server[C]. https://arxiv. org/abs/1504.00325, 2015.
[6] Liu S, Huang D, Wang Y. Learning Spatial Fusion for Single-Shot Object Detection[J/OL]. https://arxiv.org/ abs/1911.09516, 2019.

Dynamic equivalence of doubly-fed wind turbines based on parameter identification and optimization

Yaxin Li, Jie Yang*, Bainian Yi, Rengcun Fang, Dongyin Zhang

State grid hubei electric power company limited economic research institute, Wuhan, Hubei province, 430074, China

*Corresponding author's e-mail: yangjie16@hb.sgcc.com.cn

Abstract. The dynamic equivalence and modeling of large-scale wind farms is the basis for studying the problem of multi-temporal and spatial scale wind farms grid connection, and the establishment of a high-precision equivalence model is a key issue among them. This paper takes doubly-fed wind turbines as the research object. Aiming at the problem of large dynamic errors in direct aggregation of equivalents into a single unit, an equivalent model combined with particle swarm optimization algorithm is established to improve the accuracy of dynamic equivalents. The electrical parameters, reactive power reference values, active power controller parameters, and active DC voltage controller parameters are optimized using particle swarm optimization. The research results show that the equivalent model after optimizing the parameters of the active power and DC bus voltage controller is the closest to the detailed model in describing the dynamic characteristics of multiple machines.

1. Introduction

In order to deeply reveal the interaction mechanism between wind farms and power systems, and to lay a theoretical foundation for the safe and stable operation of power systems under the new energy structure, it is necessary to simulate and model large wind farms. At present, there are two main ideas for modeling large-scale wind farms. The first is to establish a detailed wind turbine model and superimpose the output results of each model as the wind farm output. The second is to use a single or fewer wind turbines to characterize the entire wind farm, then solve the model results of these several representative wind turbines and superimpose them as the overall output result. The results obtained by the first way of thinking can most accurately express the impact of the entire wind farm on the power system, but it is difficult to solve and not very adaptable to parameter changes. It requires high computer performance and has a large workload for model modification. Although the second way has some errors in describing the overall system, the model is simple and easy to modify, the physical problems can be simplified while accurate conclusions can be drawn, thus of great significance in studying the impact of wind farms on the power system.

In the past research, there are many literatures on the equivalence problem of wind farms. Literature[1-4] focuses on clustering indicators, including wind speed, active power, and wind turbine state vector[1], rotation speed during fault removal[2], three-dimensional correlation coefficient matrix of wind speed, wind direction, and wind turbine number[3], 13 state variables obtained from wind speed, nameplate parameters, and power curve[4]. Literature[5-7] focuses on parameter aggregation methods. Most parameter aggregation methods use capacity weighting, or use simplex and genetic optimization algorithms to optimize parameters on this basis[5][6]. Literature[8-10] began to try to apply the idea of parameter identification to wind farm equivalence. Literature[8] verified that the short-circuit fault disturbance data can effectively identify the electrical parameters of the wind turbine, and the gust disturbance data can effectively identify the mechanical parameters of the wind turbine. Literature[9] proposed the identification of the electrical parameters of doubly-fed wind turbines based on particle swarm optimization, and improved the algorithm's deficiencies. Literature[10] uses the particle swarm optimization algorithm to adjust the voltage/reactive power reference value of the equivalence unit, and uses a single unit to equalize the entire wind farm, which effectively solves the equivalence problem of doubly-fed generator wind farms with large input wind speed differences.

The output characteristics of multiple doubly-fed units as accurately as possible with the least number of equivalent units is described in this paper. First, multiple doubly-fed units are equalized into a single unit using the capacity weighting method, and electrical parameters, reactive power reference values, active power controller parameters, active power and DC bus voltage controller parameters are optimized using particle swarm algorithm. Then, by comparing the optimized equivalent model and the active output curve of the original detailed model in turn, it is verified that the equivalent model obtained by optimizing the active power and DC bus voltage controller parameters after the capacity-weighted equivalent is used to form a single machine is the closest to the dynamic characteristics of the original detailed model, when a three-phase short-circuit fault occurs at the grid connection point of the wind farm.

2. Mathematical model of doubly-fed wind turbine

The state equation of the doubly-fed wind turbine under the synchronous rotating coordinate axis considers the mechanical dynamics, stator and rotor dynamics at the same time. The motor voltage equation under the synchronous rotating coordinate system is as follows.

$$\begin{cases} u_{ds} = \dfrac{d\psi_{ds}}{dt} - \omega_s \psi_{qs} + R_s i_{ds} \\ u_{qs} = \dfrac{d\psi_{qs}}{dt} + \omega_s \psi_{ds} + R_s i_{qs} \\ u_{dr} = \dfrac{d\psi_{dr}}{dt} - (\omega_s - \omega_r)\psi_{qr} + R_r i_{dr} \\ u_{qr} = \dfrac{d\psi_{qr}}{dt} + (\omega_s - \omega_r)\psi_{dr} + R_r i_{qr} \end{cases} \tag{1}$$

Where u_{ds}, u_{qs}, u_{dr}, i_{ds}, i_{qs}, i_{dr}, i_{qr} are d-axis, q-axis stator and rotor winding voltage and current respectively. Ψ_{ds}, Ψ_{qs}, Ψ_{dr}, Ψ_{qr} are d-axis, q-axis stator and rotor winding flux linkage respectively. R_s, R_r are the stator and rotor winding resistance respectively. ω_s is the rotational angular velocity of the coordinate system and is equal to the synchronous speed ω_0. ω_r is the rotational angular velocity of the rotor.

The corresponding flux linkage equation is shown below.

$$\begin{cases} \psi_{ds} = (L_{ls} + L_m) i_{ds} + L_m i_{dr} \\ \psi_{qs} = (L_{ls} + L_m) i_{qs} + L_m i_{qr} \\ \psi_{dr} = (L_{lr} + L_m) i_{dr} + L_m i_{ds} \\ \psi_{qr} = (L_{lr} + L_m) i_{dr} + L_m i_{ds} \end{cases} \tag{2}$$

Where L_{ls}, L_{lr} are the leakage inductance for each phase of stator and rotor respectively. L_m is mutual inductance.

The generator is equivalent to a transient impedance $Z`=R_s+jX`$, a transient voltage source $E`=E`_d+jE_q`$ in simplified DFIG third-order model. Ignoring the transient process of stator flux, a simplified third-order dynamic model can be obtained.

$$\begin{cases} \dfrac{d\psi_{ds}}{dt} = 0 \\ \dfrac{d\psi_{qs}}{dt} = 0 \end{cases} \tag{3}$$

Finally, the DFIG third-order dynamic model composed of the stator and rotor voltage equations and the rotor motion equations is as follows.

$$\begin{cases} \dfrac{dE_d}{dt} = \dfrac{R_r}{L_r}[E_d + \omega_1 (L_s L_s^{'})(i_{qs})] + \omega_s E_q \omega_1 \dfrac{L_m}{L_r} U_{qr} \\ \dfrac{dE_q}{dt} = \dfrac{R_r}{L_r}[E_q \omega_1 (L_s L_s^{'})(i_{ds})] \omega_s E_d + \omega_1 \dfrac{L_m}{L_r} U_{dr} \\ \dfrac{T_j}{N_p} \dfrac{d\omega}{dt} = T_m T_e \end{cases} \tag{4}$$

Where L_s` is stator transient inductance. E_d, E_q are the potential dq-axis component within the DFIG equivalent. $\omega_s = \omega_1 \omega_r$.

3. Capacity weighting method parameter aggregation

The capacity weighting idea considers that a certain parameter of the equivalent unit is equal to the parameter in the single unit multiplied by the weight. But in fact, the situation that the overall characteristics of each wind turbine are completely equal to the superposition of the characteristics of a single wind turbine must be based on specific assumptions, such as the same speed.

From the perspective of the wind turbine to the grid side, the parameters that need equivalent values mainly include: moment of inertia (there is also a model corresponding to the inertia time constant), damping coefficient, stator resistance, stator leakage inductance, rotor resistance, rotor leakage inductance, stator and rotor mutual inductance, proportional and integral constants of each PI controller, DC bus capacitance connecting the rotor side and grid side converter, impedance parameter between the grid-side converter and the grid-connected point, box-type transformer parameters at the outlet of the wind turbine.

In the case where the speed of each wind turbine is different, the input wind speed needs to be equivalent. Usually, the equivalent wind speed is the wind speed of each unit multiplied by the capacity weighting coefficient of the corresponding unit. The equivalent formula of each parameter is as follows.

3.1. Equivalent moment of inertia

$$J_{eq} = \sum_{i=1}^{n} J_i \tag{5}$$

In the formula, the subscript eq represents the parameter of the equivalent unit, the subscript i (i=1,2,...n) represents the parameter of the equivalent unit, and n is the number of units required to be equivalent in the wind farm.

3.2. Equivalent inertia time constant

$$H_{eq} = \frac{J_{eq} \omega_m^2}{S_{eq}} \tag{6}$$

In the formula, ω_m is the standard value of the mechanical speed of the wind turbine, that is, the corresponding mechanical speed when the generator rotor is at the synchronous speed. S_{eq} is the equivalent wind turbine capacity, which is the algebraic sum of the unit capacity to be equivalent. For wind turbines with the same moment of inertia and the same capacity, the inertia time constant of the equivalent unit is equal to the inertia time constant of each unit. Damping coefficients mostly appear in the form of per unit value in wind turbine parameters, and the equivalent damping coefficient is the damping coefficient of each unit multiplied by the corresponding capacity weighting coefficient.

3.3. Equivalent generator impedance parameters

$$R_{eq} + jX_{eq} = \sum_{i=1}^{n} \frac{S_i}{S_{eq}} R_i + j\sum_{i=1}^{n} \frac{S_i}{S_{eq}} X_i \tag{7}$$

Where R_i, X_i, S_i are the resistance, reactance, and wind turbine capacity of the *i-th* unit.

3.4. Equivalent impedance parameter between grid-side converter and grid connection point

$$\frac{K_{eq}}{S_{eq}} = \sum_{i=1}^{n} \frac{K_i}{S_i} \tag{8}$$

The equivalence of each PI controller parameter in the rotor-side converter and the grid-side converter follows the principle of equal standard unit value, and the proportional and integral constants are proportional to the multiples of the equivalent unit and the unit to be equivalent. In the formula above, K_i is the nominal value of the proportional and integral constants of each PI controller.

3.5. Equivalent DC bus capacitance between the converters on both sides

$$C_{eq} = \sum_{i=1}^{n} C_i \tag{9}$$

The equivalent analogy of the impedance parameters of the box transformer from the outlet of the wind turbine to the grid connection point is the equivalent of the impedance parameters of the generator, and the capacity of the box transformer after the equivalent value is equal to the sum of the box transformer capacity of each wind turbine.

4. Particle swarm parameter identification equivalent method

When a three-phase symmetrical short-circuit fault occurs in a wind farm, the stator and rotor currents all contain DC transient components, AC components, and steady-state components[11]. The AC components of the stator and rotor are respectively related to the power frequency and the rotor angular frequency, and the steady-state component is related to the depth of the voltage drop. Through simulation, it is found that when a three-phase short-circuit fault occurs at the grid connection point of the wind farm, the dynamic error during the fault process is not large. That is because most of the faults set in the simulation are metallic ground faults, the voltage is close to zero, and the transient current has little effect on the active power curve. However, in the process of fault recovery, the dynamic error between the single-machine equivalent model obtained by the capacity weighting method and the detailed model is very large. The fault recovery process is similar to the instantaneous grid connection process of wind turbines, and is related to factors such as the phase angle of the wind turbine grid-connected point voltage and the speed of the wind turbine when the fault is removed.

Because the transient power of the grid connection process is difficult to quantitatively describe, the current literature mostly focuses on the grid connection control strategy of doubly-fed wind turbines. It is difficult to quantitatively express the dynamic errors of the detailed model and the equivalent model during the fault recovery process. Therefore, the output characteristics of the equivalent model and the detailed model can be similar by optimizing some parameters that affect the dynamic characteristics of the doubly-fed wind turbine.

The method of using identification methods to obtain accurate values of physical component parameters has been widely used in synchronous generators and electric loads[12]. The parameter identification method also has certain advantages in the dynamic equivalence of doubly-fed wind turbines. It only needs to know the electrical and mechanical parameters of each wind turbine when it leaves the factory as an optimal initial value. And then through the iterative optimization algorithm, it can achieve equivalent values for the dynamic characteristics of the entire wind farm. Appropriate initial value selection and appropriate algorithm are conducive to quickly find the optimal solution and improve simulation efficiency. However, the difficulty of parameter identification lies in the need to obtain disturbance data, which is difficult to obtain under normal conditions. The parameter identification method in this paper is only carried out in a simulation environment for the time being.

The particle swarm optimization algorithm has the characteristics of good overall performance and fast convergence speed, and is widely used in power systems and other engineering fields. In each iteration of the particle swarm algorithm, the particle tracks two extreme values, one is the optimal solution found so far by the particle itself, and the other is the optimal solution found so far by the entire population. The flow chart of the basic particle swarm algorithm in the parameter identification of the doubly-fed fan is as follows.

- Set the disturbance, simulate the disturbance data corresponding to the real parameters and store it.
- Set the population size, number of particles, number of iterations, initial position and speed.
- Solve the particle fitness function in the initialized population to find the global optimum.
- Update the particle position, obtain the fitness function value and update the optimal fitness of the particle.
- Update the optimal fitness of the population and the corresponding particle position.
- Determine whether the maximum number of iterations or iteration accuracy is reached. If yes, end the process. If not, go back to the fourth step until the maximum number of iterations is reached or the iteration accuracy is reached.

Among them, the fitness function for parameter identification is as follows. It is used to find the optimal one and minimize the value of the fitness function.

$$F(\theta) = \min \sum_{i=1}^{n} [\text{y\_real} - \text{y\_sim}(\theta)]^2 \tag{10}$$

5. Dynamic equivalent optimization simulation using particle swarm identification algorithm

5.1. Capacity weighted equivalence model

The following is a simulation of the doubly-fed wind turbine model based on the MATLAB/Simulink platform. The schematic diagram of three 1.5MW double-fed wind turbines connected to the 120kV system is shown in Figure 1.

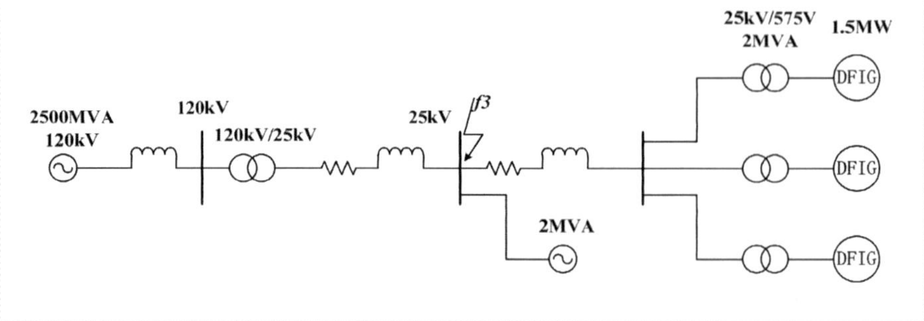

Figure 1. DFIG integrated with a 120kVpower system.

Taking the simulation start time as the timing starting point, at t=1.2s, a three-phase short-circuit fault occurs on the 25kV bus, and the fault will be removed after 0.05s. The parameters of each wind turbine and the box transformer at the outlet of the wind turbine are exactly the same, and the input wind speed of each wind turbine is 9m/s. It is assumed that the rotor crowbar loop action during the fault process is not considered, and the line protection action is also not considered. First, the capacity of three wind turbines is weighted and equalized into a single unit. The parameter aggregation method is shown in section 3. The transformer parameter aggregation means the capacity overlap. And the impedance parameter is based on the principle of consistent standard unit value. The mechanical, electrical parameters and controller parameters of the equivalent unit are shown in Tables 1 and 2 respectively. After intercepting the data within 0.6s after the fault occurs, the active and reactive dynamic curves of the capacity weighted equivalent model and the detailed model are shown in Figures 2 and 3, respectively.

Table 1. Mechanical, electrical parameters of DFIG.

Parameters	Single	Equivalent	Parameters	Single	Equivalent
rated power(MVA)	1.5/0.9	5	Stator resistance	0.00706	0.00706
Stator terminal voltage(kV)	0.575	0.575	Winding rotor resistance	0.005	0.005
Rated frequency(Hz)	60	60	Magnetizing inductance	2.9	2.9
Inertial time constant(s)	5.04	5.04	Stator leakage inductance	0.171	0.171
Mechanical damping coefficient	0.01	0.01	Rotor leakage inductance	0.156	0.156
Grid side resistance	0.0015	0.0015	DC capacitor(μ F)	1000	3000
Grid side inductance	0.15	0.15	DC bus voltage(kV)	1.2	1.2

Table 2. Controller parameters of DFIG.

	Proportional constant		Integral constant	
	Single	Equivalent	Single	Equivalent
Rotor side inner ring	0.05	0.15	1.5	4.5
Active outer ring on the rotor side	0.16667	0.5	16.667	50
Rotor side reactive power outer ring	0.00833	0.025	0.8333	2.5
Network side inner ring	0.16667	0.5	16.667	50
Outer loop of grid-side DC voltage	0.000333	0.001	0.00833	0.025

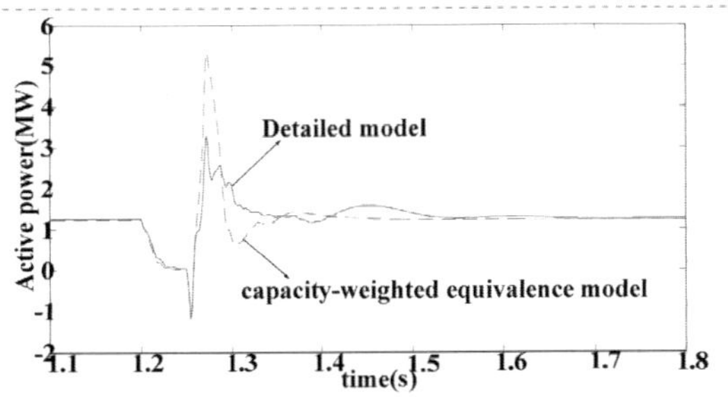

Figure 2. Dynamic active power curves of capacity weighted model and detailed model.

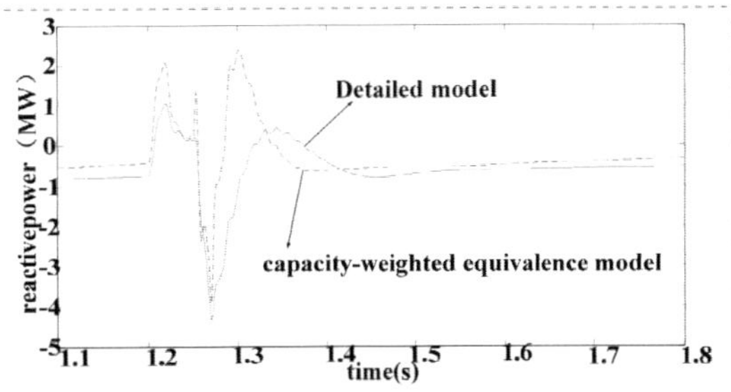

Figure 3. Dynamic reactive power curves of capacity weighted model and detailed model.

When the failure recovery process reaches the steady state again, the dynamic characteristics of the capacity-weighted equivalent model and the detailed model have large errors. Therefore, the particle swarm optimization algorithm is used to optimize the electrical parameters, reactive power reference values, active power controller parameters, and DC bus voltage controller parameters of equivalent units. Set the number of particles to 5, the number of independent variables to 5, 1, 1, 2, 2, the maximum number of iterations is 100, the learning factors 1 and 2 are both set to 2, and a constant inertia weight of 0.2 is used.

5.2. Comparison of particle swarm optimization models

5.2.1. Optimize electrical parameters. The grid-side fault can stimulate the fast dynamic mode of the electrical part of the wind farm[8], and electrical parameters also have a certain impact on the dynamic process of fault recovery. Based on the capacity-weighted aggregation parameters, the use of basic particle swarm optimization to optimize the electrical parameters of the equivalent unit can reduce the dynamic error in the fault recovery process.

5.2.2. Optimize the reactive power reference value. The reactive power characteristics of doubly-fed wind turbines affect the equivalent accuracy in the dynamic process. By superimposing a compensation quantity ΔQ on the reactive power reference quantity of the doubly-fed wind turbine, the particle swarm algorithm is used to find the optimal ΔQ to meet the minimum average relative error in the dynamic process. It is noteworthy that optimize the ΔQ can cause static error increases.

5.2.3. Optimize active controller parameters. In the converter of the doubly-fed unit, if the rotor side adopts the stator flux-oriented control strategy, the electrical quantity related to the active output characteristic is mainly I_{rq}. By optimizing the active controller parameters, the I_{rq_ref} of the equivalent unit and the detailed model is consistent, which effectively reduces the dynamic error.

5.2.4. Optimize active power and DC bus voltage controller parameters. The main parameter related to active power in the grid-side converter of the doubly-fed wind turbine is the DC bus voltage U_{dc}. After optimizing the parameters of the active controller, optimizing the parameters of the U_{dc} controller can further reduce the dynamic error.

5.2.5. Comparison. The final optimization results of each optimization model parameter are shown in Table 3. The electrical parameters and reactive power parameters in the table are in p.u., and the controller parameters are dimensionless. The optimized active power controller parameters k_p is 0.0512, and k_i is 6.011. The DC voltage controller parameter optimization value K_p is 0.0005618 and K_i is 0.00167.

Table 3. Optimization Results of model parameters.

Model 1[a]	R_s	R_r	L_{ls}	L_{lr}	L_m
	0.00270	0.0126	0.262	0.291	2.061
Model 2[b]	ΔQ				
	0.243				
Model 3[c]	k_p		k_i		
	0.0512		6.011		
Model 4[d]	P_k_p	P_k_i	$U_{dc}\_k_p$	$U_{dc}\_k_i$	
	0.0512	6.011	0.0005618	0.00167	

[a] Optimized equivalent model of electrical parameters.
[b] Optimized equivalent model of reactive power reference value.
[c] Optimized equivalent model of active power controller parameters.
[d] Optimized equivalent model of active power and DC voltage controller parameters.

The 15 iteration errors of each optimization model and the active dynamic curve within t=1.1-1.7s are shown in Figure 4 and Figure 5. It can be seen from the figure that the dynamic error between the model after optimizing the parameters of the active power and DC bus voltage controller and the detailed model is the smallest. The effect of reactive power reference value optimization and DC bus voltage controller parameter optimization is less obvious than electrical parameter optimization and active power controller parameter optimization.

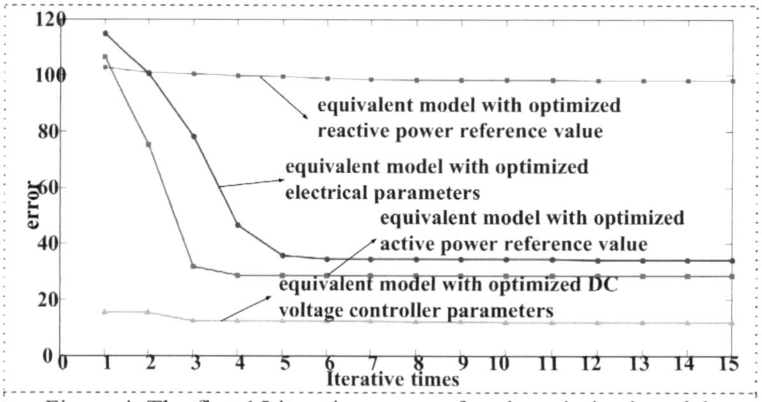

Figure 4. The first 15 iteration errors of each optimized model.

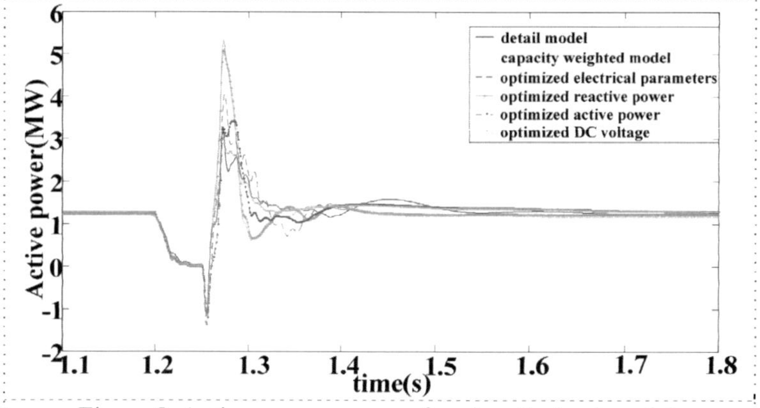

Figure 5. Active power curves of each optimized model.

In order to establish a high-precision equivalent model of wind farms, whether static or dynamic, the input wind speed of the equivalent unit can be obtained by optimizing the input wind speed in the static state firstly, then build a stand-alone equivalent model based on the capacity weighting method. Finally, by optimizing the parameters of the active power controller and the DC voltage controller, higher accuracy in the dynamic process is obtained, and the equivalent model with the overall static and dynamic performance closest to the detailed model is obtained.

6. Conclusion

This article first aggregates multiple wind farm units into one unit using the capacity weighting method. Because of the large dynamic error between the capacity weighted model and the detailed model in the case of a short-circuit fault, the period of large dynamic error is concentrated in the process from fault removal to restoration of steady state. The basic particle swarm algorithm is used to optimize the electrical parameters, reactive power reference values, active power controller parameters, and active DC voltage controller parameters. By comparing various equivalent models with detailed models, the dynamic equivalent errors of three-phase short-circuit faults are compared. The research results show that after optimizing the parameters of the active DC voltage controller, the dynamic error between the equivalent model and the detailed model is the smallest, and the dynamic process of the detailed wind farm model can be described more accurately.

Acknowledgments

This work was financially supported by the State Grid Hubei Electric Power Company Science and Technology project(52153818001F).

References

[1] Wang, X. (2008) Research on Dynamic Equivalent Model of Wind Farm. North China Electric Power University, Beijing.
[2] Mi, Z.Q., Su, X.W., Yang, Q.X., ect. (2010) Multi-Machine Representation Method for Dynamic Equivalent Model of Wind Farms. TRANSACTIONS OF CHINA ELECTROTECHNICAL SOCIETY, 25(5): 162–169.
[3] Cao, N., Yu, Q. (2012) A Grouping Method for Wind Turbines in a Grid-connected Wind Farm During Wind Speed Fluctuation. Automation of Electric Power Systems, 36(2): 42–46.
[4] Chen, S.Y., Wang, C., ect. (2012) Dynamic Equivalence for Wind Farms Based on Clustering Algorithm. Proceedings of the CSEE, 32(4): 11–19.
[5] Liu, L.Q., Yu, Y., ect. (2012) Dynamic Equivalence Method of Variable Speed Wind Turbine with Doubly-Fed Induction Generators. Proceedings of the CSU-EPSA, 24(2): 63–66.
[6] Zhang, B.H., Li, G.H., ect. (2013) Impact of wind farm integration on relay protection(2): DFIG-based wind farm electromagnetic transient equivalent model. Electric Power Automation Equipment, 33(2): 1–7.
[7] Jin, Y.Q., Huang, H., ect. (2012) Collector network transformation methods for wind farm aggregated modeling. Power System Protection and Control, 40(16): 34–41.
[8] Pan, X.P., Ju, P., ect. (2013) A Two-step Method for Estimating DFIG Parameters in a Wind Turbine and the Measurement Selection. Proceedings of the CSEE, 33(13): 116–126.
[9] Liu, Y.K., Pan, X.P., ect. (2014) Identification of DFIG parameters based on improved PSO algorithm. Journal of Hohai University(Natural Sciences), 42(3): 273–276.
[10] Jin, Y.Q., Ju, P., ect. (2014) Analysis on Controller Aggregation Method for Equivalent Modeling of DFIG-based Wind Farm. Automation of Electric Power Systems, 38(3): 19–24.
[11] Li, H., Zhao, M., ect. (2010) Evaluation and analysis of transient current of a DFIG wind generation system under grid fault. ELECTRIC MACHINES AND CONTROL, 14(8): 45–51.
[12] Ju, P. (2010) The theory and method of electrical power system modeling. Science Press, Beijing

An Intelligent Fault Diagnosis Method for CNN-SVM Circuit Breaker Based on Quantum Particle Swarm Optimization

Zhongting Huang[1], Longying Wang[2], Qiyun Ge[2], Yongyi Chen[1] and Dan Zhang[1*]

[1] School of Information Engineering, Zhejiang University of Technology, Hangzhou, Zhejiang, 310023, China

[2] Kerun Intelligent Control CO., LTD, Jiangshan, Zhejiang, 324102, China

[*]Corresponding author's e-mail: danzhang@zjut.edu.cn

Abstract. In order to make use of fewer fault data samples to diagnose the main fault types of circuit breakers accurately in real time, an intelligent fault diagnosis method for circuit breakers based on convolutional neural network (CNN) and quantum particle swarm optimization (QPSO) is proposed. Firstly, the key features of the circuit breaker operational signal are extracted through the CNN model, and the extracted feature vectors are input into the support vector machine (SVM) for fault diagnosis. In order to improve the diagnostic performance, this paper uses QPSO algorithm to optimize the parameters of the classifier, it effectively solves the local optimal problem. The experimental results show that the method presented in this paper has achieved good results in fault diagnosis of circuit breakers, and the accuracy of diagnosis is up to 100%, which highlights the superiority of this method.

1. Introduction

As an important power equipment, the working state of circuit breaker directly affects the safety of the power system. According to relevant data, more than 2/3 of circuit breaker failures are caused by abnormal operation of the operating mechanism[1]. Traditional manual patrol mode always bring unnecessary shutdown, in the context of smart grid, major power equipment manufacturers have increased research efforts on circuit breaker fault diagnosis technology. There are abundant vibration signals in the operation of circuit breakers. Taking the vibration signals during charged operation as input signals, the characteristic factors representing the state can be effectively extracted and used to analyze the fault location of circuit breakers, so as to realize state monitoring and diagnosis of circuit breakers[2].

As for the selection of fault diagnosis algorithms, most of the current artificial intelligence algorithms are machine learning. In Literature[3], Back Propagation (BP) algorithm and Radial Basis Function (RBF) were used to carry out fault diagnosis of circuit breakers, and the results showed that RBF neural network had fast training speed and good classification performance. In literature[4], wavelet packet transform was used for data preprocessing, and then Particle Swarm Optimization (PSO) algorithm and RBF were combined to build fault diagnosis network for circuit breakers. The accuracy of RBF network model after PSO was higher than that of RBF neural network. Although the diagnostic accuracy of traditional algorithms has been greatly improved, these algorithms generally have problems such as slow diagnosis speed and poor generalization ability, which limits the development of fault diagnosis technology of circuit breaker.

Because the traditional artificial intelligence algorithm still has many defects, this paper proposes an intelligent fault diagnosis method of circuit breaker based on convolutional neural network (CNN) and quantum particle swarm optimization (QPSO). This method is based on the CNN, using the circuit breaker operational signal data as the training sample to train the CNN model. The trained CNN model is combined with the QPSO algorithm to optimize the parameters of the SVM classification model, and the optimal parameters are outputted while CNN parameters remain unchanged. The SVM fault classification model was built based on the optimal parameters, and the test samples were input to the CNN-QPSO-SVM model, and the final fault diagnosis results were output.

2. Convolutional neural network model

Inspired by the human brain's processing of the external information received, LeCun proposed the convolutional neural network (CNN) model for the first time in 1994[5]. CNN model consists of convolutional layer, pooling layer and full connected layer, as shown in figure 1.

The convolutional layer has the characteristics of local perception and weight sharing, and the network parameters to be trained can be greatly reduced to reduce the amount of calculation. The operation of the convolution layer is as follows:

$$X_j^l = f(\sum_{i=M_J} x_i^{l-1} * K_{ij}^l + B^l) \tag{1}$$

where x_i^{l-1} is a feature in the filter of the upper layer, M_j is the filter corresponding to neuron j, K_{ij}^l is the Jth corresponding weight of neuron I in layer l, B^l is the unique offset of layer l, and *f(x)* is the activation function. Common activation functions are sigmoid, tanh, and ReLU, and the ReLU activation function is usually chosen.

The pooling layer is used to reduce the size of the feature map and reduce the number of parameters to be trained by reducing the dimension of the feature map. The pooling layer enhances the robustness of feature extraction through the lower sampling factor. The pooling function is expressed as downsample(). For each x_j^l, there is:

$$X_j^L = downsample(x_j^l) \tag{2}$$

After the convolution layer and the pooling layer, several full connected layers are usually connected, with the purpose of integrating the local feature information in the convolution layer or the pooling layer that can effectively distinguish the different categories. In order to improve the overall performance of CNN, the activation function of the full connected layer generally adopts the ReLU function. Its feedforward process is the same as that of the standard artificial neural network (ANN), as shown in equation (3).

$$x_j^l = f(\sum_{i \in l-1} w_{ji} x_i^{l-1} + B^l) \tag{3}$$

where w_{ji} is the weight of node j at layer 1 to node i at layer l+1.

The final layer of the neural network generally uses the softmax classifier to classify the different categories of data. The feature vectors with a fixed dimension are the output of the previous layer and serve as the input to the softmax classifier, which converts the output of the last layer into a basic probability distribution to predict the class labels of the input data.

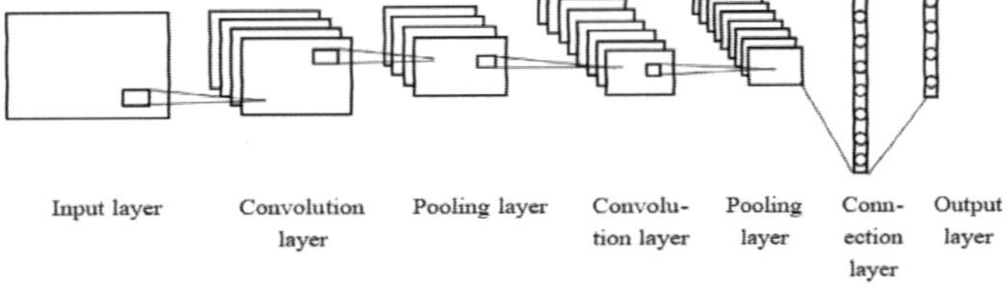

Figure 1. Convolutional neural network structure.

3. Support vector machine

Support Vector Machine (SVM) has good generalization performance and classification accuracy. The principle of SVM is to map data from a low-dimensional space to a high-dimensional space and find a hyperplane that can effectively divide samples of different categories. This hyperplane maximizes the minimum distance from a point in the data set to this surface, that is, find the maximum classification interval. Finally, the hyperplane corresponding to all the maximum classification intervals is determined to achieve the classification of the sample set[6].

By introducing kernel function to deal with various nonlinear problems, SVM shows good generalization ability in the training set and achieves high accuracy. There are four kinds of kernel functions commonly used: linear kernel function, polynomial kernel function, sigmoid kernel function and radial basis kernel function[7]. Radial basis function (RBF) was used as the kernel function of SVM in this paper, and the RBF expression is shown in equation (4).

$$k = \exp(-gamma \bullet d^2) \tag{4}$$

where d represents the distance between two points. The bigger the *gamma*, the fewer the support vectors, the smaller the *gamma*, the more support vectors. The number of support vectors influences the speed of training and prediction.

SVM selects $gamma = 0.1$ and $C = 0.8$ as parameters. Where C is the penalty coefficient, namely the tolerance to error. The higher C is, the less errors can be tolerated and the overfitting is easy. The smaller C is, the easier it is to underfit.

4. Quantum particle swarm optimization algorithm

In order to make the classification of fault diagnosis classifier more accurate, this paper uses the quantum particle swarm optimization (QPSO) to optimize the SVM parameters. Sun Jun first introduced quantum behavior into PSO algorithm and proposed QPSO algorithm in 2004[8].

QPSO algorithm introduces the quantum idea that the particle can appear at any position in the search space, and the probability of appearing at a certain position is determined by a distribution function. The specific position of particles in each generation is determined by an "observation" operation. In terms of algorithm implementation, Monte Carlo method is used to recurse the specific position of particles in the n+1 generation in the nth iteration.

The central point introduced by the QPSO algorithm is shown in equation (5).

$$m_{best} = \frac{1}{M}\sum_{i=1}^{M} P_i \tag{5}$$

where m_{best} is the average historical best position of particles. Equation (6) and equation (7) are used in the update process.

$$P_i(t+1) = \phi P_i(t) + (1-\phi)P_g(t) \tag{6}$$

$$X_i(t+1) = P_i(t+1) \pm \lambda \left| m_{best} - X_i(t) \right| \ln(1/u) \tag{7}$$

where ϕ and u are random variables and are uniformly distributed on (0,1): the probability of getting positive and negative is 0.5. λ is the only control parameter, in practical application, the value of λ is generally less than 1, which will affect the convergence rate of the population.

QPSO algorithm introduces the degree of evolution and degree of aggregation of particles to modify parameter β to avoid premature convergence.

(1) Evolution rate e of particle swarm is defined as:

$$e = \frac{\min\{p_g(T-1), p_g(T)\}}{\max\{p_g(T-1), p_g(T)\}} \tag{8}$$

where e is the evolution rate factor, and $p_g(T)$ is the global optimal solution of the T generation particle swarm.

(2) The degree of particle aggregation c is defined as:

$$c = \frac{\min\{ p_g(T), \overline{p_i(T)}\}}{\max\{ p_g(T), \overline{p_i(T)}\}} \tag{9}$$

where c is the particle aggregation factor, and $\overline{p_i(T)}$ is the average value of the current individual optimal solution of all particles.

The relation expression of β, c and e is:

$$\beta = \beta_0 + c\beta_c - e\beta_e \tag{10}$$

where β_0 is the initial value of β, usually β_0=1, β_c, β_e are random numbers between [0,1].

To sum up, if the evolution rate of particle swarm slows down,the value of β can be reduced in order to search the particle swarm in a small space. If the degree of particle swarm aggregation increases, the particle swarm should be dispersed, the search space should be enlarged, that is, the value of β should be increased, so as to avoid falling into the local optimal solution.

5. Experiment and result analysis

5.1. Experimental data set

The data set in reference[9] was used for verification in this paper. The acceleration sensor was used to record the vibration acceleration parameters during the opening/closing process of the circuit breaker. The sampling frequency was 10kHz. A total of 7 groups of vibration signal data were collected, including 4 main fault types of circuit breakers. Seven experiments were conducted under each condition. Typical signals of circuit breakers in normal and fault states are shown in figure 2, and the four fault types and their serial numbers are shown in table 1.

Table 1. Four types of mechanical fault numbers of circuit breakers.

Fault types	Fault number
Normal state	0
Tripping to close the electromagnet blocking state	1
Spindle clogging state	2
Half shaft clogging state	3

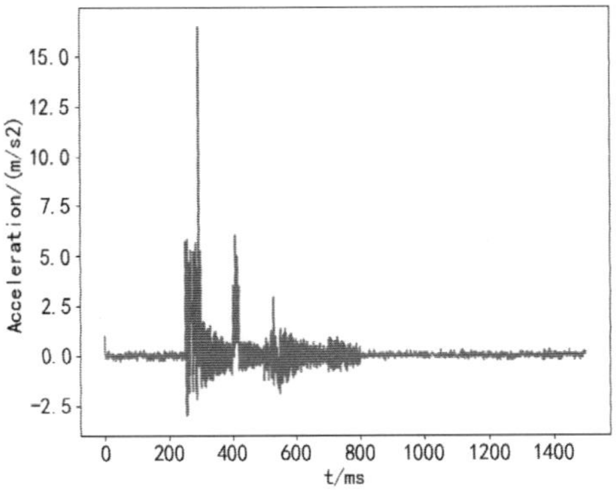

(a) Normal state.

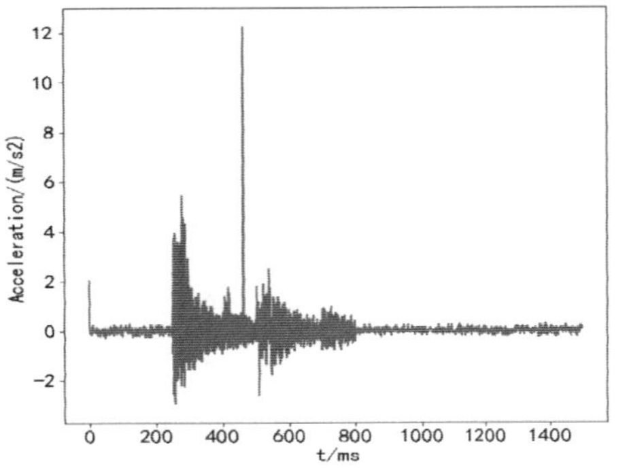

(b) Tripping to close the electromagnet blocking state.

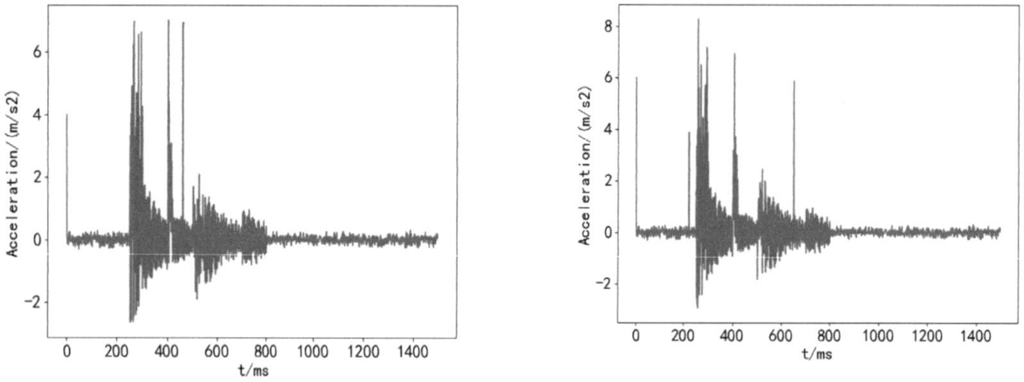

(c) Spindle clogging state. (d) Half shaft clogging state.

Figure 2. Vibration signals of four states are collected by acceleration sensor.

In seven data sets, three of them are selected as training dataset and the remaining four are selected as test dataset for each state.

5.2. CNN-QPSO-SVM model construction

The circuit breaker fault diagnosis method flow based on CNN and QPSO is shown in figure 3.

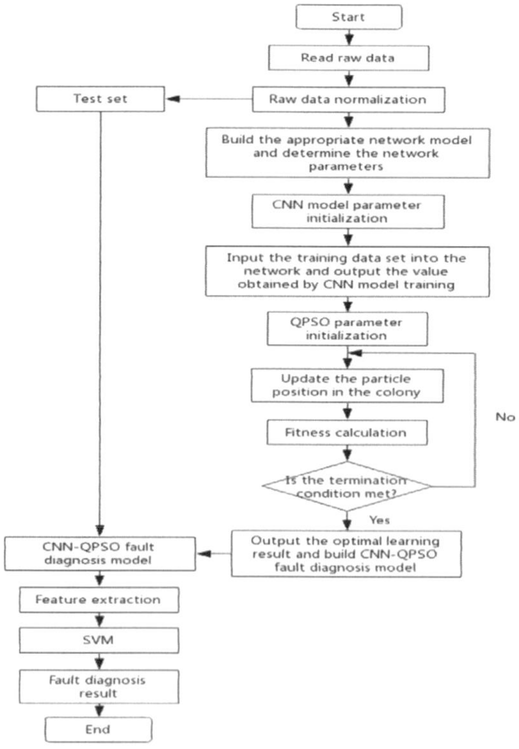

Figure 3. Flow chart of circuit breaker fault diagnosis method based on CNN and QPSO.

In this model, the fault sample data of circuit breakers are input into the corresponding sample data of 4 state types, and these sample data are divided into two groups, one group as the training sample and the other group as the test sample. The specific process is as follows:

(1) Vibration signals of circuit breakers in normal state, tripping to close the electromagnet blocking state, spindle clogging state, half shaft clogging state were collected and used as original vibration signals for model training.

(2) The original signal is normalized, as shown in equation (11).

$$\hat{x}_i = \frac{x_i - x_{\min}}{x_{\max} - x_{\min}}, i = 1,2,\cdots,n \tag{11}$$

where x_i is the original sample data, $x_{max} = max(x_i)$, $x_{min} = min(x_i)$. In order to find the optimal parameter combination of SVM, the normalized samples were classified, and part of the samples were selected as the training data set and the rest as the test data set.

(3) The training data set is used to train CNN, and the loss function of CNN is based on the cross-entropy loss function of softmax. The specific training method is Adam gradient descent method, the batch size is 2, and the number of iterations is 200. After the completion of the training, the parameters of the CNN model are kept unchanged.

(4) The trained CNN model is combined with the QPSO algorithm to train the SVM classification model. In the iteration process, the adaptive value deviation of each generation of particles is tracked in real time. If the deviation of a generation is less than the threshold value, interference operation is carried out on the average optimal position to enhance the global optimization ability of the algorithm. The algorithm outputs the optimal SVM parameters until the algorithm meets the termination condition.

(5) The SVM fault classification model was built based on the optimal parameters. The test samples were input into the CNN-QPSO-SVM model and the fault diagnosis results were output.

The neural network structure constructed by this model is shown in figure 4. It can be seen from the figure that the CNN network designed in this paper has 6 convolution layers, 4 pooling layers, 1 output layer and 1 softmax layer, among which the structural parameters of each layer are shown in table 2.

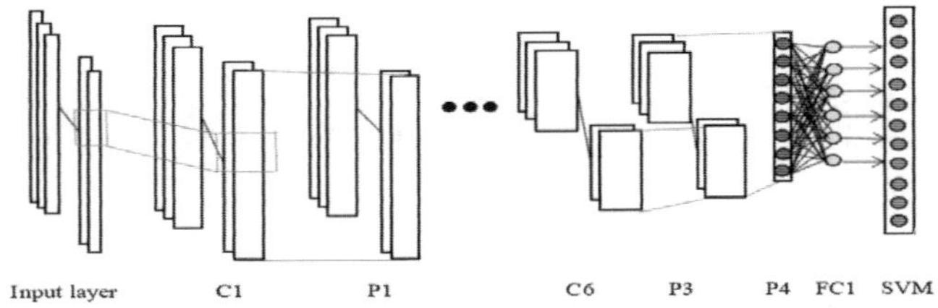

Figure 4. Structure diagram of CNN.

Table 2. Structural parameters of CNN.

Serial number	Network layer name	Size of convolution kernel	Number
C1	Convolution layer1	8×1	16
C2	Convolution layer2	8×1	16
P1	Max pooling layer1	2×1	16
C3	Convolution layer3	4×1	32
C4	Convolution layer4	4×1	32
P2	Max pooling layer2	2×1	32
C5	Convolution layer5	4×1	64
C6	Convolution layer6	4×1	64
P3	Max pooling layer3	2×1	64
P4	Global Average Pooling layer1	-	-
FC1	Softmax	4	1

5.3. Experimental analysis

The trained CNN-QPSO-SVM model is used to detect the fault data sets of the circuit breakers in four different states. Finally, the simulation graph of the accuracy of the CNN-QPSO-SVM model on the verification samples is obtained. Figure 5 shows the result of the change of accuracy of training set and test set with the number of iterations. It can be seen from the figure that when the number of iterations is 25, the accuracy tends to the maximum and then becomes stable. The curve of the verification set can completely fit the curve of the training set, and there is no over-fitting or under-fitting phenomenon, and the model has been effectively learned. At the same time, cross entropy loss function is used as the loss rate calculation method, and the loss rate of the test set is obtained as shown in figure 6. It can be seen from the figure that when the number of iterations is 25, the loss rate tends to the minimum value and then becomes stable. The curve of the verification set can fit the curve of the training set perfectly, and there is no over-fitting or under-fitting phenomenon.

In order to verify the superiority of the CNN-QPSO-SVM algorithm proposed in this paper, CNN-LSTM algorithm[9], CNN-SVM algorithm and SVM algorithm are respectively used to compare with this algorithm, and the results obtained are shown in table 3.

It can be seen from table 3 that the prediction accuracy of CNN-QPSO-SVM algorithm is 100%, while the prediction accuracy of CNN-SVM or SVM algorithm is 75% and 50% respectively, which cannot effectively identify the fault type. It shows that QPSO parameter optimization has a great influence on the improvement of prediction accuracy of CNN and SVM algorithms. In addition, although the prediction accuracy of CNN-QPSO-SVM algorithm and CNN-LSTM algorithm[9] are both close to 100%, the one-dimensional convolutional neural network is better than the two-dimensional convolutional neural network in the characterization of circuit breaker vibration signals. Therefore, CNN-QPSO-SVM algorithm has a higher recognition accuracy than CNN-LSTM algorithm. Through the above comparison, it can be concluded that CNN-QPSO-SVM algorithm has the best performance among these algorithms.

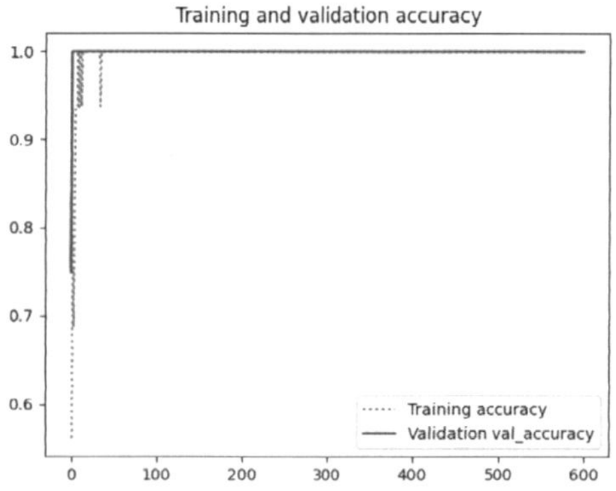

Figure 5. Diagnostic accuracy results.

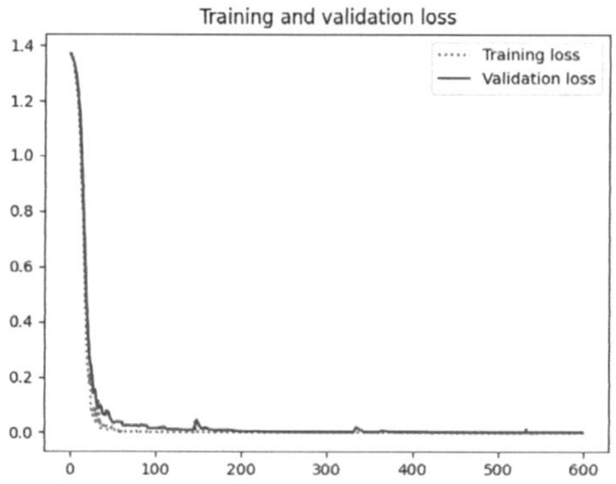

Figure 6. Diagnostic loss rate results.

Table 3. Comparison of diagnostic accuracy of different algorithms.

Diagnosis algorithm	Diagnostic accuracy
CNN-QPSO-SVM	100%
CNN-LSTM	99.8%
CNN-SVM	75%
SVM	50%

6. Conclusion

In view of the difficulties in obtaining vibration signals, the lack of fault samples and the inaccurate identification of eigenvectors during the operation of circuit breakers, a circuit breaker fault diagnosis method based on CNN and QPSO is proposed in this paper. The method uses CNN moedel to extract circuit breaker signal features effectively. Furthermore, the QPSO algorithm is used to effectively eliminate the local optimal phenomenon. Finally, SVM is used to classify the samples so that the model can diagnose the fault categories of circuit breakers effectively and accurately. The experimental results show that CNN-QPSO-SVM model can extract effective feature vectors more accurately than the other three models, and make more accurate diagnosis of feature vectors. Moreover, with the increase of the number of samples, it can still maintain a high diagnostic accuracy. In the future work, the signal preprocessing method can also be further studied, so that the model can show better performance in a more complex environment.

References

[1] GUAN Yonggang, YANG Yuanwei, ZHONG Jianying, et al. (2018) Review on mechanical fault diagnosis methods for high voltage circuit breakers. High Voltage Apparatus, 54: 10-19.

[2] MEI Fei, MEI Jun, ZHENG Jianyong, et al. (2013) Application of Particle Swarm Fused KFCM and Classification Model of SVM for Fault Diagnosis of Circuit Breaker. Proceedings of the CSEE, 33: 134-141.

[3] CHEN Li, TIAN Shu. (2013) Application of fault dignosis in high voltage circuit breaker based on BP and RBF neural networ. Colliery Mechanical & Electrical Technology, 3: 8-11.

[4] Dong Di, Li Hengzhen, Wang Yunfei, et al. (2017) Condition monitoring and fault diagnosis of high voltage circuit breaker based on morphology wavelet and support vector machine. Electrical Measurement & Instrumentation, 54: 92-98.

[5] Lecun Y, Bengio Y. (1994) Word-level training of a handwritten word recognizer based on convolutional neural networks. In: International Conference on Pattern Recognition, Montreal. pp: 88-92.

[6] Abdi M.J., Giveki D. (2013) Automatic detection of erythemato squamous diseases using PSO-SVM based on association rules. Engineering Applications of Artificial Intelligence, 26: 603-608.

[7] YUAN Jinli, LI Kui, GUO Zhitao, et al. (2011) Mechanical failure diagnosis of high voltage circuit breaker based on SVM and opening/closing coil current parameters. High Voltage Apparatus, 47: 26-30.

[8] Sun J, Feng B, Xu W.B. (2004) Particle swarm optimization with particles having quantum behavior. In: Proceeding of Congress on Evolutionary Computation, Portland. pp: 326-331.

[9] CAO Yupeng, LUO Lin, WANG Qiao, et al. (2021) Fault diagnosis of high-voltage vacuum circuit breaker with a convolutional deep network. Power System Protection and Control, 49: 39-47.

Design and research of temperature measurement system for intermediate joint of 10kV power cable

Mengting Mao [1], Zhuan Liu [2], Rui Xu [3] and Longwei Wang [4]

[1]State Grid Jiangxi Electric Power Research Institute;Nanchang, Jiangxi;

[2]State Grid Jiangxi Electric Power Research Institute; Nanchang, Jiangxi;

[3]State Grid Jiangxi Electric Power Research Institute; Nanchang, Jiangxi;

[4]Beijing University of Aeronautics and Astronautics, communication engineering; Beijing

[*]Corresponding author's e-mail: maomt@jx.sgcc.com.cn

Abstracts:Cable intermediate joint is the weak link of power cable. Its operation determines the quality of power transmission and distribution in power system. The temperature rise of the intermediate joint is the direct manifestation of a variety of faults. Therefore, monitoring the temperature of the intermediate joint of power cable and finding and eliminating cable faults in advance is of great significance to ensure the safe and reliable operation of power system. Based on the analysis of the temperature characteristics of the cable intermediate joint, the design scheme of the temperature measurement system of the power cable intermediate joint is given, and its hardware system and software system are designed. Finally, the temperature measuring device for the intermediate joint of power cable is obtained, tested on the spot, and the power consumption of the data is analyzed. The debugging results show that the system has good characteristics and meets the design requirements.

1. Introduction

The defect of cable accessories is an important cause of cable failure. For long-distance transmission cables, it is bound to need more intermediate joints to connect all parts of cables together [1]. The manufacturing process of the intermediate joint requires high requirements. Once it is unqualified, parasitic resistance, inductance and capacitance will be generated, and long-term operation will lead to the failure of the intermediate joint [2-3]. At present, the manifestation of most faults is the change of temperature. Therefore, it is very necessary to design the temperature measurement system of power cable intermediate joint. At the same time, because the output power of the intermediate joint of the power cable is not high, the low power consumption of the temperature measurement system is also a very important index. In order to solve the above problems, this paper intends to use the finite element method [4-6] to calculate the joint temperature field, and carry out preliminary simulation analysis on the design of the system. On this basis, the main system of low-power cable intermediate joint is designed [7-8]. According to the design idea,f the main system complete of hardware system and software system, and finally make the physical device for testing[9]. The test results well meet the system design requirements.

Cable joint temperature change is a complex non-linear process, using the cable skin temperature to derive the core temperature is affected by the load current, ambient temperature, etc. The derivation of

Journal of Physics: Conference Series **2113** (2021) 012048 doi:10.1088/1742-6596/2113/1/012048

the joint core temperature is a multivariate nonlinear regression problem.

2. Temperature characteristic analysis of power cable intermediate joint

The calculation of the intermediate joint temperature of power cable can determine the temperature measurement range of the temperature sensor of the temperature measurement system, estimate the conductor temperature and determine the high temperature alarm value. The temperature field of the intermediate joint will be solved by using the finite element calculation software COMSOL to calculate the temperature distribution of the intermediate joint under different operating conditions.

2.1 calculation of temperature rise of power cable

The temperature characteristic of the intermediate joint of power cable determines the initial value setting of the subsequent system. The temperature measurement system needs to monitor the surface temperature of the intermediate joint, so it is necessary to study the variation range of the surface temperature under different operating conditions, so as to provide reference for the design of the temperature monitoring system. The surface temperature of the joint fluctuates in a certain range under different operating conditions. When the joint structure and material parameters are determined, the specific value is determined by the circulating current value of the joint and the ambient temperature. The higher the current, the higher the ambient temperature and the higher the surface temperature of the joint. For the temperature measurement system, two data are mainly needed: 1) the joint surface temperature of the cable under rated operation to determine the main working temperature range of the temperature sensor; 2) The maximum working temperature on the surface of the cable joint to determine the upper limit of the temperature measurement of the temperature sensor. After these two data are determined, the temperature measurement range of the temperature sensor is determined. The calculation of temperature rise of intermediate joint of power cable under two rated boundary conditions is analyzed below.

Set the boundary condition to 25 ° C air, 415 a current, this calculation condition simulates the rated operation of the connector, and the results are shown in Fig. 1 and Fig. 2.

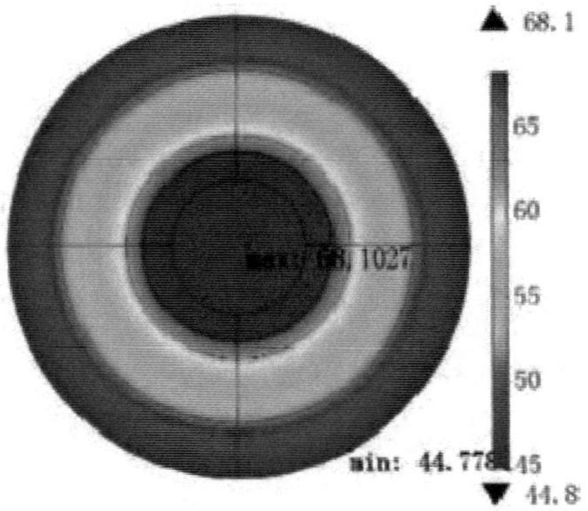

Fig. 1 steady state distribution

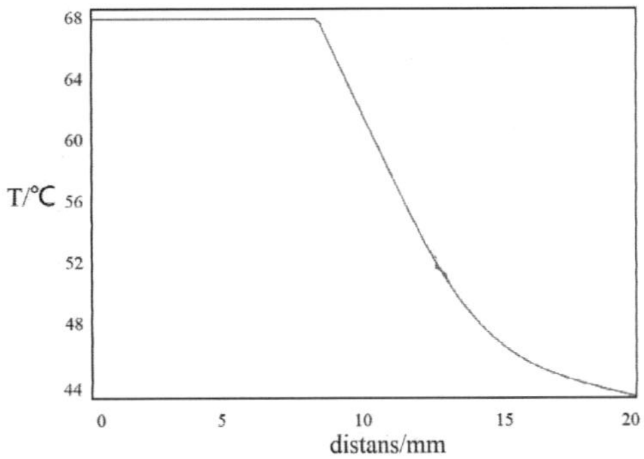

Fig. 2 radial temperature variation curve

It can be seen from the figure that under this condition, the conductor temperature of the joint is about 68.1 ° C. In the process of heat transfer from inside to outside, due to the different structural parameters of each layer, the temperature change rate of each layer is different, and the temperature finally transferred to the outer surface is 44.8 ° C. And communicate with the outside world 25 ° C natural convection of air occurs. Therefore, 45 ° C can be considered as the main working temperature of the joint surface.

Set the boundary condition to 40 ° C air, 250 ° C conductor, which simulates the maximum allowable conductor temperature in case of power cable short-circuit fault. The calculation results are shown in

Fig. 3 and Fig. 4. Figure 3 steady state distribution figure 4 radial temperature change curve shows that in case of short circuit, the outer surface temperature is about 136.4 ° C. This is the maximum value of the joint surface temperature. From the perspective of accommodating a certain margin, the upper temperature limit of the temperature sensor is 150 ° C is better. Since the temperature sensor also needs to measure the ambient temperature, the lower limit of temperature measurement can be - 30 ° C. Based on the above analysis, the temperature measurement range of the temperature sensor is - 30 ~ 150 ° C. It can ensure the full section detection of the temperature of the intermediate joint of the power cable.

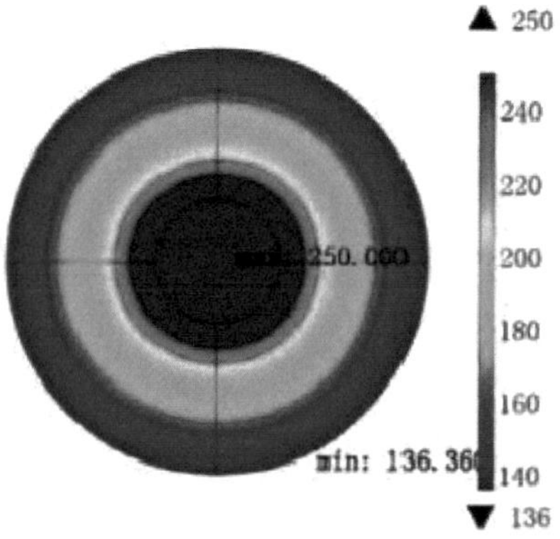

Fig. 3 steady state distribution

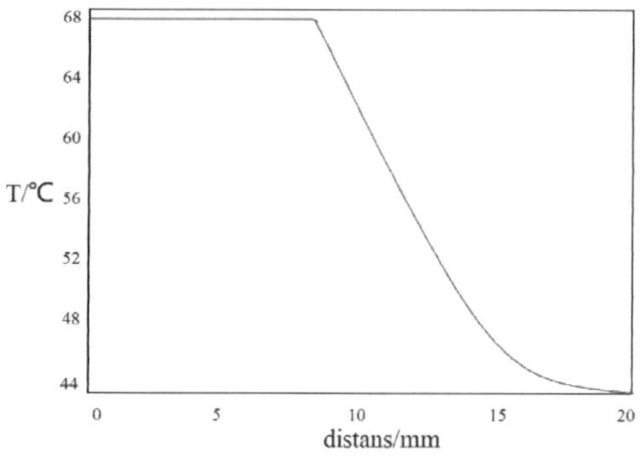

Fig. 4 radial temperature variation curve

2.2 Relationship between internal and external temperature rise of power cable and alarm value

A perfect temperature measurement system needs to establish a set of fault judgment basis in order to alarm in case of joint failure. GB / T 12706.2-2008 standard stipulates that the maximum conductor temperature of XLPE cable under normal operation is 90 °C ° C. So more than 90 ° C can be regarded as a fault. Because the conductor temperature is not easy to detect, this paper uses the method of estimating the internal conductor temperature by the external temperature rise to determine the temperature alarm value.

The surface temperature and ambient temperature of the joint are easy to detect, so the conductor temperature can be set to 90 ° The joint surface temperature at C is determined as the high temperature alarm value. When the measured surface temperature exceeds this value, the conductor temperature must be higher than 90 ° C. This is the fault state. ninety ° The conductor of C can be obtained by adjusting the current. To 40 ° C air as an example, conductor 90 ° The steady-state temperature field distribution of the joint at C is shown in Fig. 5 and Fig. 6.

As can be seen from the figure, at 40 ° C in air environment, the surface temperature of the joint is about 62.94 ° C。 When the calculation method remains unchanged, different joint surface temperatures can be calculated only by changing the ambient temperature. The high temperature alarm value can be a little higher than the corresponding joint surface temperature.

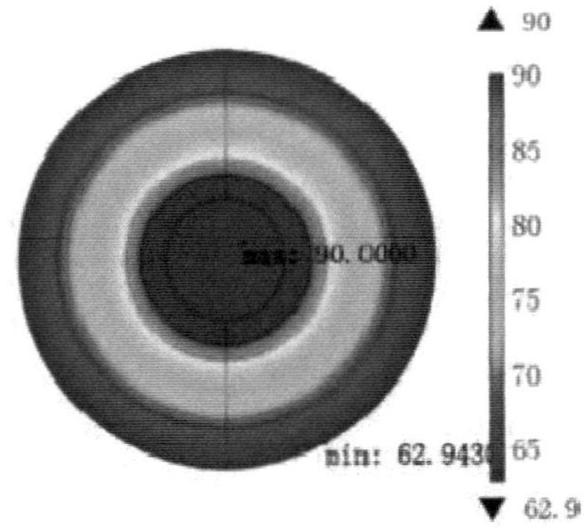

Fig. 5 steady state distribution

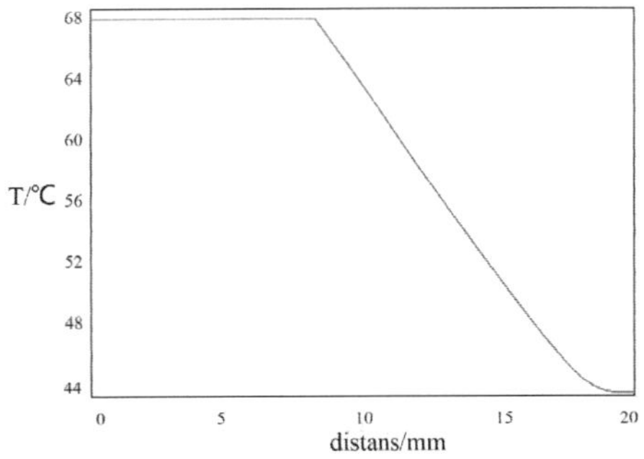

Fig. 6 radial temperature variation curve

It is difficult to measure the conductor temperature directly, so this paper uses the estimation method to estimate the temperature of the internal conductor by using the conductor surface temperature. This paper will solve the problem of conductor temperature estimation from the perspective of temperature field calculation. Let the current flowing through the conductor be I, the ambient temperature be x, the surface temperature be y, and the conductor temperature be Z. Both conductor temperature and surface temperature are determined by the internal heat source and ambient temperature, and the heat source is generated by the thermal effect of current. Therefore, when the current and ambient temperature are known, the conductor temperature and surface temperature are determined, and there must be a corresponding functional relationship between them, which is as follows:

$$(z,y)=f_0(I,x)z=f(x,y) \tag{1}$$

If the corresponding conductor temperature and surface temperature can be obtained by selecting several sample values of current and ambient temperature, and then these data can be analyzed to fit the functional relationship:

$$z=f(x,y) \tag{2}$$

Then the corresponding conductor temperature can be calculated according to the ambient temperature and surface temperature data detected by the sensor. The conductor temperature estimation process is shown in Fig 7.

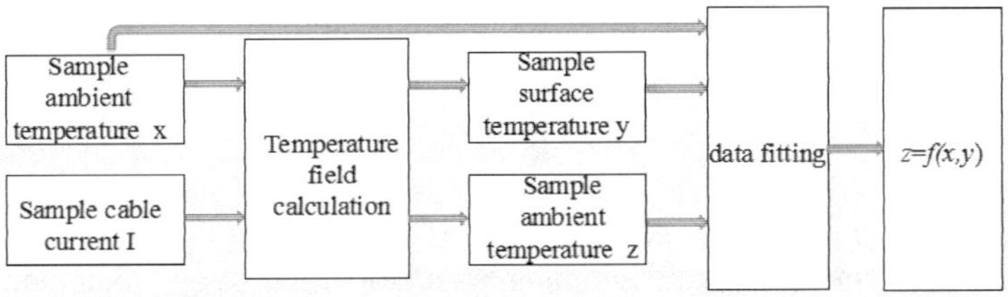

Fig. 7 Conductor temperature estimation process

3. System Design

According to IEC61850 standard of substation network and communication protocol (hereinafter referred to as IEC61850), this paper puts forward the concept of information layering in substation, and divides the communication system of substation into three parts: process layer, bay layer and station control layer. The designed temperature measurement system is mainly composed of temperature measurement terminal, aggregation terminal and monitoring station, which constitutes a hierarchical management communication system. Wireless data communication is adopted between temperature measurement terminal and aggregation terminal. Coordination between temperature measurement terminal and convergence terminal work, obey the unified dispatching of the monitoring station, so as to realize the all-round monitoring of the intermediate joint temperature. The overall structure design of the system is shown in Fig 8.

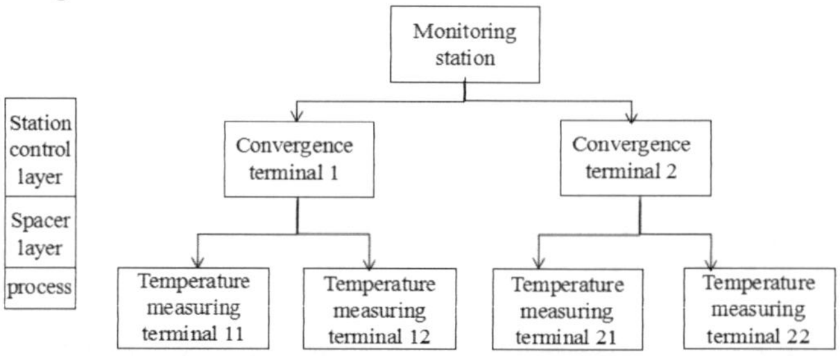

Fig.8 Conductor temperature estimation process

The temperature measurement terminal is located in the process layer and installed at the middle joint of the cable. Therefore, it is difficult to obtain power from the temperature measurement terminal, which is the part of the temperature measurement system that really needs to realize low-power design. The temperature measurement terminal needs to collect temperature data, then transmit the data to the aggregation terminal and receive the control signal of the aggregation terminal. According to this design, the temperature measurement terminal is mainly composed of controller, wireless communication circuit, temperature sensor and power supply.

The commissioning of the system is carried out in the substation of a power supply company of the State Grid. The commissioning objects are 5 temperature measurement terminals, 1 convergence terminal and 1 monitoring station. The temperature measurement terminal is installed at the middle joint of 10 kV cable, and the address is set to 0 ~ 4. The convergence terminal is installed in the weak current well, and the monitoring station is a computer. In terms of temperature measurement accuracy, due to the limitation of the operating environment, it is impossible to verify the precision temperature measurement equipment on site. Here, fluke infrared temperature measuring gun in the substation is used as the temperature measurement accuracy detection. The ambient temperature measured by infrared temperature measuring gun on site is 19 ° C. The temperature of each joint is mainly concentrated at 39 °C ° C. It is basically consistent with the temperature displayed by the aggregation terminal. In terms of communication distance, the field test results show that the wireless communication distance is about 230 m at 0 d BM transmission power and 360 m at 10 d BM transmission power, which meets the system design index

4. Conclusion

By analyzing the temperature characteristics of the intermediate joint of power cable, this paper gives the approximate relationship between the internal and external temperature at the joint, and determines the alarm temperature and related data; the design scheme of low-power cable intermediate joint temperature measurement system is given, and the hardware and software design of the system are completed. Finally, the physical test device is completed and tested on the spot. The results show that the temperature measurement system can accurately measure the temperature of power cable intermediate joint, and can feed back the corresponding information in real time to meet the system design requirements.

References

[1] Harry Orton. Overview of power cable technology (English) [J]. High voltage technology, 2015, 41 (04): 1057-1067.
[2] Wang Yani, Zhang Hongliang, Wu Jiandong, et al. Simulation calculation of temperature field and electric field of HVDC cable under different laying methods [J]. Insulating materials, 2017, 50 (07): 71-78.

[3] Zhang Lei, Zhu Lixiang, Chen Qingyun. Simulation analysis of temperature field of single core cable with different frequency harmonics [J]. Hubei electric power, 2016, 40 (03): 45-51.
[4] Tang Ke, Wen Wu, Ruan Jiangjun, et al. Simulation of temperature field of single core cable based on finite element method [J]. Journal of Wuhan University (Engineering Edition), 2018, 51 (09): 811-816.
[5] Xie guangbin, Xie Wei. Study on ampacity of single core cable based on temperature field [J]. Sichuan electric power technology, 2010, 33 (01): 86-88.
[6] Xing Ya, Hou Zhe, Yang Yuqi, et al. Numerical calculation of transient temperature field of single core cable based on distributed optical fiber temperature measurement system [J]. Wire and cable, 2013 (06): 7-11.
[7] Jiang Hua, Li Xizhong, Qin Xiaolai, et al. Power cable temperature measurement device based on passive wireless sensing technology [J]. Instrument technology and sensors, 2018431 (12): 62-66.
[8] Gao Qiang, Wang Hongli, Liu Xin, et al. Design of wireless temperature measurement and early warning system for high voltage substation [J]. Instrument technology and sensors, 2009 (01): 105-107 + 110.
[9] Luo Junhua, Zhou zuochun, Li Huachun, et al. Application Research on on-line detection technology of operating temperature of power cable [J]. High voltage technology, 2007 (01): 169-172.

Journal of Physics: Conference Series **2113** (2021) 012049 doi:10.1088/1742-6596/2113/1/012049

Linear Active Disturbance Rejection Control based on Finite Impulse Response filter for harmonic control in microgrid

Xiong Yang[1], Xuhui Zhang[2*] and Jiahao Guo[3]

[1] Distribution Network Technology Department, State Grid Jiangsu Electric Power Co., LTD. Research Institute (JSEPRI), Nanjing, Jiangsu, 211103, China

[2] Graduate School and Mechanical Engineering, Nanjing Institute of Technology, Nanjing, Jiangsu, 211167, China

[3] Distribution Network Technology Department, State Grid Jiangsu Electric Power Co., LTD. Research Institute (JSEPRI), Nanjing, Jiangsu, 211103, China

[*]e-mail: y00450200412@njit.edu.cn

Abstract: Island for micro grid is vulnerable at the run time, which is caused by the problem of load fluctuation, voltage frequency deviation, and distributed power supply power. These issues are generated by the uncertainty of three-phase voltage waveform output at the same time. All of these problems are urgently needed to solve, this paper designs a linear active disturbance rejection control to do robust estimation. This controller is based on optimal finite impulse response (FIR) filter, which is also combined with the immunity of the micro grid system harmonic method. The proposed controller is simulated. And it is compared with the classical PI controller and the linear active disturbance rejection controller respectively too. The simulation results show that the performance and robustness of the proposed model are better than those of traditional algorithm in microgrids.

1. INTRODUCTION

With the continuous progress and development of China's new energy science and technology, while the improvement of people's awareness of environmental protection, the microgrid based on new energy has been widely applied and promoted. As an important link of new energy connection load, inverter is also an important topic in the research of new energy power generation technology [1].

In the important historical process of rapid transformation from China's traditional intelligent distribution network to modern intelligent distribution network, and then to modern intelligent distribution network, micro-grid will play an important role in promoting the continuous progress and development of China's intelligent distribution network. The microgrid management system is mainly a small, medium-scale and large-scale power generation and distribution grid system that generally refers to a collection of various distributed on-grid power sources, energy storage power generation equipment, loads and other grid data systems. The autonomous power grid system can be connected to the grid internally and operate independently at the same time. It is a provincial autonomous power grid system that can completely independently realize the functions of self-operation control and management service. The extensive existence of microgrid not only changed the basic network operation structure, but also system structure of China's traditional large power transmission and distribution network. The continuous updating of network topology and structure will have a great direct impact on the development of China's traditional large power transmission and distribution network.

From a macro overall research perspective, microgrid integrates the distributed micropower supply, load, energy consumption storage devices and other technologies, and forms an overall control of micropower network system through the connection, control of inverters, which provides customers with large capacity of active and reactive power. Not only can the grid-connected inverter of the public grid be connected to the grid-connected operation of the public grid, but also can be disconnected from the inverter in the public grid in the case of accidents or maintenance of the public grid[2] .

Domestic and foreign scholars have carried out a lot of research on microgrid, it is constantly improved. In literature [3], the microgrid parallel multi-inverter is controlled, and its output power, output frequency will change with the change of active power and reactive load. So, the stable operation of the microgrid parallel multi-inverter is realized, and great progress has been made in the research on the stable operation control of the microgrid parallel multi-inverter. In literature [4], FIR filter and Kalman filter are used to design a method of robust estimation of harmonics in power system by using optimal finite impulse response (FIR) filter, which is applied to the state space representation of noisy current or voltage signals to estimate the amplitude of harmonic components. The main purpose of literature [5] is to combine the droop control and active disturbance rejection controller organically and closely in China's microgrid. By compensating disturbance in advance, the control effect is improved, at the same time, the simulation results show that it can effectively suppress the output error of the system.

2. MICROGRID MODEL BASED ON DROOP CONTROL

2.1 Droop control principle

The so-called low-power droop control, that is, droop control, also refers to the one-time low-power frequency modulation control principle of my country's traditional power system (such as wind turbines) to perform a variety of power electronic devices (such as inverters) and other equipment. A kind of low power control. In this low power control principle, the active and reactive power output by the inverter can be decoupled separately, it based on the relationship between active power and frequency, the low power based on the relationship between reactive power and voltage. The power droop characteristic is realized as a curve. Automatic regulation and control of inverter[6].

In Figure 1, the equivalent calculation of a microgrid power supply is transformed into a precise and simplified calculation model of a type containing a type of micropower, which supply to facilitate our analysis and design of the basic characteristics of various types of micropower sources , their power and signal transmission located within the microgrid.

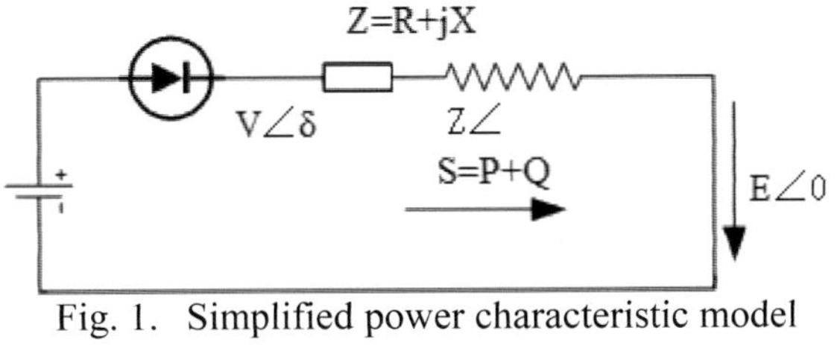

Fig. 1. Simplified power characteristic model

Z and θ respectively represent the current amplitude ($Z=\sqrt{R^2+X^2}$) and impedance Angle ($\theta=\arctan(X/R)$) of the equivalent voltage impedance of the input and output of the AC side of the bus of an inverter in a large microgrid, V is the current amplitude of the equivalent voltage impedance of the input, output of the AC side of the bus and the grid side of an inverter; E is the impedance amplitude of the equivalent AC voltage of the two inputs at the connection point of the bus of the public line. δ is the amplitude vector of the equivalent voltage impedance of two inputs and outputs on the AC side of the bus and the grid side of an inverter and the phase Angle difference of the equivalent voltage impedance of two inputs and outputs at the connection between the AC power of the common bus and the bus.

According to the electrical theory, the output complex power of inverter AC side can be expressed as:

$$\begin{aligned} S &= P + jQ = \dot{V}\dot{I} = V\angle\delta\left[\frac{V\angle\delta - E\angle 0}{Z\angle\theta}\right] = Ve^{j\delta}\left[\frac{Ve^{j\delta} - Ee^{j0}}{Ze^{j\theta}}\right] \\ &= Ve^{j\delta}\left[\frac{V}{Z}e^{j(\delta-\theta)} - \frac{E}{Z}e^{j(-\theta)}\right] = Ve^{j\delta}\left[\frac{V}{Z}e^{-j(\delta-\theta)} - \frac{E}{Z}e^{-j(-\theta)}\right] \\ &= \frac{V^2}{Z}e^{j\theta} - \frac{EV}{Z}e^{j(\theta+\delta)} \\ &= \frac{V^2}{Z}(\cos\theta + j\sin\theta) - \frac{EV}{Z}[\cos(\theta+\delta) + j\sin(\theta+\delta)] \end{aligned} \tag{1}$$

According to the formula, the active power and reactive power output on the AC side of the inverter are respectively:

$$P = \frac{V^2}{Z}\cos\theta - \frac{EV}{Z}\cos(\theta+\delta) \tag{2}$$

$$Q = \frac{V^2}{Z}\sin\theta - \frac{EV}{Z}\sin(\theta+\delta) \tag{3}$$

Due to impedance Angle $\theta = \arctan(X/R)$,so $\sin\theta = X/Z, \cos\theta = R/Z$

$$P = \frac{V}{R^2 + X^2}[R(V - E\cos\delta) + XE\sin\delta] \tag{4}$$

$$Q = \frac{V}{R^2 + X^2}[X(V - E\cos\delta) + RE\sin\delta] \tag{5}$$

In the general low frequency and high voltage AC transmission signal line, the AC impedance of the line is usually measurable and sensitive, is due to X>>R, so in the case of the value of R impedance can be fully automated to ignore, therefore, the current phase radian error between the vectors E and V that output high-frequency voltage downward at the AC side of the line and the vectors V and E that output the AC voltage upward on the side of the common line bus is very small, so in radians we're going to have approximately $\sin\delta = \delta, \cos\delta = 1$, so this simplifies to:

$$\delta = \frac{XP}{EV} \tag{6}$$

$$V - E = \frac{XQ}{V} \tag{7}$$

According to the analysis of, the transmission of active power of inverter mainly depends on the power angle δ, the transmission of reactive power depends mainly on the voltage difference（U1-U2）. The mathematical equation of droop control is obtained:

$$\begin{cases} f = f^* - m(p - p^*) \\ V = V^* - n(Q - Q^*) \end{cases} \tag{8}$$

The droop control characteristics described can be described in the figure shown:

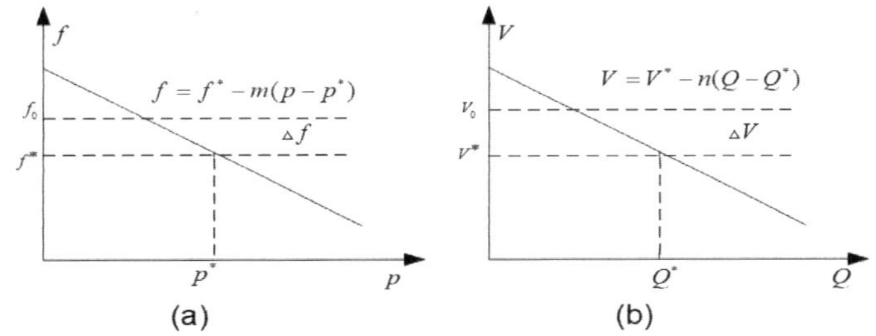

Fig. 2. Droop control characteristic diagram

Given a reference output power threshold based droop when the function of an essential meaning of the control strategy can be as follows: the AC inverter of each load according to the reference output signal based on the reference input power threshold requirements for decoupling, then according to the load of each load droop angle movement characteristics curve, to each load based on the reference output power threshold requires a reasonable power distribution, the signal of each AC inverter control circuit and so will be automatically calculated to get its computations based on reference output parameter values, and then through the control circuit of the signal conversion control module, it is concluded that a computations based on reference output power requirement signal parameter values, then hang through each stage of the inverter load movement In order to achieve the uniform balance of the output power reference requirements of each load, the ac side board of each inverter can finally make the reference power requirement of the output power signal completely reach the rated power reference required to be set. As shown in the figure, FIG. 2(a) refers to the line determined by the m in the slope and the reference power P_0 , the reference frequency f_0, while FIG. 2(b) refers to the line determined by the n in the slope and the reference power Q_0 , the reference voltage V_0, with its slope m, n and reference points can be directly used to adjust the actual operating position of each distributed universal micropower supply by a new dynamic calculation method, determine how to correctly arrange each inverter according to the characteristics of various low vertical curve and the structure characteristics of the high plane curve after decoupling.

2.2 Dq transform

In the process of experiment, because the voltage mathematical model is analyzed in the time domain, it is time varying, which makes it particularly inconvenient to design the inverter control technology. Therefore, it is necessary to transform the three-phase stationary (A-B-C) coordinate system into the synchronous rotation of the grid fundamental frequency two-phase (D-Q) coordinate system by coordinate transformation. In this way, the fundamental wave sine variable in the three-phase A-B-C coordinate system is transformed into the dc component in the synchronous rotation D-Q coordinate system by coordinate transformation, and the design of the inverter controller can be realized [7].

Before the space of the two coordinate system transformation, according to the regulation, make this condition in three phase ac circuits, the two different (d or q) axis in the coordinate system of the two d and q axis space respectively and their corresponding active component and its reactive component, among them, for the dq axis rotation angular velocity, counterclockwise as the positive direction of rotation, q axis ahead of d axis 90 degrees.

From the equation of the phasor of the transformation matrix of the two coordinates, it can be analyzed that the three phase motion static (A-B-C) coordinate system is transformed into each other in order to obtain the transformation matrix of the two phase static rotation (D-Q) coordinate system $T_{3d/3s}$ can be expressed as:

$$T_{3d/3s} = \frac{2}{3}\begin{pmatrix} \cos(\omega t) & \cos(\omega t - 2\pi/3) & \cos(\omega t + 2\pi/3) \\ -\sin(\omega t) & -\sin(\omega t - 2\pi/3) & -\sin(\omega t + 2\pi/3) \\ 1/2 & 1/2 & 1/2 \end{pmatrix} \tag{9}$$

The inverse matrix of formula 10 is

$$T^{-1}{}_{3d/3s} = \frac{2}{3}\begin{pmatrix} \cos(\omega t) & -\sin(\omega t) & 1 \\ \cos(\omega t - 2\pi/3) & -\sin(\omega t - 2\pi/3) & 1 \\ \cos(\omega t + 2\pi/3) & -\sin(\omega t + 2\pi/3) & 1 \end{pmatrix} \quad (10)$$

Due to the main working principle of three-phase inverter is mainly three-phase ac circuit of three-phase input current evenly balanced and symmetrical each other, at the same time, based on the work of three-phase loop current balance can also be used for three-phase coordinate system before and after the transformation of the working principle of power conservation, shaft three-phase current conservation vector expression can be used to get two phase current of d - q vector of conservation of shaft current Can be expressed by formula 11 and 12

$$\begin{bmatrix} i_d \\ i_q \end{bmatrix} = \frac{2}{3}\begin{bmatrix} \cos(\omega t) & \cos(\omega t - 2\pi/3) & \cos(\omega t + 2\pi/3) \\ -\sin(\omega t) & -\sin(\omega t - 2\pi/3) & -\sin(\omega t + 2\pi/3) \end{bmatrix} \quad (11)$$

$$\begin{bmatrix} u_d \\ u_q \end{bmatrix} = \frac{2}{3}\begin{bmatrix} \cos(\omega t) & \cos(\omega t - 2\pi/3) & \cos(\omega t + 2\pi/3) \\ -\sin(\omega t) & -\sin(\omega t - 2\pi/3) & -\sin(\omega t + 2\pi/3) \end{bmatrix} \quad (12)$$

Thus, the voltage and current vector equation of the three-phase inverter in the microgrid in the two-phase rotation (D-Q) coordinate system is expressed as formula 13.

$$\begin{cases} u_d = Ri_d + L\frac{di_d}{dt} + \omega Li_q + u_{od} \\ u_q = Ri_q + L\frac{di_q}{dt} - \omega Li_d + u_{oq} \\ i_d = C\frac{du_{od}}{dt} + \omega Cu_{oq} + i_{od} \\ i_q = C\frac{du_{oq}}{dt} - \omega Cu_{od} + i_{oq} \end{cases} \quad (13)$$

2.3 Design of microgrid based on droop control

The control system structure of the isolated island microgrid designed in this paper is shown in figure 3. In order to effectively simplify the circuit model, the distributed load power supply model can be directly replaced by the DC load voltage source. The DC load power supply Udc can be connected to the three-phase AC load power supply R after the three-phase AC inverter and LC filter constitute a circuit. Output voltage value of the actual work of U0, current I0 after calculate the load power values of P, Q, droop control and park transformation to obtain the voltage reference Ud *, Uq *, the actual voltage value is obtained by extended state observer (ESO voltage measurements, the voltage reference by differential tracker TD do smoothness processing compared with voltage measurements, the error signal is used as the voltage signal of the inverter through the inner ringlinar active disturbance rejection controller.

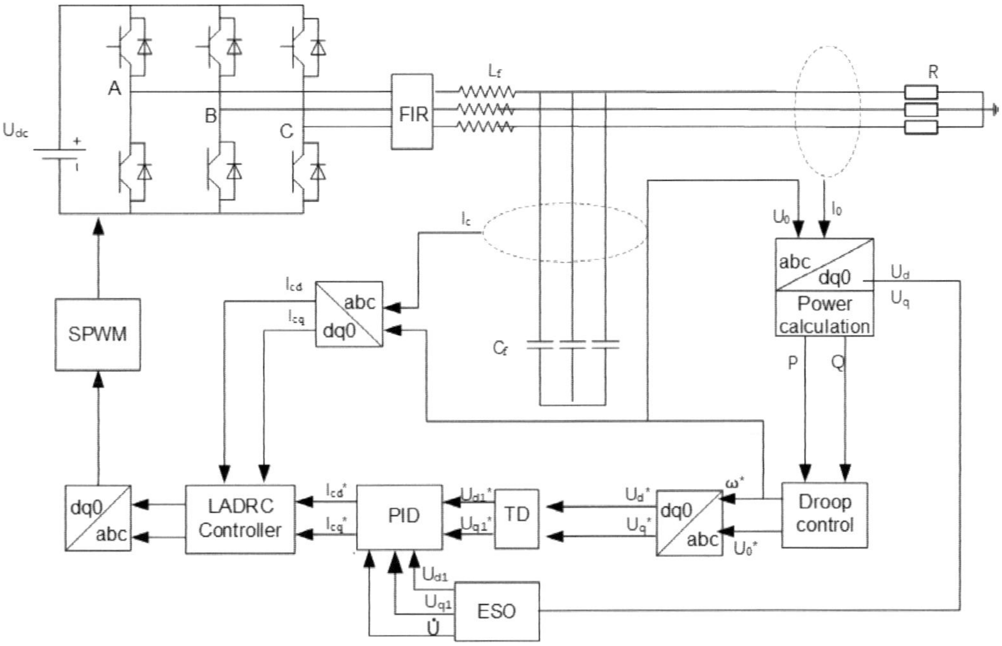

Fig. 3. System structure diagram

3. Design of controller based on active disturbance Rejection and FIR Filter

3.1 TD

In a general controller system, the error is obtained by the difference between the set value and the system input, which can cause the initial error to be very large, which can easily cause overshoot. In order to solve the problem of the speed and overshoot of the motor system, we reduce the initial error in the initial state. Since the initial output of the control system is generally zero, we arrange the transition process to the initial target value, which is what TD does. The discrete TD is shown in formula (14).

$$\begin{cases} x_1(k+1) = x_1(k) + h \cdot x_2(k) \\ x_2(k+1) = x_2(k) + h \cdot fhan(x_1(k) - x_d(k), x_2(k), r, h_0) \end{cases} \tag{14}$$

In formula (14), h Is the integral step, h_0 is a new variable independent of h, it has a certain filtering effect on noise, k is the sampling time, $x_1(k)$ is the transition process of the reference voltage, $x_2(k)$ is the derivative of the transition process, $fhan(x_1(k) - x_d(k), x_2(k), r, h0)$ is a nonlinear function, the expression is shown in formula (15).

$$fhan = -\begin{cases} r \cdot sign(a), & |a| > d \\ r \cdot \dfrac{a}{d}, & |a| \le d \end{cases} \tag{15}$$

In formula (15), r is the speed factor. Within the bearing capacity of the controlled object, the greater the value, the faster the speed of approaching the target value. The definition a of and d is shown in the following formula：

 2113 (2021) 012049 doi:10.1088/1742-6596/2113/1/012049

$$\begin{cases} d = r \cdot h_0 \\ d_0 = h_0 \cdot d \\ y_0 = x_1(k) - x_d(k) + h_0 \cdot x_2(k) \\ a_0 = \sqrt{d_2 + 8r \cdot | y_0 |} \\ a = \begin{cases} x_2(k) + \dfrac{a_0 - d}{2} \cdot sign(y_0), & | y_0 > d_0 | \\ x_2(k) + \dfrac{y_0}{h_0}, & | y_0 \leq d_0 | \end{cases} \end{cases} \quad (16)$$

3.2 ESO

Considering the inaccuracy of system modeling and the unpredictability of renewable energy output, an effective method is to treat system uncertainties as disturbances. If the sum effect of various disturbances can be estimated and their compensation is eliminated at the input end of the system, the system will be simplified as an integral series system [8]. The discrete ESO is shown in formula (17):

$$\begin{cases} e = z_1(k) - y(k) \\ z_1(k+1) = z_1(k) + h \cdot (z_2(k) - \beta_{01} \cdot e) \\ z_2(k+1) = z_2(k) + h \cdot (z_3(k) - \beta_{02} \cdot fal(e, 0.5, \delta) + b \cdot u(k)) \\ z_3(k+1) = z_3(k) - h \cdot \beta_{03} \cdot fal(e, 0.25, \delta) \end{cases} \quad (17)$$

In formula (17), h is sampling period, k is the kth sampling time, b is observer gain, u is control output value, β_{01} 、 β_{02} 、 β_{03} are observer parameters, z_1 、 z_2 、 z_3 are state variables, among them z_3 is real-time estimation of internal uncertainties (such as parameter perturbations) and external disturbances of model objects, nonlinear fal function expression is shown as follows.

$$fal(e, \alpha, \delta) = \begin{cases} e \cdot \delta^{\alpha - 1}, & | e | \leq \delta \\ | e |^{\alpha} \cdot sign(e), & | e | > \delta \end{cases} \quad (18)$$

In formula (18), α is a constant between 0 and 1, $\delta = j \cdot h$, j is a positive integer, fal function is one essential meaning is to "big error, small gain" in quality control system engineering; small error, large gain "set mathematics for analog number fitting [9].

3.3 Design of FIR filter based on MATLAB

As at present in the mobile industry in the process of the information age of digital, intelligent, and push the "mobile network construction process, the digital filtering processing also gradually mature development has become a very widespread extremely important technical subject and professional research hotspot, and digital filtering processing technology is widely used in the application in the field of the major also one by one gradually has been more local attention and wide recognition. Compared with IIR type digital phase filter, FIR filter has made great progress in China with its very high phase linearity and phase frequency advantage. The hardware design of the digital signal filter can generally be realized directly by two simple control methods of direct control hardware or software. The main technical advantage of the latter is that the user can also directly change the digital filter to manual control through the hardware by adjusting the various parameters of the signal digital filter of signal processing performance of the whole, and for the MATLAB The need to provide all kinds of digital signal filter processing control toolbox is because it is a completely is a powerful digital signal processing function control function, it not only can not only directly corresponding used for control and automated design all kinds of digital signal filter, even big enough to directly make the software design staff can do the biggest worry, so it has become to FIR and other digital signal filter applications need to design a powerful control means.

3.3.1 FIR filter

FIR digital differential filter system is a nonlinear recursive function system, in which the impulse response function H (n) can be expressed as a finite length of function sequence, and the general expression form of each sequence function in the recursive system is:

$$H(z)=\sum_{k=0}^{N-1} h(k)z^{-k} \tag{19}$$

In the above equation, h(n) is a causal sequence, H(z) is a polynomial of degree N-1 of z1, with poles of order N-1 only at Z =0, no poles elsewhere, only any pole position of any N-1 zeros in a finite region Z plane. The basic structure of FIR system is direct type and cascade type. It has been widely used in the acquisition and processing of various radar transmitted signals, as correlator and radar reverser at the same time.

From the above equation, we can get an implementation framework structure of FIR filter in FPGA, as shown in figure 4, which mainly includes three parts: delay unit Z^{-1}, multiplier and accumulator. This structure is a direct type of FIR filter structure, also known as the transverse structure.

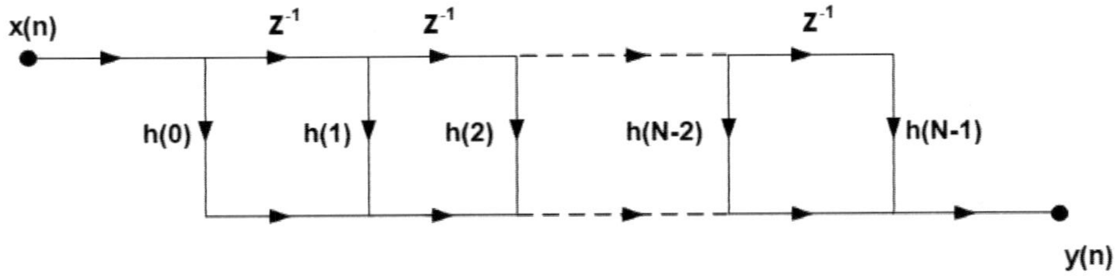

Fig. 4. FIR internal structure

The design process of digital filter is:

1) According to the actual situation, the performance requirements of various filters are determined. Usually, this method is to put forward a performance requirement for a given digital filter in the signal frequency domain, which is generally the amplitude and phase response, namely its technical index.

2) In demand in the filter need to find a can satisfy the given default filtering performance requirements of very short time discrete filter of linear system, which is a very great stability of causal mathematical system using the method of functional group and fast approaching a preset performance requirements based on the requirement of the given filter, filter coefficient of the linear order to define the demand.

3) The system is implemented with finite precision. This includes a selection operation structure that quantifies the filter's coefficients, input variables, intermediate variables, and output variables to a fixed word length.

4) Through simulation, their frequency characteristics and phase characteristics are analyzed to verify whether the system we need to design meets the given performance.

3.3.2 FIR design based on filterDesigner

After setting the key indicators, enter the FDATOOL operation interface, and set the parameters of the designed filter in the interface: select Lowpass in the Filter Type; select FIR (Window) in Design Method; in Filter Order, select the specified order and select Kaiser window. In Frequency Specification, set the Frequency unit to Hz, and set the sampling Frequency to Fs=30.72 MHz, Fc= 8M Hz. After setting the parameters, click the "Design Filter" button to obtain the amplitudefrequency characteristic curve as shown in the figure.

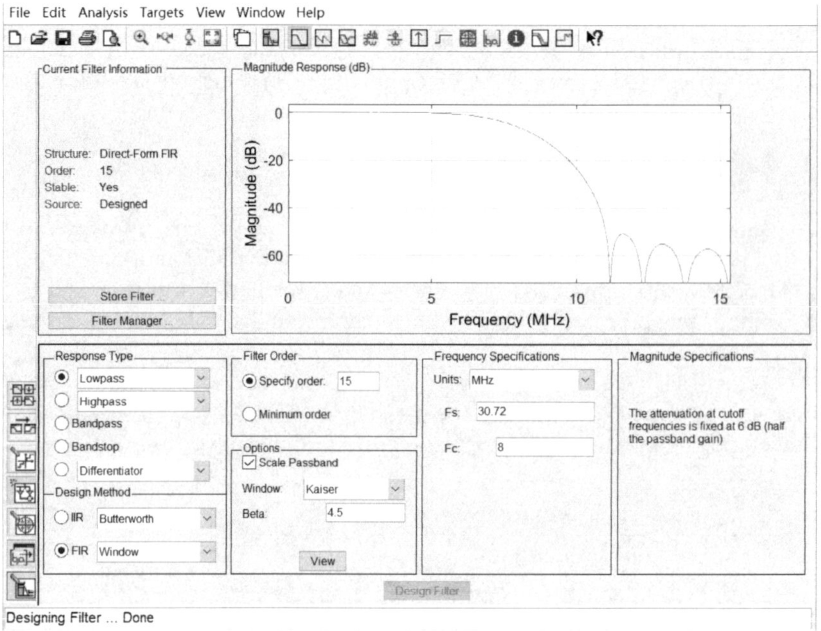

Fig. 5. FIR parameter setting

Using Matlab provided by the Simulink simulation tool box to achieve the characteristics and performance of the filter. Press the right mouse button in the Matlab command window and type Simulink to enter the simulation interface to create a new model workspace.

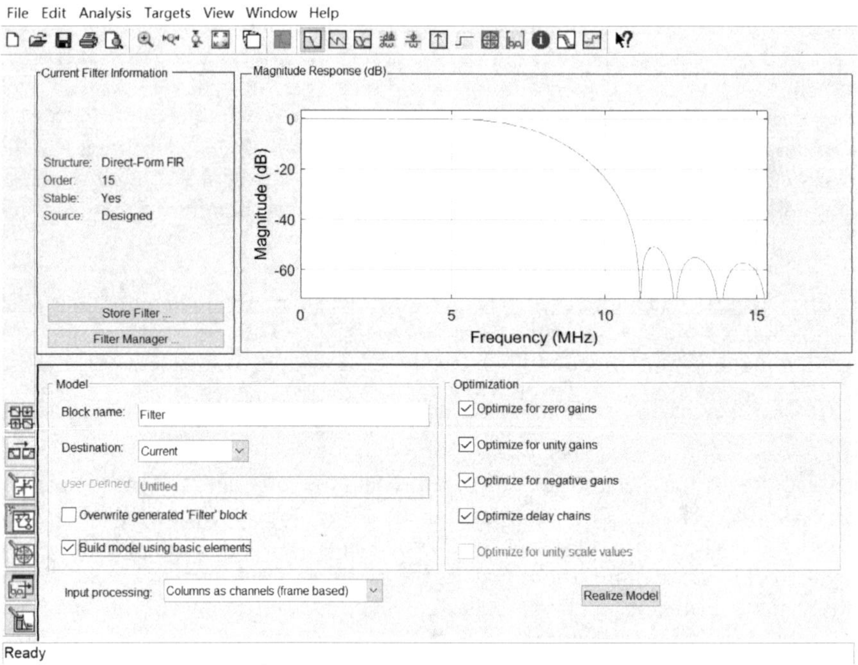

Fig. 6. Model interface to establish FIR Model

Two different simulation signal source jointly produced by the two analog signal filter frequency to 2 hz, 50 hz, respectively, in accordance with the design drawing 7 good even after two cables to filter movement simulation, the circuit using oscillograph carries on the simulation test before filtering after two basic structure and the waveform of the simulation signal source, after two of the same design analog signal source with a filter. The basic structure of the waveform of the two analog signal sources calculated after the simulation is shown in figure 8 below.

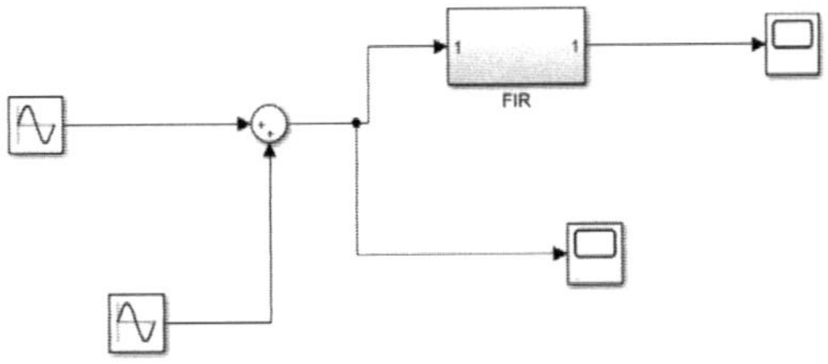

Fig. 7. FIR simulation model

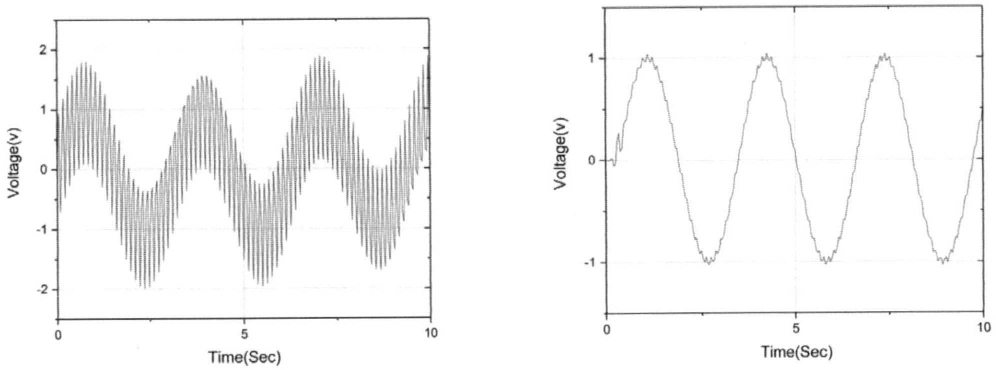

Fig. 8. Comparison of effect pictures before and after filtering

4. TIME DOMAIN SIMULATION RESULTS ANALYSIS

In this section, the isolated island microgrid model is built on the Matlab/Simulink simulation platform, the performance of the designed controller based on active disturbance rejection and FIR is studied when the load is switched off and the disturbance is added, then compared with the traditional PI controller and LADRC. Among them, the simulation parameters of PI, LADRC are shown in Table 1 and Table 2 respectively, and the FIR parameters are described in 3.3. The voltage regulator input, output filter load voltage on the DC side of the inverter is 800v, the AC output input filter load capacitor of the inverter is 1500 μ F, the filtered load inductor is 0.6mh, the droop power coefficient m is 1e-7 and n is 3e-4.

TABLE I. PI CONTROLLER PARAMETERS

parameters	value
kp	10
ki	100

TABLE II. LADRC CONTROLLER PARAMETERS

parameters	value	parameter	value
h	0.01	β_{02}	50
$h0$	1	β_{03}	100
r	10	α_1	0.5
b	19	α_2	0.25
β_{01}	150	δ	0.0025

TABLE III. FOPID CONTROLLER PARAMETERS

parameters	value
Kp	180
Ki	3.6532
Kd	8.9811
λ	0.3
μ	0.7

4.1 Comparison of switching load simulation

For the convenience of comparison, only two distributed power supplies are simulated, and the simulation time is 2s. One load is cut off at 0.3s, restored at 0.6s. Figure 9 and 10 are the comparison diagram of frequency, voltage changes during the operation of load switching and switching respectively. As can be seen from Figure 9, the FIR controller based on active disturbance rejection reduces the frequency to 50HZ with the minimum oscillation and the shortest time, while the accuracy of PI regulator is significantly worse than that of the combination of LADRC and FIR controller. However, under the condition of no interference, the FIR response based on active disturbance rejection is basically consistent with that of active disturbance rejection.

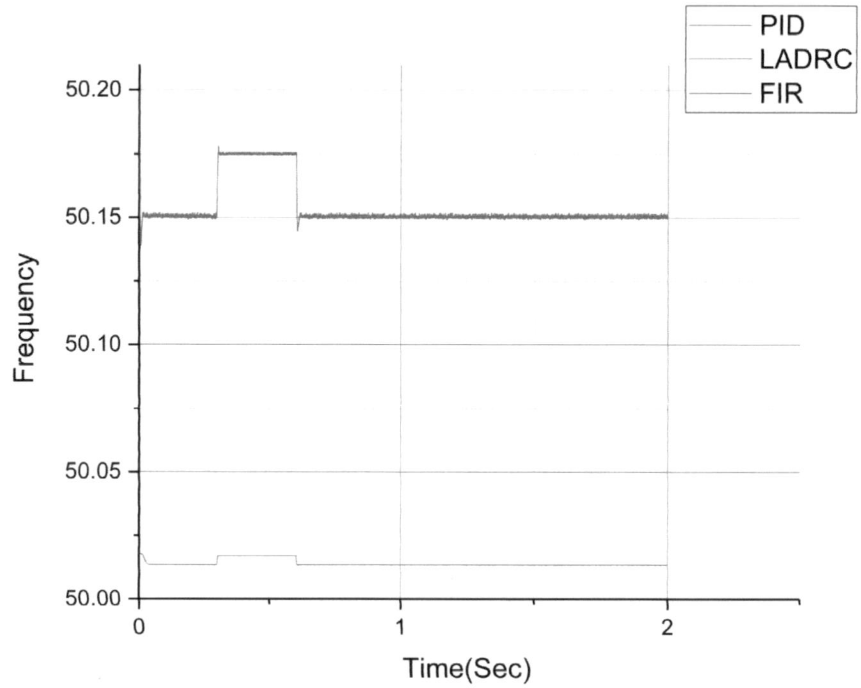

Fig. 9. Frequency response of the three controllers to load changes

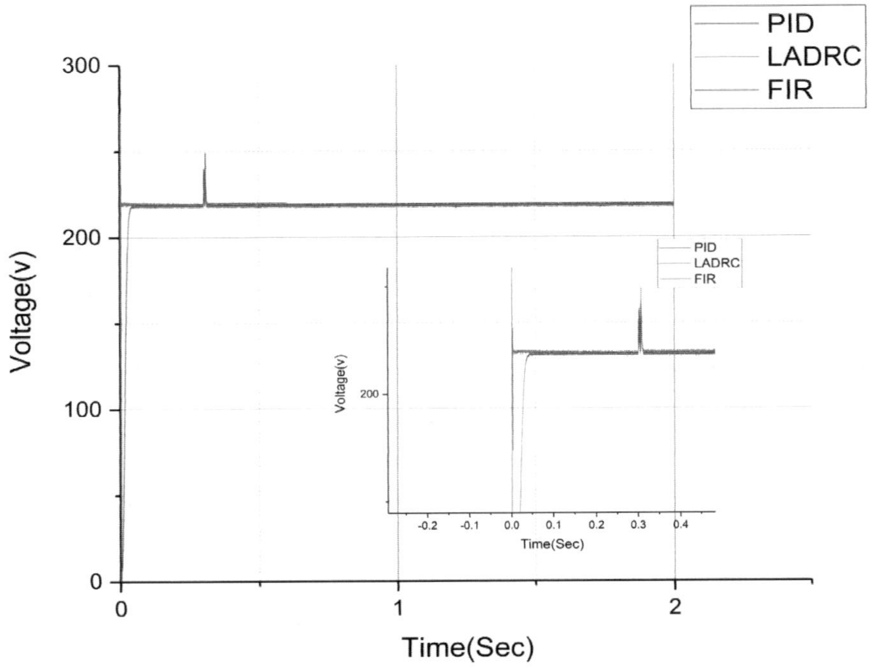

Fig. 10. Voltage response of the three controllers to load changes

4.2 Simulation and comparison of sudden interference

In order to simulate the instability of the output power of clean energy such as wind and solar power, random interference is added to the DC side of the inverter. As shown in Figure 11, sinusoidal signal with amplitude 1 and random signal interference with variance 50 were added at 1s.

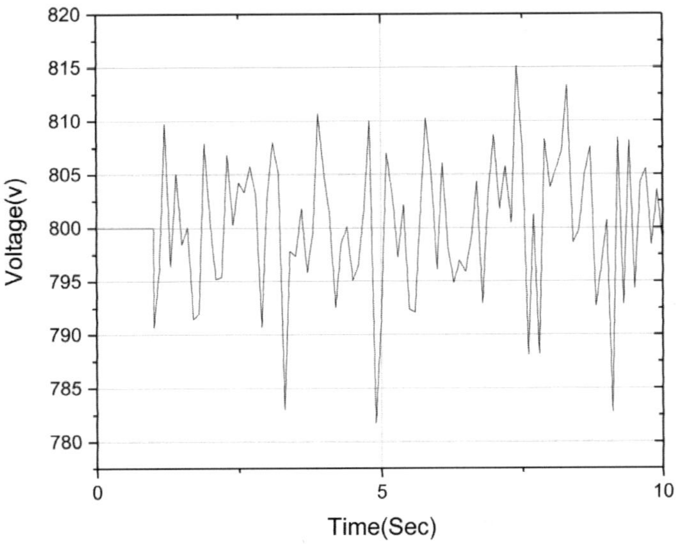

Fig. 11. Voltage waveform of the DC side-rush with interference

FIG. 12 and 13 show the response waveform comparison between frequency and voltage when sudden interference is added. It can be seen that the three controllers can ensure the frequency stability of isolated island microgrid, but PI is obviously affected by disturbance, the response waveform is very rough, and the frequency fluctuation becomes large. However, when using FIR combination, the frequency and voltage waveform change little, but the waveform appears a sudden spike when the disturbance amplitude is larger, indicating that the frequency and voltage are still affected by the disturbance. Based on the immunity of the FIR under the condition of disturbance can get very good control effect, the waveform is basic and same, in the absence of any disturbance in tends to steady state compared with PI controller and LADRC controller, the speed of response time is shorter, more responsive, adopt the immunity of FIR not only balance the LADRC smoother response, and of better on the response speed, excellent disturbance resistance. In the period from 0.3s to 0.6s, the responses of the three controllers are basically the same. Table 1 shows the frequency response indicators of the three controllers. As can be seen from FIG. 12, the FIR controller has the fastest response speed, and the voltage reaches a stable state at 6.023ms. On the basis of ensuring the reduction of overshot, the response speed performance is improved. The second is PI, the response time is 12.124ms, but the system stability is insufficient, while the response time of active disturbance rejection is 19.548ms. When one load is cut off at 0.3s, the three controllers can quickly reconverge, and the FIR is affected for the shortest time. Table 6 is the voltage response index of the three controllers. In general, the designed FIR controller based on active disturbance rejection has smoother response and shorter adjustment time when switching load.

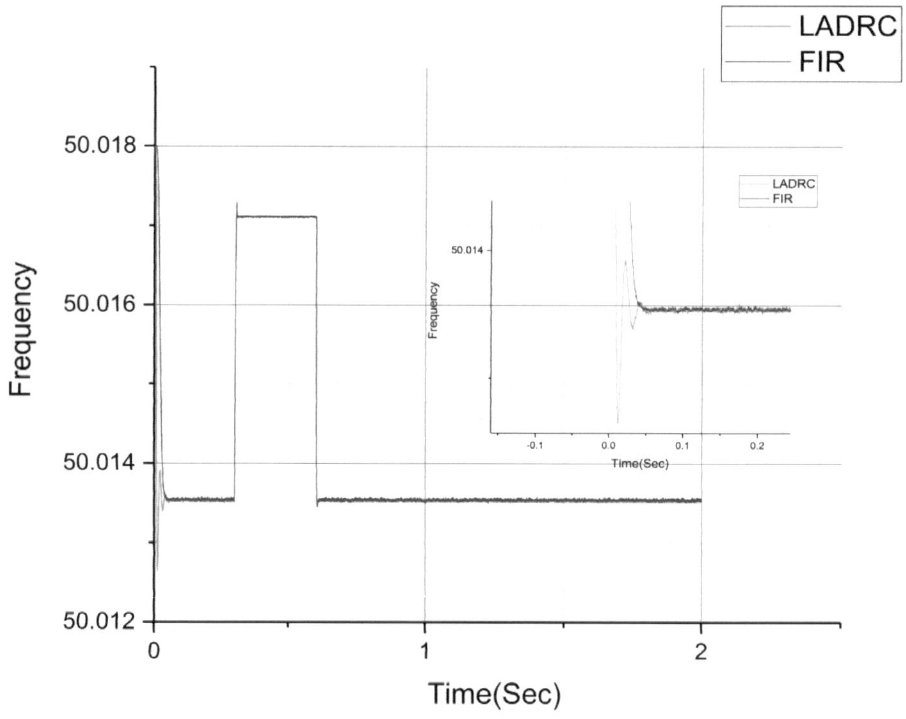

Fig. 12. Frequency response of three controllers under perturbation

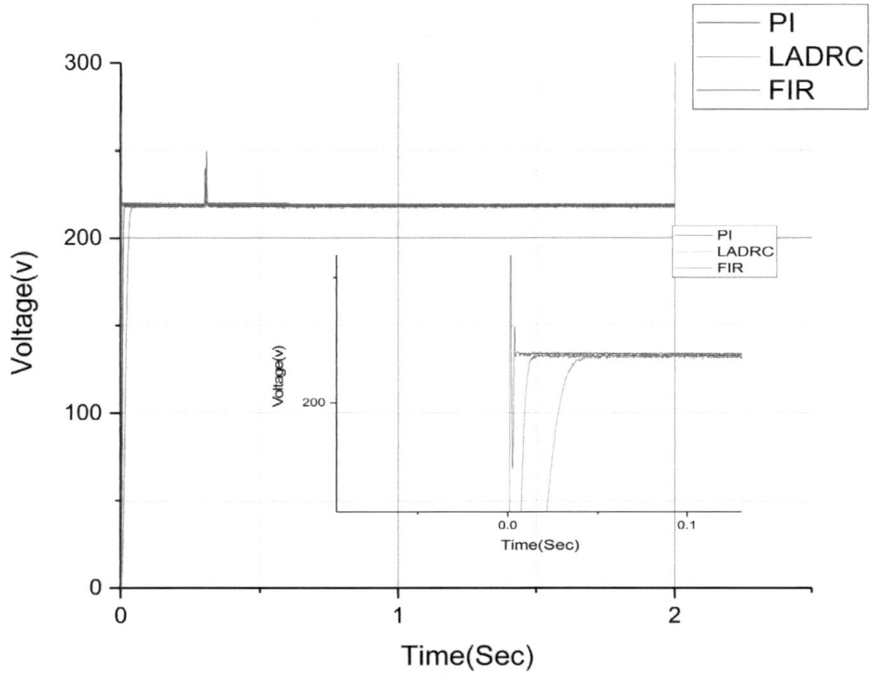

Fig. 13. Voltage response of three controllers to disturbance

TABLE IV. THE FREQUENCY RESPONSE INDEX OF THE THREE CONTROLLERS UNDER THE DISTURBANCE OF THE NEEDLE TO THE CHANGE OF LOAD

Time(s)	controller	Overshoot(%)	Setting time 2%(ms)
t=2s	FIR	4.755	5.008
	LADRC	4.875	5.015
	PI	11.135	19.968

TABLE V. THE VOLTAGE RESPONSE INDEX OF THE THREE CONTROLLERS TO THE LOAD CHANGE UNDER DISTURBANCE

Time(s)	controller	Overshoot(%)	Setting time 2%(ms)
t=2s	FIR	0.170	6.023
	LADRC	0.357	19.548
	PI	25.949	12.124

5. Conclusion

This paper studies the secondary control strategy of microgrid based on droop control, and designs a control system based on active disturbance rejection and FIR filter to adjust the parameters of the controller. The various and possible problems are analyzed, which may need to be dealt with the actual applications in the microgrid. It also simulates and tests the load shedding and additional interference on the DC side of the inverter. So, it needs to design a controller based on noise immunity. It compares the controller based on FIR-based ADRC, the classic PI controller and LADRC controller. The results of the Matlab/Simulink simulation platform show that the designed controller can estimate the internal very well. It also shows that the external disturbances of the entire system has strong anti-interference and performance. All of these can guarantee the performance of the microgrid effectively. The controller designed in this paper combines FIR and ADRC to solve the problem of three-phase voltage waveform distortion caused by the uncertainty of the output power of the distributed power supply. It has important practical significance for harmonic control and stable operation of microgrid. It can be adjusted quickly and accurately under interference conditions. The shortcoming of this research that the combination of ADRC and FIR filter has little difference when it only switches the power load. The next work is to do more research on this problem. I plan to improve the harmonic suppression effect and reliability of ADRC based on FIR.

References

[1] Zeng Zheng, Li Hui, Ran Li. Review on Control Strategy of AC Microgrid Inverter [J]. Automation of electric power systems, 2016,40 (9):142-151.

[2] Robert H. Lasseter, Paolo Piagi, Control and Design of Microgrid Components. Final Project Report, 2006.1.

[3] Jiang Feng-feng. Study on Stable Operation Control of Microgrid Parallel Multiple Inverters considering Circulation suppression [J/OL]. Journal of Power Sources :1-12[2021-07-27].

[4] Chen Biao, Fang Xufeng, Chen Shiwei. Application of Finite Impulse response filter to power System Harmonic Robust Estimation [J]. Energy and Environmental Protection, 201,43(04):217-222.

[5] Xue Yang, Shi Yufei, Kan Dongyue, Chen Mingchao, Zhang Maosheng. Sag Control of microgrid based on auto-disturbance rejection [J]. Grid and clean energy of China, 2016,32(11) : 25-30.

[6] Jin CHENG. Research on mode switching strategy of wireless parallel Inverter System from independent to grid-connected based on droop Control [D]. Hangzhou: Zhejiang University, 2013.

[7] Se-Kyo Chung. A Phase Tracking System for Three Phase Utility Interface Inverters[J]. IEEE Transactions on Power Electronics, 2000, 15(3):431-438.

[8] J. Han, "From PID to Active Disturbance Rejection Control," in IEEE Transactions on Industrial Electronics, vol. 56, no. 3, pp. 900-906, March 2009, doi: 10.1109/TIE.2008.2011621.

[9] Zhiqiang Gao, Yi Huang and Jingqing Han, "An alternative paradigm for control system design," Proceedings of the 40th IEEE Conference on Decision and Control (Cat. No.01CH37228), 2001, pp. 4578-4585 vol.5, doi: 10.1109/CDC.2001.980926.

Intrusion Detection Method of Electric Power Information Network in Cloud Computing Environment

Jiaqi Zhang [1*], Guoping Feng [2], Dexi Zhou[2] and Mingjiu Li [2]

[1] Guangzhou Power Supply Bureau of Guangdong Power Grid

[2] China Energy Engineering Group Guangdong Electric Power Design Institute Co., Ltd.

*Corresponding author's e-mail: zhangjiaqi@gzpsc.csg.cn

Abstract. With the widespread application of power grid systems, the information security problems faced by power grids have become more obvious. Various internal and external intrusion attacks that occur frequently have become an important issue affecting the normal operation of power generation and operations. The purpose of this paper is to study the intrusion detection method of electric power information(PI) network in the cloud computing environment. With the help of the cloud platform's ability to process big data, and based on the analysis of the PI network structure, a DBN optimized BP network algorithm is proposed, and the optimized BP neural network is used as a runtime classification program. Experimental results show that MR-DBN-BP has a detection rate of 96.7% for intrusion detection of PI networks, which can effectively detect intrusions and effectively protect the power dispatch system network.

1. Introduction

With the rapid development of Internet applications and distributed computing technology [1-2], cloud computing has become a mature network technology and is widely used in PI networks [3-4]. As the PI system is subject to various external and internal attacks, the information security problem of the PI system has become more obvious, which has seriously affected the normal production and stable operation of the power system [5-6]. With the continuous improvement of the integrated information network of the power supply network, the security threats from the Internet have become more complex and changeable. These threats include the information network, the collection of energy user information, smart power consumption, power trading, power distribution network automation, chargers, etc. [7-8].

In the research of intrusion detection methods for PI networks based on cloud computing environments, many scholars have studied them and achieved good results. For example, Suzuki S, Hiraoka M, Shiraishi T and others proposed the first profile based on user behavior characteristics. A real-time intrusion detection system model [9]. Wang D, Bing-Yu G E, Xuan J X and others have proposed a GPU-based high-speed network intrusion detection system for the high underreporting problem faced by traditional intrusion detection under high-speed traffic conditions [10].

Based on the analysis of the PI network structure and its security partitions, this paper proposes a PI network intrusion detection framework based on cloud computing technology in view of the intrusion attack problems in the PI network.

2. PI Network Intrusion Detection Method Under Cloud Computing Environment

2.1. PI Network Structure

When constructing a secure power system network, it is necessary to comprehensively consider how to improve the security of the entire power system network. Therefore, it is necessary to study the security between the real-time system and the local area network to ensure the most important power production real-time system security, to study the security between the real-time systems to ensure the independence between systems, and to avoid mutual interference; to study the relationship between the local area network and the Internet To ensure the safety of power application systems such as MIS, OA, and power management systems.

2.2. Structure of PI Network Intrusion Detection System

Nowadays, the emergence of high-speed networks and the continuous update of attack methods make it more and more difficult to update and maintain the knowledge base by manual methods. As a method of obtaining knowledge from massive data, cloud computing technology was first introduced by WenkeLee et al. to intrusion detection. The field subsequently developed vigorously. The system adopts misuse intrusion detection, that is, judges whether the data in the network has intrusions based on known intrusion patterns. The intrusion detection system must increase the speed of rule generation, increase the accuracy of the rules, and reduce the system's false negative and false positive rates; it must detect intrusions quickly and accurately; the rule base must be continuously updated to accurately discover new ones in a timely manner Intrusion attacks. Due to the huge number of power information networks and unknown sources, it is very important to strengthen the maintenance of their safety performance. Applying cloud computing to the security detection of power information networks, by establishing a database, it can effectively detect intrusions in the power information network, maintain the security of the power information network, and ensure its normal operation.

This article will obtain network data information flowing through the power dispatch automation system. After preprocessing the data, use data mining methods to discover the attack mode of the system data information, thereby detecting illegal behaviors and intrusion attacks.

2.3. BP Neural Network Optimization Based on DBN

Because the BP network relies heavily on the initial weight, it is easy to fall into the local minimum, and the convergence speed is slow, and the training time is long. Many improved BP algorithms have been proposed. In order to solve the problem of BP's sensitivity to initial weights, a global search strategy in the range space using a global search algorithm is proposed. Starting from the BP algorithm itself, an adaptive variable step size for the momentum item is proposed, based on LM (Levenberg-Marquardt) Improved algorithm of algorithm. The above method can effectively solve the problem of BP network in solving the global minimum and the number of iterations, but the first strategy is prone to oscillations, and the second method is prone to overfitting.

Combined with the deep belief network, a BP improved algorithm based on DBN optimization (DBN-BP) is proposed, because the multiple Boltzmann machines of DBN can obtain better model parameter values in the process of unsupervised learning, and then pass BP. The algorithm is fine-tuned, and it can quickly converge to the global minimum. For a given RMB state vector (v, h), define the energy of RBM as:

$$E(v,h|\theta) = -\sum_{i=1}^{n} a_i v_i - \sum_{j=1}^{n} b_j\, h_j - \sum_{i=1}^{n}\sum_{j=1}^{m} v_i\, w_{ij}\, h_j \quad (1)$$

Equation (1) is the parameter of RBM, and the joint probability of the visible layer and the hidden layer is:

$$P(v,h|\theta) = \frac{1}{Z(\theta)}\exp(-E(v,h|\theta)) \quad (2)$$

Where Z(θ) is the normalization factor, Z(θ)=∑_(v,h)▒〖exp (-E(v,h∣θ))〗. The distribution of observation data defined by RBM is P(v∣θ) and P(h∣θ):

$$P(v|\theta) = \frac{1}{Z(\theta)}\sum_v exp(-E(v,h|\theta)) \quad (3)$$

$$P(h|\theta) = \frac{1}{Z(\theta)}\sum_h exp(-E(v,h|\theta)) \quad (4)$$

3. Experimental Research on PI Network Intrusion Detection Method Under Cloud Computing Environment

3.1. Description of Test Data Set

The data set used in this article is the network traffic audit log data of a power company. In order to increase the data of various intrusion behaviors, a part of the security audit data set is added to the data set after analysis and processing to form an intrusion detection data set (KD). A total of 2 million network traffic is collected as training data, and a test data set is a total of 1 million data sets. It includes four types of intrusions: Dos attacks, port scanning attacks, unauthorized access by remote hosts, and unauthorized access by local users. Each record has 53-dimensional attributes, and the last attribute is its category.

3.2. Data Processing Method

Because the data contains two types of discrete and continuous data. In order to fit the input of neurons and eliminate the situation that big numbers eat decimals, the data needs to be standardized and normalized.

(1) Data standardization

The input of the neural network requires numeric input, and uniformly encodes items whose attribute fields are of character type. The encoding method uses lexicographic sorting to assign ordinal numbers to character fields in turn. Other character fields are standardized in the same way.

(2) Data normalization

The size range of the attribute is reduced to the interval of [0,1]. This article uses the formula a=(a-min)/(max-min) to normalize the data, where a represents the value of the attribute field.

4. Data Analysis of PI Network Intrusion Detection Method Under Cloud Computing Environment

4.1. MR-DBN-BP Algorithm Classification Performance Test Under Cloud Computing Applications

Test the extraction performance of the DBN-BP algorithm under different data volumes. KD dataset data is randomly generated into multiple groups of datasets, KD1 to KD8. Among them, KD1 contains 10,000 pieces of sample data, KD2 contains 30,000 pieces of data, KD3 contains 50,000 pieces of data, KD4 contains 70,000 pieces of sample data, KD5 contains 100,000 pieces of sample data, KD6 contains 130,000 pieces of sample data, and KD7 contains 16 pieces of data. There are 10,000 pieces of sample data, and KD8 has 200,000 pieces of sample data.

The training time of DBN-BP algorithm on single machine is compared with that of MR-DBN-BP in cluster environment. The comparison results are shown in Table 1.

Table 1. Single-machine DBN-BP and 4-node cloud cluster MR-DBN-BP data

Sample data (ten thousand)	Stand-alone DBN-BP (training time S)	Cluster MR-DBN-BP (training time S)
1	218	320
3	345	369
5	420	384
7	645	418
10	972	458
13	1341	489
16	1867	550

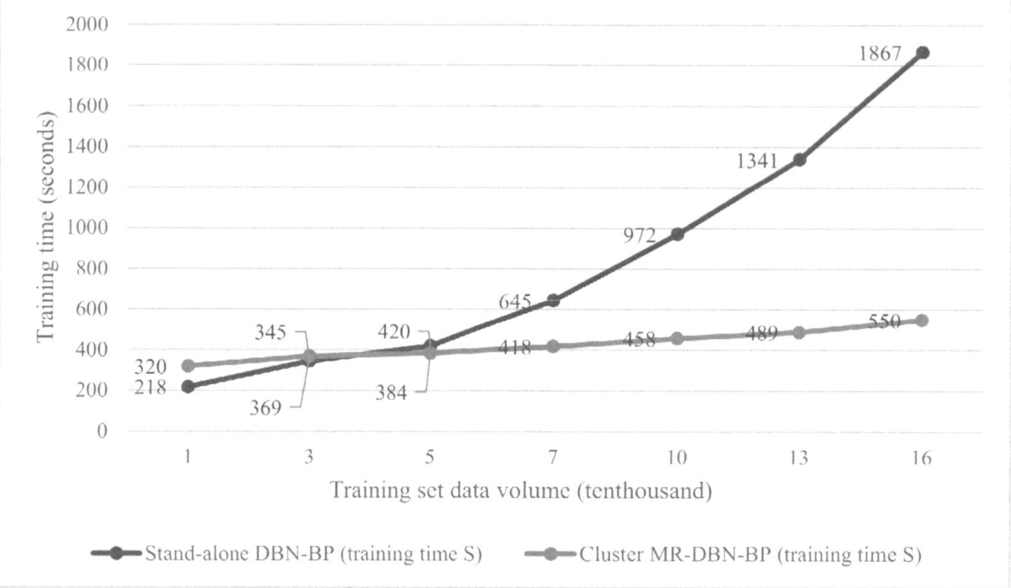

Figure 1. Single-machine DBN-BP and 4-node cloud cluster MR-DBN-BP

As shown in Figure 1, with the expansion of the training set size, the execution time of the MR-DBN-BP algorithm on a 4-node cloud computing cluster has little change. In a stand-alone environment, with the expansion of the training set size, DBN -BP algorithm execution time is also increasing. Through the analysis of the experimental results, it can also be concluded that in a 4-node cloud computing cluster environment, the training speed of the MR-DBN-BP algorithm is more than 3 times that of the single-machine DBN-BP algorithm. In practical application, the increase of the number of cloud cluster nodes and the growth of power information network connection record sample data set are conducive to the operation of MR-DBN-BP algorithm for cluster computing.

4.2. Parallel Performance Analysis Of MR-DBN-BP Algorithm in Cloud Computing Applications

In order to compare the method in this paper with the BP method, PSO-BP, and GA-BP method that introduces the steepness factor λ, now set the network parameters: set the number of RBMs constituting the DBN network to two, and the number of nodes to be 53-22 -12, 1000 iterations; BP neural network has a three-layer structure, the number of nodes is 12-6-3, 500 iterations, the learning rate is 0.01, and the error rate is 0.01. Table 2 is the comparison between MR-DBN-BP and λBP, PSO-BP, GA-BP in three aspects: detection rate, false alarm rate and false alarm rate. It can be seen that compared with λBP, MR-DBN-BP, PSO-BP and GA-BP have the highest detection rate of 96.7%.

Table 2. Comparison of BP network detection rate with other optimization methods

Detection method	Normal record	Attack record	Detection rate (%)			Time-consuming detection (S)
	850	500	R_C	R_W	R_I	
λ BP	784	466	91.3	4.51	10.8	18.7
PSO-BP	815	488	94.5	2.13	7.6	21.4
GA-BP	826	472	95.2	1.25	7.4	20.7
MR-DBN-BP	831	491	96.7	1.09	6.5	21.7

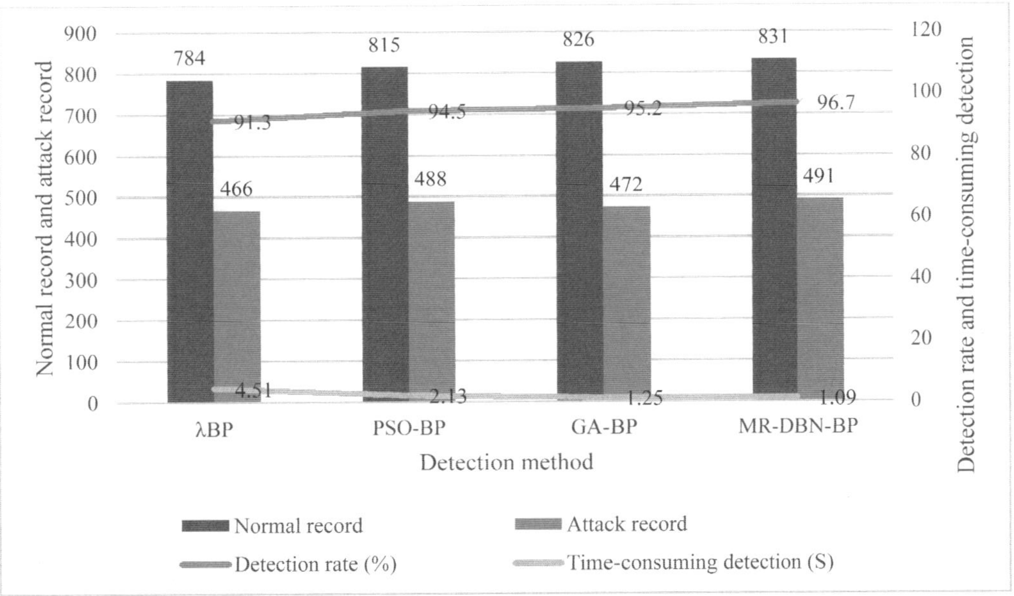

Figure 2. Comparison of network detection rates

According to Figure 2, the detection time of the DBN-optimized BP network in a single machine is slightly longer than other methods, which is 21.7 seconds. However, because this method can select the more essential features of the data during training, the detection accuracy is Compared with other methods, it also effectively reduces the false detection rate and missed detection rate.

5. Conclusions

The rapid development of big data and cloud computing has further ensured the safety and maintenance of power systems. Through the introduction of cloud computing, the detection and prevention of intrusion behaviors in the power system information network is of great significance for effectively improving power information security and ensuring the normal operation of the power system. However, it is undeniable that due to the limitations of time, energy and knowledge, the intrusion detection experiments carried out in this research have certain limitations. In order to obtain a more scientific power information network intrusion detection method, further research is needed.

References

[1] Li S , Wang X P , Wang Q D , et al. Research on intrusion detection based on SMDP reinforcement learning in electric PI network[J]. Electric Power Automation Equipment, 2006, 26(12):75-78.

[2] Li C , Kang H , Chen S . Research on the Service Innovation Path for Information Platform in the Cloud Computing Environment[J]. International Journal of Grid & Distributed Computing, 2016, 9(10):129-140.

[3] Wang T , Zhang G , Yang X , et al. A Trusted and Energy Efficient Approach for Cluster-Based Wireless Sensor Networks[J]. International Journal of Distributed Sensor Networks,2016,(2016-4-10), 2016, 2016:1-13.

[4] Xinhui D , Shuai W , Juan Z . Research on Marine Photovoltaic Power Forecasting Based on Wavelet Transform and Echo State Network[J]. Nephron Clinical Practice, 2017, 24(s2):53-59.

[5] Kb A , Bbg A . Hypothesis Test for Low-rate DDoS Attack Detection in Cloud Computing Environment - ScienceDirect[J]. Procedia Computer Science, 2018, 132:947-955.

[6] Chuang C L , Chiu W Y , Chuang Y C . Dynamic Multiobjective Approach for Power and

Spectrum Allocation in Cognitive Radio Networks[J]. IEEE Systems Journal, 2021, PP(99):1-12.
[7] Chen Y , Li L , Shen Y , et al. Group decision-making framework for site selection of coastal nuclear power plants in a linguistic environment: a sustainability perspective[J]. International Journal of Green Energy, 2021:1-12.
[8] Gulmkhan O , El-Saadany E , Youssef A , et al. Cyber Security of Market-based Congestion Management Methods in Power Distribution Systems[J]. IEEE Transactions on Industrial Informatics, 2021, PP(99):1-1.
[9] Suzuki S , Hiraoka M , Shiraishi T , et al. Power Prediction for Sustainable HPC[J]. Journal of Information Processing, 2021, 29:283-294.
[10] Wang D , Bing-Yu G E , Xuan J X , et al. Research on Operation Test Technology in PI System[J]. Electric PI and Communication Technology, 2017(6):50-55.

Research on latent defect detection technology of high voltage cable

Sanwei Liu[1*], Chao Qiu[2], Yi Xie[1], Jianjia Duan[1], Fuyong Huang[1], Xiaoli Duan[1], Zeyu Zeng[1]

[1] State grid Hunan electric power co, LTD. Electric power research institute, Changsha, Hunan, 410000, China

[2] State grid Hunan Zhuzhou electric power co, LTD, Zhuzhou, Hunan, 410002 China

[*]Corresponding author's e-mail: liusanwei@cqu.edu.com.cn

Abstract: As a component of the Internet of things, high-voltage cables are the power supply infrastructure for the modern development of cities. The operation experience shows that the high-voltage cable has been broken down many times, due to the defective operation. At present, due to the limitation of detection technology, the research on detection and identification of defects in high-voltage cables is progressing slowly. Therefore, a new DR technology based on X-ray digital imaging is proposed in this paper to realize real-time detection of defects in the semi-conductive buffer layer of high-voltage cables, and intelligent detection of DR images of high-voltage cables by using image depth processing technology to realize intelligent identification of defects in the buffer layer of power cables. The results show that using the new DR technique proposed in this paper, the accurate and intuitive DR image of high-voltage cable can be obtained quickly, and the intelligent identification of defects can be realized.

1. Introduction

High voltage cable is an important infrastructure equipment for building the Internet of things. Operation experience shows that under the influence of production process, transportation and storage, distribution defects will occur in the semi conducting buffer layer of high-voltage cable. Which is easy to cause breakdown of cable body under random insulation during operation, and the probability of breakdown increases significantly with the increase of operation time. Over the past 20 years, the defects in the high-voltage cable industry have shown multiple natures. With problems such as factory distribution, unclear defect mechanisms, imperfect defect detection methods, and difficulty in implementing status assessments[1]. Before that, the State Grid Corporation and the South Grid Corporation decided that "the existing common high-voltage cable online monitoring means are difficult to obtain such defects effectively[2].

As an effective non-destructive testing method, X-ray Digital Radiography (DR) was first used in the medical, aerospace and other industries, and has gradually been widely used in materials, machinery and other industrial fields[3-4]. In order to meet the demand of power Internet of things construction, this technology has also been applied in the field of state perception of power transmission and transformation equipment. Typical application scenarios include transmission line tension line clamping and compression quality test and structure test of substation GIS equipment[5].

Different from tension clamps and GIS equipment, due to the special structure of high-voltage cables, there is still no effective method to detect and identify the defects of the semi-conductive buffer layer of high-voltage cables. Therefore, this paper proposes a new Image Depth Intelligent processing-digital Radiography (DR) technology to realize Intelligent recognition of high-voltage cable X-ray Digital Image Processing and power cable buffer layer defects. By turning passivity into initiative, the insulation breakdown accident in the operation of high-voltage cable can be forewarned in advance, and secondary disasters such as cable fire and loss of important section of power grid can be avoided.

2. DR detection technology

X-rays of different energies have different abilities to penetrate the material being examined. Theoretical and experimental studies show that when narrow beams and single-energy x-rays pass through a layer of material of uniform thickness, their intensity decays exponentially, and the incident intensity decays with the increase of the thickness of the penetrating object, that is:

$$I = I_0 e^{-\mu \rho x} \tag{1}$$

The plate detector is used to receive the X-ray that penetrates the detected workpiece, and then the crystal circuit inside the plate detector converts it into the current signal according to the X-ray dose intensity, and finally presents it in the form of digital image on the terminal computer[6].

3. Defect generation mechanism

A large number of cable disintegration tests with buffer layer defects show that there are a large number of white powdery substances in the cable buffer layer with water resistance structure, aluminum sheath and insulation shielding layer. Under the action of operating voltage for a long time, the defect of the buffer layer gradually develops, ablates and runs through the semi-conductive layer of the cable, finally leading to the breakdown of the main insulation of the cable. According to the infrared spectrum, the white powdery substance contains sodium, aluminum and other metal elements.

The main component of the water-blocking powder in the water-belt buffer layer of cable semi-conducting resistance is low-crosslinked polyacrylate sodium salt, which has strong water absorption. When the cable is poorly controlled due to production, storage, transportation, construction and other links, the semi-conductive resistance water buffer layer will absorb water and be affected by moisture, forming free OH- and Na+ ions. When the OH- ions absorb CO2 in the air, they will form sodium carbonate (Na2CO3) or sodium bicarbonate (NaHCO3) :

Small amount of CO2:

$$CO_2 + 2OH^- = CO_3^{2-} + H_2O \tag{2}$$

$$CO_3^{2-} + 2Na^+ = Na_2CO_3 \tag{3}$$

A glut of CO2：

$$CO^2 + OH^- = HCO_3^- \tag{4}$$

$$HCO_3^- + Na^+ = NaHCO_3 \tag{5}$$

Sodium carbonate (Na2CO3) or sodium bicarbonate (NaHCO3) are white powder crystals. Therefore, the sodium element in the white powder exists in the form of sodium bicarbonate or sodium bicarbonate.In addition, the water-blocking powder forms a free OH- ion, which contacts with the buffer layer on the inside of the trough of the aluminum sheath, and a chemical reaction occurs between the OH- ion and the aluminum, resulting in AlO2-, an aluminum-biased acid radical:：

$$2Al + 2OH^- + 2H_2O = 2AlO_2^- + 3H_2 \tag{6}$$

AlO2- and Na+ ions in the water blocking powder form NaAlO2, and the reaction produces NaAlO2, which is easily soluble in water, and absorbs CO2 in the air and reacts to form Al(OH)3:：

Small amount of CO2:

$$AlO_2^- + CO_2 + 2H_2O = Al(OH)_3 + HCO_3^- \tag{7}$$

A glut of CO2：

$$2AlO_2^- + CO_2 + 3H_2O = 2Al(OH)_3 + CO_3^{2-} \tag{8}$$

Finally, the generated aluminum hydroxide Al(OH)3 is unstable in nature and is easy to decompose into aluminum oxide Al2O3:

$$2Al(OH)_3=Al_2O_3+3H_2O \qquad (9)$$

Therefore, the main components of the white powder are sodium carbonate (Na2CO3), sodium bicarbonate (NaHCO3) and alumina (Al2O3) crystals.

Most of the white powder at the defects was generated at the trough where the aluminum sheath had the closest contact with the buffer layer. The reason was that the sodium aluminate NaAlO2 produced by the above reaction was easily soluble in water, and the sodium aluminate ions would migrate around the contact point between the aluminum sheath and the buffer layer and inside the buffer layer under the effect of concentration gradient. On the other hand, the solution will produce electrophoresis phenomenon under the electric field, and the particles with high permittivity will move in the direction of the high electric field, which will cause the sodium aluminate here to accumulate towards the insulation shield of the cable core.

4. Field application

4.1 Field briefing

The distribution diagram of cable line is shown in figure 1. Line A, line B and line M of 110 kV are derived from the tea garden at intervals of 504 through the cable, and laid along the "220 kV X to 220 kV Y cable tunnel (about 2400m). The total length of the cable is about 3,100m, and it is divided into four sections, including two tunnel segment joints and one directly buried segment joint. At section III of the tunnel, it is laid with lines C and D and E of 110 kV.

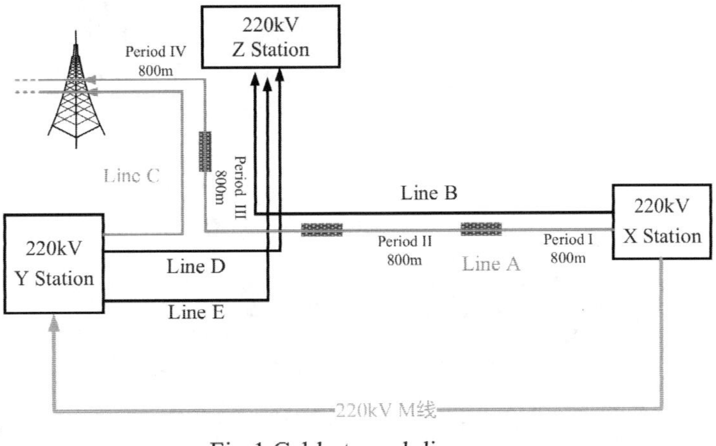

Fig.1 Cable tunnel diagram

The circuit diagram of line A of 110 kV is shown in figure 6. The detection scheme takes two joints as A section and changes from X to Y, as shown in figure 1. There are altogether 4 sections, which can be labeled as I, II, III, and IV sections, each of which is about 800m long. 8 points are detected in each section, and the interval between each point is 100m, marked as 1, 2, 3.. 8 PM. The detection points of A, B and C are set in this way.

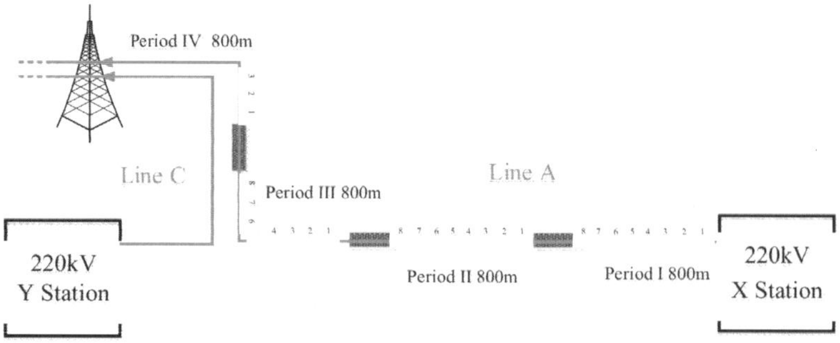

Fig.2 Check point layout

4.2 Test results and analysis

A, B and C of 110kV tea chrysanthemum steel line are cables made in lingnan, guangzhou. A total of 64 tests have been completed in sections I, II, III and IV, and 21 defect points and 43 normal points have been detected.

In this paper, full convolutional neural network FCN is used for defect identification. The DR detection image of normal cable is shown in figure 3 below::

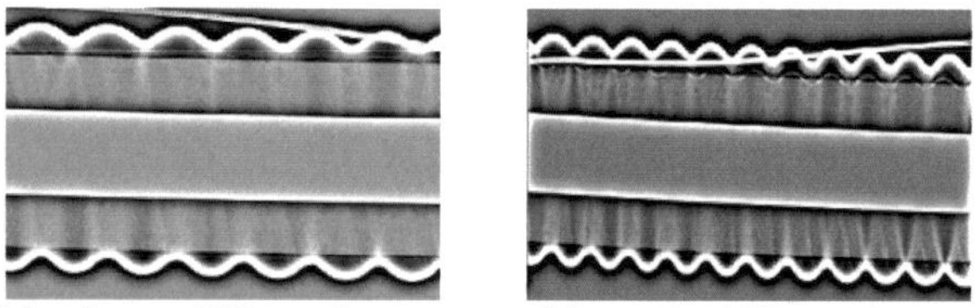

(a) horizontal X-ray image (b) vertical X-ray image

Fig.3 Line A normal cable DR detection image

Using the DR detection technology proposed in this paper, the whole line of line A was tested by DR, and the defective cable segments were found: A phase I, II, III, B phase I, II, III, IV, C phase I, II, III, IV. The DR detection image of a typical defective cable is shown in figure 4.

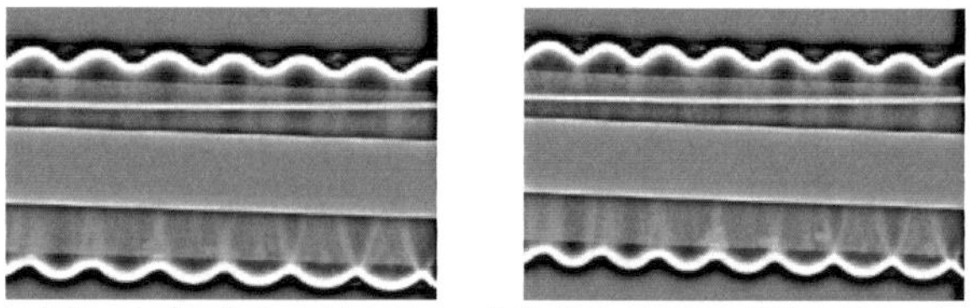

(a) original drawing (b) defect identification drawing

(1) A-phase defective cable

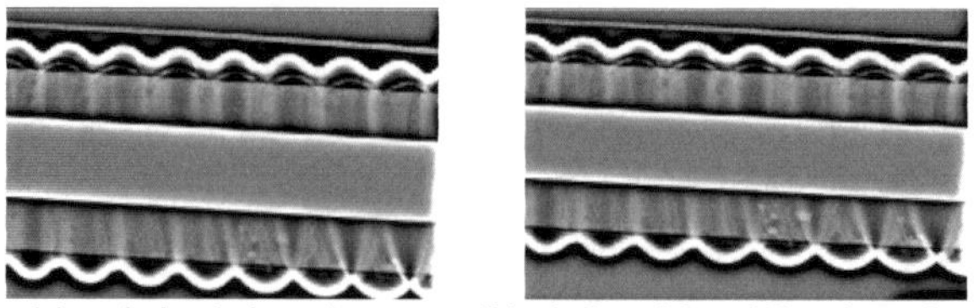

(a) original drawing (b) defect identification drawing

(2) B-phase defective cable

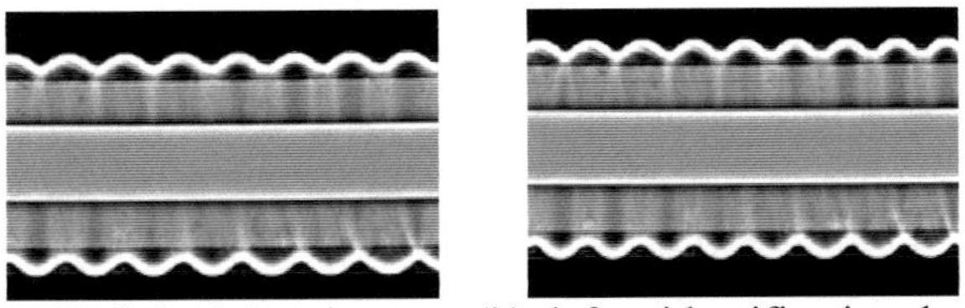

(a) original drawing (b) defect identification drawing
(3) C-phase defective cable

Fig.4 Line A defect cable DR detection image

Characteristic diagnosis and analysis are carried out for the DR image data set detected in the field. The DR image of the cable buffer layer has the following characteristics:

(1) cable buffer layer defects during imaging, if the defects are on the same section as the incident X-ray and cable core, the cable defects will be covered by the cable core. Therefore, during X-ray detection, at least two images should be taken, and the angles of the two images should cover 360° to ensure that the defects can be photographed.

(2) most of the defects occurred at the trough where the aluminum sheath and buffer layer had the closest contact, and the test results were consistent with the theoretical analysis in 1.2.

(3) the defect of the cable buffer layer has a significant position feature: when shooting in the horizontal direction, the defect is basically located at the bottom of the image, and the corresponding actual position is at the bottom of the cable; When shooting vertically, the defect is basically located at the top of the image, and the corresponding actual position is the cable near the inside of the wall. Therefore, from the image analysis, the cable buffer layer defect is basically located in the cable bottom inner region.

5. Conclusion

Aiming at the problem of defect detection and recognition in the buffer layer of high voltage cable, the validity of the DR detection method proposed in this paper is proved by the spot defect detection, and the rule of the defect detection in the buffer layer is summarized.

References

[1] Duarte MF, Davenport MA, Takhar D, Sun T, Kelly KF, Baranuik RG. Signal-pixel imaging via compressive sampling [J]. IEEE Signal Processing Magazine, 2008, 25(2): 83-91.

[2] [Gan H, Cheng Z, Yang S, et al. Circulant and toeplitz chaotic matrices in compressed sensing [J]. Journal of Computational Information Systems, 2015, 11(4): 1231-1238.

[3] Gan H, Cheng Z, Yang S, et al. Circulant and toeplitz chaotic matrices in compressed sensing [J]. Journal of Computational Information Systems, 2015, 11(4): 1231-1238.

[4] Duarte MF, Davenport MA, Takhar D, Sun T, Kelly KF, Baranuik RG. Signal-pixel imaging via compressive sampling [J]. IEEE Signal Processing Magazine, 2008, 25(2): 83-91.

[5] Sun Y, Wang X , Tang X . Deep Learning Face Representation from Predicting 10,000 Classes[C]// 2014 IEEE Conference on Computer Vision and Pattern Recognition (CVPR). IEEE Computer Society, 2014.

[6] Srinivasan V, Sankar A R, Balasubramanian V N. ADINE: An Adaptive Momentum Method for Stochastic Gradient Descent[J]. 2017.

Research progress of fault detection equipment and methods for track train's shape part

Yingqi Hao [1], Kun Xie [1]*, Zhepu Xu [2]

[1]School of Rail Transportation, Shanghai University of Technology,Shanghai, 200000, China

[2]School of Rail Transportation, Shanghai University of Technology,Shanghai, 200000, China

*Corresponding author's e-mail: 206151106@mail.sit.edu.cn

Abstract: Rail transit has many advantages such as large capacity, fast speed, energy saving and environmental protection, and has gradually become an indispensable means of transportation in people's lives. At the same time, the safety and reliability of rail transit is also facing severe challenges. The running part plays the role of load bearing, running and traction guidance, etc. It is one of the most important systems of the train, and it is also the accident-prone part. Aiming at the failures of rail vehicle wheelsets, bearings, overall frame and suspension parts, this article uses CNKI, Wanfang Data Knowledge Service Platform, IEEEXplore and ScienceDirect as the search database to investigate and analyze the research results published at home and abroad in recent years. The fault diagnosis technology of the traveling department summarizes the sensor types, installation positions and usage methods used in various diagnosis technologies, and finally puts forward the prospect of the follow-up research. The work of this paper will help researchers to systematically grasp the relevant research trends of the fault detection of the walking part, and then make clearer and deeper thinking.

1. introduction

Rail transit has many advantages such as large capacity, fast speed, energy saving and environmental protection.Has gradually become the indispensable transportation way of people's lives, the field is the national natural science foundation of the mechanical engineering discipline development strategy report (2011-2020) ", "national medium and long-term science and technology development plan outline (2006-2020), 2025 national key r&d plan and made in China and other key research direction of national science and technology strategy.At the same time, at the same time, the safety and reliability of rail transit is also facing severe challenges.A number of accidents in recent years have caused service disruptions and social panic, such as the 2011 Nagamon Line accident and the 2017 Shinkansen emergency inspection accident, in which a 44-centimeter crack was found.[1].High-speed rail systems need more reliable and thorough detection and monitoring methods.

Rail vehicle is composed of body, running part, brake system, electrical system and so on.Among them, the walking part plays the role of bearing, walking and traction guidance, which can ease the impact between the vehicle and the track, restrain the vibration of the vehicle, and ensure the smooth and safe operation of the train. It is one of the most important systems of the train.Subway train running gear mainly consists of steering device, elastic suspension device, basic brake device and driving device[2].The steering device mainly includes wheelset, axle box and frame[3]. **Figure 1** is a tree diagram of train components.

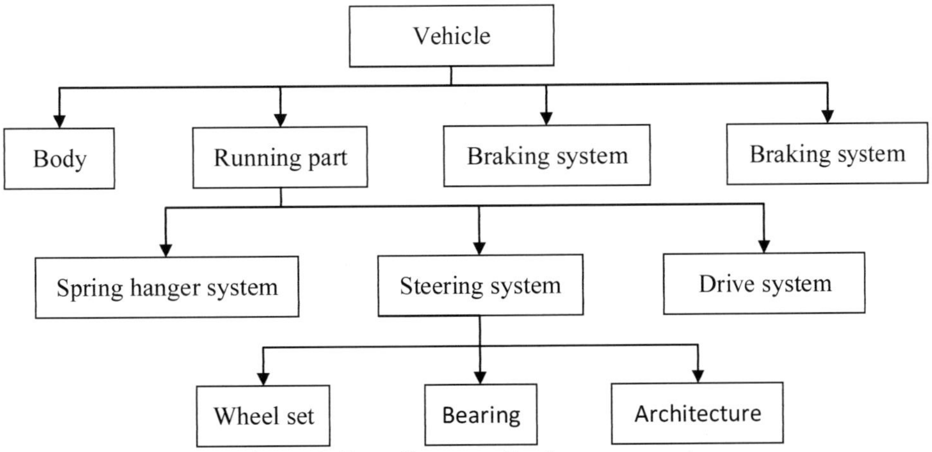

Figure 1 .Tree diagram of train components

Statistical analysis of accidents at home and abroad shows that the running part is the most frequent part of accidents, and the faults of wheelset, axle box, whole frame and suspension part are the main parts.With the development of railway fault diagnosis, more and more achievements have been accumulated, including many new and advanced theoretical methods and reliable and practical technical means.Review, collation, summary and summary of research results can help researchers understand the research status in this field as a whole, so as to help researchers conduct clearer and deeper research in this field.

Due to the complexity of the structure of rail vehicles and the wide range of technical fields involved, a large number of relevant research results have been published at home and abroad.Aiming at the wheel, bearing, the overall structure of railway vehicles, and suspension parts fault, with China hownet, ten thousand data knowledge service platform, IEEEXplore and ScienceDirect as a searchable database, to analyses research published research results at home and abroad in recent years, introduced different types of fault detection of data collected in the type and use of sensor types, including the data analysis methods.

Other contents of this paper are arranged as follows: The second part introduces the functions and typical faults of the main components in the running part;The third part reviews the types of sensors and data analysis methods used by researchers to deal with different faults in recent years.The last part summarizes the technical means mentioned in the article and puts forward suggestions for further research.

2.Typical faults of rail transit train running part

2.1 Wheel set Faults

Wheelset consists of two identical wheels and an axle. The basic structure of wheelset is shown in Figure 2.Wheel includes rim, rim, spoke plate, tread and other parts.Wheelset is an important factor for vehicle safety. With the continuous improvement of vehicle speed, the operating performance and safety requirements of vehicle tires are also improved accordingly.[4].

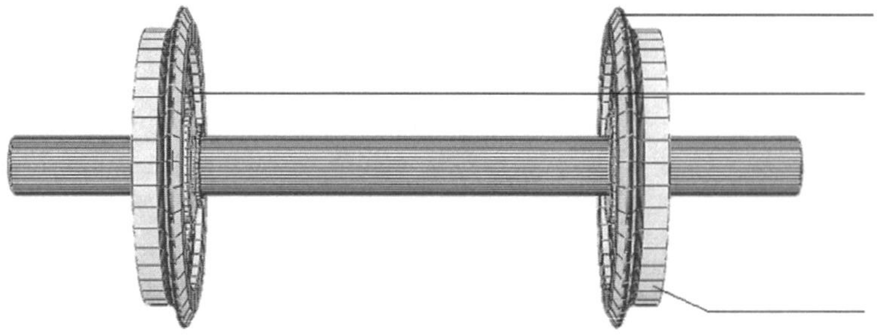

Figure 2 . Basic structure of wheelset

Table 1 lists the main wheel set failures and their mechanisms[5-8].Figure 3 shows strip and abrasion of track tread[9].

Table 1. Main faults and mechanism of wheelset

MAIN MODE OF FAILURE	THE MECHANISM OF
THE TREAD STRIPPING[10, 11]	Structural deformation, as well as surface hardening and metal fatigue occur when metal is crushed[12], and then the change of tread temperature before and after train braking produces rupture and detachment[13]
TREAD SCRATCHES	As a result of the train emergency braking, wheelset and rail sliding, so that the wheel tread scratches, cracks and so on
TREAD CIRCLE	When the wheel is moving, it will produce periodic violent impact in a larger radius, resulting in the wheel more and more deviate from the normal range of rotation
TREAD IN THE HEAP	It is mainly caused by poor material quality of wheelset, which produces plastic deformation at high temperature and forms rolling pile after cooling
BRIM ABRASION	The material at the contact point of the rim and rail is subjected to repeated shearing and extrusion, causing stress and corrosion on the metal surface[14]

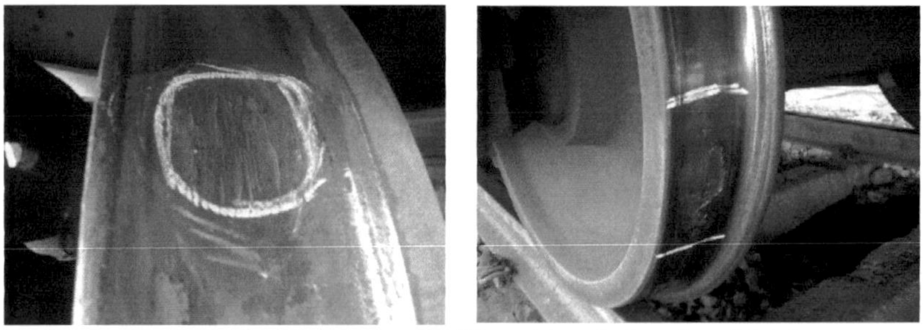

Figure 3. Tread stripping (left) and tread abrasion (right)

2.2 Bearing Failure

Axle box is the key component of running part of EMU, which plays the role of connecting wheelset and frame, bearing and transferring various loads between wheelset and running part.Bearing is the most important part in the hub box, so it is the most important to ensure the safe operation of emU.Bearing is a component with high probability of failure because it operates in high speed and high pressure environment.The bearing structure is shown in Figure 4.

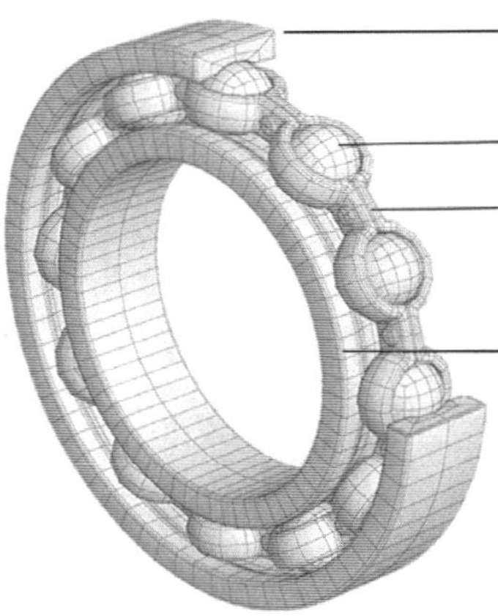

Figure 4 . Structure diagram of rolling bearing

The common faults and mechanisms of rolling bearings in rail transit are shown in Table 2[14-16]. **Figure 5** shows rolling bearing faults caused by different causes, wherein (a), (b) and (c) show fatigue shedding of inner and outer rings and rolling body, **figure** (d) show failure types of rolling body wear, **figure** (e) show inner ring corrosion faults caused by electrical corrosion, and **figure** (f) show cage fracture[17-19].

Table 2. Main bearing faults and mechanisms

MAIN MODE OF FAILURE	THE MECHANISM OF
FATIGUE PEELING	The inner and outer rings are in contact with the rolling body and bear a lot of lateral loads.At the same time, due to the lack of lubricating oil, the formation of cracks, and then the formation of irregular pits, and eventually the formation of large areas of shedding
WEAR AND TEAR	Due to poor bearing sealing performance, dust or foreign matter invasion and insufficient lubrication cause surface wear between inner and outer rings, rolling body and cage
CORROSION	There are generally three forms of expression, one is the chemical wear of lubricating oil, moisture and moisture;The other is the electric spark caused by the current through the electrical corrosion;The third is the wear caused by the relative movement of current.Put up with it for a long time.
CAGE FRACTURE	Due to the wrong operation in the process of production and use, once the cage deformation or fracture will increase the friction between the rolling body[13], thus aggravating vibration and noise, bearing damage
PLASTIC DEFORMATION	When the rolling bearing bears too much load, too much heat and powerful foreign body invasion, the bearing will not return to its original shape when the external force disappears
GLUE	When running under the condition of poor lubrication and high speed and heavy load, the metal on the surface of the bearing is attached to the surface of other metal parts because of friction heating

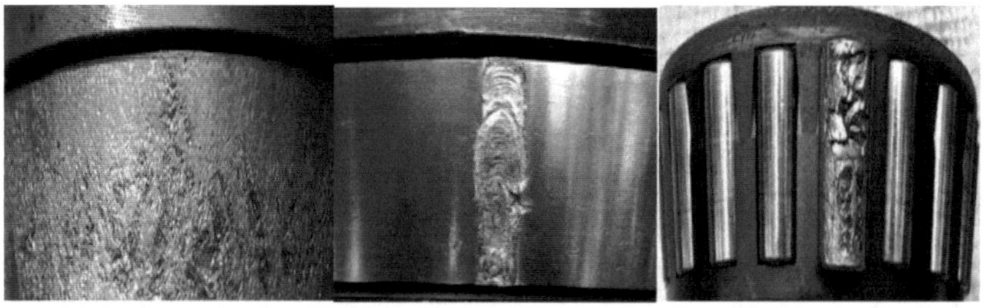

(a) the inner ring is off, (b) the outer ring is off (c) the rolling body is off

(d) rolling wear, (e) inner ring corrosion, (f) cage fracture
Figure 5 . Fault diagram of different types of rolling bearings

2.3 Frame and hanging part of walking part

The walking part, also known as the walking part or trolley, is located below the car body to guide the running vehicle along the track and transfer the weight and load of the car body to the rail.Running gear is an essential part of locomotive and rolling stock. Its function lies in carrying, pulling, running and braking.It is the most important part of the safe and stable operation and dynamic performance of the train.

The running part is mainly composed of six parts: frame, wheelset axle box device, spring suspension device, central traction connecting device, traction motor and gear variable transmission device, and basic brake device.The frame is the main part of bogie bearing and the installation base of bogie components.Figure 6 shows the basic bogie frame.

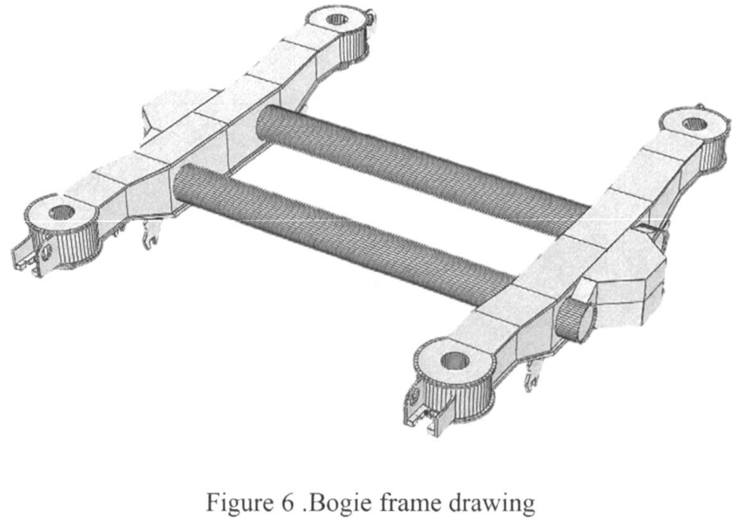

Figure 6 .Bogie frame drawing

Table 3 lists the main faults and mechanisms of the walking frame and suspension part. **Figure** 7 shows the cracks in the middle part between the motor hanger and vertical stop in the steering structure of bullet train[20, 21].

Table 3. Main faults and mechanism of the frame and suspension part of the walking part

MAIN MODE OF FAILURE	THE MECHANISM OF
STRUCTURAL DEFORMATION AND CRACKS	Fatigue caused by use
BROKEN AIR SPRING	Aging or long-term use
AIR SPRING IS OUT OF ORDER	Poor processing technology
LATERAL AND VERTICAL SHOCK ABSORBERS LEAK OIL	Pressure cylinder seal rubber ring fatigue use, high valve nut vibration loosening, anti-side roll torsion rod rubber ring crack aging and fatigue use

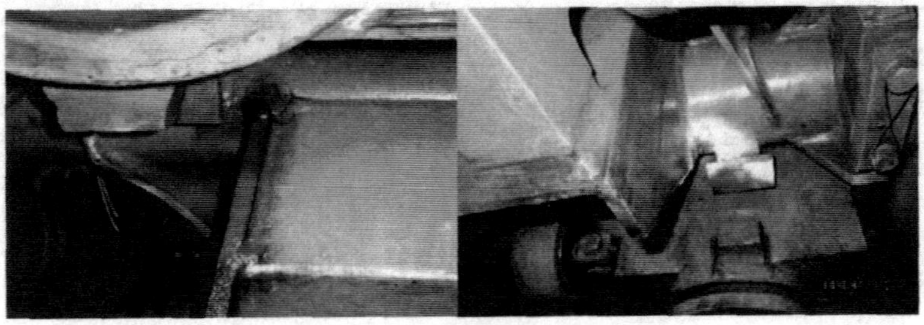

Figure 7 . Bogie frame cracks

3.Sensors and technologies used in fault detection and monitoring

3.1 Wheel set failure

Due to different types of wheelset faults, the corresponding sensors selected are also different, among which vibration acceleration sensor, fiber Bragg grating sensor, eddy current sensor, CCD camera, laser sensor and other sensors appear frequently in the literature.

(1) Vibration acceleration sensor

Vibration acceleration sensor is widely used in vibration signal acquisition.Zhu Yue installed vibration acceleration sensor on the track to collect the coupling vibration signal of wheel and track, and analyzed abnormal vibration and impact effect to detect the wheel tread fault.Four pairs of sensors are installed to prevent information interference.The average speed of the train passing through the acquisition area is measured by the wheel and axle position sensor, and the noise interference is reduced by wavelet packet signal analysis to complete the data processing[22].Wang Fengtao[9, 23]In the research and development of subway wheelset tread fault diagnosis system, the data acquisition method similar to Zhu Yue's is adopted to collect vibration signals and train running speed.Using wavelet packet algorithm and fractal dimension combined with particle swarm optimization algorithm, singleton fractal dimension and multifractal dimension are regarded as feature vectors, and fault identification of tread is accomplished by support vector machine.Installation position of vibration acceleration sensor is shown in Figure 8.

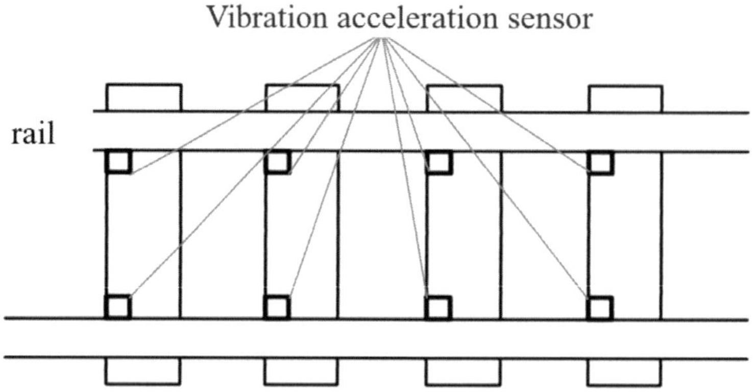

Figure 8 . Installation of vibration acceleration sensor

The depth or length of the tread fault will affect the vibration energy, so the vibration energy of the tread can be indirectly used as the basis for judging whether there is a wheel set fault and its severity.Guo Jinlong[6]In the research of wheelset tread fault detection system of locomotive and rolling stock, the resistance strain gauge is arranged on both sides of the rail to form a weighing sensor, and the weight data is collected to better calculate the vibration energy, so as to distinguish wheelset faults.Hlibert-huang transform was used to preprocess signals to reduce noise, and time-frequency entropy discrimination and fault classification were used to complete fault detection and monitoring.Xiao-hua yu[24]In the research and development of train tread abrasion monitoring system, piezoelectric acceleration sensor is used.He installed four sensors on the wheelset bearing seat to extract the vibration difference, impact, frequency signal and other parameters of the tread during operation, and installed two sensors on the brake disc to measure the working state of the brake disc.Based on the traditional FFT method, the discrete cosine envelope method and Hilbert-Huang transform method are used to identify the wheel-set tread abrasion faults.

(2) Fiber grating

Optical fiber network has the advantages of wavelength coding, small size, high precision, anti-electromagnetic interference and so on. It can be directly connected to the optical fiber system, which is incomparable to many other devices.With the development of fiber Bragg grating technology, various active and passive devices based on fiber Bragg grating technology, performance and application have been developed.

wang[25]In the research of train wheel tread defect monitoring technology based on fiber Bragg grating, fiber Bragg grating sensors are arranged on the rail.Compared with the traditional resistance strain sensor and piezoelectric sensor, the FIBER Bragg grating sensor has better measurement accuracy and long-term stability in the complex and harsh environment, and improves the accuracy of train tread detection, and makes the detection Angle resolution reach 5°.

(3) Eddy current sensor

The methods of measuring wheel geometry parameters can be divided into two categories: contact measurement and non-contact measurement.Contact measurement requires the measuring instrument to be in contact with the surface under test.The advantage is that the measurement accuracy is high, the surface roughness of the measured workpiece is more inclusive, and can adapt to the harsh detection environment[26].Eddy current sensors are often used in contact measurement.

Bear for the star[27]Using the contact measurement method, eddy current sensor is set on the rail side to measure the relative position of the measured object and the probe end face, and based on the parallelogram mechanism to achieve the purpose of online dynamic detection of wheelset tread abrasion, and then the wavelet transform method is used to analyze the data to get the tread abrasion depth.Wang Yijun[28]In the research of WTFDS vehicle wheel-set tread fault detection system, the method of measuring the change of wheel tread height relative to rim height with eddy current sensor is also used to detect the wheel set tread fault.

(4) Laser light source and its sensor

The disadvantage of contact measurement is that the probe is easy to wear, resulting in error increase, affect the measurement effect, low efficiency, non-contact measurement can effectively avoid these problems.Using the camera or the corresponding sensor to obtain the wheelset pictures or signals in a specific detection range in a non-contact way, compared with the standard database, to find out the geometric parameters of the wheel set defects, and then achieve the purpose of detection and measurement.

To wei bin[29]In the research of wheel set tread abrasion fault diagnosis based on vehicle detection, a vehicle detection method combining laser scanning sensor and vibration acceleration sensor was proposed to detect geometric parameter changes of wheel set.The accurate real-time data can be obtained by setting the sensor above the wheel tread under the carriage to measure the tread information.Wavelet packet, rough set theory and support vector machine classification are used to diagnose and detect wheel-set tread abrasion faults.The detection device of laser scanning sensor and vibration acceleration sensor is shown in Figure 9.Liu chang[30]In the research of geometric parameters detection technology of train wheelset tread, the laser displacement sensor and laser triangulation measurement principle are also used to collect wheelset tread data by using semiconductor laser, lens and laser displacement sensor.Compared with the vibration acceleration sensor, the wheel set abrasion position can be detected more intuitively.

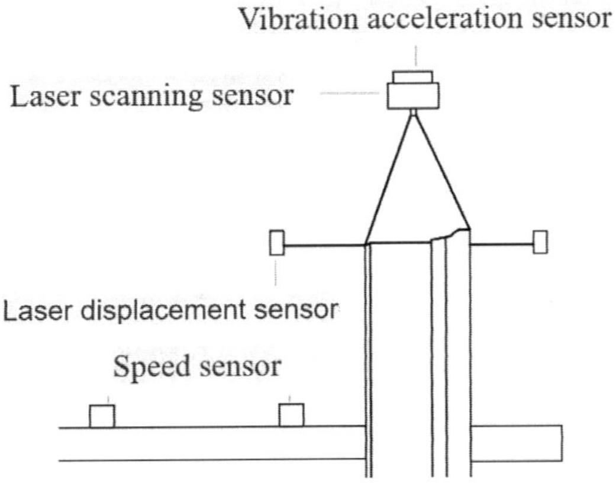

Figure 9 . Laser scanning sensor and vibration acceleration sensor detection device

(5) CCD camera

In addition to laser detection, CCD camera imaging is also used in the tread detection field.Hai-peng xiao[31]Based on the deep study of locomotive wheel tread damage detection, to reduce the influence of vibration on the camera imaging, five groups of on-orbit Settings to use CCD camera in the tread of wheel and rail, on the edge of the image, using the gaussian low-pass filter to filter the raw data, using OpenCV open-source library and YOLOv2 based model to tread the damage detection of the image.Huang Xi[32]Based on image processing and machine vision of wheel tread of vehicle detection technology in the study, using CCD camera in same tread images, according to the characteristics of wheel tread band of contour image, put forward the corresponding all kinds of image processing algorithm, get the band binary image contour, abrasion, through the difference method to get the right amount, using additive operation and define the flange height, flange thickness and other related parameters.Sun ran[33]In the on-line wheel-set tread abrasion imaging detection system, three CCD cameras are used to collect and synthesize the whole cycle image of wheel-set tread on each side, and MATLAB and C++ mixed programming are used to correct and fuse the tread image of the three cameras.

3.2 Bearing failure

Sensors suitable for bearing fault detection include temperature sensor, vibration acceleration sensor, acoustic sensor and so on.

(1) Temperature sensor

In the process of running, axle box bearings will constantly friction and heat, and the faults of axle box bearings are mainly thermal expansion and hot shaft problems caused by heat dissipation delay[34].The temperature changes of bearings can be monitored in real time by temperature sensors.

tramal[35]Install temperature sensor to collect the emu bearing in the axle box bearing temperature, the temperature data cleaning, normalization, after reduction and feature extraction of pretreatment, the improved AHP based on decision tree, the SVM method - DT - SVM algorithm for identifying the fault effectively, and realize the emu axle box bearing health assessment and trend analysis.Li Jindong[36]The temperature sensor was installed on the bogie shaft end to collect the bearing temperature data, clean and integrate the original data and divide the time window, extract and construct features in each time window, and predict the bearing faults of high-speed train axle box based on machine learning method.Yang xm[37]Nine temperature sensors were installed on each axle to collect the temperature information of the axle. Based on online learning algorithm, the bearing error prediction frame of high-speed train was generated, and an evaluation method of bearing performance degradation of high-speed train was proposed based on deep neural network and error matrix.

(2) Vibration acceleration sensor

Bearing will be accompanied by vibration in the process of operation, obtaining vibration information is also an important means to detect whether there is a bearing fault.At present, vibration detection is mainly used in early diagnosis of train rolling bearing faults in China.A variety of diagnostic systems have been successfully developed and installed on the axle inspection line at the depot.Through the measurement and analysis of bearing vibration signal, the bearing condition can be judged, the early fault can be found, and the basis for maintenance can be provided.

Liang Yu[38]Biaxial sensors are installed in different parts of the train axle box to collect vibration and impact information in both vertical and horizontal directions. The sensor signals are adaptively decomposed into a series of single component signals, and the spectrum diagram of bearing signals is calculated. The faults of subway train bearings are diagnosed on the way based on the characteristic parameters in time-frequency domain.Yellow light[39]A fault rolling body location method based on envelope spectrum and convolution average was proposed, and the relationship between maximum impact time interval and defect location was deduced.Wu Jili[40]A new VMD method based on whale optimization algorithm after vibration signals are collected by acceleration sensor solves the problem of early fault identification of high-speed train rolling bearings.by[41]Acceleration sensors are arranged in horizontal and vertical directions to collect bearing vibration signals.The early faults of gearbox bearings of high speed trains were diagnosed by using tianniu optimization algorithm.Yang jieren made[42]The MED theory is used to process vibration signals, and then the vibration signals and the original signals are used to form two-dimensional tensors in the training convolutional neural network, which improves the accuracy of train rolling bearing fault diagnosis.Zhang Jiancai[17]The original vibration signal was degraded by VMD method.PSO algorithm is used to optimize PNN network to realize fault type identification and classification.Hui-ying Yang[43]In order to extract the fault characteristic frequency, an algorithm combining double tree complex wavelet packet transform and total variation is proposed, which makes the impact feature in vibration signal significantly enhanced.Hai-ming wang[44]A defect vibration data repair model based on compressive sensing is established.According to the prior knowledge, a complete dictionary is trained by the improved KSVD algorithm, and the fault characteristic frequency is identified by single point fault diagnosis based on compressed sensing envelope detection.The envelope of filtered time-domain signal is compressed and reconstructed by subspace tracking algorithm to remove noise to complete signal defect supplement and fault detection.

(3) Acoustic sensor

The audio frequency diagnosis method can diagnose faults in the case of vehicle bearing without hot shaft, and realize early detection and early prediction of bearing faults.Zhao Jiameng[45]Pulse sound level meter and vibration sensor are set up on the orbit side to collect the sound and vibration signals of rolling bearings, and the fault is monitored online by time-domain diagnosis method, and the fault is located accurately by wavelet transform method.Fan Zhuo deep and remote[46]The method of t-Music algorithm and Angle interpolation resampling is proposed to solve the problem of signal separation from multiple sound sources of bearing.Shang-bin zhang[47]Aiming at the limitation of collecting spatial sound source by single microphone signal in traditional method, a processing scheme of analyzing spatial sound source by time-varying array is established, and the time-varying spatial position information of sound source can be captured by microphone array.economists[48]In order to solve the problem of multi-source aliasing of data collected from microphone array, a space-time filter rearrangement method and an adaptive sparse filter method are proposed.

3.3 The structure and suspension of the running part are faulty

Vibration acceleration sensor, temperature sensor, thermal imager and acoustic sensor are mainly used for fault detection of walking frame and suspension part.

(1) Vibration acceleration sensor

Zhang Li[35]An acceleration sensor and a strain sensor were installed on the bogie of Beijing Metro Line 2 to collect the bogie load information, vibration signals and the corresponding stress, establish the finite element model of the bogie frame, analyze the bogie load characteristics in time and frequency domain, and establish the load spectrum of the train bogie frame.Ren Ning[49]Acceleration sensors and angular acceleration sensors are installed on the car body, bogie and wheel set to collect the vibration information of the train and the bogie's turn Angle acceleration. The method based on FFT transformation and frequency domain calculation is adopted, and the performance parameters of the suspension system are estimated by using the principle of least square method.Zhang Yuchen[50]The vibration acceleration sensor is installed on the key parts of the car body, frame, axle box and so on, and the vibration acceleration and vibration displacement data of running gear are collected, and four working conditions of running gear of high-speed train are analyzed.According to the pattern recognition of various working conditions, the fault classification and diagnosis method of high-speed train running part is proposed, and the genetic factor of genetic algorithm is introduced to improve it, which can accurately detect the fault type.

(2) Temperature sensor, thermal imager

Li Mengchen[51]The temperature sensor is set on the important node of bogie to collect the temperature data under different running conditions of train, and the feature extraction method based on dynamic time warping can diagnose faults in time.BaoChen inscription[52]In heavy-duty train tread brake wheel brake fault diagnosis temperature field analysis and study, using set on-orbit edges of infrared thermal imager contactless temperature measurement, based on wheel under different conditions of temperature change judge the fault type and grade of the train, the possible fault bogie inspection advice of friction pair and prediction of failure.

(3) Acoustic sensor

Li Qi[53]In the study of railway side acoustic diagnosis system for freight car faults in early operation, several groups of railway side acoustic sensors are used to collect the acoustic signals of train wheel and axle assembly, and the data of several sensors are combined with magnetic method.Steel signal and processing data to parse the acoustic signal of the sensor.High-speed train coupling and high-speed train fault signal location is difficult, the follow-up fault detection and maintenance work is completed.Zhi-gang qu[54]In the train fault audio diagnosis method based on deep belief network, the audio signals collected by the audio sensors on the track side are detected based on DBN analysis and wavelet packet feature vector extraction, and the current fault types of the train are identified.

4. Conclusions and prospects

In this paper, many fault diagnosis techniques for wheelsets, bearings, frames and suspension parts of rail transit train running parts are reviewed, and the sensor types, installation positions and application methods used in these diagnosis techniques are summarized.Through literature research and sorting, it is found that the studies on the fault of running stock in rail transit mainly focus on wheelset tread, bearing, running stock frame parts and suspension part, and the following conclusions can be drawn (the summary of the concise version is shown in Table 4) :

Table 4. Summary of fault detection equipment

The fault structure	Collect data	Used sensor
Wheel tread	Wheel vibration signal	Vibration acceleration sensor, fiber Bragg grating sensor, eddy current sensor
	Wheel tread image information	CCD camera, laser light source and its sensor
bearing	Bearing vibration signal	Vibration acceleration sensor
	Temperature information	Temperature sensor
	Track side acoustic signal	Acoustic transducer
Steering frame and suspension structure	Frame vibration signal	Vibration acceleration sensor
	Temperature data of important nodes	Temperature sensor, thermal imager
	Track side acoustic signal	Acoustic transducer

(1) For wheelset tread faults, vibration acceleration sensor, fiber Bragg grating sensor and eddy current sensor are used to collect train running part vibration, and the faults are analyzed by comparing with the vibration of components under normal working conditions.Laser detection or CCD camera can be used to directly collect tread information to determine the fault location.

(2) For bearing faults, temperature sensors are often used to detect shaft temperature information to judge and predict the occurrence of faults due to obvious temperature changes under fault conditions;Vibration acceleration sensor can also be selected to collect abnormal bearing vibration to determine the type of fault.

(3) For structural and suspension faults, stress and strain gauges can be used to observe the changes of stress and strain in key parts to predict the occurrence of faults;Abnormal vibration can also be detected by vibration acceleration sensor to predict the type and location of the fault.

(4) In addition, acoustic sensors can also be set up on the rail side to detect the fault of the rail. Different faults will send out different acoustic signals, which can be used as the basis for judging the fault of the rail.With the continuous improvement of train speed, the performance requirements of high-speed train system and components are higher and higher, and the realization of intelligent train operation and maintenance is expected to be higher.In the future, we can make full use of sensor technology to build a full range of security monitoring system on the ground, the integrated use of big data technology, from design, manufacturing, operations and other key components to collect data, using intelligent analysis diagnosis algorithm to realize the automation and remote diagnosis, vehicle is constructed based on active intelligent integrated operations and emergency response platform.

reference

[1] THE JIJI PRESS L. Crack in Shinkansen Bullet Train Nearly Caused Derailment [M]. 2017-12-20.
[2] Huaixian Yin. Research on Hidden Danger Assessment System of Subway Train Running Department Based on Reliability Analysis [D]; Beijing Jiaotong University, 2019.
[3] Zhiliang Liu, Deng Pan, Mingjian Zuo, et al. Research progress of Rail Vehicle fault Diagnosis [J].Journal of Mechanical Engineering, 2016, 52(14): 134-46.
[4] Effect of rail corrugation on vertical dynamics of railway vehicle coupled with a track [J]. Acta Mechanica Sinica, 2005, 21(01): 95-102.

[5] Xin Jiang, Tai Liu, Kelei Li. Analysis of Wheel Tread Abrasion of Urban Rail Vehicle [J].Electric Drive for Locomotive, 2012, 06): 61-3.
[6] Jinlong Guo. Research on Wheel Set Tread Fault Detection System of Locomotive and Rolling Stock [D]; Shijiazhuang Tiedao University, 2016.
[7] QIBO FENG J C, YAN ZHAO, YING PI, YONGPING TENG. A dynamic and quantitative method for measuring wheel flats and abrasion of trains [M]. 15th WCNDT World Conference on Non-Destructive Testing,Rome. 2000.
[8] JOHANSSON A,NIELSEN J C O. Out of round Railway Wheels contact Forces and Track Response Based on Field Tests and Numerical Simulation [J]. Proceedings of the Institution of Mechanical Engineers,Part F: Journal of Rail and Rapid Transit, 2003, 217(2):
[9] Fengtao Wang. Research and Development of Wheel Set Tread Fault Diagnosis System for Metro Vehicle [D]; Guangdong University of Technology, 2018.
[10] Peng Jia. Research on CBTC On-board Equipment Online Fault Diagnosis System [J]. Railway Transport and Economy, 2018, 40(11): 94-8.
[11] Lei Xu, Wanming Zhai. Stochastic Analysis Model of Vehicle-track System Under Combined Effects of Wheel Tread Wear and Track Irregularity [J].Journal of the China Railway Society, 2020, 42(02): 79-85.
[12] VERNERSSON T, CAPRIOLI S, KABO E, et al. Wheel Tread Damage:A Numerical Study of Railway Wheel Tread Plasticity under Thermomechanical Loading [J]. Proceedings of the Institution of Mechanical Engineers, Part F: Journal of Rail and Rapid Transit, 2010, 224(5):
[13] ANONYMOUS. Research and Markets; Wheel/Rail Interface Handbook [J]. Engineering Business Journal, 2009,
[14] Weiyao Xie. Influence of Wheel Tread Wear on Locomotive Running Gear [J]. Railway Locomotive & Bullet Train, 2020, 07): 45-6+6.
[15] Yunqi Li. Application of Empirical Wavelet Transform and Support Vector Machine to Rolling Bearing Fault Diagnosis [D]; Southwest Jiaotong University, 2018.
[16] Energy Conversion; New Findings on Energy Conversion from J. Harmouche and Co-Authors Summarized (Improved Fault Diagnosis of Ball Bearings Based on the Global Spectrum of Vibration Signals) [J]. Energy Weekly News,In 2015,
[17] Jiancai Zhang. Research on Rolling Bearing Fault Diagnosis Method based on Rail Transit [D]; Qingdao University, 2020.
[18] CHEN B, ZHANG W, SONG D, et al. Blind deconvolution assisted with periodicity detection techniques and its application to bearing fault feature enhancement [J]. Measurement, 2020, 159(
[19] YAN H, JIANHUI L, ZECHAO L, et al. A Morphological Filtering Method Based on Particle Swarm Optimization for Railway Vehicle Bearing Fault Diagnosis [J]. Shock and Vibration, 2019, 2019(
[20] Lei Dong, Jianhui Cheng, Delong Fu. Crack Fault Analysis of Metro Vehicle Frame [J].Railway Vehicles, 2016, 54(08): 5-8+4.
[21] Li Zhang. Load Test and Research on Bogie Frame of Beijing Metro Line 2 [D]; Beijing Jiaotong University, 2017.
[22] Yue Zhu. Research on On-line Monitoring System of Wheel Set Tread Fault of Urban Rail Vehicle [D]; Nanjing University of Science and Technology, 2013.
[23] Guobiao Wang, He Zhengjia, Chen Xuefeng, et al. Basic Research on Mechanical Fault Diagnosis "Where to Go" [J].Journal of Mechanical Engineering, 2013, 49(01): 63-72.
[24] Xiaohua Yu. Research and Development of Train Tread Abrasion Monitoring System [D]; Hunan University of Science and Technology, 2007.
[25] Pengfei Wang. Research on Monitoring Technology of Wheel set Tread Based on Fiber Bragg Grating [D]; Beijing Jiaotong University, 2012.
[26] ALEMI A, CORMAN F, LODEWIJKS G. Condition monitoring approaches for the detection of railway wheel defects [J]. Proceedings of the Institution of Mechanical Engineers, Part F:Journal of Rail and Rapid Transit, 2017, 231(8):

[27] Weixing Xiong. Research on On-line Dynamic Measurement System of Wheelset Tread Abrasion Based on Parallelogram Mechanism [D]; Beijing Jiaotong University, 2015.
[28] Yijun Wang. Research on Wheel Set Tread Fault Detection System of WTFDS- ⅰ Vehicle [D]; Southwest Jiaotong University, 2007.
[29] Weibin Xiang, HE Deqiang, MIAO Jian, et al. Research on Fault Diagnosis of Wheel set tread Abrasion based on Vehicle Detection [J].Journal of Guangxi University (Natural Science Edition), 2016, 41(02): 434-42.
[30] Chang Liu. Research on Geometric Parameters Detection Technology of Train Wheelset Tread [D]; Changchun University of Science and Technology, 2014.
[31] Haipeng Xiao. Damage Detection of Locomotive Wheelset Tread Based on Deep Learning [D]; Nanjing University of Aeronautics and Astronautics, 2019.
[32] Xi Huang. Research on Wheel Set Tread Detection Technology Based on Image Processing and Machine Vision [D]; Central South University, 2008.
[33] Ran Sun. On-line Wheel-set Tread Abrasion Imaging Detection System [D]; Nanjing University of Aeronautics and Astronautics, 2014.
[34] MAKINO T, SAKAI H, KOZUKA C,et al. Overview of fatigue damage evaluation rule for railway axles in Japan and fatigue property of railway axle made of medium carbon steel [J]. International Journal of Fatigue, 2020, 132(
[35] Min Shu. Health Status Evaluation and Trend Analysis of EmU Axle Box Bearings [D]; Beijing Jiaotong University, 2018.
[36] Jindong Li. Research on Bearing Fault Prediction Method of High-speed Train Axle Box Based on Machine Learning [D]; Beijing Jiaotong University, 2020.
[37] Xiaoming Yang. Research on Real-time Fault Warning of High-speed Train Bearing Based on Shaft Temperature Characteristics [D]; Beijing Jiaotong University, 2020.
[38] Yu Liang. Research on Bearing Fault Diagnosis and In-route Diagnosis System of Metro Train [D]; Beijing Jiaotong University, 2014.
[39] Chenguang Huang, Bing Zhang , et al. Multi-fault rolling body vibration model of high-speed train axle box bearing and its defect location method [J].Journal of Vibration and Shock, 2020, 39(18): 34-43.
[40] Jili Wu, Hongyong Cao, Xu Wang. Parameter Adaptive VMD and Its Application in Early Fault Diagnosis of High-speed Train Bearings [J].Journal of Shaanxi University of Technology (Natural Science Edition), 2021, 37(02): 35-41.
[41] Dan Zhu, Yanchen Su, Qi Sun, et al. Application of BSO-MCKD in early Fault Diagnosis of gearbox bearings of high-speed trains [J].Railway Locomotive and Vehicle, 2020, 40(02): 14-9.
[42] Jieli Yang, Jianhui Lin, Liang Chen. Train Bearing Fault Diagnosis Method Based on MED Assisted Feature Extraction CNN Model [J].China Measurement and Testing, 2020, 46(10): 124-9.
[43] Huiying Yang, Chuanhui Wu, Yanping Li, et al. Application of DTCWPT-TV in High-speed train Gearbox bearing Fault Diagnosis [J].Machinery Design & Manufacture, 2020, 09): 9-13+6.
[44] Haiming Wang. Research on Fault Diagnosis Method of Train Axle Box Bearing Based on Compressive Sensing [D]; Shijiazhuang Tiedao University, 2020.
[45] Jiameng Zhao. Rolling Bearing Fault Diagnosis Based on Acoustic Signal Wavelet Transform [D]; Beijing Jiaotong University, 2009.
[46] Zhuoyou Fan. Research on Train Bearing Fault Diagnosis Based on Track Edge Acoustic Doppler Distortion Signal [D]; Southwest Jiaotong University, 2019.
[47] Shangbin Zhang. Study on Fault Acoustic Spectrum Identification of Train Bearing Rail Edge Acoustic Diagnosis based on Time-varying Array Analysis [D]; University of Science and Technology of China, 2017.
[48] Wei Xiong. Design Method of Multi-Source Aliasing Space-time Filter for Acoustic Fault Diagnosis of Train Bearing Rail Edge [D]; University of Science and Technology of China, 2019.
[49] Ning Ren. Performance Monitoring of Train Suspension System [D]; Zhejiang University, 2014.

[50] Yuchen Zhang. Fault Diagnosis of High-speed Train Running Gear Based on Improved SVM and T-SNE [D]; Southwest Jiaotong University, 2016.
[51] Mengchen Li. Research on Fault Diagnosis Method of High-speed Train Bogie Based on Data Drive [D]; Changchun University of Technology, 2020.
[52] Chenming Bao. Research on Temperature Field Analysis and Brake Fault Diagnosis of Tread Brake Wheel of Heavy Train [D]; Beijing Jiaotong University, 2020.
[53] Qi Li. Research on Acoustic Diagnosis System of Rail Side Fault in Early Operation of Freight Car [D]; Harbin Institute of Technology, 2014.
[54] Zhigang Qu, WANG Man, Li Jiqing, et al. Train fault audio diagnosis method based on deep belief network [J].Measurement and Control Technology, 2020, 39(05): 65-8.

A novel current differential protection for MMC-HVDC lines

Yu Qin[1*], Minghao Wen[1], Yu Bai[2], Yuxi Wang[1], Zeya Fang[1]

[1] School of Electrical and Electronic Engineering, Huazhong University of Science and Technology, Wuhan, Hubei, 430074, China

[2] State Grid Economic and Technological Research Institute Co., Ltd, Beijing, 102209, China

[*]Corresponding author's e-mail: d201880421@hust.edu.cn

Abstract. The present current differential protection for MMC-HVDC transmission lines has absolute selectivity and powerful ability to withstand high transition resistance, while it is easily affected by distributed capacitive current and data synchronization error. To solve the problem above, this article proposes a novel current differential protection scheme. The distributed capacitive current can be calculated by integrating the linear voltage distribution in real-time. Thus, the differential value of the midpoint currents of DC line, which are calculated based on the low-pass filtered measure voltages and currents on both sides, can be adopted to identify the fault. Besides, the data synchronization error can be eliminated based on the waveform matching of the calculated midpoint currents. This novel current differential protection has excellent performance and can solve the problems of traditional current differential protection for HVDC lines.

1. Introduction

The DC transmission line protection is important to ensure the security of the entire system [1-2]. In existing MMC-HVDC transmission projects, traveling wave protection (TWP) and differential undervoltage protection are usually used as the primary protection, and current differential protection is used as backup protection for the DC lines. However, TWP and differential undervoltage protection for HVDC lines have problems of invalid and inadaptable setting criterion, the dependency on simulation and being sensitive to transient resistance. Current differential protection has absolute selectivity and powerful ability to withstand high transition resistance. However, the long transmission distance and high voltage level in HVDC lines will produce a large distributed capacitive current (DCC) during fault transient period. Thus, large time delay has to be set in the presently applied current differential protection to prevent mal-operation [3-4].

Some improved current differential protections for HVDC lines have been proposed. Refs [5-7] propose current differential protection principles based on the distributed parameter line model and frequency-dependent parameter line model, by which the DCC can be compensated. However, the complex calculation and sensitivity to noise and parameter errors make them difficult to be applied in actual engineering. Refs [8-9] propose novel differential protection schemes for HVDC lines, in which the DCC can be calculated based on the linear distributed voltage. However, they are very likely to mal-operate for the lack of capability against data synchronization error (DSE). Thus, they have high requirements for communication real-time quality between both sides [10-11].

To solve those problems mentioned above, a novel current differential protection for MMC-HVDC lines is proposed. When the voltage and current signals at both line ends are filtered by the low-pass filter, the voltage distribution can be approximately regarded as linear distribution. The DCC can be calculated based on the linearly distributed voltage. Thus, the differential value of the midpoint currents of DC line, which are calculated with filtered measure voltages and currents, can be adopted to identify the fault. Besides, the data synchronization error can be eliminated based on the waveform matching of the calculated midpoint currents. With PSCAD/EMTDC, an MMC-HVDC transmission system is modeled and extensive simulation results demonstrate that the newly proposed protection stands out for absolute selectivity, fast response and powerful fault resistance withstand ability.

2. Compensation method of DCC

The distributed parameter model of transmission line is shown in Figure 1. According to Kirchhoff's current law, the current along the DC transmission line satisfies equation (1).

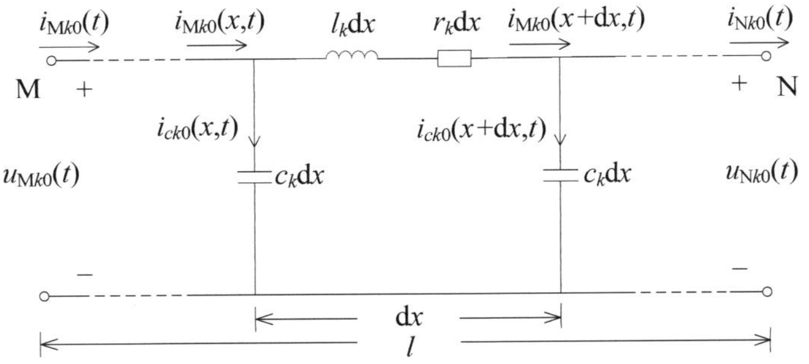

Figure 1. Distributed Parameter Model of Transmission Line

$$i_{\mathrm{M}ko}(t)-\int_0^l i_{cko}(x,t)\mathrm{d}x = i_{\mathrm{N}ko}(t) \tag{1}$$

Where $i_{Mko}(t)$ and $i_{Nko}(t)$ represent the measure currents on both sides of DC transmission line, $i_{cko}(x,t)$ represents the DCC per unit length at x, k=1 means 1-mode, k=0 means 0-mode, l means the DC line length. Since $i_{cko}(x,t)$ cannot be measured, the current differential protection in practical HVDC systems usually sets a long time delay to avoid maloperation caused by DCC during transient process. In order to improve the operating speed of current differential protection in MMC-HVDC systems, it is necessary to compensate for the DCC along the DC line.

When there is no fault or external fault on the DC line, the sum of the DCC can be calculated by

$$\int_0^l i_{cko}(x,t)\mathrm{d}x = \int_0^l c_k \frac{\partial u_{ko}(x,t)}{\partial t}\mathrm{d}x \tag{2}$$

According to [8], after voltage and current signals at both line ends are filtered by the low-pass filter, the voltage distribution can be approximately regarded as linear distribution. The cut-off frequency f_c of the low-pass filter should satisfy

$$f_c < \frac{v\arccos(1-E_{k\_max})}{180l} \tag{3}$$

Where E_{k_max} is the maximum linear fitting error. Thus, the voltage along the DC line can be given by

$$u_k(x,t) = u_{\mathrm{M}k}(t) + \frac{u_{\mathrm{N}k}(t)-u_{\mathrm{M}k}(t)}{l}x \tag{4}$$

Where $u_k(x,t)$ is the low-pass filtered voltage along the line, $u_{Mk}(t)$ and $u_{Nk}(t)$ are the low-pass filtered measure voltages on both sides. According to (2) and (4), when there is no fault or external fault on the DC line, the sum of DCC can be given by

$$\int_0^l i_{ck}(x,t)\mathrm{d}x = \int_0^l c_k \frac{\partial u_k(x,t)}{\partial t}\mathrm{d}x = \frac{c_k l}{2}\left[\frac{\mathrm{d}u_{\mathrm{M}k}(t)}{\mathrm{d}t}+\frac{\mathrm{d}u_{\mathrm{N}k}(t)}{\mathrm{d}t}\right] \tag{5}$$

Where $i_{ck}(x,t)$ represents the DCC per unit length caused by $u_k(x,t)$. According to (5), when the measure voltages and currents on both sides are processed by the low-pass filter, the original distributed parameter line model can be equivalent to Figure 2.

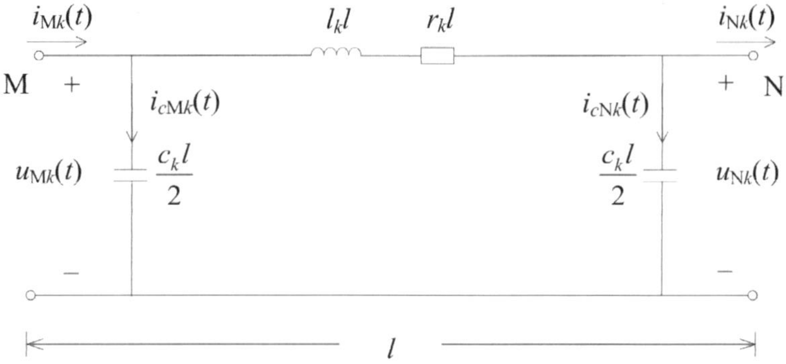

Figure 2. The equivalent model of DC line

The expression of $i_{cMk}(t)$ and $i_{cNk}(t)$ can be given by

$$\begin{cases} i_{cMk}(t) = \dfrac{c_k l}{2} \dfrac{\mathrm{d}u_{Mk}(t)}{\mathrm{d}t} \\ i_{cNk}(t) = \dfrac{c_k l}{2} \dfrac{\mathrm{d}u_{Nk}(t)}{\mathrm{d}t} \end{cases} \tag{6}$$

When there is an external fault, the low-pass filtered current along the DC line can be given by

$$i_{Mk}(t) - i_{cMk}(t) = i_{Nk}(t) + i_{cNk}(t) \tag{7}$$

Where $i_{Mk}(t)$ and $i_{Nk}(t)$ represent the low-pass filtered measure currents on both sides.

It is obvious from (3) that f_c is related to E_{k_max} and l. The larger the l is, the lower the f_c should be set under the same E_{k_max}. However, the lower cut-off frequency will lead to longer time delay. In order to realize accurate compensation to the DCC as well as fast protection response, a sectional compensation method is proposed in this article, in which the DC line is divided into N sections. Based on this, it is feasible to realize sectional compensation to DCC. For the DC line of 200km, considering calculation complexity, voltage linear fitting error and protection operating speed, N can be set as 2. Thus, the length of each section of the DC line is 100km, and cut-off frequency f_c can be set lower than 190Hz according to (3). The sectional compensation method is shown in Figure 3.

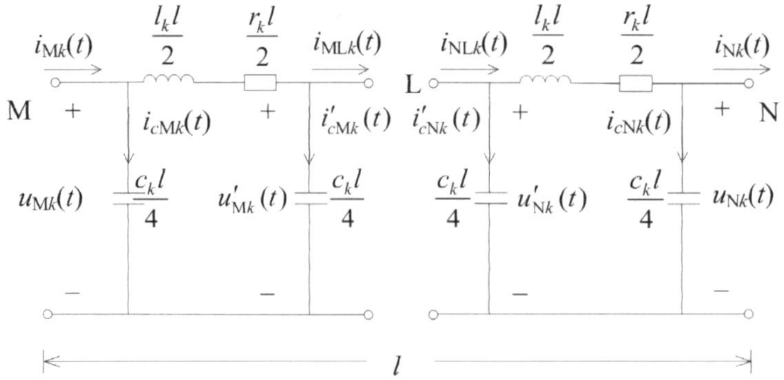

Figure 3 Sectional compensation to the DCC

When there is no fault or external fault on the DC line, the currents in Figure 3 satisfy

$$i_{MLk}(t) = i_{Mk}(t) - i_{cMk}(t) - i'_{cMk}(t) = i_{NLk}(t) = i_{Nk}(t) + i_{cNk}(t) + i'_{cNk}(t) \tag{8}$$

$$\begin{cases} i_{\mathrm{cM}k}(t)=\dfrac{c_k l}{4}\dfrac{\mathrm{d}u_{\mathrm{M}k}(t)}{\mathrm{d}t} \\ i'_{\mathrm{cM}k}(t)=\dfrac{c_k l}{4}\dfrac{\mathrm{d}u'_{\mathrm{M}k}(t)}{\mathrm{d}t} \\ i'_{\mathrm{cN}k}(t)=\dfrac{c_k l}{4}\dfrac{\mathrm{d}u'_{\mathrm{N}k}(t)}{\mathrm{d}t} \\ i_{\mathrm{cN}k}(t)=\dfrac{c_k l}{4}\dfrac{\mathrm{d}u_{\mathrm{N}k}(t)}{\mathrm{d}t} \end{cases} \quad \begin{cases} u'_{\mathrm{M}k}(t)=u_{\mathrm{M}k}(t)-\dfrac{r_k l}{2}[i_{\mathrm{M}k}(t)-i_{\mathrm{cM}k}(t)]-\dfrac{l_k l}{2}\dfrac{\mathrm{d}[i_{\mathrm{M}k}(t)-i_{\mathrm{cM}k}(t)]}{\mathrm{d}t} \\ u'_{\mathrm{N}k}(t)=u_{\mathrm{N}k}(t)+\dfrac{r_k l}{2}[i_{\mathrm{N}k}(t)+i_{\mathrm{cN}k}(t)]+\dfrac{l_k l}{2}\dfrac{\mathrm{d}[i_{\mathrm{N}k}(t)+i_{\mathrm{cN}k}(t)]}{\mathrm{d}t} \end{cases} \tag{9}$$

Where $i_{\mathrm{ML}k}(t)$ and $i_{\mathrm{NL}k}(t)$ are the midpoint currents of DC line, which are calculated with low-pass filtered measure voltages and currents on both sides. The 1-mode values are adopted and the internal faults and external faults can be discriminated by $i_{\mathrm{dL1}}(t)$.

$$i_{\mathrm{dL1}}(t)=\left|i_{\mathrm{ML1}}(t)-i_{\mathrm{NL1}}(t)\right|>I_{\mathrm{set}} \tag{10}$$

Where I_{set} is the setting value of the differential current.

3. The method to estimate DSE

The proposed protection above still requires strict synchronization. The current differential protection may mal-operate under DSE. To solve the problem mentioned above, a method to estimate DSE is proposed. The waveforms of $i_{\mathrm{ML1}}(t)$ and $i_{\mathrm{NL1}}(t)$ will be matched under external faults. Accurate DSE can be obtained by matching the waveforms of $i_{\mathrm{ML1}}(t)$ and $i_{\mathrm{NL1}}(t)$ under external faults. Thus, the differential current $i_{\mathrm{dL1X}}(t)$, in which the DCC is compensated and DSE is eliminated, can be obtained based on $i_{\mathrm{dL1}}(t)$. The differential current $i_{\mathrm{dL1X}}(t)$, which is greater than the threshold under internal faults and lower under internal faults, can be employed to establish the protection criterion. To describe the waveform similarity between $i_{\mathrm{MLk}}(t)$ and $i_{\mathrm{NL1}}(t)$, the Pearson correlation coefficient (PCC) is adopted. For any two sets of discrete signal sequences $x=\{x_1,x_2,\cdots,x_n\}$ and $y=\{y_1,y_2,\cdots,y_n\}$, the PCC $p_{x,y}$ of x, y can be given by

$$p_{x,y}=\frac{\sum_{i=1}^{\mathrm{N_d}}[x_i-\frac{1}{\mathrm{N_d}}\sum_{i=1}^{\mathrm{N_d}}x_i][y_i-\frac{1}{\mathrm{N_d}}\sum_{i=1}^{\mathrm{N_d}}y_i]}{\sqrt{\sum_{i=1}^{\mathrm{N_d}}[x_i-\frac{1}{\mathrm{N_d}}\sum_{i=1}^{\mathrm{N_d}}x_i]^2}\sqrt{\sum_{i=1}^{\mathrm{N_d}}[y_i-\frac{1}{\mathrm{N_d}}\sum_{i=1}^{\mathrm{N_d}}y_i]^2}} \tag{11}$$

Where $\mathrm{N_d}$ denotes the sampling number. The greater $p_{x,y}$ indicates the stronger correlation between x and y. When x and y are exactly matched, $p_{x,y}$=1; When x and y have opposite trends, $p_{x,y}$=-1; When x and y are unrelated, $p_{x,y}$=0.

After the fault occurs, the midpoint currents of the DC line $i_{\mathrm{ML1}}(t)$ and $i_{\mathrm{NL1}}(t)$, which are calculated based on (8)-(9), have obvious waveform characteristics. The waveforms of $i_{\mathrm{ML1}}(t)$ and $i_{\mathrm{NL1}}(t)$ are exactly the same regardless of the DSE, and can be completely matched at a certain moment during the data window sliding process. Through waveform matching, an accurate DSE can be calculated and the differential current, in which the DCC is compensated and DSE is eliminated, can be obtained.

4. Current differential protection scheme of DC line

A novel current differential protection for MMC-HVDC lines which can solve the problems of DCC and DSE is proposed. The protection scheme is composed of four parts.

(1) Protection start-up element

The maximum value of the modulus of positive, negative and 1-mode measure voltage change rate is adopted as the protection start-up element, which can be given by

$$\left|\frac{\mathrm{d}u(t)}{\mathrm{d}t}\right|_{\max}=\max[\left|\frac{\mathrm{d}u_{\mathrm{po}}(t)}{\mathrm{d}t}\right|,\left|\frac{\mathrm{d}u_{\mathrm{no}}(t)}{\mathrm{d}t}\right|,\left|\frac{\mathrm{d}u_{\mathrm{1o}}(t)}{\mathrm{d}t}\right|]>D_{\mathrm{set}} \tag{12}$$

Where $u_{\mathrm{po}}(t)$, $u_{\mathrm{no}}(t)$ and $u_{\mathrm{1o}}(t)$ are measure voltages of the positive pole, negative pole and 1-mode respectively. D_{set} is setting threshold setting value.

When the start-up element is tripped, the measure voltage and current signals will be sent to the opposite side, and calculate start time of opposite side by opposite side measured voltage. t_{Mq} is start time of side M, t_{Nq} is start time of side N.

(2) Low-pass filter

Butterworth low-pass filter is applied to filter the measure voltages and currents, the cut-off frequency of which can be set based on (3).

(3) DSE estimation

Taking the relay M for an example, the process of DSE estimation include the following steps.

a. Select the sampling data of $i_{NL1}(t)$ during $[t_{Nq}-\Delta t_L/2,\ t_{Nq}+\Delta t_L/2]$ as the scanning sample.
b. Set $t_{Mq}-\Delta t_L/2-\Delta t_H$ as the match start point and $t_{Mq}+\Delta t_L/2+\Delta t_H$ as the match terminal point.
c. Set the match start point as the calculation start point, select the sampling data of $i_{ML1}(t)$, the data window of which is equal to Δt_L, and calculate the PCC $p_{iML1,iNL1}(t)$ between it and the scanning sample.
d. If $t<t_{Mq}-\Delta t_L+\Delta t_H$, the match start point will move to next sampling point, and turn back to step c. Otherwise, move to next step.
e. Calculate the maximum $p_{iML1,iNL1}(t_{max})$ of PCC, in which t_{max} is time of the greatest PCC. If $p_{iML1,iNL1}(t_{max})$ is less than p_{set}, which is the threshold, the synchronization error $\Delta t_d=0$. If $p_{iML1,iNL1}(t_{max})$ is greater than p_{set}, the synchronization error $\Delta t_d=t_{max}-(t_{Nq}+\Delta t_L/2)$.

(4) Current differential protection criterion

The current differential protection criterion can be expressed as

$$i_{dL1X}(t)=\left|i_{ML1}(t)-i_{NL1}(t-\Delta t_d)\right|>I_{set} \tag{13}$$

Where $i_{ML1}(t)$ and $i_{NL1}(t-\Delta t_d)$ can be obtained based on (8)-(10). If there are K values satisfying (13) continuously, an internal fault is detected.

5. Simulation Analysis

For a further study and verification of the novel differential current protection scheme, a ±500kV four-terminals MMC-HVDC transmission system model is established in the transient simulation software PSCAD/ EMTDC. U_{S1}~U_{S4} denote the AC power sources, MMC1~MMC4 denote the converters, Line1~Line4 denote the DC lines, U_{dc1}~U_{dc4} denote the DC bus voltages, and L_T is the DC reactor. D_{ij} (i, j=1,2,3,4) represents the DC circuit breaker and R_{ij} (i, j=1,2,3,4) represents the relay.

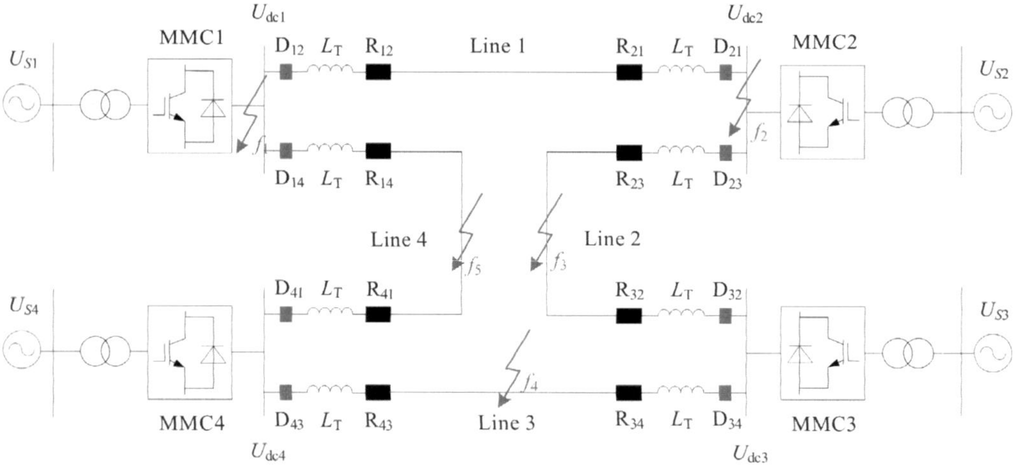

Figure 4 Structure of MMC-HVDC system

Table 1 Parameters of converters

	C_{sm}/μF	R_{sm}/Ω	L_{arm}/mH	N_{sm}
MMC1	8000	0.005	100	244
MMC2	8000	0.005	100	244
MMC3	15000	0.005	50	244
MMC4	15000	0.005	50	244

Table 2 Parameters of DC transmission lines

Projects		Parameters	Projects	Parameters
Length	Line12	200 km	Unit resistance	0.03984 Ω/km
	Line23	157 km	External radius	0.03623 m
	Line34	205 km	SAG	16.5 m
	Line41	50 km	Bundle spacing	0.5 m

The faults occurring time is set at 0.8s, sampling rate is set as 10kHz, D_{set} is set as 100kV/ms, I_{set} is set as $0.1I_N$, and K is set as 3. Δt_L is set as 4ms and Δt_H is set as 2ms. E_{k_max} is set as 2%, and f_c of the low-pass filter should satisfy

$$f_c < \frac{v \arccos(1-E_{k\_max})}{180l/2} = \frac{3\times10^5 \arccos(1-2\%)}{180\times100} = 191\text{Hz} \tag{14}$$

Thus, f_c can be set as 150Hz. To verify the performance of the novel protection scheme, the test system is simulated with numerous cases under different fault conditions. The simulation results under a synchronization error of 1ms are shown in Table 3 and Table 4.

Table 3 Simulation results under internal faults

Fault Conditions			Protection Results		
Fault Type	Position	R_f/Ω	Fault Area	Time	Action
Positive pole to ground fault	0%	0Ω	Internal	4.2ms	YES
		500Ω	Internal	4.2ms	YES
		1000Ω	Internal	4.2ms	YES
	50%	0Ω	Internal	4.5ms	YES
		500Ω	Internal	4.5ms	YES
		1000Ω	Internal	4.5ms	YES
	100%	0Ω	Internal	4.8ms	YES
		500Ω	Internal	4.8ms	YES
		1000Ω	Internal	4.8ms	YES
Negative pole to ground fault	0%	0Ω	Internal	4.2ms	YES
		500Ω	Internal	4.2ms	YES
		1000Ω	Internal	4.2ms	YES
	50%	0Ω	Internal	4.5ms	YES
		500Ω	Internal	4.5ms	YES
		1000Ω	Internal	4.5ms	YES
	100%	0Ω	Internal	4.8ms	YES
		500Ω	Internal	4.8ms	YES
		1000Ω	Internal	4.8ms	YES
Pole to pole fault	0%	0Ω	Internal	4.2ms	YES
	50%	0Ω	Internal	4.5ms	YES
	100%	0Ω	Internal	4.8ms	YES

Journal of Physics: Conference Series **2113** (2021) 012053 doi:10.1088/1742-6596/2113/1/012053

Table 4 Simulation results under external faults

Fault Conditions			Protection Results	
Fault Type	Position	R_f/Ω	Fault Area	Action
DC Bus Fault	f_1	0Ω	External	NO
DC Bus Fault	f_2	0Ω	External	NO
DC Line2 Fault	f_3	0Ω	External	NO
DC Line3 Fault	f_4	0Ω	External	NO
DC Line4 Fault	f_5	0Ω	External	NO

Table 3 shows the performance of the novel current differential protection under the internal fault. From the table, when the internal fault location points from 0% to 100% and the fault resistance varies from 0Ω to 1000Ω, the internal faults can all be correctly identified and the protection has fast response. From Table 4, with different fault locations, types and resistances, the proposed protection will not operate reliably. Consequently, after changing the fault conditions, including the fault location, fault type, and fault resistance, the proposed protection method can discriminate internal faults from external faults accurately and quickly.

6. Conclusions

A novel current differential protection scheme for MMC-HVDC lines is proposed in this paper. The distributed capacitive current can be calculated by integrating the linear voltage distribution in real-time. Thus, the differential value of the midpoint currents of DC line calculated with low-pass filtered measure voltages and currents on both sides can be adopted to identify the fault. Besides, the data synchronization error can be eliminated based on the waveform matching of the calculated midpoint currents. Numerous simulations have proved that the novel current differential protection has fast response under all test conditions and does not require strict data synchronization on both sides.

References

[1] X. Lin, Z. Li, K. Wu and H. Weng, "Principles and implementations of hierarchical region defensive systems of power grid," IEEE Trans. Power Del., vol. 24, no. 1, pp. 30-37, Jan. 2009.

[2] S. Mirhosseini and M. Akhbari, "Wide area backup protection algorithm for transmission lines based on fault component complex power," International Journal of Electrical Power & Energy Systems, vol. 83, pp. 1-6, 2016.

[3] Li A, Cai Z, Sun Q, Li X, Ren D, Yang Z. Study on the dynamic performance characteristics of HVDC control and protections for the HVDC line fault. In: Proc power energy soc gen meeting, Calgary, AB, Canada, 2009. p. 1–5.

[4] Naidoo D, Ijumba NM. HVDC line protection for the proposed future HVDC systems. In: Proc IEEE powercon conf, Singapore, 2004. p. 1327–32.

[5] Gao. S, Liu Q, Song G. Current differential protection principle of HVDC transmission system. IET Generation, Transmission & Distribution 2017;11(5):1286-1292

[6] Chu. X. Unbalanced current analysis and novel differential protection for HVDC transmission lines based on the distributed parameter model. Int J Electr Power Energy Syst 2019; 171:105-115

[7] Chu X. Transient numerical calculation and differential protection algorithm for HVDC transmission lines based on a frequency-dependent parameter model. Int J Electr Power Energy Syst 2019; 108:107-116

[8] Zheng, J, et al. A novel differential protection scheme for HVDC transmission lines. International Journal of Electrical Power & Energy Systems, 2018. 94: pp. 171-178.

[9] Zheng, J, et al. A Novel Differential Protection Scheme With Fault Line Selection Capability for HVDC Transmission Line. Proceedings of the CSEE 2018; 38(15):4350-4358.

[10] Tong N, Fan L, Lin X, et al. Waveform matching based protection strategy for VSC-MTDC independent on synchronization and boundary component. Proceedings of the CSEE 2019; 39(12):3820-3833
[11] Chen L, Lin X, Li Z, et al. Similarity comparison based high-speed pilot protection for transmission line. IEEE Trans. Power Del 2018;33(2):938-948

Journal of Physics: Conference Series **2113** (2021) 012054 doi:10.1088/1742-6596/2113/1/012054

Intelligent support and command system based on WAPI terminal equipment

Yu Wang [1*], Zongchun Wei [1], Haibo Bao [1], Gaofeng Deng [2], Jiangwei Li [1], Pei Li [1], Leigang Nie [1] and Hua Fu [1]

[1] Nanning power supply bureau of Guangxi Power Grid Co. Ltd

[2] Guangxi Power Grid Co. Ltd

*Corresponding author's e-mail: wangy0515@nng.gx.csg.cn

Abstract: In order to solve the problem that the process of WAPI terminal connection is troublesome, this paper proposes an intelligent support and command system of WAPI terminal equipment based on centralized control AC + AP architecture. The system stores WAPI certificate on AC (centralized controller), WAPI terminal equipment is associated with AP (wireless access point), and AP equipment notifies AC of the associated events of the terminal, AC equipment and terminal equipment enter authentication and complete certificate authentication. Complete access authentication through AC and terminal equipment, and store the unicast and multicast keys generated by authentication negotiation on AC. This method is more convenient to maintain the certificate. At the same time, on the premise of ensuring the security of the certificate, it also improves the encryption and decryption efficiency and improves the user's business experience.

1. Introduction

WAPI wireless LAN Authentication and privacy infrastructure is a secure transmission protocol for IEEE802.11 protocol [2]. In the gb15629.11 [3] standard of wireless local area network in China, there is a clear security solution for WLAN (wireless LAN), which adopts the certificate mechanism of public key cryptosystem and realizes the two-way authentication between mobile terminal (MT) and wireless access point (AP) in a real sense, making the information transmission between both ends more secure.

However, when all terminals supporting WLAN function need to support WAPI security protocol, the problem of secure storage of WAPI certificate also arises. At present, many WAPI certificates are directly stored on terminals, WAPI security modules or intermediate wireless access points. This method of saving WAPI certificates has great security risks. WAPI certificates are easy to be illegally copied or stolen by others, which leads to the threat of network security. Similarly, the data security of the business between AP equipment and AC equipment also depends on the tunnel defined by each manufacturer, which can not ensure sufficient security and privacy. For a lightweight AP, the performance of the device is limited. When maintaining the corresponding certificate, when multiple terminals are associated at the same time, the performance of the AP device will be seriously affected, resulting in packet loss or fragmentation during data transmission. Daily maintenance and management are also difficult, and ultimately affect the user experience.

By studying an intelligent support and command system based on WAPI terminal equipment, this paper proposes a WAPI technology system based on centralized control AC + AP architecture. In this system, the equipment certificate is installed on AC, which has high security, convenient maintenance and effectively improves the user experience.

2. Composition and implementation method of intelligent support and command system based on WAPI terminal equipment

2.1. System architecture composition and function

The intelligent support and command system based on WAPI terminal equipment is a system of WAPI technology based on centralized control AC + AP architecture. The system includes AS server (WAPI authentication server), AC (access controller), AP (wireless access point) and WAPI terminal. The system architecture is shown in Figure 1:

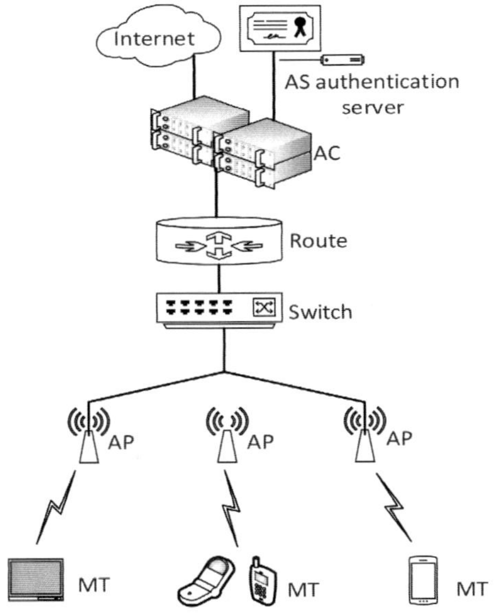

Figure 1 system architecture

AC (access controller) is mainly responsible for configuration management of all downstream AP devices, authentication, management, security, storage and installation of wireless users, generation and issuance of WAPI certificates from AS servers and other control functions. After the AC establishes a CAPWAP tunnel with the lightweight access point AP equipment, the AC equipment performs online centralized management of the AP and centralized management of the terminal services, but does not directly issue relevant WAPI certificates to the AP equipment. The WAPI terminal notifies the AC of related events through the AP, the AC and the terminal enter the authentication process, and the AC equipment and the AS server complete the authentication of the certificate.

AP (wireless access point) is mainly responsible for providing wireless access for the terminal and forwarding the encryption and decryption service data based on CAPWAP tunnel. It mainly plays the role of seamless integration and bridging of wired and wireless management frames and service frames.

As server (WAPI authentication server) is the core equipment in the system architecture based on WAPI protocol. It is mainly responsible for providing WAPI identity certificate management and authentication services. AS authentication server is divided into CISU (certificate issuing function of AS) and ASU (identity authentication function of AS). Among them, CISU supports WAPI certificate issuance service, forces the use of digital identity certificate as identity certificate, and can issue digital identity certificate for the accessed AP and STA. The system adopts SM4 symmetric cryptographic algorithm approved by the State Password Administration to encrypt and decrypt the transmitted data, which fully ensures the security of data transmission and the integrity of user information. ASU (authentication service unit) is mainly used to manage the certificates required in the information interaction of each network element, such as certificate generation, issuance, revocation and update [4] [5]. A certificate contains the public key and signature of the issuer and holder of the certificate. In addition, ECC based algorithm is adopted in the whole certificate authentication process, which can make the secure message hash algorithm more ensure the integrity of the message, and make it difficult for those who deliberately attack the system to modify and forge the authenticated information, so as to improve the security.

WAPI terminal is a WAPI module with encryption chip, including the main control chip, the encryption chip and RF chip, the connection management and configuration of the main control chip to the whole WAPI module, and the WAPI protocol stack. The most important is that the encryption chip can verify the WAPI certificate used for WAPI authentication, key negotiation, key decryption and so on.

2.2. Implementation method of intelligent support and command system based on WAPI terminal equipment

Store the AP's WAPI certificate on the AC device. After the AP and AC establish a capwap tunnel, and the terminal is associated with the AP and registered online on the AC, the terminal completes the authentication of WAPI certificate through AC and AS. WAPI protocol adopts the public key cryptography mechanism to authenticate and authenticate the terminal and AP in the system by using the issued certificate. Figure 2 shows the flow chart of the working process:

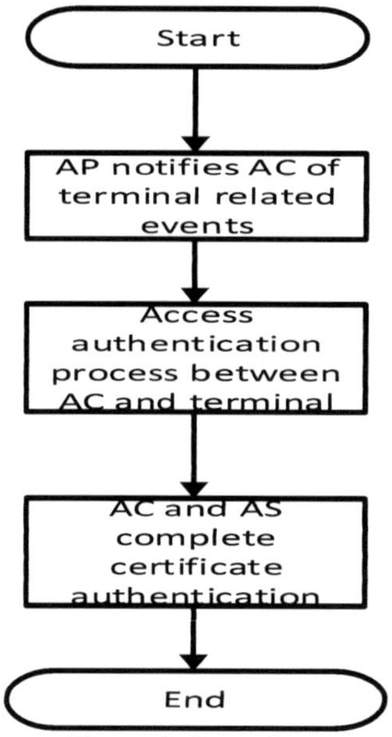

Figure 2 flow chart

The terminal is associated with the AP device through SSID, and then the AP device sends the terminal associated event notification message to the AC device to inform the AC device. After the terminal is associated with the AP, the AP notifies the AC of the associated event, and the terminal and AC start mutual identity authentication. During this period, the terminal must send its own certificate and current time to the AP, and the AP forwards the relevant information to the AC.

The AC device accesses the WAPI certificate sent by the AP device, initiates the access authentication process, and the intermediate interactive authentication message is forwarded by the AP device. The AC signs the message sent by the AP, and sends the terminal certificate, submission time and the private key of the AP device to the AS server.

The authentication of WAPI certificate is carried out interactively in AC and AS [6] [7]. The AS server completes the authentication of relevant certificates. After receiving the authentication request message sent by the AC equipment, the AS will first verify the signature of the AP in the request message and the integrity and validity of the AP certificate. The second is to verify the certificate information of the terminal [8]. Finally, the AS authentication server will sign the authentication results of the AP and the terminal with its own key, and send the signature to the AC together with the previous two authentication results. AC signs and further verifies according to the message authentication result sent by AS, and obtains a corresponding final authentication result, so as to decide whether to allow the terminal to access. The AC forwards the authentication result to the terminal through the AP device, and then the terminal device identifies and verifies the signature result of the as, so as to determine whether it can access the AP.

2.3. Specific implementation of intelligent support and command system based on WAPI terminal equipment

The system involved here adopts capwap tunnel (a special tunnel in WLAN system) on the wired link to enhance the security of data. WAPI technology is adopted in the aspect of air interface. It adopts ternary symmetric security architecture. The trusted third-party authentication server as will conduct two-way authentication between the terminal and AP, which can effectively ensure the security of air interface data transmission. The message interaction process [9] is shown in Figure 3:

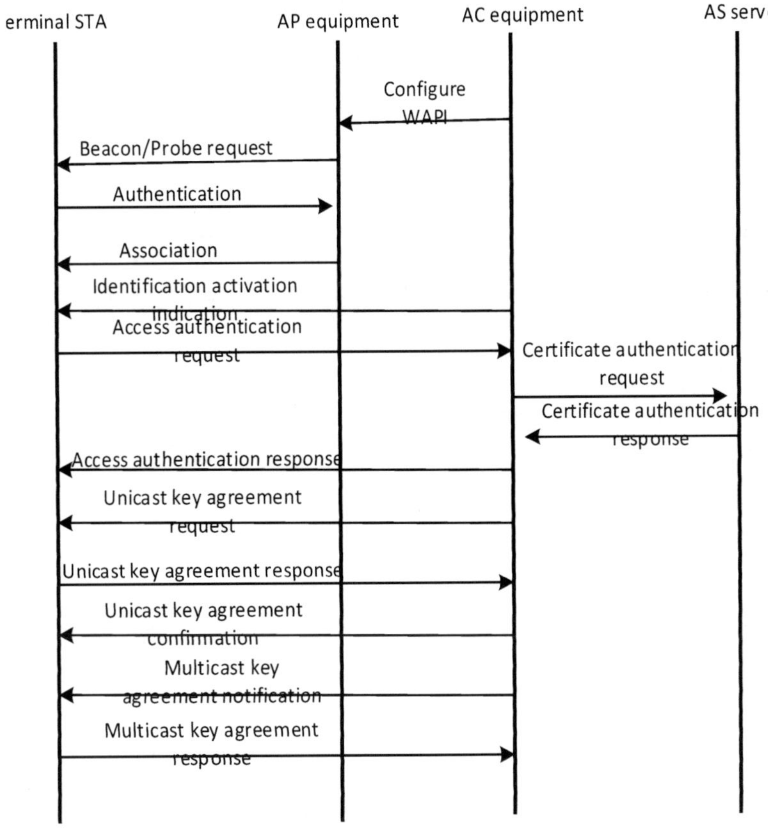

Figure 3 message interaction process

Install the WAPI certificate issued from the as authentication server on the AC, and configure the association relationship between the AC and the AP's WAPI certificate, so that the configuration capability of the AC can be connected to the AP.

The AP sends beacon / probe request to the terminal Frame message, which includes SSID and WAPI capability set information. After receiving the Becon broadcast frame message sent by the AP, the terminal will send the Authentication message frame. After receiving the Authentication message sent by the terminal, the AP sends an Association message to the terminal, which includes its associated SSID, WAPI capability set and other information. After the terminal successfully associates with the AP, the AC device sends an authentication activation instruction to the terminal, which is forwarded to the terminal through the AP. After receiving the authentication activation instruction, the terminal immediately sends an access authentication request to the AC device, and the same request is forwarded to the AC device through the AP. After receiving the corresponding request message, the AC equipment will send the certificate authentication request message to the AS and wait for the as equipment to analyze and respond to the message. After receiving the authentication request message sent by the AC, the AS performs parsing processing. If the certificate is authentic, it sends the response message of the certificate authentication result to the AC equipment. After receiving the response authentication message from the AS, the AC sends back the access authentication response message to the terminal [10] [11]. The response message is finally forwarded to the terminal through the AP equipment.

After the previous access authentication operation is completed, the subsequent key negotiation is carried out. The AC device sends a unicast key agreement request message to the terminal device, which is forwarded to the terminal through the AP device. Then, the terminal sends a unicast key agreement response message to the AC device, which is forwarded through the AP device. After receiving the key negotiation response message from the terminal, the AC device continues to send unicast key negotiation confirmation to the terminal [12].

Finally, after the terminal matches the AS key, the AC device starts to send multicast key notification to the terminal. The terminal sends the corresponding multicast key agreement response message to the AC equipment, and finally completes the whole WAPI protocol operation process.

In general, the first is the link negotiation process, the second is the unicast key negotiation process, the third is the multicast key negotiation process, and the last is the WAPI access control process.

The intelligent support and command system based on WAPI terminal equipment shares a WAPI certificate for multiple AP devices without spending too much resources. It manages multiple WAPI certificates for centralized management and maintenance. According to different services among all AP devices under the same AC, multiple WAPI certificates can be used at the same time. For example, one AP uses one WAPI certificate and multiple AP devices share one WAPI certificate.

The AP equipment mentioned in this paper has two logical ports: controlled port and uncontrolled port. Before the terminal passes the authentication, all interactive messages are transmitted through the uncontrolled port. Once the terminal passes the authentication, the service data of the terminal is transmitted through the controlled port.

3. System verification

3.1. Experimental resources

In order to show the advantages of the command system, the required software and hardware are shown in the table below:

Table 1 list of experimental resources

resources	quantity	explain
As server	1	Provide WAPI identity certificate management and authentication services
Ac server	1	It is used for centralized management of WAPI certificates. It is also convenient for centralized management of AP, and then centralized management of all details
AP	3	It is used for terminal equipment access management and forwarding interactive messages between terminal and AC.
Wireless terminal	some	For access and service test
Switch	2	The wired link communication network element between AC and AP has the function of mirror port
Wire shake	1	It is used to capture the interactive message of wired link for analysis
Omni Peek	1	It is used to grab the wireless link interaction message for analysis
WAPI negative tester (xwn1000t)	1	The abnormal message is constructed to test the judgment and processing ability of the mobile terminal to the abnormal message
WAPI protocol analysis tool	1	Capture WAPI protocol related messages, comprehensively and carefully analyze the message content, and give accurate prompts for possible error fields and locations

The AC server here can be any one of 512, 1024, 2048 and 4096, and the number in the model represents the maximum number of APS it supports.

3.2. Verification process

The test networking is shown in Figure 4. WAPI negative tester and WAPI protocol analysis tool are used in the test environment. WAPI negative tester can construct abnormal message to test the judgment and processing ability of mobile terminal to abnormal message. WAPI protocol analysis tool can efficiently capture WAPI protocol related messages, comprehensively and carefully analyze the message content, and give accurate prompts for possible error fields and locations.

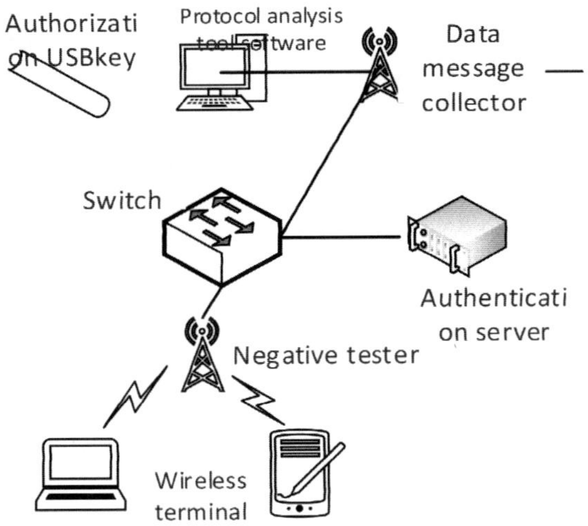

Figure 4 test networking diagram

Take the protocol integrity test sample of AE signature attribute type field in the access authentication response message as an example, analyze it, and analyze its message packet by using capture tools such as WAPI protocol analysis tool.

AE signature attribute type value should be '1'. When the AE signature attribute type is given an exception value. At this time, the tested mobile user terminal does not respond to the abnormal message, does not continue the subsequent key negotiation steps, and continues to wait for the retransmission message.

The negative tester first sends two abnormal multicast key announcement packet messages, with an interval of 100 ms. Mobile user terminals do not respond to abnormal multicast key announcement packet packets. Finally, the negative tester sends a normal multicast key announcement packet message (AE signature attribute type field value is "1"), and the interval between normal message and abnormal message is 100 ms. Then the mobile user terminal can respond to the normal multicast key announcement packet message, complete the whole process of subsequent key negotiation, and finally establish the connection.

After receiving the abnormal specified field message for the first two times, the terminal still does not disconnect, but continues to make the third attempt. When the message of normal Wai header field is received for the third time, normal processing is carried out and the connection is finally established.

4. Conclusion

In this paper, by installing the WAPI certificate on the AC device instead of sending it to the AP device, the keys generated after subsequent terminal negotiation are stored on the AC, which makes the certificate maintenance convenient and more secure. The user's air interface message encryption and decryption is carried out on the AC equipment, which improves the efficiency of encryption and decryption and the user's business experience. At the same time, it also improves the security and robustness of the business data of the whole system.

Reference

[1] SUN Hao. Design and implementation of WLAN network monitoring system [J], communication and information system, 2014.
[2] Yang Kunyuan. Planning and implementation of wireless campus network in University Park [J], computer technology, 2014.
[3] Zhu Weiwei. Research on WLAN construction and application [J], wireless interconnection technology, 2013.
[4] Chen Jigang. Security Protocol Improvement of WLAN based on management frame [b], 2008.
[5] Zhao Yujuan. Research on security mechanism of WLAN [J], computer software and theory, 2006.
[6] Deng Guoqiang, ye Zhao, Lu Yiqin. Wireless roaming between colleges and universities using WAPI [J], China education network, 2015.
[7] Nie Ming, Liu Changrui. Some problems to be solved in the application of WAPI to operator networks [J], Internet world, 2013.
[8] Hou Liqun, Mao Yuming. Comparative analysis of WLAN user security access solutions [J], national network and information security technology seminar, 2004.
[9] Cai Kai. WLAN security protocol testing technology research and system implementation [J], computer system architecture, 2011.
[10] Shi Yunzhi. Research on WLAN security mechanism and implementation of wireless access point WAPI protocol [J], communication and information system, 2010.
[11] Sun Xuan. Analysis of WAPI protocol and its implementation in WLAN integrated authentication platform [J], computer system architecture, 2006.
[12] Zhang Hao. Network security integration of WLAN and 3G based on WAPI [J], electronic and communication engineering, 2012.

Research on dynamic current sharing method of parallel connected IGBT modules for NPC three level converters

Xi Peng*, **Sheng Yin and Yingqin Zou**

Wuhan Institute of Marine Electric Propulsion, Wuhan, Hubei, 430074, China

*khl303041@163.com

Abstract. The parallel connection of IGBTs has been being applied in high power neutral point clamped (NPC) three level converters. This paper investigates the impact of gate parameters (gate resistor and capacitance) on dynamic current imbalance of parallel connected IGBT for NPC three level converter. A gate parameters calculation method is proposed in the paper, and the delay time and collector current difference can be analysed quantitatively. Experimental results have shown the effectiveness of the method.

1. Introduction

Multilevel converters have been attracting particular attention for high power applications such as marine electric propulsion. Insulated gate bipolar transistor (IGBT)-based the neutral point clamped (NPC) three level converters are widely studied topology. Although the output current ratings of these high power IGBT modules are steadily increasing, the parallel operations of IGBT modules are commonly adopted to reach higher output currents as are required by high power converter applications.

However, parallel operation of IGBT modules require a careful selection of IGBT modules and the special design of gate driver. Besides, although the parallel operation of IGBT modules expands the capacity of the high power NPC converters, it is difficult to maintain the complete consistency of the output current of parallel branches, which brings the problem of static and dynamic current imbalance due to the difference of internal parameters of each module and the asymmetry of circuit topology and driving circuit layout.

Figure 1 shows the definitions of static and dynamic current imbalance. The former means that the IGBT modules bears different current when in steady-state operation, and the latter means that the IGBT module bear unequal currents at the moment of switching action. Both of these two situations cause current imbalance and heat distribution imbalance of the device, and even damage the IGBT modules in serious cases.

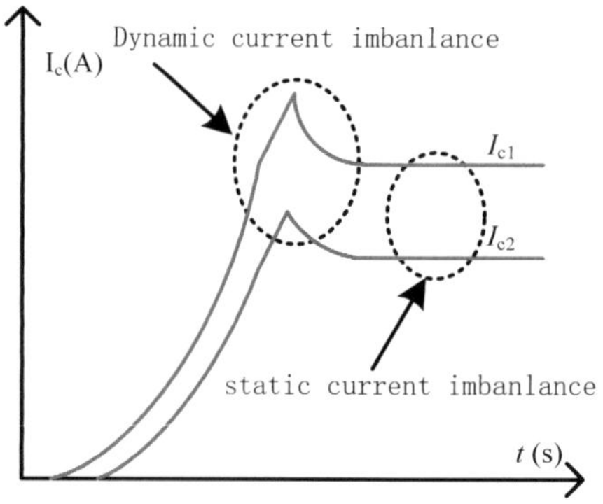

Figure 1. Current imbalances in parallel connected IGBT modules

The degree of current imbalance can be defined as

$$i_{cim} = \frac{|i_{c1} - i_{c2}|}{i_{c1} + i_{c2}} \tag{1}$$

Current balancing influence factors of parallel connected IGBT modules relate to static and dynamic cases. The former includes collector-emitter saturation voltage (V_{cesat}), collector current, junction temperature and euivalent resistance of commutation circuit, the latter includes gate threshold voltage(V_{GEth}), gate capacitance, gate resistor, junction temperature, stray inductance, gate driver, diode reverse recovery characteristics, load characteristics.

In the field of high power converter, many efforts have been made to achieve equal current sharing in the parallel connected IGBT modules. [1] proposed to control the gate conduction time and gate emitter voltage by checking the load current or emitter current to achieve current balance. The dynamic current imbalance in IGBT parallel connection is analyzed in [2]. gate resistance compensation method is proposed in [3] to realize IGBT parallel current equalization. The realization of IGBT parallel current balance by gate resistance compensation method is demonstrated in [4-5], and the selection of resistance parameters is analyzed. Generally, the dynamic current sharing methods are more important to implement. Among the methods, optimization of gate resistor and gate capacitance are most feasible for high power NPC three level converters because,

- Change of Commercial IGBT module parameters is impossible
- The overall dimension of IGBT module is large, and each IGBT module generally adopts an independent drive board
- The commutation loops in busbar are different and complex, and the commutation loops of parallel connected IGBT modules are difficult to be symmetrical, which makes the stray inductances of commutation loops to be different

This paper studies the current sharing of parallel IGBT module of high power NPC three-phase converter, focusing on determination of gate resistor and gate capacitance to improve the dynamic current balance of parallel module.

2. Gate parameters analysis for current sharing

Figure 2 shows a circuit configuration of one phase of the IGBT-based NPC three level converters. Each phase contains four IGBT units and two diode units. Each IGBT unit includes two parallel connected IGBTs. In high power converters, the IGBT modules, diode modules, absorption capacitors and temperature sensors are generally installed on the barbus, which is necessary to form a power unit.

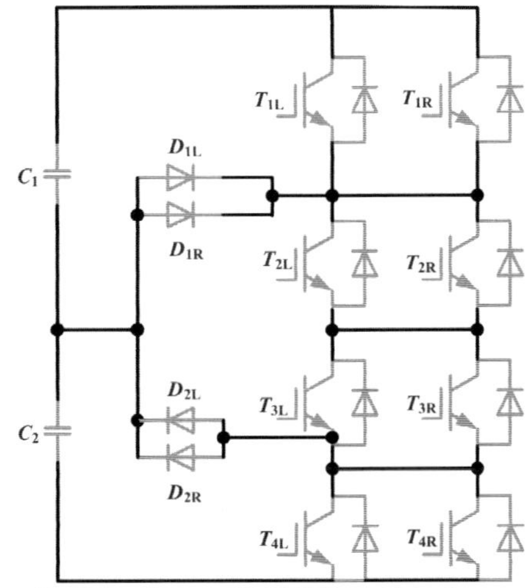

Figure 2. One phase of NPC three level converter

The commutation loops of NPC three-level converter are complex, and the commutation of parallel IGBT module are difficult to be symmetrical. As a result, the stray inductance of the parallel connected IGBT modules are different.

As shown in Figure 2, the turn-on and turn-off of IGBT can be approximately regarded as the charge and discharge process of a voltage source to a resistance-capacitance circuit. Rg and Cg are the gate resistance and gate capacitance set by the IGBT drive, respectively, and Cgi is the gate equivalent input capacitance of the IGBT module.

The amplitude of turn-on and turn-off current pulse and the rise rate of gate voltage can be limited by adjusting the gate resistance and capacitance, that is, the smaller gate resistance and capacitance are, the shorter the gate tur-on and turn-off time are, and the the switching loss is reduced. Therefore, changing the gate parameters can affect the dynamic current sharing of IGBTmodules.

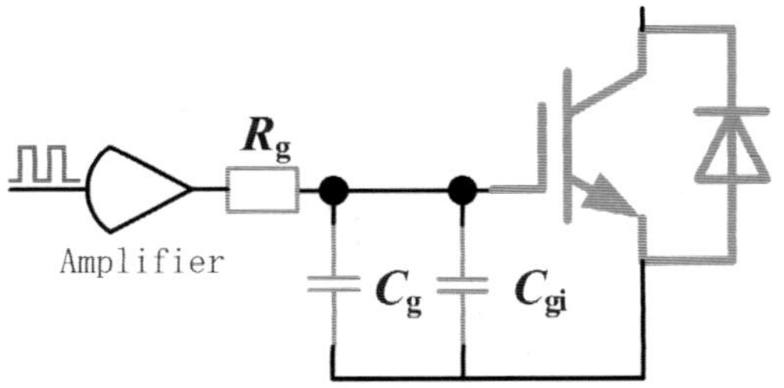

Figure 3.Typical IGBT gate drive circuit

For the IGBT-based NPC three level converters shown in Figure 3, the detailed gate drive circuit for two parallel connected IGBT modules could be shown in Figure 4, in which the inherent stray capacitors (C_{gei}, C_{gci} and C_{cei}) are considered as well. R_{g1}, R_{g2}, C_{g1}, C_{g2} are the gate resistance and gate capacitance set by the first and the second IGBT drive, respectively, $L_{\sigma 1}$, $L_{\sigma 2}$ ($L_{\sigma 1} \neq L_{\sigma 2}$) are the stray inductance of the two parallel connected IGBT modules.

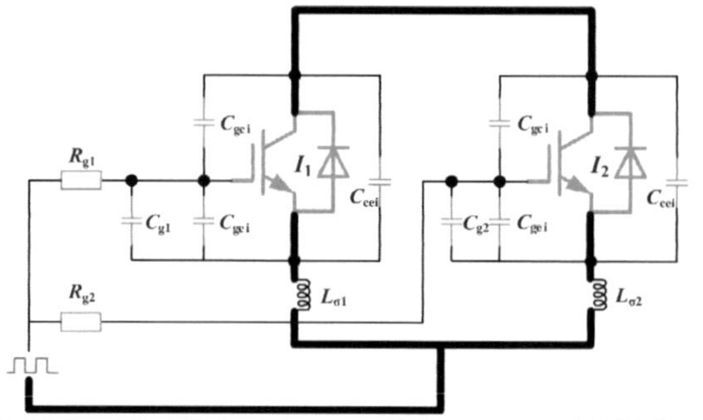

Figure 4. Drive circuit for two parallel connected IGBT modules

IGBT turn-on delay time (td) is greatly affect the transient current balancing of parallel connected IGBT modules.

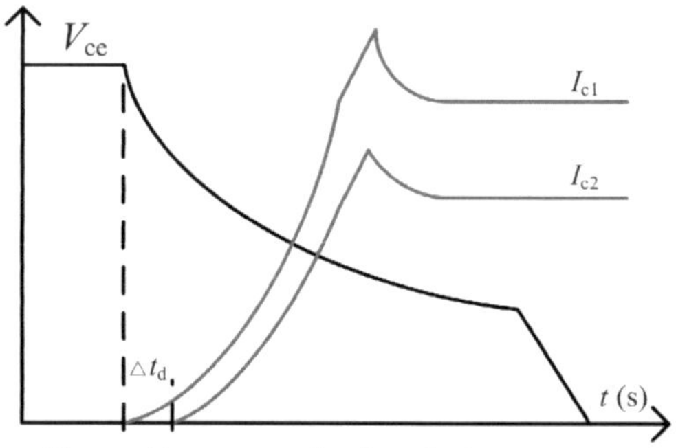

Figure 5. current sharing during turn-on

The turn-on delay time for the two parallel connected IGBTs are calculated as

$$t_{d1} = -\tau_1 \times \ln\left(1 - \frac{V_{GEth}}{V_{GE}}\right) \tag{2}$$

$$t_{d2} = -\tau_2 \times \ln\left(1 - \frac{V_{GEth}}{V_{GE}}\right) \tag{3}$$

where the time constant τ1 and τ2 are

$$\tau_1 = R_{g1} \times (C_{g1} + C_{ge1} + C_{gc1}) \tag{4}$$

$$\tau_2 = R_{g2} \times (C_{g2} + C_{gei} + C_{gci}) \tag{5}$$

The delay time difference then can be obtained

$$\triangle t_d = t_{d1} - t_{d2} = \ln\left(1 - \frac{V_{GE(th)}}{V_{GE}}\right)(\tau_1 - \tau_2) \tag{6}$$

In Figure 4, if Lσ1 > Lσ2, the difference between the collector current of the two parallel connected IGBT modules is described as

$$\triangle i_{c\max} = \begin{cases} 0 & t \leq \min(t_{d1}, t_{d2}) \\ k_1 \triangle t_d & t = \max(t_{d1}, t_{d2}) \end{cases} \tag{7}$$

where

$$k_1 = \frac{V_{GE} - V_{GE(th)}}{\tau_1 / g_{fs1} + L_{\sigma1}} \tag{8}$$

and gfs is the forward transconductance.

if Lσ1 > Lσ2, the difference between the collector current of the two parallel connected IGBT modules is described as

$$\triangle i_{c\max} = \begin{cases} 0 & t \leqslant \min(t_{d1}, t_{d2}) \\ k_2 \triangle t_d & t = \max(t_{d1}, t_{d2}) \end{cases} \tag{9}$$

where

$$k_2 = \frac{V_{GE} - V_{GE(th)}}{\tau_2 / g_{fs2} + L_{\sigma 2}} \tag{10}$$

In order to obtain the minimum Δi_c, the Lagrange multiplier method can be used to analyze the minimum of the function

$$f = \frac{(V_{GE} - V_{GE(th)}) \times \ln\left(1 - \frac{V_{GE(th)}}{V_{GE}}\right) \times (\tau_1 - \tau_2)}{\tau_1 / g_{fs1} + L_{\sigma 1}} \tag{11}$$

and leads to

$$\begin{cases} R_{g1}(C_{g1} + C_{ge1} + C_{gc1}) = g_{f1} L_{\sigma 2} \\ R_{g2}(C_{g2} + C_{ge1} + C_{gc2}) = g_{f2} L_{\sigma 1} \end{cases} \tag{12}$$

Careful selection of IGBT pairs is generally enough to ensure $C_{ge1}=C_{ge2}$, $C_{gc1}=C_{gc2}$ and $g_{f1}=g_{f2}$, and they could be obtained in manufacture materials. $L_{\sigma 1}$ and $L_{\sigma 2}$,could be obtained using test or finite element calculation method. R_{g1} and R_{g2} can not be too large in order to limit switching loss. The selection of R_g and C_g interact and R_g is suggested to be obtained first

3. Experimental verification

Experimental study presented in this paper is based on Infineon IGBT modules FZ3600R17HP4 and DIODE module DZ3600S17K3.

The test circuit configuration for the characterization of IGBT under hard switching condition is illustrated in Figure 6. A set of two pairs of IGBT modules is connected in parallel in the experimental setup (Figure 7). The middle parallel pair of IGBT modules are devices under test (T_1 location) while other pairs of IGBT modules remain under passive mode. The load inductor Lload is set to 400uH. The stray inductance $L_{\sigma 1}$ and $L_{\sigma 2}$ in the test circuit are tested to be can be 74uH and 49Uh, respectively.

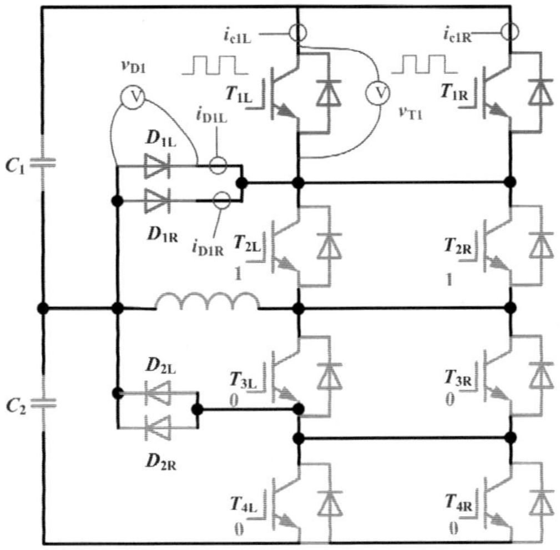

Figure 6. Test circuit

Figure 7. Test bed

The original IGBT driving board parameters are R_{g1} = 3 Ω, R_{g2} = 2 Ω, C_{g1} = 4.7uF, C_{g2} = 3.9uF. According to (12) and IGBT module parameters, and keep the gate resistance unchanged, it is calculated that C_{g1} = 5.6uF, C_{g2} = 7.1uF.

Different gate parameters are used to test the imbalance of Ic of the two parallel connected IGBT located on T1, the result is shown in Figure 8 and Table 1. It can be seen that the current imbalance is reduced from 21.76% to 11.21%, which is beneficial to the safe operation of IGBT modules.

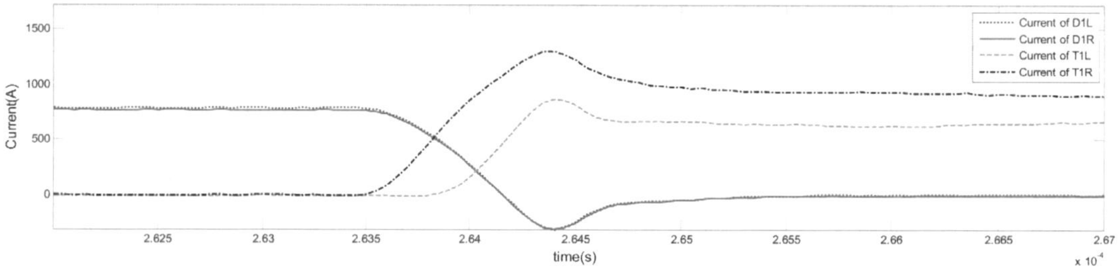

(a) R_{g1} = 3 Ω, R_{g2} = 2 Ω, C_{g1} = 4.7uF, C_{g2} = 3.9uF

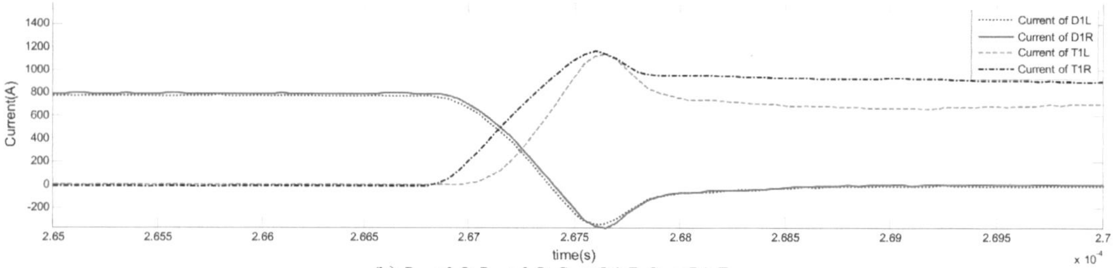

(b) R_{g1} = 3 Ω, R_{g2} = 2 Ω, C_{g1} = 5.6uF, C_{g2} = 7.1uF

Figure 8. Imbalance of Ic test results

Table 1. Imbalance of collector currents of two parallel connected IGBTs

Gate parameters	Imbalance of Ic
R_{g1} = 3 Ω, R_{g2} = 2 Ω, C_{g1} = 4.7uF, C_{g2} = 3.9uF	21.76%
R_{g1} = 3 Ω, R_{g2} = 2 Ω, C_{g1} = 5.6uF, C_{g2} = 7.1uF	11.21%

4. Conclusion

This paper investugates the impact of gate parameters (gate resistor and capacitance)on dynamic current imbalance of parallel connected IGBT for NPC three level converter.A gate parameters calculation method is proposed in the paper, and the delay time and Ic difference can be analyzed quantitatively. Experimental results have shown that the current imbalance is reduced from 21.76% to 11.21%.

References

[1] Hofer, R, Karrer, N., Gerster,C. (1996) Paralleling intelligent IGBT power modules with active gate-controlled current balancing. Proceedings of IEEE Power Electronics Specialists Conference (PESC), 2:1312-1316.

[2] Hofer-Noser P, Karrer N.(1999) Monitoring of paralleled IGBT/diode modules. IEEE Transactions on Power Electronics, 14(3):438-444.

[3] Schlapbach U. (2010)Dynamic paralleling problems in IGBT module construction and application. Integrated Power Electronics Systems (CIPS), 2010 6th International Conference on. 2010:1-7.

[4] Y. Xiao, Q. Liu, Y. Tang, L. Du and H. Ma, (2014)Current sharing model of parallel connected IGBTs during turn-on. *IECON 2014 - 40th Annual Conference of the IEEE Industrial Electronics Society*, 1350-1355,

[5] R. Letor. (1992)Static and dynamic behavior of paralleled IGBTs. *IEEE Transactions on Industry Applications*, 28(2): 395-402

Active and reactive power coordinated optimal dispatch in active distribution network based on model predictive control

Hanbing Qu [1*], Zheng Xu [2], Bo Wang [1], and Pu Zhao [1]

[1]State Grid Jinan Electric Power Company, Jinan, Shandong Province, 250100, China

[2]State Grid Shandong Electric Power Company, Jinan, Shandong Province, 250100, China

[*]Corresponding author's e-mail: 201914193@mail.sdu.edu.cn

Abstract. With the proliferation of the distributed energy resources (DERs), the scheduling and control of the distribution network have become more complicated. To cope with the uncertainty nature of distributed generation, a multi-timescale optimal dispatch method in active distribution network (ADN) based on the model predictive control (MPC) is proposed in this paper. First, based on MPC, a hierarchical scheduling framework for ADN is established, including long-timescale stage, and short-timescale stage. Then, via coordinated control of various resources in the ADN, i.e., distributed generators, energy storage, capacitor banks and OLTC transformer, the impact of intermittent renewable energy and load forecast errors can be reduced. Finally, considering the coupling characteristics of active and reactive power in the ADN, a joint active and reactive power optimization model is proposed to further reduce the network loss. Numerical simulation on a modified IEEE-33 distribution network system verifies the correctness and superiority of the proposed scheduling approach.

1. Introduction

The large-scale integration of distributed energy resources (DERs) has increased the control complexity of the distribution network operates [1], [2]. On the one hand, the prediction error of DERs may lead to the non-optimality of the scheduling results. Therefore, the network operator needs to consider the prediction errors explicitly [3], [4]. On the other hand, DERs can bring bi-directional power flow in the distribution network (DN), which may result in security issues such as line congestion and over-voltage [5]. To cope with the challenge brought by the increasing penetration of DERs, researchers proposed the concept of an active distribution network (ADN) [6], aiming to control these DERs actively and reduce their negative impact on the DN.

Based on the feature that the prediction accuracy of renewable energy sources (RESs) accelerates with the decrease of the predicted timescale [8], [9], the optimal scheduling scheme is modified step by step to diminish the dominance of the random flux of RESs [10]. In [11], a multi-timescale active power scheduling method for the large-size wind power grid connection is proposed, which realizes the step-by-step correction of wind power(WP) prediction error. According to the characteristic that the prediction error of wind power decreases gradually with timescale shortening [12], a coordinated formulation strategy of multi-stage unit commitment scheduling method considering WP generation is proposed. By coordinating and optimizing unit commitment in all timescale, the cost of power generation including wind power generation can be reduced. Considering that with the continuous subdivision of timescale, the forecast accuracy of WP and load is continuously improved, the

Journal of Physics: Conference Series **2113** (2021) 012056 doi:10.1088/1742-6596/2113/1/012056

coordinated control between wind power units and conventional units is realized to maximize the consumption of wind power [13], [14]. Based on the latest power prediction information, [15] builds closed-loop optimal control, which corrects the scheduling deviation caused by prediction errors timely, to guarentee the security and stability of the ADN and maximize its operation benefit.

Connecting a large number of DG has changed the original power flow and energy structure of the ADN [16], and have a certain impact on the system voltage stability and network loss. In [17], the multi-agent system is used in the voltage/var optimization problem for the distribution network. It shows that the usage of DERs for reactive power compensation leads to the great decrease in the electric energy losses, and proves the quickness and the effectiveness of the proposed algorithm. However, only reactive power is considered in this method, which is not complete for the ADN. Different from the transmission network, the line resistance is close to the reactance in the distribution network, which leads to the strong coupling between active and reactive power. Therefore, it is not complete and comprehensive to conduct a unilateral active/reactive power optimization analysis based on the traditional theory of active and reactive power decoupling in the ADN. For example, a coordinated optimization model of active-reactive power for comprehensive energy in a power system is established in [18], which takes the minimum total generation overall charge of the distribution system and the minimum active power losses of the network as the objective function and solves the optimal comprehensive benefits solution.

To address the above issues, based on the model predictive control (MPC) technique, this paper proposes a coordinated control method of active and reactive power dividing the control process into long-timescale and short-timescale. Optimal dispatch problem at different timescale performs separate MPC for their respective control objective and control variables. The long timescale is derived from the day-ahead 24-hour prediction of RESs and load. The multi steps rolling optimization approach is adopted to optimize the active and reactive output of each controllable device. The short timescale is derived from the optimized results of the longer timescale and ultra-short-term forecast of RESs. In the short-timescale optimal scheduling problem, the incremental power output of active/reactive devices is optimized with a rolling horizon. The overall goal of the proposed method is to reduce network loss and ensure the safety and economic operation of the ADN, which can make the distribution system operator handle the fluctuation and randomness of RESs and loads more effectively and actively.

2. MPC-based active-reactive power coordination dispatch model

A multi timescales joint optimal dispatching method is proposed in this paper. In the long timescale, 1h is taken as the time resolution, the network losses is selected as the optimization objective, and the scheduling plan of the future 24h period is optimized. In the short timescale, taking RESs short-term rolling forecast as a benchmark, and long-timescale optimal operation value as reference, distribution system operator scrolls to calculate active power increment of the next 15 min. However, only the first control instruction of the optimal control sequence can be executed.

2.1. Objective function

2.1.1. Long timescale. The optimization control variables include active and reactive output of RESs, charging/discharging power of energy storage, compensation capacitor tap position, the output of static var compensator and the OLTC tap ratio. Taking the current operation state as the initial value, RES and load demand forecast data as input variables, ΔT as the resolution and minimum network loss as the optimization goal, scroll to solve the active and reactive regulating resources' output for the coming time $M\Delta T$.

To guarantee the economy of the overall system operation reducing network losses, the optimization objective of the longer timescale is to minimize the network loss of distribution network, i.e.

$$\min F = \min \sum_{t=t_0}^{t_0+M\Delta T} \sum_{i=1}^{n} \sum_{j\in u(i)} r_{ij} \tilde{I}_{ij,t} \tag{1}$$

$$\tilde{I}_{ij,t} = F(K_{ij,t}, H^{c}_{i,t}, P^{ch}_{i,t}, P^{dis}_{i,t}, Q^{DG}_{i,t}, Q^{SVC}_{i,t}) \tag{2}$$

whereas t_0 indicates the starting time of rolling optimization under longer timescale, $u(i)$ is a set of all the line that point to node i, r_{ij} represents the resistance of the transmission line ij, $\tilde{I}_{ij,t}$ is the square of the current amplitude on the transmission line ij at time periods t, K_{ij} is the tap position of on-load tap-changer transformer, $H_{c,i,t}$ represents the current tap position of the capacitor bank, $P_{ch,i,t}$ and $P_{dis,i,t}$ are the charging and discharging power of the energy storage device respectively, $Q_{DG,i,t}$ is the reactive power output of distributed generators, and $Q_{SVC,i,t}$ is the reactive power output of the static var compensator.

2.1.2. Short timescale. The optimal objective is still to minimize the operating network losses of the distribution network, but the time horizon is shortened from 24h to period. In addition, the regulation plan of compensation capacitor and OLTC has been determined in the optimized scheduling under the long timescale, so the optimized scheduling under the short timescale will not be changed.

To address the random fluctuations of RESs output and load demand, prevent voltage from exceeding the limit and guarantee the secure operation of the system and keep consistent with the overall optimization trend of the longer timescale, the optimization control of the shorter timescale still takes the minimization of network losses as the objective.

$$\min F = \min \sum_{t=t_0}^{t_0+N\Delta t} \sum_{i=1}^{n} \sum_{j\in u(i)} r_{ij} \tilde{I}_{ij,t} \tag{3}$$

$$\tilde{I}_{ij,t} = F(P^{T1}_{ch,j,t} + \Delta P^{ch}_{j,t}, P^{dis,T1}_{j,t} + \Delta P^{dis}_{j,t} \tag{4}$$

$$Q^{DG,T1}_{j,t} + \Delta Q, Q^{SVC,T1}_{j,t} + \Delta Q^{SVC}_{j,t}) \tag{5}$$

whereas N indicates the optimization horizon, $P^{ch,T1}_{j,t}$, $P^{dis,T1}_{j,t}$, $Q^{SVC,T1}_{j,t}$ are the reference of the charging and discharging active power of the energy storage and the reactive power of the distributed generators and the static var compensator solved by the long timescale optimization at the time t respectively.

2.2. Constraints

2.2.1. Power flow model. According to the characteristics of the radial network, the power flow equation is expressed in the form of Distflow in this paper (see [19]).

$$\sum_{i\in\alpha(j)} (P_{ij,t} - \tilde{I}_{ij,t} r_{ij}) + P_{j,t} = \sum_{k\in\beta(j)} P_{jk,t} \tag{6}$$

$$\sum_{i\in\alpha(j)} (Q_{ij,t} - \tilde{I}_{ij,t} x_{ij}) + Q_{j,t} = \sum_{k\in\beta(j)} Q_{jk,t} \tag{7}$$

$$k^2_{ij,t} \tilde{U}_{j,t} = \tilde{U}_{i,t} - 2(r_{ij} P_{ij,t} + x_{ij} Q_{ij,t}) + (r^2_{ij} + x^2_{ij}) \tilde{I}_{ij,t} \tag{8}$$

$$\tilde{U}_{i,t} \tilde{I}_{ij,t} = P^2_{ij,t} + Q^2_{ij,t} \tag{9}$$

where $\alpha(j)$ indicates the set of all the nodes that point to node j, $\beta(j)$ is the set of all the nodes that point to node j, $P_{ij,t}$ and $Q_{ij,t}$ are the active and reactive output at the head point of transmission branch ij, $P_{j,t}$ and $Q_{j,t}$ are the injection of active and reactive power, r_{ij} and x_{ij} represent the resistance and reactance of the transmission line branch, $\tilde{I}_{ij,t}$ represents the current amplitude square of transmission branch ij, and $\tilde{U}_{i,t}$ is the voltage amplitude square of bus i.

2.2.2. Distribution network safety limits.

$$I_{ij,t} \le I^{\max}_{ij} \tag{10}$$

$$U^{\min}_{i} \le U_{i,t} \le U^{\max}_{i} \tag{11}$$

where $I^{\max}_{ij}$ is the maximum limitation of current amplitude on the branch ij, $U^{\min}_{i}$ and $U^{\max}_{i}$ are the upper and lower limitation of voltage amplitude at bus i respectively.

2.2.3. Distributed generators.

$$P_{DG,i,t} = P_{DG,i,t}^{pre} \tag{12}$$

$$Q_{DG,i,t}^{\min} \le Q_{DG,i,t} \le Q_{DG,i,t}^{\max} \tag{13}$$

$$Q_{DG,i,t}^{\max} = -Q_{DG,i,t}^{\min} = \sqrt{S_{DG,i}^{2} - \left(P_{DG,i,t}^{pre}\right)^{2}} \tag{14}$$

where $P_{DG,i,t}^{pre}$ represents the predicted active power output of distributed generators, $Q_{DG,i,t}^{\max}$ and $Q_{DG,i,t}^{\min}$ are the upper/lower reactive power output limits respectively, while $S_{DG,i}$ is the inverter's nominal rated power.

2.2.4. Static Var Compensator.

$$Q_{SVC,i}^{\min} \le Q_{SVC,i,t} \le Q_{SVC,i}^{\max} \tag{15}$$

whereas $Q_{SVC,i}^{\max}$ and $Q_{SVC,i}^{\min}$ represent the upper/lower limits of the adjustable reactive production of the static var compensator.

Capacitor Banks

$$H_{i,t}\Delta Q_{i,t}^{c} = Q_{i,t}^{c} \tag{16}$$

$$0 \le H_{i,t} \le H_{\max} \tag{17}$$

where $H_{i,t}$ is an integer variable that represents the switched times of the capacitor banks, $\Delta Q_{i,t}^{c}$ indicates the reactive power output for each single capacitor, and $H_{\max}$ is the total number of capacitor banks.

2.2.5. Energy Storage.

$$E_{i,t}^{SOC} + P_{i,t}^{ch}\eta_{ch}\Delta T - \frac{P_{i,t}^{dis}}{\eta_{dis}}\Delta T = E_{i,t+\Delta T}^{SOC} \tag{18}$$

$$E_{SOC,i}^{\max} \times 20\% \le E_{i,t}^{SOC} \le E_{SOC,i}^{\max} \times 80\% \tag{19}$$

$$0 \le P_{i,t}^{ch} \le P_{i,t}^{ch,\max} D_{i,t}^{ch} \tag{20}$$

$$0 \le P_{i,t}^{dis} \le P_{i,t}^{dis,\max} D_{i,t}^{dis} \tag{21}$$

$$D_{i,t}^{ch} + D_{i,t}^{dis} \le 1 \tag{22}$$

where $E_{i,t}^{SOC}$ is the stored energy, η_{ch} and η_{dis} are the charging/discharging efficiency respectively, $E_{SOC,i}^{\max}$ is the maximum stored electric energy of ES, both $D_{i,t}^{ch}$ and $D_{i,t}^{dis}$ are 0-1 variable, representing the charging and discharging state, respectively.

3. Case study

The established model is verified on a revised IEEE 33-bus distribution network which is given in Figure.1. The parameters are derived from [6]. The forecast data of WP, photovoltaic(PV) and load are shown in figure 2. The numerical simulation is carried out in MATLAB 2018b, modelled in Yalmip package and solved by commercial solver Gurobi on a PC with Intel(R) Core(TM) i7-5300 2.56Ghz, 16G RAM.

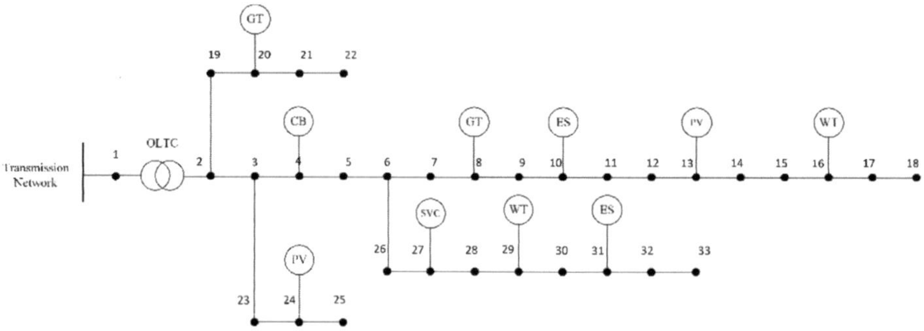

Figure 1. The modified IEEE 33-bus distribution network system

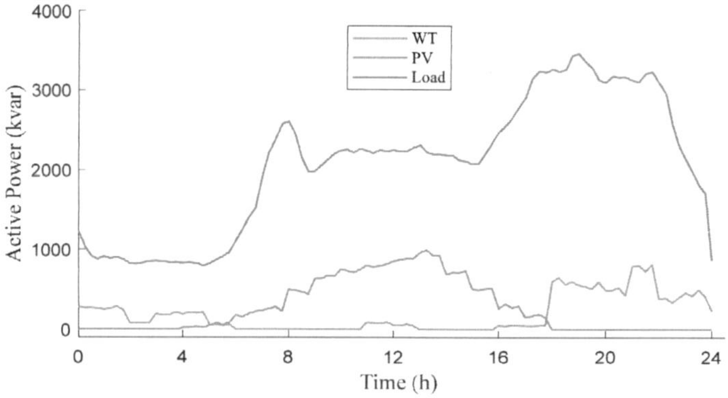

Figure 2. Load and renewable energy curve.

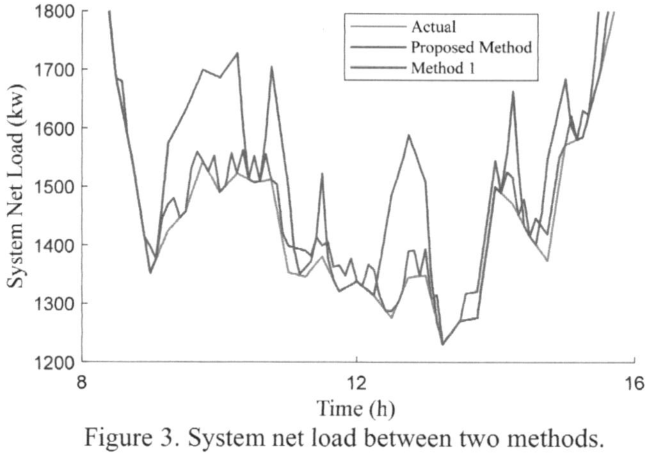

Figure 3. System net load between two methods.

For the ease of presentation, the single open-loop method is called Method 1, while the single active dispatch method is called Method 2 and the conventional single reactive power scheduling approach is called Method 3. The objective of Method 2 is the minimization of the operation cost of ADN by optimizing active power regulating resources (i.e., the control variables), which mainly includes ES and DGs, not includes CBs and OLTC. Minimzing the losses of the ADN by optimizing the reactive output of CBs ,DGs and OLTC taps is the objective function of Method 3.

3.1. System Netload Analysis

Figure 3 dsiplays the overall user demand curve of the ADN predicted by the presented method and Method 1. Comparing with the actual overall user demand of the ADN, we can easily infer that the presented method can better track the flux of renewable energy sources.

The prediction result of Method 1 is quite differed from the actual total user demand of the system at some time slot. However, the prediction result of the proposed method can be consistent with the actual total load curve of the ADN. Therefore, the presented approach in the paper is much better than Method 1 in tracking system fluctuations. The main reason is that the single open-loop optimization only predicts the total load of the system in the day-ahead, while the proposed method, based on MPC, repeatedly revises the forecast results within the day, reducing the prediction error. We can further conclude that the receding horizon control strategy can reduce the prediction error.

3.2. Distribution Network Losses Analysis

To verify the superiority of the presented method in diminishing system network losses, the proposed method is compared with Method 2 and Method 3. The network loss using these dispatch methods is obtained through numerical simulation, as shown in figure 4. Table 1 exposes the total operational system cost of these three methods.

Table 1. System Cost for Three Method

Dispatch Method	System Cost ($)
Method 2	3.8477e+04
Method 3	4.3954e+04
The proposed method	3.4853e+04

Table 2. Solution time for two case study systems.

Numerical example	Indicators	Original Model	Second-order Cone Relaxed Model
IEEE 33-bus system	Single optimization duration	18.24 s	0.61 s
IEEE 118-bus system	Single optimization duration	173 s	1.42 s

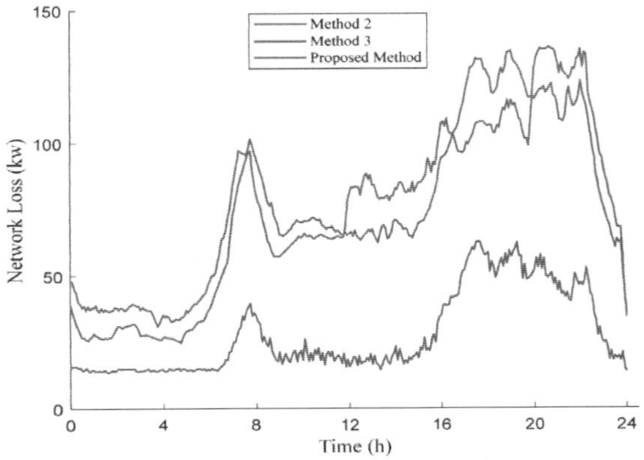

Figure 4. Network loss between three methods.

It can be watched from figure 4 and table 1 that the prosented method in this paper is much better than Method 2 and Method 3 in reducing network losses. The reason for this phenomenon is that, different from the transmission network, the branch resistance is close to the reactance in the DN, and there exists a strong coupling between active power and reactive power. Therefore, the effect of joint active and reactive power optimization approach in the ADN is much better. Real-time system operation results of the proposed method are given in figure 5 and figure 6.

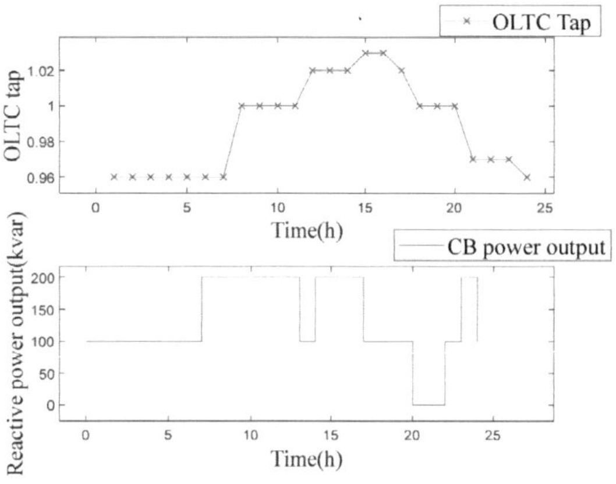

Figure 5. Reactive device actions.

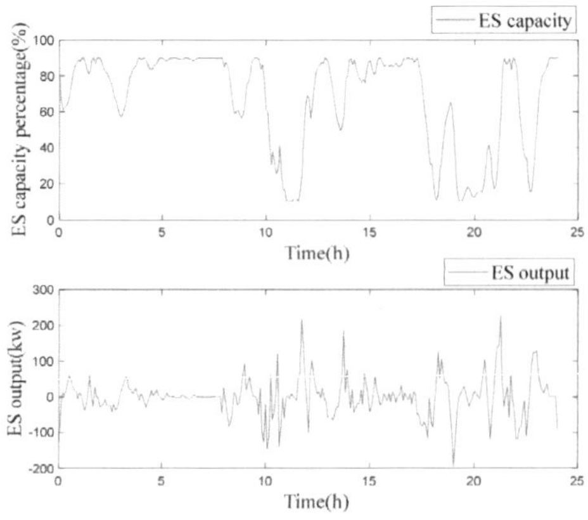

Figure 6. Energy storage status.

Taking 15:00 to 18:00 as an example for analysis, during this period, photovoltaic output decreases and load demand increases. To alleviate the shortage of active power of the system, the ES system discharges. Meanwhile, to preventing the increase of network loss brought by the long-distance reactive power transmission, the OLTC ratio drops and the reactive compensation output of CB increases.

3.3. Solution time analysis

Numerical simulations are varified on the IEEE 33-bus and IEEE 118-bus network system to further verify the effect of the second-order cone relaxation on algorithm solution time, as given in table 2.

It can be observed from table 2 that, the second-order cone relaxation technique can reduce the algorithm solution time significantly, which is more obvious in large-scale systems. Thus, the proposed MPC-based multi-timescale joint active and reactive optimal scheduling approach is effective and superior.

4. Conclusion

To effectively deal with the fluctuation and uncertainty of the renewable energy sources output and load, this paper presents a multi-timescale dispatching method coordinating active and reactive power in the ADN. In the proposed model, a variety of regulating resources are formulated and coordinated to guarantee the economic performance and security of the system. The main conclusions are summarized as follows:

1) The multi-time scale dispatching framework based on MPC technique is presented to deplete the influence of the fluctuation of RESs in the ADN, which is conducive to the stability of system operation.

2) Unlike the conventional single active or reactive power optimal dispatching, the joint optimization method of active and reactive power may further diminish the network operating losses, which improves the economic performance of the active distribution network.

3) The second-order cone relaxation technique is adopted to recast the dispatch problem, which improves the convergence speed of the algorithm markedly.

With the liberalization of electricity market, distributed energy sources within the demand side will take a vital part in the energy collaborative management of ADN and generally displace the conventional generation plants. The new operating paradigm of ADN integrating various and dispersed DERs is the focus of the authors' future work.

Acknowledgments

The authors gratefully acknowledge the financial support from Science and Technology Project of Jinan Power Supply Company, State Grid Shandong Electric Power Company Limited (Research and

Application of Distributed Generators on Cooperative Control and Optimization Using Dispatch Cloud Technology).

References

[1] H. Sheng, C. Wang, and J. Liang. "Multi-timescale active distribution network optimal scheduling considering temporal-spatial reserve coordination." International Journal of Electrical Power & Energy Systems 125: 106526.

[2] P. Li, C. Wang, Q. Wu and M. Yang, "Risk-Based Distributionally Robust Real-Time Dispatch Considering Voltage Security," in IEEE Transactions on Sustainable Energy, vol. 12, no. 1, pp. 36-45, Jan. 2021, doi: 10.1109/TSTE.2020.2964949.

[3] P. Li, M. Yang and Q. Wu, "Confidence Interval Based Distributionally Robust Real-Time Economic Dispatch Approach Considering Wind Power Accommodation Risk," in IEEE Transactions on Sustainable Energy, vol. 12, no. 1, pp. 58-69, Jan. 2021, doi: 10.1109/TSTE.2020.2978634.

[4] Z. Li, Y. Xu, "Optimal coordinated energy dispatch of a multi-energy microgrid in grid-connected and islanded modes", Applied Energy, Vol. 210, No. 1, pp. 974-986, Jan. 2018.

[5] Ren Jiayi, Gu Wei, Wang Yong, Ji Wenlu, Liu Haibo, et al, "Multi-time scale active and reactive power coordinated optimal dispatch in active distribution network based on model predictive control", in Proceedings of the CSEE, vol. 38, no. 05, pp. 1397-1407, June 2017.

[6] S. Chen, C. Wang and Z. Zhang, "Multi-time Scale Active and Reactive Power Coordinated Optimal Dispatch in Active Distribution Network Considering Multiple Correlation of Renewable Energy Sources," in IEEE Transactions on Industry Applications, doi: 10.1109/TIA.2021.31 00468.

[7] Z. Li, Y. Xu, et al. Multi-objective Coordinated Voyage Scheduling and Energy Dispatch for the Multi-energy Cruising Ship System with Renewable Energies, IEEE Transactions on Industrial Applications, Vol. 1, No. 1, pp. 1-9, Nov. 2019.

[8] Z. Li, Y. Xu, "Temporally-coordinated optimal operation of a multi-energy microgrid under diverse uncertainties", Applied Energy, Vol. 240, No. 1, pp. 719-729, Apr. 2019.

[9] Zhengmao Li, Yan Xu, Sidun Fang, Stefano Mazzoni. Optimal Placement of Heterogeneous Distributed Generators in a Grid-Connected Multi-Energy Microgrid under Uncertainties, IET Renewable Power Generation (RPG), Vol.13, No. 14, pp. 2623-2633, Sep. 2019.

[10] Zhang Boming, WU Wenchuan, ZHENG Taiyi, et al, "Design of a multi-time scale coordinated active power dispatching system for accommodating large scale wind power penetration", in Automation of Electric Power Systems, vol. 35, no. 01, pp. 1-6, Jan 2011.

[11] Z. Li, Y. Xu, L. Wu and X. Zheng, "A Risk-Averse Adaptively Stochastic Optimization Method for Multi-Energy Ship Operation Under Diverse Uncertainties," in IEEE Transactions on Power Systems, vol. 36, no. 3, pp. 2149-2161, May 2021, doi: 10.1109/TPWRS.2020.3039538.

[12] Q. Fujun, Q. Hanbing, F. Depin, W. Jun and S. Hongzhang, "Multi-time Scale Active and Reactive Power Coordinated Dispatch on Distribution Network with Distributed Generators," 2020 5th Asia Conference on Power and Electrical Engineering (ACPEE), 2020, pp. 437-441, doi: 10.1109/ACPEE48638.2020.9136208.

[13] Yu Ji, Wanxing Sheng, Wei Jin, Ming Wu, Haitao Liu, Feng Chen. "Coordinated control of active and reactive power of distribution network with distributed PV cluster via model predictive control", IOP Conference Series: Earth and Environmental Science, 2018.

[14] Z. Li, Y. Xu, et al. Robust Operation of a Hybrid AC/DC Multi-energy Ship Microgrid with Flexible Voyage and Thermal Loads, IEEE Transactions on Smart Grid, Early access, 2020.

[15] Zhang Na, LV Quan, ZHANG Xian, et al, "A multi-level unit commitment coordinated strategy for power system with wind power", in Automation of Electric Power Systems, vol. 37, no. 11, pp. 39-45, June 2013.

[16] Zheng Taiyi, FAN Guoying, SUN Yong, et al, "A coordination control strategy for multiple power source under grid-integration of large-scale wind farm", in Power System Technology, vol. 37, no. 11, pp. 3091-3095, Nov 2013.
[17] Hanbing Qu, Fu Yang, Qirong Lin, Yujian Jia. Active and Reactive Power Coordinated Optimal Dispatch in Active Distribution Network Considering Spatial-temporal Correlation of Wind Power[J]. IOP Conference Series: Earth and Environmental Science, 2021, 687(1):012093 (8pp).
[18] Masoud Farivar, Steven H. Low, "Branch flow model: Relaxations and convexification—Part I", in IEEE Transactions on Power Systems, vol. 28, no. 3, pp. 2554-2564, Aug. 2013

Design of Smoke Alarm Disperser Based on Single-chip Microcomputer

Yang Wang[1], Lichun Han[2],Yonghui Zhao[3] ,Yuquan Xue[4] and Liming Wang[1*]

[1,1*,2,3,4] College of Electrical Engineering,Naval University of Engineering,Wuhan,Hubei,430033,China

[*]Corresponding author's e-mail: licesoar@163.com

Abstract. With the rapid development of society and economy, various laboratories have become more complete in number and equipment, but subsequent laboratory accidents frequently occur. Therefore, it is imperative to construct a laboratory safety management system to ensure the safety of laboratory personnel and complete equipment. This article intends to design a smoke alarm, which mainly includes measuring the sensitivity of the smoke sensor to the concentration of smoke in the air, and the linear change of the sensor converting the corresponding analog signal into a digital signal. By designing the fan speed to be controlled by the change of smoke density, the speed of the motor is adjusted. And the alarm module can sound an alarm when the smoke concentration is certain, and at the same time, it can automatically cancel the alarm when the smoke concentration decreases. Finally, a simulation experiment is designed to simulate the effect of the sound alarm of the smoke alarm disperser on the dispersal of smoke under different working environments.

1. Introduction

With the development of science and technology, laboratory equipment and facilities have become more and more complete, but the safety performance of the facilities is still a major hidden danger in the laboratory, especially the safety accidents caused by improper operation of the laboratory, especially the laboratory fire and Explosion accidents are particularly frequent[1]. This project hopes to design a new type of smoke alarm device that can give out sound and light alarms when a fire occurs in the laboratory[2]. This research can inform the personnel to put out fire and evacuate the dangerous area in the early stage of fire, which can effectively reduce the casualties caused by fire.

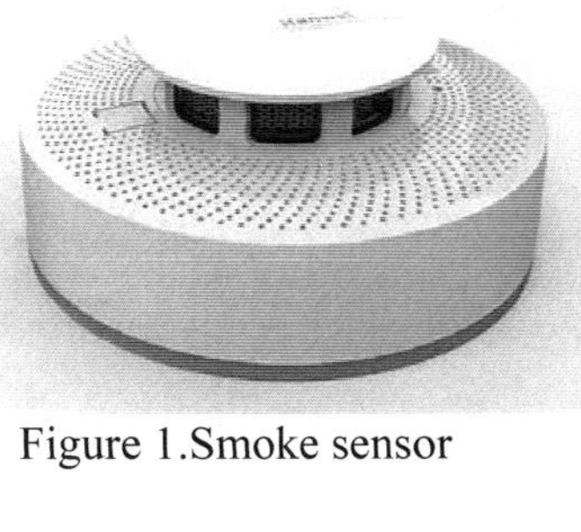

Figure 1.Smoke sensor

The Design of smoke alarm disperser

The smoke alarm disperser is based on the smoke concentration signal collected by the smoke sensor and input into the single-chip microcomputer for analysis. When the smoke concentration is higher than the alarm threshold, it has an alarm and smoke dispersal function. Its components include: smoke signal collection circuit, Motor drive module circuit, analog-to-digital conversion circuit, Arduino microcontroller control circuit[3].

1.1. The Principle of MQ-2 Smoke Sensor

The main semiconductor component of the MQ-2 smoke sensor is tin dioxide. When tin dioxide comes in contact with smoke, it causes a change in surface conductivity[4]. When the smoke concentration becomes larger, the conductivity of the sensor increases, the output resistance becomes smaller, and the output analog voltage signal increases. Therefore, the output voltage analog signal can be used to indicate the level of smoke concentration in the environment.

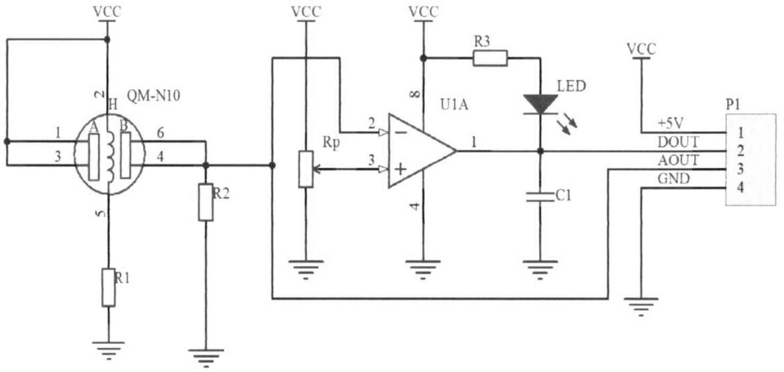

Figure 2.Schematic diagram of mq-2 smoke sensor circuit

This experiment scheme adopts the comparator circuit, in which the MQ-2 smoke sensor outputs analog signals into Pin 6 of the Arduino control board through the LM393 comparator digital tube. Pin 6 has an analog-to-digital conversion function, which can convert the input analog signals into digital signals[5].

1.2. The Working Principle of Smoke Alarm Disperser

In this paper, based on The Arduino microcontroller, MQ-2 smoke sensor is used to collect smoke concentration information and transmit it to the Arduino. The microcontroller itself is the voltage signal amplification module. Analog signals are converted into digital signals for analysis through analog-to-digital conversion, and the concentration of smoke is calculated[6]. When the smoke concentration value is higher than the alarm threshold, the buzzer gives an alarm, and the single-chip microcomputer controls the motor speed to drive the fan to rotate, so as to achieve the effect of dispersing the smoke. When the smoke concentration continues to increase and reaches a certain limit, the microcontroller controls the motor acceleration, so the fan speed increases, if the smoke concentration is still rising, the microcontroller controls the motor speed to reach the maximum, in order to achieve the effect of dispersive smoke; When the smoke concentration falls to the alarm threshold, the buzzer and motor stop working and return to normal[7].

2. The Hardware Setting of Smoke Alarm Disperser

This chapter mainly introduces the hardware settings of the smoke alarm disperser, especially the circuit analysis and characteristics of each functional module, and explains the working principles and related technologies of the buzzer, MQ-2 smoke sensor and motor fan module in the smoke alarm in more detail index[8].

2.1.Selection of MCU

In the design of this paper, the single-chip microcomputer is the core component of the entire smoke alarm disperser. First of all, it must meet the computing power and the number of external interfaces required by the smoke alarm disperser. At the same time, the overall model of the alarm is small. Reduce costs as much as possible in terms of calculation accuracy, stability, reliability and anti-interference.The single-chip microcomputer used in this paper is Arduino and software Arduino IDE.

2.2.Hardware Circuit Design of Smoke Alarm Disperser

Buzzers are generally divided into two types: active buzzers and passive buzzers. Compared with passive buzzers, active buzzers are easier and more convenient to operate. At the same time, they can send out sound source signals with prompt frequencies. , So it can meet the requirements of the design scheme. Since the output of the Arduino UNO control board is a DC signal, an active buzzer is selected to sound an alarm when the smoke concentration is higher than the alarm threshold.

The motor speed is controlled by the PWM signal, so as long as the speed range of 0~1023 is input to the IA pin, the speed can be controlled. In the design of this paper, the smoke density analog signal in the environment is read through the Arduino single-chip microcomputer, and the smoke density is converted into an electrical signal through the A/D analog-to-digital conversion circuit. The corresponding interval of the smoke density is [0,1023].

The MQ-2 smoke sensor outputs an analog voltage signal through the LM393 comparator, and inputs the signal into the Arduino single-chip microcomputer. Especially the ADC analog-to-digital conversion module of the single-chip microcomputer converts the analog signal into a digital signal. For the single chip microcomputer to analyze the data to determine whether the existing alarm conditions are met.

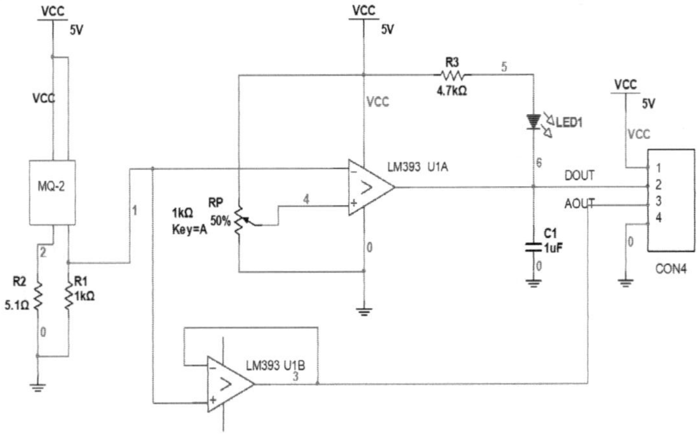

Figure 3. MQ-2 smoke sensor comparator circuit

For the schematic diagram of the MQ-2 smoke sensor circuit, LM393U1A is used as a comparator to compare the analog voltage input by the Rp sliding rheostat and the analog voltage input by MQ-2, U1A outputs high and low levels; LM393U1B is used as an operational amplifier, adding In the negative feedback link, the magnification is 1, which is used to follow the voltage and play a certain buffering effect[9].

3. Software Design of Smoke Alarm Disperser

This chapter mainly describes the software design of the smoke alarm disperser, mainly the introduction of the development environment ArduinoIDE and the design of the program flow chart. When the smoke alarm disperser responds to different smoke concentrations, the single-chip microcomputer adjusts the PWM wave to control the motor speed through data analysis, so as to achieve the purpose of controlling the speed of the electric fan, ensuring that the speed of the electric fan in the environment can be adjusted independently according to the ambient smoke concentration[10].

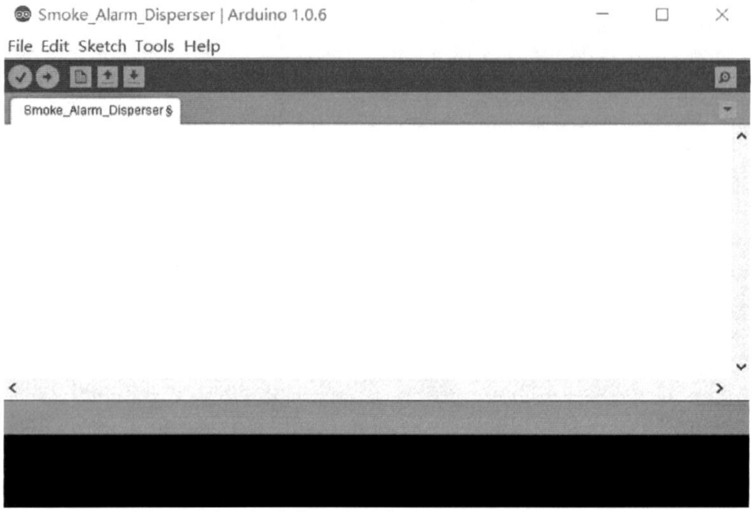

Figure 4. Arduino development environment

Arduino series MCU debugging and R&D tools: The software programming of this system adopts ArduinoIDE software, built-in open source Simple I/O interface version, and has a Processing/Wiring development environment similar to Java and C language. The main language of the Arduino core library is C language, and the latest Arduino core library is written in a mixture of C/C++.

Through the program initialization, MQ start work - 2 type smoke sensor, acquisition environment of smoke concentration and enter into the Arduino microcontroller, through the ADC modulus conversion, the smoke concentration corresponds to the voltage analog signal is converted to facilitate analysis of digital signal, through processing compared with alarm threshold, in order to judge whether the smoke concentration exceeds the alarm threshold, If the threshold is exceeded, an acoustic alarm should be issued and the motor should be started to control the fan to disperse the smoke, so as to achieve the purpose of reducing the smoke concentration[11].

4. Experimental Verification

This experiment mainly uses the Arduino control board, MQ-2 smoke sensor, buzzer and motor fan dispersing module to realize the simulation of the smoke alarm disperser[12]. For the smoke produced by the proposed cigarette, the speed threshold of the fan, the concentration threshold of the MQ-2 smoke sensor and the change of the fan speed at different concentrations are measured.

4.1. The relationship between fan speed and Arduino input voltage

The experiment mainly verified the sensitivity threshold of the MQ-2 smoke alarm, the minimum speed and maximum speed that the fan can reach, and the relationship between the fan speed and the digital input of the Arduino control board.

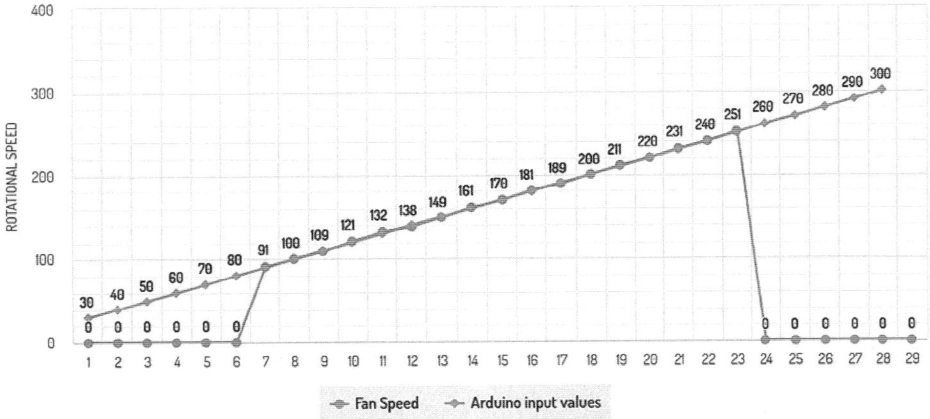

Figure 5. The relationship between the fan and Arduino input voltage.

Through the above data, we carry out corresponding mathematical analysis and draw the following conclusions:

1. There is a certain area for the speed of the fan driven by the motor. The minimum and maximum speeds of the motor correspond to 90 and 270 of the digital signal respectively.

2. The fan speed and the motor speed change in a linear relationship. At the same time, the higher the motor speed, the faster the fan speed.

4.2. Smoke concentration signal collection

Through the design, the smoke alarm disperser is measured in a smoke-free environment. As a calibration value, the experimental results can tell that, under normal circumstances, the smoke concentration collected by the smoke sensor is 255~275.When the paper is ignited, the concentration detected by the smoke sensor is indefinite between 285 and 310. When the toilet paper is gradually burned sufficiently, the concentration of smoke gradually increases. At this time, the concentration detected by the smoke sensor is indefinite between 350 and 400.

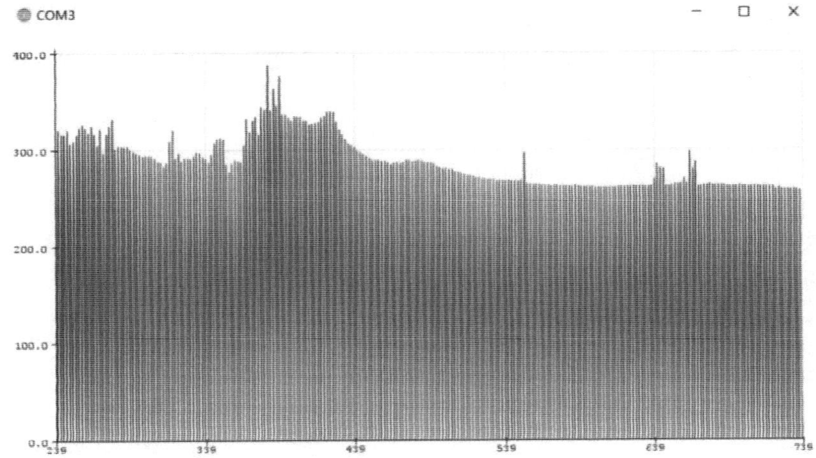

Figure 6. Smoke sensors measure the concentration of paper when it ignites

According to the chart drawn by the serial port plotter, it can be seen that when the concentration continues to rise, the motor speed increases, and the dispersing ability is strengthened. After a certain period of time, the concentration gradually decreases, which has a certain effect of dispelling smoke.

5. Conclusion

The research and design content of this paper is based on the Arduino microcontroller to add smoke sensor module, motor fan module, sound and light alarm module design, so that in a certain sealed environment, when the smoke reaches a certain concentration, it can timely send sound and light alarm, start the steering gear, control the fan wind speed, and achieve the alarm and disperse function. For the smoke alarm dispeller, as long as the smoke concentration reaches a certain concentration value, the alarm will be issued, can not have according to the different gas to set the corresponding alarm threshold for some low concentration harmful to the human body can not be detected in time. Next, we hope to combine fire extinguishing devices with wireless communication modules.

References

[1] Allison,R.S , Johnston,J.M , Wooster,M.J, (2021)Sensors for Fire and Smoke Monitoring.[J]. Sensors , 21(16), Switzerland.
[2] XIE,X.H, (2018) An Intelligent Smoke Detector System Based on LoRa and Indoor Positioning. In:2018 International Conference on Computer,Communication and Network Technology （CCNT 2018）.Tongxiang, Zhejiang Province, China.pp.6.
[3] Liu,P.S, Fang,J.L, Huang,H.J. (2021) Node deployment method of Intelligent smoke sensors across high space using many-objective optimization algorithm[J]. Journal of Physics: Conference Series, 1883(1).
[4] Huang,Y , Wang,X.G , Gan,S.C.(2021) Design and Evaluation Method of Wireless Fire Detection Node Based on Multi-source Sensor Data Fusion[J]. International Journal of Sensors and Sensor Networks, 9(1).
[5] Ahmad,S.and Kim,D.H.(2021) A task orchestration approach for efficient mountain fire detection based on microservice and predictive analysis in IoT environment[J].Journal of Intelligent & Fuzzy Systems, 40(3).
[6] Lule,E, Mikeka,C, Ngenzi,A. (2020) Design of an IoT-Based Fuzzy Approximation Prediction Model for Early Fire Detection to Aid Public Safety and Control in the Local Urban Markets[J] Symmetry,12(9).
[7] Hassan,S,A, Ingram,M.A,(2014) Analysis of an opportunistic large array line network with Bernoulli node deployment[J] .IET Communications , 8(1).
[8] Miao,K. (2021) Research on multi feature fusion perception technology of mine fire based on inspection robot[J]. Journal of Physics: Conference Series , 1955(1).
[9] Rowland,J.H, Yuan,L, Thomas,R.A (2021) Evaluation of Carbon Monoxide and Smoke Sensors at a Low Ventilation Velocity[J]. Mining Metallurgy & Exploration, 38(1) : 603-608.
[10] Wang,L, (2021) Research on the development and testing of remote automatic fire alarms based on the internet of things[J]. Academic Journal of Engineering and Technology Science, 4(3).
[11] Bai,X.T, Lu,H.R, (2021) Research on Fire Alarm Monitoring System Based on Computer[J]. Journal of Physics: Conference Series, 1744(2) : 022088-.
[12] Chris,G, Tim,F, (2018) Smoke from the Fire Alarm Control Room: How Your Risk Assessment Must Change[J]. Fire Engineering, 171(1).

Design of entertainment drone with rotating LED display technology

Anhua You[1*], Jiahao Li[2], Zifang Xu[1], Wencong Liu[1]

[1]Nanjing University of Science and Technology, Nanjing, Jiangsu, 210094, China

[2]Nanjing Foreign Language School Xianlin Campus, Nanjing, Jiangsu, 210046, China

[*]Corresponding author's e-mail: youanhua@njust.edu.cn

Abstract. The drone light show is a popular form of performance. This paper designs a new type of entertainment drone with rotating LED display technology. This drone is equipped with a rotating LED display, a single drone can display a complete picture, and has a good display effect. The application of the new entertainment drone can reduce the cost and technical difficulty of the drone light show, and expand the applicable scenarios of the drone light show.

1. Introduction

Drone light show is an emerging form of performance, often used in festival performances or large-scale celebrations[1-2]. Compared with the traditional firework show, the drone show does not produce air pollution, is more environmentally friendly, and can flexibly display text, patterns and even videos. At present, the drone light show is dominated by drone formation performances[3-4]. The formation performance is composed of hundreds of drones equipped with LED lights in the air to form a dot matrix, and the position and LED lights of each drone are controlled by programs. Formation performances require a large number of drones, and the control of drone formations is difficult and prone to malfunctions, which affects the overall performance. The cost of the formation performance is positively related to the number of drones used in the performance. The more complex the images displayed in the light show, the more drones are needed and the higher the cost of the light show. However, a drone usually only carries one LED light for performance, which does not fully utilize the performance of a drone. This paper proposes to apply rotating LED display technology[5] to drone performances and design a new type of entertainment drone. A drone equipped with a rotating LED display can display a complete pattern in the air, and the number of drones required to display a large pattern is lower than that of commonly used solutions. The application of the new entertainment drone can reduce the cost and technical difficulty of the drone light show, and expand the applicable scenarios of the drone light show. The new type of entertainment drone designed in this paper is mainly composed of two parts, a rotating LED display and a quadrotor.

2. Rotating LED display

The drone used in drone light show usually carries only one LED light, forms a pixel of the overall pattern in the performance. The application of LED display makes a drone to display more content. Ordinary LED displays are composed of semiconductor LED dot matrix, and display different pictures by controlling the colour of LED lights in different positions on the screen. The rotating LED display is a display device that works according to the principle of persistence of vision. Persistence of vision traditionally refers to the optical illusion, which occurs when visual perception of an object does not cease for some time, after the rays of light proceeding from it have ceased to enter the eye.

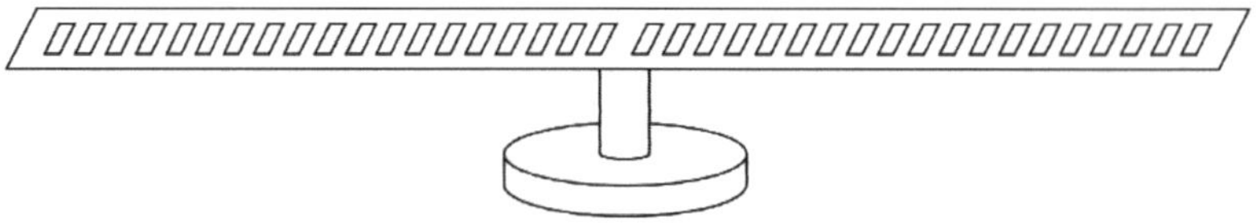

Figure 1. Rotating LED display structure

The rotating LED display is mainly composed of two parts: the base and the LED dot matrix. The motor on the base makes the LED dot matrix to rotate at a high speed, and the single-chip microcomputer controls the LED light to light up in an appropriate position. The persistence of vision phenomenon causes the brain to visually overlap and see a complete picture.

Ordinary LED displays are opaque. In the application scenario of the drone light show, the opaque display will block the background of the performance scene and negatively affect the visual effect of the drone light show. When the rotating LED display is working, the display will not block the background, and the observer feels that the picture is floating in the air. Especially in the case of displaying a three-dimensional picture, the display effect of the rotating LED display is better than that of the ordinary LED display, and the picture is more three-dimensional and realistic, giving people an immersive feeling[6]. And the viewing angle of the rotating LED display is larger than that of the ordinary LED display, allowing more audiences to see the performance under the same conditions. The weight of the display will have an important impact on the flight time of the drone. The greater the weight of the display is, the shorter the flight time of the drone is. Because of the difference in principle and structure, to display the same number of pixels, the number of LED lights required to rotating LED display is much lower than that of the ordinary LED display. Therefore, under the same conditions, the weight of the rotating LED display is smaller, and the power consumption is lower.

After analysing the above situation, this article chose a rotating LED display with a diameter of 43cm. The display is equipped with 224 LED lights and weighs about 200g. The rotating LED display can play the video stored in the SD card, and can also get the picture on the player in real time via WiFi. Multiple rotating LED displays can be spliced into a larger display to display more content. The physical image and display effect of the rotating LED display are shown in the figure.

Figure 2. Rotating LED display and display effect

3. Design and realization of the quadrotor

The drone is the carrier of the display, and a stable and reliable drone is the prerequisite for the performance. As one of the multi-rotor aircraft, the quadrotor has the advantages of small size, simple structure, flexible operation, vertical take-off and landing[7]. In this paper, a quadrotor is selected as the carrier. The quadrotor is mainly composed of a flight platform and a flight control system.

As the basis of drone flight, the flying platform of the quadrotor needs to have the characteristics of light weight and high strength. Carbon fibre material is a light-weight and strong material, often used in drones. We designed and processed the frame made of carbon fibre materials to meet the above requirements. The structure of the entire drone platform is basically symmetrical, and the centre of gravity is on its geometric centre. The battery provides power for the flight control system and drive system. In order to improve the load capacity of the quadrotor flight platform, a lithium polymer battery with high energy density and high power is selected; the brushless DC motor drives the propeller to rotate and provides lift for the drone. The structure of the drone is shown in the figure 3, and the wheelbase of the quadrotor is 650mm.

Figure 3. Quadrotor flying platform

The flight control system collects data from various sensors in real time and calculates the attitude, position, speed and other information of the drone. According to the control instructions, it calculates the control instructions of the motor, and converts the control instructions into PWM signals to drive the motor to rotate, and finally achieve flying manoeuvres. Pixhawk is an open-source flight control project that aims to provide a high-end autopilot with low cost, good stability, high safety, and multiple functions for academic researchers, drone enthusiasts and industrial teams. In this paper, Pixhawk is used as the flight control of the quadrotor. The drone equipped with GPS positioning modules can complete automatic flight missions in outdoor environments.

The positioning accuracy of ordinary GPS is at the meter level, which does not meet the position accuracy requirements of the drone light show. Real-time kinematic positioning (RTK) is the application of surveying to correct for common errors in current satellite navigation (GNSS) systems. It uses measurements of the phase of the signal's carrier wave in addition to the information content of the signal and relies on a single reference station or interpolated virtual station to provide real-time corrections, providing up to centimetre-level accuracy. In order to achieve high-precision position control of the quadrotor, the RTK positioning system[8] is used in this article. The base station is placed in an open area to continuously observe GPS satellites, and the observation data and base station information are sent to the rover in real time through radio transmission equipment. The rover calculates its position in real time based on its own observation data and data from base station. The high-precision position data of the rover is transmitted to Pixhawk through the serial port, and the quadrotor can achieve centimetre-level positioning accuracy, which meets the requirements of the drone light show.

4. Entertainment drone performance experiment

An entertainment drone, the controller and other auxiliary devices are required to realize the performance experiment of the drone light show. In this paper, the rotating LED display is installed vertically on a quadrotor to form an entertainment drone. In order to achieve the desired performance, we need to realize the control of the drone's position and the rotating LED display. In order to realize the position control of the entertainment drone, this article uses the Mission Planner software running on the computer to edit and generate flight missions, and upload them to the flight controller via wireless data transmission. The flight mission specifies the position, speed and hovering time of the drone, and the drone completes the flight action according to the prescribed route. The RTK positioning system provides high-precision positioning information for the drone. High-precision entertainment drones can more finely control the images presented in different positions and get better performance effects. In order to achieve the control of the picture on the rotating LED display, this article uses a high-power outdoor AP to achieve WiFi coverage of the performance venue. The rotating LED display and the player are connected to the same WiFi local area network to achieve a stable link. Figure 4 shows the entertainment drone designed in this article. The drone takes off and landed on the landing gear fixed on the ground, this can reduce the weight of the drone and increase the endurance.

Figure 4. The entertainment drone with rotating LED display technology

5. Conclusion

In this paper, the rotating LED display technology is applied to the drone performance, and a quadrotor equipped with a rotating LED display is designed. The drone can display complex patterns, which greatly reduces the cost and technical difficulty of the drone light show. At the same time, multiple drones can form a flight formation, which can display more complex pictures. The drone uses an open-source flight control solution, which is highly expandable. The performance experiment of the drone has achieved a good display effect, which proves the feasibility of this design. However, the current picture played on the drone and the position of the drone are still independently controlled. Further work is needed to realize the synchronous control of the picture and the position, which can achieve better performance effects.

References

[1] Trask, M. (2019) Intel's Drone Tech Illuminates Night Skies. Https://www.commercialdroneprofessional. com/intels-drone-tech-illuminates-night-skies
[2] Austin, P.L. (2018) A Record-Breaking Drone Show Ended With Quadcopters Falling from the Sky. https://gizmodo.com/a-record-breaking-drone-show-ended-with-quadcopters-fal-1825745673
[3] Dong, X., Zhou, Y., Ren, Z., & Zhong, Y. (2016). Time-varying formation tracking for second-order multi-agent systems subjected to switching topologies with application to quadrotor formation flying. IEEE Transactions on Industrial Electronics, 64(6), 5014-5024.
[4] Dong, X., Yu, B., Shi, Z., & Zhong, Y. (2014). Time-varying formation control for unmanned aerial vehicles: Theories and applications. IEEE Transactions on Control Systems Technology, 23(1), 340-348.
[5] LIANG, G. S., QIN, J., CHEN, S. H., & CHEN, Z. Y. (2014). Design and implementation of colourful rotating LED display. Chin. J. Liq. Cryst. Disp, 29(5), 850-855.
[6] Ortega, M. X. P., Ramírez, J. C. D., Castro, J. W. V., Carrión, J. L. A., Carvajal, A. A. Á., & Segarra, A. N. G. (2020). Application of the technical-pedagogical resource 3D holographic LED-fan display in the classroom. Smart Learning Environments, 7(1), 1-13.
[7] Bouabdallah, S., & Siegwart, R. (2007, October). Full control of a quadrotor. In 2007 IEEE/RSJ international conference on intelligent robots and systems (pp. 153-158). Ieee.
[8] Eling, C., Klingbeil, L., & Kuhlmann, H. (2014). Development of an RTK-GPS system for precise real-time positioning of lightweight UAVs. In Proc. of.

A design of laser array harp based on multi-dimensional wavelet transform and audio signal reconstruction

Bin Liu[1],Yan Ren[1,*]

[1]Henan Forestry Vocational College,Luoyang, Henan, 471000,China

[*]Corresponding author's e-mail:271804648@ hnfvc.edu.cn

Abstract. This paper introduces a design scheme of laser array harp based on multi-dimensional wavelet transform and audio signal reconstruction. The green light beams from multiple high-power lasers simulate harp strings, use photoresistors as the signal receiving end, and use a signal conditioning system composed of analog circuits and LM393 comparators to collect and adjust the resistance signal of the laser sensor[1], and finally it is adjusted to a level signal that can be recognized by the CPU. After receiving the signal, the CPU core board analyzes the string signal, and sends control commands to the audio processing system through the industrial bus according to the analyzed digital signal. After receiving the control command, the audio processing system uses the audio signal reconstruction technology composed of multi-dimensional wavelet packets, deep learning and other algorithms to simulate the audio signals of various string music, so as to achieve the purposes of using the lasers as virtual strings and imitating musical instruments for musical performance.[2]

1. Introduction

The traditional harp is a plucked string instrument with rich connotation and beautiful sound quality, while the laser harp uses laser to simulate the string, processes the collected signals, and sends out the corresponding sound, and can choose different timbres according to the music. The most important thing in the laser harp is the signal acquisition, processing and timbre restoration. In this paper, the acquisition and processing circuit of laser string signal is designed, and the corresponding audio signal is decoded and restored by multi-dimensional wavelet transform [3] [4] and reconstruction.

2. Hardware circuit

The stm32f407 processor is used as the CPU of this system. The resistance value of the photoresistor is adjusted to the high and low level that the single-chip microcomputer can recognize through the signal conditioning board, and then the single-chip microcomputer is used to conduct real-time inspections on the signal conditioning board to determine which photoresistor changes, and then the single-chip microcomputer sends control commands to the voice chip. After receiving the signal, the CPU core board analyzes the string signal, and sends control commands to the audio processing system through the industrial bus according to the analyzed digital signal. After receiving the control command, the audio processing system uses the audio signal reconstruction technology composed of multi-dimensional wavelet packets, deep learning and other algorithms to simulate the audio signals of various string music, so as to achieve the purposes that using the lasers as virtual strings and imitating musical instruments for musical performance.

The hardware circuit mainly includes a sensor composed of a photoresistor array, a signal acquisition and conditioning circuit composed of an LM393 comparator, a potentiometer, etc., and a signal processing and control circuit composed of a CPU and an audio processing chip. Fig. 1 is the overall block diagram of the design hardware:

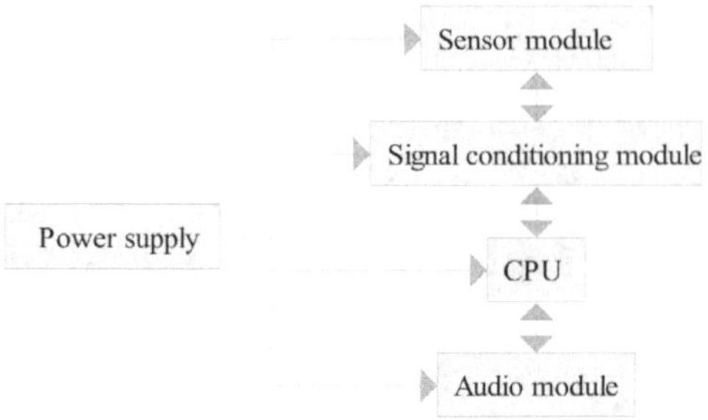

Figure 1. Overall block diagram of hardware

2.1. Selection of laser head and sensor

This design uses a green laser head with a wavelength of 532nm. The power supply of the laser head is 3.3V and the power is 200mW. In this design, a 10KΩ photoresistor is selected as the sensor. Considering that the laser spot and the area of the photoresistor are relatively small, for more precise positioning, two photoresistors in parallel are used as a sensor, which enlarges the area of the photoresistor receiving the light source. The resistance of the photoresistor after parallel connection is about 9KΩ to 11KΩ under normal indoor lighting, and about 1KΩ under laser beam irradiation.

2.2. Design of string signal acquisition and conditioning system

The signal acquisition circuit is composed of 4 comparators, a photoresistor sensor array and peripheral circuits, as shown in Fig. 2 below. Among them, the comparator uses the LM393 chip produced by Texas Instruments. LM393 is a dual-voltage comparator integrated circuit. There are two internal comparators with low current consumption and a TTL level signal output.

The photoresistor is connected to the in-phase input terminal of LM393. When the light is blocked between the laser head and the photoresistor, the resistance of the photoresistor will decrease. LM393 compares the voltage at the in-phase input terminal with the reference voltage, and outputs signals of different levels to the I/O port of the microcontroller. When the laser irradiates the photoresistor, the output port of the comparator will output low level; when the light between the laser head and the photoresistor is blocked, the output port will output high level. The 8 output ports corresponding to the 8 photoresistors are connected to the P0 port on the single-chip microcomputer. The state value of the P0 port is read to determine which laser string is being played, and the corresponding sound is emitted.

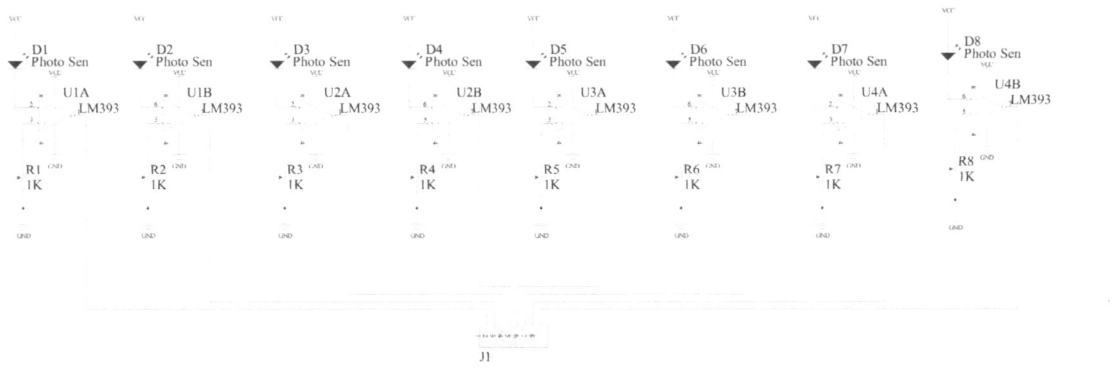

Figure 2. Signal acquisition and conditioning circuit

2.3. Audio reconstruction system

This design selects stm32f407 processor as the core board. It contains the IO ports and resources required by the laser harp, and its performance can also meet the system requirements.

This module is mainly composed of audio processing chip WT588D-U and external SPI Flash two circuits. WT588D-U requires external SPI bus Flash module to store the tone color sample library. In order to enable the laser harp to play a variety of musical instruments, different sound libraries can be created in advance, so that the laser harp can control the sound of a variety of musical instruments by pressing the keys, such as piano, violin, harp, etc., so as the function can be enabled Richer.

The audio module circuit diagram is shown in Fig.3 In the figure, U6 is a 32M external SPI Flash GD25Q32SIP, which is used to build a sample library of deep learning algorithms, and U7 is an audio chip WT588D, which is used to reconstruct audio signals of various musical instruments.

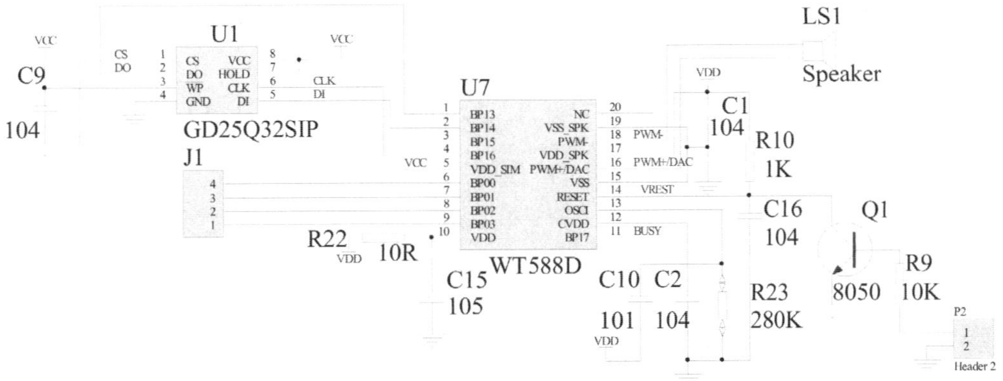

Figure 3. Voice module circuit

3. software program

3.1. Main program design

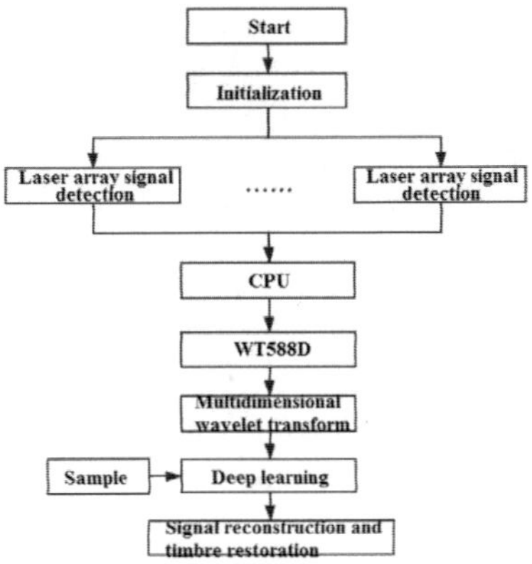

Figure 4. Main program flow chart

3.2. Multi-dimensional wavelet analytic transform

A wave composed of multiple frequencies is decomposed by a multi-dimensional wavelet, and all frequencies are decomposed, and then these sub-frequencies are weighted and reconstructed through the inverse operation to obtain the final result of restoring the timbre. In the discrete wavelet transform (DWT), the space $V_j=V_{j-1}+W_{j-1}$ represents the audio signal, and each audio signal X(t) represented on V_j can be represented by the basis functions in the above two spaces.

$$x(t) = \sum_{b} cA_0^{(k)\phi} Jb(t)$$
$$= \sum_{b} cA_1(k)\phi_J - 1, k(t) + \sum_{b} c^D, (k) u_J - 1, k(t) \quad (1)$$

When we decompose the coefficient A0(k) in the scale metric space J, two coefficients D1(k) and A1(k) can be obtained in the scale metric space, and we can also reconstruct the two coefficients A1(k) and D1(k) to get the coefficient A0(k).

$$A_0(k) \rightarrow \boxed{Analysis} \begin{matrix} \rightarrow D1(K) \rightarrow \\ \rightarrow A1(K) \rightarrow \end{matrix} \boxed{Analysis} \rightarrow A_0(k) \quad (2)$$

The specific coefficient calculation process is as follows:

$$cA_1(k) = \sum_{m} h_0(n-2k) cA_0(n) \quad (3)$$

$$CD_1(k) = \{y(t), uv-1, k(t)\}$$
$$= \sum_{R} cA_0(n)\phi_J, n(t), Uv-1, (t) \quad (4)$$
$$= \sum_{R} lA_0(n)\{\phi_J, n(t), w-1, n(t)\}$$

$$CD_1(k) = \sum_{R} h_1(n-2) cA_0(n) \quad (5)$$

4. Result

Fig.5 shows the laser harp prototype. Through experimental analysis, the effectiveness of the system is verified. The debugging experiment of hardware and software is carried out on MATLAB platform. The operating system is Microsoft's windows 10 system, and the running memory of the system environment is 8GB.

The experimental results are realized in the laser harp.

Figure 5. Finished drawing of laser harp

Fig5. Fig6. and Fig7. shows the reconstruction results of signals in different frequency bands of harp by audio reconstruction system [8].

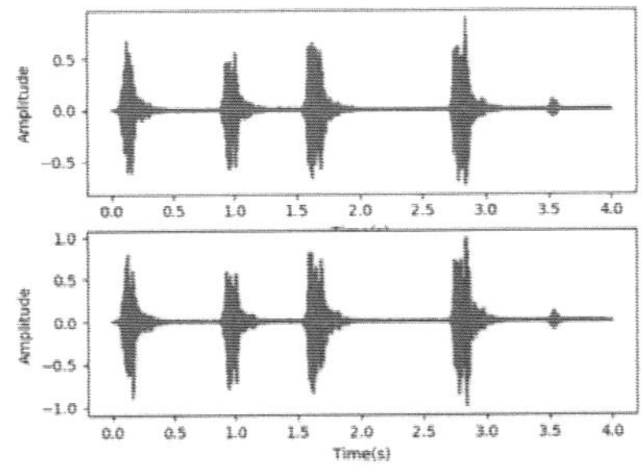

Figure 6. Reconstruction results of guitar signal (frequency band 95 Hz ~ 4.3 kHz) by audio reconstruction system

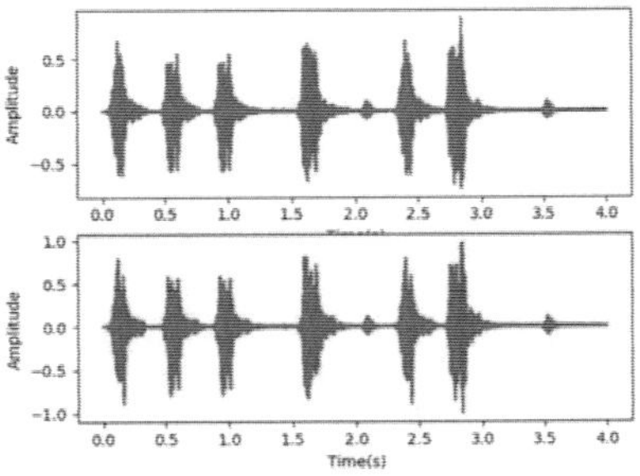

Figure 7. Reconstruction results of guitar signal (frequency band 120 Hz ~ 8 kHz) by audio reconstruction system

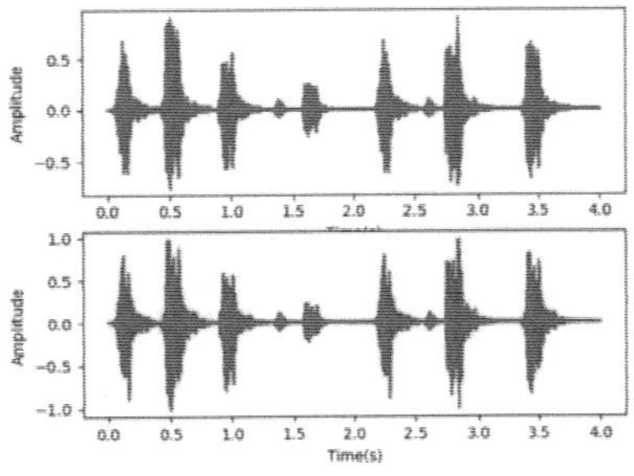

Figure 8. Reconstruction results of guitar signal (frequency band 65 Hz ~ 1.7 kHz) by audio reconstruction system

The experiment reconstructs the timbre information of the three frequency bands respectively. With the change of the number of wavelet packet iterations, the error of the system to the three audio processing shows a downward trend. Among them, the processing error of piano timbre is the lowest, up to about 0.1%, while the lowest errors of the other two kinds of audio are about 1.3% and 1.8% respectively. In contrast, the processing error of the proposed system is reduced by about 1.2% and 1.7% respectively. This is because the internal circuit of the chip designed by the proposed system has Zener regulated reference voltage, which can ensure that the output voltage is constant at 3.3 V and the output accuracy error is less than 1%, so as to improve the processing accuracy of the proposed system. The experimental results show that the multi-dimensional wavelet transform has good restoration characteristics for low, medium and high frequency bands of various timbres, and the mean square value of error is less than 0.97, especially in medium and high frequency bands, the mean square value of error can reach 0.99.

5. Conclusion .

In this paper, stm32f407 processor is used as the core, and the signal acquisition system is composed of signal processing module composed of high-power laser head, photoresist array and LM393. Many technologies are applied, such as signal acquisition, signal processing, data storage and transmission, SPI communication, audio decoding, audio output and so on. Through the patrol inspection of the signal end, the state of the signal end can be judged in real time. When a signal end is triggered, the processor analyzes the audio signal by using the methods of multi-dimensional small transformation and audio signal reconstruction, and uses the signal reconstruction technology to restore the real timbre.

References

[1] Chen Xin, Zhang Lihua. Design of a laser harp circuit [J]. Journal of electrical and electronics teaching, 2011,33 (05)
[2] Wan Dongqin, Qing Chaojin, Yang Qingyao, Cai bin, Yu Wang. Superimposed feature assisted 1-bit CS audio transmission [J]. Computer engineering and applications, 2020,56 (16). 1211-1213
[3] Cheng Zhengxing, Yang Shouzhi, Feng Xiaoxia. Theory, algorithm, progress and application of wavelet analysis [M]. Beijing: National Defense Industry Press, 2007
[4] Liu Mingcai. Wavelet analysis and application [M]. Beijing: Tsinghua University Press, 2005
[5] Yu Zhenjiang. Design of a voice signal acquisition and analysis system based on MATLAB [J]. Scientific and technological information development and economy, 2012,22 (12): 112-113
[6] Guan Huihui, Zhou Xiaohui, Lu Weina. Matlab technology implementation of multi wavelet audio signal processing [J]. Gansu Journal of science, 2017,12 (6): 11-14
[7] Zhang Dengqi, Jiang min. implementation of audio signal processing technology based on MATLAB [J]. Computer knowledge and technology, 2009,16 (3)
[8] Wang Yingxue, Liu Yu, Duan Jizhong, Zhai Li. Multichannel signal reconstruction and improvement based on compressed sensing [J]. Journal of Nankai University (NATURAL SCIENCE EDITION), 2020,53 (04)

Research on 5G positioning technology for urban canyon

Xin Xiong[1*], Shumin Liu[2]

[1]School of Software Engineering, Jiangxi University of Science and Technology, Nanchang, Jiangxi, 330013, China

[2]School of Software Engineering, Jiangxi University of Science and Technology, Nanchang, Jiangxi, 330013, China

[*]Corresponding author's e-mail: 6720190970@mail.jxust.edu.cn

Abstract. Aiming at the problem of poor positioning accuracy of GNSS for urban canyon environment, this paper proposes a TDOA/AOA combined positioning technology based on 5G.Firstly, the AOA estimation algorithm based on conventional beam forming is compared with that based on MUSIC algorithm. Then, the weighted matrix is smoothed forward and backward.Finally, according to the measurement data of AOA and TDOA, Chan and Taylor combined localization algorithm is used to obtain the estimated location of the user.

1. Introduction

With the development of economy, urban building clusters gradually become dense, resulting in urban canyon, and the traditional GNSS positioning effect is not good.Cellular wireless Location service (LCS) can address such problems[1]. 5G positioning technology includes AOA (Angle of Arrival), TDOA (Time Difference of Arrival), RSSI (Received Signal Strength Indication)[2].

For the problem of signal interference, the traditional beam forming algorithm has low accuracy[3]. Schmidt et al. proposed the MUSIC algorithm[4]. Evans proposed a spatial smoothing algorithm to achieve decoherence[5].Literature improves the spatial smoothing algorithm to deal with coherent sources more effectively[6-8]. In this paper, based on MUSIC algorithm and weighted forward and backward space smoothing algorithm, AOA estimation effect is better.

The positioning algorithm of TDOA will lead to more than one registration point in the case of fewer base stations or unreasonable distribution of base stations[9]. Common algorithms for TDOA position calculation include Fang algorithm[10], Chan algorithm and Taylor algorithm[11-12]. Fang algorithm can not use redundant information to improve location accuracy. Chan algorithm can improve its accuracy by increasing the number of base stations.

Many scholars have studied the combined positioning method based on the above algorithms[13]. In this paper, the improved MUSIC algorithm is used to complete AOA estimation. Combined with TDOA measurement data of 5G base station, TDOA/AOA combination positioning combined with Chan algorithm and Taylor algorithm is adopted to obtain more accurate THREE-DIMENSIONAL location information.

2. AOA Location

2.1. AOA positioning principle

The positioning algorithm of AOA (Angle of Arrival) uses the built-in antenna array at the receiving end of the base station to receive the azimuth of the electromagnetic wave sent by the user's equipment to construct the tangent function equation. The three-dimensional position principle of AOA in the urban environment is shown in Figure 1:

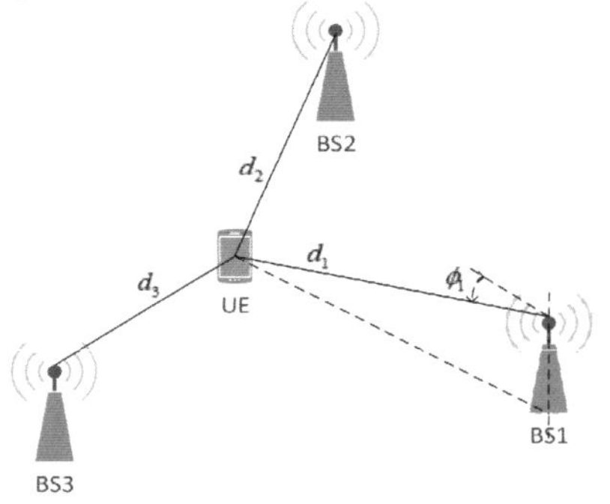

Figure 1. AOA location principle

According to the figure, the relationship between ϕ_i, UE and BS_i is:

$$\tan(\phi_i) = \frac{z_i - z}{\sqrt{(x_i - x)^2 + (y_i - y)^2}} \tag{1}$$

Coordinate $[x, y, z]$ of UE can be achieved by combining three or more equations.

2.2. AOA estimation by MUSIC algorithm

The principle of MUSIC (Multiple Signal Classification) is to decompose the covariance matrix of the output data of any array by eigenvalue, and the eigenvectors of different eigenvalues form orthogonal signal subspace and noise subspace.

The auto correlation matrix can be expressed as:

$$\begin{aligned} R_{xx} &= E\left[x(t)x^H(t)\right] \\ &= AE\left[s(t)s^H(t)\right]A^H + E\left[n(t)n^H(t)\right] \\ &= AR_{ss}A^H + R_{nn} \end{aligned} \tag{2}$$

Spatial spectrum of MUSIC algorithm $P_{MUSIC}(\theta)$:

$$P_{MUSIC}(\theta) = \frac{1}{a^H(\theta)E_N E_N^H a(\theta)} \tag{3}$$

2.3. MUSIC algorithm for backward and forward space smoothing

The forward space smoothing algorithm divides M array signals into L overlapping sub-arrays, as shown in the figure below.

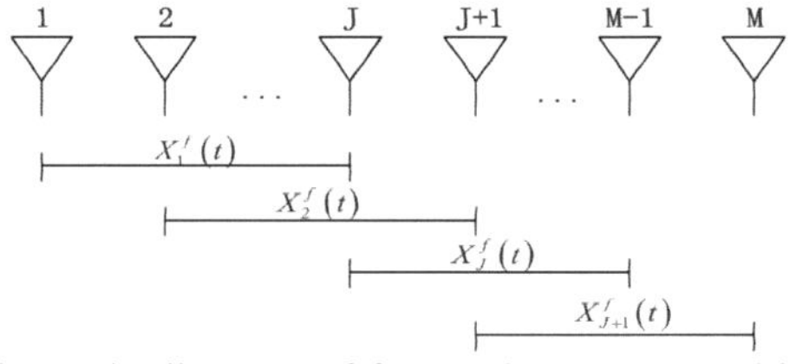

Figure 2. Schematic diagram of forward space smoothing algorithm

Each sub-array has $N = M - L + 1$ elements. The spatial smoothing matrix of the J sub array is:

$$R_N^{jj} = A_1 D^{j-1} R_s \left(D^{j-1}\right)^H A_1^H + \sigma^2 I_N \qquad j = 1, 2, ..., L \tag{4}$$

The covariance moment after smoothing forward space can be obtained by averaging the covariance of all sub arrays, which can be expressed as:

$$\tilde{R}^{f} = \frac{1}{L}\sum_{j=1}^{L} R_{N}^{jj} \tag{5}$$

The backward space smoothing algorithm smooths the covariance matrix of the backward sub array conjugate received data, and improves the number of smoothed sub arrays, as shown below.

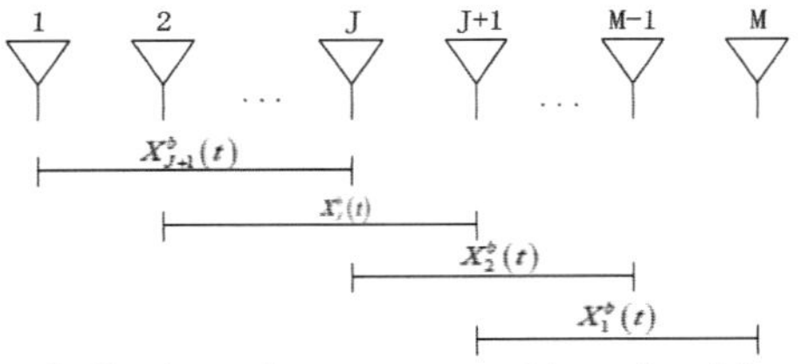

Figure 3. Backward space smoothing algorithm

Each sub-array has $L = M - P + 1$ elements. The relationship between backward space smooth covariance matrix $\tilde{R}^{b}$ and forward space smooth covariance $\tilde{R}^{f}$ is:

$$\tilde{R}^{b} = J * \left(\tilde{R}^{f}\right)^{*} * J \tag{6}$$

The covariance after the forward and backward space smoothing is expressed as:

$$\bar{R} = \frac{1}{2}\left(\tilde{R}^{f} + \tilde{R}^{b}\right) \tag{7}$$

2.4. Weighted forward and backward space smoothing MUSIC algorithm

Weighted forward-backward space smoothing algorithm is a weighted average of forward-backward space smoothing and forward-backward space smoothing. The estimated value of the source $\left[\hat{\theta}_1, \hat{\theta}_2, \ldots, \hat{\theta}_K\right]$ is obtained by using the forward-backward space smoothing MUSIC algorithm. A weighted matrix W, W, can be expressed as:

$$W = \left(B \cdot B^{H}\right)^{+} \tag{8}$$

The optimal weight matrix W_P is obtained by smoothing W.

$$W_P = \frac{1}{L}\sum_{k=1}^{L} F_k \cdot \left(W + J \cdot W^{*} \cdot J\right) \cdot F_k^{T} \tag{9}$$

The weighted matrix W_P is combined with the covariance matrix of forward-backward space smoothing, and the obtained weighted forward-backward space smoothing covariance matrix is:

$$\tilde{R}^{fb} = \frac{1}{L}\sum_{j=1}^{L} \left(R_{N}^{jj} + J * \left(R_{N}^{jj}\right)^{*} * J\right) \cdot W_P \tag{10}$$

2.5. Simulation and analysis of improved MUSIC algorithm

The simulation uses a uniform linear array with 16 array elements and half wavelength spacing, and the two signal angles are 0° and 5° respectively. In the case of the SNR of -5dB, two groups of experiments are carried out on the forward-forward-backward space smoothing MUSIC algorithm and the improved weighted forward-forward-backward space smoothing MUSIC algorithm. The spatial spectrum obtained is shown in the figure below.

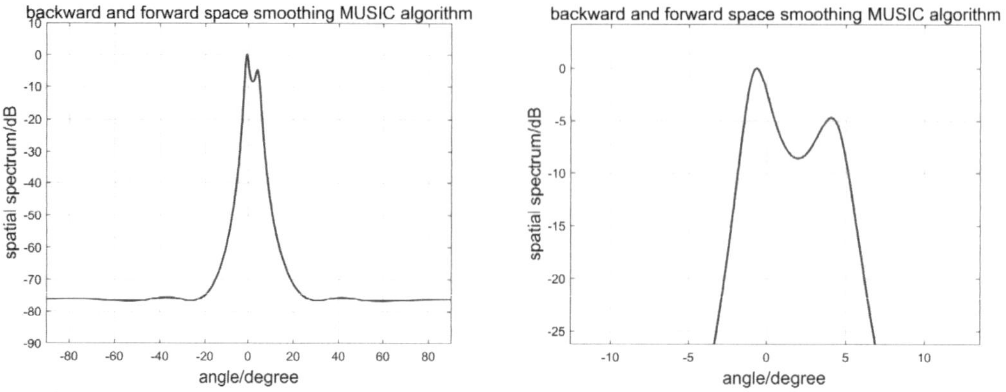

Figure 4. MUSIC algorithm and details of backward and forward space smoothing

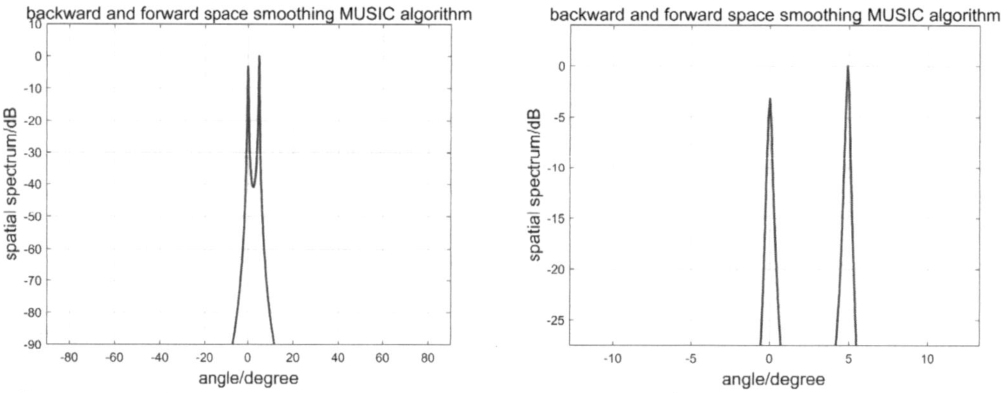

Figure 5. Weighted forward-backward space smoothing MUSIC algorithm and its details

By analyzing the above simulation results and details, the weighted forward-backward space smoothing MUSIC algorithm is superior to the forward-backward space smoothing MUSIC algorithm under the condition of low SNR.

3. TDOA Location

TDOA (Time Difference of Arrival): To locate by measuring the Time Difference between user device signals and different base stations. The positioning principle of TDOA is shown in Figure 3-1. To determine a point in three-dimensional space, at least three distance differences and four base stations are required. The coordinate of the base station is $BS_i = [x_i, y_i, z_i], i = 1, 2, 3, 4$. Figure 3-1 shows the TDOA locating principle.

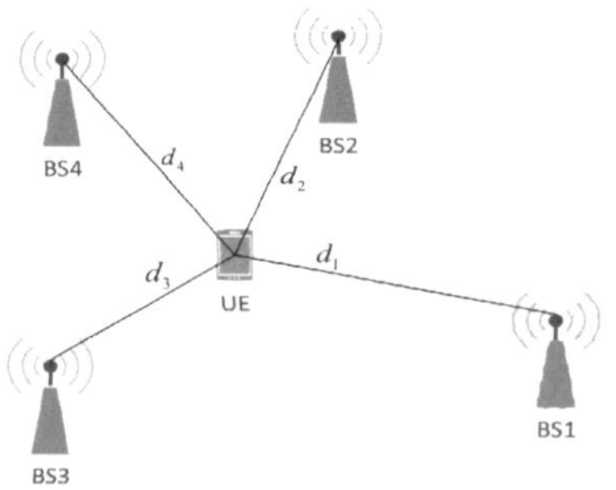

Figure 6. TDOA locating principle

Taking BS_1 as the reference base station, it is known that the time difference between the arrival of UE at each secondary station and the arrival of the main station is $\tau_{i1}(i=2,3\ldots)$. Assuming that the signal propagation velocity is the speed of light c, the distance difference between UE and the primary station and each secondary station is:

$$\Delta d_{i1} = c\tau_{i1} \tag{11}$$

Can also be expressed as:

$$\Delta d_{i1} = d_i - d_1 = \sqrt{(x_i - x)^2 + (y_i - y)^2 + (z_i - z)^2} - \sqrt{(x_1 - x)^2 + (y_1 - y)^2 + (z_1 - z)^2} \tag{12}$$

set $q_i^2 = x_i^2 + y_i^2 + z_i^2$,

$$x(x_1 - x_i) + y(y_1 - y_i) + z(z_1 - z_i) = \Delta d_i d_1 + \frac{\Delta d_{i1}^2 + q_0^2 - q_i^2}{2} \tag{13}$$

The above equations should have three, which can be rewritten as a matrix:

$$\begin{bmatrix} x_2 - x_1 & y_2 - y_1 & z_2 - z_1 \\ x_3 - x_1 & y_3 - y_1 & z_3 - z_1 \\ x_4 - x_1 & y_4 - y_1 & z_4 - z_1 \end{bmatrix} \begin{bmatrix} x \\ y \\ z \end{bmatrix} = \begin{bmatrix} -\Delta d_2 \\ -\Delta d_3 \\ -\Delta d_4 \end{bmatrix} d_1 + \begin{bmatrix} l_2 \\ l_3 \\ l_4 \end{bmatrix} \tag{14}$$

UE coordinate $[x, y, z]$ can be obtained.

4. TDOA/AOA combination localization

4.1. TDOA/AOA positioning principle

In this section, the positioning method combined with TDOA and AOA is adopted to establish a three dimensional positioning system model by using the measured values of AOA and TDOA. Chan algorithm is a non-recursive positioning method based on TDOA positioning technology, and generally coordinates positioning with other algorithms to provide initial values for other algorithms. Chan algorithm was used to preliminary estimate the coordinates of x and y of UE, and then Taylor algorithm was used to iterate and revise the initial estimated value until it met a certain threshold value, and the position of UE was solved.

The relationship between ϕ_1 and d_1 is:

$$\cos(\phi_1) = \frac{\sqrt{(x_1 - x)^2 + (y_1 - y)^2}}{d_1} \tag{15}$$

The z coordinate of UE can be obtained from equation (1):

$$z = z_1 - \tan(\phi_1)\sqrt{(x_1 - x)^2 + (y_1 - y)^2} \tag{16}$$

The distance between UE and each base station can be expressed as:

$$d_i = \sqrt{(x_i - x)^2 + (y_i - y)^2 + \left(z_i - \left(z_1 - \tan(\phi_1)\sqrt{(x_1 - x)^2 + (y_1 - y)^2}\right)\right)^2} \tag{17}$$

It can be obtained from the above formula:

$$y = ax + b$$

$$\begin{cases} a = \dfrac{d_{31}x_2 - d_{21}x_3}{d_{21}y_3} \\ b = \dfrac{(1-\beta^2)d_{31}(d_{21} - d_{31})}{2y_3} - \dfrac{d_{31}x_2^2 - d_{21}(x_3^2 + y_3^2)}{2d_{21}y_3} \\ \beta = \sin(\phi_1) \end{cases} \tag{18}$$

The obtained $[x, y]$ is used as the preliminary estimate of Taylor algorithm for the next iteration correction. The distance difference between UE and the reference base station and other base stations is $d_{i1} = d_i - d_1$. The distance differences g and d are the form of the coordinates x and y of UE and are functions of ϕ_1. Define $\rho = [x, y]^T$, and the cost function of about is expressed as ρ:

$$J(\rho) = \|g - d\|^2 \tag{19}$$

The objective function of ρ estimation is:

$$\hat{\rho} = \arg\min_{\rho} J(\rho) \tag{20}$$

The objective function of ρ_{k+1} estimation is:

$$J(\rho_{k+1}) \cong \|\hat{g} - H\rho_{k+1}\|^2 \tag{21}$$

The least squares solution of x and y can be obtained:

$$\rho_{k+1} = \rho_k + \left(H^T H\right)^{-1} H^T (g - d_k) \tag{22}$$

If $|\rho_{k+1} - \rho_k| \leq \varepsilon$, the estimated value of coordinates x and y can be determined by equation (18), and finally substituted into Equation (16) to obtain the position coordinate $[x, y, z]$ of UE.

4.2. Simulation and analysis
This section uses MATLAB simulation software to simulate and compare the improved AOA positioning algorithm, TDOA positioning algorithm and combined TDOA/AOA positioning algorithm in this paper. Assuming that four base stations are involved in positioning and the real positions of base stations and UE are known, the relationship between SNR and estimated standard deviation under different SNR is shown in figure 7.

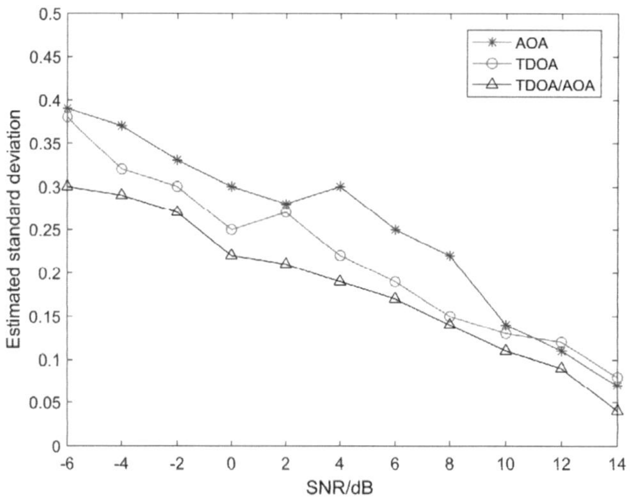

Figure. 7 Relationship between estimated standard deviation and SNR

The simulation results show that the estimated standard deviation of the three methods decreases with the increase of SNR, but the TDOA/AOA positioning algorithm has the lowest estimated standard deviation and good stability.

5. Conclusion
In this paper, weighted spatial smoothing MUSIC algorithm is used to improve the AOA estimation accuracy. Chan algorithm was used to preliminary locate UE position, and Taylor algorithm was used for iterative correction, which effectively improved the positioning accuracy. 5G is a hot technology at

present. Next, we should make full use of the characteristics of 5G signal to improve positioning accuracy and make life more convenient for people in urban valleys in the future.

Acknowledgments
This work was supported in part by National Natural Science Foundation of China (No. 61761021), Natural Science Foundation of Jiangxi Province (Grant No. 20181bab202018, 20202BAB212003), Special Fund for Postgraduate Innovation of Jiangxi Province(No. YC2020-S481).

References
[1] Peral-Rosado J D, Raulefs R, Lopez-Salcedo J A, et al. Survey of Cellular Mobile Radio Localization Methods: From 1G to 5G[J]. IEEE Communications Surveys & Tutorials, 2018:1124-1148.
[2] Dammann A, Raulefs R, Zhang S. On prospects of positioning in 5G[C]// IEEE International Conference on Communication Workshop. IEEE, 2015.
[3] Mortaheb M , Abbasfar A. Beamforming - based random access protocol for massive MIMO systems[J]. Transactions on Emerging Telecommunications Technologies, 2020(8).
[4] Schmidt R O. Multiple Emitter Location and Signal Parameters Estimation[C]// Proc Radc Spectrum Estimation Workshop. 1979.
[5] High resolution angular spectrum estimation techniques for terrain scattering analysis and angle of arrival estimation. 1981.
[6] Kamiya Y, Kikuma N, Sakakibara K. DOA estimation of desired signals using in-phase combining of multiple cyclic correlations and spatial smoothing processing. IEEE, 2017.
[7] Ray K P, Kulkarni R K , Ramkrishnan B K. DOA estimation in a multipath environment using covariance differencing and iterative Forward and Backward spatial smoothing[C]// International Conference on Recent Advances in Microwave Theory & Applications. IEEE, 2008.
[8] Iwai T, Hirose N, Kikuma N, et al. DOA estimation by MUSIC algorithm using forward-backward spatial smoothing with overlapped and augmented arrays[J]. IEEE, 2015.
[9] Yang ZHEN. TDOA/AOA localization algorithm based on EKF to solve TDOA Multi-solution Problem [J]. Industrial control computer, 2019, 32(09):104-106.
[10] Chen Sihan. TDOA indoor Location Technology Based on Fang Algorithm [J]. Journal of Terahertz Science and Electronic Information, 2017, 15(05):752-755.
[11] Chan Y T, Ho K C. A simple and efficient estimator for hyperbolic location[J]. IEEE Transactions on Signal Processing, 2002, 42(8):1905-1915.
[12] Foy W H. Position-Location Solutions by Taylor-Series Estimation[J]. IEEE Transactions on Aerospace and Electronic Systems, 2007, AES-12(2):187-194.
[13] Yan Lei-bing, Lu Yin, Zhang Ye-rong. TDOA/AOA Localization Method Based on Improved Least Squares Algorithm [J]. Chinese Journal of Radio Science, 2016, 31(02):394-400.

Analysis and Testing of the Dynamic Characteristics of Rotor Worktable

Jianmin Wang[1, a], Dong LIU[1, b*] and Yi Bian[2, a]

[1, a]College of mechanical and material engineering, North China University of Technology, Beijing, Beijing, 100144, China

[1, b*] College of mechanical and material engineering, North China University of Technology, Beijing, Beijing, 100144, China

[2, a] China Precision Forming Industry Technology Research Institute (Anhui) Co., Ltd, Wuhu, Anhui, 241080, China

*Corresponding author's e-mail: liudong@ncut.edu.cn

Abstract. The finite element modelling, mesh generation and modal analysis of the rotor worktable are carried out by combining the analysis technology of finite element software in this paper. Then, the software analysis results and the actual experimental results are compared and analyzed, so as to get the causes of error, which can provide good basic data for the use of the rotor table in the future that could better meet the needs of scientific research and teaching.

1. Introduction

As a kind of power equipment, rotating machinery is widely used in production and life [1]. As an important part of rotating machinery, the normal operation[2] of the whole equipment will be affected or even determined by the performance of Flexible rotor. In the process of the actual research on the performance of the rotor, it is usually necessary to use the experimental instrument such as the rotor workbench that facilitate the various experiments on the rotor[3][4]. But in the actual experiment, the results of rotor experiment are affected by the service life, the degree of wear, the vibration of rotor rotation[5]-[6] and other factors of rotor table. Therefore, it is particularly important to be familiar with the structure and performance of rotor worktable.

Modal analysis is a modern method to study the dynamic characteristics of structures[7], which is mainly used to determine the vibration characteristics of structures or machine components. For the modal analysis of rotor vibration characteristics, a lot of research have done by domestic scholars. The modal analysis error of flexible rotor system of Maglev motor under nonlinear contact are analyzed and verified by Ziyuan Huang[8]. The modal analysis of magnetically suspended rotor that verified by relevant experiments is carried out by Kelei Li[9]. The safe area of working speed of flexible rotor of high-power high-speed permanent magnet motor is analyzed by Guanghui Du[10] through modal analysis.

In this subject, the results of the modal analysis of the rotor table that is carried out by combining the analysis technology of the finite element software are compared and analyzed with the calculation results of the actual experiment that would obtain the cause of the error, which would provide good basic data for the use of the rotor workbench in the future and some reference for the design and analysis of the rotor workbench.

2. Simulation model modeling of rotor working table

2.1. Rotor table modeling

The rotor table is basically composed of base, oil cup, oil bearing support, oil cup, eddy current sensor mounting support, counterweight plate and DC motor. In this paper, SolidWorks software is used to model inv 1612 multifunctional flexible rotor worktable. The assembly drawing of parts is shown in Figure 1.

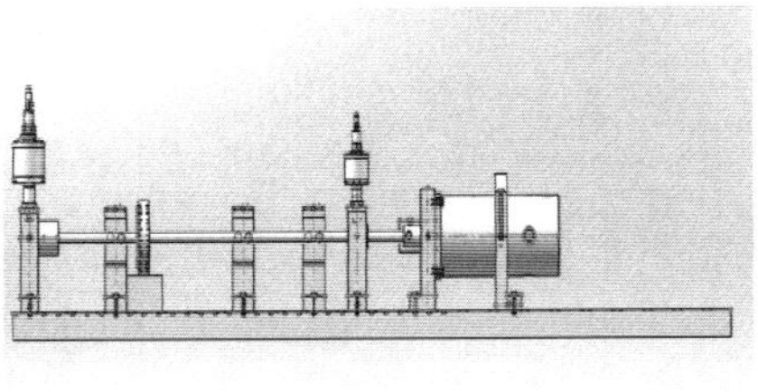

Fig.1 Overall assembly drawing of rotor table.

2.2. Finite element modeling of rotor table

First, the model built in Solidworks must be saved as .x_t file format, and then the file must be imported into ANSYS finite element software. According to the actual situation of the rotor worktable, the rotor worktable is fixed and restrained (fixed support is added on the base to limit all degrees of freedom of points, edges or faces,but no load is added. The whole rotor table is simplified to be composed of one material, and the material parameter settings are shown in Table 1:

Tab 1. Material properties.

Materials	45 Steel
Elastic modulus	210GPa
Damping ratio	0.3
Density	7850kg/m

The finite element mesh generation of the rotor workbench is carried out in the modal module. The mesh generation is automatic, and the mesh size is appropriate. The finite element mesh model is shown in Figure 2.

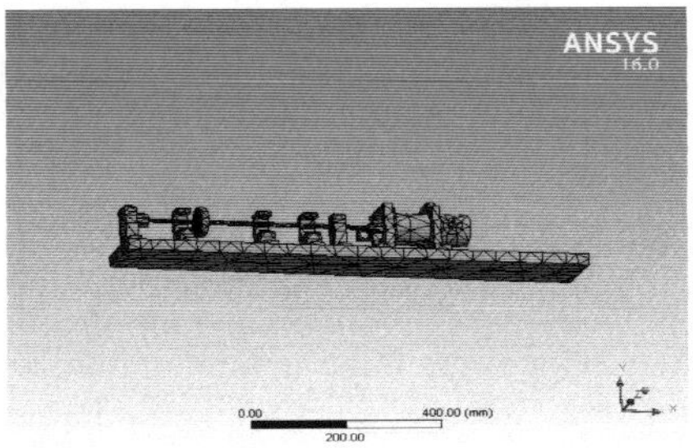

Fig.2 Finite element mesh model of rotor table.

2.3. Analysis of post-processing results

Since the low-order modes of the rotor worktable have practical reference value for the dynamic characteristic analysis of the rotor worktable, the first six (n = 6) modes of the rotor worktable are obtained by using ANSYS 16.0 software, and the natural frequencies of each order are shown in Table 2. In the table: n - Order of mode; ω-Natural frequency of rotor table.

Tab 2. The first six natural frequency parameters (unit: Hz).

n	1	2	3	4	5	6
w	140.28	140.43	262.73	494.44	495.7	909.3

The first six modes of the rotor can be clearly seen in the modal analysis module of ANSYS, and its vibration mode diagram can be analyzed. In the vibration mode of the first-order natural frequency of the rotor table, it can be seen that it has longitudinal bending vibration, which is mainly shown that the downward bending of the counterweight shaft, as shown in Figure 3; in the second-order mode, it also shows longitudinal bending vibration, with large deformation and upward bending of the heavy shaft, as shown in Fig. 4; in the third mode, it has torsional bending, as shown in Figure 5; in the fourth mode, it bends longitudinally, which is mainly manifested in the large deformation and downward bending of the shaft of the eddy current sensor on the right side of the counterweight plate, as shown in Figure 6; in the fifth mode, it has longitudinal bending vibration, which is mainly manifested in large eddy current deformation and upward bending of the counterweight plate on the right side of the sensor shaft, as shown in Fig. 7; it has transverse bending vibration in the sixth mode, as shown in Figure 8.

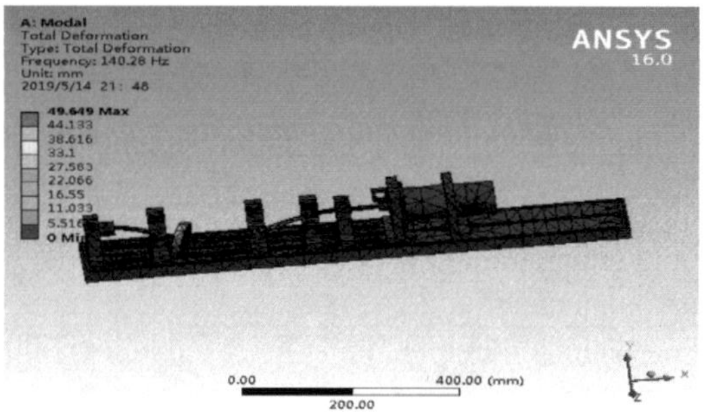

Fig.3 First order mode.

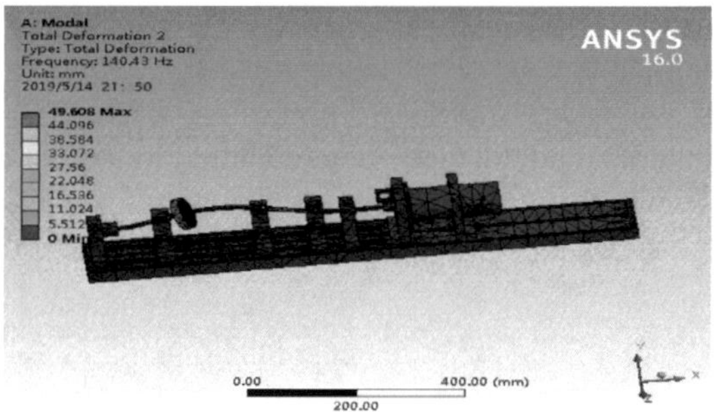

Fig.4 Second order mode.

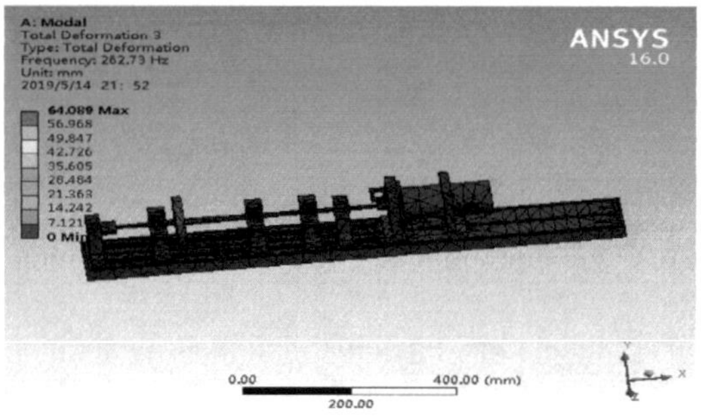

Fig.5 Third order mode.

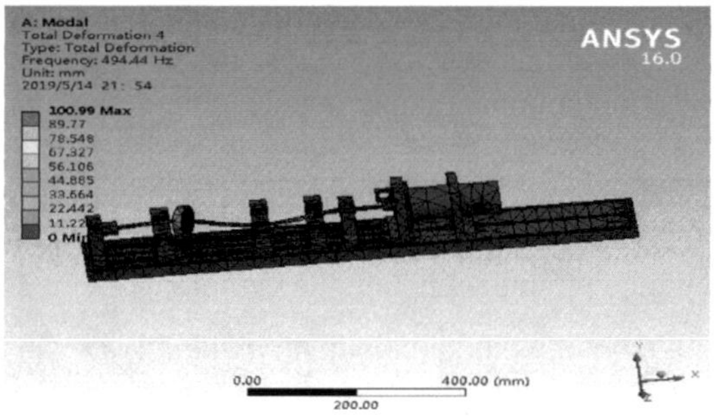

Fig.6 Fourth order mode.

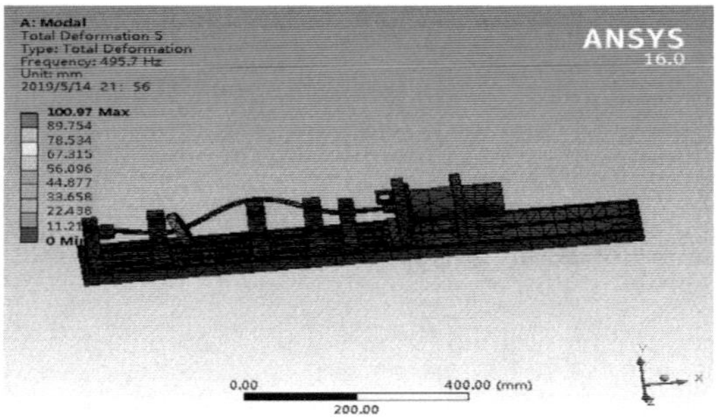

Fig.7 Fifth order mode.

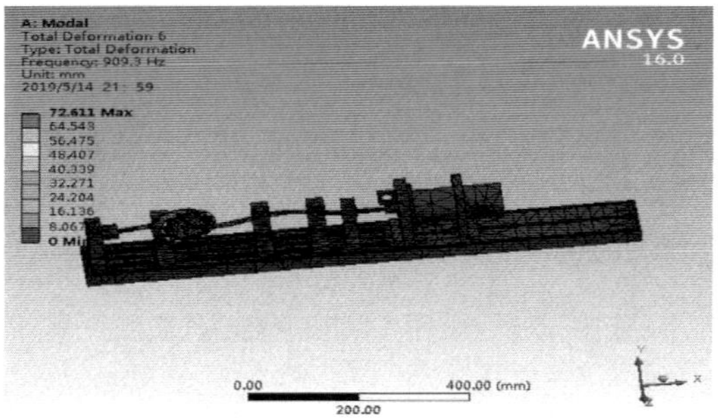

Fig.8 Sixth order mode.

3. Experimental analysis

3.1. Experiment Scheme

This test is mainly to test the dynamic characteristics of rotor worktable by hammering method. Compared with other methods, hammering method has obvious advantages, simple operation and easy control.

Firstly, according to the experimental instructions of the rotor workbench, the rotor workbench is connected with computers that need to set the parameters. Then the rotor workbench should be hit by hammer in turn and check the response attenuation signal should be checked. Finally the frequency value and damping value of the current waveform in the spectrum diagram should be observed and recorded.

3.2. Composition of experimental system

The system is mainly composed of rotor worktable, force hammer, INV1601B vibration teaching, experimental instrument and so on. The INV1601B vibration teaching experimental instrument and hammer used in the experiment are shown in Fig. 9 and Fig. 10 respectively.

Fig.9 Vibration teaching experiment instrument.

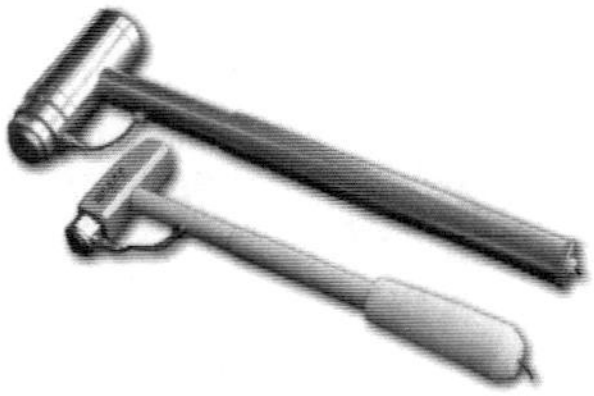

Fig.10 Force hammer.

3.3. Experimental results and analysis

Tab 3. Spectrum peak list.

No.	Frequency (Hz)	Amplitude spectrum (Peak)	Damping ratio
1	19.04297	15.79974	----
2	49.80469	1309.072	1.21%
3	99.60938	19.10921	----
4	107.4219	16.856	----
5	109.375	16.51903	----
6	111.8164	16.04238	----
7	149.9023	102.9708	0.34%
8	250	75.21978	0.27%
9	349.6094	34.44186	0.18%
10	449.707	82.59216	0.12%

In the experimental data in Table 3, the 10th order natural frequencies during rotor bench test are listed. Their trend is also gradually increasing, while the damping ratio is gradually decreasing, which is in line with the actual situation. However, due to the limitations of experimental operation, experimental equipment and professional level, only five damping ratios are measured, which is a pity.

4. Conclusion

(1) Through the modal analysis of the dynamic characteristics of the rotor table, the vibration and deformation of the first six modes are different. With the increase of frequency, the deformation is basically bending from left to right.

(2) Through comparison, it can be found that there is a deviation between the simulation experimental results and the actual results, which can be seen from the difference between the natural frequency of the simulation results and the natural frequency of the experimental results. Although the simulation has tried to imitate the actual situation, it still deviates from the actual situation. For example, the simulation is based on the integration of rotor worktable, which is different from the actual situation. In addition, the rotor workbench used in the laboratory is a device that has been used for some time, which has been worn and will also affect the experimental results. But in general, this provides good basic data for the future use of the rotor worktable, and also provides a certain reference for the research of the rotor worktable.

Reference

[1] Xiaojing Ma, Mai Timin Aini, Yaling Yan, et al. (2012) Experimental study on vibration characteristics of single disk flexible rotor. Machine Tool & Hydraulics, 40: 34-36.

[2] Yan Wang, Jisheng Ma, Haiqi Zheng, et al. (2012) Dynamic characteristic analysis of gear bearing system with flexible rotor. Vibration. Journal of Vibration, Measurement & Diagnosis, 32: 51-55.

[3] Changsheng Zhu, Chuan Mao. (2019) Experimental study on vibration reduction characteristics of concentric squeeze film damper with partial contact between inner and outer rings of oil film. Journal of Vibration Engineering, 32: 116-122

[4] Shengxi Jia, Longxi Zheng, Wangqun Deng, et al. (2018) Dynamic balance test system of flexible rotor without test weight mode. Journal of Vibration, Measurement & Diagnosis, 38: 1108-1113+1287-1288.

[5] Jie Hong, Cheng Chen, Yongfeng Wang, et al. (2018) Vibration characteristics test of high speed flexible rotor system under sudden unbalanced excitation. Journal of Aerospace Power, 33: 15-23.
[6] Yanlei Gao, Yong Li, Deyou Wang. (2001) Experimental study on vibration characteristics of flexible rotor. Aeroengine, :21-24.
[7] Bin Yang, Jianping Tan, Jianxiong Yao, et al. (2013) Dynamic characteristic analysis of fault flexible rotor system. Journal of Mechanical Transmission, 37: 106-109.
[8] Ziyuan Huang, Bangcheng Han, Yinfeng Zhou, et al. (2014) Modal analysis of flexible rotor system of Maglev motor under nonlinear contact. Proceedings of the CSEE, 34: 2438-2444.
[9] Kelei Li, Zhenyu Xie. (2008) Modal analysis of magnetic levitation rotor based on ANSYS. Journal of Mechanical & Electrical Engineering, :1-3.
[10] Guanghui Du, Na Huang, Fengge Zhang, et al. (2017) Modal analysis of flexible rotor system of high power and high speed permanent magnet motor. Transactions of China Electrotechnical Society, 32: 101-107.

Journal of Physics: Conference Series **2113** (2021) 012062 doi:10.1088/1742-6596/2113/1/012062

Anomaly detection in multi-class time series

Weihong Wang[1], Zhuolin Wu[2*], Xuan Liu[3], Lei Jia[4] and Xiaoguang Wang[5]

[1] College of Computer Science and Technology, Zhejiang University of Technology, Hangzhou, Zhejiang, 310000, China

[2] College of Computer Science and Technology, Zhejiang University of Technology, Hangzhou, Zhejiang, 310000, China

[3] Faculty of Computer Science, Dalhousie University, Halifax, Nova Scotia, B3H 4R2, Canada

[4] Research and Development Center of Agricultural Bank of China, Hangzhou, Zhejiang, 310000, China

[5] Beijing Broada Software Ltd, Hangzhou, Zhejiang, 310000, China

[*]Corresponding author's e-mail: 2112012132@zjut.edu.cn

Abstract. For modern operation and maintenance systems, they are usually required to monitor multiple types and large quantities of machine's key performance indicators (KPIs) at the same time with limited resources. In this paper, to tackle these problems, we propose a highly compatible time series anomaly detection model based on K-means clustering algorithm with a new Wavelet Feature Distance (WFD). Our work is inspired by some ideas from image processing and signal processing domain. Our model detects abnormalities in the time series datasets which are first clustered by K-means to boost the accuracy. Our experiments show significant accuracy improvements compared with traditional algorithms, and excellent compatibilities and operating efficiencies compared with algorithms based on deep learning.

1. Introduction

Huge amounts of data are generated daily for modern enterprises like financial securities companies. Generally, these data are arranged in strict accordance with the time sequence based on a statistical index. Based on our extensive experiences in real businesses, we summarize the following several challenges facing current time series anomaly detection services:

- Wide variety of data types.
  When monitoring tens of thousands of machine metrics simultaneously, different types of anomalies are associated with different tasks. For instance, innovational outlier, additive outlier, level shift, temporary change. Many models in the past have good detection accuracy only for a certain type of data, like DONUT [4] model.
- Limited computing resources.
  There are some good approaches for time series anomaly detection based on neural network like LSTM-based VAE-GAN [5]. However, these methods generally consume a lot of computing resources.
- Category change of data.

This is a point that many studies have missed. In reality, the type of data produced by a machine is not always the same, it changes with the services provided by the machine. Therefore, a successful model should possess the ability to detect these changes in time and respond accordingly.

With these challenges in mind, we decided to design a more competitive model which combines accuracy, time efficiency, and compatibility all together. Figure 1 shows the difference between our solution and traditional solutions. Our main design is to cluster the data first, then use different detectors to handle different types of data.

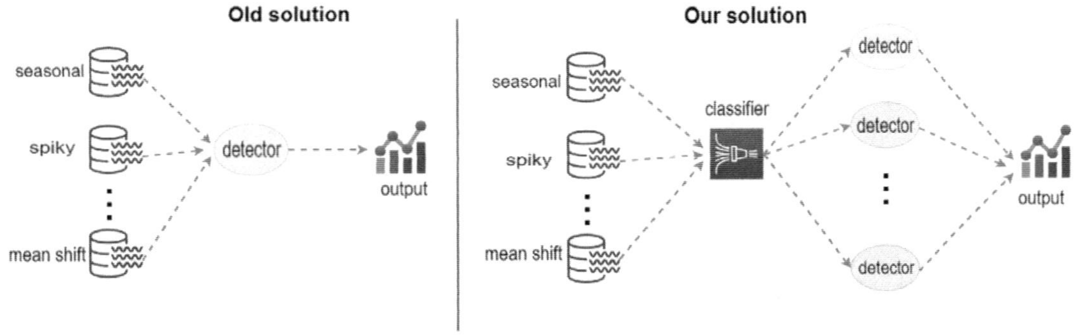

Figure 1. Our solution compared to traditional solutions.

Our contributions for this paper are as follows:

- We designed four high-efficiency and high-accuracy anomaly detection algorithms named R detector, T detector, $SR-\theta$ detector, SG detector. R and T detector are completely based on our own theory.
- We propose a self-selection algorithm model based on k-means clustering algorithm. We define the WFD distance which was adopted to measure the similarity between the sequences.
- We created and open sourced a well-labeled time series dataset: Robin. This data comes from the actual production data of an anonymous commercial bank.

2. Theory

Below we explain the basic theories used in our algorithm.

2.1. Periodic analysis

Regarding whether a sequence has periodicity, we can use contour plot of the square of wavelet coefficients' modulus to describe it more vividly, or use the wavelet variance curve to visually determine its periodic intensity. The wavelet variance can be deduced as:

$$w(a,b)=\int_R f(t)|a|^{-\frac{1}{2}}\overline{\varphi}\left(\frac{t-b}{a}\right)\mathrm{d}x \tag{1}$$

$$V(a)=\int_{-\infty}^{\infty} w(a,b)^2\,\mathrm{d}b \tag{2}$$

$w(a,b)$ represents wavelet coefficients, a is the scale factor, b is the translation factor, $|a|^{-\frac{1}{2}}\varphi(\frac{t-b}{a})$ represents subwavelet generated by basis wavelet function $\varphi(t)$, $\overline{\varphi}$ is the complex conjugate function of φ. $f(t)$ requires to be a sequence exists in a square integrable real number space.

$V(a)$ represents the wavelet variance, which integrates the wavelet coefficients in the b domain. The change curve of V with a, which is what we call the wavelet variance curve.

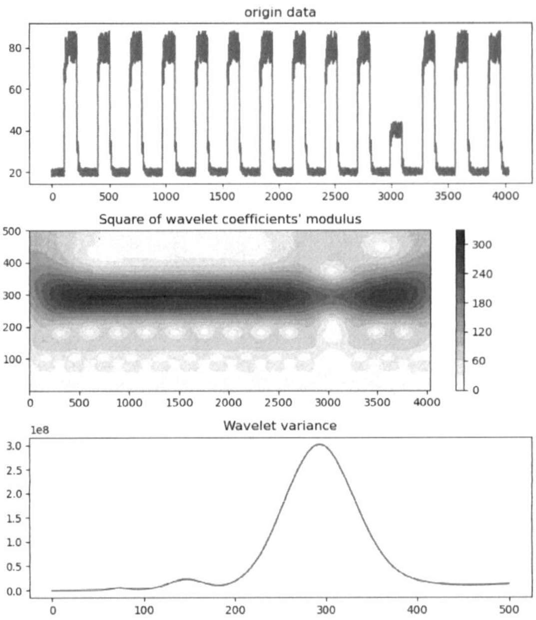

Figure 2. It is clearly that the periodicity of this sequence is relatively strong at about 300 on the whole, but the periodicity is relatively weak near 3000 on the time scale, due to the fact that there is a descending anomaly near this position in the original sequence.

2.2. SR

Spectral Residual is an algorithm that uses Fast Fourier Transform (FFT) to calculate the salient part of an image. [6] has proved that SR also has great efficiency in time series anomaly detection. It uses the log amplitude spectrum of the input to subtract the average log amplitude spectrum to get the salient part. Given a sequence $x = (f_0, f_1, \ldots, f_n)$:

$$A(f) = Amplitude(\mathcal{F}(x)) \quad (3)$$

$$P(f) = Phrase(\mathcal{F}(x)) \quad (4)$$

$$L(f) = log(A(f)) \quad (5)$$

$$AL(f) = h_q f \cdot L(f) \quad (6)$$

$$R(f) = L(f) - AL(f) \quad (7)$$

$$S(x) = \| \mathcal{F}^{-1}(exp(R(f) + iP(f))) \| \quad (8)$$

Where $\mathcal{F}$ denote Fourier Transform, $\mathcal{F}^{-1}$ denote Inverse Fourier Transform. The $S(x)$ we calculated above is called Saliency Map.

2.3. SG

Savitzky-Golay filter is widely used in data stream smoothing and denoising. It is a filtering method based on local polynomial least square fitting in time domain. When filtering the signal, the low-frequency components are fitted and the high-frequency components are smoothed out.

3. Method

In this section, we elaborate on our proposed model and framework.

3.1. Algorithms

Based on the theoretical basis of section 2, we now develop our own WFD distance and four detection algorithms.

3.1.1. WFD.

In our model, we mapped each sequence to a point (x, y, z) in the three-dimensional space. Among them, x reflecting the relative strength of the overall fluctuation of the data. The coordinate y represents the standard deviation of the sequence, and z represents the periodic intensity of the sequence. Given a sequence $f=(f_0,f_1,\ldots,f_n)$, this can be calculated as:

$$x = 15 + tan^{-1}\left(\frac{\overline{A}}{\sum_{i=1}^{5}\frac{\overline{D_i}}{5}} - 14\right) \tag{9}$$

$$y = std(f) = \sqrt{\frac{\sum_{i=0}^{n}(f_i-\overline{f})^2}{n}} \tag{10}$$

$$z = \max\left(V(a)\right),\ a \in (1\ldots K) \tag{11}$$

$\overline{A}$ represents the mean value of the approximate coefficients after 5th order wavelet decomposition of the sequence, $\overline{D_i}$ represents the detail coefficients obtained by decomposition of each order. The value of K depends on the specific situation. $V(a)$ is in (2). When the three coordinate values of a sequence are obtained, we use the Euclidean distance of the three-dimensional space coordinate points to determine the finally distance between the time series.

3.1.2. T detector.

This detection algorithm is mainly used to detect periodic abnormalities. Given a sequence $f=(f_0,f_1,\ldots,f_n)$, it will be deduced as follows:

$$yhalf = Half\left(\frac{\mathcal{F}(f)}{len(f)}\right) \tag{12}$$

$$peak = Maxpeak(yhalf) \tag{13}$$

$$T = \frac{len(f)}{peak} \tag{14}$$

where $Half(a)$ means the first half of the sequence a, $\mathcal{F}$ denote Fourier Transform, $Maxpeak(a)$ means the maximum crest of the sequence a, $len(a)$ means the length of sequence a. T will be considered as the main oscillation period of the input sequence. Then, for each point f_i:

$$G_1 = \left(i-T-10\ldots i-T+10\right) \tag{15}$$

$$G_2 = \left(i-2\cdot T-10\ldots i-2\cdot T+10\right) \tag{16}$$

$$K_1 = \min\left(|f_i - f_j|\right), j \in G_1 \tag{17}$$

$$K_2 = \min\left(|f_i - f_j|\right), j \in G_2 \tag{18}$$

$$Label(i)=\begin{cases}1, & if\ K_1+K_2>K\\ 0, & else\end{cases} \tag{19}$$

The value of K depends on the specific situation. Finally, $Label(i)=1$ means that f_i was detected as an anomaly point.

3.1.3. R detector

A very simple and effective way to locate mean drift abnormalities is to perform r transformation on the sequence $f=(f_0,f_1,\ldots,f_n)$, which can be formulated as:

$$r_i=\frac{\sum_{i-w}^{i} f_i}{\sum_{i}^{i+w} f_i} \tag{20}$$

$$Label(i)=\begin{cases}1, & if\ r_i-\overline{r}>3*std(r)\\ 0, & else\end{cases} \tag{21}$$

Where $std(r)$ denotes the standard deviation of r. $Label(i)=1$ means that a mean shift abnormal occurred at f_i, w depends on the specific situation. As shown in figure 3, in the r-space, the abnormal data can be observed clearly.

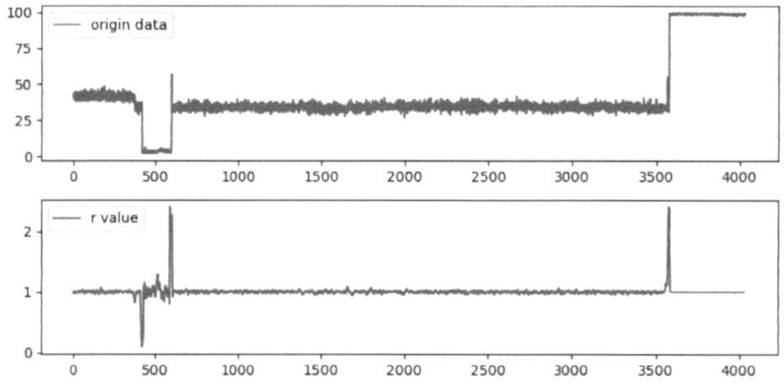

Figure 3. Example of r transform.

3.1.4. SR $-\theta$ detector

After getting the Saliency Map, we defined the following rules to locate and identify outliers:

$$Label(i)=\begin{cases}1, & if\ S_i-\overline{S}>\theta*std(S)\\ 0, & else\end{cases} \tag{22}$$

Where S denotes $S(x)$ in equation (8). θ is a parameter that depends on the actual situation. $std(S)$ denotes the standard deviation of S. $Label(i)=1$ means that f_i was considered to be an anomaly point.

3.1.5. SG detector

Given a sequence $f=(f_0,f_1,\ldots,f_n)$:

$$Label(i)=\begin{cases}1, & if\ f_i>(1+r)*smooth(i)\\ 1, & if\ f_i<(1-r)*smooth(i)\\ 0, & else\end{cases} \tag{23}$$

Where $smooth$ denotes the smoothed value. r is a parameter depends on the actual situation. $Label(i)=1$ means that f_i was considered to be an anomaly point.

3.2. Framework

Our model structure is mainly divided into two parts: the ***offline computing module*** responsible for time series clustering and the ***online detection module*** responsible for real-time detection. Different detectors will be saved in our detectors pool. The offline module automatically reads data periodically to select the cluster centers and finds a best detector for each cluster center, then writes it into the correspondence table to help provide better detection accuracy for the online detection module. When a new sequence to be detected reaches the detection module through the queue or pipeline, the program will quickly calculate the similarity between this sequence and the cluster centers. Then it will look up the correspondence table to obtain the corresponding detector.

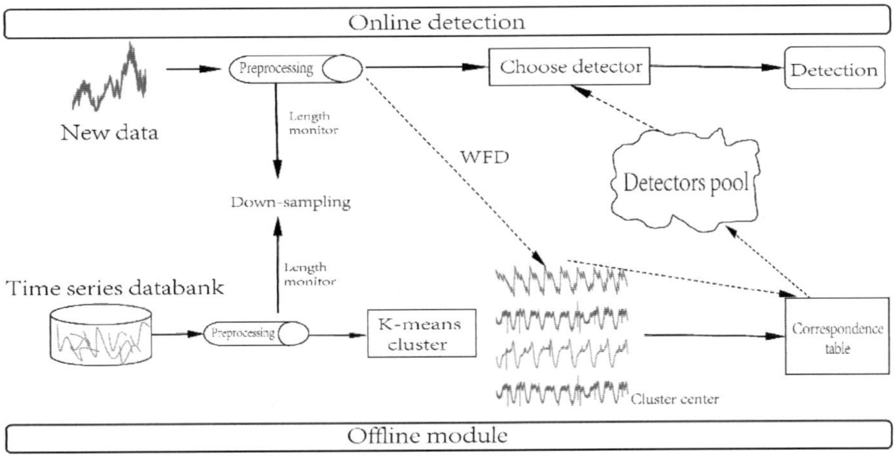

Figure 4. Model structure.

4. EXPERIMENTS

We contributed a dataset named ***Robin time-series-dataset*** to the open-source community. Data provided by an anonymous commercial bank, published at https://github.com/Voce-lin/Robin-time-series-dataset.

4.1. Datasets

- *Our Robin*: The dataset contains a total of 22 time series, each time series contains 2400 sampling points, the sampling interval is 5 minutes.
- *AIOps' KPI-final*: This dataset is released by the 1st match for AIOps (2018AIOps). It contains 58 curves, 5922913 data points and 134114 tagged abnormal points.
- *Yahoo's Webscope S5*: This dataset is provided by Yahoo to benchmark anomaly detection algorithms. The dataset consists of real time series and artificial time series with marked outliers. It contains 58 curves, 572966 data points and 3915 tagged abnormal points.
- *Numenta's NAB*: It is a novel benchmark for evaluating algorithms for anomaly detection in streaming, real-time applications. It is composed of over 50 labeled real-world and artificial timeseries data files. 365558 data points and 33495 tagged abnormal points.

4.2. Metrics.

In some anomaly detection work like [4][6], researchers aggregate multiple consecutive abnormal points in a time series into an abnormal event. As long as the algorithm detects one point in an event, it is deemed to have correctly detected the abnormal event. This measurement strategy is relatively loose and is suitable for tasks that do not require high detection accuracy. In our work, we focus more on the accuracy, so we adopt a more rigorous measurement strategy, that is, when the algorithm detects an abnormal point, only this single point will be considered as abnormal. To determine an abnormal event, all abnormal points have to be detected. In our measurement strategy, we use 4 indicators, namely **F1**, **recall**, **precision** and **CPU time**.

4.3. Comparison of results.

For the experiment of this article, we used seven operators in the detectors pool with different parameters according to the mentioned four kinds of detection algorithm to meet our actual needs. We compared our model MCD with DONUT [4], LSTMAD[7], Banpei's Hotelling[1] based on Hotelling's theory, luminol[2] and iqr_ad[3]. Among them, Hotelling, luminol, and iqr_ad do not require training data, so when comparing these three algorithms, we treat all the data as both the test set and the training set. For DONUT and LSTMAD, they are based on deep learning and require additional training data sets. Therefore, our strategy is to treat the first half of each series as the training set and the second half as the test set. In our algorithm model, this strategy is also adopted.

Table 1. Results of cold-start comparison.

	KPI				Yahoo				NAB				ABC			
model	F1	recall	precision	time	F1	recall	precision	time	F1	recall	precision	time	F1	recall	precision	time
Hotelling	0.36	0.40	0.33	19.44	0.38	0.52	0.30	12.82	0.26	0.19	0.40	4.59	0.16	0.40	0.10	0.20
luminol	0.12	0.23	0.08	932.82	0.46	0.66	0.36	149.65	0.31	0.29	0.34	15.25	0.18	0.45	0.11	1.01
iqr_ad	0.38	0.31	0.49	15.40	0.45	0.31	0.78	10.43	0.31	0.25	0.39	2.24	0.37	0.31	0.46	0.23
MCD	**0.51**	**0.48**	**0.54**	**9.62**	**0.71**	**0.79**	0.65	32.71	**0.32**	0.21	**0.64**	3.78	**0.62**	**0.73**	**0.54**	0.26

Table 2. Results of train-test split comparison

	KPI				Yahoo				NAB				ABC			
model	F1	recall	precision	time	F1	recall	precision	time	F1	recall	precision	time	F1	recall	precision	time
LSTMAD	0.52	0.51	0.54	1042	0.47	0.68	0.36	84.47	0.36	0.79	0.23	33.18	0.31	0.70	0.20	6.60
DONUT	0.44	0.42	0.47	7600	0.30	0.77	0.18	2431	0.30	0.78	0.18	406.15	0.57	0.54	0.60	52.80
MCD	0.50	0.51	0.49	**84.83**	**0.73**	**0.77**	**0.69**	136.09	0.31	0.25	**0.42**	53.89	**0.74**	**0.73**	**0.74**	12.21

It can be seen from table 1 that compared with these three traditional algorithms, the accuracy of our proposed algorithm model is the best. In terms of time-consuming performance, our performance is relatively stable. It can be seen from table 2 that the overall effect of the algorithm based on deep learning is higher than that of the traditional algorithm. The performance of our algorithm model still surpasses them in the Yahoo dataset and the Robin dataset. In addition, the detection time of these two deep learning algorithms rises sharply with the length of the time series (the length of the KPI data set is the highest), while our algorithm model does not have this phenomenon.

5. RELATED WORKS

STL [8], Based on loess, the data at a certain time is decomposed into trend component, seasonal component and residual. [9] Focus on the detection of time series up and down translation. In 2018, [4] proposed DONUT, they implement unsupervised anomaly detection using VAE, [6] Combined SR with

CNN neural network. [10] proposed a robust anomaly detection algorithm (MEDIFF) to monitor online business metrics in real time, using robust statistical metric--median--of the time series to decouple the trend and seasonal components. The above methods can achieve satisfied detection results on specific types of time series, but our model can deal with more types of anomalies at the same time.

6. CONCLUSIONS & FUTURE PLAN

Time series anomaly detection is still an extremely challenging topic. According to the experimental data obtained from our proposed model, what we can foresee is that pre-classifying a large number of time series and then detecting them is a path worth exploring. Our experiments showed significant improvements. Compared with both traditional algorithms and algorithms based on deep learning, our F1-score has improved by nearly 40% on average, our CPU-time consumption are relatively stable on different datasets, showing excellent efficiency and compatibility. Our model provided well satisfied intelligent operation and maintenance experience for our customers in the actual business.

In the future, we will work closely with engineers to find better feature engineering methods and get better clustering effect. We will also consider using decision tree in the model and some lightweight deep learning methods based on neural network to deploy high-precision anomaly detection for core services and services with higher priority. And we will provide fast and real-time detection services for massive low priority business simultaneously.

References

[1] GitHub. (2020) Banpei. https://github.com/tsurubee/banpei.
[2] GitHub. (2018) luminol. https://github.com/linkedin/luminol.
[3] GitHub. (2020) Anomaly Detection Toolkit (ADTK). https://github.com/arundo/adtk.
[4] Xu, H. , Feng, Y. , Chen, J. , Wang, Z. , Qiao, H. , and Chen, W. , et al. (2018). Unsupervised anomaly detection via variational auto-encoder for seasonal kpis in web applications. 187-196.
[5] Niu, Z., Yu, K., and Wu, X. (2020). LSTM-based VAE-GAN for time-series anomaly detection. Sensors, 20(13), 3738. doi:10.3390/s20133738
[6] Hansheng Ren, Bixiong Xu, Yujing Wang, Chao Yi, Congrui Huang, and Xiaoyu Kou, et al. 2019. Time-series anomaly detection service at Microsoft. In Proceedings of the 25th ACM SIGKDD International Conference on Knowledge Discovery and Data Mining (KDD '19). 3009–3017. DOI:https://doi.org/10.1145/3292500.3330680
[7] Pankaj Malhotra, Lovekesh Vig, Gautam Shroff, and Puneet Agarwal. 2015. Long short term memory networks for anomaly detection in time series. In 23rd European Symposium on Artificial Neural Networks, Computational Intelligence and Machine Learning, ESANN 2015.
[8] Cleveland RB, Cleveland WS, McRae JE, Terpenning I (1990) STL: a seasonal-trend decomposition procedure based on Loess. J Off Stat 6: 3–73.
[9] Ma, M. , Zhang, S. , Dan, P. , Xin, H. , and Dai, H. . (2018). Robust and Rapid Adaption for Concept Drift in Software System Anomaly Detection. 2018 IEEE 29th International Symposium on Software Reliability Engineering (ISSRE). IEEE.
[10] Li, T. , Geng, Y. , and H Jiang. (2020). Anomaly detection on seasonal metrics via robust time series decomposition.

Journal of Physics: Conference Series **2113** (2021) 012063 doi:10.1088/1742-6596/2113/1/012063

GAN Image Generation and Detection Technology Based on Fractal Dimension

Yuheng Guo*

Henry Samueli School of Engineering, University of California, Los Angeles, Los Angeles, 90095, USA

Corresponding Authpr's Email: yuhengguo@g.ucla.edu

Abstract. With the development of modern technology today, image analysis technology has become an important research direction in the computer field. Computer image processing technology is a cross and fusion involving multiple disciplines, which includes multiple research directions such as physics, chemistry, and computer science. Starting from the fractal dimension, this article studies the image generation and detection technology. The purpose of the research topic in this article is to have a deep understanding of image generation technology so as to master image processing methods proficiently. The methods used in this article include case analysis, data, comparison, and experimental methods, etc., and carry out related researches on image generation detection technology and fractal dimension. The experimental results show that the algorithm accuracy of GAN in the detection technology of computer-generated images is higher than other algorithms, and the improvement space is about 0.06.

1. Introduction

Modern machine vision systems based on fractal dimensions have been widely used in many fields. The fractal dimension of an image is a very important and generally valued technology. It has a lot to do with our requirements for information processing speed, data storage capacity, and image resolution under different objects and environments in practical applications. Therefore, people have been committed to studying how to efficiently use existing equipment to achieve high-quality, high-precision conversion of two-dimensional images into digital images and generate fractal dimensions that can be used in engineering practice.

There are many researches on image generation and detection technology. For example, Liu Jun proposed a composite edge detection algorithm based on multifractal theory and mathematical morphology in response to the problem of license plate image positioning and processing in the existing license plate recognition technology [1]. Zhou Linna said that the introduction of generative confrontation networks into steganography can reduce the traces of carrier modification and improve the concealment of steganography [2]. Wang Wang said that image super-resolution is a technology that uses software algorithms to improve image spatial resolution [3]. In order to improve the stability of GAN, Zhang Guanghua proposed an improved neuron abandonment method [4]. Therefore, the research of GAN's image generation and detection technology in this paper is closely following the pace of scholars and has the significance of the times.

This dissertation first studies the related knowledge of fractal and fractal dimension, and secondly studies the image desiccation and image super-resolution reconstruction. Then the research is on the edge detection technology of digital image, and the algorithm research is carried out. Finally, analyzing and experimenting around computer-generated images and detection, and get the final result.

2. GAN Image Generation And Detection Technology Based on Fractal Dimension

2.1. Fractal Dimension

(1) Fractal Features

Fractal sets all have proportional details at arbitrarily small scales. The fractal set has a self-similar form. The method of defining fractal sets is very simple [5-6].

(2) Fractal dimension

Dimension is the number of coordinates of a point in a geometric object. The fractal dimension reflects the efficiency of the space occupied by complex shapes. The formula can be expressed as:

$$c = l\,\underset{\partial \to 0}{im}\left[\log M(\partial) / \log(\frac{1}{\partial})\right] \quad (1)$$

Where ∂ is the length of one side of the small cube, M_{∂} is the number obtained.

Generally speaking, when calculating the fractal dimension of an image, the calculation method must first be selected according to a specific fractal pattern. The image fractal model is compatible with different image fractal dimension calculation methods [7-8].

(3) "Box" dimension

The set X in s dimension, if X can be expressed as a subset S_a that does not cover each other, then X is self-similar. The fractal dimension F is as follows:

$$F = \frac{\log(S_a)}{\log(1/a)} \quad (2)$$

Among them, a is the scale factor and the number of subsets.

Assuming that the highest and lowest gray values in the (m,n)th grid respectively fall in the Ith and hth small boxes, then the distribution of S_a in the (m,n)th grid is :

$$S_{\mathrm{a}}(m,n) = I - h + 1 \quad (3)$$

(4) Application of fractal dimension in image field

1) Fractal coding

Fractal image coding is to divide the graph into w blocks, and then search for p blocks that match it in the image, and perform affine transformations such as spatial shrinkage, rotation, reflection, and brightness transformation on the p blocks [9-10].

2) Image segmentation

In image processing, in order to describe and analyze an image, it is often necessary to subdivide the image into several sensitive areas. This technique of dividing an image into multiple hot spots for effective post-processing is called image segmentation.

3) Image texture analysis

Fractal dimension is the main measurement tool used in other fractal image processing techniques. As a measure of the unevenness of the image surface, the fractal dimension can not only measure the complexity, but also show the invariance of multi-scale and multi-resolution changes [11-12].

2.2. Image Denoising Method Based on Generative Adversarial Network

(1)GAN network

Generative adversarial network is an unsupervised learning method, which is widely used in various fields of computer vision. The original generative confrontation network is used to learn the sample images in the image library, so as to generate images similar to the samples in the library. The advantage of generating an adversarial network is that it does not memorize image sample blocks through the network and piece the sample image blocks together into output images, but learns the image characteristics of the sample image blocks in the database and uses the features to reconstruct the image. In order to avoid the loss of more image details when denoising the image, the multi-scale features containing redundant information are extracted for subsequent denoising. In addition, other channel features still retain the structural information, and images with clear texture details can still be restored during image fusion. The convolutional network with weight sharing is used in the generation network, because the shared weight structure is closer to the biological visual system, which reduces the complexity and parameter amount of the network, and has stronger adaptability and robustness [13-14]. The specific block diagram is shown in Figure 1:

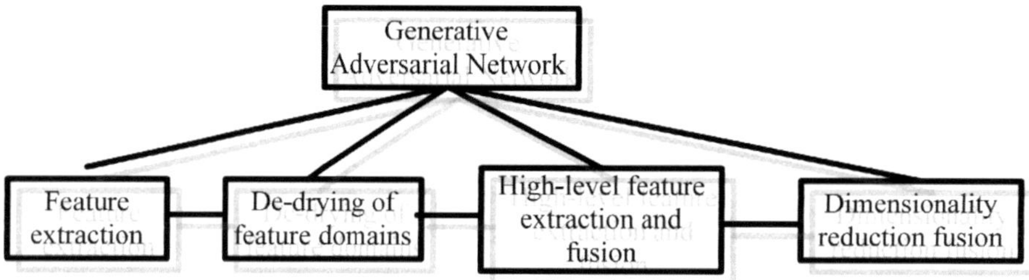

Figure 1. Generate the overall structure of the network

(2) Feature extraction

Using multi-scale convolution kernel to process the image, you can extract the information of different scales of the image, and convert the image into a feature domain. Processing images in the feature domain can not only retain more image information, but also has more room for optimization when the generating network and the discriminant network are used for confrontation training.

(3) Domain denoising

Since the image input to the generation network is a noisy image, the multi-scale features output by the feature extraction layer are also features contaminated by noise. For these noisy features, this paper uses a stacked convolutional layer structure to extract the noise in the feature, and through cross-layer connection, the noisy feature is subtracted from the extracted noise to obtain the denoised image feature for subsequent use Network processing.

(4) High-level feature extraction

After obtaining the denoising results of the feature domain, it is necessary to perform multi-dimensional fusion of the input denoising features, and use the abstraction of high-dimensional features and the information retention ability of low-dimensional features to strengthen the network's fitting ability.

(5) Dimensionality reduction fusion

The multi-layer network is used to filter the features in the image, and the features that contain noise or loss of information are removed from the feature domain. And gradually merge the features suitable for the restored image, and use the non-linearity of the multi-layer network to construct the correct denoising image.

2.3. Image Super-resolution Reconstruction of Generative Adversarial Network

With the increase in the number of layers of convolutional neural networks, the image quality of super-resolution reconstruction is also improving. Generative confrontation networks have achieved great success in the field of computer vision. The image super-resolution method based on generative confrontation network is a network model that is more in line with subjective visual effects.

The generation of the confrontation network does not require repeated sampling of the Markov chain. It is mainly to learn the probability distribution of the real data through the training data of the sample, so that the generated sample distributes the data in the same way as the actual data. In the training process of generating the adversary network, first repair the generator, train the discriminator to distinguish between real data and fake data, and then repair the discriminator, train the generator to generate deceptive data current discriminator, repeat the above process, optimize the creation of the grid and distinguish grid. The loss function can be expressed as follows:

$$\min_{F}\max_{G} W(G,F) = \mathrm{P}_{a\sim s\,data}(a)\left[\log G(a)\right] + \mathrm{P}_{c\sim sc}(c)\left[\log(1-G(F(\mathrm{c})))\right] \quad (4)$$

Generating the loss function of the adversarial network is a minimum-maximum optimization process, which is essentially two optimization problems.

The discriminant network takes the super-resolution image and the original high-resolution image as input, and outputs the discrimination result. The generative network and the discriminant network confront each other. When the discriminant grating cannot be distinguished strongly, SRGAN has completed the task of reconstructing super-resolution images. The super-resolution image reconstructed by the SRGAN model will appear blurry at higher magnifications, and the generation of the adversarial network itself is also unstable.

2.4. Digital Image Edge Detection Technology

Edge detection is the process of identifying sudden changes in brightness and discontinuous points in the image through edge detection operators and positioning the edges according to the positions of these points. Search-based edge detection uses the first derivative of the gray curve to obtain the brightness gradient of the original data to find the peak value in the image brightness gradient. The geometric shape of the peak point is the edge. Based on the zero-crossing method, the derivative is obtained from the image to determine the edge.

Image acquisition is the first prerequisite for edge detection and subsequent analysis. Discretize the collected images, and convert continuous gray values into discrete values through quantization. The collected images are preprocessed before processing and analysis to remove redundant information and reduce the amount of calculation.

2.5. Algorithm Research

Since the communication signal has certain modulation characteristics, its fractal dimension and noise have different randomness, so this difference can be used to construct a spectrum sensing model. Among them, the randomness of the communication signal mainly depends on the modulation type and the phase of the carrier, and noise has little effect on its randomness, that is, within a certain range of signal-to-noise ratio, the fractal dimension is not sensitive to noise.

(1) Katz fractal dimension

The method of calculating the fractal dimension according to Katz is derived from the fractal dimension of the plane curve. The fractal dimension of the curve can be expressed as:

$$H = \frac{\lg(G)}{\lg(h)} \quad (5)$$

The total length and distance of the curve are accumulated to get:

$$S = \sum_{\mathrm{i}=1}^{M-1} (g_{i+}1 - g_i)^2 + (h_{i+1} - h_i)^2 \quad (6)$$

(2) Sevcik fractal dimension

Suppose the signal waveform length is L, which is composed of a series of points. First, normalize the signal waveform:

$$A_i^* = \begin{cases} A_i \\ A_{\max} \end{cases}, B_i^* = \begin{cases} B_i - B_{\min} \\ B_{\max} - B_{\min} \end{cases} \quad (7)$$

2.6. Computer-generated images and their detection

Computer-generated images are mainly generated by 3D software such as 3D Max and Maya, which are mainly based on the principles of computer graphics.

(1) Blanking technology

The use of computers to generate realistic images of three-dimensional objects is an important part of computer graphics research. Projective transformation usually causes blurred images. In order to resolve the ambiguity, invisible lines or areas need to be eliminated. In line blanking, the most basic operation is to determine the occlusal relationship with the line. The body must also be decomposed into curved surfaces, and then the occlusal relationship between the curved surfaces and the lines must be evaluated. When evaluating the occlusion, the cutting operation between the lines needs to be repeated.

(2) Eliminate hidden surfaces

Ray tracing technology. The most basic ray tracing algorithm is to follow specular reflection and refraction.

(3) Texture mapping technology

During the mapping process, the texture is converted to the surface of the 3D object to create the final image.

(4) Shadow generation technology.

The shaded area is a semi-open three-dimensional area, with the illuminated area as the top area. Each object area contained in the shadow box must be a shadow box. The field of view The screen field usually represents a square pyramid. Use this square cone to crop the shadow field of the object.

3. Computer-Generated Image Detection Algorithm Experiment

3.1. Experimental Environment

The experiment uses the Python interface of the deep learning framework to build the SRWGAN algorithm model.

Processor: Intel, Core, CPU

Memory: 64GB

Graphics card: NVIDIA

The experiment uses the SRGAN model as a benchmark, and uses subjective visual evaluation and objective quantitative data as evaluation indicators. Three experiments are set up: First, the comparison between different interpolation algorithms and the experimental analysis of the improvement of the residual structure. Second, the comparison between loss functions. Third, the experimental analysis of the entire improved model.

3.2. Experimental Comparison of Improved Generative Model

Mainly introduce the experiment and result analysis of the SRWGAN generative model. The parameters of each layer of the improved generative network are shown in Table 1. Name the convolutional layers 1, 2, 3, 4, and 5 from left to right.

Table 1. Generate Network Layer Parameters

	Convolution kernel size	Convolution kernel number	Step size
1	4*4	32	1
2	4*4	32	1
3	9*9	4	1
4	4*4	32	1
5	4*4	128	1

4. Analysis of Experimental Results

The algorithms used in this article include generative adversarial networks, recursive algorithms, residual networks and loss functions. The specific experimental results are shown in Table 2:

Table 2. Results of Different Super-Resolution Algorithms

	Average time	PSNR/SSIM	Promote
SRGAN	0.187	31.56/0.845	+0.08/+0.007
VDSR	0.157	31.65/0.875	+0.07/+0.005
SRWGAN	0.165	31.54/0.869	+0.05/+0.003
DRCN	0.135	31.95/0.898	+0.02/+0.001

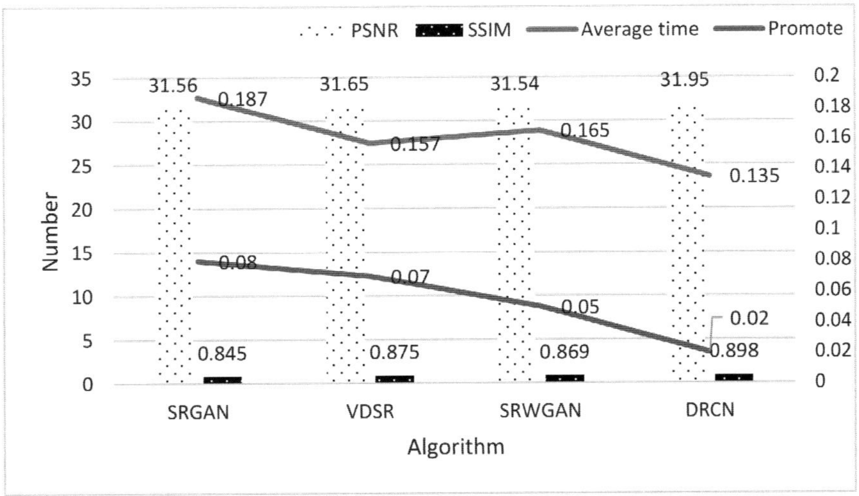

Figure 2. Results of Different Super-Resolution Algorithms

As shown in Figure 2, we can see that among the four algorithms, the RSNP value recursive algorithm has the highest data, while the SSIM data has the highest residual network. The one with the most value increase uses the generative adversarial network algorithm. This shows that the accuracy of the generative adversarial network algorithm in discrimination is higher.

5. Conclusion

Fractal dimension is an important method in solid geometry. It has extensive and in-depth applications in solving complex problems and optimizing design. It can not only provide important information as digital signal processing, radar system and remote sensing technology. It also has many functions that are not involved in other disciplines or have not yet been developed. GAN is an optical network analysis system based on fractal dimension. This technology is mainly used to detect changes in the shape and size of objects. The principle of GAN scanning is to analyze the pixel positions and topographic features of each point in the image in different regions, so as to obtain the image classification. The research results of this paper show that the role of generative adversarial networks in image generation is relatively large, and its accuracy is relatively high and it is worth a try.

References

[1] Liu Jun, Liu Jinglong. Research on image edge anti-noise detection algorithm based on fractal dimension and mathematical morphology[J]. Journal of Shenyang Ligong University, 2017, 36(005): 14-17.

[2] Zhou Linna, Lu Xinyi. Overview of Information Hiding Technology Based on GAN Image Generation[J]. Information Security Research, 2019, 005(009): P.771-777.

[3] Wang Wang, Xu Junwu. Research on image super-resolution method based on GAN[J]. Software Guide, 2019(6):181-183.

[4] Zhang Guanghua, Wang Fubao, Duan Weijun. Research on Astronomical Image Generation Based on GAN[J]. Journal of Northwestern Polytechnical University, 2019, 037(002):315-322.

[5] Zhang Wenkai, Sun Hao, Sun Xian, et al. Visual summary of image collection based on MFF-GAN[J]. Computer Engineering, 2019, 045(002):202-206.
[6] Zhang Guanghua, Wang Fubao, Duan Weijun. Research on an astronomical image evaluation method generated by GAN[J]. Computer Simulation, 2020, v.37(01):470-474.
[7] Huang Luyao, Ye Shaozhen. Research on low-light image enhancement algorithm based on GAN[J]. Journal of Fuzhou University: Natural Science Edition, 2020(5):551-557.
[8] Cheng Huanxin, Liu Wenhan. Research on Image Super-resolution Based on Generative Countermeasure Network[J]. Electronic Measurement Technology, 2020, v.43; No.346(14):137-140.
[9] Qi Yan. Research on image recognition and generation based on artificial intelligence algorithms [J]. Electronic Components and Information Technology, 2019, 003(011): P.45-47.
[10] Sun Xiyin, Feng Huamin, Liu Biao, etc. Research on adversarial sample generation based on GAN[J]. Computer Applications and Software, 2019, 036(007):202-207,248.
[11] Liu Weitao. Image generation model based on SinGAN[J]. Journal of Qufu Normal University (Natural Science Edition), 2020, v.46; No.176(02):73-77.
[12] Liang Weipeng, Hong Shengwei. Research on image denoising and denoising principles based on GAN generative confrontation network [J]. Science & Technology Information, 2019, 017(013): 26-27.
[13] Jiang Daihong, Zhang Sanyou, Liu Qikai. Image classification algorithm based on feature recalibration generative adversarial network [J]. Journal of Computer Application Research, 2020, 037(003):932-935.
[14] Wang Tongping. Improved heterogeneous face image synthesis algorithm based on CycleGAN[J]. Modern Computer, 2020, No.686(14):50-54.

Journal of Physics: Conference Series **2113** (2021) 012064 doi:10.1088/1742-6596/2113/1/012064

The High-Speed Low-Power Dynamic Comparator

Menghua Cao[1, a, *, †], **Weixun Tang**[2, b, *, †]

[1]Taiyuan University of Technology, Taiyuan, China

[2]Nanjing Normal University, Nanjing, China

[*]Corresponding author email: [a]caomenghua2193@link.tyut.edu.cn, [b]21180438@njnu.edu.cn

[†]These authors contributed equally.

Abstract. This paper comments on four works for the optimization of comparator design. Today, with the development of integrated circuits, the requirements for comparators about low power, low delay, few offset voltage, and low noise are highly desirable. Specifically, these works made progress in the conventional comparator, which comprises a preamplifier and a latch. They also solved some problems, such as decreasing power and delay. Some works employ a positive feedback cross-coupled pares to provide a larger gain in the preamplifier, use PMOS switch transistors to accelerate the definition phase, or a double-tail architecture to increase the latch regeneration speed. Other work designs a charge pump to improve speed.

1. Introduction

In two high-speed parallel ADCs, a high-speed comparator is the main factor limiting the max samples of ADCs and the power consumptions. The performance of the comparator is based on speed, precision, and power consumption, in the ideal condition. However, the delay between the output and the input of the comparator is inevitable due to response time and settling time which support the comparator to work properly. So, reducing the delay is the main purpose of improving the performance of ADCs.

Some works made progress among the conventional comparators. The first is that Samaneh Babayan-Mashhadi and his team members substitute the classic structure of single-tail with a kind of double-tail structure to increase the latch regeneration speed. As to the second one, Ata Khorami and Mohammad Sharifkhani demonstrate a low-power comparator. PMOS is provided as input of the latch stage and the cross-coupled circuit is used for the preamplifier. The third one, YAO WANG et al., present a novel comparator with the latching stage using divided gate-biasing, providing a much faster speed and lower energy consumption.

Moreover, a novel innovation is designed by Haoyu Zhuang and his teammates. They add a charge pump to the conventional Miyahara's comparator, which is used in the bootstrap switch, increasing the actual voltage of the latch, thus increasing the regeneration speed. It decreases the delay compared to conventional Miyahara', but it can still be improved since the power consumption increases.

This paper describes the design, progress, and innovation based on the conventional comparator. The paper is organized as follows. Section II introduces the conventional comparator. The innovations are discussed in Section III, and finally, Section IV concludes the paper.

2. Conventional dynamic comparator

The schematic diagram of the conventional dynamic comparator is shown in Fig.1.The operation of the comparator is as follows [1]. During the reset phase, the CLK equals 0, and Mtail is off. Both Outn and Outp start to work and are charged. In the comparison phase, Outp and Outn, which have been pre-charged, will start to discharge with different discharging rates depending on the VINN and VINP

The delay of the comparator is comprised of two times delays, which are t0 and tlatch. As given in Fig1. The CL is the comparator load, Itail is the function of input common-mode voltage (Vcm) and VDD. Also, β1,2 is the input transistors' current factor, so the delay is obtained as:

$$t_{delay} = 2\frac{C_L|V_{thp}|}{I_{tail}} + \frac{C_L}{g_{m,eff}} \cdot \ln\left(\frac{V_{DD}}{4|V_{thp}|\,\Delta V_{in}}\sqrt{\frac{I_{tail}}{\beta_{1,2}}}\right) \tag{1}$$

This structure has the advantages of high input impedance, rail-to-rail output swing, no static power consumption, and good robustness against noise and mismatch. However, a needed sufficiently high supply voltage, and only one current path are the disadvantages.

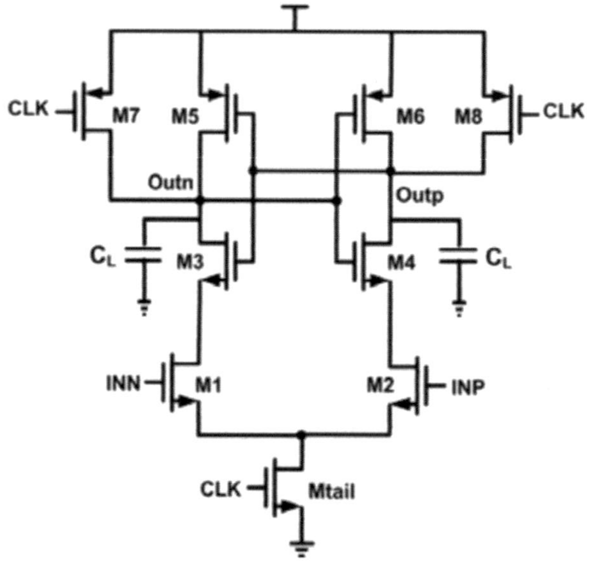

Fig.1 Conventional comparator

3. Proposed comparator

3.1. double-tail dynamic comparator

Samaneh Babayan-Mashhadi and his teammates design a low-power double-tail comparator in 2014. The schematic of the conventional double-tail dynamic comparator is shown in Fig.2(a). This comparator can operate at lower supply voltages due to its topology. The double tail structure can enable a large current in the latching stage and a wider Mtail2, which leads to the fast latching independent of the input common-mode voltage and a small current in the input stage used for low offset [2]. Also, a shielding between input and output is provided by the intermediate stage results in the reduced value of kickback noise. Similar to the conventional dynamic comparator, the total delay of the comparator is achieved as follows:

$$t_{delay} = 2\frac{C_{Lout}V_{thn}}{I_{tail2}} + \frac{C_{Lout}}{g_{m,eff}}\ln\left(\frac{V_{DD}\cdot I_{tail2}^2\cdot C_{L,fn(p)}}{8\,V_{thn}^2\cdot C_{Lout}\,g_{mR1,2}g_{m1,2}\,\Delta V_{in}}\right) \tag{2}$$

However, the performance of the conventional double-tail comparator can still be improved since it can obtain a more optimal delay. To optimize the conventional comparator, a novel comparator is designed, and Fig.2(b). demonstrates the schematic diagram of the proposed novel dynamic double-tail comparator. It is based on the double-tail structure due to the better performance in low-voltage applications. Increasing ΔVfn/fp to improve the latch regeneration is the main design idea. To achieve this goal, two measures are operated. The first is that two control transistors are added in a cross-coupled manner. The second is two added nMOS switches, Msw1 and Msw2, which can avoid static power consumption because they can emulate the operation of the latch and increase the fn and fp's voltage difference.

The operation of the proposed comparator is as follows. During the reset phase, fn and fp nodes are pulled to VDD by M3 and M4, and the intermediate stage transistors MR1, MR2 reset both latch outputs to the ground. In the decision-making phase, fn and fp start to drop with different rates according to the voltages. Unlike in conventional double-tail dynamic comparator that ΔVfn/fp is just the function of input transistor transconductance and input voltage difference, in the proposed comparator when one fn/fp nodes is discharging faster, the other one will be pulled back to VDD. By doing this, ΔVfn/fp increases exponentially, and the latch regeneration time will be reduced.

In summary, the delay of the proposed comparator is reduced for two factors. Firstly, it increases the initial output voltage difference at the beginning of the regeneration, defined as ΔV0. What's more, the effective transconductance gmeff is enhanced because one of fn and fp will charge up back to the VDD and turn on one of the intermediate stage transistors. So the total delay of the proposed comparator follows the formula:

$$t_{delay} = 2\frac{V_{thn}C_{Lout}}{I_{tail2}} + \frac{C_{Lout}}{g_{m,eff} + g_{mR1,2}} \times \ln\left(\frac{V_{DD}/2}{4V_{thn}|V_{thp}|\frac{g_{mR1,2}}{I_{tail2}}\frac{g_{m1,2}\Delta V_{in}}{I_{tail1}}\exp\left(\frac{G_{m,eff1}\cdot t_0}{C_{L,fn(p)}}\right)}\right) \quad (3)$$

The proposed comparator utilizes the inner positive feedback in double-tail operation and finally strengthens the latch regeneration, improving low supply voltages.

The proposed comparator and the conventional comparator are all designed in 0.18um CMOS technology with VDD=1.2V. Fig.3 displays the post-layout simulation results of the delay and the energy per conversion of the mentioned dynamic comparators versus supply voltage variation [3]. As seen from the simulation, the delay is significantly reduced in low-voltage supplies, which means the proposed comparator can be used in the lower supply voltages to consume the same power as the conventional one.

Fig.4 presents the simulated performance as a function of input common voltage. It's apparent that the double-tail comparator is less affected by the variation of the input common-mode voltage than the conventional one. Fig.5 illustrates the dependence of the comparator delay on power supply level at various differential input voltages. The larger the differential input voltage, the smaller the comparator delay. Fig.6 shows the simulated delay of the comparator versus differential input voltage under different conditions of input common-mode voltage at VDD=1.2V. When the Vcm is a fixed value, the delay decreased as differential input voltage increased, and the delay is dependent on the variation of common-mode voltage.

In summary, the method used to decide important parameters and emulate them provides new ideas for us to optimize the comparator.

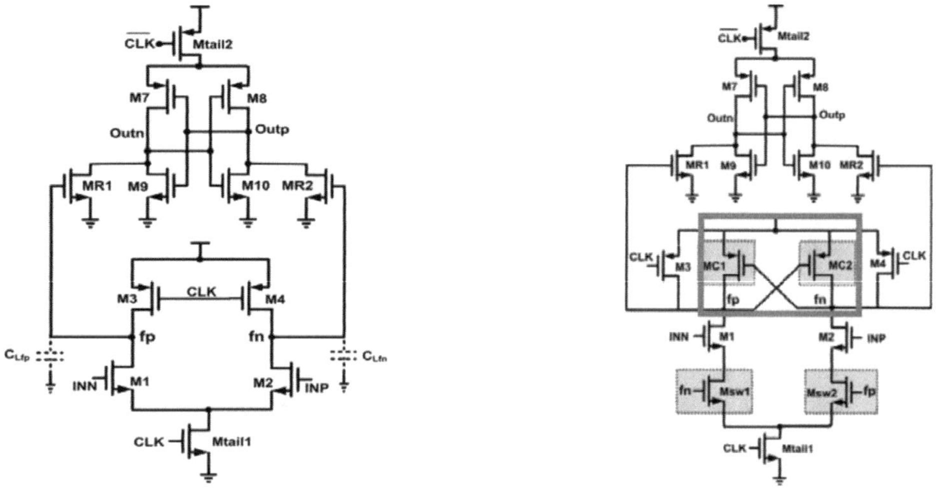

(a) (b)

Fig.2 Double-tail dynamic comparator; (a) conventional double-tail dynamic comparator (b) proposed double-tail dynamic comparator

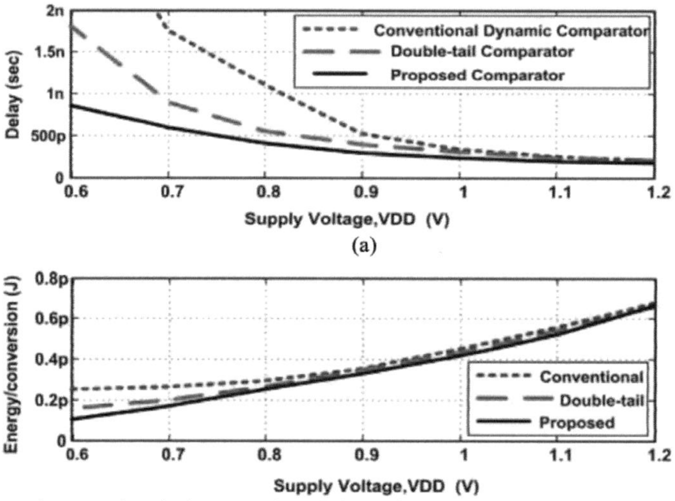

Fig.3 The post-layout simulation results of the delay and the energy per conversion

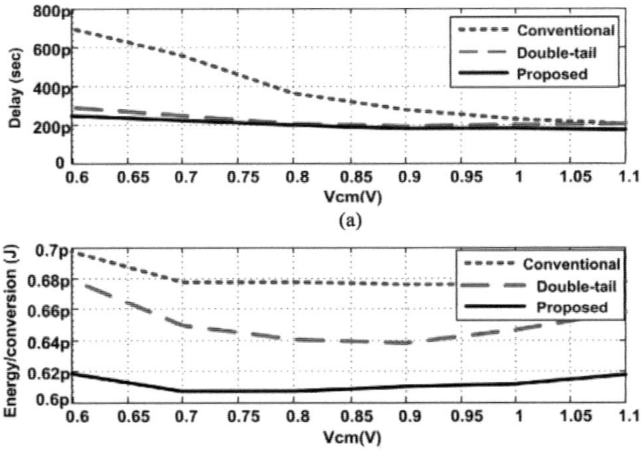

Fig.4 The simulated performance as a function of input common-voltage

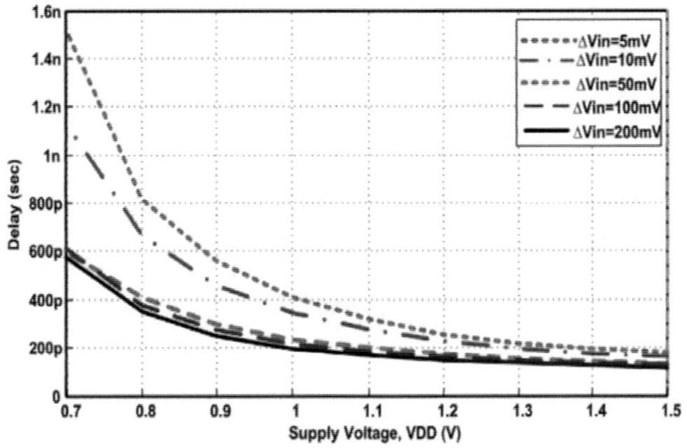

Fig.5 The dependence of the comparator delay on power supply level at various differential input voltages

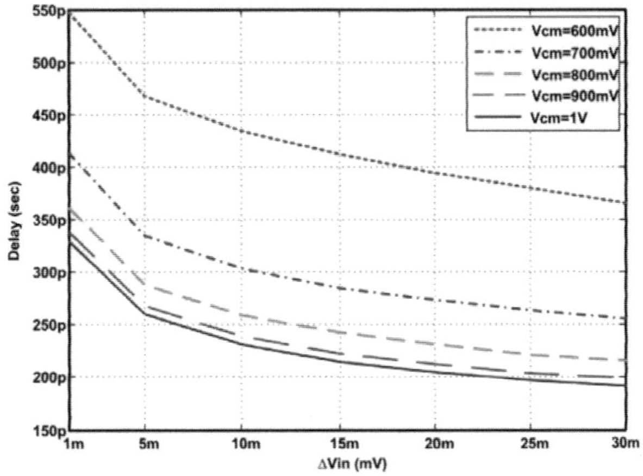

Fig.6 The simulated delay of the comparator versus differential input voltage under different conditions of input common-mode voltage at VDD=1.2V

3.2. a low-power high-speed comparator for precise applications

Ata Khorami and Mohammad Sharifkhani demonstrated a low-power comparator in which PMOS is provided as input of the latch stage. A cross-coupled circuit is used for the preamplifier.

Fig.7 is the circuit of the proposed comparator. During the reset phase, the clk, clkb1, and clkb2 are kept logic '1' to provide GND with output voltage which belongs to the preamplifier and latch [3]. During the evaluation phase, first, clk and clkb1 switch to logic '0' to open the preamplifier, which charges parasitic capacitors of O1+ and O1- nodes variously.

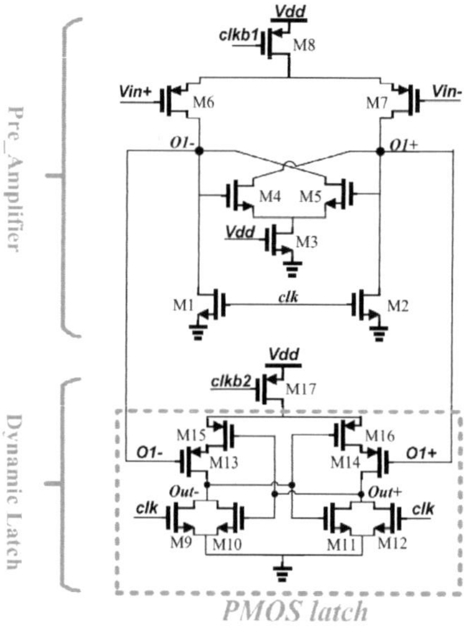

Fig.7 The circuit of the proposed comparator.

Referred to the proposed comparator, two innovations are a cross-coupled circuit (Fig.8) and a latch with input PMOS transistors(Fig.9). The following is a detailed description of these innovations.

Firstly, the innovation about the cross-coupled circuit is below. During the evaluation phase, due to M4 and M5 are largely in the subthreshold region, the differential voltage is slightly increasing by the cross-coupled circuit. And the common-mode voltage has a decline providing a significant drive to the input PMOS latch stage. The impact of the latch on the input-referred offset voltage is removed by growing Vidl [3]. Therefore, the cross-coupled circuit achieves enough delay to provide the minimum required preamplification to reach a given speed and latch offset elimination. Detailed data analysis of offset is below.

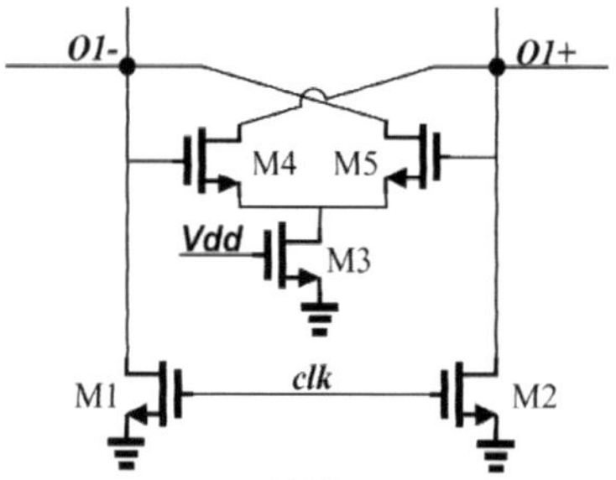

Fig.8 Failure characteristics of specimen SJ2

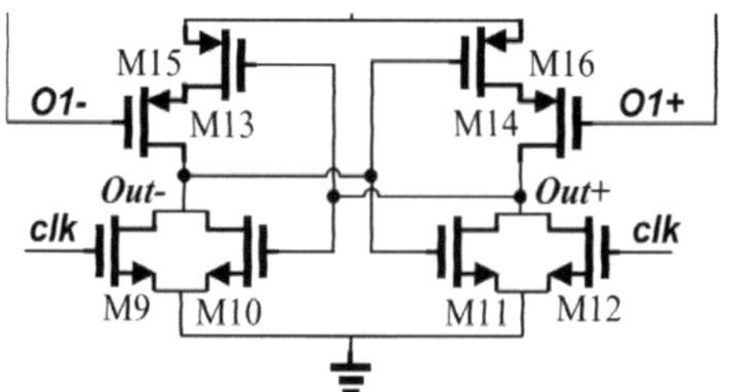

Fig.9 A latch with input PMOS transistors.

The latch delay is described as:

$$t_{latch} = \tau_{inv} \times \ln\left(\frac{VDD-GND}{V_{idl}}\right) + \frac{K_{latch}}{(VDD-V_{cml}-V_{thp})^2} \quad (4)$$

Vcml is the input common-mode voltage of latch (Vcml =(VO1++VO1-)/2). Vidl is the differential voltage of the latch (Vidl =VO1++VO1-). The two figures are all from the preamplifier stage. Then they can be figured in the following formulas

$$\begin{cases} V_{cml} = \frac{V_{O1+}+V_{O1-}}{2} = \frac{I\times\frac{t_{amp}}{C}}{1+\alpha\left(\frac{t_{amp}}{C}\right)}, I = I_1 + I_2 \\ V_{cml} = V_{O1+} - V_{O1-} = \frac{I\times\frac{t_{amp}}{C}}{1-\alpha\left(\frac{t_{amp}}{C}\right)}, I = I_2 - I_1 \end{cases} \quad (5)$$

Because the α term of the conventional comparator is equal to 0, the proposed method is bigger. Consequently, a lower Vcml and higher Vidl can be achieved. So, a fast speed is realistic according to formula (4).

To prove the results of derivations, this paper designs a sample of the proposed comparator to achieve an offset voltage with 2mV. Using the formula about the delay, Vcml and Vidl get an illustration (Fig.10) about simulation and analytical derivations in the delay of the proposed comparator versus Vcm, which verifies the analytical derivations doping out an exact delay.And the delay is stable in Vcm = 0.2v - 0.5v and Vcm = 1.0v - 1.4v ,in which the former is lower than latter. After that, a decreasing trend about the delay changing with Vid about simulation and analytical derivations when Vcm is equal to 1.1v can be concluded in Fig.11. According to the definition of offset voltage, if an input difference voltage of up to VOS is applied to the comparator, after amplified, VO1++VO1-. Therefore, Fig.12 illustrates the offset voltage versus Vcm about simulation and analytical derivations. The result is that the two lines almost coincide.

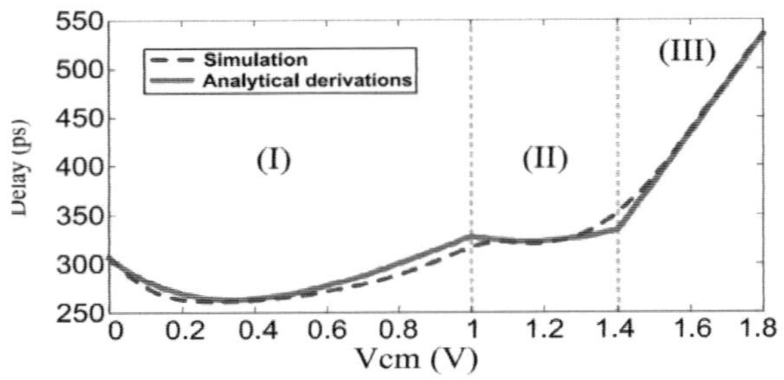

Fig.10 The delay of the proposed comparator changing with Vcm.

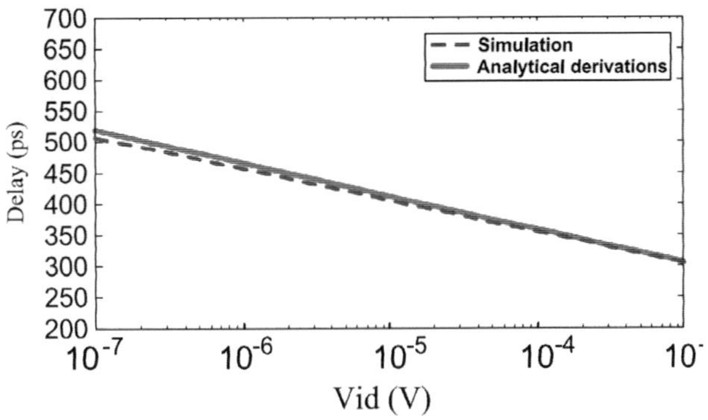

Fig.11 The delay changing with Vid.

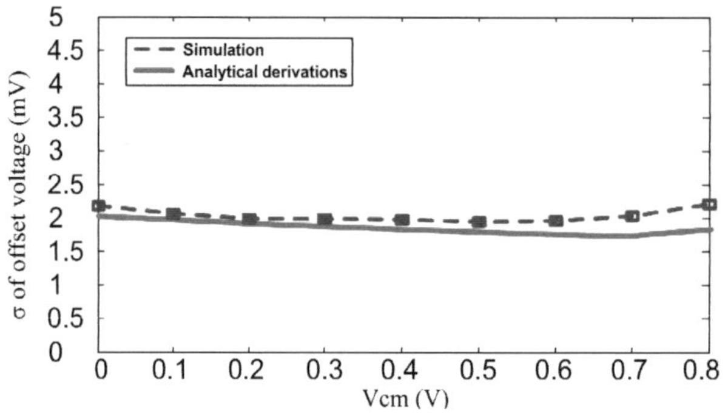

Fig.12 The offset voltage versus Vcm.

3.3. A Low-Power High-Speed Dynamic Comparator with a Transconductance-Enhanced Latching Stage

YAO WANG et al. have presented a novel comparator with a latching stage using divided gate-biasing to provide a much faster speed and lower energy consumption.

Referred to the proposed comparator (as Fig.13 shown), in the reset phase, M0, M1, M4, and M5 are in the strong-inversion region compared with other cross-coupled structures that have two transistors in the cut-off region and the other two in the strong-inverter region.

A more effective transconductance can be concluded in the comparison phase, which has a higher regeneration speed. What's more, lower energy consumption is made because of a shorter metastable period in the inverters provided by a speed-up comparison.

According to the proposed comparator [4], a significant decline in delay and energy consumption will be achieved with the new latching stage. In fact, it can be achieved by improving the effective transconductance at the start of the comparison phase.

The latching delay of two cross-coupled inverters is given by

$$t_{latch} = \frac{C_{OUT}}{g_{m,eff}} \cdot \ln \frac{\Delta V_{out}}{\Delta V_0} = \frac{C_{OUT}}{g_{m,eff}} \cdot \ln \frac{V_{DD}/2}{\Delta V_0} \tag{6}$$

The total delay is described as

$$t_{latch} = \frac{V_{THN} C_{OUT} \cdot 2\mu_P C_{OX} \frac{W_1}{L_1}}{g_{m,eff}^2} = \frac{C_{OUT}}{g_{m,eff}} \cdot \ln \frac{V_{DD}/2}{\Delta V_0} \tag{7}$$

According to this formula, to lower tdelay, the higher gm and lower COUT should be chosen. The effect of VDD can't be concluded directly, so the result of the simulation would be represented later.

This paper cited postlayout simulated delay of comparators of [5] and [6] in contrast with proposed and conventional comparators and illustrated the relationship of postlayout simulated delay and Vcm (VDD=1.2v; Vid=50mv) in Fig.14.Additionally, Fig.15 shows the relationship between energy consumption and Vcm. Evidently, the delay, which is the minimum in the Vcm=1.0v, is slower than the proposed comparator by 150 ps. And the energy consumption, which is the minimum in the Vcm between 1.0v and 1.2v, is considerable to [6] but clearly lower than the conventional comparator and [5].

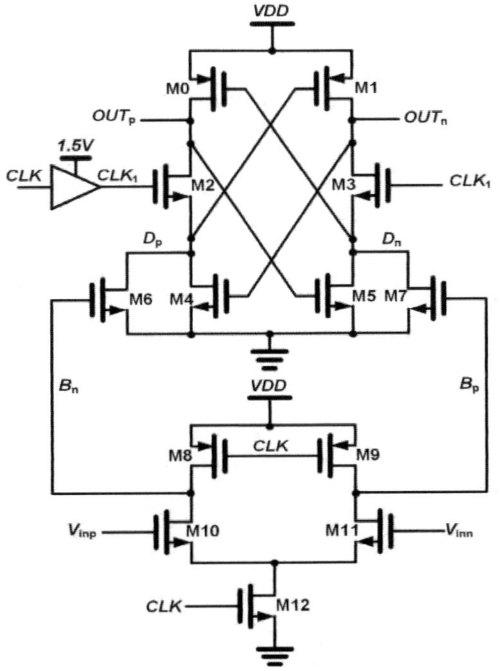

Fig.13. The proposed comparator

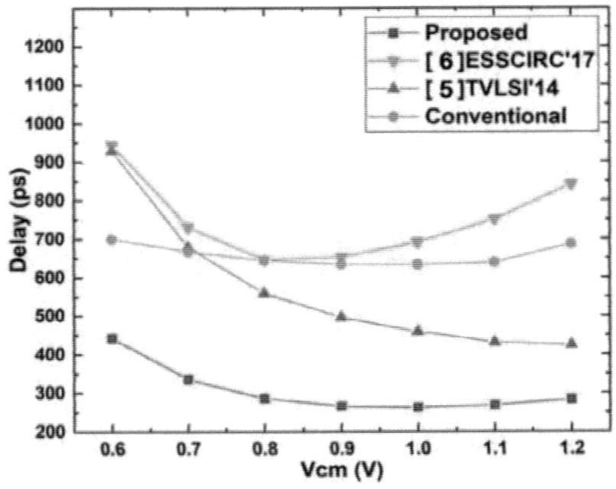

Fig.14. The relationship of postlayout simulated delay.

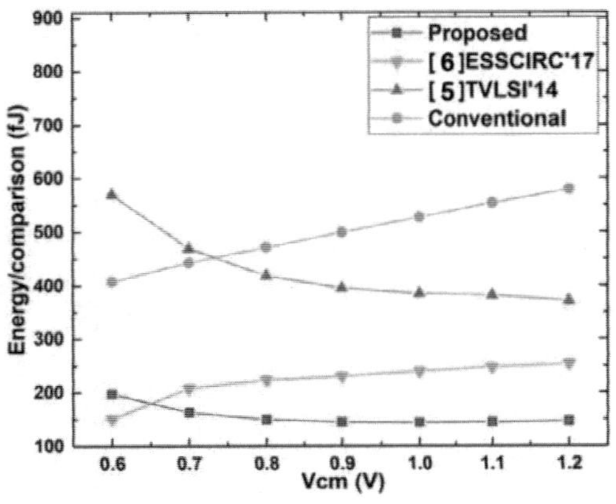

Fig.15. The relationship of energy consumption and Vcm

Simulation results in Fig.16 and Fig.17 show the delay and energy consumption versus VDD(Vid=50mv; Vcm=VDD-0.1v).Fig.16 shows that the proposed comparator is a better fit for low-voltage and low-power applications due to the greatly lower delay and energy consumption than other comparators. In this paper, the proposed chooses VDD=1.2v.

Based on the above, low delay and energy consumption can be achieved when VDD=1.2V. Therefore, Fig.4 illustrates the dependence of delay in the difference in input voltage (VDD=1.2v) and the multiple input common-mode voltage. For the multiple of input common-mode voltage ranging 0.7-1.1v, the delay remains stable and declines along with the decrease of the differential voltage.

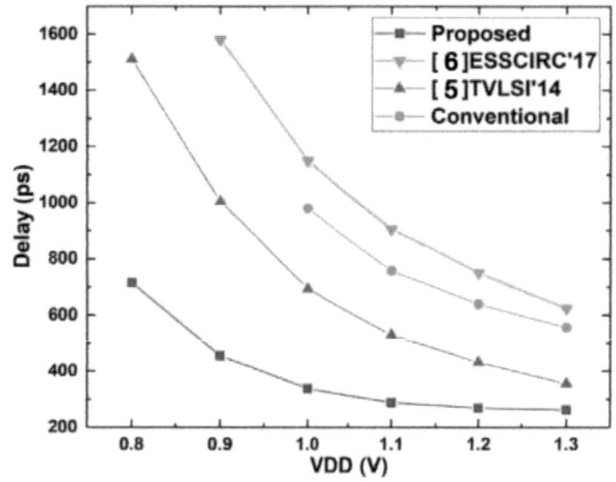

Fig.16 The delay versus VDD

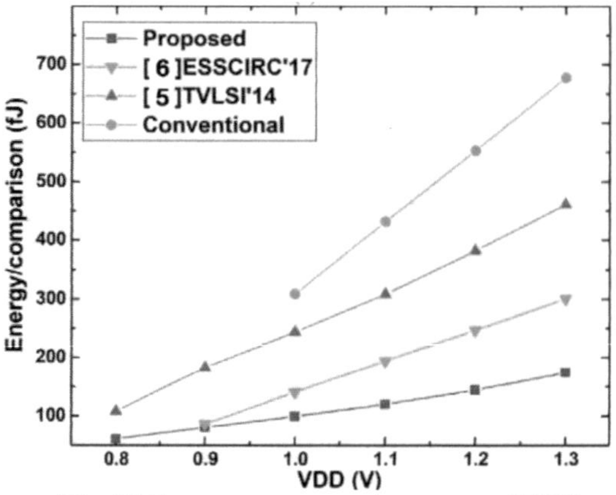

Fig.17 Energy consumption versus VDD

Based on the above, low delay and energy consumption can be achieved when VDD=1.2V. Therefore, Fig.18 illustrates the dependence of delay in the difference in input voltage (VDD=1.2v) and the multiple input common-mode voltage. For the multiple of input common-mode voltage ranging 0.7-1.1v, the delay remains stable and declines along with the decrease of the differential voltage.

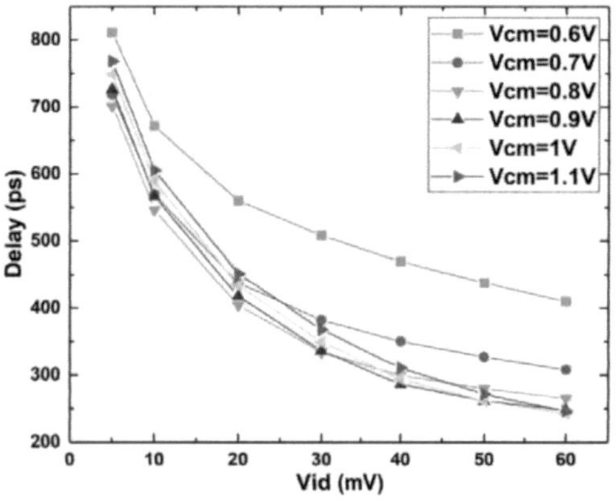

Fig.18 Post layout simulated delay versus Vid.

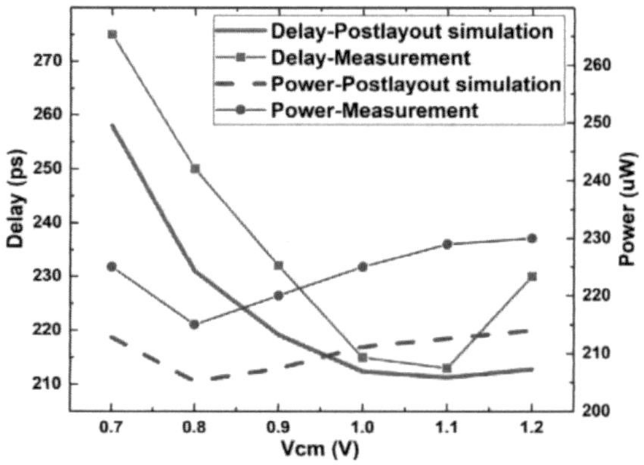

Fig.19 The delay and power versus Vcm.

The demonstration about the simulation and measurement of delay and power versus Vcm(Vid=0.1v, VDD=1.2v,f=2GHz) is represented in Fig.19. It can be concluded that the delay is under 275ps, and the consumption of power is about 225μW with the input common-mode voltage ranging 0.7-1.2v. The trend of experimental results is basically the same as the simulated results. But there is a constant error because of the methods and devices of measurement. In conclusion, such a high speed and low power consumption are realistic in the proposed comparator.

3.4. a novel comparator with a charge pump based on the conventional Miyahara's comparator

Haoyu Zhuang and his teammates design a voltage comparator 60% faster than the Miyahara's by using a charge pump in 2020.

The schematic of the conventional Miyahara's comparator is shown in Fig.20. It has three phases: the reset phase, the amplification phase, and the regeneration phase. The operation of the comparator is described as follows. DIP and DIN are reset to VDD in the reset phase, while OUTP, OUTN, QP, and QN are reset to GND. And in the amplification phase, DIP and DIN fall, while OUTP, OUTN, QP, and

QN rise. M10 and M11 are turned on in the regeneration phase so that the comparison result can be generated.

Based on the analysis, we can conclude that the input-referred noise can be reduced by increasing the input pair transconductance and the integration time [7]. So, new technology is applied to the conventional Miyahara's comparator, which can greatly improve the comparison speed and will not degrade the noise performance.

The proposed comparator based on the comparator above-mentioned is shown in Fig. 21. It just has one modification compared to the classic one, a charge pump connected to the VTOP. So, the VTOP is equal to VDD at CLK=0 and is slightly larger than VDD at CLK=1. Besides, the bulk of M6~M9 is also connected to the VTOP to make sure their bulk-source voltage is not negative. The operation of the proposed comparator is similar to Miyahara's comparator. In the reset phase, all the sources are reset. In the amplification phase, CLK's rise contributes to the rising of VTOP, which makes the second-stage amplification phase turn on earlier than conventional Miyahara's comparator, which helps accelerate the comparison speed of the proposed comparator.

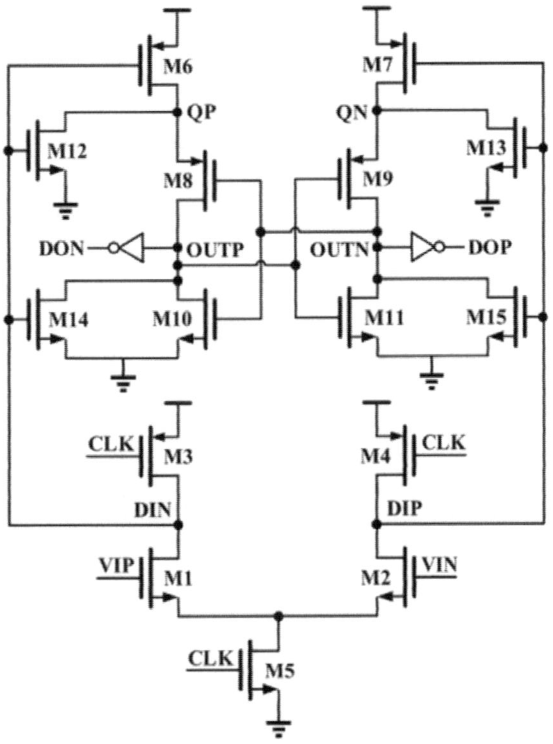

Fig.20 Miyahara's comparator

Journal of Physics: Conference Series **2113** (2021) 012064 doi:10.1088/1742-6596/2113/1/012064

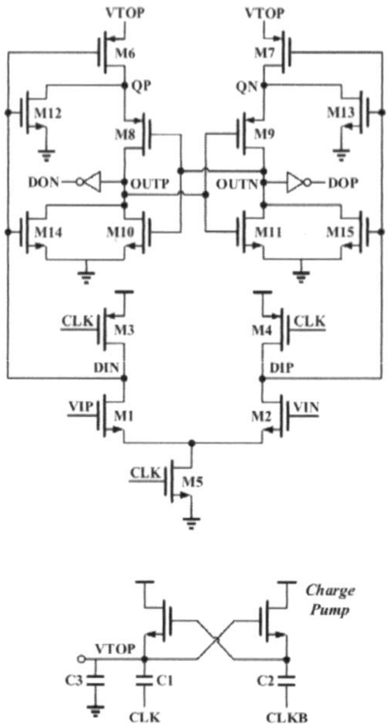

Fig.21 The novel comparator based on Miyahara's comparator

Meanwhile, the current at the beginning of the second-stage amplification is much larger than Miyahara's comparator. Fig. 22 displays the comparison of the current in both comparators. As can be seen, the current in the proposed comparator is about 3 times larger, which is caused by the earlier start of the second-stage amplification and the larger current in M12 and M13. The faster rise of OUTP and OUTN in the proposed comparator accelerates the comparison speed.

The larger current also increases the latch transconductance, which cancels out the penalty in the noise performance due to the reduced integration time caused by the earlier start of the second-stage amplification.

The proposed comparator is designed in the same as conventional Miyahara's comparator. Fig.23 shows the measured power consumption versus the differential input voltage. The conclusion is that the power of the proposed comparator is larger by 47% than Miyahara's comparator due to the larger voltage the proposed comparator works at.

Fig.24 shows the measured delay versus the differential input voltage and figures out that the delay of the proposed comparator is 60% smaller than Miyahara's comparator.

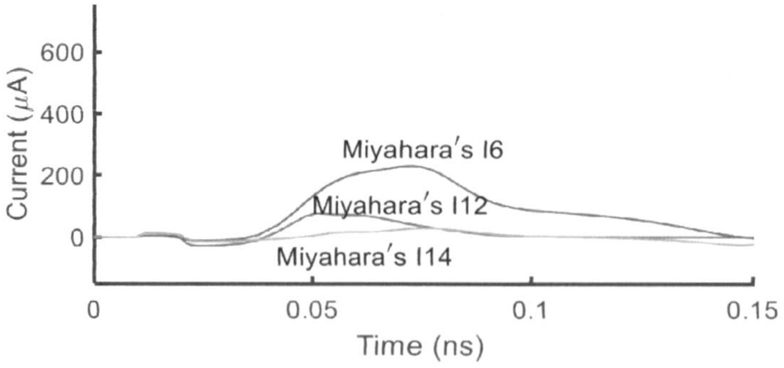

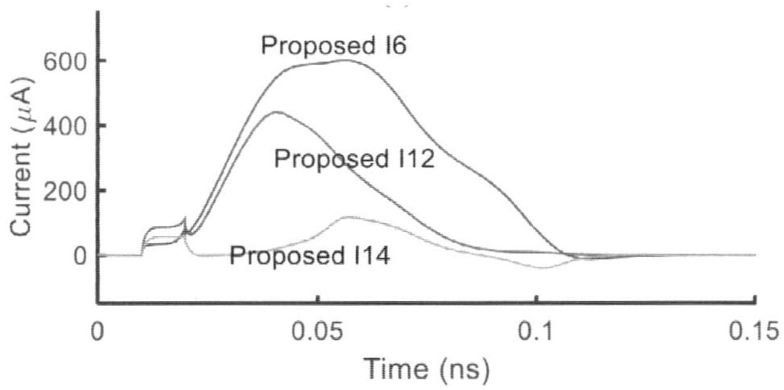

Fig.22 The comparison of the current in both of the comparators

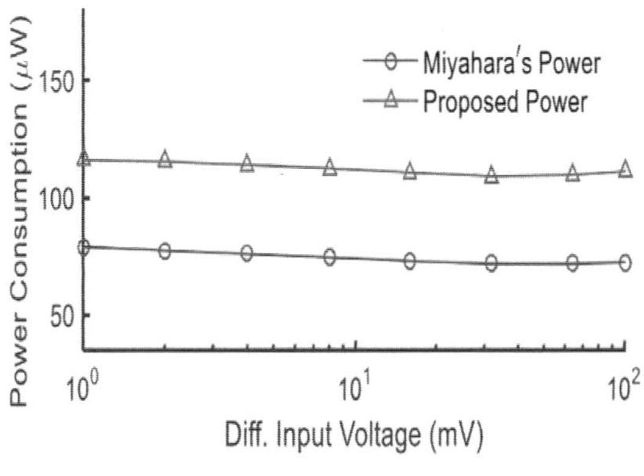

Fig.23Measured power consumption versus the differential input voltage

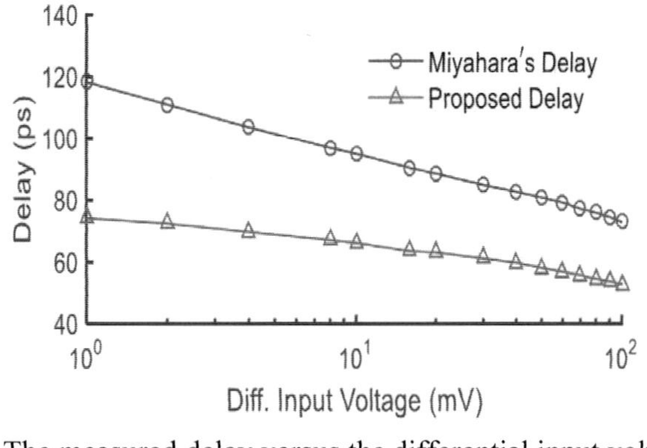

Fig.24 The measured delay versus the differential input voltage

Table.1 lists the process of the performance of the comparator. The delay and power consumption are greatly reduced, but the offset and kickback noise are almost remained unchanged. Under the 180nm technique, there are some different technologies. The new double-tail dynamic comparator designed in 2014 can reduce the power consumption and delay. The comparator in 2018 uses the input of the latch and the cross-coupled circuit while the comparator in 2019 uses the separated gate-biasing cross-coupled transistors to substitute the conventional cross-coupled inverter structure. It can be concluded that the delay [8] of 2014 is greater than that of 2019, the power consumption is much higher than that of 2019. Under the 40nm technique, it greatly increases the speed without needing extra calibration for comparator offset. Through Table I, it can be found that the power consumption has been greatly reduced compared to the previous design. However, the optimization of the circuit still needs to consider some factors involved in the noise, delay, power, and offset voltage [9, 10].

Table.1 The performance of dynamic comparator

Parameter	CDC	CDDC	2014	2018	2019	2020
CMOS technology （nm）	180	180	180	180	180	40
Supply voltage (V)	0.8	0.6	1.2	1.1	1.2	0.5
clock frequency (GHz)	0.9	1.8	0.5	0.5	0.5	1
Delay time (ps)	940	358	550	370	268.6	84
Offset (mv)	—	7.91	7.8	2	7.3	—
power consumption per conversion (μw)	—	—	329	230	72.2	115
kickback noise （mv）	51.3	5.3	43	—	—	—

*CDC and CDD stand for Conventional Dynamic Comparator, Conventional Double-tail Dynamic Comparator, respectively

4. Conclusion

In summary, four dynamic comparators are given to improve circuit performance.

From the analytical expressions, a new dynamic comparator is proposed by Samaneh Babayan-Mashhadi, and the power consumption and delay time are significantly reduced. These are improvements to the circuit structure, thereby optimizing the performance of the comparators to a certain extent. The main idea of Ata Khorami and Mohammad Sharifkhani is PMOS transistors used at the input of the latch and the cross-coupled circuit to keep the common-mode voltage of the preamplifier outputs at a low level. Yao Wang put forwards separated gate-biasing cross-coupled transistors instead of the conventional cross-coupled inverter structure in the latching stage, significantly decreasing the delay and energy consumption.

And there is an innovation to optimize the comparator. Haoyu Zhuang designs an extra charge pump based on the classic Miyahara's two-stage comparator to improve voltage, decrease the delay, and not degrade the noise performance. But its comparator offset causes the offset of the entire SAR ADC.

Generally, through the study of four papers, we can have more ways to deal with design problems in future design work. However, the noise is still a problem. Therefore, many critical factors are considered when designing dynamic comparators.

References

[1] S. S. Chiwande and S. P. Akarte, "Performance analysis of low voltage, low power dynamic double tail comparator for data converter application," 2016 World Conference on Futuristic Trends in Research and Innovation for Social Welfare (Startup Conclave), 2016, pp. 1-4, doi: 10.1109/STARTUP.2016.7583918.

[2] Samaneh Babayan-Mashhadi, et al. "Analysis and Design of a Low-Voltage Low-Power Double-Tail Comparator." IEEE TRANSACTIONS ON VERY LARGE SCALE INTEGRATION (VLSI) SYSTEMS, VOL. 22, NO. 2, FEBRUARY 2014.

[3] Ata Khorami, Mohammad Sharifkhani. "A Low-Power High-Speed Comparator for Precise Applications." IEEE TRANSACTIONS ON VERY LARGE SCALE INTEGRATION (VLSI) SYSTEMS, VOL. 26, NO. 10, OCTOBER 2018.

[4] YAOWANG, MENGMENGYAO, et al. "A Low-Power High-Speed Dynamic Comparator With a Transconductance-Enhanced Latching Stage " Received May 4, 2019, accepted June 27, 2019, date of publication July 9, 2019, date of current version July 29, 2019. Digital Object Identifier 10.1109/ACCESS.2019.2927514.
[5] S.Babayan-MashhadiandR.Lotfi,''Analysisanddesignofalow-voltage low-power double-tail comparator,'' IEEE Trans. Very Large Scale Integr. (VLSI) Syst., vol. 22, no. 2, pp. 343–352, Feb. 2014.
[6] H. S. Bindra, C. E. Lokin, A.-J. Annema, and B. Nauta, ''A 30fJ/comparison dynamic bias comparator,'' in Proc. 43rd IEEE Eur. Solid State Circuits Conf. (ESSCIRC), Sep. 2017, pp. 71–74.
[7] Haoyu Zhuang, He Tang, and Xiaoxian Liu. "Voltage Comparator With 60% Faster Speed by Using Charge Pump." IEEE TRANSACTIONS ON CIRCUITS AND SYSTEMS—II: EXPRESS BRIEFS, VOL. 67, NO. 12, DECEMBER 2020.
[8] D. Schinkel, E. Mensink, E. Klumperink, E. van Tuijl and B. Nauta, "A Double-Tail Latch-Type Voltage Sense Amplifier with 18ps Setup+Hold Time," 2007 IEEE International Solid-State Circuits Conference. Digest of Technical Papers, 2007, pp. 314-605, doi: 10.1109/ISSCC.2007.373420.
[9] M. Shim et al., "Edge-Pursuit Comparator: An Energy-Scalable Oscillator Collapse-Based Comparator With Application in a 74.1 dB SNDR and 20 kS/s 15 b SAR ADC," in IEEE Journal of Solid-State Circuits, vol. 52, no. 4, pp. 1077-1090, April 2017, doi: 10.1109/JSSC.2016.2631299.
[10] A. K. Dubey and R. K. Nagaria, "Low-power high-speed CMOS double tail dynamic comparator using self-biased amplification stage and novel latch stage," Analog Integrated Circuits and Signal Processing, vol. 101, (2), pp. 307-317, 2019.

Progress in Data Transmission and Storage for Long Pulse Data Acquisition System

Yu Gu[1], Mengqi Fan[1], Chuanqian Tang[1], Guojing Zhang[1], Xiaodan Zhang[1*]

[1]Department of Computer Technology and Applications, Qinghai University, Xining, Qinghai, 810016, China

[*]Corresponding author's e-mail: xdzhang@qhu.edu.cn

ABSTRACT. Since the study in the field of fusion has gradually developed toward the long-pulse experiment mode, long-pulse data has gradually become one of the main data types for pulsed experiments in the field of fusion. For long-pulse data, which is a kind of pulse-type data, it will be more difficult to transmit and store than short-pulse data because of its significant characteristics. In addition, in the design of data acquisition and control system (DACS) in fusion field, Experimental Physics and Industrial Control System (EPICS) has now gradually become the main framework of experimental control system to meet the diversity of devices and complexity of subsystems in large experimental system. However, due to the limitation of EPICS, its effectiveness in handling data transmission and storage under high speed data acquisition is not satisfactory. To solve the data transmission and storage under high-speed sampling, this paper proposes a data transmission and storage solution based on TCP/IP protocol and MDSplus database, which is designed with the concept of segmentation, i.e., data generated from experiments longer than 100 seconds are uploaded and stored in a segmented form. Currently, this system has been tested and applied, and the test result shows that the solution is feasible and the overall test system operates stably and reliably.

1. Introduction

At present, long pulse experiments have become an important study direction in the field of fusion. Through long pulse experiments, experimenters can more accurately analyse the operating state of each subsystem and the behaviour of large experimental devices during the experiments, which will help to study the effects of different experimental parameters on the experimental results and further upgrade the experimental devices. Compared with short-pulse experiments, the duration of long-pulse experiments usually exceeds 100s, so the experimenter usually needs to analyse the data only after the experiments are finished. Therefore, it is necessary to explore a method to synchronize the transmission, storage and visualization of long pulse data. With the study development of the experimental system, the control system of large fusion hosts is now usually developed using EPICS[1], which is first applied to the KSTAR system in Korea [2], and then the ITER organization and Huazhong University of Science and Technology have adopted EPICS to improve their own systems[3-4]. Compared to some data acquisition frameworks, EPICS provides input output controller (IOC) and channel access (CA) protocol, making it possible to support most current bus architectures on the one hand, and on the other hand, those control systems developed based on EPICS can precisely control subsystems through reliable CA protocol. However, nothing is perfect, the characteristics of EPICS make it ineffective in terms of data storage and data transmission for high speed data acquisition. In fact, EPICS does not provide reliable data retrieval and storage, so for data storage processing, an additional database is usually required. For data transmission, EPICS can achieve data transmission at low acquisition rate through channel archiver [5], which creates a predefined buffer and archives the collected data in real time according to the scanning rate of EPICS. However, when the data acquisition rate is higher than 1kHz, EPICS will have a certain delay in scanning the data. Therefore, based on the above data transmission mechanism, it is easy to conclude that it is difficult to achieve better data transmission through channel archiver, which means that the system designed based on EPICS needs to consider data transmission at high speed data acquisition to avoid data loss.

Inspired by the ITER CODAC system, the current DACS has been upgraded based on the EPICS framework [6], and the architecture of it is shown in Figure 1. DACS is composed of three layers, namely the central control system, the server control system, and the sub-control system. The central control system is the human-machine interface (HMI), which includes both remote and local HMIs; the server control layer is mainly composed of a data server and a network server, the data server is used for long-term storage for experimental data, while the network server is used to monitor the data server and provide remote services for users; the sub-control system runs on an industrial personal computer (IPC), which can be composed of multiple IPCs to form a workgroup. The overall workflow of DACS can be briefly described as follows: the experimenter issues a trigger command from the HMI, the relevant configuration data will be obtained by the sub-control system, and at the same time, the sub-control system will start the data acquisition task, and the collected data will be asynchronously uploaded to the data server for long-term storage and visualization on the HMI. This paper focuses on the design of the DACS for data transmission and data storage in the case of high-speed sampling.

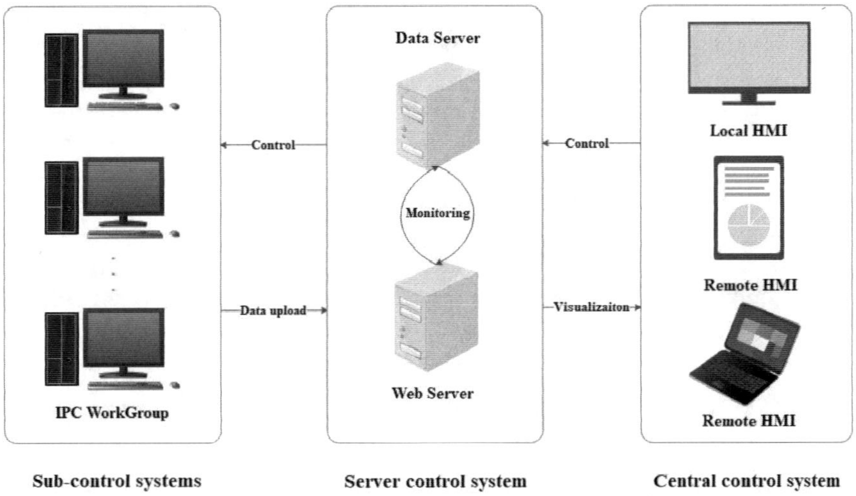

Figure 1. The architecture of DACS

2. The overall design of data transmission and storage

As mentioned above, EPICS is deficient in data transmission and storage. Therefore, MDSplus [7] is used for data storage. MDSplus database is a software that integrates data storage, data processing, and data visualization which is specially designed for long pulse experiments by MIT. It has been widely used for long pulse data processing and storage due to its easy operation and user-friendly interface. As for data transmission, the data transmission process here refers to the real-time transmission of the experimental data from the IOC to the data server for long-term storage. As mentioned earlier, MDSplus provides channel archiver to archive data from the EPICS IOC to MDSplus. However, limited by the working mechanism of the channel archiver, this data transmission method is not suitable for high speed data acquisition. In view of the fact that data transmission under low speed data acquisition can be completed by channel archiver, this paper only discusses data transmission under high speed data acquisition. The collected data is finally transferred to MDSplus in the data server for long-term storage for later analysis and utilization by the researchers. The process of data transmission and storage under high speed data acquisition is shown in Figure 2.

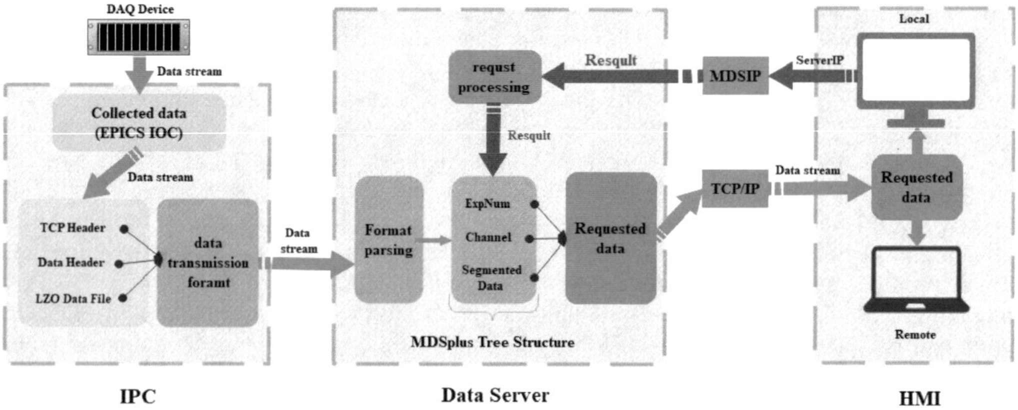

Figure 2. The process of data transmission and storage under high speed data acquisition

In the design and implementation about the scheme, Data acquisition tasks at high sampling speeds are performed by IPC and then the collected data will be scanned by EPICS IOC and read in the process variable (PV). In order to reduce the network transmission load and save storage space, the raw experimental data will be compressed by a data compression algorithm. After that, the compressed data will be integrated into a defined and standardized data structure, which is then transmitted to the data server via TCP/IP based data transmission structure [8]. The packet will then be parsed and the data will be stored in MDSplus using data segmentation techniques. When the experimenter needs to access the experimental data, it can be obtained based on the specific sampling time and channel number of a particular experiment. Finally, the waveform of the raw data will be visualized on the HMI.

It is worth noting that the storage format needs to be specified in advance. Data obtained by different data acquisition channels in the same experiment number will be stored in different data files to prevent data confusion; while experimental data from different experiment numbers but under the same channel should be stored by experiment number to prevent data overwriting. In order to easily display data waveforms to the user, the data segmentation requirements must be met in a specific data format. The requirements of this system under high-speed sampling can be summarized as follows.

- Implementation of communication between the IPC and the data server for data transmission and reception at high speed data acquisition.
- Correct parsing of the data structure to facilitate access to the original data and prevent data loss.
- Long pulse data segmentation processing.
- Experimental data segmentation storage.
- Accessing segmented data from the MDSplus database according to pre-set parameters.
- A clear, simple and user-friendly data visualization interface.

3. The detailed design and implementation

3.1. High speed data acquisition

Given that the overall framework of EPICS is implemented through C and C++, the data acquisition program is also implemented using the C programming language in order to maximize compatibility between the data acquisition program and EPICS. In addition, since EPICS provides a variety of PVs for implementing different functions, we establish a generic device model based on 'ai (analog input)', 'bo (binary output)', and 'calc (calculation)' in order to enhance the compatibility of the system with different DAQ cards. The 'ai' is used to read the collected data, the 'bo' is used to control the experimental devices, and the 'calc' is used to monitor the status of those devices. This device model unifies the abstract programming standard of experimental devices, especially for DAQ cards, so it provides great convenience for the subsequent maintenance, update and expansion of DACS. Before the data acquisition task starts, the experimenter needs to configure the experimental devices in detail through HMI, such as the number of acquisition channels, the selection of voltage range, the setting of acquisition frequency, and the acquisition time. At the same time, the program will retrieve the specified directory to determine the current experiment number to prevent the new experiment data from overwriting the original experiment data. When the experimental system is on standby, the IPC will be in a waiting state, ready to receive the 'START' command from the central control system, after which the data acquisition task will be started with the obtained configuration. Considering the need for real-time access in long pulse experiments, the steps of data acquisition, data transmission and data segment storage will be performed synchronously. Through the high speed data acquisition program, the data collected by the DAQ devices will be continuously scanned into the EPICS IOC, and the next step is to transfer and store the collected data for further processing.

In this study, PCI-9112 DAQ card is used on the IPC for data acquisition [9]. The card has 12-bit sampling accuracy and 16 channels, the highest sampling rate can reach 110 kHz for each independent channel. In addition, the minimum hardware and software environment required for the overall system operation should meet the following requirements: CPU performance of 2.0 GHz or above; memory performance of 4 GB or above; hard disk storage capacity of at least 500 GB. The software platform is a C/C++ programming environment based on the Linux operating system. Moreover, MDSplus database and Linux shell programming language are required for the successful implementation of the system.

3.2. Data transmission at high speed sampling

Once data acquisition is finished, one thing that needs to be considered is data transmission. Data acquisition at high speed sampling temporarily stores the collected data in a file named after the channel number that is used for data acquisition in this experiment. In order to reduce network transmission load and save data storage space, this data file will be compressed at this time using a data compression algorithm. The LZO (Lempel-Ziv-Oberhumer) data compression algorithm [10] is chosen for data compression after considering the support for data segmentation and the rate of compression and decompression and space size. After testing, the size of original data file processed by LZO can be reduced by 30%.

Lossless transmission and reception of raw data is crucial, which is why we adopt the TCP/IP protocol to achieve reliable data transmission under high-speed sampling. The data transmission structure based on TCP/IP protocol is shown in Figure 3. As can be seen from the figure, the data structure contains three parts, which are TCP header, data header and compressed data file. The TCP header includes the IP address of the IPC host, the IP address of the data server, the size of this segment of the data structure and CRC checksum. In this design, the IP of IPC is used to mark the source address of data, and the IP of the data server is used to indicate the address to which this data will be sent, and finally, the CRC is used to verify the correctness of the packet after transmission. The data header mainly includes some configuration information about the acquisition of this data, including the acquisition channel number, signal name, sampling frequency, sampling duration, sampling accuracy and voltage range. This information is used to accurately parse the data structure in the data parsing phase and store the data in the correct node in the MDSplus tree. And the last part is the compressed data file. After comparison, the collected data can be transmitted correctly in this way.

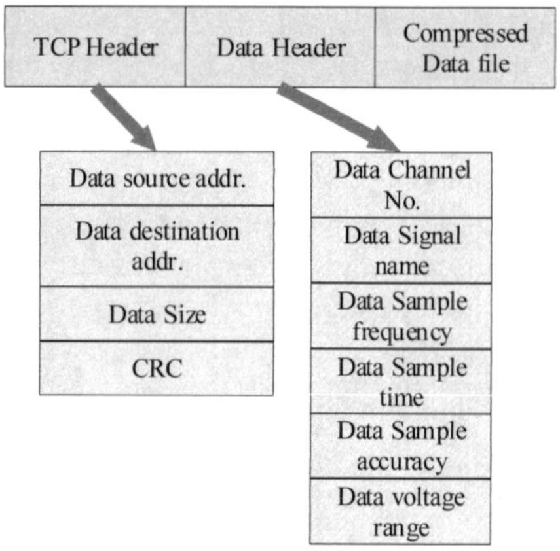

Figure 3. The data transmission structure based on TCP/IP Protocol

3.3. MDSplus-based data segmentation storage

Now, the collected data has been transmitted to the data server through the above data structure. As mentioned before, the collected data needs to be compressed by LZO, and the original data file will be changed to a lzo file after being compressed. Since MDSplus does not support LZO files for storage, in order to ensure that the data files can be stored successfully, it is necessary to perform decompression operation after the data has been transmitted to the data server. As for decompression, LZO's library provides decompression function that can be called directly at the time of use. After decompression, the compressed file will be converted to a txt file and the txt file is the one that will be stored in MDSplus.

The next step is data segmentation and storing the segmented data into the MDSplus. MDSplus stores all experimental data with a tree structure [11]. On the one hand, it unifies the data storage format; on the other hand, it can clearly express the logical relationship between tree nodes. There are two types of trees in the MDSplus: the model tree and the pulse tree. The model tree mainly stores the structure information of the tree, such as the node name and other specific tree structures. The pulse tree will be built after the experimental data was stored in the model tree. After this, if the structure of the model tree is changed again afterwards, the logical structure of the created pulse tree will not change again [12]. In other words, the change of the logical structure of the model tree does not affect the structure of the pulse tree. The relationship between them is shown in Figure 4.

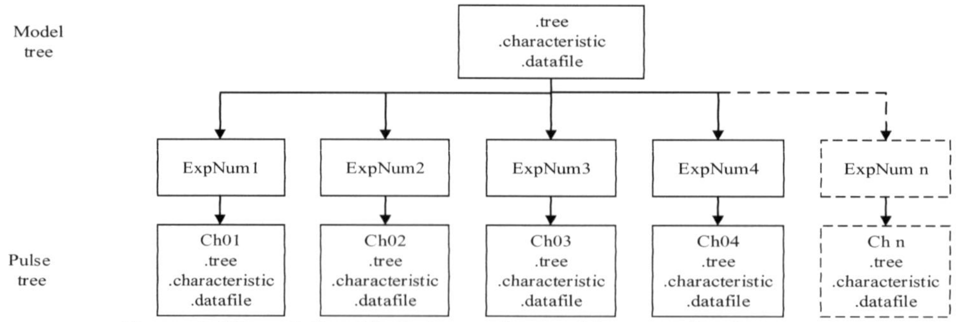

Figure 4. The relationship between the model tree and pulse tree

3.3.1 MDSplus Model Tree and pulse tree. Although the model tree specifies the logical structure of the MDSplus tree, no data is actually stored in it. The overall framework of model tree can be built using the TCL language, and then the jTraverser provided by MDSplus can be used to help optimize the structure of the model tree, which can save a lot of development time. In this paper, in order to better present the data storage structure of the MDSplus after data transmission, a four-level structure of the MDSplus tree was created in the test server, consisting of the experiment number, device number, channel number and data. The final MDSplus tree is shown in Figure 5.

In this MDSplus tree, 'EXPNUM X' represents the experiment number, 'DEVICE X' represents the device number, 'CH X' represents the channel number and the 'DATA X' represents the segmented data file. As mentioned before, the DAQ card has 16 data acquisition channels, so, there are 16 one-to-one corresponding subtree nodes named after the channel number of the DAQ card in this tree structure. In this design, the model tree of this MDSplus consists of four branches, including parameter, timing, raw and process. The parameter branch stores some constant data, such as amplification, data type, etc. The timing branch is used to store the trigger time, data acquisition time and the frequency of this experiment. The raw branch stores the raw experimental data. The processing branch stores the corresponding signal information of the transmitted packets after processing. As mentioned before, the pulse tree will be built after store the data into the model tree. Therefore, if the experiment starts, the data server needs to stay connected to each IPC all the time when the data server receives a data packet from the IPC, the server program will first parse the packet and write the data to the corresponding MDSplus tree node according to the parsed result. In fact, the parsed data is saved in three files according to the file type in the pulse tree, including tree file, data file and characteristic file. The tree file stores information about the structure of the MDSplus tree; the data file stores the real data. In theory, all data from a long pulse experiment will be written to the node of the pulse tree, while the actual data will only be stored in the data file of the tree node, and some information about this experiment, such as the offset address, will be stored in the access log of the node. MDSplus only maps the '*.tree' file to memory [13], and as for the actual data, it is stored in the hard disk, in other words, the data and the tree information are stored separately, which has the advantage of greatly increasing the data transfer speed.

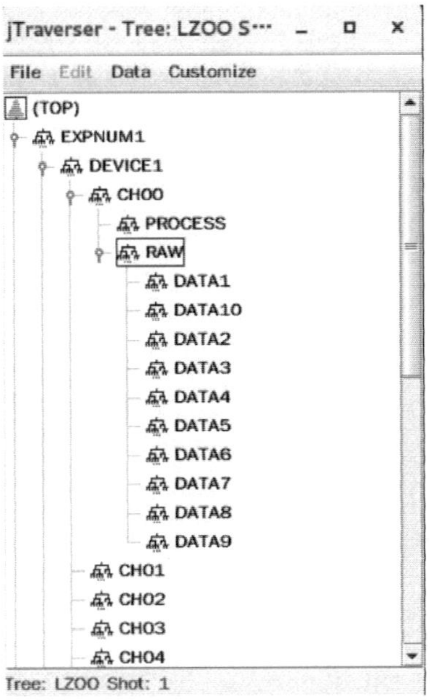

Figure 5. The MDSplus four-level tree structure

3.3.2 Data segmentation storage and accessing. In the specific MDSplus data storage format, segmented storage is used to store data. The core design concept of segmented storage is the "data segment", by which MDSplus can achieve continuous data reads and writes. The key technology of data segmentation is the segmented reading and segmented writing of the collected data. Segmented write means that each segment data is appended to the data node, instead of overwriting the original data, as storage space allows; segmented read means that each segment can be read from MDSplus in a separate form, without being read only with the overall data. By reading and writing in segments, the collected data can be stored in MDSplus in the form of segments without waiting for the end of the experiment. In addition, the experimenter can read the stored experimental data in time during the experiment. The principle of this technique is that when retrieving data from segmented records, MDSplus will select the corresponding segment based on the timestamp, then read all segments that match the retrieval time and join these segments to finally construct the final result that satisfies the retrieval criteria and return it to the user. Through data segmentation storage, the experimenter can read data from MDSplus that is in the experimental period but has been collected and stored. The associated program can then visualize the target data on HMI. The process of accessing data from MDSplus is shown in Figure 6.

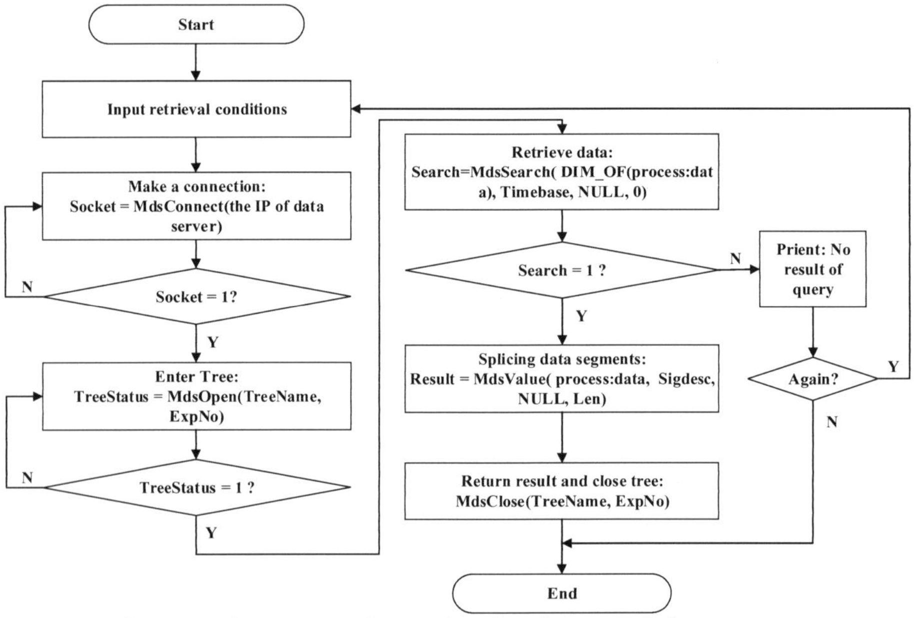

Figure 6. The process of accessing data from MDSplus

4. Testing

In order to verify the integrity of the experimental data and the reliability of the segmented storage after processing by this system at high speed sampling, after the segmented experimental data is transmitted and stored in MDSplus, NBWave [14] is used for accessing the same experimental data in segment and in whole respectively, and the segmented data is compared with the original data to check whether the data is correctly stored in MDSplus. The test experiment is a long pulse experiment of up to 100 seconds with a square wave and 100 kHz data sampling frequency. In this test, 10 seconds is chosen as the segmentation basis. Due to the high speed data acquisition, for better display of the comparison results, a segment of data is taken here, whose time length is 10s, and then it is divided into 10 segments on the basis of 1 second. The raw data waveform for 10 seconds under this configuration is shown in Figure 7. Then the 2nd to 3rd seconds data stored in MDSplus is randomly selected and then visualized in the other data visualization window and compared with the original data waveform. The result of comparison between them is shown in Figure 8. In this figure, above the middle red line is the original data waveform, and below is the data of the 2nd to the 3th seconds. It can be seen that the upper part and the lower part are exactly the same in terms of the y-axis value that corresponds to any value of the x-axis. According to the comparative analysis of the waveform chart, the results are consistent with the expectations.

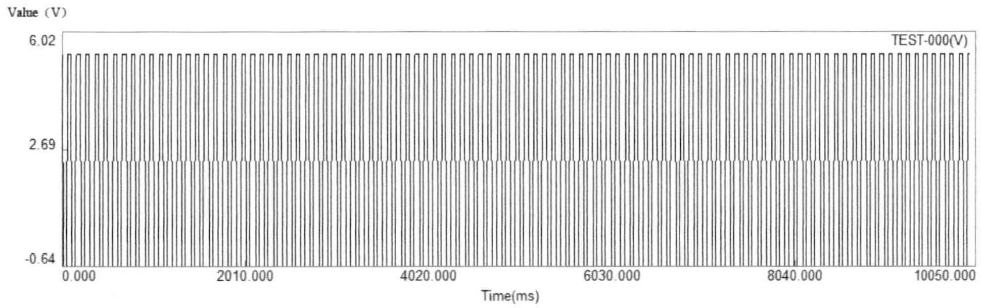

Figure 7. The raw data waveform for 10 seconds

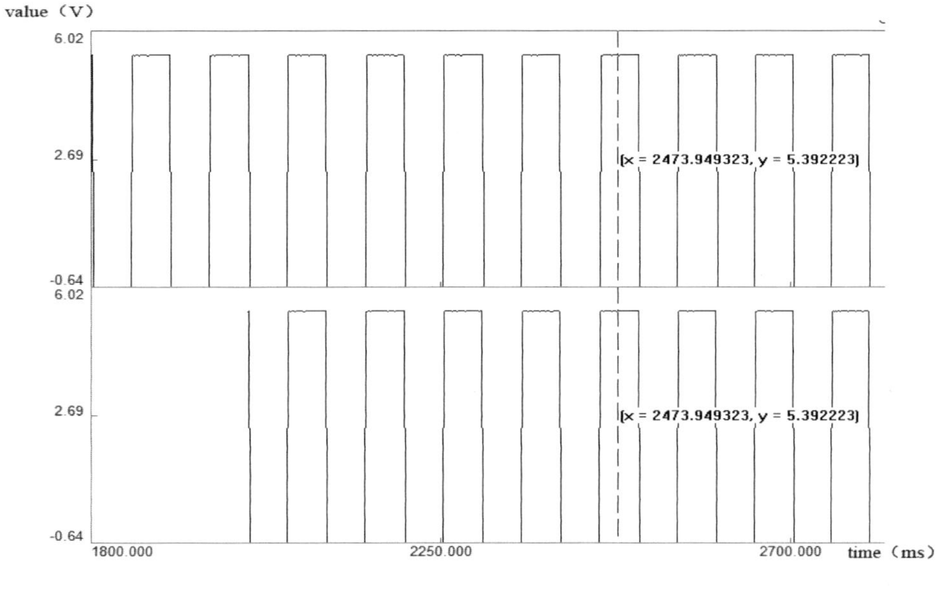

Figure 8. Comparison results between the data stored using segmentation and the original data.

5. Conclusion

This paper presents the progress of data transmission and storage for long pulse data acquisition system. It proposes a scheme to cope with high speed data acquisition based on EPICS and designs a reasonable data transmission structure based on TCP/IP protocol, and also proposes a segmented storage scheme based on MDSplus for the data storage in long-pulse experimental mode. It integrates many tools, including EPICS system, MDSplus database, LZO data compression algorithm, TCP/IP network transmission protocol and data segmentation processing technology. The application of the program makes it possible to access real-time data for high speed data acquisition in long-pulse experimental mode. After testing, the results show that the method is effective and fully meets the requirements of practical applications.

Acknowledgments
The authors gratefully acknowledge the supports of the open project of state key laboratory of plateau ecology and agriculture, Qinghai university (2021-ZZ-05), the experimental education teaching research project of Qinghai university under No.SYJG202012, the scientific research cooperation project (IPP-ZC-19071806).

References
[1] Experimental Physics and Industrial Control System Website. Available from: http://www.aps.anl.gov/epics/
[2] Kwon M, Choi IS, Choi JW, et al. The control system of KSTAR. Fusion Engineering and Design. 2004;71(1).
[3] Lister JB, Farthing JW, Greenwald M, et al. The ITER CODAC conceptual design. Fusion Engineering and Design. 2007;82(5).
[4] W. Zheng et al, "The J-TEXT CODAC system design and implementation," Fusion Engineering and Design, 89, 5, (2014); 10.1016/j.fusengdes.2014.03.048.
[5] MDSplus Channel Archiver for EPICS. Home Page. Available from: https://www.mdsplus.org/index.php/Documentation:Users:EPICS.
[6] Y Gu, X Zhang, Y Zhao, et al., The redesign and implementation of data acquisition and control system for EAST NBI[J], Fusion Eng. Des. 164 (10) (2021), 112219.
[7] MDSplus Home Page. Available from: http://www.mdsplus.org/.
[8] X. Xie. Computer Networks. Publishing House of Electronics Industry, Beijing. (2017).
[9] ADLINK Official website [Online]. Available: https://www.adlinktech.com/Products/Data_Acquisition/Multi-FunctionDAQ/PCI-9112?lang=zh-cn, Accessed on: may. 21, 2020.
[10] LZO [Online] http://www.oberhumer.com/opensource/lzo/lzofaq.php.
[11] D. T. Lin, B. J. Xiao, and Q. P. Yuan, "MDSplus Data Storage Application," Microcomputer Information, 2010, 26(34): 111-113.
[12] T. Fredian, J. Stillerman, G. Manduchi. "MDSplus extensions for long pulse experiments". Fusion Engineering and Design,2007,83(2).
[13] W. J. Huang, H. Z. Wang, F. Yang, S. Li, and B. J. Xiao, "MDSplus Real-Time Data Acquisition System," Measurement and Control Technology, 2014, 33(02): 16-19.
[14] L.J. Xiao, J.R. Luo, P. Sheng, Z.M. Liu, The design and implementation of data publishing system for NBI [J], Microcomputer Inf. 27 (05) (2011), 123-125+179. (In Chinese).

Panoramic image mosaic method based on image segmentation and Improved SIFT algorithm

Lei Zhuang, Jiyan Yu, Yang Song*

Mechanical Engineering, NJUST, Nan Jing, Jiangsu Province, 210000, China

*Email: SongY126@njust.edu.cn.

Abstract. Aiming at the problem of large amount of calculation in extracting image feature points in panoramic image mosaic by SIFT algorithm, a panoramic image mosaic algorithm based on image segmentation and Improved SIFT is proposed in this paper. The algorithm fully considers the characteristics of panoramic image stitching. Firstly, the stitched image is divided into blocks, and the maximum overlapping block of image pairs is extracted by using mutual information. The SIFT key points are extracted by SIFT algorithm, and the dog is filtered before the spatial extreme value detection of SIFT algorithm to eliminate the feature points with small intensity value; When establishing the feature descriptor, the 128 dimension of the original algorithm is reduced to 64 dimensions to reduce the amount of calculation. In the feature point registration process, the feature descriptor is reduced to 32 dimensions, the feature point pairs are roughly extracted by the optimal node first BBF algorithm, and the feature point pairs are registered and screened by RANSAC; Finally, the image transformation matrix is obtained to realize panoramic image mosaic. The experimental results show that the proposed algorithm not only ensures the panoramic mosaic effect, but also extracts the feature points in 11% of the time of the traditional SIFT algorithm, and the feature point registration speed is 27.17% of the traditional SIFT algorithm.

1. Introduction

Panoramic images are spliced from actual scene images, which can display a wide range of information. At present, panoramic images have been widely used in remote sensing image processing, military applications, medical image analysis, machine vision and other fields. Two or more images taken from different positions and angles are spliced into a panoramic image with wide viewing angle, which has the characteristics of large field of view and high resolution [1]. Image registration is a key technology in panoramic image Mosaic and it is a process of finding geometric relations between different images [2]. Image registration methods are mainly divided into region-based and feature-based registration methods [3]. The image registration method mainly uses the feature-based registration method [4], which uses the feature information in the image to establish the geometric relationship between the image and the image, and has feature invariance under the conditions of image scale transformation, translation, rotation and so on.

SIFT first proposed SIFT algorithm in 1999 and optimized in 2004.As the algorithm has good robustness under the conditions of image affine transformation, perspective change, noise and illumination, it has gradually become a panoramic image Mosaic based on feature points Research hotspot of algorithm. Li Yufeng et al. [1] proposed an image Mosaic algorithm based on region segmentation and SIFT features, which only divided the image into 4 equal parts and extracted feature points of similar regions using SIFT algorithm. Li Jia et al. [5] proposed a video panorama Mosaic method based on image block matching, which divides images into blocks according to the number of processors, uses multi-thread to match and search sub-images, and establishes the relationship between the sub-images. The speed of this method is too dependent on the performance of hardware. Zhao Yan et al. [6] divided the overlapping area into 128×128 image blocks and only selected the two image blocks with large standard deviation and distributed in different positions to detect their SIFT feature points. This method needs to adjust the size of sub-image blocks according to different images and has poor adaptability. Li Liang et al. [7] reduced the DIMENSION of SIFT feature descriptor from 128 to 64, and changed the direction of feature descriptor to 0, $1/4\pi$, $1/2\pi$ and $3/4\pi$, but did not include the direction between $3/4\pi$ and 2π. Li Dan et al. [8] reduced the dimension of the feature vector by reducing the size of the window centered on the feature point, from the original 128 dimensions to 24 dimensions. In this method, the reduction of the window centered on the feature point is easy to make the feature point descriptor affected by illumination and Angle of view changes.

Based on the above deficiencies, this paper proposes a panoramic image Mosaic method based on image segmentation and improved SIFT, which extracts the overlapping areas of images and reduces the feature descriptor from 128 dimensions to 32 dimensions. This method can obtain the panoramic image quickly and stably while retaining the CORE SIFT algorithm.

2. The general process of stitching

SIFT has a large amount of calculation when establishing feature point descriptor, and the real-time performance of panoramic image mosaic is not strong. According to the mosaic characteristics of panoramic images, only the feature points of overlapping parts are effective. Therefore, the images to be spliced are processed in blocks to extract the overlapping area of the image, which will greatly reduce the running time of the algorithm.

2.1. Regional block

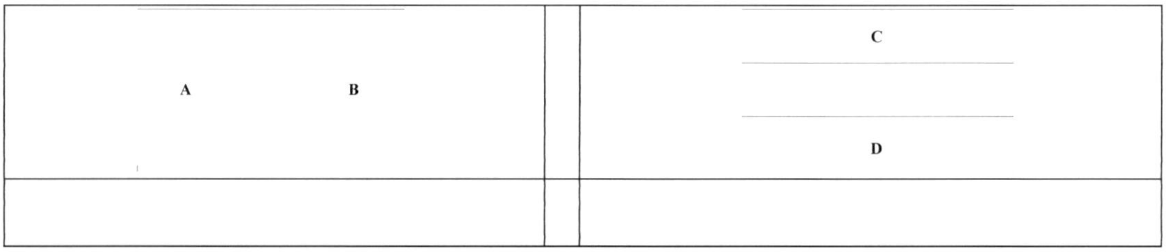

The two blocking methods in Figure 1、2 are selected to extract the feature points of four sub images A, B, C and D. It is concluded that the blocking strategy reduces the time used to search the feature points of similar areas by 1 / 3 compared with the traditional SIFT algorithm, makes full use of the mosaic characteristics of panoramic images, eliminates a large number of invalid feature points, and fully retains the feature points of overlapping areas.

2.2. Analysis of regional similarity

In this paper, the image after blocking is analyzed with mutual information as a similarity measure. Mutual information measures the similarity of two random variables, which is commonly expressed by joint entropy and individual entropy. H (a) and H (b) represent the entropy of image a and image B respectively, and H (a, b) represents the joint entropy of image a and image B. Then the joint entropy H (a, b) and individual entropy H (a) of the image are expressed as:

$$H(A)=-\sum_a P_A(a)\log P_A(a) \quad (1)$$

$$H(A,B)=-\sum_a P_{AB}(a,b)\log P_{AB}(a,b) \quad (2)$$

Where, $P_A(a)$ represents the probability distribution of A, $P_{AB}(a,b)$ represents the joint probability distribution of A and B.

Then the mutual information of image A and image B can be expressed as:

$$I(A,B)=H(A)+H(B)-H(A,B) \tag{3}$$

$$I(A,B)=\sum_{a,b} P_{AB}(a,b)\log\frac{P_{AB}(a,b)}{P_A(a)\cdot P_B(b)} \tag{4}$$

In practical problems, the entropy values of different images vary greatly. NMI is used as a similarity measure. The normalized mutual information can be expressed by the following formula:

$$NMI=2\frac{I(A,B)}{H(A)+H(B)} \tag{5}$$

When the two images are completely consistent, the normalized mutual information is 1; When the two images are completely inconsistent, the normalized mutual information is 0.

3. Improved SIFT algorithm

The steps of generating image feature point feature descriptor by Improved SIFT algorithm include: 1) improved scale space extreme value detection; 2) Determination of the principal direction of feature points; 3) Improved feature point feature descriptor.

3.1. Improved scale space extreme value detection

The Gaussian pyramid of the image is used to realize the scale space of the image, establish the dog of the image, set the threshold, propose the points with low intensity in the image, and finally find the local extreme points in the dog scale space.

The image scale space $L(x,y,\sigma)$ is defined as the convolution of the two-dimensional Gaussian $G(x,y,\sigma)$ function with the original image $I(x,y)$, σ is the image scale parameter. As follows:

$$L(x,y,\sigma)=G(x,y,\sigma)*I(x,y) \tag{6}$$

$$G(x,y,\sigma)=\frac{1}{2\pi\sigma^2}e^{-\frac{x^2+y^2}{2\sigma^2}} \tag{7}$$

Subtract two Gaussian images of the same group of adjacent scales to obtain the Gaussian difference scale space of the image, namely:

$$D(x,y,\sigma)=(G(x,y,k\sigma)-G(x,y,\sigma))*I(x,y)=L(x,y,k\sigma)-L(x,y,\sigma) \tag{8}$$

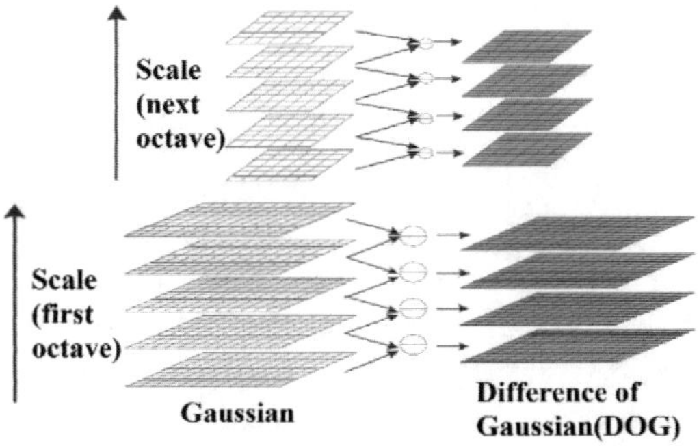

Figure 3. Dog pyramid

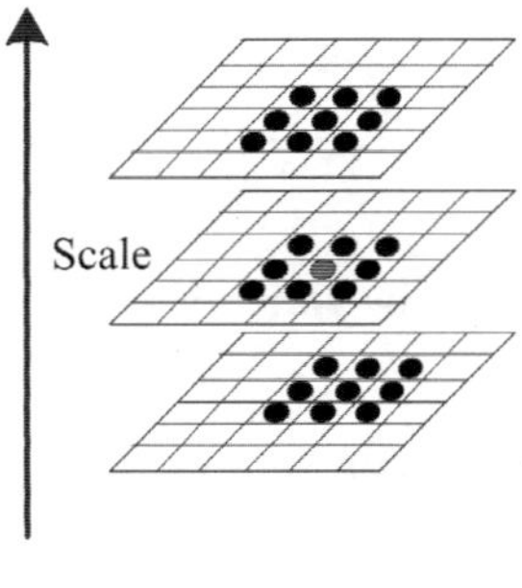

Figure 4. Extreme point detection

Remove the points with small intensity value in Gaussian difference scale space, namely:

$|D(x,y,\sigma)|<K$， K is the threshold

If the above conditions are true, the pixel is updated to 0. As shown in Figure 4, the detection point is compared with 26 points in the field. If the detection point is a local extreme point, the point is tentatively regarded as a feature point. Finally, feature points with poor stability and feature points under edge effect are eliminated to reduce the error of image matching.

3.2. Determination of principal direction of feature points

In order to realize the rotation of feature points without deformation, the gradient of the neighborhood of feature points is used to describe the reference direction of feature points.

$$m(x,y)=\sqrt{(L(x+1,y)-L(x-1,y))^2+(L(x,y+1)-L(x,y-1))^2} \tag{9}$$

$$\theta(x,y)=\tan^{-1}\left(\frac{L(x,y+1)-L(x,y-1)}{L(x+1,y)-L(x-1,y)}\right) \tag{10}$$

Calculate the gradient amplitude and direction of each pixel in the neighborhood of feature points, and use histogram to represent the gradient and direction of pixels in the neighborhood of feature points. The gradient histogram divides 360 ° into 36 histograms on average, and each histogram represents the range of 10 °. The direction with the largest histogram peak is selected as the main direction of the feature point, and the direction with the histogram peak greater than 80% of the maximum peak is reserved as the auxiliary direction of the feature point.

3.3. Determination of principal direction of feature points

SIFT algorithm takes a lot of time to establish the feature descriptor of feature points. In this paper, the 128 dimensional feature vector is simplified to 64 as the feature vector, and the four directions of 0, 1 / 2 π, π and 3 / 2 π are selected to describe the feature points. The four directions of 1 / 4 π, 3 / 4 π, 5 / 4 π and 7 / 4 π in SIFT algorithm can be projected in the four directions of 0, 1 / 2 π, π and 3 / 2 π, which can not only reduce the dimension of feature vector, but also represent the characteristics of feature points in 360 ° direction. As shown in the figure 5:

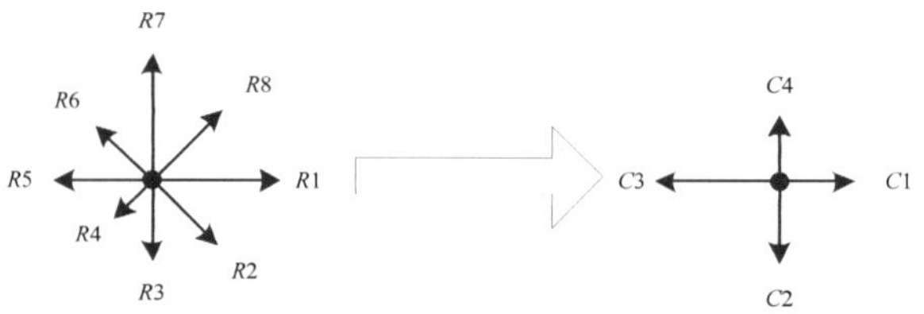

Figure 5. Feature descriptor dimensionality reduction graph

The description of the sub region in the neighborhood of each feature point is changed from 8 directions to 4 directions. This method not only reduces the running time of the algorithm, but also retains the rotation invariance of the feature points, which can better realize the mosaic of panoramic images.

4. Feature point matching and registration

In this paper, an algorithm combining average dimensionality reduction with BBF (best bin first) is proposed. The feature points are roughly matched by this algorithm, and then the transformation matrix of the image to be spliced is obtained by RANSAC algorithm, so as to obtain the panoramic image through geometric transformation.

4.1. The average dimensionality reduction method and BBF algorithm in this paper

BBF algorithm is an improvement on the k-dimension tree nearest neighbor search algorithm. It is based on KNN algorithm , and is mainly used for the search of key data in multi-dimensional space. However, the dimension of the feature descriptor of the Improved SIFT algorithm in this paper is as high as 64 dimensions. When processing high-dimensional data, the performance of BBF algorithm will be seriously degraded due to backtracking search, so this paper first reduces the dimension of the feature descriptor.

In this paper, an average dimensionality reduction method is adopted. Let the feature point descriptor be represented by matrix A of M * n (n is an even number). M is the number of feature points and N is the dimension of feature points. In this paper, n = 64, and each feature point has 64 dimensions. After the dimensionality reduction of the feature descriptor, the BBF algorithm is used to roughly match the feature points of the stitched image, eliminate the wrong matching points and improve the registration accuracy.

The average of odd and even dimensions of each feature point is obtained as the new first dimensional feature descriptor of the feature point, that is, the feature descriptor after dimension reduction can be expressed as:

$$A_{m*k}=\frac{A_{m*i}+A_{m*(i+1)}}{2},\ \text{i is 1, 3, 5...n-1, k=(i+1)/2} \tag{11}$$

4.2. RANSAC obtains the transformation matrix

In this paper, RANSAC algorithm is used to further optimize the filtered feature point pairs. Through this algorithm, the transformation matrix of the feature point pairs of the two images can be obtained, so as to transform and splice the images with the transformation matrix to obtain the panoramic image.

Randomly select three pairs of feature points x of two images to be spliced $X_1=(x_1,y_1)^T$、$X_2=(x_2,y_2)^T$、$X_3=(x_3,y_3)^T$ and $X_1'=(x_1',y_1')^T$、$X_2'=(x_2',y_2')^T$、$X_3'=(x_3',y_3')^T$, then the feature point pair meets:

$$\begin{bmatrix} x_1 & x_2 & x_3 \\ y_1 & y_2 & y_3 \end{bmatrix}=H\begin{bmatrix} x_1' & x_2' & x_3' \\ y_1' & y_2' & y_3' \\ 1 & 1 & 1 \end{bmatrix} \tag{12}$$

$$H=\begin{bmatrix} a_{11} & b_{12} & c_{13} \\ a_{21} & b_{22} & c_{23} \end{bmatrix} \tag{13}$$

Where, H is the affine transformation matrix of the image, a_{11}、a_{21}、b_{12}、b_{22} is the image rotation and scale transformation parameter, c_{13}、c_{13} is the offset parameter in the horizontal and vertical directions of the image.

5. Result and analysis

5.1. Feasibility verification of image segmentation

In this paper, two images to be spliced are selected for experimental analysis. The original size of the images is 574 * 324. According to the blocking method proposed in this paper, the spliced image is processed in blocks. The blocking results are shown in Fig. 4. The blocks a, B, C and D of the image are extracted for verification test.

Figure 6. Image to be spliced 1

Figure 7. Image to be spliced 2

Table 1. NMI values between block images

	A'	B'	C'	D'
A	0.170500	0.093356696	0.082555360	0.07765363
B	0.3124302	0.1160261831	0.128396	0.097717
C	0.11396	0.0522	0.137880508	0.08615464410
D	0.10215183	0.069312	0.108013	0.098617

Mutual information is used to measure the similarity of images. The greater the normalized mutual information NMI value, the higher the similarity of images. As can be seen from table 1, block B in Fig. 6 and block A' in Fig. 7 have the largest NMI and have the highest similarity. The panoramic image is spliced by extracting the feature points of the two blocks.

5.2. Data dimensionality reduction efficiency analysis

This algorithm reduces the feature descriptor dimension of the traditional SIFT algorithm, tests 13 images, and obtains the time distribution of this algorithm and the traditional algorithm in Fig. 8.

For a single image, the time used by this algorithm to extract feature points is much less than that of the traditional SIFT algorithm, which is about 11% of that of the traditional algorithm.

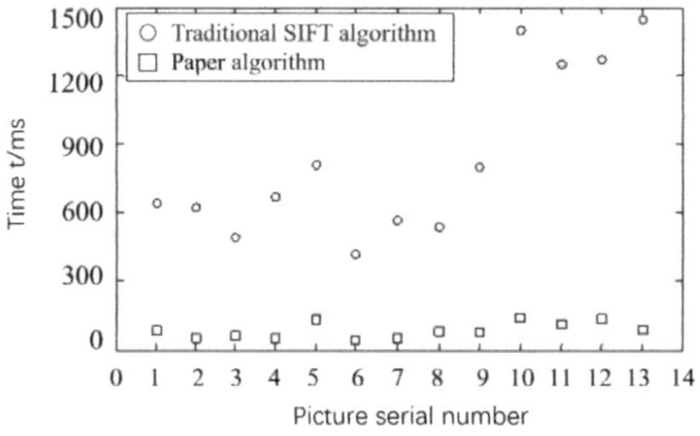

Figure 8. Time curve used to extract feature points

5.3. Registration efficiency comparison

In order to further reduce the time of panoramic image mosaic, the feature descriptor is reduced to 32 dimensions in the image registration stage, and 5 images to be spliced in different scenes are selected for comparative analysis.

Table 2. Comparison of two algorithms

Picture number	Number of feature points to be spliced		Matching time		Number of correct matches	
	Figure 1 to be spliced	Figure 2 to be spliced	Paper algorithm (reduced to 32 dimensions)	Traditional SIFT algorithm	Paper algorithm 128 → 32	Traditional SIFT algorithm
1	1006	959	0.089	0.313	258	300
2	757	1136	0.077	0.285	98	122
3	763	958	0.067	0.242	128	143
4	997	1225	0.105	0.388	79	88
5	2101	1967	0.370	1.442	40	60

It can be seen from table 2 that the feature point pairs successfully registered by the dimensionality reduction algorithm in this paper are about 82.46% of that of the traditional algorithm, but the registration time is only 27.17% of that of the traditional algorithm, which greatly reduces the time cost. The registration results of the traditional SIFT algorithm are shown in Figure 9, and a total of 51 pairs of feature points are successfully registered. The registration results of the algorithm in this paper are shown in Figure 11, and a total of 42 pairs of feature points are successfully registered.

Figure 9. Feature points of traditional BIFT algorithm

Figure 10. Feature point of

5.4. Panoramic image mosaic effect
Two images with the same scene are experimentally analyzed by using this algorithm and the traditional SIFT algorithm, and figures 11 and 12 are obtained. There is no obvious difference in the image quality of the two algorithms, and there is an obvious transition stage at the splicing.

Figure 11. Paper algorithm

Figure 12. Traditional SIFT algorithm

6. Conclusion
Aiming at the problems of large amount of calculation in extracting image feature points and high dimension of feature point registration in panoramic image mosaic, this paper proposes an improved panoramic image mosaic algorithm, which blocks the image to be spliced, extracts the feature points of the image block with the largest coincidence degree, reduces the dimension of the feature points before registration. This method avoids the problems of slow speed and high dimension of the traditional SIFT algorithm, and reduces the time of the algorithm. Experimental results show that the proposed algorithm reduces the registration time by about 73% on the premise that the feature point pair of successful registration is 82.46% of that of the traditional algorithm. At the same time, there is no obvious difference between the panoramic image stitched by this algorithm and the panoramic image stitched by traditional SIFT algorithm.

References
[1] Li, Y.F., Li G.Z., Gu, S.H. (2016) Image mosaic algorithm based on region segmentation and scale invariant feature transformation. J. Optics and Precision Engineering., 5: 1197-1205.
[2] Xiao, M., Bao, Y.L., Yan, Z.X. (2015) Overview of image registration methods based on point features. J. Acta Armamentarii., S2.
[3] Barbara, Z., Jan F. (2003) Image registration methods: A Survey. J. Image and Vision Computing., 11: 977-1000.
[4] Lowe, D.G. (2004) Distinctive Image Features from Scale-Invariant Keypoints. J. International Journal of Computer Vision., 60(2): 91---110.
[5] Li, J., Duan, P., Zhang, C. (2018) Video panoramic mosaic method under image block matching. J. Journal of applied basic and Engineering Sciences., 26(4): 697-708.

[6] Zhao, Y., Chen. Y., Wang, S.G. (2017) Fast sift image mosaic combined with projection error correction. J. Optical precision engineering., 25(6): 1645-1651.
[7] Li, L., Gu, G.H., Qian, W.X. (2013) Infrared sequence image mosaic based on feature point and Poisson fusion. J. Infrared and laser engineering., 42(9): 2584-2588.
[8] Li, D., Sun, H.T, Wang, H.L. (2015) An Improved SIFT image stereo matching algorithm. J. Journal of Southwest Jiao Tong University., 50(03): 490-496.

Comparative analysis of dynamic state recognition between permutation entropy and weighted permutation entropy

Wenxiang Luo[1,2], Li Wan[2,*] and Hui Liu[2]

[1]Software Engineering Institute of Guangzhou, Guangzhou City, Guangdong Province, 510990, China

[2]Guangzhou University, Guangzhou City, Guangdong Province, 510006, China

*Corresponding author's e-mail: wanli@gzhu.edu.cn

Abstract. Permutation entropy (PE) and weighted permutation entropy (WPE) are indexes of the complex system. In this paper, the nonlinear sequence of two chaotic states is simulated by Logistic mapping, and the effectiveness of the PE and WPE algorithms in chaotic state recognition is compared and analyzed with different sliding step sizes. The results show that the value of WPE changes more obviously than PE in different chaotic states, and the recognition effect of WPE is better than PE in dynamic states, which can provide theoretical support for further analyzing the complexity characteristics of actual data.

1. Introduction

Entropy is a measure of system complexity, which can be used for time series dynamic state recognition. After the establishment of information theory, the theory and application of entropy have been developed rapidly. In recent years, new entropies, such as approximate entropy [1], sample entropy [2], and permutation entropy [3], have been proposed based on information entropy. Weighted permutation entropy [4] is an improved method proposed by Fadlallah in 2013 for the calculation of permutation entropy without considering the amplitude information in time series, which has good anti-noise and anti-interference ability. The WPE algorithm is simple and operable, and can effectively magnify the small changes of time series. It is widely used in signal mutation detection and random signal analysis, which has achieved good application effects in mechanical fault diagnosis [5], medicine [6], and other fields [7-10].

This paper uses Logistic mapping to simulate chaotic sequences to test the effectiveness of the permutation entropy and weighted permutation entropy algorithms in identifying different dynamics, providing theoretical support for further complexity analysis of actual data.

2. Methods

2.1 permutation entropy

Given a time series $X = \{x(n), n = 1, 2, \cdots, N\}$, reconstructed the phase space of X and obtained the matrix $\boldsymbol{X}_K$ is in the following:

$$\begin{bmatrix} x(1) & x(1+\tau) & \cdots & x(1+(m-1)\tau) \\ \vdots & \vdots & & \vdots \\ x(i) & x(i+\tau) & \cdots & x(i+(m-1)\tau) \\ \vdots & \vdots & & \vdots \\ x(K) & x(K+\tau) & \cdots & x(K+(m-1)\tau) \end{bmatrix}, i = 1, 2, \cdots, K, \tag{1}$$

Where $K = N - (m - 1)\tau$, m is the embedding dimension, and τ is the delay time. Rearranges $X_i = (x(i), x(i + \tau), \cdots, x(i + (m - 1)\tau)$ in ascending order can get

$$x(i + (j_1(i) - 1)\tau) \leq x(i + (j_2(i) - 1)\tau) \leq \cdots \leq x(i + (j_m(i) - 1)\tau), \tag{2}$$

where $i = 1, 2, \cdots, K$, $\pi_i = (j_1(i), j_2(i), \cdots, j_m(i))$ is some permutation of $(1, 2, \cdots, m)$ which represents the index of the column of the elements in X_i. If the elements are equal, they are sorted by the size of the element index. Obviously, it has at most $m!$ different arrangement of species in m elements. For instance, Given a time series {7, 6, 5, 8, 6}, when $m = 3$ and $\tau = 1$, the first reconstructed component is (7, 6, 5). The index of (7, 6, 5) is (2, 1, 0) since $x_{t+2} < x_{t+1} < x_t$, Similarly, the index of the second reconstructed component (6, 5, 8) is (2, 1, 0), and the index of the third reconstructed component (5, 8, 6) is (0, 2, 1) [11].

Counting all the permutations of π_i, the probability of each permutation is

$$p(\pi_r) = \frac{f(\pi_r)}{K}, \quad K = N - (m-1)\tau, \tag{3}$$

Where $0 < r \leq m!$, $0 < f(\pi_r) \leq N - (m - 1)\tau$. The permutation entropy (PE) is defined as:

$$\mathrm{PE}(m) = -\sum_{r=1}^{m!} p(\pi_r)\log(p(\pi_r)), \tag{4}$$

where log is with base 2. PE(m) gets the maximum value log($m!$) when $p(\pi_r) = 1/(m!)$. In practical applications, log($m!$) is used to normalize PE(m), that is:

$$0 \leq \mathrm{PE}(m) = \mathrm{PE}(m)/\log(m!) \leq 1. \tag{5}$$

The value of PE(m) is consistent with the randomness of the time series. The larger the value of PE(m) is, the higher the randomness of the time series is.

2.2 weighted permutation entropy

The permutation entropy algorithm does not consider the specific value of the time series but shows its complexity by examining the permutation pattern of the time series. This algorithm is simple and easy to understand. In the calculation of PE, because the amplitude information in the time series is not considered, the amplitude difference between the sequences with the same permutation is ignored, which will inevitably lead to the reconstructed components with different amplitude to obtain the same PE(m) value. To solve this problem, Fadlallah (2013) proposed an improved algorithm of permutation entropy --weighted permutation entropy (WPE). WPE considers the amplitude information of time series when calculating the probability of a permutation. In order to clarify the difference with PE, formula (3) in the PE algorithm is rewritten as follows:

$$p(\pi_r) = \frac{\sum_{i=1}^{K} 1_{u:\text{type}(u)=\pi_r}(X_i)}{\sum_{i=1}^{K} 1_{u:\text{type}(u)\in\Pi}(X_i)}, \tag{6}$$

where $1_u(v) = \begin{cases} 1, & v \in u \\ 0, & v \notin u \end{cases}$, $\Pi = \{\pi_r\}_{r=1}^{m!}$.

The probability of each permutation pattern in the WPE algorithm is defined as:

$$p(w\pi_r)=\frac{\sum_{i=1}^{K}1_{u:\text{type}(u)=\pi_r}(X_i)w_r}{\sum_{i=1}^{K}1_{u:\text{type}(u)\in\Pi}(X_i)w_r}, \quad (7)$$

where w_r is the weight of the reconstructed component X_i, which is represented by the variance of X_i:

$$w_r=\frac{1}{m}\sum_{q=1}^{m}\left[x(i+(q-1)\tau)-\overline{X}_i\right]^2, \quad (8)$$

$\overline{X}_i$ is the mean of the reconstructed component X_i, which is defined by

$$\overline{X}_i=\frac{1}{m}\sum_{q=1}^{m}x(i+(q-1)\tau). \quad (9)$$

WPE is an improved algorithm of PE, which is defined as:

$$\text{WPE}(m)=-\sum_{r=1}^{m!}p(w\pi_r)\log(p(w\pi_r)), \quad (10)$$

where log is with base 2. It can be obtained the normalized form by

$$0\le \text{WPE}(m)=\text{WPE}(m)/\log(m!)\le 1. \quad (11)$$

Weighted permutation entropy considers the amplitude information of reconstructed components when calculating the probability of a certain permutation, which can fully reflect the detailed characteristics of time series, and has a good effect on the applications that need to describe the detailed characteristics of time series.

3. Dynamic state recognition comparison

A chaotic ideal time series (IS) with length 1000 is simulated by Logistic mapping, and its iterative equation is as follows:

$$x_{n+1}=\lambda x_n(1-x_n),\quad x_n\in[0,1]. \quad (12)$$

λ represents the growth rate of the Pest Model where $\lambda\in[0, 4]$. For $x_0 = 0.7$, IS consists of two parts, and the parameters of the 500 data before and after are $\lambda = 3.6$ and $\lambda = 3.7$ respectively, both of which are in a chaotic state. The evolution curve of IS is shown in figure 1.

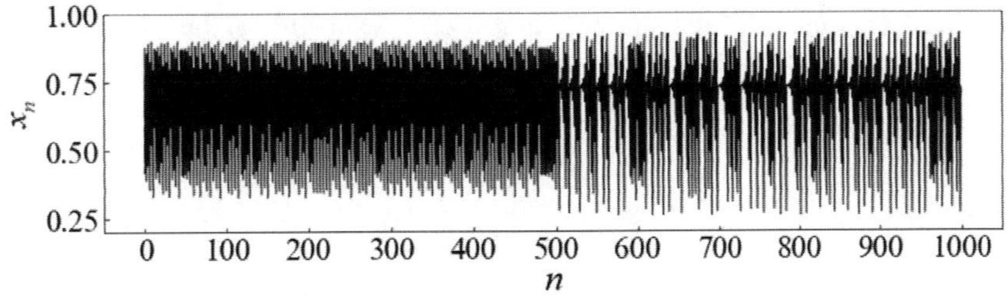

Figure 1. Chaotic ideal time series IS

According to the structure of IS, the evolution of the system changes from a low growth rate of 3.6 to a high growth rate of 3.7 at $n = 501$, and the complexity of the first 500 points of IS sequence is lower than that of the last 500 points.

A sliding window with length $M = 150$ is used to select the data of IS to form a subsequence. The PE and WPE algorithms are used to calculate the entropy value of the subsequence for $m = 3$, $\tau = 2$ respectively. Selecting a new subsequence with the step size of $S = 10$ and $S = 20$, respectively, and calculating the corresponding entropy value of the new subsequence until the end of IS, which can be obtained a sequence of entropy values that changes gradually with the step size, as shown in figure 2.

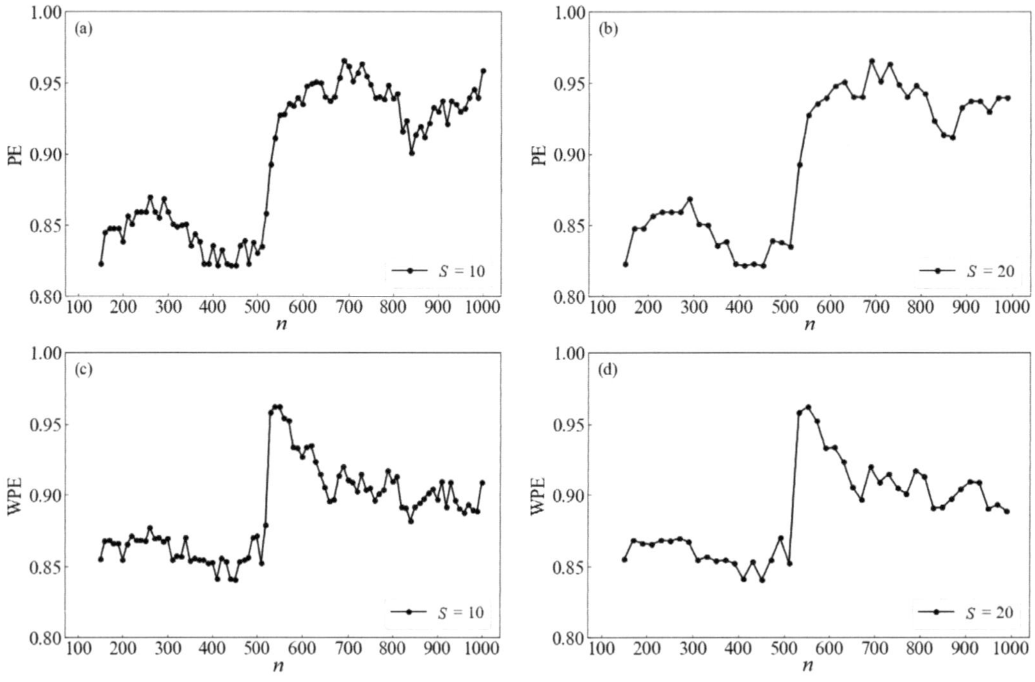

(a) PE sequence, $S = 10$ (b) PE sequence, $S = 20$
(c) WPE sequence, $S = 10$ (d) WPE sequence, $S = 20$
Figure 2. Comparison of PE and WPE of IS

As shown in figure 2, the entropy value calculated by the two algorithms presents two different stable states before and after $n = 500$, indicating that IS belongs to different chaotic states. According to the process of sliding data, the edge of the two kinds of chaotic data should be at the beginning of the entropy rising interval, about at $n = 501$, which is consistent with the structure of IS. The entropy increase interval obtained by the WPE algorithm is steeper than that obtained by the PE algorithm with better differentiation. On the other hand, according to the analysis of the sliding data process, when the two kinds of chaotic data occupy half of the sliding window size respectively, the entropy value reaches the maximum and then decreases to a similar level with the latter kind of data. The entropy obtained by WPE decreases after reaching the maximum value, which is consistent with the process of sliding data. However, the entropy obtained by PE does not reach the maximum value in the interval (500, 650), and then decreases to a certain extent. Therefore, WPE performs better than PE in dynamic recognition of chaotic time series.

4. Conclusion

The nonlinear sequences of different chaotic states are simulated by Logistic mapping, and the effectiveness of PE and WPE in chaotic state recognition is compared and analyzed with different sliding step sizes. The results show that the change of entropy value of WPE is more evident than that of PE in different chaotic states, indicating that WPE can effectively magnify the slight change of time series. Furthermore, WPE is better than PE in dynamic state identification, which can provide theoretical support for further analyzing the complexity characteristics of actual data.

Acknowledgments

This work was financially supported by the National Natural Science Foundation of China (41872246); Scientific Research Project of the Software Engineering Institute of Guangzhou (ky201909).

References

[1] Pincus, S. M. (1991) Approximate entropy as a measure of system complexity. Proceedings of the National Academy of Sciences, 88: 2297-2301.

[2] Richman, J. S., Moorman, J. R. (2000) Physiological time-series analysis using approximate entropy and sample entropy. American Journal of Physiology-Heart and Circulatory Physiology, 278: 2039-2049.

[3] Bandt, C., Pompe, B. (2002) Permutation entropy: A natural complexity measure for time series. Physical Review Letters, 88: 174102.

[4] Fadlallah, B., Chen, B. D., Keil, A., et al. (2013) Weighted-permutation entropy: A complexity measure for time series incorporating amplitude information. Physical Review E, 87: 022911.

[5] Ding, J. X., Wang, Z. Y., Yao, L. G., et al. (2021) Rolling bearing fault diagnosis based on GCMWPE and parameter optimization SVM. China Mechanical Engineering, 32: 147-155.

[6] Lee, C. H., Chen, S. H., Jiang, B. C., et al. (2020) Estimating postural stability using improved permutation entropy via TUG accelerometer data for community-dwelling elderly people. Entropy, 20, 1097.

[7] Chen S. J., Shang P. J., Wu Y. (2018) Multivariate multiscale fractional order weighted permutation entropy of nonlinear time series. Physica A: Statistical Mechanics and its Applications, 515: 217-231.

[8] Wu P., Guo L. L., Duan Y. Y., et al. (2019) Control loop performance monitoring based on weighted permutation entropy and control charts. The Canadian Journal of Chemical Engineering, 97 : 1488-1495.

[9] He C., Wu T., Liu C. C., et al. (2020) A novel method of composite multiscale weighted permutation entropy and machine learning for fault complex system fault diagnosis. Measurement, 158, 107748.

[10] Li, Y. X., Geng, B., Jiao, S. B. (2021) Refined composite multi-scale reverse weighted permutation entropy and its applications in ship-radiated noise. Entropy, 23, 476.

[11] Luo, W. X., Wan, L., Lai, S. M. (2018) Comparison of sequence mutational detection of two kinds of moving permutation entropy. Journal of Xiamen University of Technology, 26: 91-96.

Fault-tolerant architecture for Cache : Summaries, Assessments and Trends

Xuru Wang[1,2], Xin Gao[1,2], Zongnan Liang[2,3], Jiawei Nian[2,3] and Hongjin Liu[1,2*]

[1] Beijing Institute of Control Engineering, Beijing ,100090, China

[2] Beijing SunWise Space Technology Ltd, Beijing 100090, China

[3] School of computer science and technology, Xidian University, Xi'an, Shanxi, 710071, China.

[*]Corresponding author's e-mail: wangxr0321@cast.cn

Abstract: Fault-tolerant design of cache is a key aspect of highly reliable processor design. In this paper, based on the key metrics in Cache architecture design: reliability, power consumption, latency and area, we divided the related research into two categories: one is to maximize reliability with guaranteed latency, power consumption and area, the other is to minimize latency, power consumption and area loss while ensuring fault tolerance reliability. Based on the classification, by analyzing different studies of Data and Tag in Cache, this paper gives the characteristics of these methods and the future development trend.

1. Introduction

Accuracy of data and instructions is an important guarantee in computer processes However, in some specific environments such as space environment, the correctness of processor execution is seriously harmed because of the exposure to extreme conditions like radiation [5][8][12]. In particular, single and multiple bit upsets can cause transient or permanent changes in data bits, which can lead to catastrophic failures. In the past decades, Single Event Upset (SEU) was considered as the main influencing factor for Cache reliability, but with the gradual reduction of the feature size of semiconductor devices, the chance of Multiple-Bits Upset (MBU) is increasing and needs the same attention in the design [15][17][21]. Therefore, in processor design, the reliability design of the Cache, the key part in data processing, should be mainly concerned. In the design of fault-tolerant computer architecture, four metrics should be focused: reliability, latency, power consumption, and area. Therefore, recent studies can be divided into the following two categories according to the indicators of concern: the first one is to maximize fault tolerance with guaranteed latency, power consumption and area, the other is to minimize latency, power consumption and area loss while ensuring fault tolerance reliability [22]. In this paper, we analyze and evaluate these two types of research, summarize their technical points, and give the development trend of fault-tolerant technology for Cache architecture.

The data in Cache is divided into Data and Tag, where the differences are the way they are stored. Therefore, the related research can be mainly divided into fault-tolerant protection for Data and fault-tolerant protection for Tag. This paper is organized as follows: The section2 provides an overview of the basic fault-tolerance methods and their principles. In Section 3 and Section 4, the relevant researches on the protection of Data and Tag is analyzed, respectively. Section 4 summarizes the whole paper and gives the development trend of Cache fault-tolerance protection technology.

2. Methods and basic principles

The main form of fault-tolerant design for Cache architecture in common use today is redundant design, which is divided into three types: temporal redundancy, spatial redundancy, and information redundancy. Temporal redundancy means using the same hardware to repeat instruction operations to ensure reliability. However, this is a sacrifice of time for reliability, and the execution speed of each instruction will become slower, which is contradictory to the goal of pursuing high performance. Thus, temporal redundancy is rarely used in design. Spatial redundancy, on the other hand, uses multiple backups on hardware circuits, and instructions are executed in these circuits simultaneously to ensure reliability by comparison of results, but it will cause an increase in power consumption and area. Information redundancy uses additional EDC (Error Detection Code) and ECC (Error Correction Code) to protect the data, and the area power consumption is relatively smaller to spatial redundancy. However, the delay problem could arise due to the extra coding and decoding logic. The following are some classic and commonly used fault tolerance methods:

2.1. TMR(Triple Modular Redundancy)

Triple Modular Redundancy (TMR), for short, is one of the most commonly used spatial redundancy. Three same modules will perform the instruction simultaneously, taking the majority through voting system as correct output, which is often referred to as triple take two. Therefore, the correct output of system can be guaranteed unless the same error occurs in at least two modules., the error of the faulty module can be masked and the correct output of the system can be guaranteed. The structure is illustrated in figure.1. Since the three modules are independent of each other, the probability that two modules have errors simultaneously is quite small., so the reliability of the system can be greatly improved.

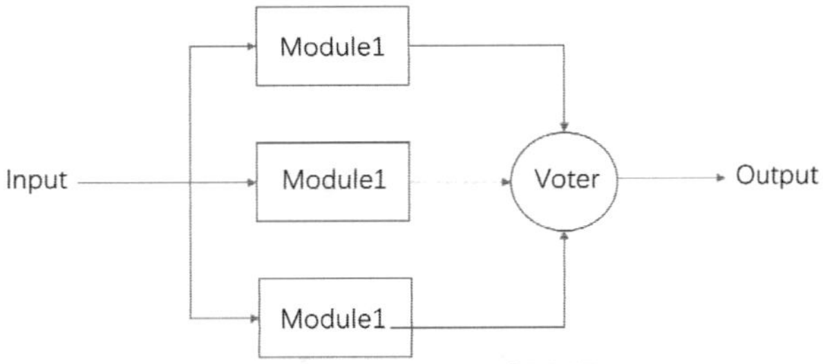

Figure1. The structure of TMR

The reliability of each module is assumed to be equal and is denoted by Rm. The reliability of the table decider is denoted by Rv. Then the system reliability is:

$$R = (3R_m^2 - 2R_M^3) \cdot R_v$$

The key to this structure is how to implement the design of the three-take two, which is the voting machine.

2.2. Parity Check

Parity checking is a traditional method of information redundancy for fault tolerance [10].

There are two types of parity bits: even parity bits and odd parity bits. If the number of 1 bit in a given set of data bits is odd, then the even parity bit is set to 1, so that the total number of 1 bit is even. If the number of 1's in a given set of data bits is even, then the odd parity bit is set to 1, making the total number of 1's odd. Even parity is actually a special case of cyclic redundancy, where a 1-bit CRC is obtained by the polynomial $x + 1$.

If an odd number of data bits including the parity bit are flipped, the parity bit will be changed, indicating an error during transmission. Therefore, the parity bit is an error detection code, but since there is no way to determine which bits are flipped, it cannot be error corrected. Once an error is detected, the received data have to be discarded and then transmission reruns.

2.3. Hamming Code

Hamming codes can be used to detect and correct single bit errors when a processor stores or moves data that may generate data bit errors. Because of the simplicity of Hamming codes, they are widely used in storage units [1][7].

Hamming code utilizes the concept of parity bits by adding some bits after the data bits, the validity of the data can be verified. By using more than one parity bit, Hamming Code not only verifies the validity of the data, but also indicates the error location in case of data errors.

The purpose of a Hamming code is to correct a single bit of error. It can be expressed as Hamming(n,r), where the information code has n bits and the Hamming code has r bits, the total code length is $n + r$ bits. In order to detect an error in one of the $n + r$ bits, the Hamming code must be able to represent at least $n + r + 1$ states, where $n + r$ means that there is an error in one of the $n + r$ bits, and the extra one means that the whole code is correct. Then the length of the Hamming code needs to satisfy the following relationship:

$$2^r > n + 1 + r$$

The encoding algorithm for the Hamming code is as follows.

1. fill the new code as $2^{(r-1)}, (r \geq 0)$ bits with zeros (i.e. parity bits)
2. fill the remaining bits of the new code with the source code in the original order
3. encode the rth check digit starts from then the $2^{(r-1)}$ bit of the new code, and for every $2^{(r-1)}$ bits of the iso-or calculated, jump $2^{(r-1)}$ bits, and then calculate the next set of 2^(k - 1) bits of the iso-or, fill in $2^{(r-1)}$ bits. The rth check digit is then computed from the $2^{(r-1)}, (r = 1,2,3 \ldots)$th bit of the new code. Each time it needs to calculate the$2^{(r-1)}$ iso-or, jump $2^{(r-1)}$ bits, then calculate the next set of $2^{(r-1)}$ bits iso-or, fill in the $2^{(r-1)}$position.

If the checksum bits generated by the receiver and the received checksum bits are the same, the data is correct, otherwise they are different indicating an error.

3. Fault-Tolerant Protection of Cache Data

Since the method of spatial redundancy is mainly designed for fault tolerance through the method of backup, which is relatively simple and has rarely studies on. Therefore, the research on the fault-tolerant architecture of Cache data mainly focuses on information redundancy.

When designing information redundant architecture, we mainly focus on four metrics: reliability, power consumption, area and latency, but these four metrics interact with each other and need to be considered in a compromise. Therefore, the research related to information redundancy can be divided into two main classes:

1. Maximize fault tolerance reliability with guaranteed latency, power consumption and area;
2. Minimize latency, power consumption and area loss while ensuring fault tolerance reliability.

In the first group of studies, researchers primarily focused on reliability improvements, for example, improving the data detection and correction range of EDAC codes. This will inevitably result in complex logic circuits, increasing power consumption. The second group of studies primarily focuses on optimizing the overall performance of the processor after adding a fault-tolerant design. In these studies, the loss of power and latency is alleviated by adjusting the structure of the Cache read/write path and the read/write and replacement strategies.

3.1. Maximize fault tolerance reliability with guaranteed latency, power consumption and area

The Common approaches to information redundancy are Hamming code, RM code [2] (Reed Muller code), RS code [3] (Reed-Solomon code), BCH code [4] (Bose-Chaudhuri-Hocquenghem), etc.

Hamming code is simple in logic [1], easy to implement, and is the most commonly used EDAC code for Cache information redundancy. However, its weak error correction capability (only one bit error correction) cannot satisfy the growing reliability requirements, because the size of devices decreases and the probability of MBU gradually increases with the rapid development of semiconductors. RS, RM, BCH codes have more complex logic and higher error correction and detection capability, for example, RM (2, 5) can achieve three error correction, but the complex circuit logic will lead to more delay and power loss. In a word, although the implementation of Hamming code is simple, its reliability is not high; while RS, RM, BCH code is more reliable, but its coding and decoding logic is complex. Therefore, in this part, relevant researches are mainly divided into two kinds: one is to optimize the coding and decoding logic to improve the error tolerance with the same amount of redundancy; the other is to use basic ECC and EDC (such as Hamming code, parity check code) to improve the error tolerance by increasing the amount of redundancy.

3.1.1. Optimize code logic

This type of research mainly adjusts and optimizes the encoding and decoding logic of parity and Hamming codes to improve error correction capability. Some researchers added one parity check code on the basis of Hamming code and ensured its different forms by modifying the same adjacent correction factors to achieve error location, which could realize detecting two and three adjacent errors [5]. [6] added the unit matrix optimized coding and decoding matrix on this basis. It could achieve two-bit error detection as well as multi-digit adjacent error detection while the number of codes remains unchanged. [7] used the measure of solving the Boolean-function satisfaction condition problem (SAT) to design the check matrix H, which can detect 2-3-bit errors and correcting one-bit errors when the amount of redundancy is the same as that of ordinary Hamming codes. In [8], the data to be protected were divided into information bits and key bits and decoded separately, while the decoding H matrix was optimized to merge the unused row data. Two-bit error correction can be realized with the same amount of redundancy as the Hamming code.

3.1.2. Increase the amount of redundancy

In addition to optimizing the codec logic, some studies have improved the error tolerance by increasing the amount of EDAC redundancy. For error detection capability, [9] proposed to group the parity check matrix, part of which was used to detect the type of error occurred (number of error bits, whether it is an adjacent error, etc.), and the other part was used to search the location of the error, causing up to 7 bits of error could be detected. [10] used OLS code to replace the parity check code, and also grouped the code into two parts for error type and error location separately. [11] proposed to divide the multi-bit data into several groups and group them for Hamming code coding, which increases the error correction and error detection capability of the same protected data by reducing the coding granularity, but causes an increase in the amount of redundancy. To detect more errors, [12] proposed the method of PBD (per bit detection), which interleaves a data into groups (i.e., non-adjacent data into one group) and assigns error detection codes to each group, improving the error detection efficiency. In order to improve the error correction capability, [13] proposed Matrix code, which treated the data to be protected as a matrix form, and used Hamming code to protect the row data and parity check code to protect the column data. It can realize the detection of three errors and correction of three errors. On this basis, CLC [14] and CLC-E [15] used parity check code plus Hamming code for the data in the same matrix form. Column data (including error correction code) were protected by parity code, which could detected more errors and correct up to 5-bit errors, and in addition, [15] proposed that an additional error correction process can be performed to achieve higher error correction capability. [16] adopted the same idea and proposes the PCoSA approach, which protected column data (including error correction code) also by Hamming code and parity, achieving 7-bit error correction but the area, power consumption, and delay overhead are large. Using the codec idea of 3 bit code, [17] continued the data protection form of Matrix code, which can achieve 6-bit or 12-bit error correction according to the form of data matrix.

Table 1 Comparison of metrics of some fault-tolerant methods (16bit)

Method	Reliability	Area (μm^2)	Delay (ns)	Power consumption (mw)	Number of codes
RM	Correction: 3	4816	2.86	0.774	16
Matrix	Dection:3 Correction:3	1388	1.16	0.06	16
CLC	Dection:4 Correction:5	1786	1.68	0.100	24
CLC-E	Dection:4 Correction:7	3752	1.40	0.913	24
PCoSA	Dection:n Correction:7	10579	2.34	0.891	48
PBD	Dection:7 Correction:1	569	0.73	0.041	4
[17](2)	Correction:6			0.79	24

Since these studies used different evaluation criteria, only some of the studies with the same evaluation criteria were selected for comparative analysis in this paper. These methods (for 16bit data protection) are compared here for reliability, area, latency, and power consumption in table 1. It can be seen that PCoSA performs better in terms of reliability, but has the largest area, power consumption and latency loss compared with other approaches; CLC-E adds two error corrections relative to CLC, which can correct more errors but consumes more power than CLC; and PBD can only perform one bit error correction, but has the smallest redundancy and the lowest area, power and latency metrics. This also shows that improving the error detection capability has less power consumption, latency penalty than improving the error correction capability.

3.2 Minimize latency, power consumption and area loss while ensuring fault tolerance reliability

Area, latency and power consumption are mainly influenced by the following factors:

- The coding redundancy in memory space. More redundancy will result in increased storage usage.
- The complex encoding and decoding logic. It is generally located on the critical path of the Load/Store instructions. This causes increased latency in instruction execution
- Encoding and decoding require sophisticated circuit support. Frequent coding and decoding operations will cause power cost.

Therefore, to resolve the above problems, a large amount of research has been conducted to achieve optimization of power consumption, area and latency loss mainly by optimizing the structure of Cache, such as simplifying the read/write path logic and storage strategy, while guaranteeing the fault tolerance reliability (codec logic).

To address the storage problem of EDAC, [18] proposed to separate ECC code from EDC code. EDC was stored in Cache for error detection, while ECC was stored in main memory for error correction when an error occurred, FIFO policy was applied to store ECC. Continuing the idea of separating ECC and EDC, [19] allocated storage blocks for ECC in Cache along with data, which could be stored according to coded data to determine the location and state, thus reducing the area overhead.

Besides, to optimize the latency problem caused by decoding on the critical path, [20][21] took advantage of the parallel reading of data in the Cache and decode before the Tag comparison selection data, which saved the wating time for Tag comparison of data and also enhances the detecting probability of useless data. PSC-Cache [22] stepped further by placing both encoding and decoding logic in Tag comparison before selecting data, which could encode and decode. multiple data simultaneously, multiplex the coding and decoding logic, and reduce the overhead. EUC [23] combined the above two approaches, separating ECC and EDC, while increasing the coding granularity as well as the amount of read data decoding data, and classifying them into three coding modes according to the different granularity of coding: minimum area, minimum latency, and highest reliability. [24] added two registers to the original Cache structure, one was to store the iso-or of the data in the Cache and the other was to store the iso-or of the dirty data replaced out, optimizing the write read logic of the Cache. [25] proposed two methods of using parity, the first was to use parity only on the read path to improve reliability while reducing power consumption, and the second was to protect the smaller the granularity of data with higher the reliability.

Table 2 Summary of metrics of some fault-tolerant methods

Method	Configuration	Power consumption* (%)	Reliability[a] (%	Area (%)
SEA Cache	L1Caches:Split I/D cache, 32 KB, 4-way, 32B line, L2Cache 1MB cache, 8-way, 64B line,	19.4	1	12.5
REAP-cache	L1Caches:Split I/D cache, 32 KB, 4-way, 64B line, L2Cache 1MB cache, 8-way, 64B line,	2.7	171×	≤1
CLEAR	L1Caches:Split I/D cache, 32 KB, 4-way, 64B line, L2Cache:256KB cache, 8-way, 64B line,	1.52	4.52 ×	≤2
EUC-MinDelay	L1Caches:Split I/D cache, 32 KB, 4-way, 64B line, L2Cache:256KB cache, 8-way, 64B line,	16.1	1	8.1
CPPC	L1 D cache 32KB, 2-way, 32 byte lines, L1 I cache 16KB, 1-way, 32 byte lines, L2 cache 1MB unified, 4-way, 32 byte lines,	7	1	

([a]Compare to the baseline)

As can be seen from the table 2 above, these methods differ distinctly in power consumption, area and reliability due to the various configurations and EDAC codes used.

There are also several works that have been designed for other storage medium. For the STT-SRAM Cache, [26] used orthogonal Latin square code (OLSC) to encode the data, and LCLL and VnC are used for read and write strategies respectively, while the storage space of ECC was optimized to reduce the latency and area. For the storage on FPGAs, [27] used Column Mux to encode the data separately, while the correction factor was multiplexed to reduce the reliability problems caused by adjacent errors.

4. Fault-Tolerant Protection of Cache Tag

The data stored in Cache is divided into data and Tag, and most of the studies focus on the fault-tolerant design of data in Cache. Since an error in the Tag comparison process will lead to fault in reading or writing data, and even affect the data in main memory, which will seriously affect the normal execution of instructions, several studies also focus on the fault-tolerant design of Tag. In addition, Tag is located in the critical read and write path of Cache, complex coding and decoding logic will bring delay problems, will affect the speed of the processor. Hence, Tag-based fault tolerance studies mainly start from simple EDAC codes or spatial redundancy.

Based on the idea of spatial redundancy, TRB [28] provided a small capacity buffer for the Cache Tag to provide a copy for the Tag of dirty rows, so as to prevent Tag comparison errors when the data was replaced. To extend the range of protectable Tag [29], the memory space in the cache-adjacent group was expanded to store Tag copies. In group-adjacent Cache, the above strategy requires index bits to locate Tag copy locations, so [30] applied this strategy in direct mapping Cache to reduce the storage Tag copy space. RAW-Tag [31] adopted TRB active replication to ensure that copies of dirty row Tag all existed and were stored in different ways of adjacent sets, using a complex replacement strategy to prevent simultaneous errors in the master copy. In addition to spatial redundancy, Tag could also be protected by simple EDAC. FastTag [32] chose to directly encode the Tag in the address and then compare it with the Tag in the Cache in terms of code distance, which reduced the overhead problem caused by repeated encoding and decoding, compared with the common one of decoding the EDAC-protected Cache Tag and then comparing it. Smartag [33], on the other hand, took advantage of the feature that the high bits of the address are basically the same, so the high bits of the Tag were designed to be shared by the ECC and the address, saving the storage space of the ECC.

Table 3 Comparison of metrics of some fault-tolerant methods in Tag

Method	Category	Power consumption[a] (%)	Correction capability	Area[a] (%)
TRB	Spatial redundancy	19.9	91.5	16.3
SimTag	Spatial redundancy	11		
RAW-Tag	Information Redundancy	6.9	97.9	
FastTag	Information Redundancy	40.8**		25.6**
SMARTag	Information Redundancy	7	99.6	

([a]Compare to the baseline)
([b]Reduction compared to baseline)

From the relevant data in the table 3, it shows that the fault tolerance study for Cache is more concerned with the protection rate of Tag data.

5. Conclusion

As a key memory unit in processor, cache plays a decisive role in processor performance. It is located near the pipeline, and data can be easily and quickly forwarded to other units in case of errors. So the reliability design of Cache is critically required for processors running in radiation environment. The problem of increasing area, power and latency due to reliable design also needs to be considered as a compromise. Therefore, the current work focuses on either improving reliability or alleviating the area, power, and latency problems. In terms of reliability, it is now possible to achieve multi-bit correction like seven-bit fault tolerance; in terms of reducing the area, power and latency, there are various pragmatic methods such as separate EDC/ECC, the coding and decoding logic simplification, and the cache read/write path optimization. Typically, these approaches are applied to embedded processors or SoCs because they are more concerned with power consumption and real-time data. However, there are still some shortcomings in these researches, such as the inability to detect autonomously, the imbalance of the four indicators, etc.

As the semiconductor technology evolves, some new storage media emerge, such as STT-RAM type Cache, storage cells on FPGAs, etc. Future fault-tolerant technologies will probably be more flexible and adaptable to different storage media.

References

[1] R. W. Hamming, "Error detecting and error correcting codes," Bell Syst. Tech. J., vol. 26, no. 2, pp. 147-160 , Apr. 1950.
[2] Tam S . Multiple Bit Error Correction[C]// IEEE Conference on Information & Communication Technologies. IEEE, 2004.
[3] S. Reed, "A class of multiple error correcting codes and decoding scheme" IRE Trans, vol. U-4, pp. 38-49, Sep 1954.
[4] DE Muller, "Application of Boolean algebra to switching circuit design and to error detection, . IRE Trans, EC-3,pp,6-12, Sep 1954.
[5] A Sánchez-Macián, Reviriego P , Maestro J A . Hamming SEC-DAED and Extended Hamming SEC-DED-TAED Codes Through Selective Shortening and Bit Placement[J]. IEEE Transactions on Device and Materials Reliability, 2012.
[6] Neale, A, Sachdev, et al. A New SEC-DED Error Correction Code Subclass for Adjacent MBU Tolerance in Embedded Memory[J]. Device & Materials Reliability IEEE Transactions on, 2013.
[7] Cui Y , Lou M , Xiao J , et al. Research and implementation of SEC-DED Hamming code algorithm[C]// Tencon IEEE Region 10 Conference. IEEE, 2014.
[8] Namba K , Lombardi F . A Single and Adjacent Error Correction Code for Fast Decoding of Critical Bits[J]. IEEE Transactions on Computers, 2018.
[9] Das A , Touba N A . Low Complexity Burst Error Correcting Codes to Correct MBUs in SRAMs[C]// The. 2018:219-224.
[10] Das A , Touba N A . Layered-ECC: A Class of Double Error Correcting Codes for High Density Memory Systems[C]// 2019 IEEE 37th VLSI Test Symposium (VTS). IEEE, 2019.
[11] Antony B . Modified Hamming Codes with Double Adjacent Error Correction along with Enhanced Adjacent Error Detection.
[12] Goerl R C , Villa P , Poehls L B , et al. An efficient EDAC approach for handling multiple bit upsets in memory array [J]. Microelectronics Reliability, 2018, 88.
[13] Argyrides, C.,Pradhan, D. K.,Kocak, T. Matrix Codes for Reliable and Cost Efficient Memory Chips[J]. IEEE Transactions on Very Large Scale Integration Systems, 2011, 19(3):420-428.
[14] Castro H , Silveira J , Coelho A , et al. A correction code for multiple cells upsets in memory devices for space applications [C]// New Circuits & Systems Conference. IEEE, 2016.
[15] Silveira F, Silveira J, et al.. An Extensible Code for Correcting Multiple Cell Upset in Memory Arrays[J]. Journal of Electronic Testing, 2018, 34:417-433.
[16] Df A , Dm A , Rg B , et al. PCoSA: A product error correction code for use in memory devices targeting space applications - ScienceDirect[J]. Integration, 2020, 74:71-80.
[17] Li J Q , Xiao L Y , Guo J , et al. Efficient Implementations of Multiple Bit Burst Error Correction for Memories [C]// 2018 14th IEEE International Conference on Solid-State and Integrated Circuit Technology (ICSICT). IEEE, 2018.
[18] Yoon D H , Erez M . Flexible cache error protection using an ECC FIFO[C]// Proceedings of the ACM/IEEE Conference on High Performance Computing, SC 2009, November 14-20, ACM, 2009.
[19] Hong J , Kim S . Smart ECC Allocation Cache Utilizing Cache Data Space[J]. IEEE Transactions on Computers, 2017, 66(2):368-374.
[20] Cheshmikhani E , Farbeh H , Asadi H . Enhancing Reliability of STT-MRAM Caches by Eliminating Read Disturbance Accumulation[C]// 2019 Design, Automation & Test in Europe Conference & amp; Exhibition (DATE). 2019.
[21] Farbeh H , Monazzah A . CLEAR: Cache Lines Error Accumulation Reduction by exploiting invisible accesses[J]. Microelectronics Journal, 2019, 90(AUG.):123-132.
[22] H Farbe h , Miremadi S G . PSP-Cache: A low-cost fault-tolerant cache memory architecture[C]// Design Automation and Test in Europe. 2014.
[23] Farbeh H , Delshadtehrani L , Kim H , et al. ECC-United Cache: Maximizing Efficiency of Error

Detection/Correction Codes in Associative Cache Memories[J]. IEEE Transactions on Computers, 2020, PP(99):1-1.
[24] [24] Manoochehri M , Annavaram M , Dubois M . CPPC: Correctable parity protected cache[C]// International Symposium on Computer Architecture. IEEE Computer Society, 2011.
[25] Ko Y , Jeyapaul R , Kim Y , et al. Guidelines to design parity protected write-back L1 data cache. IEEE, 2015.
[26] Qureshi M A , J Park, Kim S . SALE: Smartly Allocating Low-Cost Many-Bit ECC for Mitigating Read and Write Errors in STT-RAM Caches[J]. IEEE Transactions on Very Large Scale Integration (VLSI) Systems, 2020, PP(99):1-14.
[27] Rahul K , Yachareni S . Low Power and Area efficient Forward Error correction code for SRAM in FPGAs [C]// 2020 8th International Electrical Engineering Congress (iEECON). 2020.
[28] Hu J , Wang S . REPLICATING TAG ENTRIES FOR RELIABILITY ENHANCEMENT IN CACHE TAG ARRAYS:, US20140325138 [P]. 2014.
[29] Hong J , Kim J , Kim S . Exploiting Same Tag Bits to Improve the Reliability of the Cache Memories[J]. IEEE Transactions on Very Large Scale Integration Systems, 2015, 23(2):254-265.
[30] Archana M , Kashwan K R . 2016 International Conference on Recent Trends in Information Technology (ICRTIT) - An efficient architecture for improved r [J]. 2016:1-4.
[31] Farbeh H , Mozafari F , Zabihi M , et al. RAW-Tag: Replicating in Altered Cache Ways for Correcting Multiple-Bit Errors in Tag Array [J]. IEEE Transactions on Dependable and Secure Computing, 2017, PP(99):1-1.
[32] Reviriego P , Pontarelli S , Ottavi M , et al. FastTag: A Technique to Protect Cache Tags Against Soft Errors[J]. IEEE Transactions on Device & Materials Reliability, 2014, 14(3):935-937.
[33] Ghaemi S G , Ahmadpour I , Ardebili M , et al. SMARTag: Error Correction in Cache Tag Array by Exploiting Address Locality[C]// 2018:1658-1663.

Stiffness modelling of flexible support module for Large-aperture laser transmission unit

Jiabin Pan [1], Yanan Zhang [1*], Zhongtao Fu [2] and Linyong Shen [1]

[1] School of Mechatronic Engineering and Automation, Shanghai University, Shanghai 200444, China

[2] School of Mechanical and Electronic Engineering, Wuhan Institute of Technology, Wuhan 430205, China

*Corresponding author's e-mail: ynzhang@shu.edu.cn

Abstract. Large-aperture laser transmission unit (LLTU) is a device that focuses the laser beam to the center of the target, which are often designed as compliant mechanisms to achieve micro displacement adjustment. In the traditional mechanisms, they are designed as integrated micromanipulation systems, and driven by piezoelectric ceramics. However, most of these researches only focuses on motion accuracy, due to the lack of consideration of large load problems, the application is greatly limited. To this end, a flexible support module (FSM), as well as its stiffness model, was presented in this paper. Combined with finite element method (FEM) of FSM, structure size optimization was also completed, successfully solved the problems of stress concentration and load of FSM in engineering application. Moreover, a dual vision-based measurement method was introduced, to verify the stiffness model and analyze the repetitive error of FSM. From this result, the prototype enabled 5 mm and 0.007 rad of working area with average error of 0.3192 mm and -0.0036 rad. The repeatable error is within 7%, and will decreased to 4% with internal stress released in 5~15min.

1. Introduction

Large-aperture laser transmission unit (LLTU) is a device that focuses laser beams into the target centre with high precision, it is widely used in high-energy laser facilities, such as ShenGuang III (SG-III) [1] in China, National Ignition Facility (NIF) [2] in USA, and laser megajoule (LMJ) [3] in France. Generally, considering the requirements of high precision and high load, LLTU is designed as a compliant parallel mechanism (CPM). However, in most existing research cases, CPMs are designed as integrated mechanisms, and most of them only focus on motion accuracy, such as cell microinjector [4], fast steering mirror [5], micro positioning stage [6-8], which are difficult to transplant to high load compliant engineering.

Compared with integrated CPM, flexible support module (FSM) has more advantages in precision engineering applications, such as higher DOF and replaceability, which makes it more applicable for CPM research of LLTU. As the passive deformation unit of CPM, FSM usually provide the micro displacement via parallel installation. In the LLTU [9] designed by laser megajoule (LMJ) in France, the optic component was connected on both sides by elastic stems, and supported by stepping actuators. Different from LMJ, the optical frame of NIF [10] is constrained by three FSM, which were mounted on the back of the optical assembly, and axially point to the rotation centre of the optic component. The FSM developed in the NIF is a common structure, it can achieve high DOF by staggered series of multiple flexible hinges. This kind of structure has perpendicular DOF, so it is also called as flexure hook hinge. However, due to the need of confidentiality, these studies did not analyse the modelling and control process of the mechanism.

FSM were usually designed for higher degrees of freedom, such as U-shaped flexible joints [11], and spatial joint compliant module (SJCM) [12,13] developed by SIMTech, which are composed of several flexure hinges. According to different flexibility requirements of CPM, flexure hinges are designed into various shapes, including circular hinge [14], filled leaf hinge [4,15] and elliptical hinge [16]. To model these hinges, two most methods are widely used, namely finite element method (FEM) [17] and analytical modelling method [18-19], Pham et al. [20] presented an analytical stiffness modelling of CPM in six directions, and used the described model to establish the stiffness matrixes of a double linear spring and a three-degree-of freedom (DOF) translational CPM. Byung et al. [14] designed a planar 3-DOF parallel mechanism based on circular hinge, and analysed inverse kinematics, internal kinematics, and analytic stiffness modelling referenced to the task coordinate. Chen et al. [21] established the closed-loop solution for single elliptical flexure hinge, and analysed the influence of different parameters on the compliance of flexure hinge in different directions. Su et al. [22] proposed a general framework with arbitrary topology via screw algebra, which provided another mathematical method for studying the mobility of compliant mechanisms. However, the work mentioned above rarely optimizes the structural size parameters according to the engineering application requirements, which makes their non-working direction stiffness insufficient, and cannot be normally applied in heavy load precision engineering, In some follow-up studies, many novel flexure hinges have been proposed, such as bionic flexible joints [23,24] and corrugated flexure elements [25], which were originally designed to improve the working range of flexure hinges, and did not take the aforementioned issues into consideration.

2. Design of flexible support module

The traditional flexure hinge with the filled leaf can achieve the bending deformation to a certain extent, however, under the conditions with the encountered heavy load, the stress concentration will occur, and most of these stresses are distributed at both ends of the filled leaf, as shown in Fig.1 (a), the maximum concentrated stress is 3518.4MPa, which may cause the fatigue failure of the flexure hinge. For the case of stress concentration problem, an elliptical reinforced flexible hinge (ERFH) was designed, with the profile shown in Fig.4. The results of finite element analysis in Fig.1 (b) show that it can disperse the concentrated stress to the filled leaf, effectively reduce maximum stress to 1754.7MPa.

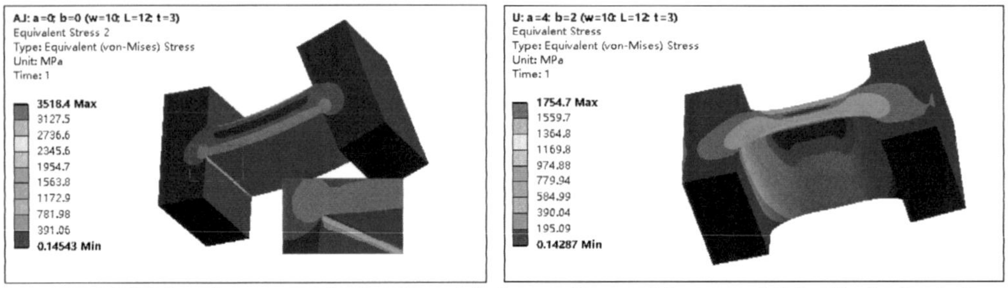

Figure 1. (a) Finite element analysis of straight beam flexure hinge, and (b) elliptical reinforced flexure hinge

Flexible support module (FSM) is designed based on ERFH, as shown in Fig.2, it is composed of four groups of flexure hinges in staggered series. For each group of flexure hinges, two ERFHs was in parallel and in the same direction, which can transform the moment in X-direction into normal stress acted on each ERFH, to improve the stiffness in the non-working direction. Different from most existing integrated CPM, the design of FSM also needs to meet the requirements of load in some precision engineering. Therefore, factors such as working flexibility, non-working direction stiffness and stress concentration will be fully considered in structural size parameters optimization process, corresponding detailed analysis will be given in Section 3

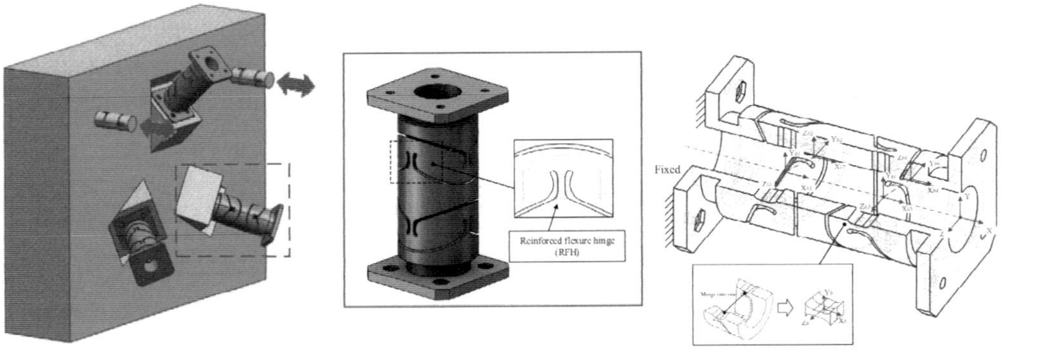

Figure 2. Model of flexible replaceable module optimized with ERFH

3. FSM stiffness analysis

Stiffness modelling is the basis of realizing the dynamic control of compliant mechanism, which gives the linear mapping between the driving force and small displacement, the flexible support module (FSM) is modelled analytically. Elliptical reinforced flexible hinge (ERFH), as the basic deformation unit of FSM, its compliance matrix is a basis to calculate structural stiffness, which will be analysed in detail in section 3.2.

3.1. FSM stiffness model

Assume a flexure hinge as a six degree-of-freedom beam with one end fixed, and subjected to bending, axial loading, shearing, and torsion. By defining the force actuating on the hinge by

$$\mathbf{F}=\begin{bmatrix}F_x & F_y & F_z & M_x & M_y & M_z\end{bmatrix}^T \tag{1}$$

and the corresponding deformations of the hinge by

$$\mathbf{X}=\begin{bmatrix}\Delta_x & \Delta_y & \Delta_z & \alpha_x & \alpha_y & \alpha_z\end{bmatrix}^T \tag{2}$$

according to the Hooke's law, the relationship between the displacement and applied force of the flexure hinge can be expressed by

$$\mathbf{F}=\mathbf{K}\cdot\mathbf{X} \tag{3}$$

where **K** represent the stiffness matrix of compliant mechanism, and is the inverse of the compliance matrix, **C**.

As shown in Fig.2, the calculation of the compliance matrix of FSM can be regarded as four groups of flexure hinges in staggered series. In order to simplify the calculation process, each group of flexure hinges can be simplified and merged into one. Therefore, the overall compliance matrix of FSM can be calculated by series formula as [26]:

$$\mathbf{C}_e=\sum_{i=1}^{n} Ad_{ie}{}^T\cdot\mathbf{C}_i\cdot Ad_{ie} \tag{4}$$

where $Ad_i C_i Ad_i^T$ is the adjoint transformation of compliance matrix, it represents the transformation of reference coordinate system. Assuming that the compliance matrix of the structural member in coordinate system $\{\mathbf{A}\}$ is $\mathbf{C}_A$, and the adjoint matrix from coordinate system $\{\mathbf{A}\}$ to $\{\mathbf{B}\}$ is Ad_1, then the compliance matrix $\{\mathbf{C}\}$ in coordinate system $\{\mathbf{B}\}$ can be defined as:

$$\mathbf{C}_B = Ad_1{}^T\mathbf{C}_A Ad_1 \tag{5}$$

with adjoint transformation and its transpose:

$$Ad(\xi)=\begin{bmatrix}\mathbf{R} & \mathbf{0}\\ \hat{\mathbf{t}}\mathbf{R} & \mathbf{R}\end{bmatrix},\ Ad^T(\xi)=\begin{bmatrix}\mathbf{R}^T & -\mathbf{R}^T\hat{\mathbf{t}}\\ \mathbf{0} & \mathbf{R}^T\end{bmatrix} \tag{6}$$

where **R** and **t** represent the rotation and translation transformation from {**A**} to {**B**} respectively, the skew matrix of the translation vector, $\hat{\mathbf{t}}$, defined as:

$$\hat{\mathbf{t}} = \begin{bmatrix} 0 & -t_3 & t_2 \\ t_3 & 0 & -t_1 \\ -t_2 & t_1 & 0 \end{bmatrix} \tag{7}$$

Therefore, for the FSM coordinate system shown in Fig.2, the kinematic relationship between them can be established as shown in Tab.1.

Table 1. Kinematics of flexure hinge in FSM

Flexure hinge	Kinematic transformation		Parameter values (mm)
1	$R_1 = [0; 0; 0]$,	$P_1 = [p; 0; 0]$	p =88.18
2	$R_2 = [90°; 0; 0]$,	$P_2 = [p; 0; 0]$	p =88.18
3	$R_3 = [0; 0; 0]$,	$P_3 = [q; 0; 0]$	q =36.82
4	$R_4 = [90°; 0; 0]$,	$P_4 = [q; 0; 0]$	q =36.82

As aforementioned, each set of ERFH is considered as a combined flexure hinge in the simplified calculation. The combined flexure hinges will have the same structure. Accordingly, the compliance matrix of ERFH in their respective reference coordinate systems, $\mathbf{C}_i$, will be the same, and have the following form [27]:

$$\mathbf{C}_h = \begin{bmatrix} \frac{\Delta_x}{F_x} & 0 & 0 & 0 & 0 & 0 \\ 0 & \frac{\Delta_y}{F_y} & 0 & 0 & 0 & \frac{\Delta_y}{M_z} \\ 0 & 0 & \frac{\Delta_z}{F_z} & 0 & \frac{\Delta_z}{M_y} & 0 \\ 0 & 0 & 0 & \frac{\alpha_x}{M_x} & 0 & 0 \\ 0 & 0 & \frac{\alpha_y}{F_z} & 0 & \frac{\alpha_y}{M_y} & 0 \\ 0 & \frac{\alpha_z}{F_y} & 0 & 0 & 0 & \frac{\alpha_z}{M_z} \end{bmatrix} \tag{8}$$

Each element in the compliance matrix is called a compliance factor, which represents the influence coefficient of displacement and angular deformation in different directions on force and moment.

3.2. Compliance Equations of ERFH

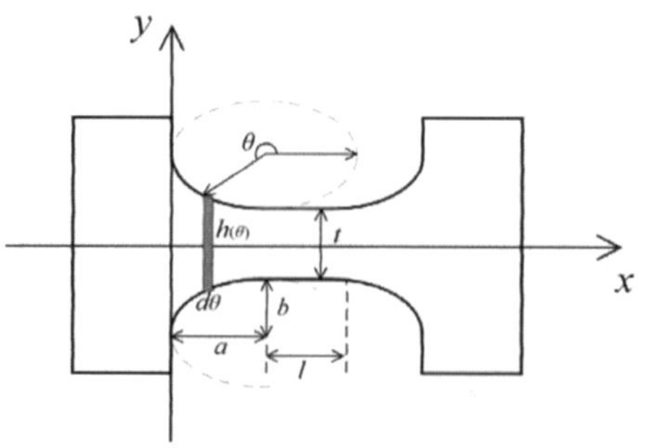

Figure 3. Analysis of the elliptical reinforced flexible hinge profile

The closed-form equation for compliance factors can be derived based on the Euler–Bernoulli beam theory. In order to deduce the compliance factor of the closed-form equation, the profile curve of ERFH should be formulated first, as shown in Fig.3, it can be described by piecewise equation,

$$y(x)=\begin{cases}\sqrt{b^2-\frac{b^2}{a^2}(x-a)^2}-b-\frac{t}{2} & (0<x<b)\\ -t/2 & (b\le x<a+l)\\ \sqrt{b^2-\frac{b^2}{a^2}(x-a-l)^2}-b-\frac{t}{2} & (a+l\le x<2a+l)\end{cases} \tag{9}$$

By parameterizing the independent variables, and applied Euler–Bernoulli beam theory, the closed form equations of compliance factors can be derived, its corresponding detailed derivation process can be seen in the appendix website (http://jiabinpan.work/Publications/AnlsRFH.html), here, the results are given directly,

$$\frac{\alpha_z}{M_z}=\frac{12l}{Ewt^3}-\frac{3a}{Ew}\cdot N_1 \tag{10}$$

$$\frac{\alpha_z}{F_y}=\frac{\Delta_y}{M_z}=\frac{6l(2a+l)}{Ewt^3}-\frac{3a(2a+l)}{2Ew}\cdot N_1 \tag{11}$$

$$\frac{\alpha_y}{M_y}=\frac{12l}{Ew^3t}-\frac{12a}{Ew^3}\cdot N_2 \tag{12}$$

$$\frac{\alpha_y}{F_z}=\frac{\Delta_z}{M_y}=\frac{6l(2a+l)}{Ew^3t}-\frac{6l(2a+l)}{Ew^3}\cdot N_2 \tag{13}$$

$$\frac{\alpha_x}{M_x}=-\frac{7a}{8Gw}\cdot N_1-\frac{7a}{2Gw^3}\cdot N_2+\frac{7l}{2Gwt^3}+\frac{7l}{2Gw^3t} \tag{14}$$

$$\frac{\Delta_z}{F_z}=\left[\frac{2l}{Ew^3t}(2a+l)(3a+l)+\frac{kl}{Gwt}\right]-\left[\frac{6a}{Ew^3}(2a+l)(3a+l)+\frac{ka}{Gw}\right]\cdot N_2+\frac{12a^2l}{Ew^3}\cdot N_3+\frac{12a^3}{Ew^3}\cdot N_4 \tag{15}$$

$$\frac{\Delta_y}{F_y}=\left[\frac{2l}{Ewt^3}(l^2+6al+6a^2)+\frac{kl}{Gwt}\right]-\frac{3a^2l}{Ew}\cdot N_1-\frac{ka}{Gw}\cdot N_2+\frac{3a^2l}{Ew}\cdot N_5-\frac{3a^3}{Ew}\cdot N_6 \tag{16}$$

$$\frac{\Delta_x}{F_x}=\frac{l}{Ewt}-\frac{a}{Ew}\cdot N_2 \tag{17}$$

where $N_1 \sim N_6$ is complex integral terms, which can be obtained via software, **Mathematica**,

$$N_1=\int_0^{\frac{\pi}{2}}\frac{\cos\theta}{(b\cos\theta-b-t/2)^3}d\theta=-8\left(\frac{t^2+4bt+6b^2}{t^2(4b+t)^2(2b+t)}+\frac{6b(2b+t)\arctan\sqrt{4b/t+1}}{t^{5/2}(4b+t)^{5/2}}\right) \tag{18}$$

$$N_2=\int_0^{\frac{\pi}{2}}\frac{\cos\theta}{b\cos\theta-b-t/2}d\theta=\frac{\pi}{2b}-\frac{2(2b+t)}{b\sqrt{t}\sqrt{4b+t}}\arctan\sqrt{4b/t+1} \tag{19}$$

$$N_3=\int_0^{\frac{\pi}{2}}\frac{\sin\theta\cos\theta}{b\cos\theta-b-t/2}d\theta=\frac{1}{b}-\frac{2b+t}{2b^2}\log\left(\frac{2b}{t}+1\right) \tag{20}$$

$$N_4=\int_0^{\frac{\pi}{2}}\frac{\cos^3\theta}{b\cos\theta-b-t/2}d\theta=\frac{3\pi+4}{4b}+\frac{(\pi+1)t}{2b^2}+\frac{\pi t^2}{8b^3}-\frac{(2b+t)^3}{2b^3\sqrt{t}\sqrt{4b+t}}\arctan\sqrt{4b/t+1} \tag{21}$$

$$N_5=\int_0^{\frac{\pi}{2}}\frac{\sin\theta\cos\theta}{(b\cos\theta-b-t/2)^3}d\theta=\frac{t-2b}{b^2t^2}-\frac{1}{b^2(t+2b)} \tag{22}$$

$$
\begin{aligned}
N_6 = \int_0^{\frac{\pi}{2}} \frac{\cos^3\theta}{\left(b\cos\theta - b - t/2\right)^3} d\theta = \frac{\pi}{2b^3} - \frac{2(2b+t)\left(6b^2 - 4bt - t^2\right)}{b^2 t^2 (4b+t)^2} \\
- \frac{2(2b+t)\left(24b^4 - 8b^3 t + 14b^2 t^2 + 8bt^3 + t^4\right)}{b^3 t^{5/2} (4b+t)^{5/2}} \arctan\sqrt{4b/t+1}
\end{aligned}
\tag{23}
$$

and E, G, μ are the modulus of elasticity, the modulus of rigidity and Poisson ratio, respectively. The shear coefficient, k, for the micro-beams with rectangular cross-section, is employed as [29]:

$$
k = \frac{12 + 11\mu}{10 + 10\mu} \tag{24}
$$

4. Optimal RFH structural parameters

There are two kinds of structural parameters for reinforced flexure hinges. The first is related to the elliptical profile, Category I (a and b), which is regarded as the parameters that determine the degree of stress concentration, and the second is main size parameters of flexure hinges, Category II (l, w and t), which is used to optimized to make the FSM meet the design goal, that is, when the compliance of the ERFH in the working direction meets the requirements, the compliance in the non-working direction need to be as large as possible. To optimized Category I and II, the parameters were limited in a certain range, as shown in Tab.2.

Table 2. Range of structural parameters to be optimized

type	symbol	range
edge profile parameters (I)	a	1 ~ 4
	b	1 ~ 3
main size parameters (II)	l	3 ~ 6
	w	8 ~ 12
	t	2 ~ 4

For parameters Category I, the optimization can be completed via finite element models (FEMs). ANSYS Workbench 19.2 is used to analyse the maximum stress for different flexure hinges, which are subjected to the external load (Mx = 19.5 N/m, My = 45 N/m), and have the same material properties ($\rho = 2770\ kg/m^3$, $E = 7.1\mathrm{e}+10$, $\mu = 0.33$). The stress analysis results of ERFH are shown in Tab.3. The increase of parameters a and b will both reduce the concentrated stress. Compared with parameter b, the effect of parameter a is more significant, especially when a=4, the influence of b on the maximum stress concentration is already very small. Therefore, a=4, b=2 (moderate value) are chosen as the parameters of the elliptical profile. At the same time, by observing the data in Tab.3, it is found that when $a \geq 3$, the maximum stress change is not so significant, which is due to the concentrated stress is dispersed to the straight beam section, and is distributed in strips (Fig.1 b).

Table 3. The maximum stress (MPa) of flexure hinge with different ellipse parameters

b \ a	1	1.5	2	2.5	3	3.5	4
1	2111	1881.7	1786.5	1750.4	1735.6	1730.2	1728.6
1.5	2088.2	1974.7	1847.3	1787.7	1760.0	1746.5	1741.6
2	2187.1	2040.6	1906.3	1823.3	1784.6	1763.9	1754.7
2.5	2293.5	2076.1	1947.4	1862.5	1808.7	1781.6	1768.2
3	2365.9	2127.7	1991.8	1890.5	1837.4	1800.4	1780.2

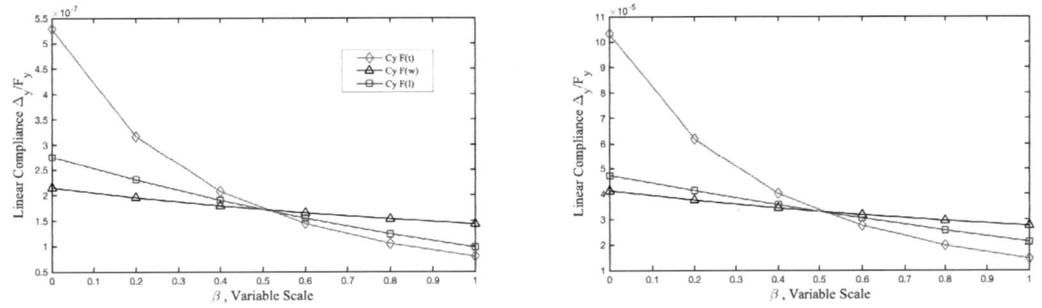

Figure 4. Variation trend of compliance factors, C_y^F and C_y^M, in terms of t, w and l.

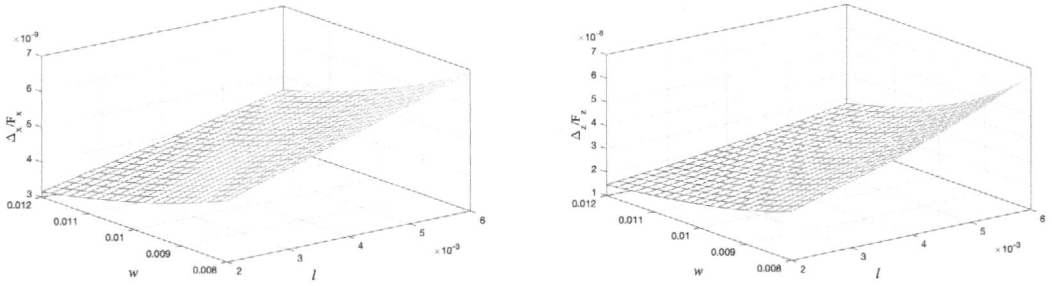

Figure 5. Compliance factors, C_x^F and C_z^F, in terms of w and l.

Among the Category II (l, w and t), thickness $\boldsymbol{t}$ has a much higher influence on the compliance of ERFH in working direction, as shown in Fig.4. In order to ensure the overall flexibility of FSM to meet the design requirements, the parameter $\boldsymbol{t}$ needs to be determined first. In this process, the other two parameters $\boldsymbol{l}$ and $\boldsymbol{w}$ of ERFH are tentatively determined as the median value (β= 0.5) within the range, that is, $\boldsymbol{l}$ = 4.5 and $\boldsymbol{w}$ = 10. In the case of this paper, the designed FSM is needed to provide a displacement of 5mm at Y-direction under external load Mx = 19.5 N/m, My = 45 N/m, which requires that the compliance factor of each ERFH should meet $C_y^M = \frac{\Delta y}{M_z} \geq 7.39 \times 10^{-6}$, therefore, we determine $\boldsymbol{t}$ = 3 as the thickness of ERFH. Under this condition, parameters $\boldsymbol{l}$ and $\boldsymbol{w}$ should ensure that the stiffness of ERFH in the non-working direction is as high as possible. From the trend diagram of the compliance factor in the x and z directions with respect to $\boldsymbol{l}$ and $\boldsymbol{w}$, as shown in Fig.5, the values of $\boldsymbol{l}$ and $\boldsymbol{w}$ are selected as 3mm and 12mm respectively.

5. Experimental Verification

In order to further verify the stiffness model, and analyse the repetitive working error of FSM, an experimental platform based on the visual measurement was designed. As shown in Fig.6, an aluminium alloy plate, attached with 10mm×10mm chessboard, is fixed at the end of FSM, and its displacement is measured by two fixed 2D cameras. In terms of the external varied force and moment required by FSM, the force gauge is placed on the back of the aluminium alloy plate, and driven by a 3-DOF slider module. By adjusting the position of the driving point and the force of gauge, the equivalent moment acting on the centre of FSM will be changed. In this experiment, the force acting point is set at a distance of Dy = -40mm and Dz = 160.75mm from the centre of FSM, and the force gauge acts on the aluminium alloy plate with a driving force of 15N in steps to obtain chessboard images of FSM under different external forces.

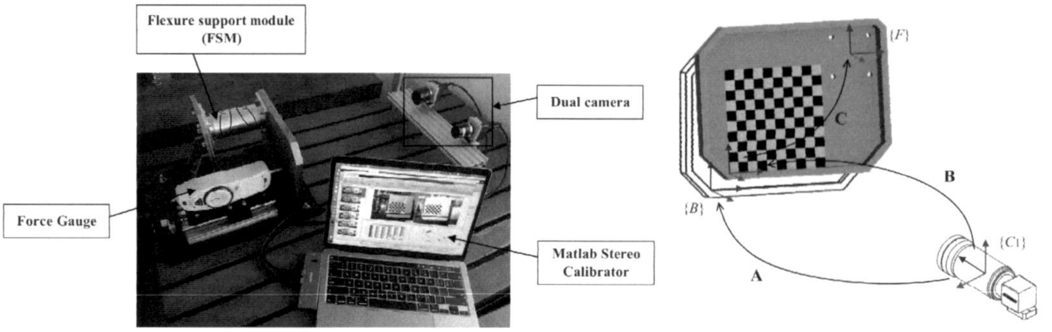

Figure 6. Vision-based displacement measurement of compliant mechanism

As mentioned before, the displacement of FSM is measured by camera calibration [30], which has two advantages: one is to avoid the error caused by the installation of the measuring device; the other is that camera calibration is non-contact measurement, which can eliminate the influence of contact error. Fig.6 show the relationship between camera calibration and displacement measurement of compliant mechanism. The deformation displacement of FSM can be expressed by HTM between flange frames $\{F\}$,

$$\mathbf{H}_F = (\mathbf{A}\cdot\mathbf{C})^{-1}\cdot(\mathbf{B}\cdot\mathbf{C}) \tag{25}$$

By transforming the $\mathbf{H}_F$ into vector form, the deformation displacement of FSM can be obtained. In these results, we mainly focus on the deformation in the direction of its working DOFs, Y and Z axis. For the direction of X axis, however, it is mainly caused by the parasitic motion rather than the deformation, which will not be considered when verifying the FSM flexibility matrix.

The expermental results, displacement of FSM under different driving forces, are shown in Fig.7, which indicate that FSM has good repetitive working accurency in the linear direction of Y axis and angular direction of Z axis, its mean error is 0.0077 mm and -2.2067e-04 rad respectively, which is better than the linear direction of Z and angular direction of Y, 0.3192 mm and -0.0036 rad. For the deviation between the calculated and measured values, we can reflect it by the corresponding compliance factor. The average measured linear compliance in Y axis, $\frac{\Delta_y}{M_z}$, is 2.249694e-01, and -2.651412e-01 in Z axis, its calculated value are both (-) 2.748678e-01. Correspondingly, the angular compliance, $\frac{\alpha_y}{M_y}$ and $\frac{\alpha_z}{M_z}$, the average measured value are 3.708488e-03 and 4.124847e-03 respectly, calculated value is 4.133350e-03. Above results show that the compliance matrix of FSM is well in line with the stiffness test of FSM.

Why is the repeatability in the linear direction of Y axis and angular direction of Z axis much lower? It is not difficult to find that the moment *My* acting on FSM is much larger than *Mz*, which leads to a greater stress in the corresponding ERFH. After the withdrawal of the external force, FSM is likely to cause errors due to the internal stress is not completely released. To verify this interpretation, we continue to measure the position and orientation variation of the FSM in these directions, as shown in Fig.10, t=0 represents the repetitive error of FSM final state without external load in Fig.8 experiment, that is, the linear displacement of Z-axis, -0.5707 mm, and the anglar displacement of Y-axis, 0.0054 rad. After 5 min, the repeatability decreased rapidly, and its value reached to -0.1805 mm and 0.0025 rad at 15 min, which confirms the previous explanation of the results in Fig.9.

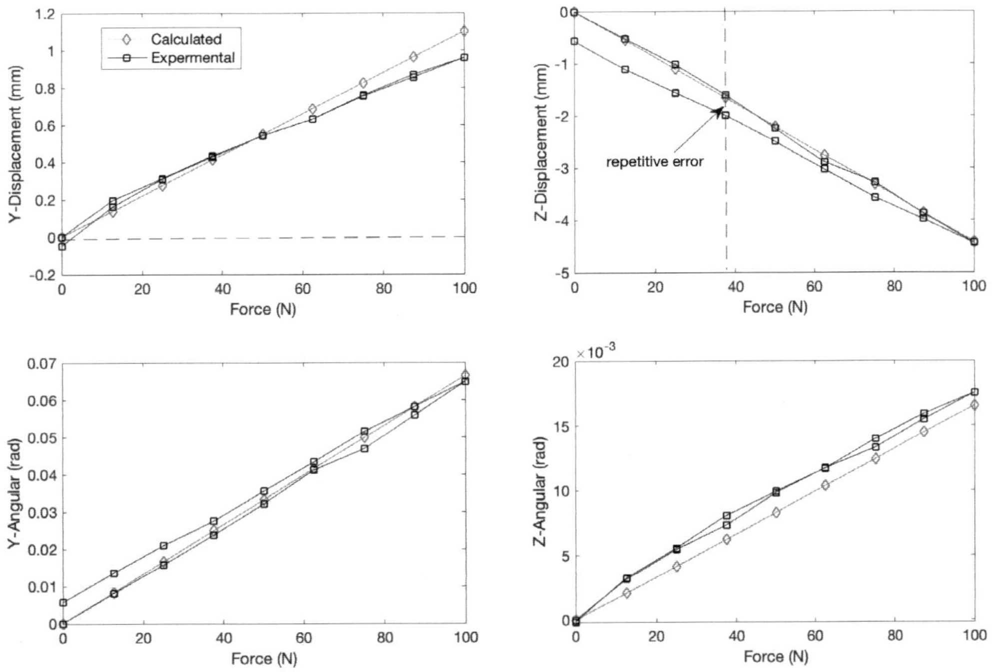

Figure 7. Calculated and expermental displacement of FSM in Y and Z direction

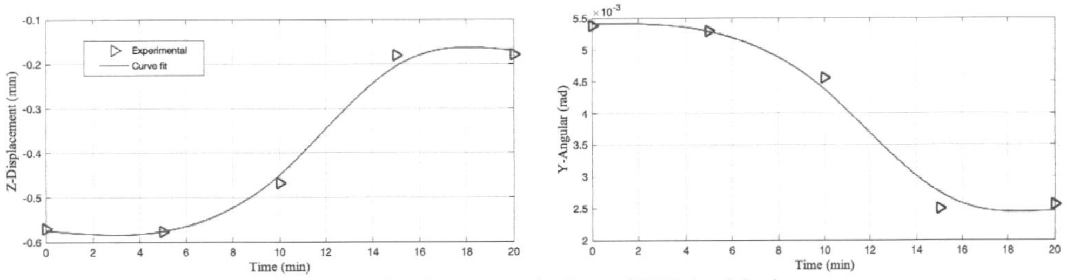

Figure 8. Final pose variation of FSM with time.

6. Conclusions

This paper presents a flexible support module (FSM) for compliant parallel mechanism, which can be used in precision engineering with high precision and large load, such as large aperture laser transmission unit. In this mechanism, the stiffness model of FSM is established based on the analysis of the elliptical enhanced flexure hinge, and its size parameters are optimized based on the finite element analysis. Moreover, to verify the stiffness model and analyse the repetitive error of FSM, a vision-based measurement method is introduced in this paper. The results show that the prototype can satisfy the application of the average error of 0.3192 mm and -0.0036 rad in the working space of 5mm and 0.07 rad, which is mainly caused by the lag of internal stress release, after the external force is removed, the repeatability error can be reduced to 4% rapidly in 5-15 minutes

Acknowledgments

This work was partially supported by the Natural Science Foundation of China (51805380, 51875415).

References

[1] Zheng, W., Wei, X., Zhu, Q., Jing, F., Hu, D., Su, J., ... Deng, X. (2016). Laser performance of the SG-III laser facility. High Power Laser Science and Engineering, 4, E21. doi:10.1017/hpl.2016.20

[2] Moses, E. I. (2009). Ignition on the National Ignition Facility: a path towards inertial fusion energy. Nuclear Fusion, 49(10), 104022. doi:10.1088/0029-5515/49/10/104022

[3] Caillaud, T., Alozy, E., Briat, M., Cornet, P., Darbon, S., Dizière, A., … Miquel, J. L. (2016). Recent advance in target diagnostics on the Laser MégaJoule (LMJ). Target Diagnostics Physics and Engineering for Inertial Confinement Fusion V. doi:10.1117/12.2238017

[4] Wei, Y., & Xu, Q. (2020). Design and Testing of a New Force-Sensing Cell Microinjector Based on Small-Stiffness Compliant Mechanism. IEEE/ASME Transactions on Mechatronics, 1–1. doi:10.1109/tmech.2020.3003992

[5] Yuan, G., Wang, D. H., & Li, S. D. (2015). Single piezoelectric ceramic stack actuator based fast steering mirror with fixed rotation axis and large excursion angle. Sensors and Actuators A: Physical, 235, 292–299. doi:10.1016/j.sna.2015.10.017

[6] Wang, N., Zhang, Z., Zhang, X., Cui, C., 2018. Optimization of a 2-DOF micro-positioning stage using corrugated flexure units. Mechanism and Machine Theory 121, 683–696.. doi:10.1016/j.mechmachtheory.2017.11.021

[7] Wang, P., Xu, Q., 2017. Design of a flexure-based constant-force XY precision positioning stage. Mechanism and Machine Theory 108, 1–13. doi:10.1016/j.mechmachtheory.2016.10.007

[8] Yao, Q., Dong, J., & Ferreira, P. M. (2007). Design, analysis, fabrication and testing of a parallel-kinematic micropositioning XY stage. International Journal of Machine Tools and Manufacture, 47(6), 946–961. doi:10.1016/j.ijmachtools.2006.07.007

[9] Hugget, A., Journot, E., Ferbos, R., Macias, F., & Fayollas, P. (2007). New design of the laser megajoule final optics assembly. New Developments in Optomechanics. doi:10.1117/12.731568

[10] Galbraith, J. Design of a Large Diffraction Grating Optical Assembly. United States: N. p., 2014. Web. doi:10.2172/1182706.

[11] Xu, Q., Li, Y., 2006. Kinematic Analysis and Optimization of a New Compliant Parallel Micromanipulator. International Journal of Advanced Robotic Systems 3, 47.. doi:10.5772/5721

[12] Teo, T.J., Yang, G., Chen, I.-M., 2014. Compliant Manipulators, in: . pp. 1–63.. doi:10.1007/978-1-4471-4976-7_102-1

[13] Yang, G., Teo, T.J., Chen, I.-M., Lin, W., 2011. Analysis and design of a 3-DOF flexure-based zero-torsion parallel manipulator for nano-alignment applications, in: .. doi:10.1109/icra.2011.5979911

[14] Byung-Ju Yi, Goo Bong Chung, Heung Yeol Na, Whee Kuk Kim, Il Hong Suh, 2003. Design and experiment of a 3-dof parallel micromechanism utilizing flexure hinges. IEEE Transactions on Robotics and Automation 19, 604–612.. doi:10.1109/tra.2003.814511

[15] Xu, Q., 2014. A novel compliant micropositioning stage with dual ranges and resolutions. Sensors and Actuators A: Physical 205, 6–14.. doi:10.1016/j.sna.2013.10.010

[16] Chen, G., Howell, L.L., 2009. Two general solutions of torsional compliance for variable rectangular cross-section hinges in compliant mechanisms. Precision Engineering 33, 268–274.. doi:10.1016/j.precisioneng.2008.08.001

[17] Zhu, Z., Zhou, X., Wang, R. et al .A simple compliance modeling method for flexure hinges. Sci. China Technol. Sci. 58, 56–63 (2015). https://doi.org/10.1007/s11431-014-5667-1

[18] Lobontiu, N., Cullin, M., Petersen, T., Alcazar, Noveanu, S., 2014. Planar Compliances of Symmetric Notch Flexure Hinges: The Right Circularly Corner-Filleted Parabolic Design. IEEE Transactions on Automation Science and Engineering 11, 169–176. doi:10.1109/tase.2012.2228853

[19] Chen, G., Liu, X., Du, Y., 2011. Elliptical-Arc-Fillet Flexure Hinges: Toward a Generalized

Model for Commonly Used Flexure Hinges. Journal of Mechanical Design 133, 081002.. doi:10.1115/1.4004441
[20] Pham, H.-H., Chen, I.-M., 2005. Stiffness modeling of flexure parallel mechanism. Precision Engineering 29, 467–478. doi:10.1016/j.precisioneng.2004.12.006
[21] Chen, G., Shao, X., Huang, X., 2008. A new generalized model for elliptical arc flexure hinges. Review of Scientific Instruments 79, 095103.. doi:10.1063/1.2976756
[22] Su, H.-J. (2011). Mobility Analysis of Flexure Mechanisms via Screw Algebra. Volume 6: 35th Mechanisms and Robotics Conference, Parts A and B. doi:10.1115/detc2011-48012
[23] Kozuka, H., Arata, J., Okuda, K., Onaga, A., Ohno, M., Sano, A., Fujimoto, H., 2012. A bio-inspired compliant parallel mechanism for high-precision robots, in: .. doi:10.1109/icra.2012.6224873
[24] Kozuka, H., Arata, J., Okuda, K., Onaga, A., Ohno, M., Sano, A., Fujimoto, H., 2012. Compliant-parallel mechanism for high precision machine with a wide range of working area, in: .. doi:10.1109/iros.2012.6386108
[25] Wang, N., Zhang, Z., Zhang, X., Cui, C., 2018. Optimization of a 2-DOF micro-positioning stage using corrugated flexure units. Mechanism and Machine Theory 121, 683–696.. doi:10.1016/j.mechmachtheory.2017.11.021
[26] Qiu, C., & Dai, J. S. (2020). Analysis and Synthesis of Compliant Parallel Mechanisms--screw Theory Approach. Springer.
[27] Cai, K., Tian, Y., Wang, F., Zhang, D., Shirinzadeh, B., 2016. Development of a piezo-driven 3-DOF stage with T-shape flexible hinge mechanism. Robotics and Computer-Integrated Manufacturing 37, 125–138.. doi:10.1016/j.rcim.2015.07.004
[28] Lobontiu, N. (2020).Compliant mechanisms: design of flexure hinges. CRC press.
[29] Zhu, Z., Zhou, X., Wang, R. et al. A simple compliance modeling method for flexure hinges. Sci. China Technol. Sci. 58, 56–63 (2015). doi:10.1007/s11431-014-5667-1
[30] Zhang, Z., 2000. A flexible new technique for camera calibration. IEEE Transactions on Pattern Analysis and Machine Intelligence 22, 1330–1334.. doi:10.1109/34.888718

Journal of Physics: Conference Series **2113** (2021) 012070 doi:10.1088/1742-6596/2113/1/012070

A new method for solving the linear recurrence relation with nonhomogeneous constant coefficients in some cases

Ben-chao Yang[1], Xue-feng Han[2,*]

[1]State Key Laboratory of Mathematical Engineering and advanced computing, Zhengzhou, Henan, China

[2] School of mathematics and Information Science, Henan Polytechnic University, Jiaozuo, Henan, China

*Corresponding author email: hanxuefeng@hpu.edu.cn

Abstract. Recursive relation mainly describes the unique law satisfied by a sequence, so it plays an important role in almost all branches of mathematics. It is also one of the main algorithms commonly used in computer programming. This paper first introduces the concept of recursive relation and two common basic forms, then starts with the solution of linear recursive relation with non-homogeneous constant coefficients, gives a new solution idea, and gives a general proof. Finally, through an example, the general method and the new method given in this paper are compared and verified.

1. Introduction

Recurrence relation, which mainly characterizes the unique laws satisfied by a sequence, plays an important role in almost all branches of mathematics. This is especially true for combinatorial mathematics, because each combinatorial counting problem has its combinatorial characteristics and laws. In many cases, recursive relationship is one of the most appropriate tools to describe this combinatorial law. Because of its importance, it has been deeply studied and applied in many aspects of engineering and life [1-4].

How to establish the recurrence relation, what are the properties of the known recurrence relation and how to solve the recurrence relation are several basic problems in the recurrence relation. Consider a sequence $\{a_n : n \geq 0\}$, where n is the number of schemes to be calculated in a specific mathematical model. Actually, it's a function of n. For such an unknown sequence, in order to calculate the number of schemes, a natural idea is to directly find its general term expression and then find all its terms. However, when the general term a_n cannot be obtained directly, if we can find a relationship between a_n and its several adjacent terms, and this relationship is always true when n is greater than or equal to a fixed positive integer, we have the possibility of finding the sequence a_n by using this relationship.

Depending on whether there is a nonhomogeneous term, recursive relations are generally divided into two types: r-order homogeneous constant coefficient linear recursive relations and r-order non-homogeneous constant coefficient linear recursive relations. The solution of r-order homogeneous constant coefficient linear recursive relation is relatively simple. As long as the characteristic polynomial is listed to solve the characteristic root, and then the formula is applied to solve the undetermined coefficient, the solution of the recursive relation can be obtained. This paper mainly studies a special case of non-homogeneous recurrence relation, gives a new solution method from another point of view, and gives a general proof. Finally, an example is given to verify the correctness of the method.

2. Common solving methods and new ideas of recursive relations

The general form of r-order non-homogeneous linear recursive relation with constant coefficients is:

$$u_n = \sum_{j=1}^{r} c_j u_{n-j} + g(n) \tag{1}$$

For the r-order linear recursive relation with non-homogeneous constant coefficients, the general solution is divided into three steps. The first step is to find the special solution of the non-homogeneous linear recursive relation; The second part is to solve the general solution of its corresponding derived recurrence relationship; Finally, the third step is to add the two solutions, and the result of the addition is the general solution of the original r-order non-homogeneous linear recurrence relationship with constant coefficients. The general solution of deducing the recurrence relation is actually to solve an r-order homogeneous linear recurrence relation. There are specific solutions in the corresponding textbooks [5,6]. Next, we mainly study the solution method of the special solution of the non-homogeneous recurrence relation. However, in the common solution, this method can only find the special solution for some special $g(n)$, which refers to the part of the linear recursive relationship with non-homogeneous constant coefficients excluding the derived recursive relationship.

In order to facilitate the solution, formula (1) is transformed into r-order homogeneous constant coefficient linear recursive relationship, and the transformed form is

$$u'_n = \sum_{j=1}^{r} c_j u'_{n-j} \tag{2}$$

where $u'_{n-i} = u_{n-i} + g'(n-i), (i = 1, 2, \cdots, r)$ .Through observation and analysis, it is known that polynomials $g(n)$ and $g'(n)$ have the same degree. It can be obtained from the combination of formulas (1) and (2) that

$$g'(n) - \sum_{j=1}^{r} c_j g'(n-j) = -g(n) \tag{3}$$

Next, a brief proof of the feasibility of this method is given.

Case 1: $g(n)$ is polynomial

Suppose if $g(n) = A_m n^m + A_{m-1} n^{m-1} + \cdots + A_1 n + A_0$, $g'(n) = a_m n^m + a_{m-1} n^{m-1} + \cdots + a_1 n + a_0$.

In this case, it is required to prove the existence of $a_i\,(i = 1, 2, \cdots,$ m). It can be obtained from equation (1-3),

$$(a_m n^m + a_{m-1} n^{m-1} + \cdots + a_1 n + a_0) - \sum_{j=1}^{r} c_j [a_m (n-j)^m + a_{m-1} (n-j)^{m-1} + \cdots + a_1 (n-j) + a_0]$$

$$= A_m n^m + A_{m-1} n^{m-1} + \cdots + A_1 \mathrm{n} + A_0$$

Comparing the coefficients of n^m, we can get $a_m - \sum_{j=1}^{r} c_j a_m = A_m$, then

$$a_m = \frac{A_m}{1 - \sum_{j-1}^{r} c_j} \tag{4}$$

Comparing the coefficients of n^{m-1}, we can get $a_{m-1}-\sum_{j=1}^{r}c_j[a_m\mathbb{C}_m^1(-j)+a_{m-1}]=A_{m-1}$.

This is a first-order equation containing two unknowns a_m and a_{m-1}. if we substitute a solved a_m by formula (4) into it, we can solve a_{m-1}. Similarly, by comparing the coefficients of all n^{m-i}, we can get

$$a_{m-i}-\sum_{j=1}^{r}c_j[a_m\mathbb{C}_m^{m-i}(-j)^i+a_{m-1}\mathbb{C}_{m-1}^{m-i}(-j)^{i-1}+\cdots+a_{m-i}]=A_{m-i}$$

The value of $a_m,a_{m-1},\cdots,\ a_{m-i+1}$ can be solved by substituting a_{m-i} solved by the previous equation into this system of equations, which contains $i+1$ univariate equations.

From the above, we can know the value of $a_m,a_{m-1},\cdots,\ a_1,a_0$, that is, we can prove that these m+1 coefficients do exist, so this new method is feasible when g(n) is a polynomial.

Case 2: $g(n)=\alpha^n$

Similar to the above idea, we can draw the conclusion of $g^{'}(n)=k\alpha^n$. Furthermore, the following conclusions can be obtained

$$g^{'}(n)=-\frac{\alpha^n}{\alpha^r-\sum_{j=1}^{r}c_j\alpha^{r-j}} \tag{5}$$

Case 3: $g'(n)=n^t\alpha^n$

First, through the analysis, it can be seen that the form of $g'(n)$ probably has the appearance of $g'(n)=kn^t\alpha^n$. Furthermore, the following conclusions can be obtained

$$k=\frac{-\alpha^r}{\alpha^r-\sum_{i=1}^{r}c_i\alpha^{r-i}} \tag{6}$$

3. Specific examples

Example 1: find the special solution of the following recursive relationship

$$u_{n+2}+u_{n+1}+u_n=n^2+n+1$$

Solution 1 (classical solution method):

It is easy to find that the characteristic equation corresponding to the recursive relationship is $c(x)=x^2+x+1$, so 1 is not the root of the characteristic equation, and its special solution can be solved directly by the difference method.

$$\begin{aligned}u_n&=(3\Delta^2+3\Delta+3I)(n^2+n+1)=3^{-1}(3^{-1}\Delta^2+\Delta+I)^{-1}(n^2+n+1)\\&=3^{-1}[I-(\Delta+3^{-1}\Delta^2)+(\Delta+3^{-1}\Delta^2)^2-\cdots](n^2+n+1)\\&=3^{-1}[I-\Delta+2\cdot3^{-1}\Delta^2+\cdots](n^2+n+1)\\&=3^{-1}[I-\Delta+2\cdot3^{-1}\Delta^2+\cdots]((n)_2+2(n)_1+1)\\&=3^{-1}[n(n-1)+2n+1-2n-2+4\cdot3^{-1}]\\&=3^{-1}(n^2-n+3^{-1})\end{aligned}$$

So $3^{-1}(n^2-n+3^{-1})$ is the special solution of this recursive relation.

Solution 2 (new solution):

Because $g(n)=n^2+n+1$, it can be assumed that $g^{'}(n)=an^2+bn+c$, and then the equations can be listed and solved to obtain $a=-\frac{1}{3},b=\frac{1}{3},c=-\frac{1}{9}$.Then

$$[u_{n+2}+(-\frac{1}{3}(n+2)^2+\frac{1}{3}(n+2)-\frac{1}{9})]+[u_{n+1}+(-\frac{1}{3}(n+1)^2+\frac{1}{3}(n+1)-\frac{1}{9})]+[u_n+(-\frac{1}{3}n^2+\frac{1}{3}n-\frac{1}{9})]=0$$

Suppose if $b_n=u_n+(-\frac{1}{3}n^2+\frac{1}{3}n-\frac{1}{9})$, Then the recursive relationship of order 2 homogeneous constant coefficients holds: $b_{n+2}+b_{n+1}+b_n=0$,of which $b_n=0$ is a special solution. Therefore, the special solution of the above linear recursive relationship with non-homogeneous constant coefficients is $u_n=\frac{1}{3}n^2-\frac{1}{3}n+\frac{1}{9}=3^{-1}(n^2-n+3^{-1})$.

Obviously, the special solutions obtained by the two methods are the same. The new method avoids the calculation of difference. In fact, in other cases, it can also be solved by this method.

4. Conclusion

It can be found that the new method is a little troublesome in determining the undetermined coefficients at the beginning, but there is no too much difference operation in the later steps of finding the special solution, and the calculation process is simplified, which is obviously simpler than the conventional method. But here, for some special functions, it is also difficult to determine the undetermined coefficients to solve, so this method still has room for improvement.

References

[1] SEYMOUR, P. D.; WELSH, D. J. A. Combinatorial applications of an inequality from statistical mechanics. In: Mathematical Proceedings of the Cambridge Philosophical Society. Cambridge University Press, 1975. p. 485-495.

[2] RAO, C. Radhakrishna. Some combinatorial problems of arrays and applications to design of experiments. In: A Survey of Combinatorial Theory. North-Holland, 1973. p. 349-359.

[3] ERDÖS, Paul. Some new problems and results in graph theory and other branches of combinatorial mathematics. In: Combinatorics and Graph Theory. Springer, Berlin, Heidelberg, 1981. p. 9-17.

[4] SMITH, Kate; PALANISWAMI, Marimuthu; KRISHNAMOORTHY, Mohan. Neural techniques for combinatorial optimization with applications. IEEE Transactions on Neural Networks, 1998, 9.6: 1301-1318.

[5] WEST, Douglas B. Combinatorial mathematics. Cambridge University Press, 2020.

[6] ROSEN, Kenneth H. (ed.). Handbook of discrete and combinatorial mathematics. CRC press, 2017.

Simulation of two dimensional OCDMA-PON system based on FBGs

Mingzhu Zheng[1], Qiwu Wu[2*], Yang Zhou[1], Lingzhi Jiang[1] and Jiaqi Liu[1]

[1]College of Information Engineering,Engineering University of PAP,Shaanxi, 710086, China

[2]College of Equipment Management and Security, Engineering University of PAP, Shaanxi, 710086, China

[*]Corresponding author's e-mail: cli@hunnu.edu.cn

Abstract. In order to study the effect of transmission distance on the performance of optical code division multiplexing passive optical network systems based on fiber Bragg Grating (FBG), a two-dimensional OCDMA-PON system based on cascaded FBGs is designed and constructed by using the VPI optical communication software. The spectrum, waveform and eye diagram of the output signal are analyzed by setting different fiber lengths in the system. The experimental results show that different fiber lengths have different effects on the system. Within a certain transmission distance, the longer the fiber length, the smaller the error generated by the output signal and the better the signal quality; beyond this certain distance, the quality of the output signal deteriorates sharply and the error generated becomes larger.

1. Introduction

Due to the rapid development of information technology, people's demand for data traffic of communication services has gradually climbed. Therefore, in order to fully develop and reasonably use the huge bandwidth resources of optical fiber, various multiplexing technologies have emerged, such as the most popular passive optical networks (PON) multiplexing technology. It mainly includes passive optical network technology based on time division multiplexing (TDM), orthogonal frequency division multiplexing (FDM), wavelength division multiplexing (WDM) or optical code division multiplexing (OCDMA) technology.Among them, OCDMA-PON technology, which combines the advantages of both optical code division multiplexing and passive optical networks, becomes one of the best candidates for future high-bandwidth reliable optical communication systems and a research hotspot [1].

Chuanqi Li studied and analyzed the effect of signal power spectrum on the BER of FBG-based code-division multiplexing optical networks.Manghu Cao gave the maximum allowable error of FBG parameters based on the study of the effects of grating center wavelength and relative position and different number of users on BER performance in FFH-OCDMA systems. It provides a basis for reducing the production accuracy as well as the cost of codecs. The impact of the degree of FBG codec mismatch on the OCDMA-PON system is analyzed by Chen Jiaming using VPI platform simulation. Yanyan Zhou studied the performance change of codecs when the data transmission rate in a two-dimensional OCDMA-PON system based on Bragg fiber grating exceeds the limiting condition and found that the system can still recover the original signal correctly even when the rate exceeds a certain limit. The relationship between multi-user interference and Bragg grating reflection bands in OCDMA-PON systems was investigated by Guimaraes.

It is clear from the above information that most of the studies related to OCDMA-PON systems based on FBGs have not analyzed the effect of fiber length on transmission performance. Therefore, in this paper, VPI is used to design and simulate a two-dimensional OCDMA-PON system based on cascade FBG. By changing the length of optical fiber, the frequency spectrum, waveform and eye image results of the output signal are observed and analyzed, and relevant conclusions are finally drawn.

2. Optical code division multiplexing passive optical network system

The OCDMA-PON system, as the name suggests, is a system that includes two key technologies, which are the optical code division multiplexing (OCDMA) and the passive optical network technology. Among them, the optical code division multiplexing (OCDMA) technology is combined with channel multiplexing and multiple access technologies in the optical domain. The user data source, laser source, optical coder/decoder, optical power combiner, transmission fiber and optical receiver are the basic components to realize this technology. In the optical code division multiplexing system, firstly, the address code of each channel is assigned to each signal, which is optically encoded by the encoder and then multiplexed by the coupler to become a one-way signal. Then the optical fiber is transmitted to the receiver by the decoder for matching optical decoding. Finally, the decoded signal is input to the receiver for processing and then output. It can be seen that the OCDMA technology's signal multiplexing and exchanging are different from other multiplexing technologies, so it has the advantages of high-speed all-optical processing, flexible address allocation, full utilization of spectrum resources, high confidentiality and support for asynchronous transmission [2]. The passive optical network (PON), on the other hand, is a future-oriented broadband access technology with good service transparency, high economic value, and ease of management. It is characterized by passive components between optical line terminals and optical network units, thus providing huge transmission bandwidth and reducing the energy consumption of system operation. Because of its unique advantages, PON technology has become the most popular access network solution today.

Since the optical codec/decoder is one of the most core components in the OCDMA-PON system [3], its performance is one of the decisive factors for good or bad system performance. That's why its design is also crucial to the overall performance of the system. In OCDMA systems, there are many devices that can be used as optical coder/decoder, including fiber Bragg grating (FBG), fiber delay line (FODL), array waveguide grating (AWG), spatial light modulator (SLM), etc [4-6]. Among these devices, FBG is the most suitable codec for use in commercialized OCDMA-PON systems due to its superior cost performance, unique filtering characteristics, and flexible codecs. Therefore, OCDMA-PON systems based on FBGs have become a more popular research content in the field of optical communication today.

3. Simulation experiments and results

3.1 Introduction to the simulation platform

In this paper, VPI, a professional optical communication software which commonly used in the field of optical network research, is used for simulation experiments. Since this simulation software has a large number and variety of optical network simulation modules, the user can build the system conceived by the user independently with any module according to the demand, thus realizing the efficient model construction of the optical transmission system. It is also possible to set or optimize the optical communication system by modifying the parameter settings of each device in the model. Using various types of analyzers, the simulation system can be operated with performance indicators such as spectrum, waveform, eye diagram and BER. Users can select the appropriate analyzer to study the system operation data according to the actual needs.

3.2 System model diagram

The experiments are based on the VPI simulation platform to build and simulate a two-dimensional single-user OCDMA-PON system based on cascaded FBGs. The structure is shown in Figure 1. The transmitter side of the system includes a laser source, data sequence generator, pulse transmitter, Mach-Zehnder modulator, FBG encoder, optical power combiner, and SOA amplifier, while the transmission module is an ideal loss-free fiber. The receiver side includes SOA amplifier, FBG decoder, optical power combiner, direct receiver, and signal analyzer. The main optical encoder/ decoder has 5 fiber Bragg gratings and 4 fiber delay lines in each, and its parameters are set according to the address code selection.

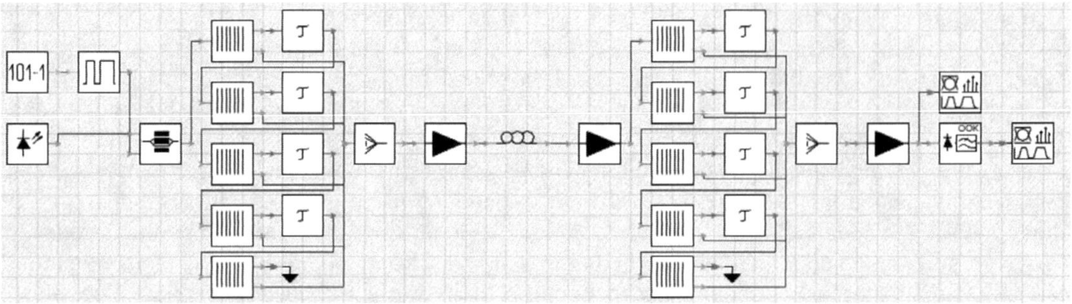

Figure 1. A two-dimensional single-user OCDMA-PON system based on cascaded FBGs.

3.3 Simulation parameter setting

The parameters of the center wavelength of each grating in the coder/decoder and the fiber delay line are set as in Table 1. And the basic parameter settings of the system are shown in Table 2.

Table 1. Codec parameter setting.

Encoder/ nm					Decoder/ nm				
FBG1	FBG2	FBG3	FBG4	FBG5	FBG6	FBG7	FBG8	FBG9	FBG10
1548.8	1549.4	1550	1550.6	1551.2	1551.2	1550.6	1550	1549.4	1548.8

Fiber delay line/ ns							
FODL1	FODL2	FODL 3	FODL 4	FODL 5	FODL 6	FODL 7	FODL 8
0.24	0.24	0.24	0.24	0.24	0.24	0.24	0.24

Table 2. System parameter setting.

Name	Value
Bitrate	1e9bit/ s
Light source center frequency	1550nm
Light source spectrum range	1548.8nm~1551.2nm
Light source power	-115dBm
Dispersion coefficient	0ps/ nm/ km
Sampling rate	128e9Hz
Sampling center frequency	193.548e12Hz
Time window	16e-9s

3.4 Simulation results and analysis

3.4.1 Spectrum analysis of the output optical signal at different fiber lengths.

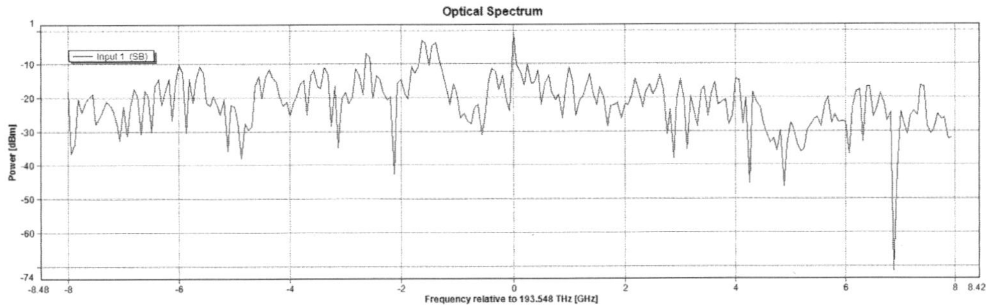

Figure 2. Spectrum of the signal at 10km length.

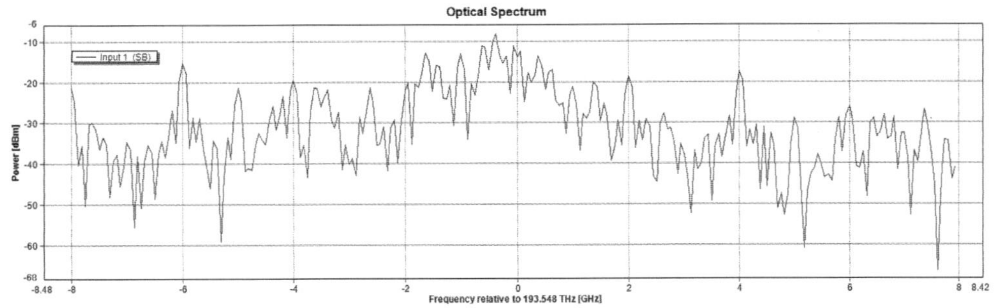

Figure 3. Spectrum of the signal at 100km length.

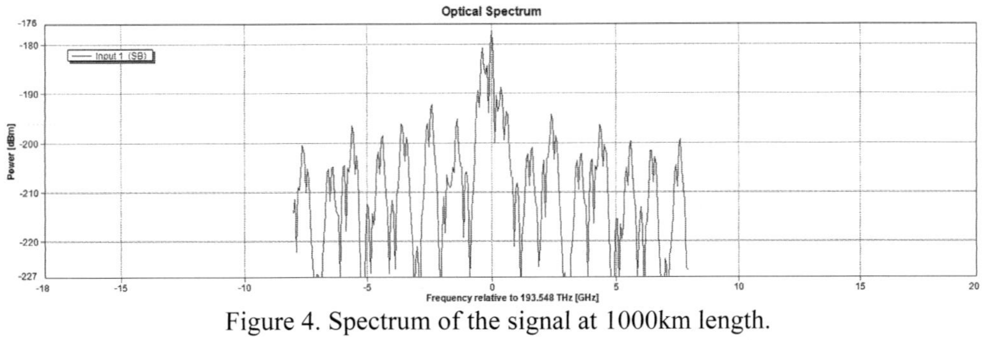

Figure 4. Spectrum of the signal at 1000km length.

The results of the integrated spectrogram show that there is no significant change in the power of the signal as well as the spectrum within 10 km; the power of the signal decreases when it exceeds 100 km, while it decreases more obviously when it reaches 1000 km. But in general, no matter how the transmission distance changes, the frequency of the signal always remains at the center frequency of 193.548 THz, and no spectral shift occurs.

3.4.2 Waveform analysis of output electrical signal at different fiber lengths.

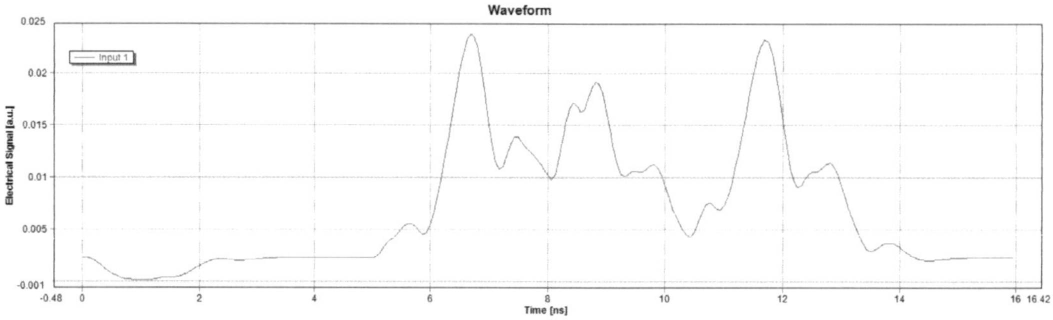

Figure 5. Decoded signal at a length of 10km

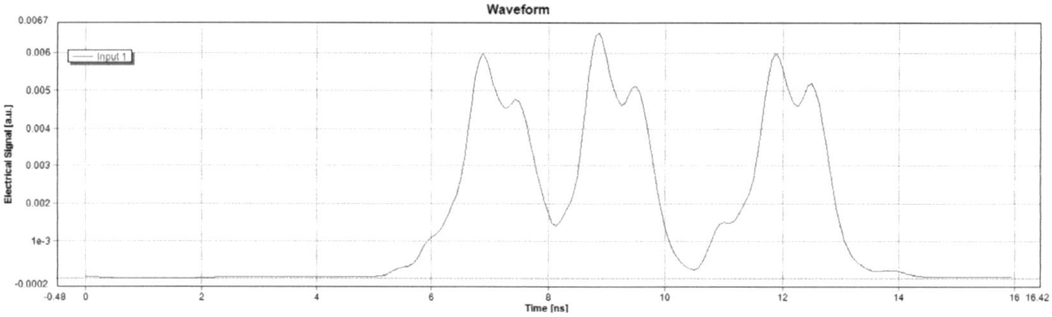

Figure 6. Decoded signal at a length of 100km

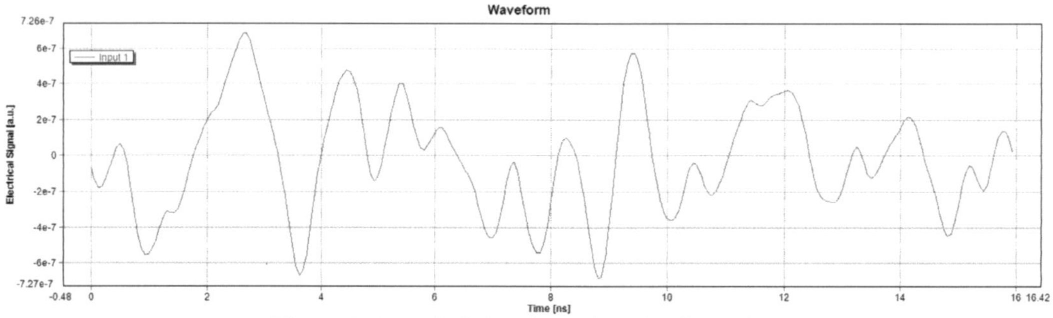

Figure 7. Decoded signal at a length of 1000km

Analyzing these waveforms, we can know that within 100km, the decoded signal can still be restored to the original user signal despite partial distortion, while beyond 1000km, the waveform changes drastically and the signal distortion is larger. This indicates that the system cannot decode normally when the transmission distance exceeds a certain length, which makes the signal have a large error.

3.4.3 Eye diagram analysis of the output electrical signal at different fiber lengths.

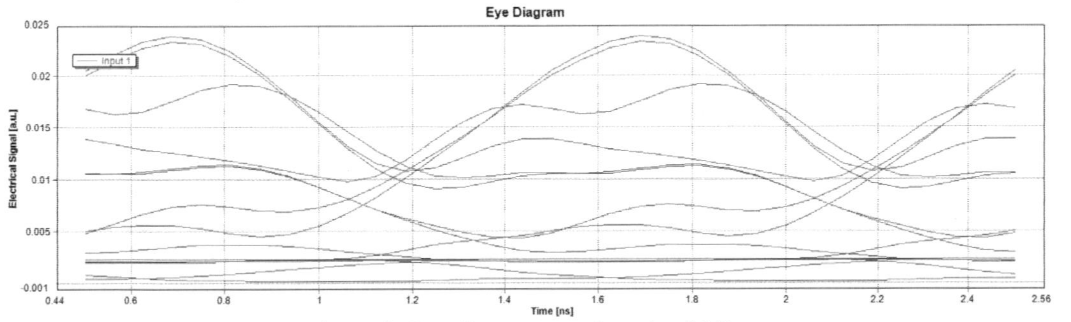

Figure 8. Eye diagram at a length of 10km

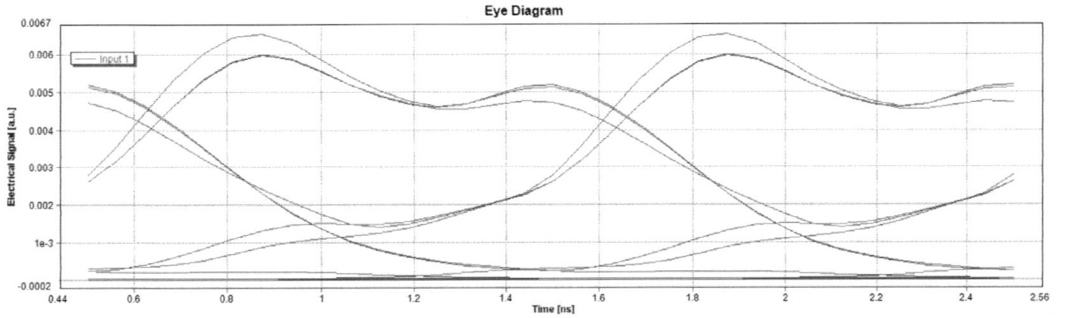

Figure 9. Eye diagram at a length of 100km

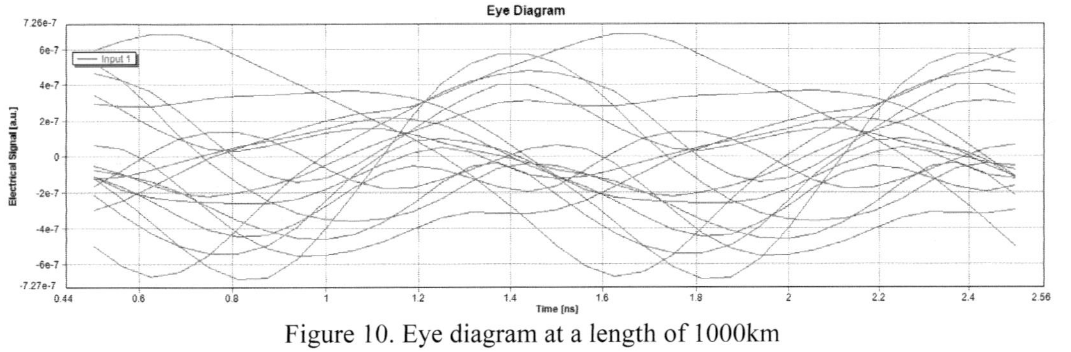

Figure 10. Eye diagram at a length of 1000km

Analyzing the eye diagram results, we can know that within the transmission distance of 100km, the eye diagram of the output signal is still clear on the whole, and some messy lines gradually decrease with the increase of transmission distance, which indicates that the anti-noise performance and signal decoding ability of the system is relatively strong, and the reliability of transmission is high. After the fiber length exceeds 1000km, the regularity of the eye diagram decreases and the lines are messy, and the signal distortion is large, which indicates that the decoding ability and anti-noise performance of the system are weak and the transmission reliability is low.

4. Conclusion

In this paper, a two-dimensional single-user OCDMA-PON system based on cascaded FBGs is constructed by VPI, and the signal quality of the system output at various fiber lengths is simulated and analyzed. From the experiments, it is clear that the noise immunity and signal decoding ability of the system are relatively affected by the variation of fiber length. Within a certain range, the longer transmission distance is, the better the quality of the output signal is instead; when the upper limit is exceeded, the signal quality decreases sharply and generates larger errors. This paper provides a certain reference value for the research of subsequent code division multiplexing type passive optical network system.

Acknowledge
This work is supported by the natural science basic research plan in Shanxi Province of China (No.2020JM-361), the young and middle-aged scientific research backbone projects of Engineering University of PAP (No.KYGG201905) , the basic research foundation project of Engineering University of PAP (No.WJY201920, No.WJY202019, No.WJY202144), the military theory research project of engineering university of PAP (No.JLY2020085), the education and teaching fund project of Engineering University of PAP (No.WJJ202039, No.WJX2021101), the research capacity improvement plan of engineering university of PAP (No.WKY202127), the PAP's Military Scientific Research Mandatory Project (No.WJ2020A020048), the basic research foundation project of college of Information Engineering of Engineering University of PAP (No. XYJC202112).

References

[1] Ahn, B. G, Park, Y. L. (2002) A symmetric-structure CDMA-PON system and its implementation. IEEE Photonics Technology Letters, 14(9): 1381-1383.
[2] XU W. (2008) Recent Progresses in OCDMA. Piscataway, N. J.
[3] Yi, C. Q. (2009) Application study of OCDMA-PON technology. Optical Communication Technology, 33(4): 18-20.
[4] Wang, X. Wang, X. H. Sun, Y. B. Li, Q. J. (2007) Advances in OCDMA systems and related devices. Semiconductor Optoelectronics, 28(2), 451-457.
[5] Li, C .Q. Sun, X. H. Zhang, M. D. Ding, D. (2005) Analysis and research of OCDMA system networking technology and core components. Journal of Quantum Electronics, 22(3): 326-333.
[6] Li, C. Q. Zhu, Y. C. Zhou, W. (2008) Research on tunable coder/ decoder for optical code division multiple access system. China Laser, 35(12): 1901-1905.

Ship classification based on random forest using static information from AIS data

Yitao Wang[1*] , Lei Yang[1], Xin Song[1] and Xuan Li[1]

[1]College of Aerospace Science and Engineering, National University of Defense Technology, Changsha, Hunan, 410073, China

[*]Corresponding author's e-mail: ytwang_15@nudt.edu.cn

Abstract. With the wide use of automatic identification system (AIS), a large amount of ship-related data has been provided for marine transportation analysis. Generally, AIS reports the type information of ships, but there are still many ships with type unknown in AIS data. It is necessary to develop algorithms which can identify ship type from AIS data. In this paper, we employ random forest to classify ships according to the static information from AIS messages. Moreover, the importance of static features is discussed, which explains the reason why some classes of ships are misclassified. The method of this paper is proved to be effective in ship classification using static information.

1. Introduction

Automatic identification system (AIS) is a navigation aid system which broadcasts the static, dynamic, voyage and safety information of ships [1]. In order to strength maritime safety, the International Maritime Organization (IMO) requires ships to install AIS equipment since 2004 [2]. The wide application of AIS promotes the development of intelligent shipping, and also plays an important role in route extraction [3], pollution prevention [4], maritime traffic feature discovery [5] and ship abnormal behavior detection [6-7].

AIS has 27 kinds of messages, in which message 5 contains the static and type information of ships. The ships type may be missing or mislabelled for various reasons [8], which probably implies the occurrence of illegal activities such as smuggling and illegal fishing [9]. It is necessary to develop the ship classification methods based on AIS data. Some researches have been done to solve this problem. Random forest [10] and KNN [11] is used to recognize ships type from static characteristics (such as ship size and draft). However, the static features considered in these methods are inadequate [10-11] and the AIS data are collected from a small region [11]. Kraus [12] combines static, geographical and dynamic features from AIS data in German Bight to identify six types of ships, while this method has the problem of data leakage.

In this paper, random forest is applied to classify ships by learning the potential pattern of static features from AIS data (message 5) received by the ocean satellite HY-1C/D [13] and HY-2B/C. The rest of this paper is organized as follows. In section 2, we firstly preprocess the static information and extract the static features, then the principle of random forest algorithm is briefly introduced. In section 3, the performance of random forest is evaluated by experiments on five types of ships. In section 4, conclusions and future works are presented.

2. Methodology

2.1. Data preprocessing and feature extraction

Figure 1 shows the quantities of the top20 kinds of ships in message 1 received by HY-2C. The abscissa in figure 1 is the codes of ship type, and the codes of passenger ships, cargo ships, tankers, fishing boats and tugs are 60~69, 70~ 79, 80 ~89, 30 and 52 respectively. The red curve is the accumulated proportion of ship quantity. According to the statistical results in figure 1, the five categories of ships mentioned above account for 93.44% of the total number of ships, and we select them to train a random forest.

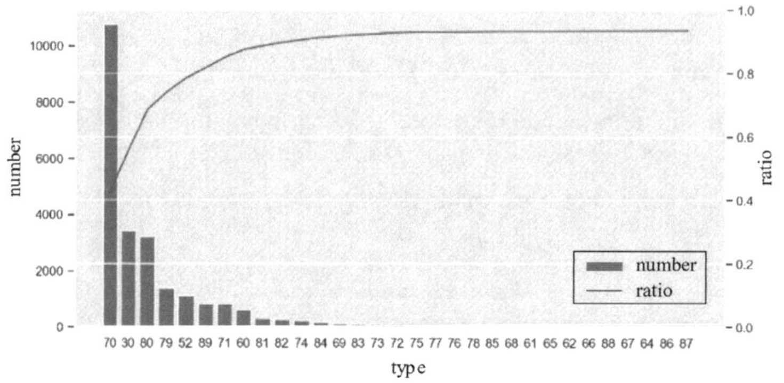

Figure 1. Quantities of different types of ships.

The original AIS static messages contain five fields named A, B, C, D and draught which can reflect the size information of ships. A, B, C and D are the distances from the reference point O used for reporting position to bow, stern, port and starboard respectively. The length and width of a ship can be calculated by equation (1).

$$\begin{cases} length = A + B \\ width = C + D \end{cases} \tag{1}$$

Except for the above 7 ship size features, other geometric features are extracted and they are defined as equation (2):

$$\begin{cases} ldivw = length / width \\ area = length * width \\ grith = length + width \end{cases} \tag{2}$$

Considering that ships static information can be incorrect, the obviously abnormal data should be removed in pretreatment process. Since the distributions of static data are various in different types of ships, the data of each type of ships need to be processed separately. Taking tankers as an example, its standardized features distribution is shown in figure 2(a). We use box plot to identify and remove outliers, and this method has no requirement on data distribution and is more conservative when filtering outliers. The specific methods are as follows. Firstly, calculate the upper quartile Q_u , lower quartile Q_l and the interquartile range IQR of each feature for one type of ships. Secondly, delete the data whose features are outside $[Q_l - 3IQR, Q_u + 3IQR]$. Figure 2(b) shows the data distribution of tankers after removing outliers. Figure 2(c) shows the distribution of original data of five types of ships, and figure 2(d) shows the data distribution of all ships after removing outliers. Table 1 shows the number of static data used in this paper.

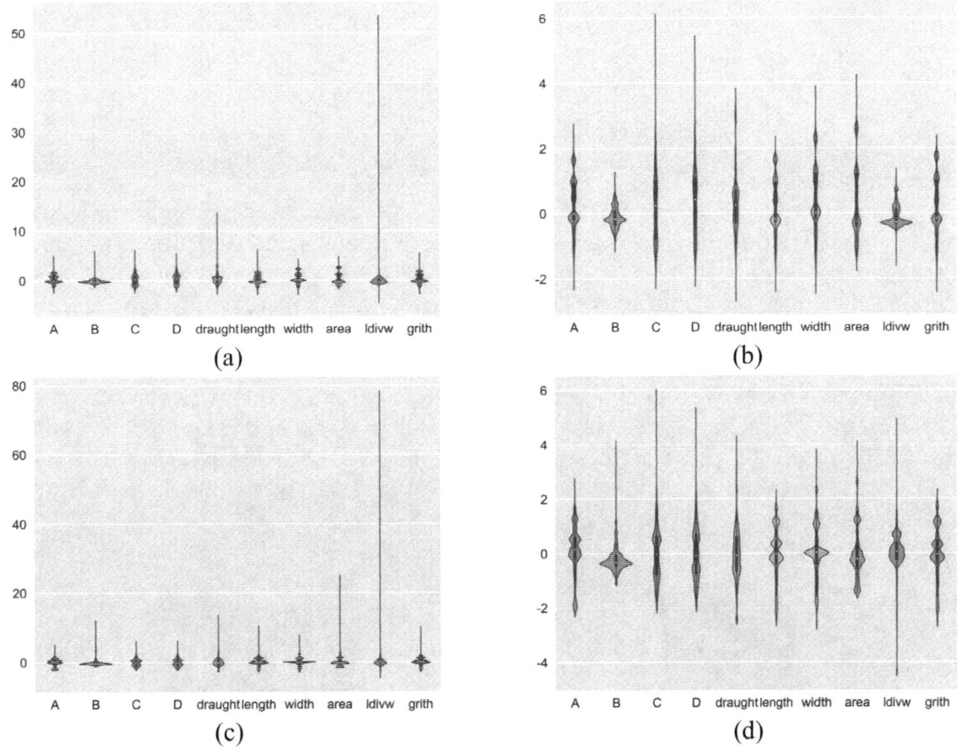

Figure 2. Distribution of original static data and static data after removing outliers.

Table 1. Quantities of static messages.

Type	Passenger ships	Cargo ships	Tankers	Fishing boats	Tugs
Number	1,285	61,477	21,094	5,968	2,046

2.2. Random Forest

Random forest is an integrated learning algorithm based on decision tree, which has the advantages of low deviation, low variance and strong generalization ability. The method for creating random forest is as follows:

- Randomly select n samples from the training samples, and use these samples to create a decision tree. The decision tree is trained by CART algorithm: m features are randomly selected at each node to be split, and then the feature k and threshold t_k that can minimize the cost function (equation (3)) are selected from the features, using which the sample set at the node is divided. $G_{left/right}$ is the impure of the left and right subsets (equation (3)), $m_{left/right}$ is the number of samples divided into left and right subsets, and $p_{i,n}$ is the proportion of class n samples in the i-th node;
- When the decision tree reaches the maximum depth or the cost function cannot be reduced, stop node splitting and terminate the creation of the decision tree;
- Repeat the first step to create a large number of decision trees to form a random forest. The prediction category of the sample is determined by the voting of the decision tree in random forest.

$$\begin{cases} J(k,t_k) = (m_{left}/m) * G_{left} + (m_{right}/m) * G_{right} \\ G_i = 1 - \sum_{n=1}^{q} p_{i,n}^2 \end{cases} \tag{3}$$

3. Results and discussion

In order to optimize the paraments of random forest and objectively evaluate the performance of 1D-CNN, we use stratified sampling to divide the data set into training set, validation set and testing set according to the ratio of 6:2:2.

The best parameters of the classifier are obtained by grid search. Considering the imbalance of the number of samples, the sample loss is weighted in training process. The weight of class I samples is the ratio of the total number of samples to the number of class I samples. Table 2 shows the best paraments for random forest. The total accuracy and F1 score of the classifier on testing set 0.8614 and 0.8693. Figure 3 shows the confusion matrixes of random forest on the validation set and testing set, where class 0 to class 5 represent passenger ships, tugs, tankers, fishing ships and cargo ships respectively. The model tends to confuse tankers with cargo ships, as well as misclassify passenger ships, tugs and fishing boats.

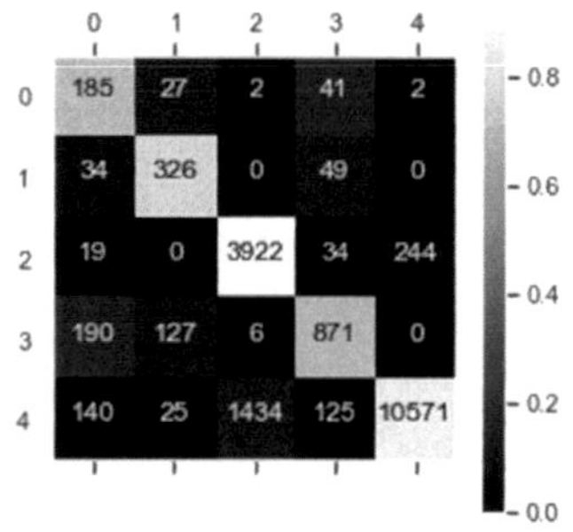

(a)

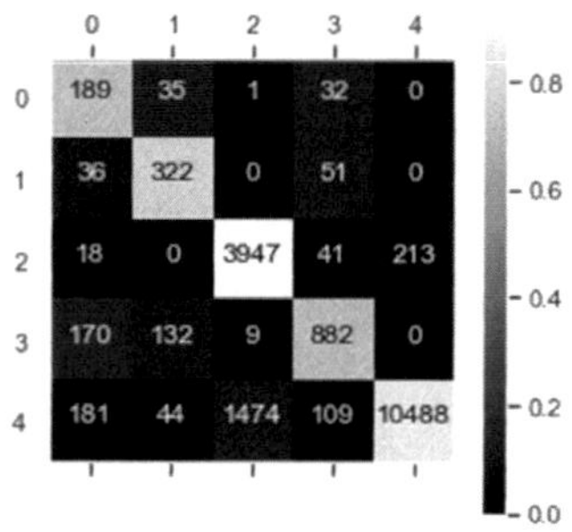

(b)

Figure 3. (a). confusion matrix on validation set, (b). confusion matrix on testing set.

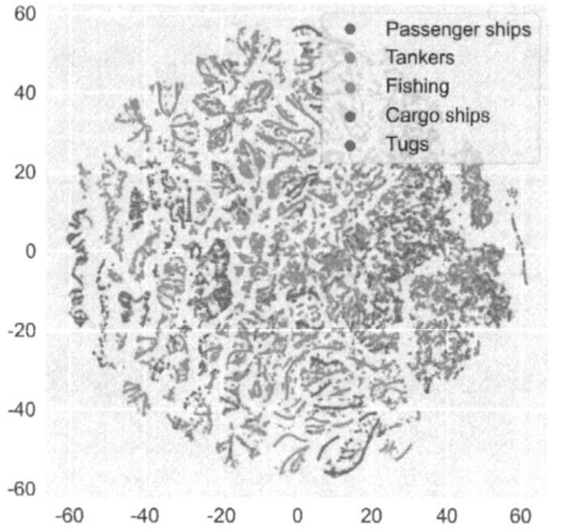

Figure 4. Visualization of static features.

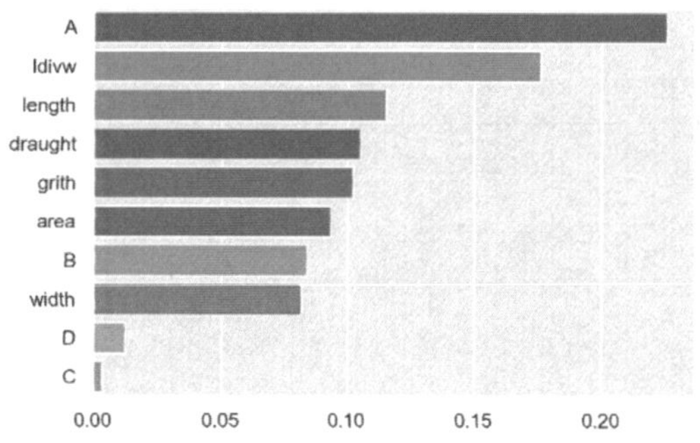

Figure 5. Static features importance.

To explain the confusion matrixes, we use t-SNE [14] to visualize the static features. t-SNE is a method of data visualization, it can map data from high-dimensional space to low-dimensional space. If two samples are similar in high-dimensional space, the distance between their maps in low-dimensional space will be close. Figure 4 shows the visual results of static features. The data distributions of passenger ships, tugs and fishing boats after dimensionality reduction overlaps partially, which also occurs on cargo ships and oil tankers. This shows that the static information of the misclassified ships is similar. Figure 5 shows the importance of static features. It can be seen that the ship's dimensional characteristics such as A, length and length-width ratio can better describe the ships features than draught. In addition, although the features C and D contribute little to the classifier, these two parameters have been reflected in the length-width ratio, girth, area and width, which proved the static features constructed in this paper are effective in the ship classification task.

4. Conclusion

In this paper, we use random forest to classify ships according to the static information in AIS data. Firstly, the static features are defined and extracted from AIS data, then experiments are conducted to get the best paraments of random forest, finally we attempt to explain the confusion matrixes by data visualization. Moreover, the importance of features is analysed, which proved the static features proposed in this paper are effective in ship type identification.

Experiments show that the static features may be similar between some kinds of ships. Using only static characteristics is not enough to distinguish five types of ships. In future work, we plan to combine static and dynamic information from AIS data to improve the performance of ship type classification.

Acknowledgments

HY-1C/D and HY-2B/C data were obtained from https://osdds.nsoas.org.cn. The authors would like to thank NSOAS for providing the data free of charge.

References

[1] Technical characteristics for an automatic identification system using time-division multiple access in the VHF maritime mobile band, Recommendation ITU-R M.1371-4. 2010. 3-4, 101-107. Available online: Template BR_Rec_2005.dot (itu.int).

[2] Tichavska M., Cabrera F., Tovar B., et al. (2015) Use of the Automatic Identification System in Academic Research. In: International Conference on Computer Aided Systems Theory. In: Computer Aided Systems Theory – EUROCAST 2015. Switzerland. pp. 33-40.

[3] Sheng, P., Yin, J. (2018) Extracting Shipping Route Patterns by Trajectory Clustering Model Based on Automatic Identification System Data. Sustainability, 10 (7): 1-13

[4] Ambjorn C. (2008) Seatrack Web forecasts and backtracking of oil spills - an efficient tool to find illegal spills using AIS. In: 2008 IEEE/OES US/EU-Baltic International Symposium. Tallinn, Estonia. pp. 1-9.

[5] Fournier, Mélanie, Casey H.R., Rezaee S., et al. (2018) Past, present, and future of the satellite-based automatic identification system: areas of applications (2004–2016). WMU Journal of Maritime Affairs, 17(3): 311-345.

[6] Rong H., Teixeira A.P., Soares C.G. (2020) Data mining approach to shipping route characterization and anomaly detection based on AIS data. Ocean Engineering, 198:106936.

[7] Castaldo F., Bastani V., Marcenaro L., et al. (2014) Abnormal vessel behavior detection in port areas based on Dynamic Bayesian Networks. In: International Conference on Information Fusion. Salamanca, Spain. pp. 1-7.

[8] Iphar, C., Napoli, A., Ray, C. (2015) Detection of false AIS messages for the improvement of maritime situational awareness. In: OCEANS 2015 - MTS/IEEE, Washington, DC. pp. 1-7.

[9] Longépé, N., Hajduch, G., Ardianto, R., Joux, R. et al. (2018) Completing fishing monitoring with spaceborne Vessel Detection System (VDS) and Automatic Identification System (AIS) to assess illegal fishing in Indonesia. Mar. Pollut. Bull, 131: 33-39.

[10] Zhong H., Song X., Yang L. (2019) Vessel Classification from Space-based AIS Data Using Random Forest. In: 2019 5th International Conference on Big Data and Information Analytics (BigDIA). Kunming. pp. 9-12.
[11] Damastuti, N., Siti Aisjah, A. (2019) Classification of Ship-Based Automatic Identification Systems Using K-Nearest Neighbors. In: 2019 International Seminar on Application for Technology of Information and Communication (iSemantic). Semarang. pp. 331-335.
[12] Kraus, P., Mohrdieck, C., Schwenker, F. (2018) Ship classification based on trajectory data with machine-learning methods. In: 2018 19th International Radar Symposium (IRS). Bonn. pp. 1-10.
[13] Zhang K., Ding Z., Gao J., et al. (2019) Design and Application of AIS on HY-1C/D Satellite. Spacecraft Engineering, 28(05): 26-31.
[14] Maaten, L.V.D., Geoffrey, E.H. (2008) Visualizing Data using t-SNE. Journal of Machine Learning Research, 2579-2605.

Journal of Physics: Conference Series **2113** (2021) 012073 doi:10.1088/1742-6596/2113/1/012073

Research on Accurate Location Algorithm of Optimized Multi-source Data Fusion Based on Improved GRU Network

Zhongzhi Li[1], Xiaoxue Li[2*] and Yufeng Huang[2]

[1] College of Computer Science, Shenyang Aerospace University, Shenyang, Liaoning, 110136, China

[2] College of Electronic and Information Engineering, Shenyang Aerospace University, Shenyang, Liaoning, 110136, China

*Corresponding author's e-mail: lixiaoxue@stu.sau.edu.cn

Abstract. The uncertainty of positioning trajectory and the existence of abnormal trajectory points are major challenges in the field of navigation and positioning. The method that only relies on global positioning system (GPS) has a large error, and the traditional location algorithms still have shortcomings in location efficiency and accuracy. Due to the successful exploration of deep learning in the field of navigation and positioning, an accurate positioning algorithm based on improved gated recurrent unit (GRU) and multi-source data fusion was proposed in this paper. Firstly, interpolation was used to fuse GPS data and inertial measurement unit (IMU) data to overcome the defect of single data. Then, an improved gated recurrent unit was introduced to process the data fusion. On this basis, the attention mechanism (AM) was added to reduce the loss of historical information and strengthen the influence of important information, so as to achieve accurate positioning. Compared with other advanced algorithms with regard to L2 norm and time of training iteration, the training time of this algorithm is reduced by more than 27%, and the highest optimization rate is 12.08%. Therefore, this algorithm can obtain higher positioning accuracy in a shorter time.

1. Introduction

The engineering application research and practice of navigation and positioning system are widely used in various living fields such as water conservancy, bridge and tunnel, water transportation, municipal administration and so on [1]. GPS positioning equipment needs to receive satellite signals in real time to get the moving track of moving objects. If there is a serious sheltered area in the positioning process, the noise, low-frequency GPS data and other factors will directly lead to the receiving equipment unable to receive satellite signals and data drift [2].

In order to reduce these errors, earlier studies mainly improved the accuracy by simply combining IMU and GPS. For example, Wang Ailin and others combined IMU, GPS and DSP processors to improve the accuracy of navigation [3]. In recent years, the research of integrated navigation has gradually matured, and the integration of various technologies has further improved the positioning accuracy. For example, Li Qingli and others proposed an integrated navigation positioning algorithm combined with Kalman filter, which effectively improved the positioning accuracy of the integrated navigation system [4]. Although the combination of IMU, GPS and other technologies has brought dawn to pedestrian navigation, its accuracy is still far from enough for low and medium precision inertial navigation. In order to improve this situation, some studies used deep learning to modify the positioning results of integrated navigation. For example, Deng Tianmin and others designed an IMU and GPS integrated navigation based on recurrent neural network, which greatly improved the positioning accuracy [5].

As a kind of deep learning algorithm, GRU has the advantages of few parameters and high efficiency, it covers a wide range of application fields and has developed well in some fields. Although some articles have studied the application of GRU in the field of positioning, there are few studies on the improvement of pedestrian navigation by GRU. Therefore, this paper used the IMU sensors in Android smartphone, collected and exported data through sensor stream IMU GPS, and finally used GRU algorithm combined with attention mechanism for real-time data correction to improve positioning accuracy. The main contributions of this paper are as follows:

1. Based on the research of traditional positioning data, this paper obtains pedestrian motion data by multi-source sensors in inertial measurement unit (IMU) on smart phone.

2. In order to deal with the problem of the inconsistency of data frequency collected by the low-frequency GPS sensor and the high-frequency IMU sensor, the interpolation algorithm is used to fuse GPS data and IMU data.

3. Based on the GRU algorithm, the attention mechanism (AM) is introduced to give different weights to the implicit state of GRU through mapping weighting and learning parameter matrix. The experimental results show that the proposed algorithm can effectively improve the positioning accuracy and get the path trajectory closer to the real route.

2. Gated recurrent unit network structure

2.1. Gated recurrent unit

The GRU network, as one of the variants of LSTM, adds gate functions to the network to deal with the problem of gradient disappearance and gradient explosion in RNN [6], such as adding gate structures and memory units [7] [8]. Compared with LSTM, its structure is simpler and the amount of calculation is much smaller. Therefore, GRU can be regarded as the simplification and improvement of LSTM, which is comparable to LSTM in performance and has been proved to be effective [9] [10].

The improvement of GRU is mainly reflected in two aspects: (1) different position states in the network structure have different effects on the current hidden layer nodes: the farther the distance, the smaller the weight; (2) If the error is generated by one or more information, the hidden layer only updates the information weight of the corresponding sequence.

The function definition of hidden layer state h of gated recurrent unit at time t is shown in formula (1):

$$h_t = Z_t h_{t-1} + (1 - Z_t)\widetilde{h}_{t-1} \tag{1}$$

where Z_t is the weight of the update gate, h_{t-1} is the previous state of the hidden layer, and the current state of the hidden layer h_t is generated by the update area of the storage unit. The update door can calculate the information retained in the previous memory. The definition of the update door is shown in formula (2):

$$Z_t = \sigma(b_i^z + \sum_j W_i^z x_t + \sum_j U_i^z h_{t-1}) \tag{2}$$

Among them, σ is the activation function, x_t is the input variable, w_i^z is the parameter of the input variable, and U_i^z is the parameter of the hidden layer. The update gate determines the weight of information that h_{t-1} passes to the next state. When $Z_t \approx 1$, h_{t-1} is almost completely transferred to h_t; When $Z_t \approx 0$, the new hidden layer state h_t is transferred to the next layer.

h_t is a memory unit. A new storage unit can be obtained through the previous hidden state h_{t-1} and the current new input, that is, the new information and historical information h_t can be integrated together, and the fusion of sequences can be determined according to the sequence vector. The definition of memory unit is shown in formula (3):

$$\widetilde{h}_t = \tanh(Wx_t + r_t U h_{t-1}) \tag{3}$$

Where *tanh* is the activation function and r_t is the weight of the reset door. Reset gate is a unique gate in GRU, which determines the influence of h_{t-1} on output h_t. If h_{t-1} is not related to the storage unit, reset door will eliminate the previous hidden layer state. The definition of this function is shown in formula (4):

$$r_t = \sigma(b_i^r + \sum_j W_i^r x_t + \sum_j U_i^r h_{t-1}) \tag{4}$$

GRU uses a special gate mechanism to control gradient propagation. Back propagation and gradient descent are used to train and update the weight of the gate unit structure to alleviate the problem of gradient disappearance and explosion. At the same time, the computer memory resources occupied is relatively small, and the efficiency is high.

2.2. Attention mechanism

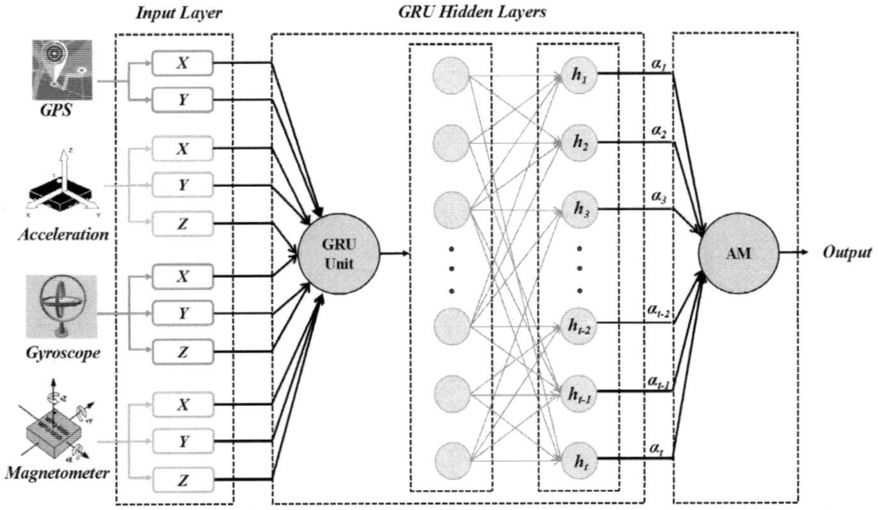

Figure 1. Framework of improved GRU model based on attention mechanism

When processing a long input sequence, since the output h_t of GRU neural network is used as the information representation of the whole input sequence, it means that all the information of the input sequence is compressed into a fixed length vector. With the increasing length of input sequence, the ability of the whole model to process information will be limited and weakened. In order to solve the problem that the data sequence is long, the attention mechanism is introduced in the decoding stage of neural network. Attention mechanism can be regarded as a simple three-layer neural network, including input layer, hidden layer and output layer. The output vector of multi-layer GRU is used as the input of the attention mechanism.

The overall framework of the improved GRU model based on attention mechanism proposed in this paper is shown in Figure 1.

3. Experimental results and analysis

3.1. Parameter configuration and experimental data set

The experimental environment of this paper is configured as Intel Core i5-8300h 2.30Ghz processor and NVIDIA GTX 1050Ti graphics card. The software environment is Python 3.6 and tensorflow 1.14.0. The experiment is implemented in the application of jupyter notebook interactive computing.

In order to verify the performance of the improved GRU network optimized multi-source data fusion precision positioning algorithm proposed in this paper, the "IMU + GPS stream" professional software is used to collect motion data. The application records the measurement results of the sensor in the format of "CSV", and the functions implemented by Python are used for analysis and data preprocessing.

Table 1. Sample data set

GPS		Acceleration			Gyroscope			Magnetometer			Delta Time
X	Y	X	Y	Z	X	Y	Z	X	Y	Z	
0.0	0.0	-0.2313	0.0514	0.1254	-0.022	-0.075	0.011	-10.7716	2.9626	-54.9108	0.0000
0.0	0.0	0.2821	0.1724	-0.5415	0.019	-0.055	-0.018	-10.0406	4.3480	-54.7535	0.03281
0.0	0.0	0.2686	0.1623	-0.5089	0.002	-0.070	-0.010	-10.2197	4.4332	-54.8432	0.0025
0.0	0.0	0.2942	0.0598	-0.3914	-0.035	-0.093	-0.005	-10.5829	4.6058	-55.0253	0.0051
0.0	0.0	-0.1882	0.0211	0.2422	-0.224	-0.307	0.095	-11.1195	3.9945	-54.3870	0.0378

Table 1 shows sample data for activity records. In this paper, the tag value of motion record is completed using the online Ordnance Survey map service [11]. This service consists of a route creation application that allows users to draw path points on satellite images. Finally, the tag value data of user motion is aligned with multi-source sensor data according to interpolation algorithm. The results before and after interpolation are shown in Figure 2. The corresponding sequence has the same color in "ground truth" and GPS, where, (a) represents the misaligned sequence and (b) represents the aligned sequence.

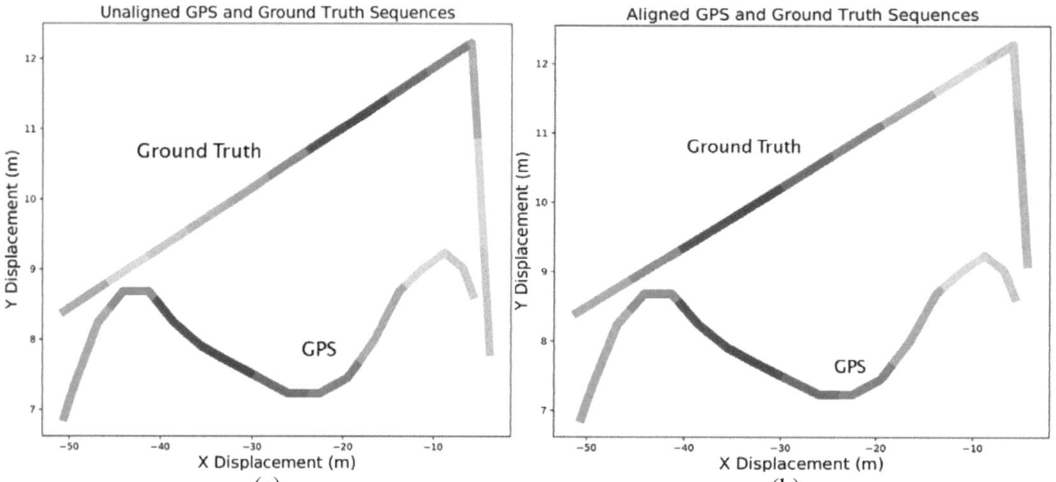

(a) (b)

Figure 2. Sequence of the "Ground truth" and GPS

3.2. Evaluating indicator

In this paper, L2 norm evaluation index is used to compare the accuracy differences between models. L2 norm is a function space composed of square integrable function on measure space, which is defined as the square root of the integral of the square of the absolute value of the function. The calculation formula of L2 norm is shown in the following formula.

$$\|x\|_2 = \left(\sum_{i=1}^{N} |x_i|^2\right)^{\frac{1}{2}} \tag{5}$$

Where, x is the absolute value of the difference between the model predicted value and the real label value y. The smaller the difference between the model predicted value and the real label value, the closer the model predicted value is to the real value, that is, the higher the accuracy of the model is, the smaller the prediction error is.

3.3. Optimized hyperparameter set

In order to obtain the optimal experimental results of the model, this paper makes a comparative experiment on the data sequence length of the input neural network, which is the most important super parameters. The sequence lengths used in the comparative experiment are 200, 400, 800, 1500 and 3500 respectively. For the test of all training processes, one tenth of the length of different sequences is selected as the test set. Figure 3 (a) shows the change of loss value of each sequence length data set during training.

It can be seen from this experiment that the sequence length of the lowest loss is 400. Therefore, a sequence length of 400 is used thereafter.

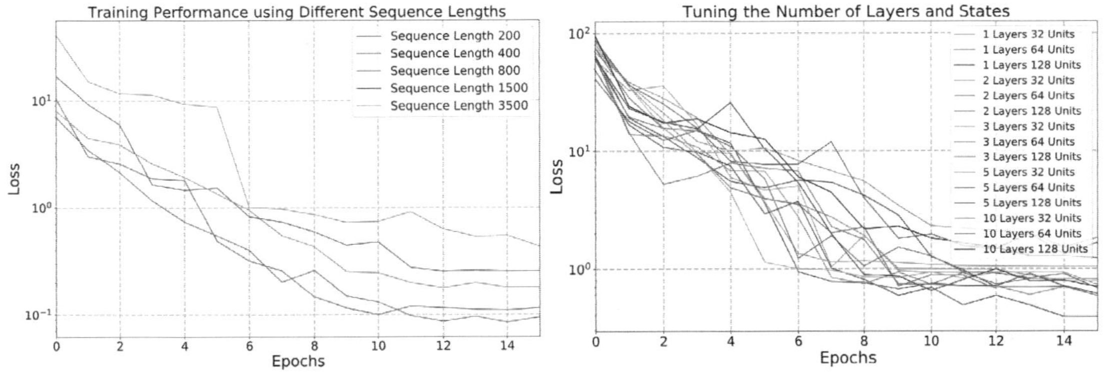

(a) (b)

Figure 3. Comparison of different data sequence and model structure

In order to reduce the training time, the structure of the neural network unit in the model is changed. 1, 2, 3, 5 and 10 neural network layers are tried, and different number of state tests on each neural network layer are carried out. The results are shown in Figure 3 (b).

Figure 3 (b) shows the network converging at the lowest loss value often has 2 or 3 layers of neural network. Although there is little difference in the loss value between them, it have an impact on the time to complete an iteration. Three-layer neural network makes the iteration time of the model about 1.5 times longer than two-layer neural network, and the convergence value of three-layer network and two-layer network tends to be consistent. Therefore, this paper selects a two-layer network with 128 hidden units as the final model structure. The detailed parameters of the model in this paper are shown in Table 2.

Table 2. Setting of super parameters of the model

Parameter	Values
Sequence length	400
Batch amount	128
Neural network layer	Two layers of GRU neural network with 128 neurons, the last being a fully connected layer
Activation function	Linear activation function

3.4. Comparative experiment

Table 3. Comparative experiments of different algorithms

Contrast algorithm	Test error	Optimization rate
Original GPS	15.40	N/A
Kalman filtering	15.75	-2.27%
CNN	15.21	+1.23%
LSTM	15.03	+2.4%
GRU	14.58	+5.32%
The improved GRU	13.54	+12.08%

It can be seen from table 3 that during the comparative test, the model proposed in this paper has the smallest test error and the highest optimization rate, which effectively improves the positioning accuracy of users in the process of movement.

4. Conclusion

The existing positioning algorithms have large deviation and low positioning accuracy. The positioning data obtained by single GPS often leads to large errors due to noise and low-frequency GPS data, which affects the accuracy of navigation and positioning. This paper presents an improved GRU positioning method based on multi-source data fusion of GPS and IMU. By introducing inertial navigation data, the problem of low quality of single GPS data itself is made up. The deep learning GRU combined with the attention mechanism is introduced to optimize the location algorithm. Experiments show that the proposed algorithm is better than LSTM, RNN and other algorithms, can effectively shorten the positioning time, and the optimization rate is 12.08%. The network structure of GRU further improves the performance of the model to a certain extent, which can better solve the problem of low accuracy of GPS navigation and positioning.

Acknowledge

This research was supported by 2020 Shenyang Aerospace University Students' Innovative Training Pr ogram (Project Number: 202010143003、202010143104、202010143135).

References

[1] Shen bin, Jiang Changjun, Zhang Zhaohui, Zheng Chunlei, Liu Haitao. Design and implementation of a distributed map matching system based on massive GPS data [J]. Small microcomputer system, 2007 (03): 479-481

[2] Zhong Haili, Tong Ruihua, Li Jun, Chen Hongsheng. Research on GPS positioning and map matching methods [J]. Small microcomputer system, 2003 (01): 109-113

[3] Wang Ailin, Chen yingtu, Xu Wenjin. Design of a GPS + IMU integrated navigation module based on TMS320F206 [J]. Silicon Valley, 2014,7 (09): 27-28

[4] Li Qingli, Xue Yongqi, Shi Pengfei. Beidou / DR Integrated Navigation fusion algorithm and its application in logistics [J]. Small microcomputer system, 2005 (12): 2150-2153

[5] Deng Tianmin, Yang Qizhi, Fang Fang, Yue Yunxia. GPS / IMU integrated navigation method based on nonlinear adaptive regression neural network [J]. Science, technology and engineering, 2019,19 (24): 274-280

[6] J. Wang, Q. Gu, J. Wu, G. Liu, and Z. Xiong, Traffific speed prediction and congestion source exploration: A deep learning method, in Proc. IEEEICDM, Barcelona, Spain, Dec. 2016, pp. 499–508.

[7] K. Cho et al. (2014). Learning phrase representations using RNN encoder-decoder for statistical machine translation. [Online]. Available: https://arxiv.org/abs/1406.1078

[8] S. Hochreiter and J. Schmidhuber, Long short-term memory,Neural Comput., vol. 9, no. 8, pp. 1735–1780, 1997. doi: 10.1162/neco.1997.9.8.1735.
[9] K. Cho, B. Van Merriënboer, D. Bahdanau, and Y. Bengio, On the properties of neural machine translation: Encoder-decoder approaches, in Proc. 8th Workshop Syntax, Semantics Struct. Stat. Transl., 2014, p. 103.
[10] J. Chung, C. Gulcehre, K. Cho, and Y. Bengio, Empirical evaluation of gated recurrent neural networks on sequence modeling,in Proc. NIPS Deep Learn. Workshop, Dec. 2014, p. 47.
[11] Ordnance Survey Maps. url: https://osmaps.ordnancesurvey.co.uk/.

Research on threat detection in cyber security based on machine learning

Qiwei Ke[1]

[1] Beijing University of Technology, Beijing, 100124, China

*Corresponding author's e-mail: keqiwei@emails.bjut.edu.cn

Abstract. The volume of the data has been rocketed since the new information era arrives. How to protect information privacy and detect the threat whenever the intrusion happens has become a hot topic. In this essay, we are going to look into the latest machine learning techniques (including deep learning) which are applicable in intrusion detection, malware detection, and vulnerability detection. And the comparison between the traditional methods and novel methods will be demonstrated in detail. Specially, we would examine the whole experiment process of representative examples from recent research projects to give a better insight into how the models function and cooperate. In addition, some potential problems and improvements would be illustrated at the end of each section.

1. Introduction

In recent years, 5G technology has established a solid stage for the internet of things (IoT). The widespread of IoT devices strengthens our connection with technical equipment around us. However, it also brings problematic challenges about cyber security at the same time as the information exchange becomes more frequent than before. Upgrading our threat detection approaches seems extremely necessary. In the previous literature, there were a few articles about the overall review of threat detection which covered diverse related fields. In this paper, it is aimed to make a survey research, which gives an outline of various aspects of machine learning-based threat detection systems by explaining the model used in each experiment and evaluate the corresponding results. At the same time, a comparative study is presented for depicting the differences between the older technique and the latest technique.

The rest of the paper is organized as follows.

Related works about threat detection are demonstrated in section 2. In the next section examples of intrusion detection are given. In section 4, recent researches on the measures of malware detection are depicted. In the following section, we will represent the achievement within the area of vulnerability detection. Finally, the paper is concluded with a discussion of potential improvements and problems.

2. Related work

This section discusses the surveys and review articles on threat detection. Al-Mhiqani et al. [1] described the insider threat detection and made a classification that depended on the insider. Arslan et al. [2] investigated the common attacks on mobile phones and analyzed the results from the captured data packets. Shaukat et al. [3] surveyed three machine-learning detection methods, namely decision tree, deep belief network, and support vector machine. Yuan et al. [4] covered the reason why the deep learning method was needed in insider threat detection and listed some existing examples to better emphasize the conclusion. Salloum et al. [5] represented the applications to intrusion detection based on machine learning, deep learning, and data mining techniques. Sahoo et al. [6] reviewed the contributions of feature presentation and learning algorithms for URL detection.

The main contributions of this review can be summarized as follows.

Firstly, most of the previous reviews mainly focused on one specific domain in threat detection while intrusion detection, malware detection, and vulnerability detection are discussed in this essay. Secondly, comprehensive detection procedures and detailed results are given in each topic rather than merely providing theoretical support. Finally, both traditional machine learning techniques and deep learning algorithms are covered in the essay.

3. Network intrusion detection

3.1 Background

With the booming usage of the Internet, it has a deeper influence on daily life, including work and study. However, a gradual rise of threats in the network has been detected in recent years. So lots of research efforts have been made to develop an Intrusion Detection System (IDS) to monitor security events and make the judgment whether they are intrusive or not. IDS can be categorized into signature-based and anomaly detection-based [7]. The former system is effective when the intrusion is known because it relies on the pre-installed signatures in it. The latter one is proven to be more practical in reality since most of the attacks are novel. To build the above system, machine learning methods, like Support Vector Machines (SVM) and Random Forests, and deep learning methods which depend on the calculation results from stacked deep networks can be utilized [10]. The crucial process behind intrusion detection is feature selection which can greatly reduce the error rate [8, 9]. In order to improve the accuracy of the detection model, deep learning techniques are now dominant in this area instead of pure machine learning techniques. The main reason is that deep learners can independently extract more concrete features from the samples while the majority of machine learning techniques require expert interaction to distinguish the features. The differences can be found in the following experiments.

3.2 Previous solution

In 2016, Javaid et al. [7] adopted a Self-taught Learning (STL) model based on a deep learning technique to solve the problem of intrusion detection. Two steps were implemented to classify the samples. Firstly, a clear feature representation could be extracted from a large collection of unlabeled data. The sparse auto encoder was chosen because it could be put into practice easily and had a relatively outstanding performance. Secondly, this representation was used for testing with labeled data.

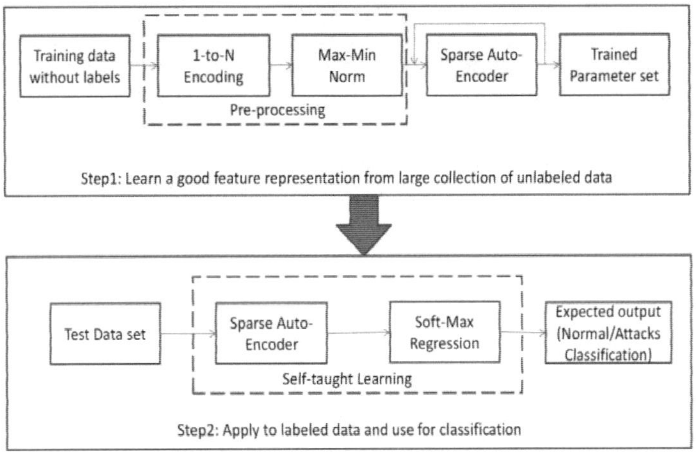

Figure 1. The detailed procedure of approaches involved in Javaid's team.

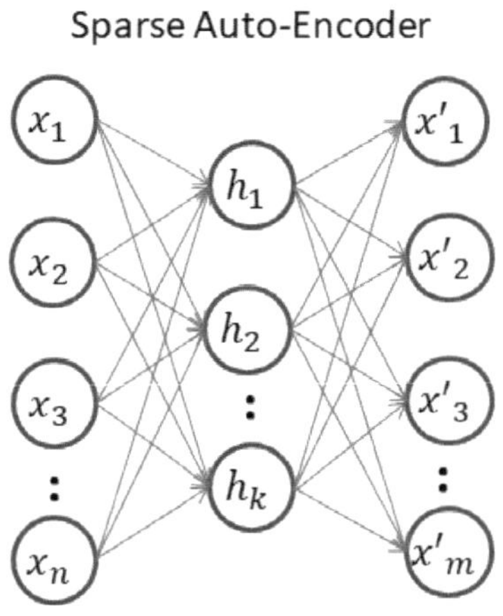

Figure 2. Structure of sparse auto-encoder. Note that bias (b) and weight (w) are not shown in the graph for clarity.

3.2.1 NSL-KDD Data Set

NSL-KDD Data Set is an updated version of the previous KDD Cup 99 Data set. Each sample contains 41 characteristics and is categorized into a specific group. It removes the redundancy of the old data set which may result in poor classification and less realism. A brief explanation of the attack type is given below.

- Denial of Service Attack (DOS): the attackers make the machine crush by sending loads of requests to occupy all the usable resources provided.
- User to Root Attack (U2R): attackers managed to obtain the root access by taking advantage of vulnerabilities based on a normal user account.
- Remote to Local Attack (R2L): attackers are able to send packets to the user's server from a remote server.
- Probing attack: attackers collect private information about computers without notice from security controls.

Table 1. Data set used in Javaid's team.

NSL-KDD	Training	Test
Normal	67343	9711
DoS Attack	45927	7458
U2R Attack	52	67
R2L Attack	995	2887
Probe	11656	2421

3.2.2 Evaluation

True Positive: the number of positive examples correctly classified

True Negative: the number of negative examples correctly classified as negative

False Positive: the number of negative examples wrongly classified as positive

False Negative: the number of positive examples wrongly classified as negative

Accuracy: the percentage of correctly classified records over the total number of records.

Precision: the ratio of the number of true positives samples divided by the sum of true positives and false positives samples.

Recall: the ratio of the number of true positives samples divided by the sum of true positives and false negatives samples.

F-Measure: the harmonic mean of precision and recall and represents a balance between them.

Table 2. Evaluation when using self-taught learning and soft-max regression (apply directly without feature learning).

2-class training data	Soft-max regression	Self-taught learning
Accuracy	97.0%	98.2%
Precision	97.2%	98.8%
Recall	96.2%	97.6%
F-measure	96.8%	98.2%

From the table, the conclusion can be drawn that with the assistance of feature learning, all four criteria for STL are at least 1% higher than the pure soft-max regression. This kind of difference can be considerable if the samples are large enough.

3.3 Update solution

In 2018, Shone et al. [11] proposed a model that combined the Non-Symmetric Deep Auto Encoder (NDAE) (deep-learning model) with the Random Forest (RF) (machine learning model). The combination showed better or at least match results from similar research and it dramatically shortened the training time. The auto encoder was an unsupervised feature extraction algorithm based on a neural network that using backpropagation to minimize the difference between the inputs and expected output. Significantly, the subsequent higher layers extracted higher-order features. As for the reason why the NDAE was applied, it could decrease the overheads of computations and time without a large impact on accuracy and efficiency. RF could be considered as an algorithm to boost the cooperation between the weak learners. In the experiment, the processed feature pattern was sent to the RF after two stacked NDAE structures. The details of the NDAE and classification model are shown below, including the parameters for the input layer and hidden layers.

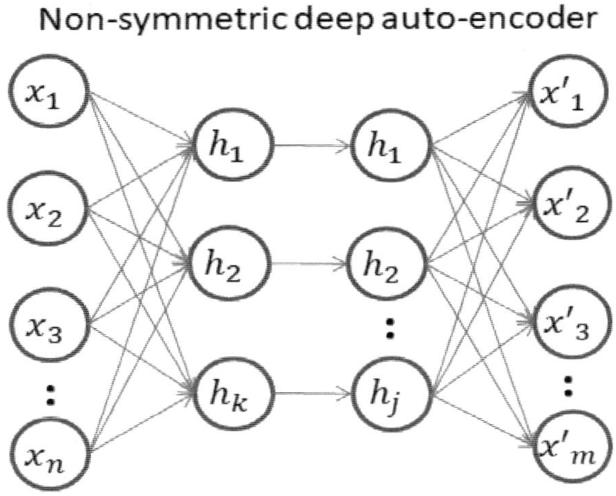

Figure 3. Structure of non-symmetric deep auto encoder. Note that bias (b) and weight (w) are not shown in the graph for clarity.

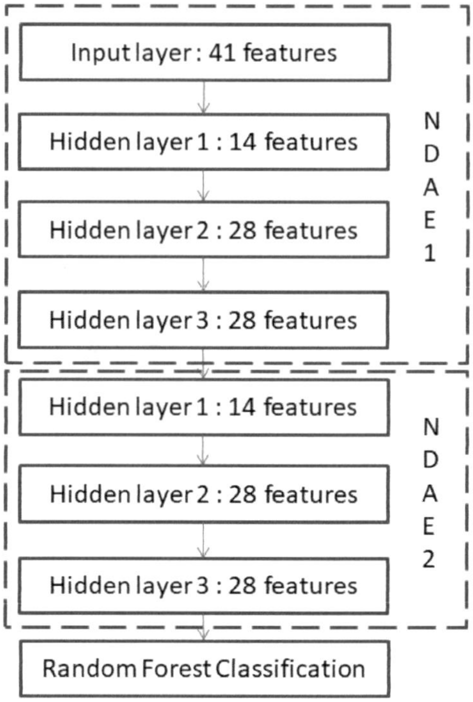

Figure 4. Stacked NDAE Classification Model proposed by Shone's team

Table 3. Data set used by Shone's team.

NSL-KDD	Training	Test
Normal	67343	9711
DoS attack	45,927	5,741
U2R attack	52	37
R2L attack	995	2199
Probe	11,656	1,106

3.3.1 Evaluation
When it came to the comparison between deep belief networks and Stacked Non-Symmetric Deep Auto-Encoder, it was evident that S-NDAE took a lead in all five criteria. By maximizing the benefit of both the deep learning method and traditional machine learning method, more practical features could be obtained from the samples which significantly increased the effects of S-NDAE.

Table 4. Evaluation for 5-class performance based on Deep Belief Networks and Stacked Non-Symmetric Deep Auto-Encoder.

5-class	DBN	S-NDAE
Accuracy	80.58%	85.42%
Precision	88.10%	100.00%
Recall	80.58%	85.42%
F-measure	84.08%	87.37%
False Alarm	19.42%	14.58%

3.4 Discussion

About the better choice between the STL model and S-NDAE, the further experiment should be carried out. Although it is noticeable that the accuracy of STL (98%) is much higher than the S-NDAE (85%), some details should be emphasized.

- To what extent may the number of kinds of classification influences the accuracy. (2-class in STL model and 5-class in S-NDAE model)
- To what extent may the pre-processing process (1-to-N Encoding and Max-Min Norm) in the STL model influences the accuracy.

Also, we can not ignore the challenges that may encounter in the future.

- How to perfect our feature selections from samples.
- More labeled samples should be created for further network training.
- How to prepare for the drastic growth in the volume of network data.
- How to design a proper norm to cooperate with diverse network protocols.

4. Malware detection

4.1 Background

With the development of mobile technology, Android devices have become a popular choice for users. However, devices with Android applications installed are exposed to diverse security threats due to the openness [12]. Third-party applications from the Android market which provide various choices for users may hide suspicious programs at the same time. Attackers take advantage of this trend and attempt to obtain user's private information including personal photos and records like address books via tempting victims to install malware. It is the malware used by attackers to infiltrate the system like spyware and worms [13]. By simply warning the users of required permissions, the success is limited since it needs domain-level knowledge to distinguish [12]. With the assistance of newly-designed malware detection techniques, this threat started to be mitigated.

4.2 Previous solution

In 2014, Yuan et al. [14] put forward a machine learning-based method that specific features could be extracted from the static and dynamic analysis of applications. Significantly, these methods could be time-saving because there was no need to design and upgrade detection rules manually. By fully making use of these, Yuan's team managed to reach 96% accuracy under real-world simulation. The detailed detection procedure is illustrated below.

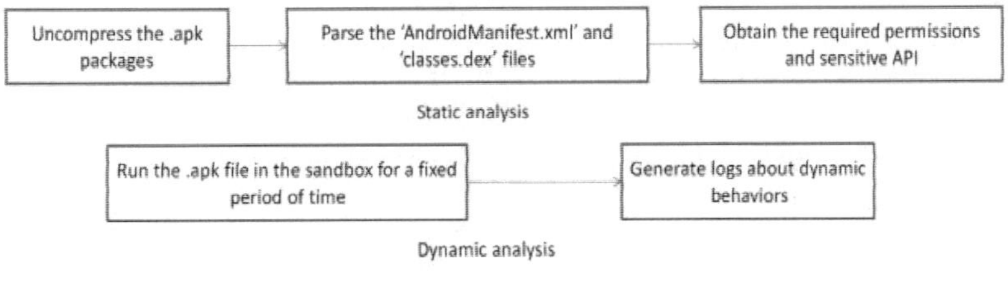

Figure 5. Detailed procedure of static and dynamic analysis

After uncompressing and parsing the corresponding apk files, required permissions from this application and sensitive application program interface (API) can be revealed. In addition, through simulating the application in the sandbox, researchers can understand the dynamic behaviors with the help of logs. These kinds of samples (more than 200 features) facilitate the later training process in the deep learning model provided below.

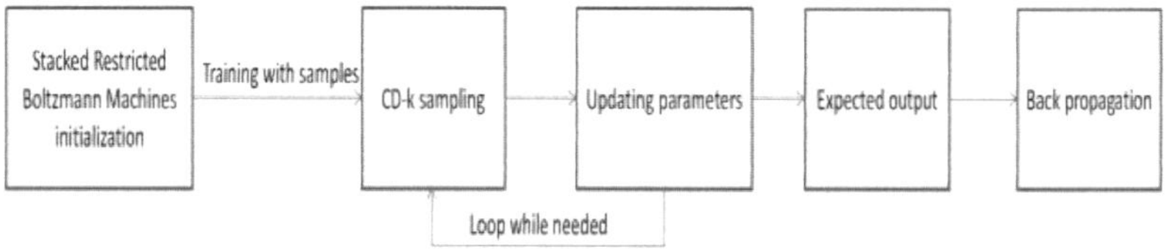

Figure 6. Deep learning model implemented in Yuan's team.

In the pre-training phase, a deep belief network that contains stacked restricted Boltzmann machines is utilized to characterize Android apps. Parameters can be updated continuously while the features input into the model. After a certain period of training section, the model can be used to distinguish the unknown samples from test sets. CD-k sampling not only increases the accuracy but also optimizes the training efficiency. For the back-propagation, the network is fine-tuned by labeled samples in a supervised manner. Yuan's team tried multiple traditional methods, namely logistic regression (LR), multi-layer perception (MLP), and support vector machine (SVM), and compared the actual effect with the deep learning method. The test result is given below for further analysis.

4.2.1 Evaluation

Under the same training and testing environment, those who took traditional methods as their model received unsatisfying results (78.0%,79.5%,80.0%). Conversely, the deep learning method achieved 96.5% accuracy among 200 testing samples which seemed to be applicable in real circumstances.

Table 5. Accuracy comparison between traditional method (LR&MLP&SVM) and deep learning method.

Model	Training / testing	Accuracy
Logistic Regression	300/200	78.0%
Multi-layer Perception	300/200	79.5%
Support Vector Machine	300/200	80.0%
Deep learning	300/200	**96.5%**

4.3 Update solution

In 2018, Kim et al. [15] designed an advanced approach to detect malware and got a more promising result. Comparing to Yuan's method, they made some essential changes to the previous model. The whole procedure is shown below for reference.

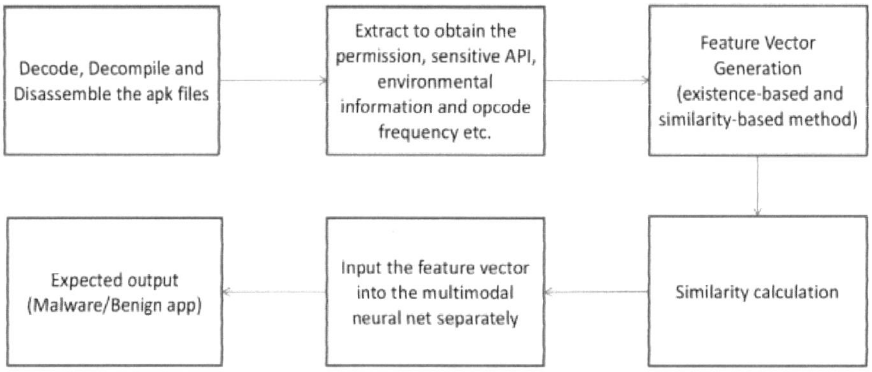

Figure 7. The overall architecture of framework proposed by Kim's team

Existence-based and similarity-based feature extraction methods were cooperated to represent the characteristics. Due to this, the advantages of encompassing multiple features could be strengthened to handle complex tasks when many properties were extremely similar between normal and benign apps. Different initial neural networks processed different types of features. The similarity-based feature (method opcode, method API, and shared library function feature vectors) focused on the elements which were similar to the malware representatives in the malicious feature database. The existence-based feature (string, permission, component, and environmental feature vectors) focused on the elements which only represented the existence of features in the malicious feature database.

Table 6. The Parameters of multimodal neural network.

Network	Layer	Number of neurons	Activation function
Initial Deep Neural Network	Input	5000	ReLU
	Hidden	2500	ReLU
	Hidden	1000	ReLU
Final Deep Neural Network	Merging	500	ReLU
	Hidden	100	ReLU
	Hidden	10	ReLU
	Output	1	Sigmoid

The initial network was built by five deep neural networks with separate input of five feature vectors (the permission/component/ environmental feature, String feature, method opcode feature, method API feature, shared library function opcode feature). The final network was one deep neural network and could output final results. The ReLU was able to make the model free of vanishing gradient issues.

4.3.1 Evaluation

The naïve binary and frequencies of the words were not comparable with the multi-feature-based algorithm, especially the formal one, which even had less than half the accuracy of the discussed method above. The substantial number of the samples contributed to the success of Kim's team.

Table 7. The evaluation among different chosen feature vectors.

Model	Malware samples/benign samples	Accuracy
uses a naïve binary as a feature vector	21260/20000	43%
uses the frequencies of the words	21260/20000	62%
Kim's team	21260/20000	**98%**

4.4 Discussion

One thing worth to be demonstrated is that the second method is purely based on static observation. Better results may probably be achieved by adding dynamic feature patterns into the design. The first method has already witnessed the beauty of dynamic analysis and it gives us an insight into how to take the first step. However, recent researches also ring the alarm that we should consolidate our processing power in order to meet the challenge from polymorphic and metamorphic as they can alter their behaviors in a short period of time [20].

5. Vulnerability detection

5.1 Background

High quality and security codes lay a solid foundation for a secure system. Tracing back to 2011, the code audit still depended on manual review which required high-level knowledge [16]. The common strategy at that time was shown below. They extracted the API symbols to help the auditor accelerate the discovery process but most of the work still demanded human interactions. Recent years have seen huge progress on vulnerability detection that the multiple software representations and multimodal deep learning ease the pressure of the auditors. These kinds of implementation can be divided into two approaches, which are similarity-based static analysis and pattern-based static analysis (it can be further divided into rule-based and machine learning-based analysis) [17]. The former is appropriate for code-cloning-based vulnerabilities while the latter requires human-defined rules or vulnerability patterns.

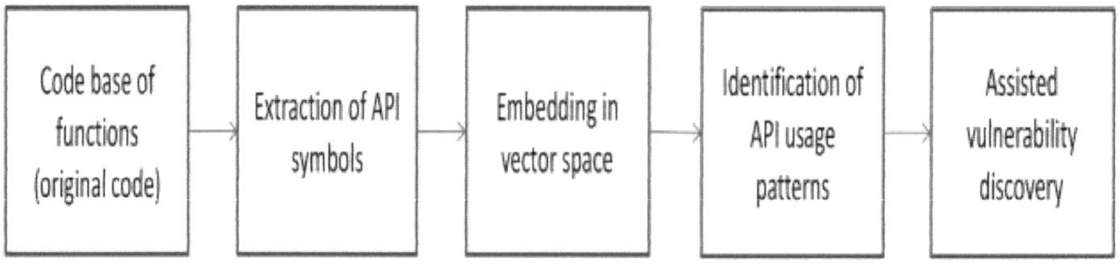

Figure 8. Initial schematic of vulnerability detection.

5.2 One possible solution

In 2020, Heidbrink et al. [18] devised to use three common multimodal deep learning models (namely correlation neural network, joint auto encoder, bidirectional deep neural network) to detect vulnerabilities and test their effects. The correlation neural network contained two (or more) inputs/outputs and a loss function term that maximized correlation across the different modalities. The joint auto encoder had extra encoder and decoder layers for each modality. It was called private branches which strikes a balance between the contributions of the modality and cross-modal mixing layers. The bidirectional deep neural network equipped the ability of multimodal representational learning via two separate networks. In the experiment, after the data normalizations, they used 5-fold cross-validation to analyze samples. The same settings were used for each architecture to test the effect. Regarding the database, the Juliet Test Suite is a well-known data set for vulnerability detection and covers common flaws within the field.

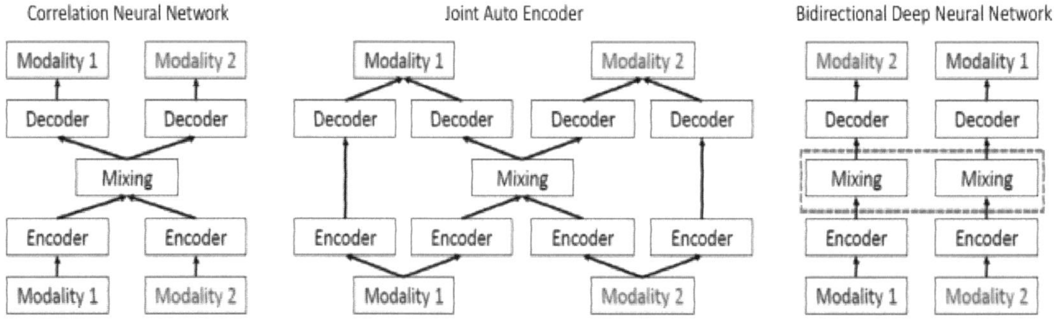

Figure 9. The common multimodal deep learning models used by Heidbrink's team

Table 8. Juliet Test Suite Data Summary.

CWE	Flaw Description	Flaw/Not Flaw
78	OS Command Injection	6102/15602
121	Stack Based Buffer Overflow	6346/16868
190	Integer Over flow	3296/12422
369	Divide by Zero	1100/4142
377	Insecure Temporary File	146/554
416	Use After Free	152/779
476	NULL Pointer Dereference	398/1517
590	Free Memory Not on Heap	956/2450
680	Integer to Buffer Over flow	368/938
789	Uncontrolled Memory Allocation	612/2302
78	OS Command Injection	6102/15602

5.2.1 Evaluation

Table 9. Flaw Detection Results on Juliet Test Suite based on Correlation Neural Network, Joint Auto Encoder, and Bidirectional Deep Neural Network (the best method among them is shown in boldfaced).

CWE	Correlation Neural Network	Joint Auto Encoder	Bidirectional Deep Neural Network
190		97.00	
369		94.00	
377	**96.00**	94.00	**96.00**
416	**72.00**	66.00	70.00
476		97.00	
680		94.00	
78	93.00	**94.00**	93.00
789	90.00	**92.00**	**92.00**
121 590		1.00	

Among the three models, the bidirectional deep neural network gave the best results but the differences were relatively small in all the test cases. The worst performance on CWE416 may attribute to the limited samples of the training set as the deep learning mechanism expected a large number of corresponding features and the difficulties of processing the specific flaw type (Use After Free).

5.3 Alternative solution

In 2020, another group of researchers demonstrated a new structure of an updated cell-based joint auto encoder that combined the neural architecture search (NAS) and multimodal learning models [19]. Specifically, the cell represented a portion of the architecture and was defined using a densely connected directed acyclic graph. NAS performed an automated search across diverse neural network architectures to find the best solutions. In the updated joint auto encoder, they replaced the decoders with a linear layer to merge the previous outputs. Among all the NAS techniques, GDAS was adopted to improve the cell search which was a gradient-based search algorithm through a differentiable architecture sampler. 10 cross-validation offered a pessimistic estimate of the generalization error.

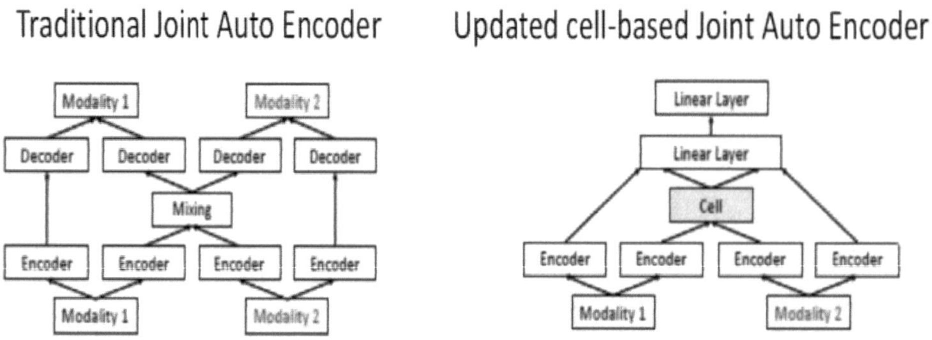

Figure 10. The comparison between the traditional joint auto encoder architecture and the updated cell-based joint auto encoder architecture

5.3.1 Evaluation

As is described in the table, there is not an obvious difference between the traditional JAE and cell-based JAE. Further work needs to be done to figure out the reason for the slight improvement. Future research can concentrate on the proper structure and functionality of the cell. Whether the effect of the removal of the decode module and the additional linear layers brings negative outputs should be investigated.

Table 10. Averaged accuracy using 10 cross-validation (better results between traditional JAE and cell-based JAE are shown in boldfaced).

CWE	Traditional JAE	Cell-based JAE
121	**0.9975±0.0008**	0.9970±0.0012
190	**0.9907±0.0059**	0.9884±0.0067
369	0.9500±0.0203	**0.9703±0.0220**
377	0.9285±0.0420	**0.9514±0.0410**
416	0.9359±0.0471	**0.9468±0.0400**
680	0.9356±0.0115	**0.9417±0.0197**
78	0.9360±0.0143	**0.9427±0.0155**
789	0.9630±0.0183	**0.9683±0.0215**
476 590	1.0000±0.0000	

5.4 Discussion

Besides the exception of CWE416, similar accuracy has been found in all four architectures under the same test suite. Some details are worth to be explored in the next stage.

- To what extent may the different cross-validation methods have influences on the accuracy. (5-fold cross-validation and 10 cross-validation)
- The potential improvement of other NAS techniques comparing to the GDAS.

6. Conclusion

In this paper, several recent approaches towards intrusion detection, malware detection, and vulnerability have been discussed. Among all the fields, deep learning techniques considerably optimize the accuracy comparing to the conventional method by introducing feature extraction and representation. At the cost of acceptable extra complexity and training periods, the large number of expert interactions can be reduced. The combination of machine learning and deep learning also brings new possibilities to threat detection in web security. More and more classification methods and neural network architecture have been designed and put into practice. Some of them have achieved productive results and are ready for additional examination in the future. The others may need to be re-designed based on their drawbacks but the overall trend is optimistic. Similar measures can be taken into account to help overcome the obstacles. First, about the feature generation method, more mechanisms should be developed to boost up precision and efficiency. Improper feature representations may lead to a huge rise in the cost of the training section. Second, about the training and testing database, the extremely limited dataset is available to the researchers in the open environment. This burning issue results from a large number of demands in the feature pattern generation. I hope most of the researchers can share their private datasets for investigation use to push the development of machine learning in all kinds of application fields, besides threat detection.

Reference

[1] Al-Mhiqani, M. N., Ahmad, R., Zainal Abidin, Z., Yassin, W., Hassan, A., Abdulkareem, K. H., ... & Yunos, Z. (2020). A review of insider threat detection: Classification, machine learning techniques, datasets, open challenges, and recommendations. Applied Sciences, 10(15), 5208.

[2] Arslan, B., Gunduz, S., & Sagiroglu, S. (2016, April). A review on mobile threats and machine learning based detection approaches. In 2016 4th International Symposium on Digital Forensic and Security (ISDFS) (pp. 7-13). IEEE.

[3] Shaukat, K., Luo, S., Chen, S., & Liu, D. (2020, October). Cyber Threat Detection Using Machine Learning Techniques: A Performance Evaluation Perspective. In 2020 International Conference on Cyber Warfare and Security (ICCWS) (pp. 1-6). IEEE.

[4] Yuan, S., & Wu, X. (2021). Deep learning for insider threat detection: Review, challenges and opportunities. Computers & Security, 102221.

[5] Salloum, S. A., Alshurideh, M., Elnagar, A., & Shaalan, K. (2020, March). Machine Learning and Deep Learning Techniques for Cybersecurity: A Review. In AICV (pp. 50-57).

[6] Sahoo, D., Liu, C., & Hoi, S. C. (2017). Malicious URL detection using machine learning: A survey. arXiv preprint arXiv:1701.07179.

[7] Javaid, A., Niyaz, Q., Sun, W., & Alam, M. (2016). A deep learning approach for network intrusion detection system. Eai Endorsed Transactions on Security and Safety, 3(9), e2.

[8] Yin, C., Zhu, Y., Fei, J., & He, X. (2017). A deep learning approach for intrusion detection using recurrent neural networks. Ieee Access, 5, 21954-21961.

[9] Karatas, G., Demir, O., & Sahingoz, O. K. (2018, December). Deep learning in intrusion detection systems. In 2018 International Congress on Big Data, Deep Learning and Fighting Cyber Terrorism (IBIGDELFT) (pp. 113-116). IEEE.

[10] Dong, B., & Wang, X. (2016, June). Comparison deep learning method to traditional methods using for network intrusion detection. In 2016 8th IEEE International Conference on Communication Software and Networks (ICCSN) (pp. 581-585). IEEE.

[11] Shone, N., Ngoc, T. N., Phai, V. D., & Shi, Q. (2018). A deep learning approach to network intrusion detection. IEEE transactions on emerging topics in computational intelligence, 2(1), 41-50.

[12] Yuan, Z., Lu, Y., & Xue, Y. (2016). Droiddetector: android malware characterization and detection using deep learning. Tsinghua Science and Technology, 21(1), 114-123.

[13] Hardy, W., Chen, L., Hou, S., Ye, Y., & Li, X. (2016). DL4MD: A deep learning framework for intelligent malware detection. In Proceedings of the International Conference on Data Science (ICDATA) (p. 61). The Steering Committee of The World Congress in Computer Science, Computer Engineering and Applied Computing (WorldComp).

[14] Yuan, Z., Lu, Y., Wang, Z., & Xue, Y. (2014, August). Droid-sec: deep learning in android malware detection. In Proceedings of the 2014 ACM conference on SIGCOMM (pp. 371-372).

[15] Kim, T., Kang, B., Rho, M., Sezer, S., & Im, E. G. (2018). A multimodal deep learning method

for android malware detection using various features. IEEE Transactions on Information Forensics and Security, 14(3), 773-788.
[16] Yamaguchi, F., Lindner, F., & Rieck, K. (2011, August). Vulnerability extrapolation: Assisted discovery of vulnerabilities using machine learning. In Proceedings of the 5th USENIX conference on Offensive technologies (pp. 13-13).
[17] Li, Z., Zou, D., Xu, S., Chen, Z., Zhu, Y., & Jin, H. (2021). Vuldeelocator: a deep learning-based fine-grained vulnerability detector. IEEE Transactions on Dependable and Secure Computing.
[18] Heidbrink, S., Rodhouse, K. N., & Dunlavy, D. M. (2020). Multimodal deep learning for flaw detection in software programs. arXiv preprint arXiv:2009.04549.
[19] Cooper, A., Zhou, X., Heidbrink, S., & Dunlavy, D. M. (2020). Using Neural Architecture Search for Improving Software Flaw Detection in Multimodal Deep Learning Models. arXiv preprint arXiv:2009.10644.
[20] Vinayakumar, R., Alazab, M., Soman, K. P., Poornachandran, P., & Venkatraman, S. (2019). Robust intelligent malware detection using deep learning. IEEE Access, 7, 46717-46738.

Optimization of operation sequencing based on GA-Jaya algorithm

Kai Zheng*, **Rui Zhang**, **Zhen-wei Zhu and Hua-dong Zhao**

College of Mechanical and Power Engineering, Zhengzhou University, Zhengzhou 450001, China

Corresponding author's e-mail ： zk1996@gs.zzu.edu.cn

Abstract: To solve the operation sequencing problem in CAPP that is a difficult problem, combining the idea of genetic algorithm, an GA-Jaya algorithm is proposed to minimize the total cost. In the GA-Jaya, the population is initialized according to the procedure priority adjacency matrix which makes the population all meet the process priority relationship. Mutation iteration operator and two kinds of crossover iteration operator are designed for process sequence and processing resource evolution. The GA-Jaya algorithm is applied to a typical case, and compared with the existing genetic algorithm, particle swarm optimization algorithm and ant colony optimization algorithm. The results show that the average quality of the solution obtained by the GA-Jaya algorithm is better than the existing genetic algorithm, particle swarm optimization algorithm and ant colony optimization algorithm.

1. Introduction

Computer Aided Process Design (CAPP) is an important part of computer integrated manufacturing system (CIMS) and an important link between COMPUTER aided design (CAD) and computer aided manufacturing (CAM). Its essence is to match the manufacturing capacity of an enterprise with the design information of parts, automatically determine and optimize the process route of designing parts [1], and convert the part design information obtained by CAD system into a series of processing operations that can be used by CAM system. CAPP is a multi-level, multi-task complex system that integrates selection, calculation, planning and drawing. It generally includes the following continuous processes: firstly, manufacturing features are identified from CAD system; Secondly, according to the manufacturing resources of the workshop, the manufacturing feature processing scheme is generated, and the processing steps and applicable manufacturing resources of each manufacturing feature are determined. Finally, priority constraints are considered and appropriate manufacturing resources are selected to prioritize processing processes to achieve an optimal plan, such as the lowest processing cost or the shortest processing time, so as to improve production efficiency and reduce manufacturing costs [2].

One of the two key technologies and difficulties in CAPP system is the process sequencing optimization of the whole parts. The essence of the process sequencing optimization of parts is the process of matching the design information of parts with the manufacturing resources of enterprises. In the past decade, many optimization algorithms have been used in process sequencing optimization, such as genetic algorithm (GA) [1-3], simulated annealing algorithm (SA) [4], ant colony algorithm (ACO) [5,6], particle swarm optimization (PSO) [7,8] and various hybrid algorithms [9,10]. However, process sequencing optimization has the potential to be further improved to improve efficiency. Jaya algorithm is a method for solving continuous optimization problems proposed by Rao in 2016 [11]. Like genetic algorithm and particle swarm optimization algorithm, Jaya algorithm belongs to swarm intelligence algorithm. Its advantages are that it does not need to adjust algorithm parameters, and it has good robustness and high efficiency [12]. Recently, there are some researches based on Jaya algorithm, such as Bayesian network structure learning [13], flexible Job Shop Scheduling [14].

In view of the characteristics of process sequencing optimization problem, this paper studies the use of this new optimization algorithm to solve the process sequencing problem, on the Jaya algorithm idea fusion of genetic algorithm variation and crossover, developed three new iteration operators to realize the iterative process of GA-Jaya algorithm, formed a GA-Jaya algorithm to solve the process sequencing optimization problem. The proposed algorithm is used to solve typical examples to verify the feasibility and superiority of the proposed algorithm.

2. Process sequencing optimization in CAPP

2.1 Problem description

The process sequencing of parts is to sort and optimize the processing schemes of each feature of parts under certain technological constraints and the condition of meeting optimization objectives, so as to obtain the optimal or approximate optimal processing sequence.

When a manufacturing feature of a part is machined, a path allowing the Tool to enter the feature is required, which is called TAD (Tool Approach Direction). Geometric reasoning is used to identify effective TADs in feature recognition[15,16]. Process sequencing involves machining sequence, machine tool selection, cutter selection and TAD selection. A good part process plan is based on two elements[7]:(1) optimal selection of machine tools, cutters and TAD in each process; (2) The optimal processing sequence of parts. Therefore, the algorithm developed needs to address these two aspects.

Among all the process constraints, there are mainly priority relationship constraints and process clustering constraints [17]. Priority relationship constraints are mandatory constraints, and priority relationship constraints must be first met in process planning. Clustering constraints generally do not need to be fully met, but the cost of violating such constraints is high. The priority relationship mainly includes datum first, coarse first, fine first, principal first, second order first, surface first and hole first, etc. The priority constraint relationship can be described by Operation Precedence Graph (OPG), which is a directed acyclic Graph, as shown in figure 1. It consists of two sets: vertex (procedure) set and directed edge (priority constraint) set. In the figure: the node represents a procedure, and the direction of the directed edge represents the sequential relationship between the two procedures.

Because computers are better at processing numbers rather than recognizing images, OPG[17] is represented by adjacency matrix $P_{n\times n}$ to realize the transfer of graphics to computers. The sum of elements in each row of the matrix is the number of constraints imposed by the procedure, and the sum of elements in each column is the number of constraints imposed by the procedure. The adjacency matrix $P_{8\times 8}$ transformed from Figure 1 is shown in figure 2.

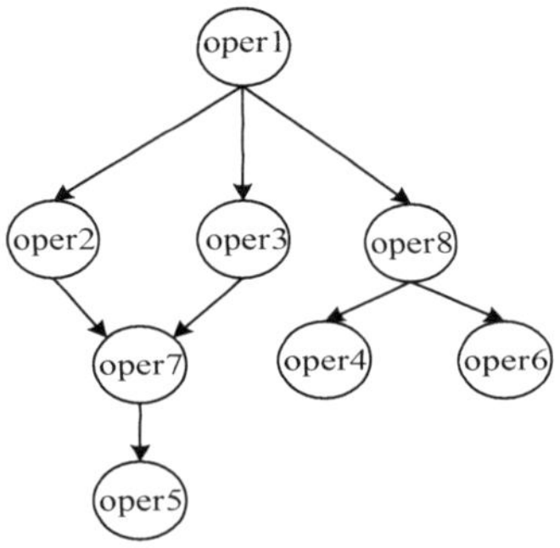

Figure 1. Process priority diagram

After the parts are clambed once, a series of processing of the same TAD on the same machine tool is regarded as a process, and each process corresponds to a number of machine tools, tools and TAD combinations [3]. Therefore, the process sequencing optimization problem referred to in this paper is a single objective multi-constraint combinatorial optimization problem. The objective of optimization is process sequence, machine tool, cutter and TAD corresponding to each process. Constraints include priority constraint and machine tool function constraint, and the objective function is to minimize the total processing cost.

	oper1	oper2	oper3	oper4	oper5	oper6	oper7	oper8
oper1	0	1	1	0	0	0	0	1
oper2	0	0	0	0	0	0	1	0
oper3	0	0	0	0	0	0	1	0
oper4	0	0	0	0	0	0	0	0
oper5	0	0	0	0	0	0	0	0
oper6	0	0	0	0	0	0	0	0
oper7	0	0	0	0	1	0	0	0
oper8	0	0	0	1	0	1	0	0

Figure 2. Adjacency matrix diagram

2.2 Objective function and calculation

The objective function of process sequencing optimization in this paper is to minimize the total machining cost (TC), including the following five types: total machine tool cost (TMC), total tool cost (TTC), total machine tool replacement cost (TMCC), total tooling replacement cost (TSCC) and total tool replacement cost (TTCC). Refer to reference [15] for specific calculation formula.

3. Ga-Jaya algorithm is used to solve the process sequencing optimization problem

3.1. Basic theory of Jaya

Jaya algorithm is Rao, a meta heuristic algorithm is put forward in 2016 came [11], the algorithm also considers the optimal individual and the worst, the effect of the individual to search the optimal solution is a combination of evolutionary algorithm and the characteristics of swarm intelligence, and do not like other evolutionary algorithms need to adjust the parameters of the structure is simple, adaptive, flexibility, improve and completeness, It is a typical swarm intelligence optimization algorithm. The iterative formula of the algorithm is as follows:

$$X'_{j,k,i} = X_{j,k,i} + r_{1,j,i}\left(X_{j,best,i} - \left|X_{j,k,i}\right|\right) - r_{2,j,i}\left(X_{j,worst,i} - \left|X_{j,k,i}\right|\right) \tag{1}$$

In the formula, $X_{j,k,i}$ is the jth variable of the kth individual in the ith iteration, $X_{j,best,i}$ is the variable corresponding to the fittest individual in iteration i^{th}, $X_{j,worst,i}$ is the variable corresponding to the individual with the worst fitness in iteration i^{th}, $r_{1,j,i}$ and $r_{2,j,i}$ are two random numbers in the range [0,1], $X'_{j,k,i}$ is the updated value of $X_{j,k,i}$, If the fitness of $X'_{j,k,i}$ is superior to that of $X_{j,k,i}$, $X'_{j,k,i}$ is substituted for $X_{j,k,i}$.

In the formula, the second term "$r_{1,j,i}\left(X_{j,best,i} - \left|X_{j,k,i}\right|\right)$" represents the tendency of the solution to converge towards the optimal solution, And the third term "$-r_{2,j,i}\left(X_{j,worst,i} - \left|X_{j,k,i}\right|\right)$" indicates the tendency of the solution to avoid the worst solution. Therefore, Jaya algorithm always moves towards the best individual while keeping away from the worst individual in the process of individual iteration update.

3.2. Solution of the coding

The solution of process sequencing is expressed as chromosomes composed of ordered genes. All candidate solutions form a solution space. In the chromosome, a gene needs to contain the process ID, machine tool ID, tool ID, and TAD ID, as well as the candidate machine tool sequence, tool sequence, and TAD sequence. Figure 3 shows the encoding of a solution for the case part.

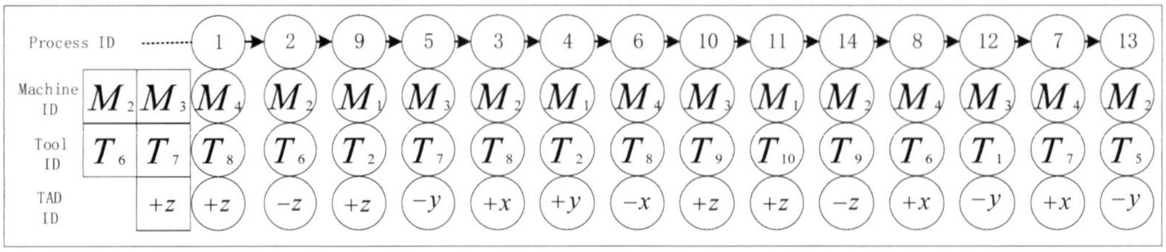

Figure 3. Chromosome coding schematic of a solution of a case part

3.3. Population initialization

Initialization of the population to the quality and speed of the algorithm has an important influence, in order to solve the speed, avoid the process sequence of random against preference constraint and not feasible, therefore, when making the population initialization, will preference constraint into consideration, through an initialization method, to ensure that the initial population contains process sequences are feasible.

In this paper, the OPG graph information of parts is transformed into adjacency matrix $P_{n\times n}$, and the population initialization is realized by using topological sorting method.

3.4. GA - Jaya iteration

The traditional Jaya algorithm is suitable for solving continuous optimization problems, while process sequencing is a typical discrete optimization problem. This paper makes full use of the influence X_{best} and X_{worst} on the individual iteration. Combining the variation and crossover ideas of genetic algorithm, a discrete GA-Jaya iterative method applied to process sequencing is proposed. The iterative method contains three iterators:

3.4.1 Mutation iteration operator

The mutation iteration operator changes the partial order of the process and reduces the number of types of processing resources (machine tools, cutters and TAD) by means of topological ordering. It includes two processes, process variation and processing resource variation, and the specific steps are as follows:
Process variation: In the process of process variation, the processing resources under the process change with the process.

Step 1: Select a random position according to the number of procedures and divide the procedure sequence into A and B parts;

Step 2: generate a random number r in the interval of (0,1). If $r \leq 0.5$, copy part A of the procedure sequence to an initially empty array C. If $r > 0.5$, copy part of the procedure sequence to an initially empty array, as shown in Figure 4(a);

Step 3: Generate a zero matrix $R_{j\times j}$ (or $R_{(n-j)\times(n-j)}$) according to the number of elements j (or $n-j$) in array C;

Step 4: Access to an array of two elements of $C, C[d]$ and $C[e]$, and obtain its process ID ($oper_u$ and $oper_v$), the two process ID combination $(oper_u, oper_v)$, as access to bottom corner of the element in the adjacency matrix $P_{n\times n}$, obtain the values in the matrix, and copies the value to the local adjacency matrix $R_{j\times j}$ (or $R_{(n-j)\times(n-j)}$), until the combination of any two elements in the array are complete, as shown in figure 4 (b);

Step 5: Compute the sum of the elements of each column of the local adjacency matrix $R_{j\times j}$ (or $R_{(n-j)\times(n-j)}$) to generate an array $D=\left[NOR(oper_i),\ldots\right]$;

Step 6: Select a random element which value is 0 from array D, assuming $NOR(oper_g)$, take $oper_g$ into the initially empty array E, $E=\left[oper_g\right]$, and remove $oper_g$ from array C;

Step 7: Delete row $oper_g$ and column $oper_g$ from local adjacency matrix $R_{j\times j}$ (or $R_{(n-j)\times(n-j)}$) to generate a new matrix $R_{(j-1)\times(j-1)}$ (or $R_{(n-j-1)\times(n-j-1)}$), as shown in figure 4(c);

Step 8: Calculate the sum of each column of the new matrix $R_{(j-1)\times(j-1)}$ (or $R_{(n-j-1)\times(n-j-1)}$) and put it back into array D;

Step 9: Select a random element which value is 0 from array D, assuming $NOR(oper_h)$, put $oper_h$ into array E, that is, $E=\left[oper_g, oper_h\right]$, and delete it from the array C.

Step 10: Repeat steps 6, 7, 8 and 9 until there is only one element in array C, as shown in figure 4(d). Add the last element in array C to the last bit of array E, and the sort in array E is the mutated sort.

Step 11: Combine the sorting in array E with A (or B) to form A complete operation sequence, as shown in figure 4(e);

Variation of processing resources: The variation process of machine tool, tool and TAD is similar, the variation of machine tool is first, the variation of tool is second, and the variation of TAD is last. Take machine tool variation as an example.

Step 1: Find the largest number of machine tools $M_{\max}$ and the smallest number of machine tools $M_{\min}$ in the sequence;

Step 2: find out the process set using the machine tool $M_{\min}$;

Step 3: Judge successively whether $M_{\max}$ is included in the candidate machine tool sequence of process $opers$ in process set $oper_x$. If so, set the machine tool in process $opers$ to $oper_x$; if not, select a machine tool randomly from the candidate machine tool sequence of process $opers$.

3.4.2. Best cross iteration operator

The sequence of the current solution is compared with that of the best solution, and the same genes in the sequence are preserved. Part of the sequence is in the order of the best solution, and part of the sequence is in the order of the current solution. The processing resources change with the process, as shown in figure 5. The specific steps are as follows:

Step 1: Select a position j randomly according to the length of the procedure sequence and divide the procedure sequence into left and right parts;

Step 2: Copy the left part of X_{best} to become the left part of X_{k+1};

Step 3: find the genes with the same sequence in the right part of X_k and X_{best}, and store them in the corresponding position of X_{k+1};

Step 4: Find out the sequence of other genes in X_{best} in X_k, and generate the right part of X_{k+1} according to the sequence of X_k;

Step 5: Output X_{k+1};

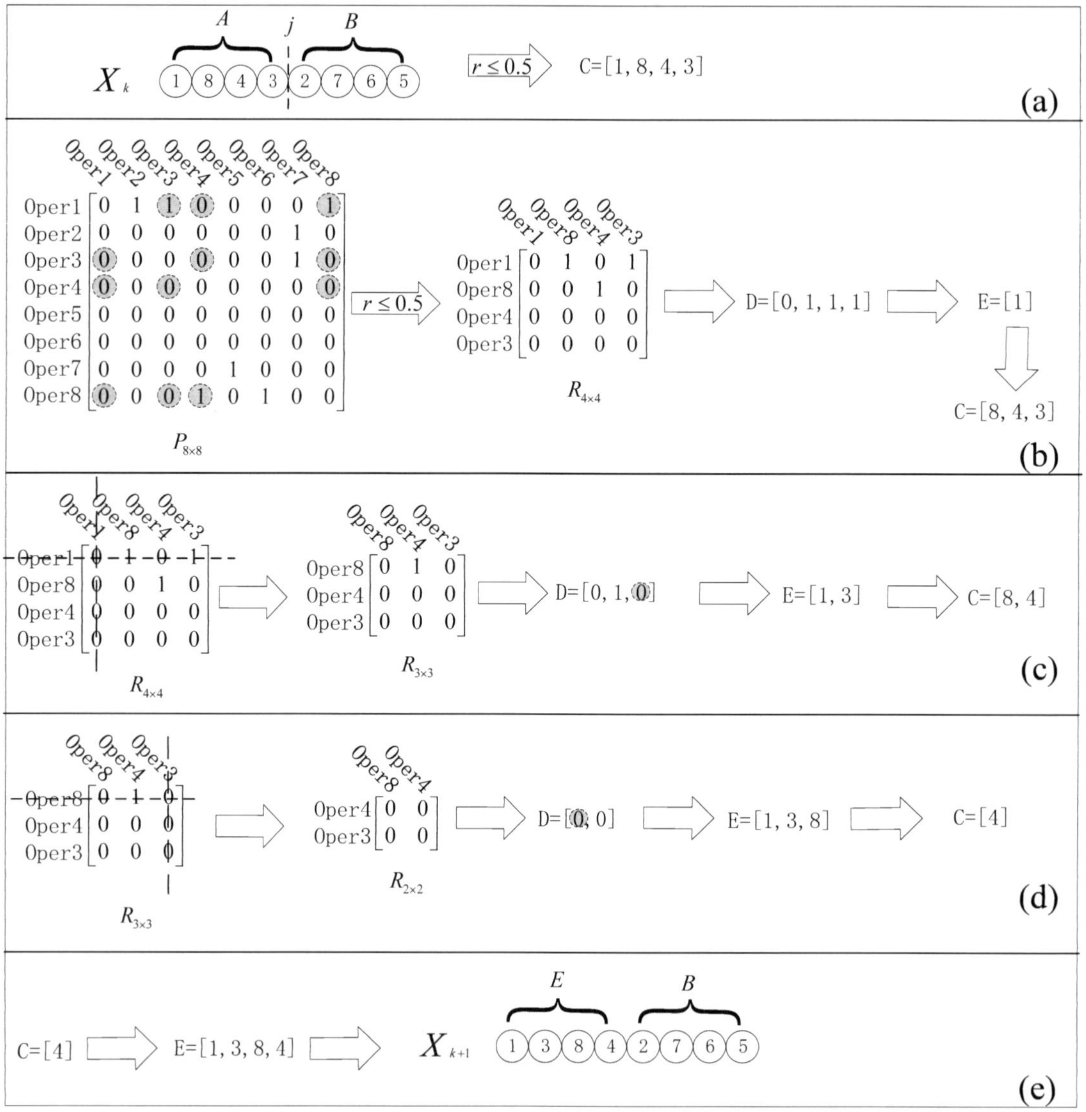

Figure 4. Schematic diagram of process variation

3.4.3. Worst cross iteration operator

The current process and worst process sequences comparison, marking the position of the sequence of the same process, choose the best process sequence in the same location of the gene as a new individual, the remaining part of the order of the sequence with the current solution, the other part of the order of the sequence with the worst, as shown in figure 6, specific steps are as follows:

Step 1: Identify genes with the same sequence in X_k and X_{worst} and mark their locations.

Step 2: Copy the gene located at the step 1 marker in X X_{best} to the corresponding location in X_{k+1}.

Step 3: Select a position j randomly according to the length of the procedure sequence and divide the procedure sequence into left and right parts.

Step 4: Save the remaining genes of the left part of X_k into X_{k+1} according to the sequence in X_k.

Step 5: Find out the sequence of genes in the right part of X_k in X_{worst}, and form the right part of X_{k+1} according to the sequence.

Step 6: Output X_{k+1};

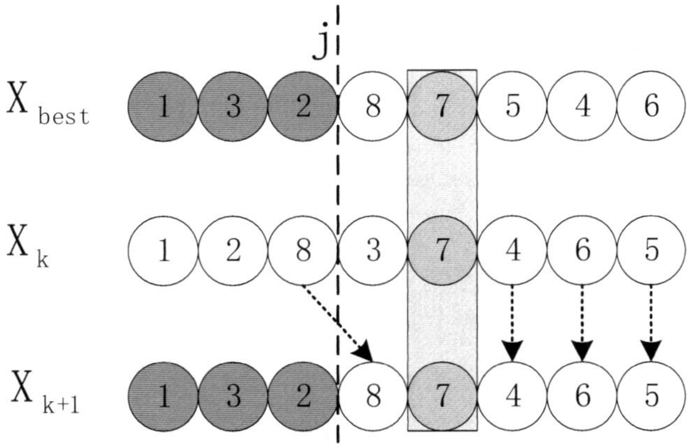

Figure 5. Schematic diagram of Best cross iteration

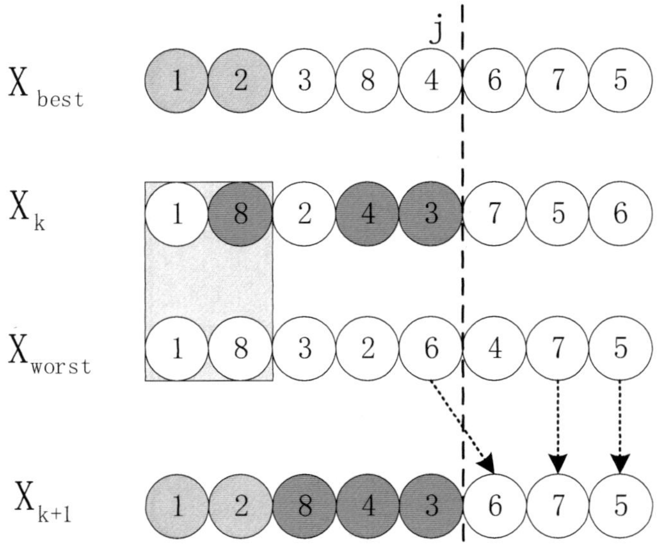

Figure 6. Schematic diagram of Worst cross iteration

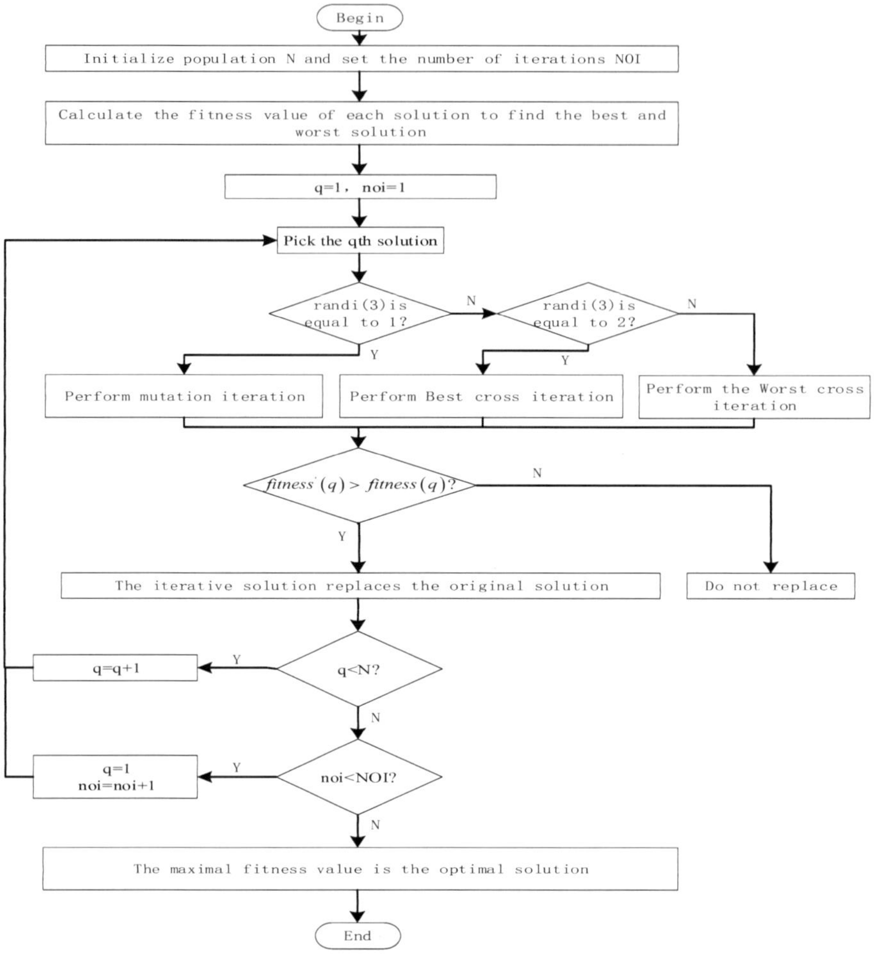

Figure 7. Algorithm flow chart

4. The case study

In order to verify the effectiveness of the algorithm, a case jointly adopted with literature [3,7,17] is adopted to verify the effectiveness and advantages of the algorithm through calculation comparison. The case part shown in figure 8 is composed of 11 manufacturing features that can be divided into 14 machining processes.

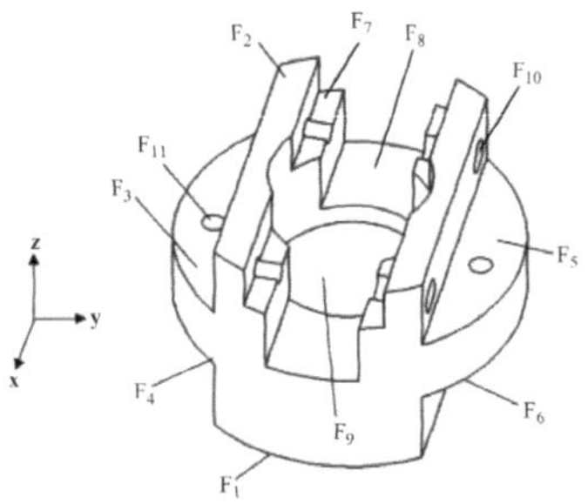

Figure 8. 11 manufacturing features of the case part

The algorithm in this paper is written by Matlab. The number of population in the algorithm is $N = 150$, and the maximum number of iterations is $NOI = 200$. Table 1 shows the comparison of results of GA-Jaya algorithm, improved genetic algorithm [3], genetic algorithm [7], particle swarm optimization algorithm [7] and ant colony algorithm [17]. The improved Jaya algorithm has been run for 10 times. figure 9 and 10 are convergence curves for obtaining the best solution.

Table 1. Comparison of calculation results of case parts

Algorithm	Optimal solution cost	Solving cost mean	Worst solution cost	Earliest algebra of convergence
GA-Jaya	1364	**1390.8**	1437	**22**
FOSOGAF	1357	**1424.7**	1459	**73**
GA	1381	**1459.4**	——	——
PSO	1361	**1430.0**	——	——
ACO	1357	**1419.0**	——	——

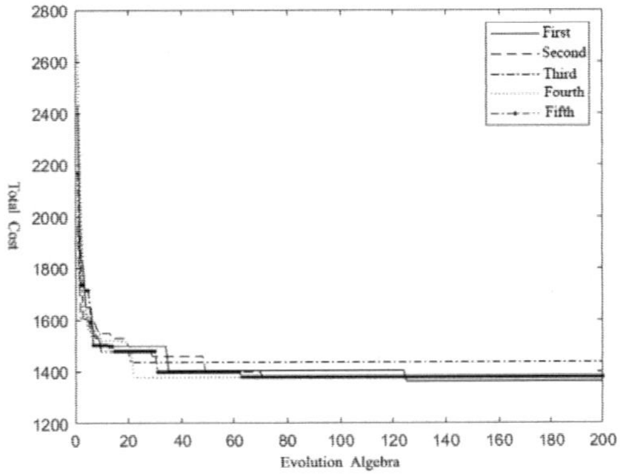

Figure 9. Convergence curves of the best solution obtained by GA-Jaya algorithm for the first five times

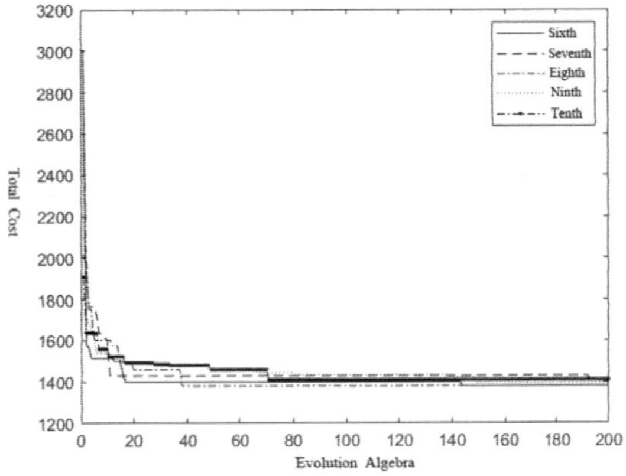

Figure 10. Convergence curves of the optimal solution obtained by ga-Jaya algorithm for the last five times

The optimal process planning scheme obtained by the algorithm in this paper is shown in Table 2, and its total cost is $TC = 1364$. The cost of the optimal solution obtained by GA in literature [7] is

1381 (see Table 16 in literature [7]). The optimal solution obtained by GA-Jaya algorithm is better than that obtained by GA. Although the best solutions obtained by SOFOGAF, PSO and ACO are slightly superior to GA-Jaya algorithm, ga-Jaya algorithm is far superior to SOFOGAF, GA, PSO and ACO in terms of average quality of solutions. And the earliest convergence algebra of GA-Jaya algorithm is better than the improved GA algorithm. By comparing the results of the five algorithms, the ga-Jaya algorithm in this paper is superior to the existing GA and PSO algorithms in terms of solution quality.

Table 2. a process planning scheme for case parts

	1	2	3	4	5	6	7	8	9	10	11	12	13	14
Process ID	1	2	3	9	10	11	6	5	8	7	4	14	12	13
Machine ID	3	3	3	3	3	3	3	3	3	3	3	3	3	3
Tool ID	6	6	6	2	9	10	7	7	7	7	2	9	1	5
TAD ID	+z	-z	-z	-z	-z	-z	-z	-z	-z	-z	+z	+z	+y	+y
$NMC=0, NSC=4, NTC=8; TMC=560, TTC=164, TMCC=0, TTCC=160, TSCC=480, TC=1364$														

5. Conclusion

In this paper, GA-Jaya algorithm is proposed to optimize the process sequencing problem which is difficult to solve in CAPP by improving Jaya algorithm by combining the crossover and mutation ideas of genetic algorithm. The algorithm generates feasible solutions according to the procedure priority adjacency matrix, and the solution is always in the feasible domain regardless of mutation iteration or crossover iteration. Thus, the search space of the algorithm is reduced and the efficiency of the optimization algorithm is improved. This algorithm is applied to the diamond roller CAPP system developed in practice. The case comparison results show that the average quality of solutions obtained by the improved Jaya algorithm proposed in this paper is better than SOFOGAF, GA, PSO and ACO.

Although ga-Jaya algorithm proposed in this paper obtains high-quality solutions for a given case, with more complex geometric shapes, the number and types of features to be produced will increase exponentially, and the constraint relationship between each feature will become more complex, which will be a challenge to the search ability of the algorithm. Therefore, in the future, new sequence adjustment methods are needed to reduce computing time and improve the performance of the algorithm in more complex process sequencing problems.

References

[1] Gao Gang. Research on decision Technology in Intelligent CAPP [D]. Yanshan University, 2018.
[2] Hua Guangru, Wang Huifeng, Zhang Zhenhuan, et al. Comprehensive Optimization Method of Machining Scheme Selection and Operation Sequencing Based on Genetic Algorithm %J journal of Shanghai Jiaotong University, 2006(02): 195-200.
[3] Dou Jianping, Li Jun, Su Chun. Process Sequencing Optimization based on Feasible Process Sequence Genetic Algorithm for %J Computer Integrated Manufacturing System [J], 2019, 25(08): 1981-1990.
[4] Zhai Dehui. Research on Key Technology of 3d Process Design of Box Parts Based on Feature [D]. Beijing Institute of Technology, 2015.
[5] Gao Tianwei. Research on Key Technology of Intelligent Process of Large Marine Engine Cylinder Based on MBD Feature [D]. Chongqing University of Technology, 2020.
[6] Pan Qinggu. Research and Application of Process Planning Method for Typical Automobile Parts Based on PLM Platform [D]. Wuhan University of Technology, 2019.
[7] Guo Y W, Mileham A R, Owen G W, et al. Operation sequencing optimization using a particle swarm optimization approach %J Proceedings of the Institution of Mechanical Engineers, Part B: Journal of Engineering Manufacture[J], 2006, 220(12).
[8] Particle Swarm Optimization; Reports from North China Electric Power University Provide New Insights into Particle Swarm Optimization (A simulation approach to the process planning problem using a modified particle swarm optimization) %J Journal of Technology &

Science[J], 2016.
[9] Huang Weijun. Research on Key Technology of Process Plan Optimization of Complex Parts [D]. Huazhong University of Science and Technology, 2012.
[10] Li W D, Ong S K, Nee A Y C. Hybrid genetic algorithm and simulated annealing approach for the optimization of process plans for prismatic parts %J International Journal of Production Research[J], 2002, 40(8).
[11] Venkata Rao R. Jaya: A simple and new optimization algorithm for solving constrained and unconstrained optimization problems[J]. International Journal of Industrial Engineering Computations, 2016: 19-34.
[12] Abu Z R, Azmi A M, A A M, et al. An Intensive and Comprehensive Overview of JAYA Algorithm, its Versions and Applications. %J Archives of computational methods in engineering: state of the art reviews[J], 2021.
[13] Yan Zhi, Zhang Peng, Xie Chuan. Research on Bayesian Network Structure Learning Algorithm based on Jaya %J Computer Engineering and Applications, 2019, 55(19): 173-177+184.
[14] Lian Yuxiang, Zhang Chaoyong, Meng Leilei, et al. Flexible Job-shop Scheduling Problem %J Computer Integrated Manufacturing System Based on Improved Jaya Algorithm [J]: 1-19.
[15] Huang W, Hu Y, Cai L. An effective hybrid graph and genetic algorithm approach to process planning optimization for prismatic parts %J The International Journal of Advanced Manufacturing Technology[J], 2012, 62(9-12).
[16] Su Y, Chu X, Chen D, et al. A genetic algorithm for operation sequencing in CAPP using edge selection based encoding strategy %J Journal of Intelligent Manufacturing[J], 2018, 29(2).
[17] Hu Q, Qiao L, Peng G. An ant colony approach to operation sequencing optimization in process planning %J Proceedings of the Institution of Mechanical Engineers, Part B: Journal of Engineering Manufacture[J], 2017, 231(3).

Robot physical modeling based on matlab/simulink

Zhiyong Wu[1],Hongping Wang[1*],Yiwen Zhang[1],Minglu Ai[1]

[1]Mechanical Engineering, Changchun University of Science and Technology, Changchun, Jilin, 130022, China

* Corresponding author: *wanghongping@cust.edu.cn

ABSTRACT: Robots play an irreplaceable role in the manufacturing industry. The use of machines instead of humans'activing.It is a long-term development trend, but it is difficult to realize the research of real robots, and the data that the robot can collect is less. In response to the above problems, this article mainly proposes a method of physical modeling using matlab/simulink. This method can realize research without robots, and can obtain more robot-related data. This method includes robot physical modeling, importing into simulink and packaging in simulink. This article is mainly aimed at IRB6700 robot to complete the simulation platform construction.

1.Introduction

ABB's IRB6700 robot is an industrial robot that can engage in welding (including cutting and spraying). According to the definition of the International Organization for Standardization (ISO) industrial robot terminology standard welding robot, an industrial robot is a multi-purpose, reprogrammable automatic control manipulator (Manipulator), with three or more programmable axes, with producting activing, it has the advantages of high efficiency, stable product quality, low labor cost, and good operating environment.[1]This article builds the robot simulation operation process as follows: 1. Robot structure analysis. Including the setting of each link, the production of joints and the establishment of the relative coordinate system of each joint. 2. Establishment of simulation model: Use solidworks to complete the robot assembly, import and use matlab/Simulink to build the robot simulation model. Since the intermediate file format is an urdf file, the model needs to be modified before importing into matlab. 3. Establish sensors for the position, velocity and acceleration of each joint, and moment of inertia. 4. Motion simulation and data processing: output and collect the motion parameter curves of each joint and verify the correctness of the simulation model.

2.Robot 3D modeling creation

2.1 IRB6700 robot assembly

The assembly process of the robot mainly needs solidworks to complete. First, you need to download the step file of the IRB6700-235 robot part on the ABB robot official website, and import the setp file into solidworks.[2] After the import is completed, you need to assemble each part. Note that when importing the part base When the ground origin coordinates need to coincide with the space coordinate system. After the import is complete, you first need to make reference coordinates on each part so that you can use these reference coordinates to complete the robot assembly.[3]

Use the established coordinate system to assemble each part. First select the part to be equipped, click Mate, and in the fit selection, click the center line of the base to coincide with the center line of the ground under one axis, that is, the line coincides with the line, and continue Add that the upper surface of the base coincides with the ground under one axis, that is, the surface matches the surface. Follow the same method as above to complete the assembly of the six-axis robot.[4]

Complete the assembly in solidworks and set the material properties of each component. The main body of the IRB6700 is made of aluminum alloy. After the assembly is completed, the model built is 3% different from the actual mass of the IRB6700 at 1205kg. It can be approximated that the built physical model meets the actual situation. After parameter setting, it is as shown in the figure.

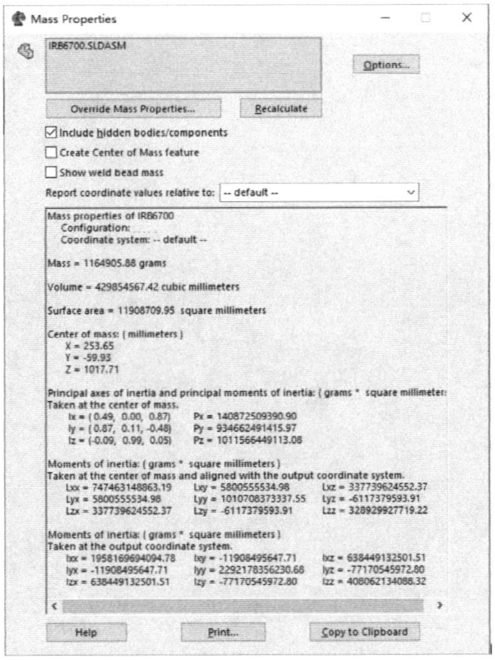

Figure 1. Robot parameters

2.2 Export the urdf file and modify it

After the assembly is completed, the coordinate system in the reference geometry needs to be set at each joint. This coordinate system is used to determine the rotation axis and the reference coordinate system in the process of exporting the urdf file. Download and install the sw2urdf plug-in on the solidworks official website to export the assembled IRB6700-235 robot model to an urdf file. During the export process, set the limit axis value and speed with reference to the robot official website data.[5]

Table 1. Rotation parameters of each axis of the robot

Axis	Type of motion	Range of movement-IRB 6700	Range of movement-IRB 6700Inv
Axis1	Rotation motion	$\pm 170° or \pm 220°$	$\pm 170°$
Axis2	Arm motion	$-65°/+85°$	$\pm 65°$
Axis3	Arm motion	$-180°/+70°$	$-180°/+70°$
Axis4	Wrist motion	$\pm 300°$	$\pm 300°$
Axis5	Bend motion	$\pm 360°$	$\pm 130°$
Axis6	Turn motion	$\pm 360°$	$\pm 360°$
		$\pm 93.7 revolution$	$\pm 93.7 revolution$

After completing the plug-in installation, restart solidworks to display the SW2URDF function in the plug-in, select Export as URDF.[6] To set in the pop-up dialog box, you first need to select the base coordinates, add the base parts, and click Load Configuration to add the next level link is the previously built reference coordinate system, and the joint activity type is set to rotate and the parts of the next level link are added, and the joint name of the next level is set, and the rotation is set Refer to the coordinate system and the rotation reference axis, select the joint motion mode as rotation, and the part section dog, and complete the addition of parts at each level as shown in the figure.

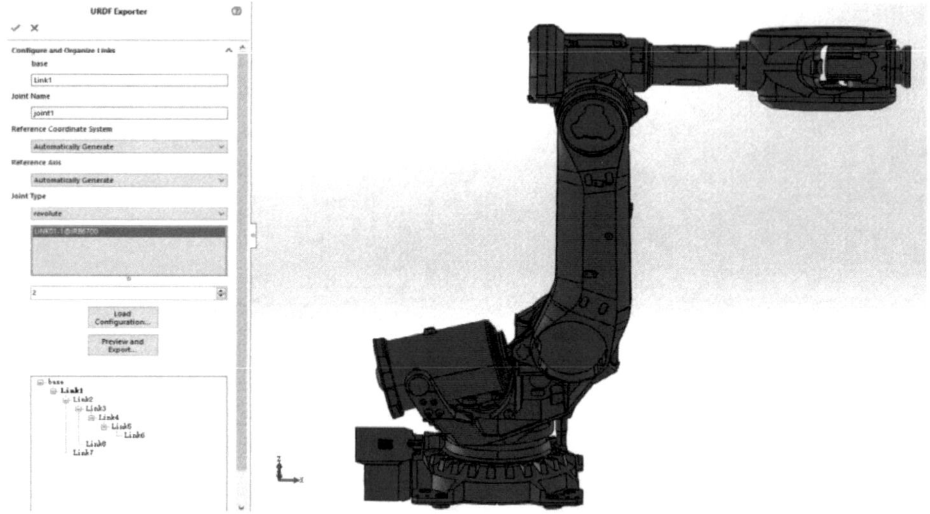

Figure 2. Complete urdf file configuration

After completing the above configuration, click the Preview and Export URDF button to export the urdf file, which corresponds to the IRB6700-235 robot axis rotation parameter table. Set the rotation parameters of each axis of the robot. When setting the rotation parameters, you need to convert the angle system to the radian system, as shown in the figure show.

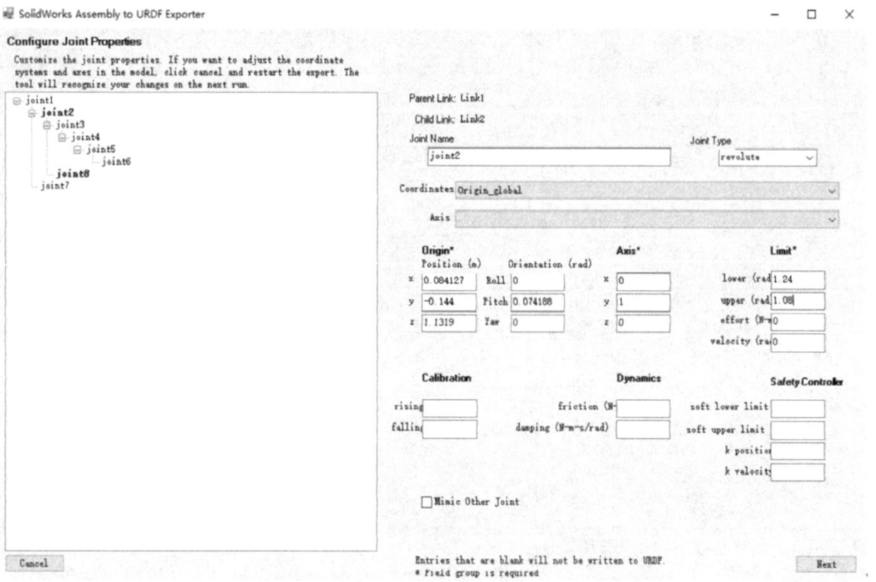

Figure 3. Set rotation limit

3.Build the model in simulink

3.1 Import urdf file into simulink

We need to import the model containing the physical parameters of the IRB6700 robot into simulink. First, we need to download the corresponding version of simscape Multibody from the matlab official website, add the path to matlab, and complete the installation.After the plug-in is installed, the exported URDF file Adjust the price to the matlab path, use the smimport() command in the command window to import the model into simulink.

3.2 Encapsulate the simulink model

This robot model mainly has a driving module, an execution module and a data acquisition module. The driving mode of the driving module is mainly driven by the rotation angle of the six axes. The angle data is intercepted in units of 0.1 seconds. The intercepted data is very close and can be approximated. The authenticity of the data is reliable, and compared to the design of the motor drive, it can achieve simple and accurate results. The input module is shown in the figure.

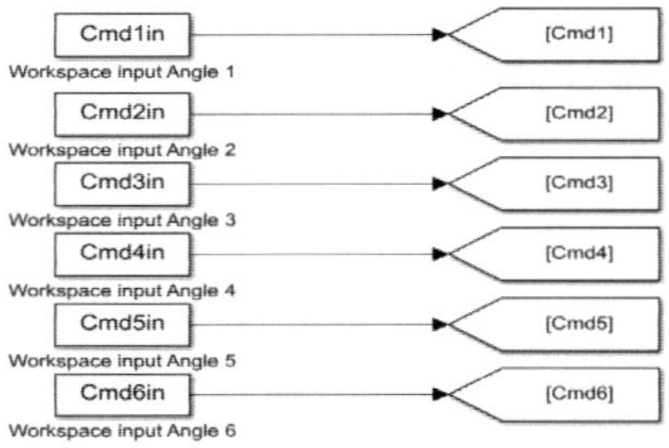

Figure 4. Corner drive module

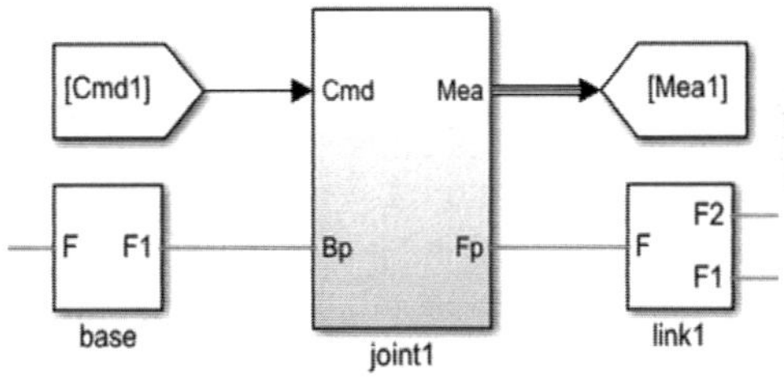

Figure 5.Package execution module

After encapsulating the input module, we need to encapsulate the execution module of the imported IRB6700 robot, as shown in the figure is the execution module, Cmd1 is the input angle of the input drive module, base is the robot base, link1 is the arm of the robot 1, and Mean1 Represents the output data of the robot.

The execution module mainly encapsulates the joint connection, which contains corner drive, joints and output data. As shown in the figure, the inside of the encapsulated execution module, joint is each axis of the robot, and m represents a set of output data, which contains each The axis speed signal, position signal, torque signal, etc., cmd corresponding to cp is the input angle data drive, p corresponding to Mea is the data output.

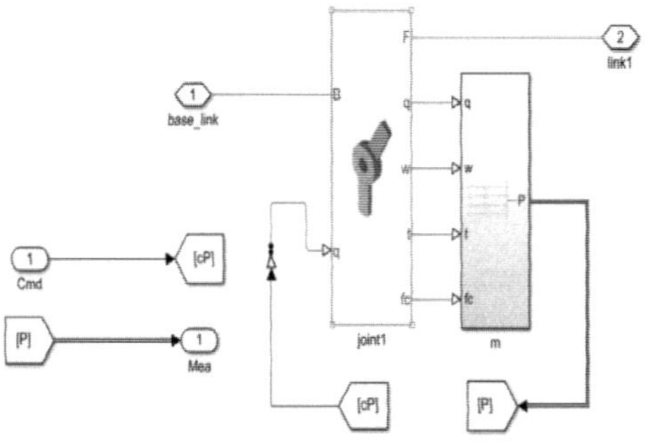

Figure 6. Inside the execution module

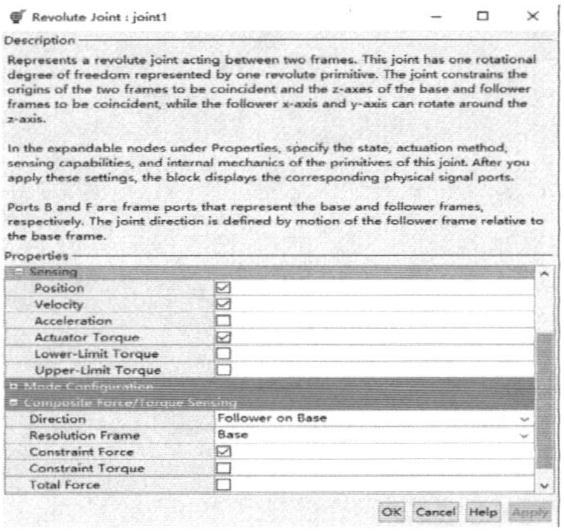

Figure 7. Joint output data type

After completing the joint encapsulation, research on the joint output data. Double-click to open the joint control. The joint control can output a variety of data.

For the parallel mechanism of the robot piston connecting rod, cylinder and robot body, triangle equations are used to solve the kinematic relationship of the mechanism. As shown in the figure, the function is written according to the kinematic relationship of the mechanism.After the completion of the drive module and the actuator, the data output module needs to be packaged.

After completing all the configurations, the overall package model of the IRB6700 robot in simulink and the robot matlab robot human model are shown in the figure.

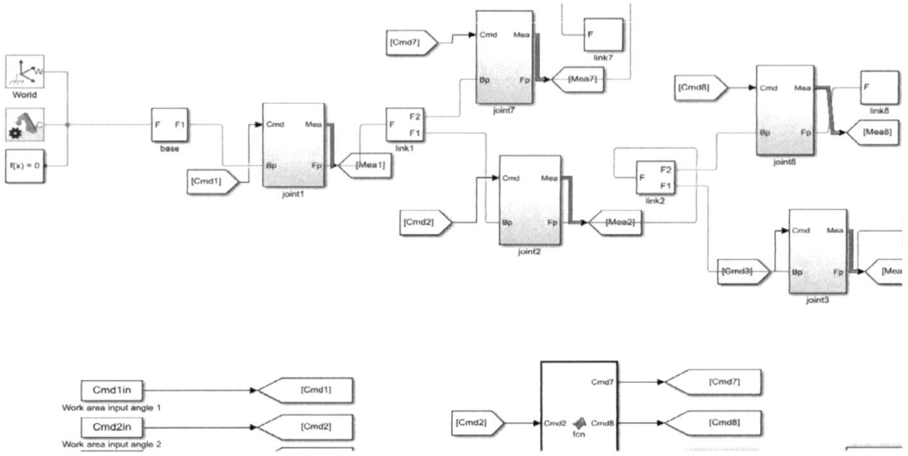

Figure 8. Simulink robot model

4.Robot movement data

Write the function to read the data in the file in the matlab script file, and set it to read in 0.1 seconds. As shown in the figure, it is part of the running data result.

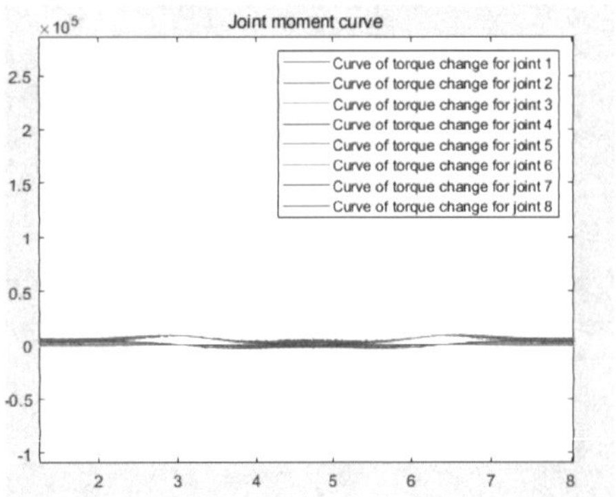

Figure 9. Partial curve of torque change of each join

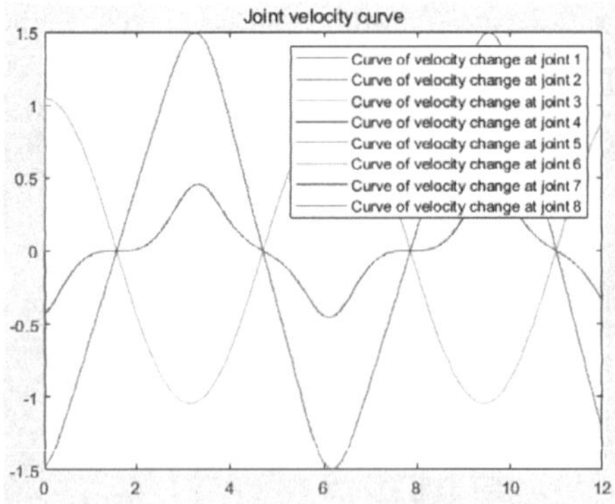

Figure 10. Partial curve of speed change of each joint

5.Conclusions

This paper realizes that the 3D model of the robot is created in solidworks, converted into a URDF file after assembly, and then the optimized model file is imported into the simulink simulation environment to complete the package and output data. Compared with the actual data of the real robot, the difference is small.This model meeting the requirements of the simulation model. The paper designs the parameters of the IRB6700-235 robot, analyzes the kinematics and dynamics of the robot, and uses the model to simulate and obtain good results.

References

[1] Lu Shifeng. Modeling and Simulation of Electro-hydraulic Stepping Cylinder [D]. Northeastern University, 2008.
[2] Cao Junkai. Improved design and simulation of built-in servo-valve electro-hydraulic stepping cylinder [D]. Northeastern University, 2012.
[3] Zeljko Situm, Tihomir Zilic, Mario Essert. High Speed Solenoid Valves in Pneumatic Servo Applications , Proceedings of the 15th Mediterranean conference on Control Automation,July 27-29,2007,Athens-Greece
[4] Xiaolong Tong, Liguang Zhang, Nadan Wang, "Research on model predictive control in purely lagging objects,"Electronic measurement technology,vol.41,n23,pp.12-16,2018.
[5] Guangxiang Yang, Hua Liang, and Xiaoli Cao, "Design of experimental teaching of combination of simulation and practice in process control system,"Experimental Technology and Management, vol.30, n1,pp. 91– 93,2013.
[6] Min Li, Tao Zou, Maying Yang, "Design and teaching pracice of synthetic experiments of process control system," Experimental Technology and Management, vol.28, n6,pp.100-104,2011.

ABB robot data collection based on dynamic link library

Zhiyong Wu[1],Hongping Wang*[1],Yiwen Zhang[1],Minglu Ai[1]

[1]Mechanical Engineering, Changchun University of Science and Technology, Changchun, Jilin, 130022, China

* Corresponding author: [1]wanghongping@cust.edu.cn

ABSTRACT: At present, with the development of the times and technology, in order to meet the needs of industrial production, robots have begun to be widely used in factories. Although the production efficiency of the robot is very high and relatively cheap compared to the labor cost, when an industrial robot fails, it will have a serious impact on the production capacity of the entire production line. Therefore, it is necessary to monitor various data of the robot. If some abnormal data can be found , it can perform effective predictive maintenance on the robot. This article mainly proposes a method for various data acquisition of ABB robots. The data acquisition interface is developed using C# and the dynamic link library is called. The data of ABB robots can be read in real time. It can realize real-time monitoring of each data of the robot.

1. Introduction

At present, the world has entered the era of Industry 4.0. In order to improve production efficiency, industry and manufacturing have begun to use robots to complete repetitive and even dangerous production operations in colleges and universities instead of manual labor.[1] Although robots can complete many tasks that replace humans, if the robot fails, it will not only lead to production stagnation, affect corporate efficiency, and even threaten the recognized life safety. Therefore, robot maintenance, life prediction, and fault diagnosis become more and more important. The ABB robot has been developed in C# and completed the data collected.[2]

At present, C# and C++ mixed programming is a relatively common method mixed programming. C# is a programming language customized by Microsoft for the Visual Studio.net framework. C# has the same powerful functions as C/C++, and the syntax of C# is simpler than C/C++, making C# one of the most popular development languages. But C# also has some shortcomings.[3] It can only be used for the bottom layer by calling Win32 dynamic link library (DLL, dynamic link library) or dynamic link library written in C++. operate. Moreover, it is widely believed that the confidentiality of C# is weaker than that of C/C++, which can easily cause its source code to be decompiled and cracked. Therefore, it is currently popular to adopt mixed programming to increase the confidentiality of C#. We can encrypt the DLL file with a certain algorithm or compress it and add it to the C# program. [4]

2. Principle and characteristics of DLL

The dynamic link library is a modular program based on Windows. The DLL file not only contains executable programs,data,and resources required for various execution programs. In the interface development program design, the DLL file can be used to integrate the entire system The program is decomposed into multiple main programs and DLLs, thereby further reducing workload during development. Due to the reduction of program reuse modules, the access speed has been greatly improved, and if it is necessary to modify a certain part of the underlying program, only The program in the corresponding DLL file can be modified, and the upper-level program does not need to be modified.

In the development process of the data acquisition system, the dynamic link library technology is applied to the functional modules of the system, and each functional module of the system module is made into DLL files, which simplifies the software development and management, and can save more memory space.

3. Dynamic link library

3.1. Import dynamic link library

First, you need to download and install Robot Studio and PCSDK on the ABB official website. ABB's software development kit PCSDK is a PC software tool that enables programmers to develop customized operator interfaces for the robot's controller and use the network provided by the SDK. Communication, establish Socket communication (TCP/IP protocol) with Robot Studio. After the installation is complete, there will be three dynamic link libraries referenced by Visual Studio with the suffix dll in the path.

It is relatively simple to use C# to call the managed dynamic link library. You only need to right-click the "reference" in the "solution" and select "add reference", select the listed dynamic link library DLL file, or browse to add the dynamic link library.[5]

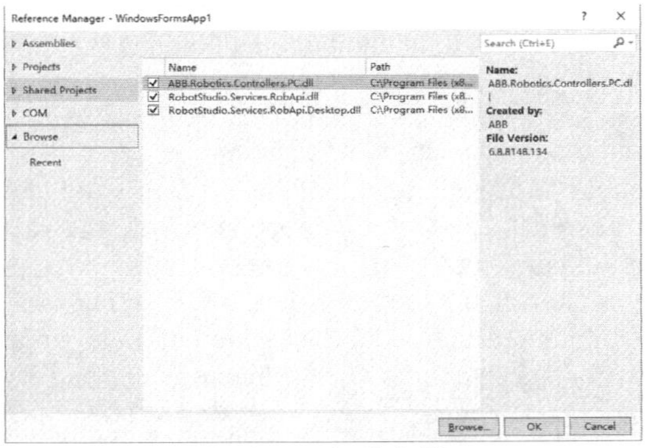

Figure 1. Add dynamic link library

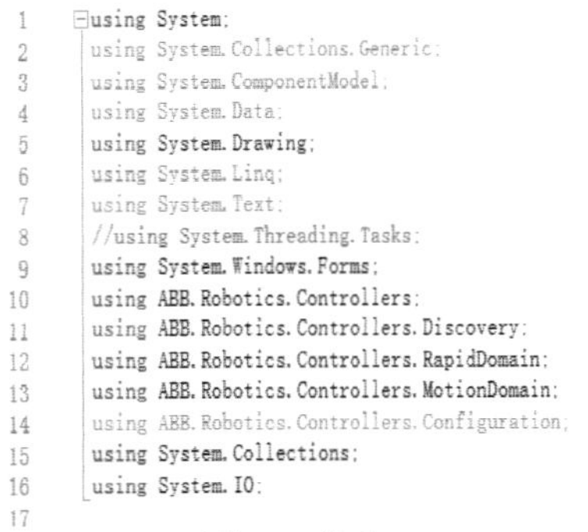

Figure 2. Import namespace

After adding the dynamic link library, use using to import the namespace you want to use, and you can call the functions in the managed dynamic link library just like using your own class.

3.2 The content of the dynamic link library

The classes used to access the functions of the robot controller constitute the controller API.

ABB.Robotics.Controllers.Discover: Scan the domains required by the controller method.

ABB.Robotics.Controllers.RapidDomain: The domain required to access and modify the Rapid program.

ABB.Robotics.Controllers.MotionDomain: The domain required to obtain the robot pose.

The ConfigurationDomain control contains classes for handling the configuration of the robot controller. This namespace allows access to the configuration database of the controller. Using this namespace, the value of the configuration parameter can be read or written to the designated database of the controller.[6] When logging in to the control It can be read-only or write-accessed. Read-only is the default access permission. In order to obtain write access, the client needs to request the master control of the controller resource it wants to operate.

ConfigurationDatabase is to access the configuration database, allowing access to the configuration database to read and write the configuration type and properties of the robot controller.

Reading and writing configuration parameters can be read from the configuration database of the controller. The figure shows how to read the unitmap value of signal TestDI, and how to read the cal offset value of rob1 l (moc.cfg), and set signal TestDI The unitmap is set to 17.

```
using ABB.Robotics;
using ABB.Robotics.Controllers;
using ABB.Robotics.Controllers.ConfigurationDomain;
 ...
 try
 {
     Controller controller = new Controller();
     Mastership master;
     ConfigurationDatabase cfg = controller.Configuration;

     using (master = Mastership.Request(controller.Configuration))
     {
         string[] path = { "EIO", "EIO_SIGNAL", "TestDI", "UnitMap" };
         cfg.Write("17", path);
     }
 }
 catch (System.InvalidOperationException ex)
 {
      MessageBox.Show("Mastership is held by another client.");
 }
 catch (System.Exception ex)
 {
     // TODO: Add error handling
 }
 finally
```

Figure 3. Read and write configuration parameters

Use DomainCollection to get a list of all domains available in the configuration database.

The ConfiguratioDomain.Attributn object contains the description of the configuration domain attribute and cannot be used to read or write the value of the attribute. To read and write the value of the attribute, you should use the GetAttribute() and SetAttribute() methods respectively.

The ConfigurationDomian.Instance object defines an instance of a specific type. This object is a set of disconnected data sets from the configuration database, as shown in the figure below, to create an instance object, use the SetAttribute() and GetAttribute() methods. Create an instance, and set the attribute entry to static, use GetAttribute() to get the same attribute.

When a network scan is executed, a collection of ControllerInfo objects will be returned. This class contains "simple" information about the controller and does not need to be connected to it. Note that when the controller state changes, the ControllerInfo object will not be updated. If you need to know whether the controller is available again, you need to execute a new Scan the network or use an existing controller object and check the status directly.

4. Acquisition system development of Data

First click on the toolbox on the design interface, find the Button-button in the common controls of the toolbox, drag the button into the design interface, and adjust the size to a reasonable size, click on the Button control, press F4 to call up the properties of the Button, and change the text to "Scan the controller", click the event, select and double-click the "Click" option, you can write the program to scan the controller.Select the public control TextBox and drag it into the Form design interface, which is mainly used to receive controller scan information. To edit the Click event of the Button control, you need to first create a network scan object, use the Scan ABB.Robotics.Controllers.Discovery namespace to scan the robot, load all controllers into the memory, and read the end of the scan The information in the returned ControllerInfo object collection is displayed in TextBox1, as shown in Figure.

```
private void Button1_Click(object sender, EventArgs e)
{
    this.Scanner = new NetworkScanner();//Create a new network scanner instance.
    this.Scanner.Scan();//Start a scan to load all controllers into memory.
    ControllerInfoCollection controllers = this.Scanner.Controllers;//Get the collection of found controllers
    foreach (ControllerInfo info in controllers)
    {
        this.TextBox1.Text += info.ControllerName + "\r\n";
        this.TextBox1.Text += info.IPAddress + "\r\n";
        this.TextBox1.Text += info.Version + "\r\n";
        Controller_Info = info;
    }
}
```

Figure 4. Scan Controller Button

Also add a button control to the Form design interface, and set the text of buton to "connect to the controller". When the controller is scanned, the controller will be loaded into the memory. Click the button to connect to the controller. . Drag the textbox again to record the connection, and add a Click event to the button. When writing this event, you must first determine whether the controller is successful. If the controller is loaded, first log off the current user, release or reset the application-defined tasks related to unmanaged resources, so as to achieve the effect of releasing the previous resources . After the resource is released, the controller instance object needs to be created first, and the controller.Logon() method is used to specify the user identity to log in. After the login is successful, the login information will be displayed in the TextBox. The event program code is shown in the figure.

```
private void Button2_Click(object sender, EventArgs e)
{
    if(this.controller!=null)
    {
        this.controller.Logoff();//Log out the current user
        this.controller.Dispose();
        //Release or reset application-defined tasks related to unmanaged resources
        this.controller = null;

    }
        //Create controller instance
        this.controller = ControllerFactory.CreateFrom(Controller_Info);
        this.controller.Logon(UserInfo.DefaultUser);
        //Log in as the specified user
        this.task = controller.Rapid.GetTask("T_ROB1");
        this.TextBox1.Text += Controller_Info.SystemName + "Logged";
        ArrayList module_ary = GetModuleList();
        comboBox1.Text = "";
        comboBox1.Items.Clear();
        for (int i = 0; i < module_ary.Count; i++)
            {
                comboBox1.Items.Add(module_ary[i]);
            }
}
```

Figure 5. Start the robot program

After the start command code is written, add the Timer control. The control can execute the code at regular intervals. According to the actual situation, when setting the Timer control, in the Timer control Tick event, first determine whether the controller is connected. When the controller is connected, "Robot is connected" is displayed, the control mode of the robot is obtained through the Getoperation_mode() method, and the control state is obtained through the Getcontroller_states() method, and displayed. Extract the robot motion data and add 6 TextBoxes, which are mainly used to obtain the current posture of the robot. First, you need to create MotionSystem and JointTarget objects, and call MidpointRounding.AwayFromZero to display the rotation angle of each axis.

After completing the page design and programming, you can run the data acquisition system and read the data of the robot in real time, as shown in the figure for the interface of the robot acquisition system.

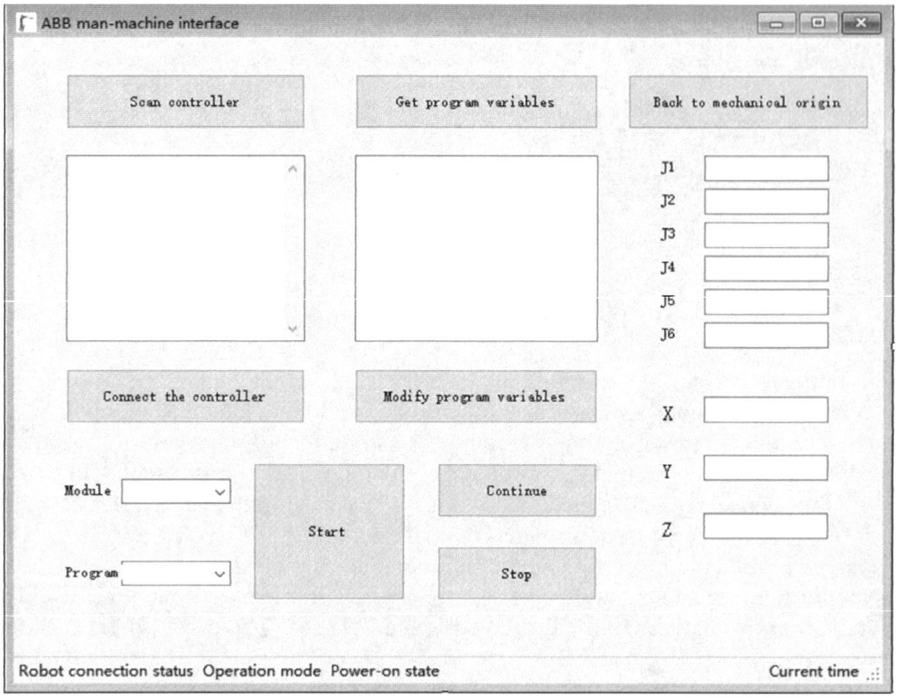

Figure 6. ABB Robot Data Acquisition System

Complete the path programming of the ABB robot through the teach pendant of the control cabinet. As shown in the figure, program the robot teach pendant. Set the written program with the Test name.

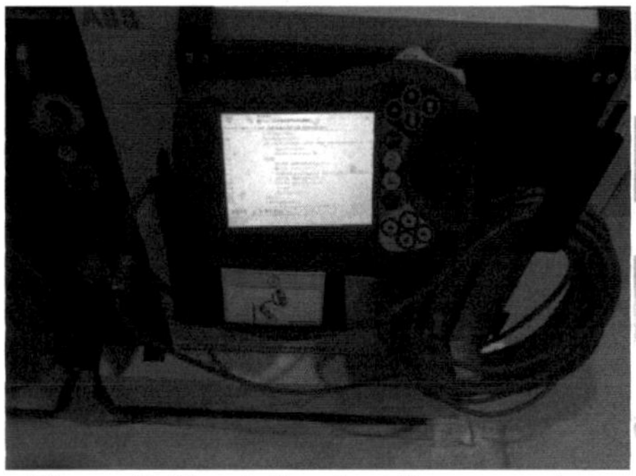
Figure 7. Teach Pendant Programming

Figure 8. Robot PLC

Connect the PLC of the robot control cabinet and the main box through a network cable, as shown in the figure.

Then open the data acquisition system developed in C#, as shown in the figure, you can collect robot data.

Figure 9. Connect to the host to collect robot data

5. Conclusion

The system can connect to the robot in real time, read the relevant data of the robot, monitor various data of the robot, and achieve the effect of predicting the failure of the robot for maintenance through the monitoring, so as to avoid unnecessary economic losses and personal injuries.

References

[1]. Nathan Wallace with, learning to use Visual C++6.0 develop- ment of Active X Template Library, University of Electronic Science and Technology Press, 2000.
[2]. Stephen D. Gilbert was, and I learn Visual C++6.0, Machinery Industry Press, 1999.
[3]. Tang Tongyi, Electrical Measurement & Instrumentation Tech- nology in Retrospect and Prospect, Electrical Measurement and Instrumentation, VOLUME 37, No. 1 2000.
[4]. Wang Xiaofeng Li Yunfei Yupeng edited, Visual C++ practice and improve, China Railway Press, 2001.
[5]. Cheng Zhiguang, "Electromagnetic Field Analysis and Verification of Power Transformers", Ph.D. Thesis, Tsinghua University, 1994.
[6]. Zhou Jianming, finite element integrated simulation method of electromagnetic field and research on leakage magnetic field of large transformer, PhD thesis of Huazhong University of Science and Technology, 1990.

Application of Fuzzy Neural Network PID In Laser-guided Mortar Projectile

Yang Song, Fangxiu Jia, Xiaoming Wang, Dingming Meng, Lei Zhuang

School of Mechanical Engineering, Nanjing University of Science and Technology, Nanjing, Jiangsu, China

email: songyangemail95@163.com

Abstract. Based on the high control performance requirement of laser-guided mortar control system, the permanent magnet synchronous motor (PMSM) is adopted in this paper as the electromechanical actuator of the system, the mathematical model of the motor is analyzed, and the vector control technology is adopted to achieve precise control of position, speed and torque of the electromechanical actuator. Aiming at the characteristics of non-linearity, strong coupling and large parameter changes of the system in flight, an improved fuzzy neural network PID control method is proposed by combining the classical PID control algorithm with fuzzy control and neural network control algorithm to realize the real-time tuning and optimization of PID parameters. The mathematical model of the electromechanical actuator control system is established and simulated. The results show that the fuzzy neural network PID control has good tracking performance, small amplitude error, and strong adaptability to load changes.

1. Introduction

In recent years, the proportion of precision-guided weapons used in war has increased. Many countries have improved the precision strike capability of artillery through the development of laser-guided projectiles. Under such demand, low-cost laser terminal guided mortars with good accuracy have emerged. The electromechanical actuator control system outputs a correction signal according to the measured deviation value, controls the deflection of the actuator surface, and realizes the control of the projectile's flight attitude. Thanks to the development of rare earth materials, microelectronics and computer technology, permanent magnet synchronous motors (PMSM) have been widely used in guided weapon systems with their simple structure, high efficiency, high power density, and good controllability. PMSM vector control can decouple the current torque component and the excitation component through coordinate system transformation, thereby simulating the DC motor torque control under the rotating magnetic field. Then simple control of PMSM is achieved by controlling armature current and excitation current[1].

The mortar flight end seeker transmits the calculated actuator surface angle to the control system according to the deviation. Based on the three closed loops of the position loop, the speed loop and the current loop, the position of the actuator is precisely controlled by the PID controller. The classic PID control algorithm has a history of more than a hundred years and is widely used in various industrial productions. It is simple, reliable, and highly robust, and is suitable for stable linear systems with precise mathematical models. But artillery shells are often affected by airflow disturbances in flight, and traditional PID control is often difficult to control this nonlinear and complex system, and it cannot achieve the control effect required by the guidance accuracy. Fuzzy control can express control experience and knowledge in language rules, and has a good control effect on nonlinear control systems with unknown mathematical models, but it is relatively lack of self-learning ability and mainly relies on expert experience. Neural network control builds models by simulating the human brain nervous system from the structure and function. It has strong ability to approximate nonlinear functions and self-learning, but it is not good at expressing knowledge explicitly.

Based on the respective advantages and disadvantages of fuzzy control and neural network control, this paper combines fuzzy control, neural network control and classic PID, and proposes a fuzzy neural network PID control technology applied to guided mortar rudder system, combined with the use of fuzzy control Experience knowledge ability and neural network self-learning ability, so as to meet the control system's requirements for accuracy and sensitivity[2].

2. Methods and theories

2.1. Design of electromechanical actuator control system

2.1.1. Three closed loop control system

The electromechanical actuator control system receives the actuator surface angle position signal calculated from the seeker. The projectile rolls at low speed during the projectile flight, and the steering gear rotates in the opposite direction relative to the projectile to achieve zero speed relative to the inertial coordinate system, and it is in position control. Control the angle of the rudder surface under the action. The system is controlled by three closed loops. The current loop is the innermost loop of the control system. The control system provides the control torque for the electromechanical actuator through the current value collected by the current sensor and the current regulator (ACR). The speed loop is in the middle ring of the system, and the speed signal collected by the Hall sensor and the speed regulator (ASR) generate the reverse rotation speed relative to the projectile to realize that the rudder surface is stationary relative to the inertial system. The position loop is in the outermost loop of the system, and the position signal collected by the Hall sensor and the position regulator (APR) output the expected phase angle of the actuator surface relative to the geodetic coordinate system.

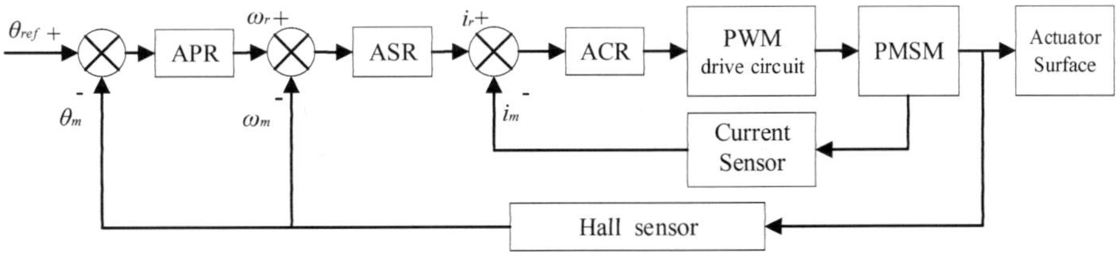

Figure 1. Steering gear control system.

2.1.2. Vector control

There are three types of coordinate systems in vector control (FOC): The ABC coordinate system formed by the three windings of the PMSM three-phase stator, the coordinate axes are separated by 120°; A static coordinate system composed of the α-axis that coincides with the A-axis and the β-axis that intersects with it; A rotating coordinate system composed of a d-axis coincident with the rotor axis and a q-axis perpendicular to it.

From the PMSM magnetic torque formula:

$$T_e = p_n\left[\psi_f i_q + \left(L_d - L_q\right)i_d i_q\right] \quad (1)$$

It can be seen that when $i_d = 0$, the electromagnetic torque T_e is proportional to the torque current component i_q.

The three closed-loop current command values are obtained through PID regulators to obtain the control voltages u_{dref} and u_{qref} of the d and q axes, and then the voltage is transformed to the

α and β axes of the static coordinate system through the inverse Park transformation, and then the vector pulse width modulation technology (SVPWM) is adopted. Control the three-phase inverter to realize the control of PMSM. The FOC control system of PMSM is as follows:

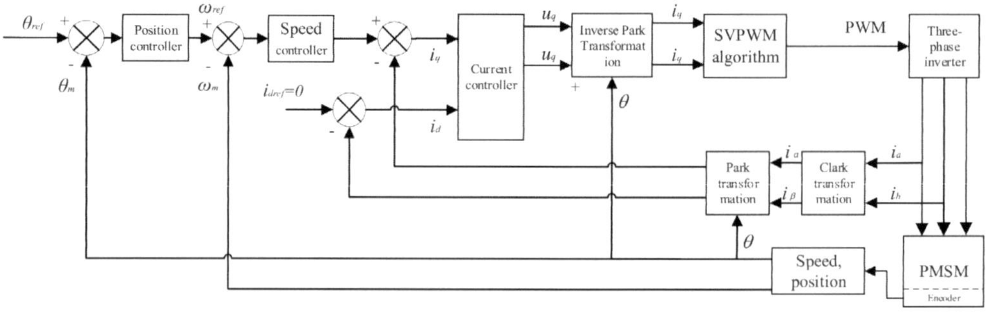

Figure 2. Vector control.

2.2. PID Controller of Fuzzy Neural Network

2.2.1. Fuzzy BP neural network PID controller

The reaction sensitivity and position accuracy of the phase angle of the actuator surface of the projectile in the terminal guidance flight stage directly determine the accuracy of the projectile[3]. Therefore, this article improves the PID controller of the position loop in the steering gear vector control system. In the fuzzy control, the self-learning ability of the neural network is integrated, and the error e of the actual feedback value of the phase angle of the steering gear and the ideal output value and the rate of change of the error e_c are used as input, and the fuzzy neural network outputs K_p, K_i, and K_d values in real time. Set the P(proportional), I(integral) and D(differential) parameters in PID[4].

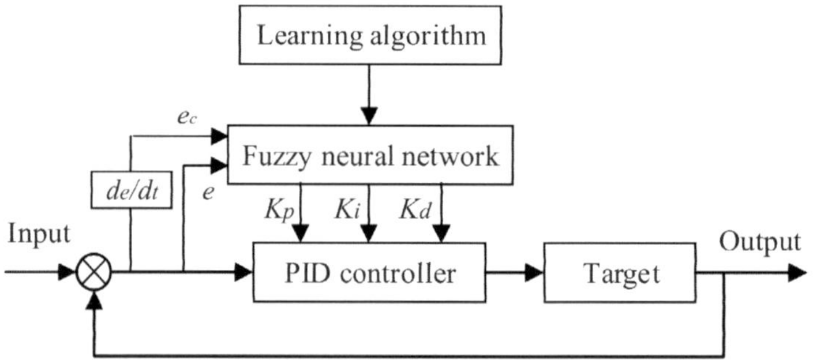

Figure 3. Fuzzy Neural Network PID.

In the figure K_p, K_i, K_d are PID controller parameters.

The control system uses BP neural network. The algorithm learning is divided into two processes of forward propagation and backward propagation. If the input value is processed layer by layer to the output layer, the actual output and expected output error exceeds a certain threshold, then the error Backward transmission layer by layer and apportion to all units of each layer, thereby correcting the weight of each unit. Repeatedly adjust the weights of each layer until the output result reaches the expected error range. According to the above principles, this paper designs a fuzzy neural network with an input layer, a fuzzy layer, a fuzzy rule layer, a normalization layer, and a clarification layer.

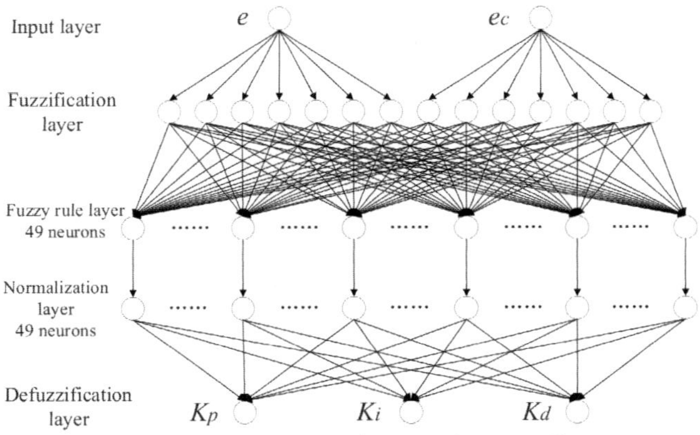

Figure 4. Fuzzy neural network controller.

2.2.2. Neural network learning algorithm

The adjustment process of the weights of each layer of the BP neural network is also the learning and training process of the neural network. The learning and training are stopped when the output error is reduced to a preset threshold or the preset number of learning times. In this paper, the direct and effective gradient method is used to find the minimum error, that is, the negative gradient descent method is used to find the optimal error function, and the inertia term is introduced to accelerate the convergence [5].

The error E between the network output and the expected value is defined as follows:

$$E = \frac{1}{2}\left[r_{in}(k) - y_{out}(k)\right]^2 \tag{2}$$

Among them, r_{in} is the expected input value of the system, and y_{out} is the actual output value of the system. The negative gradient relationship between weight ω_{kp} and its modification $\Delta\omega_{kp}$ and error E is as follows:

$$\frac{\partial E}{\partial \omega_{kp}} = \frac{\partial E}{\partial f_k^{(5)}} \frac{\partial f_k^{(5)}}{\partial \omega_{kp}} = -\delta_k^{(5)} y_p^{(4)} \tag{3}$$

The correction formula of the weighting coefficient is as follows:

$$\omega_{kp}(k+1) = \omega_{kp}(k) - \eta_1 \frac{\partial E}{\partial \omega_{wk}} + \alpha\left(\Delta\omega_{wk}(k)\right) \tag{4}$$

The relationship between the central value c_{ij} and the width σ_{ij} of the membership function and the negative gradient of the error E is as follows:

$$\frac{\partial E}{\partial c_{ij}} = \frac{\partial E}{\partial f_{ij}^{(2)}} \frac{\partial f_{ij}^{(2)}}{\partial c_{ij}} = -\delta_{ij}^{(2)} \frac{2\left(x_i - c_{ij}\right)}{\sigma_{ij}^2} \tag{5}$$

$$\frac{\partial E}{\partial \sigma_{ij}} = \frac{\partial E}{\partial f_{ij}^{(2)}} \frac{\partial f_{ij}^{(2)}}{\partial \sigma_{ij}} = -\delta_{ij}^{(2)} \frac{2\left(x_i - c_{ij}\right)^2}{\sigma_{ij}^3} \tag{6}$$

The correction formula for the central value and width of the membership function is as follows:

$$c_{ij}(k+1) = c_{ij}(k) - \eta_2 \frac{\partial E}{\partial c_{ij}} + \alpha\left(\Delta c_{ij}(k)\right) \tag{7}$$

$$\sigma_{ij}(k+1) = \sigma_{ij}(k) - \eta_3 \frac{\partial E}{\partial \sigma_{ij}} + \alpha\left(\Delta\sigma_{ij}(k)\right) \tag{8}$$

Among them η_1, η_2, η_3 represent the learning rate; k represents the iterative step, and α is the inertia factor.

2.3. System Simulation

In order to verify the control effect of the fuzzy neural network PID in the mortar servo system through simulation, Simulink is used in Matlab to establish the permanent magnet synchronous motor vector control model based on the traditional PID controller and the fuzzy neural network PID controller. Because the fuzzy neural network PID is difficult to build directly from the toolbox in Simulink, the S-function program is written to express the fuzzy neural network (FNN) sub-module.

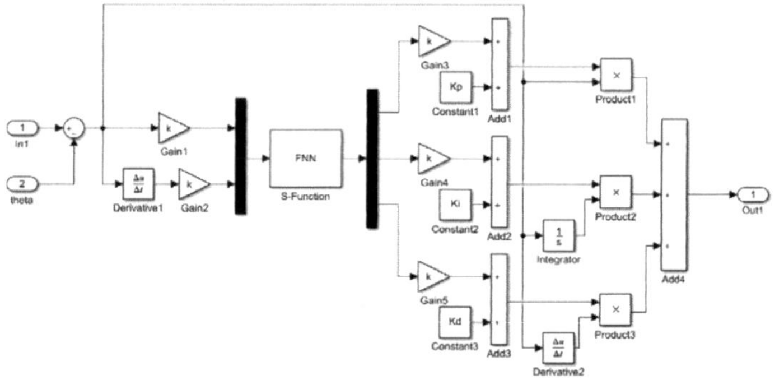

Figure 5. Fuzzy neural network PID simulation model.

The permanent magnet synchronous motor selected for this system has a rated voltage of 24V, a rated power of 64W, a rated speed of 3000 r/min, a rated armature current of 3.13A, a rated torque of 0.2NM, a torque coefficient of 0.057NM/A, and a back-EMF coefficient of 4.13V/RPM. The number of magnetic poles is 8, the phase resistance is 0.89Ω, and the phase inductance is 0.62mH. The moment of inertia of the rudder blade converted to the motor rotor is 001g.cm^2.

The initial value of the learning rate of the fuzzy neural network PID algorithm is set as $\eta_1 = \eta_2 = \eta_3 = 0.15$, The inertia factor is set to $\alpha = 0.04$.

3. Results & discussion

3.1. Step signal performance simulation

A step signal with a actuator surface angle of 1° is input into the control system, and the simulation results of Fig. 6 and Fig. 7 are obtained.

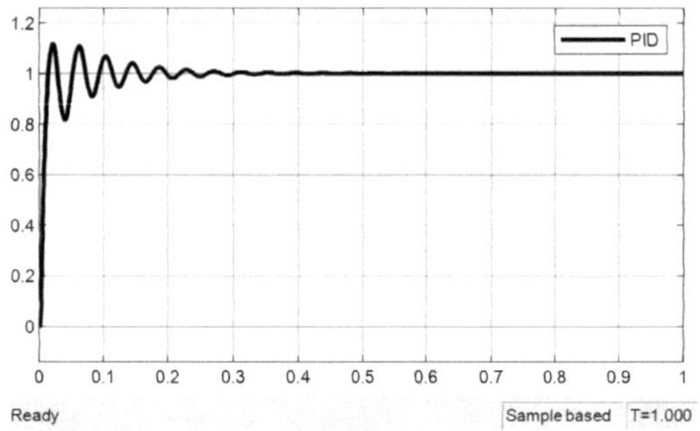

Figure 6. Traditional PID step characteristic curve.

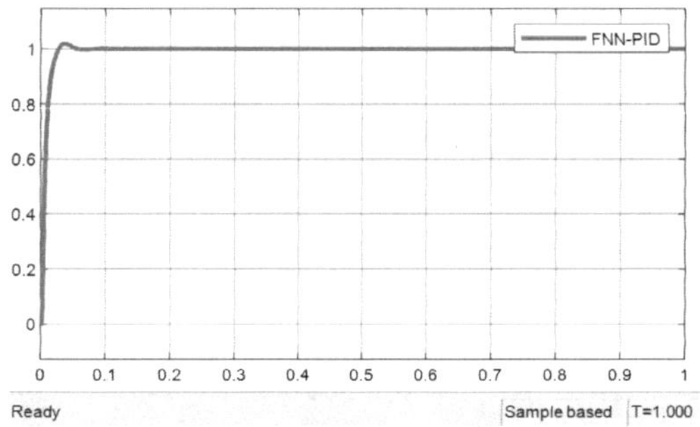

Figure 7. PID step characteristic curve of fuzzy neural network.

It can be seen that the traditional PID has an overshoot of about 10%, and the system stabilization time takes about 0.4s. The fuzzy neural network PID has only about 1% overshoot, and the signal stabilization time is about 0.1 seconds. Compared with the traditional PID, the control performance of the step signal is much better.

3.2. Follow signal performance simulation

Input a sine signal with an amplitude of 2 and a period of 50rad/s in the steering gear control system, and the simulation results of Fig. 8 and Fig. 9 are obtained.

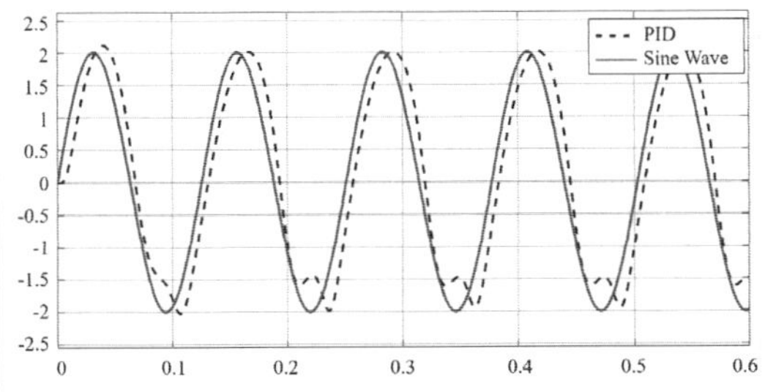

Figure 8. Classic PID followability curve.

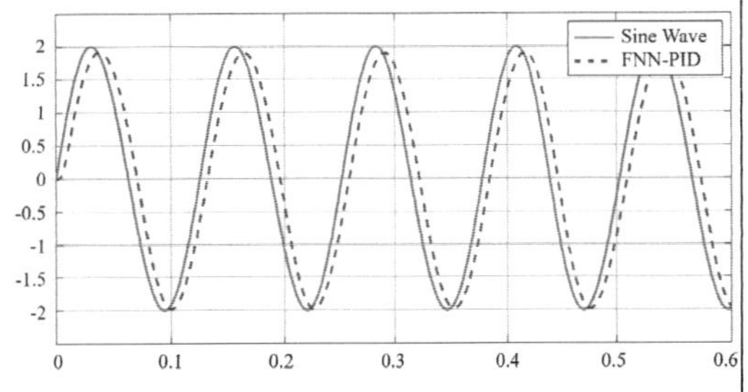

Figure 9. PID following characteristic curve of fuzzy neural network.

It can be seen that the traditional PID has signal distortion. The fuzzy neural network PID does not have the distortion problem. Compared with the traditional PID, the follow-up control performance of the continuous signal is better.

4. Conclusions

This paper studies the control system of terminal guided mortar rudder, selects permanent magnet synchronous motor and establishes three closed-loop control system, analyzes the characteristics of PMSM, selects vector control method with better control performance, and applies fuzzy neural network control to the position of vector control algorithm to improve the performance of actuator control. The steering gear control system model is established, and the position loop is controlled by traditional PID and fuzzy neural network PID respectively. Through the comparison of simulation results, the following conclusions can be drawn: Compared with traditional PID, fuzzy neural network PID has a small overshoot and system response time Short, better follow-up for continuous signals.

References

[1] Xia Y., Huang M G. (2019) Control system simulation of permanent magnet synchronous motor based on MATLAB. Electric Drive Automation., 41(05):6-9.
[2] Li X H. (2010) Application of fuzzy neural PID controller based on PSO for high precision servo system. ISCID., 2011: 291-294.
[3] Liu J K. (2013) Advanced PID control MATLAB simulation. Publishing House of Electronics Industry. Beijing.
[4] Li G Y., Yang L J. (2015) Neural· Fuzzy· Predictive Control and Its MATLAB Implementation. Publishing House of Electronics Industry. Beijing.
[5] Xu L., Guo L., Liang Y., et al. (2015) Study on intelligent PID control based on fuzzy RBF neural network. Industrial Instrumentation and Automation. 2015(06):67-69+75.

Research on intelligent early warning system of rail transit line trip based on pscada

Yuqing Zhang[1,a], Xiaohong Zhang[2,b*]

[1]School of Business, Shanghai DianJi University, Shanghai, 201306, China

[2]School of Business, Shanghai DianJi University, Shanghai, 201306, China

[a]email: 18755108238@163.com

*Corresponding author: [b]email: zhangxiaohong@sdju.edu.cn

Abstract. In this paper, an intelligent early warning scheme for rail transit line trip based on PSCADA system is proposed. The scheme takes into account the defects of low prediction accuracy and real-time prediction caused by the lack of power data in the traditional line trip prediction method. At the same time, a large number of power data generated by PSCADA system in the long-term application process are ignored in the field of rail transit[1]. Based on this situation, the prediction data set is constructed by combining the historical power data collected by PSCADA system in rail transit and the lightning weather data in traditional prediction methods. On this basis, the lightgbm machine learning intelligent algorithm is used to compare the similar support vector machine (SVM) and logistic regression algorithm to obtain a model with good prediction effect. In practical application, the real-time data set is constructed by using the real-time power data and real-time weather data collected by PSCADA system to predict, and an intelligent early warning system with the dual advantages of real-time and high accuracy is obtained.

1. Introduction

The traditional line trip prediction method is to predict the line trip probability by studying and locating the lightning occurrence area and scale and predicting the lightning weather or mountain fire disaster[2-4]. The application of this method to predict line tripping is mainly based on the important conditions that weather factors such as lightning or mountain fire affect line tripping. It is mainly to indirectly predict the tripping probability through the status and probability of lightning and mountain fire disasters. At the same time, this method ignores the power parameters generated in the process of line operation, which is of great significance, and also leads to the problem of low prediction accuracy. The PSCADA system of rail transit integrated monitoring system can collect the data of rail transit power supply system in real time[5]. If the real-time power data is used to predict the line trip fault, it will not only improve the real-time and accuracy of the prediction method, but also improve the utilization of power operation data.

2. Materials and Methods

This chapter mainly describes how the rail transit line trip intelligent early warning system proposed in this study is composed and applied to the actual operation of rail transit. This paper describes in detail how the intelligent prediction model constructs the data set, how to apply it in daily life and how to fault early warning. The flow chart of line trip intelligent early warning system is shown in Figure 1.

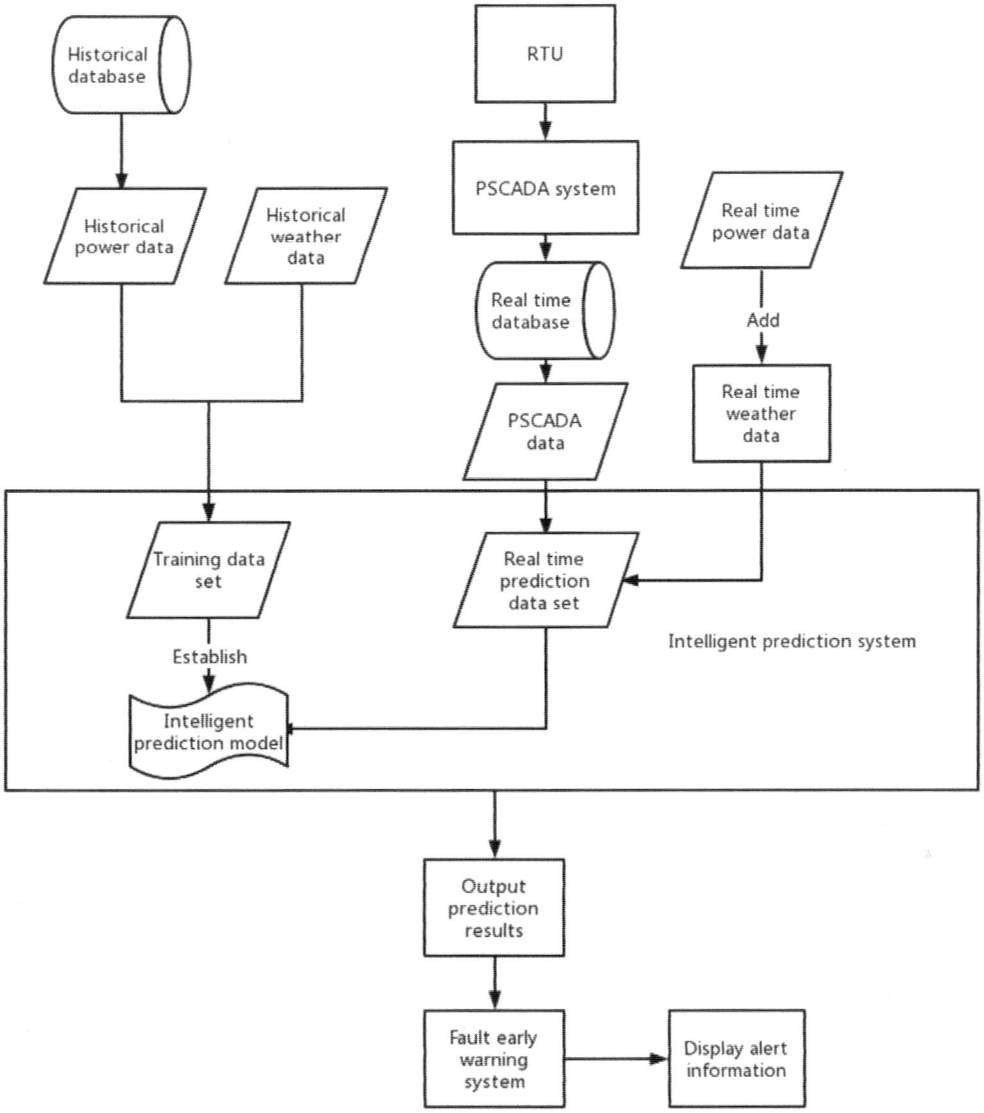

Figure 1. Flow chart of intelligent early warning scheme for line tripping.

2.1. System introduction

In this paper, an intelligent early warning scheme of line trip based on PSCADA system is proposed. In order not to lose the characteristics of relying on Lightning characteristic prediction in the classical line trip probability prediction method, the real-time power data information and weather information obtained by power monitoring system (PSCADA system) are combined in the scheme. In this way, the line trip probability can be predicted in real time to improve the real-time performance and the utilization of power data. Combined with machine learning algorithm to improve the prediction accuracy, so as to achieve more timely and accurate early warning effect, facilitate the operation and maintenance of rail transit power supply system, reduce the loss caused by line trip fault, and improve the safety of rail transit system.

2.2. Model building

The data set used for the establishment of the model comes from the relevant historical data of Metro Transmission Lines in 2021 recorded by the system in a metro station of Shanghai line 15. The data set specifically includes the operation and maintenance data of power supply lines, the relevant data of grid line tripping faults, the status of switch blades, remote voltage and current data, etc. In order not to lose the characteristics of relying on Lightning characteristic prediction in the classical line trip probability prediction method, we add the weather data to the processed operation and maintenance data set to form a prediction model data set containing lightning and other weather characteristics. The current data, weather, wind force, high and low temperature, lightning and line tripping at the current time and the first 50 recording points are combined into the same data set. The processed data set is imported into the line trip early warning model based on lightgbm for training, and the relevant parameters of the algorithm are adjusted appropriately to obtain higher accuracy and form the final optimization model. This model is compared with the model based on traditional logistic regression and SVM (support vector machine) algorithm to prove its prediction advantage.

2.3. Application Introduction

During daily operation, we combine the real-time current data and real-time weather data collected by PSCADA system as the prediction data set, predict according to the determined line trip prediction model based on lightgbm, and output the prediction results to the fault early warning system in real time to inform the dispatcher.

The fault early warning system can realize the fault early warning function based on the HMI interface of PSCADA system, timely push and inform the dispatcher of the fault predicted by the intelligent prediction system, so as to stand out from the massive alarm information and provide efficient and intuitive alarm information and auxiliary decision-making basis for the dispatching and monitoring personnel. The original alarm information in the system shall be divided into levels and display colors according to the degree of urgency. Judge according to the fault predicted by the intelligent prediction system. If the fault is predicted to occur, push it to the dispatcher in time. It is displayed in the interface with corresponding fault level and conspicuous color. If no fault is predicted, the predicted current will also be judged. If the current exceeds the set limit, it will also give corresponding alarm, play a corresponding warning role and assist the dispatcher in decision-making.

3. Results & Presentation

3.1. Result analysis

The processed data set is imported into the line trip early warning model based on lightgbm for training, and the relevant parameters of the algorithm are adjusted appropriately to obtain higher accuracy and form the final optimization model. Compared with the traditional logistic regression and support vector machine (SVM) models, this model has obvious advantages in prediction accuracy and prediction speed. The comparison of model prediction accuracy and time consumption is shown in Table 1 and figure 2.

Table 1. Accuracy and time-consuming of each model in predicting fault conditions.

	LightGBM	*SVM*	LR
Accuracy	0.9167	0.75	0.6667
Time consumption	0.018	0.032	0.054

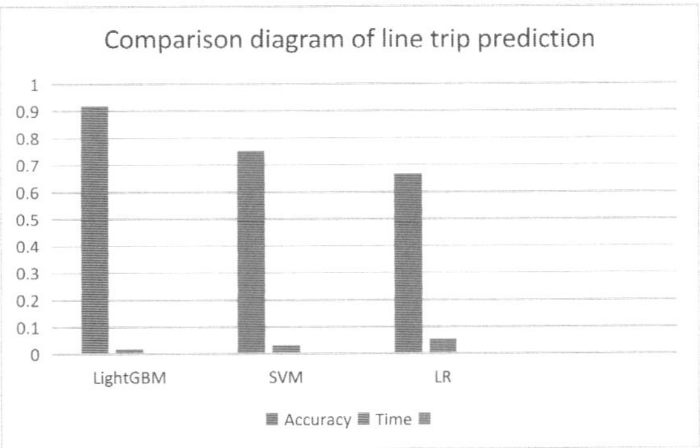

Figure 2. Comparison of predicted line tripping.

At the same time, when there is no tripping fault in the prediction result, the intelligent prediction model can be further used to predict the telemetry current. If the predicted current exceeds the set limit, the corresponding alarm will be given. Table 2 and Figure 3 show the comparison of measurement accuracy and time-consuming of the model.

Table 2. Accuracy and time-consuming of fault current prediction by each model.

	LightGBM	*SVM*	LR
Accuracy	0.833	0.75	0.833
Time consumption	0.036	0.07	0.099

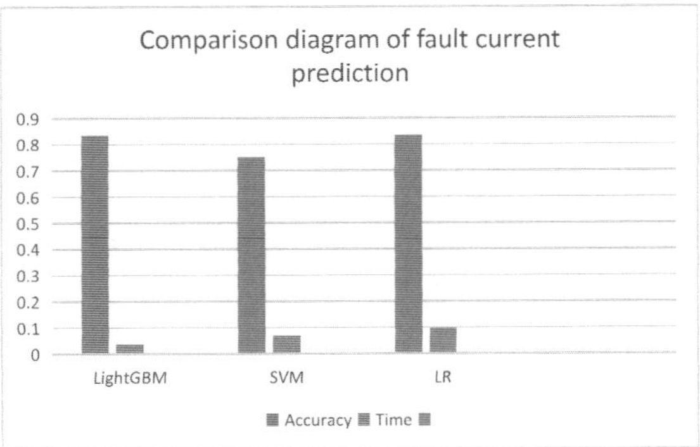

Figure 3. Accuracy and time consumption of fault current prediction by each mode.

The above chart shows that this prediction model has certain advantages in prediction accuracy and prediction speed, whether in predicting line trip or fault current.

3.2. Application result display

The results predicted by the intelligent prediction model are output to the fault early warning subsystem. The alarm information will be divided into levels and display colors according to the set emergency degree, and displayed in the interface with the corresponding fault levels and colors. The effect of early warning system is shown in Figure 4.

Journal of Physics: Conference Series **2113** (2021) 012079 doi:10.1088/1742-6596/2113/1/012079

Alarm level			Fault point	Detailed information	Operation
Frist level	2021/5/14 15:29	T3 terminal	Fault prediction - 10kV incoming line second switch trip	Switch opening	confirm
Second level	2021/5/14 15:27	Server	Relationship server - A	Insufficient capacity of historical database	confirm
Frist level	2021/5/14 15:19	T3 terminal	Fault prediction - 10kV incoming line I switching current	Over one level upper limit value:150.00	confirm

Figure 4. Effect diagram of early warning system.

4. Conclusions

This paper studies the rail transit line trip intelligent early warning system based on PSCADA system, and draws the following conclusions:

1) The intelligent early warning system of rail transit line trip based on PSCADA system effectively makes up for the defect that the traditional method can not predict in real time, and provides an idea for improving the real-time performance of line trip prediction method;

2) The prediction model of rail transit line trip intelligent early warning system based on PSCADA system has certain advantages in predicting line trip and fault current.

References

[1] Zhang,Y.X.,Zhang,B.S.(2020)Design scheme of intelligent operation and maintenance system for rail transit network.Railway Signalling & Communication Engineering,17:58-62.

[2] Gong,J.Y.,Tong,H.W.(2007)Evaluation of lightning trip rate of Zhejiang power grid line based on Lightning Location System.East China Electric Power,35:73-75.

[3] Zhao,Z.,Shi,J.Y.,Yuan,Q.H.(2015)A new method for predicting lightning trip probability of transmission lines.Automation of Electric Power Systems,3:51-58.

[4] Yang,L.,Suo,J.,Jiang,S.B.(2017)Application of deep learning technology in power grid.Technology and Economic Guide,22.

[5] Ma,L.(2020)Technical analysis of PSCADA power monitoring system for Urban Rail Transit.Microcomputer Applications,36:59-61.

Pavement State Recognition Method Considering Slope

Xiuhao Xi[1] , Jun Xiao[1*] , Qiang Zhang[1] and Yanchao Wang[1]

[1] School of Mechanical and Electronic Engineering, Wuhan University of Technology, Wuhan, Hubei, 430070, China

[*]Corresponding author's e-mail: xjun@whut. edu. cn

Abstract. For the problem of road surface condition recognition, this paper proposes a real-time tracking method to estimate road surface slope and adhesion coefficient. Based on the fusion of dynamics and kinematics, the current road slope of the vehicle which correct vertical load is estimated. The effect of the noise from dynamic and kinematic methods on the estimation results is removed by designing a filter. The normalized longitudinal force and lateral force are calculated by Dugoff tire model, and the Jacobian matrix of the vector function of the process equation is obtained by combining the relevant theory of EKF algorithm. The road adhesion coefficient is estimated finally. The effectiveness of the algorithm is demonstrated by analyzing the results under different operating conditions, such as docking road and bisectional road, using a joint simulation of Matlab/Simulink and Carsim.

1. Introduction

As an import section of the human-vehicle-road control system, the road surface adhesion has a significant impact on driver safety and vehicle stability. The methods for estimating road adhesion coefficient are mainly divided into two categories: Cause-Based and Effect-Based[1-2]. The Effect-Based method uses the existing sensors on the vehicle to measure and estimate the changes in the dynamic response of the vehicle caused by the changes in the road adhesion conditions, and indirectly estimate the road adhesion coefficient[3]. The Cause-Based method directly measures the factors related to the road adhesion coefficient (such as road unevenness[4] and tire noise[5]) by installing corresponding sensors on the vehicle, and then calculates the road adhesion coefficient. However, the Effect-Based method relies on an accurate mathematical model, while the Cause-Based method requires the installation of additional sensors, which increases economic costs.

Slope, another key information of the road surface, is an important parameter that affects the longitudinal dynamics control of the vehicle. The estimation of road slope is mainly divided into two methods: kinematics and dynamics. The kinematics method uses GPS[6]and longitudinal acceleration sensor[7] to calculate the slope. However, the low frequency of GPS can cause large slope estimation errors, and the static error of the longitudinal acceleration sensor will also affect the accuracy of the estimation result. The dynamic method is based on the longitudinal dynamics model of the vehicle and uses methods such as least square method [8] and Kalman filter [9] to estimate the road slope. However, this method is affected by the change of measurement noise during the estimation process, and the estimation accuracy is reduced.

In this paper, we estimate the road slope based on the method of fusion of dynamics and kinematics, then revise the tire model and the vehicle model by the estimation results. The road adhesion coefficient in the current driving state of the vehicle in real time is estimated and revised by using the EKF algorithm.

Finally, we use the simulation to prove the effectiveness of the algorithm. The road surface recognition algorithm framework is shown in Figure 1.

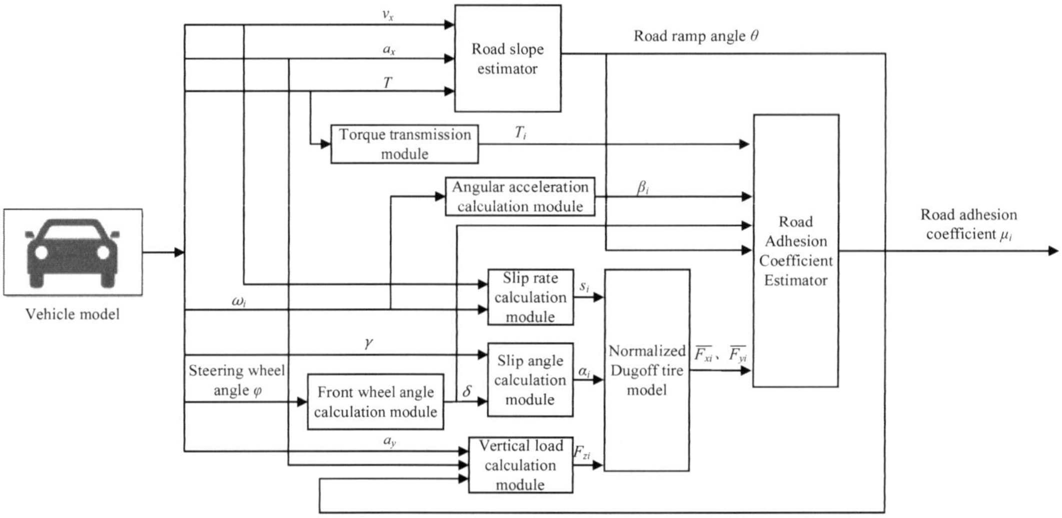

Figure 1. Pavement recognition algorithm framework

2. Slope estimation based on the fusion method of dynamics and kinematics

2.1. Slope estimation method based on dynamics

It can be seen from Newton's second law that when a vehicle moves longitudinally on a ramp, its dynamic equation is

$$m\dot{v} = \frac{T_t i_g i_0 \eta_T}{r} - mg\sin\theta - \frac{C_D A\rho}{2}v^2 - mgf\cos\theta \tag{1}$$

Where m is the vehicle mass. $\dot{v}$ is the longitudinal acceleration of the vehicle. T_t is the torque input from the engine to the gearbox. i_g is the transmission ratio of the gearbox. i_0 is the transmission ratio of the main reducer. η_T is the mechanical efficiency of the drive train. r is the effective rolling radius of the wheel. g is the acceleration of gravity. θ is the ramp angle. C_D is the air resistance coefficient. A is the windward area. ρ is the air density. v is the longitudinal speed of the vehicle. f is the rolling resistance coefficient.

From equation (1), the road ramp angle θ_a calculated based on the dynamic method can be obtained as

$$\theta_a = \arcsin\frac{2T_t i_g i_0 \eta_T - C_D A\rho v^2 - 2rm\dot{v}}{2rmg\sqrt{1+f^2}} - \arcsin\frac{f}{\sqrt{1+f^2}} \tag{2}$$

2.2. Slope estimation method based on kinematics

The value measured by the a_x longitudinal acceleration inertial sensor installed in the vehicle is affected by the actual acceleration and slope of the vehicle itself, and its value is

$$a_x = \dot{v}_x + g\sin\theta \tag{3}$$

From equation (3), the road ramp angle θ_b calculated based on kinematics method can be obtained as

$$\theta_b = \arcsin(\frac{a_x - \dot{v}_x}{g}) = \arcsin(\frac{a_x - \frac{v_x(t) - v_x(t-1)}{\Delta t}}{g}) \tag{4}$$

2.3. Slope estimation method based on the fusion of dynamics and kinematics

During the driving process of the vehicle, the road slope can be regarded as consisting of high-frequency signals and low-frequency signals. The dynamics-based slope estimation method relies on an accurate vehicle model, and the parameters in the vehicle model are affected by high-frequency noise. Thus the high-frequency part of the method is filtered through a low-pass filter, and the low-frequency part is reserved. The a_x heavily influenced by static bias is due to low frequency noise, so the low-frequency part of this method is filtered by a high-pass filter, and the high-frequency part is retained. From equations (2) and (4), the slope value θ calculated based on the fusion method of dynamics and kinematics is:

$$\theta = \frac{1}{Ts+1}(\arcsin\frac{2T_t i_g i_0 \eta_T - C_D A\rho v^2 - 2rm\dot{v}}{2rmg\sqrt{1+f^2}} - \arcsin\frac{f}{\sqrt{1+f^2}}) + \frac{s}{Ts+1}\arcsin(\frac{a_x - \dot{v}_x}{g}) \tag{5}$$

Where T is the time constant.

When the frequency of the road gradient is low, the value is greatly affected by the calculation result of the dynamic method, and it can better reflect the road gradient value under steady-state conditions. When the frequency of the road slope changes relatively high, the value is greatly affected by the calculation results of the kinematics method, and it can quickly follow the transient changes of the road slope.

3. Estimation of pavement adhesion coefficient based on EKF

3.1. Nonlinear vehicle dynamics model

When establishing the dynamics model, the vehicle is simplified as follows:

- Assuming that the centre of vehicle' mass is the origin of the moving coordinate system.
- Assuming that the vehicle is moving parallel to the ground. The influence of suspension deformation on it is ignored.
- Ignoring the influence of the steering system and assuming a linear relationship between the steering wheel angle and the front wheel angle.
- The mechanical properties of the four tires of the vehicle are the same.

The simplified vehicle dynamics model is shown in Figure 2.

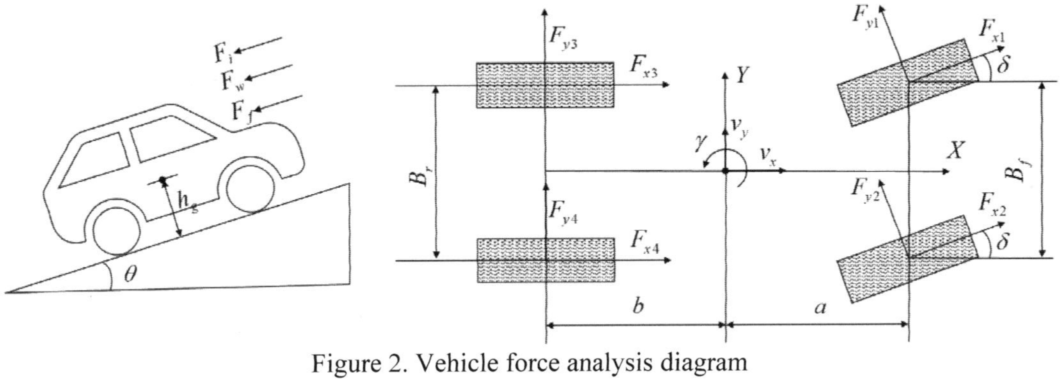

Figure 2. Vehicle force analysis diagram

Figure 2 shows that the longitudinal and lateral motion equations of the vehicle are

$$\begin{aligned} ma_x &= (F_{x1}+F_{x2})\cos\delta + F_{x3} + F_{x4} - (F_{y1}+F_{y2})\sin\delta - C_D A\rho v^2 - mg\sin\theta - mgf\cos\theta \\ ma_y &= (F_{x1}+F_{x2})\sin\delta + (F_{y1}+F_{y2})\cos\delta + F_{y3} + F_{y4} \end{aligned} \quad (6)$$

Where a_x, a_y are the longitudinal and lateral accelerations of the vehicle. F_{x1}, F_{x2}, F_{x3} and F_{x4} are the longitudinal forces of the four wheels. F_{y1}, F_{y2}, F_{y3} and F_{y4} are the lateral force of the four wheels. δ is the front wheel turning angle. θ is the road ramp angle, which can be calculated in Chapter 2.

3.2. Tire model

This paper uses Dugoff model to model and analyse the whole car tire. The force coordinate system of the whole car tire is shown in Figure 3.

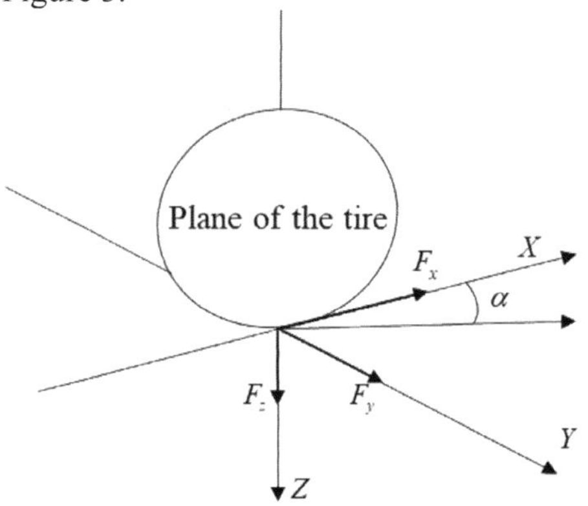

Figure 3. Tire force coordinate system

The wheel dynamic equation is

$$J_i\dot{\omega}_i = T_{di} - T_{bi} - fF_{zi}r - F_{xi}r \quad (7)$$

Where J_i (i=1, 2, 3, 4) is the rotational inertia of the four wheels. $\dot{\omega}_i$ is the angular acceleration of the four wheels. T_d, T_b is the driving and braking torque of the four wheels.

The vertical load of four wheels are

$$\begin{aligned} F_{z1,2} &= \frac{m}{2L}(gb\cos\theta - h_g a_x - h_g g\sin\theta) \mp \frac{ma_x h_g}{B_f} \\ F_{z3,4} &= \frac{m}{2L}(ga\cos\theta + h_g a_x + h_g g\sin\theta) \mp \frac{ma_x h_g}{B_r} \end{aligned} \quad (8)$$

Where L is the wheelbase. a and b are the distances from the centre of mass to the front and rear axles. h_g is the height of the centre of mass. B_f and B_r are the length of the front and rear wheels.

For a single wheel, its force equation are

$$\begin{aligned} F_x &= \mu F_z C_s \frac{s}{1-s} f(\lambda) \\ F_y &= \mu F_z C_\alpha \frac{\tan\alpha}{1-s} f(\lambda) \end{aligned} \quad (9)$$

Where

$$f(\lambda)=\begin{cases}\lambda(2-\lambda) & ,\lambda<1\\ 1 & ,\lambda\geq 1\end{cases}$$
$$\lambda=\frac{1-s}{2\sqrt{C_s^2 s^2+C_\alpha^2\tan^2\alpha}}(1-\varepsilon v_x\sqrt{s^2+\tan^2\alpha}) \tag{10}$$

Where μ is the road surface adhesion coefficient. F_z is the vertical load of the wheel. C_s and C_α are the tire longitudinal slip and cornering stiffness. s is the slip rate. α is the cornering angle. ε is the speed influence factor.

According to the expression form of the Dugoff tire model, it can be normalized to

$$F_x=\mu_0 F_z C_s\frac{s}{1-s}f(L)=\mu_0\overline{F_x}$$
$$F_y=\mu_0 F_z C_\alpha\frac{\tan\alpha}{1-s}f(L)=\mu_0\overline{F_y} \tag{11}$$

In the equation, $\overline{F_x}$ and $\overline{F_y}$ are the normalized longitudinal and lateral forces, which have nothing to do with the road adhesion coefficient. The equation (6)-(7) can be reduced to

$$a_x=\mu_1\frac{\overline{F_{x1}}\cos\delta-\overline{F_{y1}}\sin\delta}{m}+\mu_2\frac{\overline{F_{x2}}\cos\delta-\overline{F_{y2}}\sin\delta}{m}+\mu_3\frac{\overline{F_{x3}}}{m}+\mu_4\frac{\overline{F_{x4}}}{m}$$
$$-\frac{C_D A\rho v^2}{m}-g\sin\theta-gf\cos\theta$$
$$a_y=\mu_1\frac{\overline{F_{x1}}\sin\delta+\overline{F_{y1}}\cos\delta}{m}+\mu_2\frac{\overline{F_{x2}}\sin\delta+\overline{F_{y2}}\cos\delta}{m}+\mu_3\frac{\overline{F_{y3}}}{m}+\mu_4\frac{\overline{F_{y4}}}{m}$$
$$\dot{\omega}_i=-\mu_i\overline{F_{xi}}r+T_{di}-T_{bi}-fF_{zi}r \tag{12}$$

3.3. Implementation of EKF algorithm

3.3.1. Establishment of system state equation and measurement equation. For the vehicle nonlinear system, the state space expression established by EKF filtering is

$$\dot{x}(t)=f(x(t),u(t),w(t))$$
$$z(t)=h(x(t),u(t),v(t)) \tag{13}$$

Where $x(t)$ is the system state variable that needs to be estimated. $z(t)$ is the measurement output variable. $u(t)$ is the control input variable. $w(t)$ and $v(t)$ are Gaussian interference signals with a covariance of $R(t)$, representing process noise and measurement noise, respectively.

In the establishment of this model, the selected state variable is the four-wheel road adhesion coefficient μ_i, then the system state variable $x(t)=\left[\mu_1,\mu_2,\mu_3,\mu_4\right]^{\mathrm{T}}$, the selected observed variables are the vehicle longitudinal acceleration a_x, lateral acceleration a_y and four-wheel angular acceleration β_i, then the system observes the variable $z(t)=\left[a_x,a_y,\dot{\omega}_1,\dot{\omega}_2,\dot{\omega}_3,\dot{\omega}_4\right]^{\mathrm{T}}$. The control input variable is expressed as

$$u(t)=[\delta,\theta,T_{di},F_{zi},\overline{F_{xi}},\overline{F_{yi}}]^{\mathrm{T}} \tag{14}$$

3.3.2. Linearized model. After the observation equation is locally linearized, the equation (15) can be expressed as

$$z(t)=H(t)x(t)+y(t)+v(t) \tag{15}$$

Where $H(t)$ is the Jacobian matrix of the nonlinear function. $h(x(t),u(t),v(t))$ is the parameter to obtain partial derivatives of each parameter and its expression is

$$H(t)=\begin{bmatrix} \frac{\partial h_1}{\partial x_1} & \frac{\partial h_1}{\partial x_2} & \cdots & \frac{\partial h_1}{\partial x_n} \\ \frac{\partial h_2}{\partial x_1} & \frac{\partial h_2}{\partial x_2} & \cdots & \frac{\partial h_2}{\partial x_n} \\ \cdots & \cdots & \cdots & \cdots \\ \frac{\partial h_m}{\partial x_1} & \frac{\partial h_m}{\partial x_2} & \cdots & \frac{\partial h_m}{\partial x_n} \end{bmatrix} \tag{16}$$

By equation (12) can be seen

$$H(t)=\begin{bmatrix} \frac{\overline{F_{x1}}\cos\delta-\overline{F_{y1}}\sin\delta}{m} & \frac{\overline{F_{x2}}\cos\delta-\overline{F_{y2}}\sin\delta}{m} & \frac{\overline{F_{x3}}}{m} & \frac{\overline{F_{x4}}}{m} \\ \frac{\overline{F_{x1}}\sin\delta+\overline{F_{y1}}\cos\delta}{m} & \frac{\overline{F_{x2}}\sin\delta+\overline{F_{y2}}\cos\delta}{m} & \frac{\overline{F_{y3}}}{m} & \frac{\overline{F_{y4}}}{m} \\ -\overline{F_{x1}} & 0 & 0 & 0 \\ 0 & -\overline{F_{x2}} & 0 & 0 \\ 0 & 0 & -\overline{F_{x3}} & 0 \\ 0 & 0 & 0 & -\overline{F_{x4}} \end{bmatrix} \tag{17}$$

3.3.3. Initial value of filter to realize recursive algorithm. The specific workflow of the EKF filtering algorithm is shown in Figure 4. In the figure, Q and R are the covariance matrices of process noise and measurement noise. Among them, the process noise covariance matrix $Q=I_4$, and the measurement noise covariance matrix $R=I_6*0.1$. The initial value of the error covariance matrix $P^-(t_0)=I_4$, and initial value of the system state variable $\hat{x}(t_0)=[1,1,1,1]^T$.

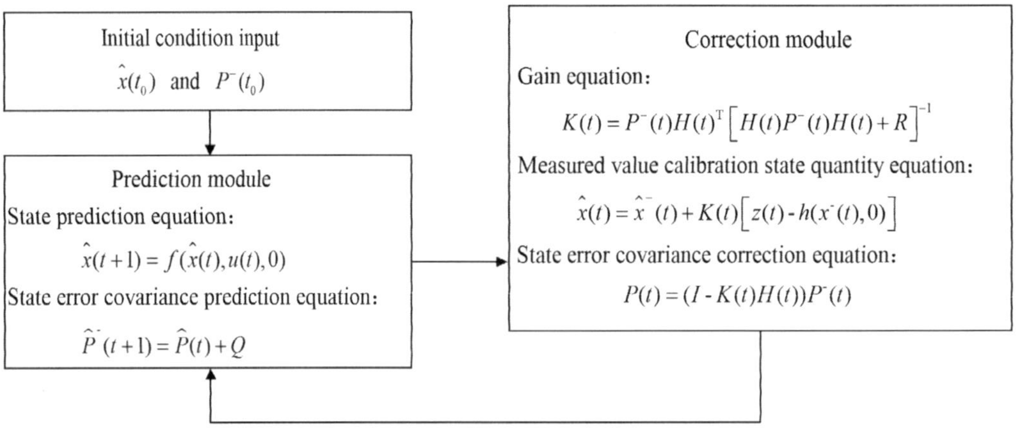

Figure 4. EKF algorithm flow chart

4. Simulation verification

The above algorithm for pavement gradient estimation of joints and adhesion coefficients was certified by Carsim/Simulink simulation. The E-class four-wheel drive SUV in the Carsim database is used as the vehicle model, and the relevant parameters are shown in Table 1.

Table 1. Vehicle parameters

Name	Symbol	Value	Name	Symbol	Value
weight	m	1860kg	Centroid height	h_g	0. 72m
Centroid distance of F axle	a	1. 18m	Effective tire rolling radius	r	0. 393m
Centroid distance of R axle	b	1. 77m	Coefficient of air resistance	C_D	0. 3
Wheelbase	L	2. 95m	Frontal area	A	3 m^2
The distance of the F wheel	B_f	1. 575m	Air density	ρ	1. 206 kg/m^3
The distance of the R wheel	B_r	1. 575m	Rolling resistance coefficient	f	0. 01

4.1. Simulation verification of single road adhesion coefficient

Considering the high and the low adhesion coefficient of the influence of the traveling state of the vehicle, two single-road driving conditions provided with CarSim are shown as follows: 1) adhesion coefficient is 0. 8 and the driver keeps a straight line at a constant speed of 60km/h; 2) adhesion coefficient is 0. 3, and the speed is maintained at a constant speed of 30km/h. The road slope of the two working conditions is set to 15%. The EKF algorithm's recognition of the road adhesion coefficients is shown in Figure 5-6.

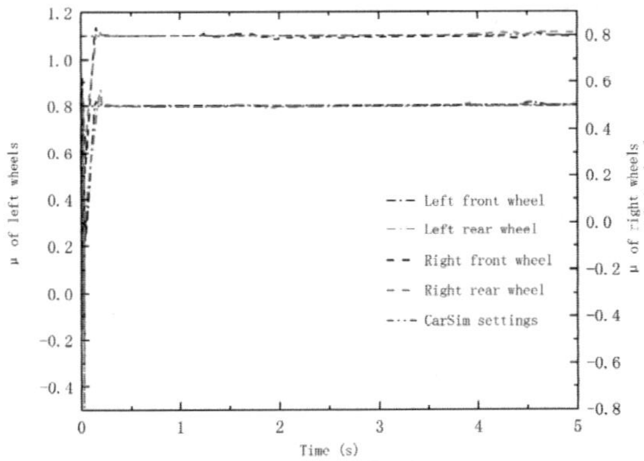

Figure 5. High-attached pavement recognition result

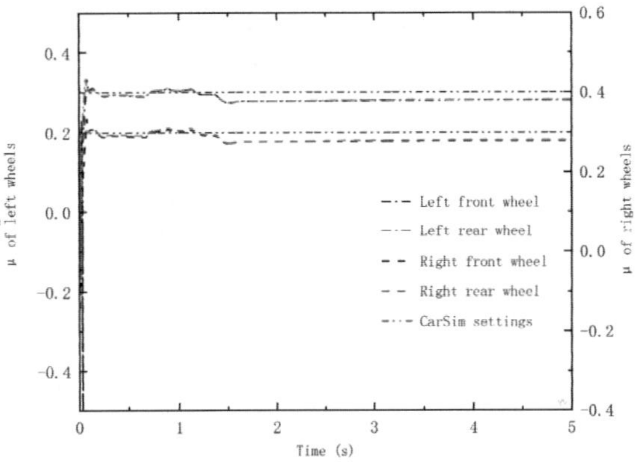

Figure 6. Low-attached pavement recognition results

4.2. Simulation verification of docking road adhesion coefficient

Considering the driving situation of the docking road, the adhesion coefficient of the road set in CarSim is changed from 0. 8 to 0. 3, and the road gradient is 3%. The vehicle keeps the speed of 40km/h and travels in a straight line.

The model is simulated, and the relationship between the calculated value of the road adhesion coefficient and the set value is shown in Figure 7. It can be seen from the simulation results that the calculated road adhesion coefficients of the four tires tend to set high value in 0. 3s. When the road adhesion coefficient suddenly drops, the algorithm can quickly calculate the low adhesion state at this time.

 2113 (2021) 012080 doi:10.1088/1742-6596/2113/1/012080

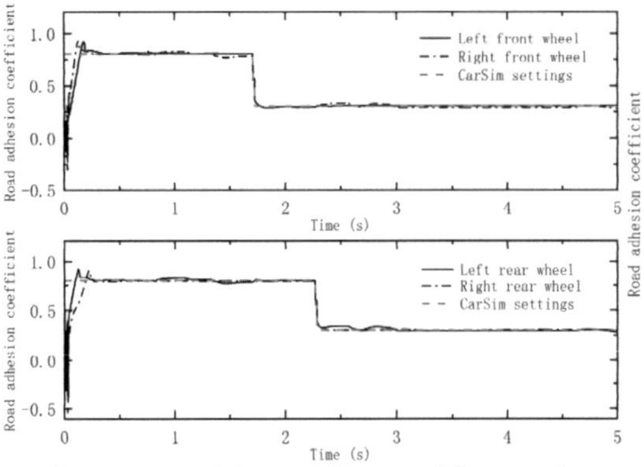

Figure 7. Docking road recognition results

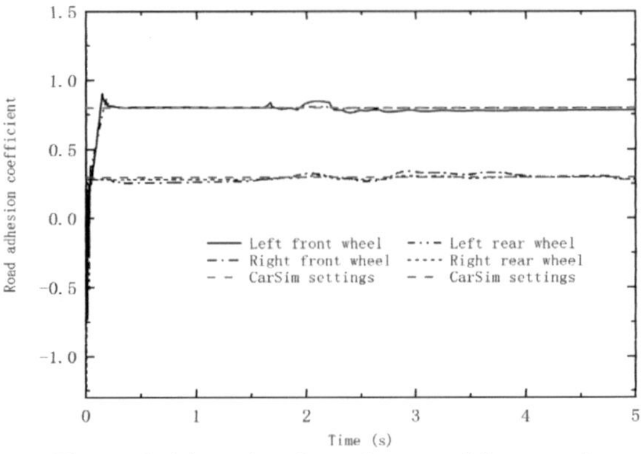

Figure 8. bisectional road recognition results

4.3. Simulation verification of bisectional road adhesion coefficient

The vehicle is set to drive on opposite roads. The left front wheel and the left rear wheel are on the high road surface, and the road adhesion coefficient is 0. 8. The right front wheel and right rear wheel are on the low-attachment road, and the road adhesion coefficient is 0.3. The vehicle drives at the speed of 60km/h, and the road gradient is 3%. Figure 8 shows the results of the simulation on the bisectional road.

From the simulation results, the adhesion coefficient is set to 0. 8 for the left wheel and 0. 3 for the right wheel in CarSim. For the two wheels on the left, although there were slight fluctuations before 0. 3 seconds, it did not affect the subsequent convergence results and the effectiveness of the estimation algorithm. For the two wheels on the right, it can quickly converge to around 0. 8 in 0. 1 second, which means that the estimation algorithm can estimate the road adhesion coefficient of the four wheels very well when the vehicle is driving on the bisectional road, which fully proves the algorithm's effectiveness.

5. Conclusion

- In view of the influence of slope on the road surface recognition accuracy, the slope is recognized by the method based on kinematics, dynamics and the fusion of kinematics and dynamics, and the recognition result is used as the input of the road adhesion coefficient recognition model to eliminate the influence of slope.
- Combined with Dugoff tire model and EKF algorithm, the state equation, measurement equation and Jacobian matrix of vehicles under different conditions are established, and finally the road adhesion coefficient of vehicles under different conditions is identified.
- The simulation verifies the recognition of the adhesion coefficient of the algorithm on the low-attached road, high-attached road, butt road, and off road. The simulation data reflects that the algorithm can accurately caculate the state and changes of the road surface under various driving conditions.

References

[1] MEYER W E, WALTER J D. Frictional interaction of tire and pavement[M]. Philadelphia: ASTM International, 1983.
[2] KHALEGHIAN S, EMAMI A, TAHERI S. A technical survey on tire-road friction estimation[J]. Friction, 2017, 5(2): 123-146.
[3] KHALEGHIAN S. The application of intelligent tires and model based estimation algorithms in tire-road contact characterization[D]. Blacksburg: Virginia Polytechnic Institute and State University, 2017.
[4] Brgeson J. Sensor data fusion based estimation of tire-road friction to enhance collision avoidance. Tampere University of Technology, 2010.

[5] EICHHORN U, ROTH J. Prediction and monitoring of tyre/road friction[J]. Proceedings Fisita, 1992: 67-74.
[6] SILVA A L D, CRUZ Z J J D, Fuzzy adaptive extended Kalman filter for UAV INS/GPS data fusion[J]. Journal of the Brazilian Society of Mechanical Sciences and Engineering, 2016, 38(6): 1-18.
[7] MASSEL T, DING E L, ARNDT M. Investigation of different techniques for determining the road uphill gradient and the pitch angle of vehicles[C]//American Control Conference. Boston, Massach-usetts: [s, n.], 2014
[8] Winstead V, Kolmanovsky I V. Estimation of road grade and vehicle mass via model predictive control ［C］ // Proceedings of 2005 IEEE Conference on Control Applications, 2005: 1588-1593.
[9] LEI Yulong, FU Yao, LIU Ke. Vehicle quality androad gradientestimation based on extended Kalman filter [J]. Journal of Agricultural Machinery, 2014, 11(11): 13-18. (in Chinese)
[10] DUGOFF H. Tire performance characteristics affecting vehicle response to steering and braking control inputs. Final report[R]. Washington D. C. : National Bureau of Standards, 1969.

Finger joint angle estimation based on sEMG signals by Attention-MLP

Zhebin Yu[1*], Hui Wang[2 3] and Wenlong Yu[4]

[1] School of artificial intelligence, Hebei University of Technology, Tianjin, 300131, China

[2] Shenzhen Key Laboratory for Innovative Technology in Orthopaedic Trauma, Department of Orthopaedics and Traumatology, The University of Hong Kong-Shenzhen Hospital, Shenzhen, China.

[3] CAS Key Laboratory of Human-Machine Intelligence-Synergy Systems, Shenzhen Institutes of Advanced Technology (SIAT), Chinese Academy of Sciences (CAS), Shenzhen, China.

[4] Shenzhen RunYiTaiYi Technology, Shenzhen, China.

*Corresponding author's e-mail: yzm9696@163.com

Abstract. sEMG(Surface electromyography) signal was widely applied in human-machine interactive field, especially in robotic arm control. In this paper, we built the Attention-MLP (Multilayer Perceptron) model to implement a type of continuous joint angle estimation method based on sEMG for six grasp movements, we tested this model on Ninapro dataset and the average Pearson correlation coefficient (CC) and the average root mean square error (RMSE) of the proposed Attention-MLP method achieved 0.812±0.02 and 10.51±1.98; the average CC and RMSE of this method are better than Sparse Pseudo-input Gaussian processes (SPGP), its average CC and RMSE are 12.14±2.30 and 0.727±0.07. Compared with the traditional method SPGP, our model performed better on continuously estimation of ten main hand joint angles under 6 grip movements.

1.Introduction

sEMG signal is a kind of bioelectrical signal that records muscle activities. Many machine learning classification algorithms were proposed for recognize motions by sEMG. Recently, in order to realize a natural and precise control methods, continuous movement estimation by sEMG has been a popular research topic[1]; the corresponding relationship between the sEMG signals and the real-time joint angles can be well excavated by deep learning models to realize fine-grained control systems such as myoelectric manipulator.

Deep learning was successfully used in signal processing for its great generalization ability; attention mechanism of this paper is an effective deep learning model for extracting important features information. The concept of attention mechanism was inspired by attention in biology and it was first proposed in Computer Vision to extract image features[2]. After 2014, attention mechanism has been successfully applied in text classification, machine translation and other tasks.[3] Recent years, attention mechanism has been used to replace Recurrent Neural Network in Sequence-to-Sequence tasks[4]; because position encoding enables attention mechanism to perceived sequence timing information. In this study, we used the Multi-Head Self-Attention mechanism processed by positional encoding to further extract the sEMG signals information; then we use Multilayer Perceptron (MLP) to generate real-time joint angles of six movements[5]. Both sEMG and joint angles comes from the public dataset, NinaPro[6]. The feature of sEMG we chose is RMS[7], for its low computational complexity and sufficient information before putting signals into the network.

2.Method

2.1 Database

Ninapro, a public dataset [6], included 7 databases. These databases include sEMG, hand kinematics when subjects doing hand motions. The requirement of each action is to continue for 5S, rest for 3S and repeat for 6 times. In this study, we selected the sEMG and 10 main joint angle sensors which is located on metacarpal and proximal of every finger before and after completing six grasping movements in the dataset for complete the training and testing of continuous joint angle estimation model, The grasp motions is shown in figure 1.

Ninapro placed 12 sEMG signal wireless sensors on the forearm and upper arm [8]. The sEMG signals with 12 channels were recorded by Trigno™ wireless sEMG recording system (Delsys, Inc,

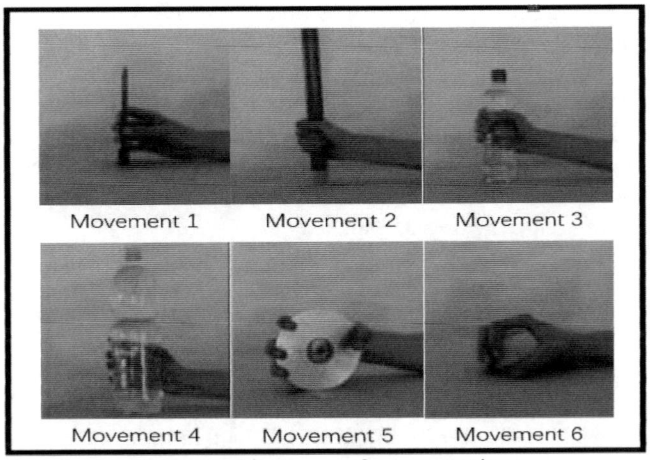

figure 1. Six types of grasp motions

Natick, Massachusetts, USA) through twelve channels separately at the sampling frequency of 2000H.

The hand joint angles with 10 channels were recorded by CyberGlove II data-glove, shown in Fig.2. The yellow dot represents our estimate of ten angles. The sensor of glove was resampled to 2000 Hz to synchronize with the sensor of Delsys.

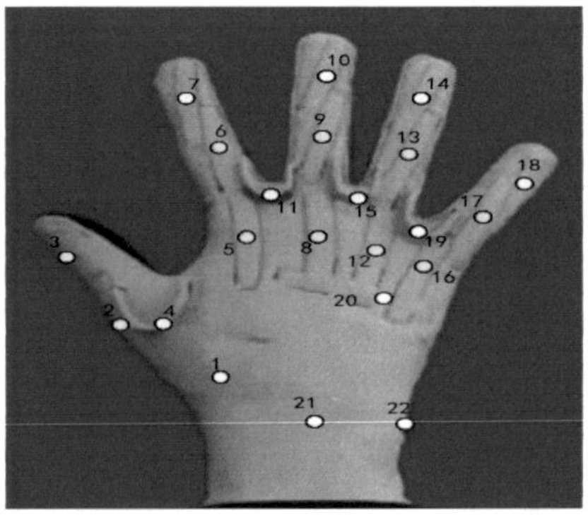

figure 2. CyberGlove II and its sensors

2.2. Data Preprocessing

Before using the data, sEMG signals and hand joint angles signals should be preprocessed. Firstly the Butterworth filter is used to remove the noise; Secondly the sEMG signals were extracted root mean square (RMS) features of every sampling point; the window of 100 ms duration and a stride length of 0.5 ms was used in features extraction[9]. Then, zero-phase lowpass filter was used to smooth the glove angle signals, which making it more consistent with the normal movement of the human body. Finally in order to speed up finding the optimal solution and improve the accuracy of the model, Min-Max normalization is used to map sEMG features and hand angles to [-1,1]. After the feature extraction, feature sequence and joint angle sequence were divided into 100 ms shifting windows with 50ms-long overlap as the inputs of the neural network[10]. In the Ninapro database, each movement have six trials; after completing the above data processing, we selected four trials of each movement as training dataset and two trials of each movement as testing dataset.

2.3. Models

2.3.1Attention-MLP

In this study, a deep network model combining Multi-Head Attention and Multilayer Perceptron (MLP) is adopted, shown as figure 4. The multi-head attention module used in this paper is a stacked multi-head attention for further feature processing [4]; multi-head attention is a special attention model which consists of multiple Scaled Dot-Product Attention models (Single head attention) shown as figure 3.

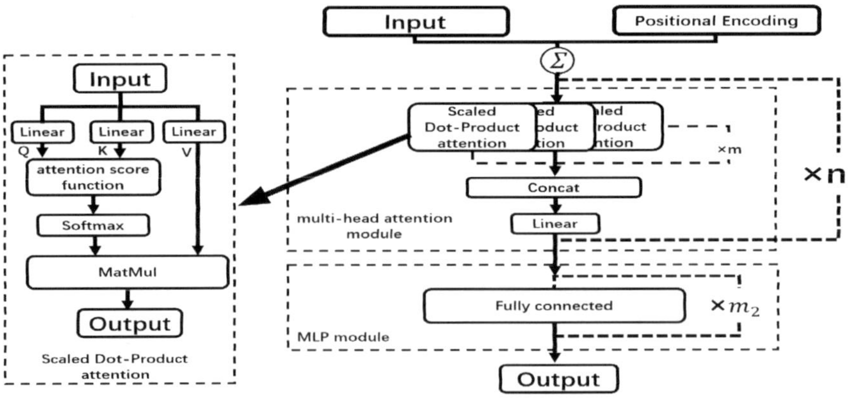

figure 3. Attention-MLP neural network(right) and Scaled Dot-Product Attention model(left).

Assuming that the input vector of sEMG feature is defined as $X = \{x_0, x_1, \ldots, x_n\} \in R^{n\times l}$ and the output of single head attenion defined as H= $\{h_0, h_1, \ldots. h_n\} \in R^{n\times d_v}$; The calculation of single head attention for a specific point x_t in the sequence can be described as the following steps : firstly it linearly map x_t to three different spaces $W_Q \in R^{l\times d_q}$, $W_K \in R^{l\times d_k}$, $W_V \in R^{l\times d_v}$ and calculate attention score function $s(x_t, x_o)$ to express weight coefficient of time t and time o, the function $S(x_t, x_o)$ is:

$$S(x_t, x_o)= \frac{(x_i W_Q)(x_o W_K)^T}{\sqrt{d_z}} \tag{1}$$

then output point h_t of input point x_t can be expressed as:

$$h_t = att(x_t) = \sum_{o=1}^{n} \frac{s(x_t, x_o) x_o W_V}{\sum_{j=1}^{n} s(x_t, x_j)} \tag{2}$$

Due to the single head attention mechanism cannot directly process the position information of timing signals, we add position encoding to each input vector x_t ,which enabling the attention model to capture the position information of the input sequence. We use the sinusoidal Positional encoding $P = \{p_0, p_1, \dots, p_n\} \in R^{n\times l}$ proposed by Google in 2017[4].

$$p_t(t, 2i) = sin(t/1000^{\frac{2i}{l}}) \tag{3}$$

$$p_t(t, 2i) = cos\left(t/1000^{\frac{2i}{l}}\right) \tag{4}$$

For the new input X+P, attention score of time t and time o is: $\left((x_i + p_i)W_Q\right)\left((x_o + p_o)W_K\right)^T$, due to the existence of $p_i W_Q W_K^T p_o^T$, the attention score function can capture the distance information between time o and time i [11]. When the actual model is used, input of this model calculates a complete sequence signal segment simultaneously, packed together into a query Vector matrix $Q = \{q_0, q_1, \dots, q_n\} \in R^{n\times d_q}$, $q_i = c_i W_Q$、 key vector matrix $K = \{k_0, k_1, \dots, k_n\} \in R^{n\times d_k}$, $k_i = c_i W_K$ and value vector matrix $V = \{v_0, v_1, \dots, v_n\} \in R^{n\times d_v}$, $v_i = c_i W_V$; The output vector sequence can be described as [4]:

$$Z = \text{Attention} = \text{softmax}\left(\frac{QK^T}{\sqrt{d_k}}\right)V \tag{5}$$

In this study, the multi head attention module consists of m single head attentions, which do not share parameters for capture different interactive information in multiple different projection spaces [12]. Concat module concatenates the single head attention's output $Z_1 \sim Z_m$ together, so, the output of Multi-Head Self-Attention can be described as

$$H = Concat(Z_1, Z_2, \dots Z_m)W_O \tag{6}$$

$W_O \in R^{md_v\times O}$, in our network, m is three;

the output of multi-head attention module will be used as the input of MLP to complete the regression task. It is composed of three-layer fully connected feedforward neural network, the activation function enhances the fitting ability of the model through.

The formula can be expressed by the following formula:

$$H_1 = \sigma(W_1 H + b_1) \tag{7}$$

$$H_2 = \sigma(W_2 H_1 + b_2) \tag{8}$$

$$O = W_3 H_2 + b_3 \tag{9}$$

W_1, W_2, W_3 and b_1, b_2, b_3 represent the weight matrixes and bias vectors of each layer respectively, and σ is the activation function described as:

$$\sigma = \tanh(x) = \frac{e^x - e^{-x}}{e^x + e^{-x}} \tag{10}$$

MLP module completes three dimensional transformations, 128, 64 and 10 dimensions respectively. In this study, the Mean Square Error between the predicted and real angle is chosen as loss function. Adam Optimizer was applied to minimize the loss of entire networks [13].

2.3.2 Sparse Pseudo-input Gaussian processes

Gaussian regression can perform regression analysis on data through prior information. According to the assumption of Gaussian processes, if we have a set of training data D={X,Y}, supposing Y obeys a normal distribution with a mean of 0 and has Gaussian noise $\varepsilon \sim N(0,\sigma_\varepsilon^2)$,the function Y = f (X) can be described as $Y \sim N(0, K(\boldsymbol{X},\boldsymbol{X}) + \sigma_\varepsilon^2 I_n)$[14],$\boldsymbol{K}(\boldsymbol{X},\boldsymbol{X})$ is the covariance matrix commonly calculated by RBF kernel function. The likelihood function of a single sample point can be expressed by:

$$P\{y|x,\boldsymbol{X},\boldsymbol{Y},\theta\} = N\left(y\middle|\begin{pmatrix} k^T(x,\boldsymbol{X})[\boldsymbol{K}(\boldsymbol{X},\boldsymbol{X}) + \sigma_\varepsilon^2 I_n]^{-1}y, \\ k_*(\mathrm{x},\mathrm{x}) - k^T(x,\boldsymbol{X})[\boldsymbol{K}(\boldsymbol{X},\boldsymbol{X}) + \sigma_\varepsilon^2 I_n]^{-1}k(\boldsymbol{X},x) + \sigma^2 \end{pmatrix}\right) \quad (11)$$

θ, σ^2, $k(\mathbf{X},x)$, $k_*(\mathrm{x},\mathrm{x})$ are hyperparameters. GPs are prohibitive for large data sets, because time complexity is O(N^3) due to the inversion of the covariance matrix. The computational complexity of predicting a sample is O (N^2), and the time required for prediction increases gradually with the increase of training set N. In order to solve this problem, SPGP was presented by Edward Selsdon et al. in 2006 [15]. In SPGP, a pseudo data set $\bar{D} = (\bar{X},\bar{f})$ replaces the real data set,$\bar{X} = \{\bar{\boldsymbol{x}}_i\}_{i=1}^M$is pseudo-inputs,$\bar{\mathbf{f}} = \{\bar{f}_m\}_{m=1}^M$is pseudo targets, this prediction distribution obtained from the pseudo data set is taken as a parameterized likelihood function.[15] Thus, the complexity of computation in training process and predicting process will reduce to O(M^2N) andO(M^2). In this study the pseudo-input M = 30. The code of this algorithm comes from Sheffield ML toolbox which contains MATLAB codes for Gaussian processes.

2.4 Post processing

In order to prevent excessive sudden changes in the output by the model from damaging the user and the controlled mechanical equipment, we analysed the absolute value of the maximum difference between two adjacent angles in the training data, when the output joint angle is greater than the threshold, we set the output joint angle as the threshold itself. At the same time, we take the average of the two adjacent joint angles of 50ms as the output of the primary joint angle, so as to enhance the smoothness and stability of the output hand joint angles and reduce the noise.

3. RESULTS

After filtering and extracting features of the raw sEMG signals, we constructed the Attention-MLP model for training and generating real-time hand joint angles under grasping movement. The Attention-MLP model was implemented using Pytorch, and SPGP was built on MATLAB; To evaluate the performance of Attention-MLP and SPGP, we used the correlation coefficient (CC) and root mean square error (RMSE) to evaluate the prediction's accuracy. The experimental results were shown as in figure. 5 and figure 6. Compared with the model of SPGP, the Attention-MLP results showed that the accuracy of the Attention-MLP model was significantly better than the Traditional machine learning model.

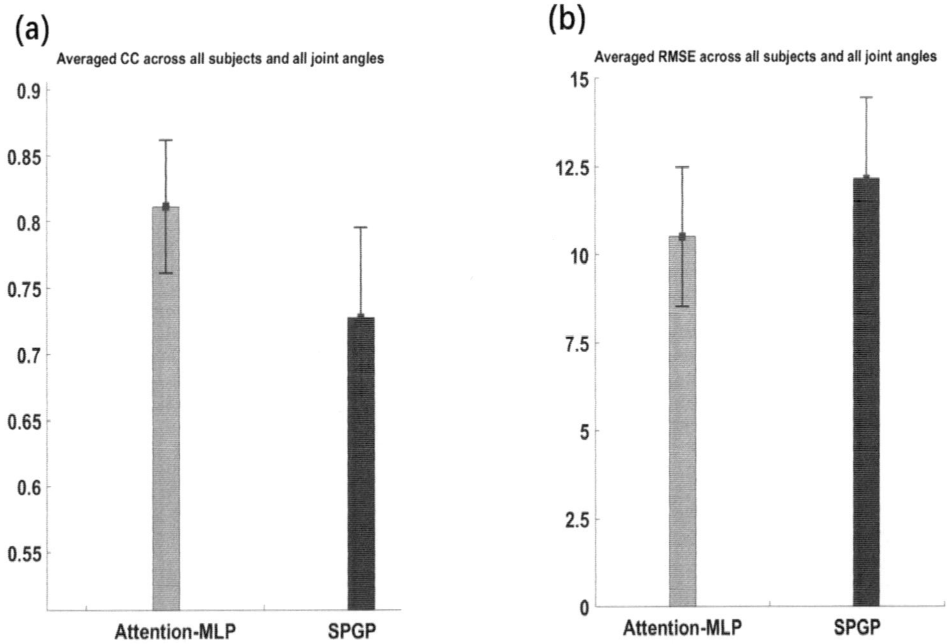

figure 4. The average accuracy of Attention-MLP and SPGP for 8 subjects. Subfigure (a) compared the correlation coefficients(CC) of SPGP and Attention-MLP, The CC of Attention-MLP and SPGP were 0.812 ± 0.02 and 0.727±0.07, respectively; subfigure (b) compares the root mean square error (RMSE) of two models, The RMSE of two models were 10.51±1.98 and 12.14±2.30 respectively.

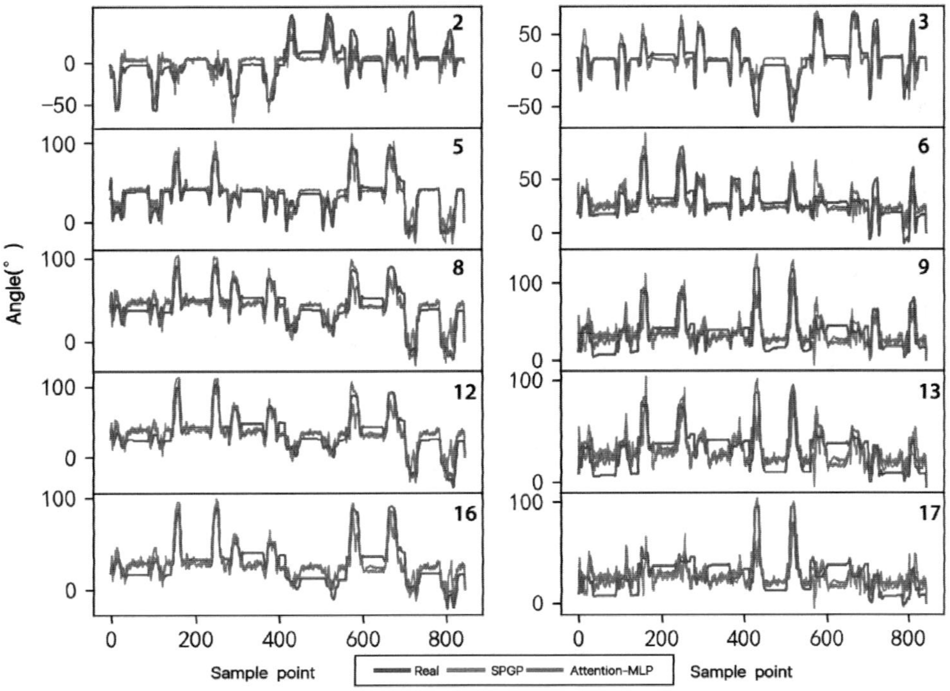

figure 5. The predicted and actual values of Attention-MLP and SPGP. The number in the upper right corner of the subgraph represents the label corresponding to the sensor on the glove

4.Conclusion

In this paper, we proposed an attention-MLP model to estimate finger joint angle during grasping movement simultaneously and proportionally. We can see this model has achieved good accuracy between 10 hand joint angles, because through attention mechanism and pre-processing, the model can extract the temporal and spatial characteristics of sEMG information, MLP can better fit the joint angles. Then we chose traditionary pattern recognition method SPGP to compare with attention-MLP under the same task, results showed that Attention-MLP is more suitable. In the future, We will further improve the attention mechanism in order to develop the potential of attention mechanism on the estimation of multi degrees of freedom continuous motion and other human–computer cooperation method based on sEMG.

References

[1] Han J, Ding Q. (2015) A State-Space EMG Model for the Estimation of Continuous Joint Movements. J. IEEE Transactions on Industrial Electronics,62:4267-4275.
[2] D Bahdanau, K Cho, Y Bengio (2014). Neural machine translation by jointly learning to align and translate. In Computer Science. pp:1409-1473.
[3] Yang Z, Yang D. (2016) Hierarchical attention networks for document classification. In NAACL, San Diego, USA. pp 1480–1489.
[4] Vaswani A, Shazeer N, Parmar N, et al. (2017). Attention is all you need. In Advances in neural information processing systems, California, USA. pp. 5998-6008.
[5] Hornik K, Stinchcombe M, White H (1989) Multilayer feedforward networks are universal approximators. J. Pergamon,2:359-366
[6] Atzori M, Gijsberts A, Castellini C, et al. (2014) Electromyography data for non-invasive naturally-controlled robotic hand prostheses. J. Scientific Data. 1:140053
[7] Phinyomark A, Phukpattaranont P, Limsakul C. (2012) Feature reduction and selection for EMG signal classification. J. Expert Systems with Applications, 39:7420-7431.
[8] Gijsberts A, Atzori M, Castellini C. (2014) Movement error rate for evaluation of machine learning methods for sEMG-based hand movement classification. IEEE transactions on neural systems and rehabilitation engineering: a publication of the IEEE Engineering in Medicine and Biology Society. 22: 735-44
[9] Guo W, Ma C, Wang Z, Zhang H, Farina D, Jiang N. (2021) Long exposure convolutional memory network for accurate estimation of finger kinematics from surface electromyographic signals. Journal of Neural Engineering, 18.
[10] Englehart K, Hudgins B. (2003) A robust, real-time control scheme for multifunction myoelectric control. J. IEEE transactions on bio-medical engineering,50: 848-854.
[11] Su J. (2021) Transformer upgrade path Sinusoidal Explore the root of Positional Encoding. https://spaces.ac.cn/archives/8231.
[12] Bishop C (2007) Pattern recognition and machine learning[M]. Springer, Berlin.
[13] Kingma D, Ba J. (2014) Adam: A Method for Stochastic Optimization, Computer Science, pp. 1–15.
[14] Xiloyannis M, Gavriel C. (2017) Gaussian process autoregression for simultaneous proportional multi-modal prosthetic control with natural hand kinematics IEEE Transactions on Neural Systems and Rehabilitation Engineering 25: 1785-1801
[15] Ahsan, Md (2009) EMG signal classification for human computer interaction: a review. J. European, Journal of Scientific Research,33:.480-501.

Identifying important nodes from content-associated heterogeneous graph by LeaderRank

Yulong Dai [1*], Qiyou Shen [2], Xiangqian Xu [3] and Jun Yang [4]

[1] College of Military Basic Education, National University of Defense Technology, Changsha, Hunan, P. R. China

[2] College of Information Communication, National University of Defense Technology, Wuhan, Hubei, P. R. China

[3, 4] College of Systems Engineering, National University of Defense Technology, Changsha, Hunan, P. R. China

*2018111436@stu.gzhmu.edu.cn

Abstract. Most real-world systems consist of a large number of interacting entities of many types. However, most of the current researches on systems are based on the assumption that the type of node or link in the network is unique. In other words, the network is homogeneous, containing the same type of nodes and links. Based on this assumption, differential information between nodes and edges is ignored. This paper firstly introduces the research background, challenges and significance of this research. Secondly, the basic concepts of the model are introduced. Thirdly, a novel type-sensitive LeaderRank algorithm is proposed and combined with distance rule to solve the importance ranking problem of content-associated heterogeneous graph nodes. Finally, the writer influence data set is used for experimental analysis to further prove the validity of the model.

1. Introduction

Information network[1] is a structured text way of knowledge representation. It contains a series of nodes and the links between nodes. Its structure reflects the structure of information stored in nodes, so it is called information network. Classic examples of information networks include academic networks[2], social media networks[3], social networks, etc. The traditional network model uses the homogeneous network to model the real system. This model ignores the difference information between entities, which leads to incomplete or missing information. Content-associated heterogeneous graphs[4] can accurately distinguish different meanings in information network and dig out more meaningful knowledge by analyzing various types of nodes in the network and various links between different types of nodes. Thus, it is of great significance to study content-associated heterogeneous graphs.

Discovering the core nodes in content-associated heterogeneous graph plays an important role in understanding the operating mode of the system and grasping the key factors of the system. It has important research value in epidemic prevention and control, network security and many other fields. At present, the research methods of node importance of homogeneous network have been relatively mature, such as PageRank algorithm[5] based on feature vector, HITS[6] algorithm, K-Shell method[7] based on node neighbor, degree centrality[8], etc. Compared with homogeneous networks, content-associated heterogeneous graphs contain richer topological structure and semantic information, which also poses new problems and challenges for the identification of important nodes. Therefore, the purpose of this paper is to propose a node importance ranking method suitable for content-associated heterogeneous graphs.

In 2015, Shi *et al.*[9] summarized the heterogeneous information network analysis, introduced the basic concepts of heterogeneous information network analysis, discussed the development of heterogeneous information network analysis in different data mining tasks, and pointed out the future research direction. In 2019, Mohammad MehdiKeikha *et al.*[10] used deep learning to learn feature vectors of nodes in the network. In 2020, Soheila Molae *et al.*[11] proposed a method to measure the importance of nodes by integrating local and global information of network.

By adding a ground node to replace the page jump probability setting in PageRank, Lü *et al.*[12] proposed the LeaderRank algorithm, which effectively solves the problem of inconsistent ranking of different subnets in PageRank algorithm. However, the original LeaderRank algorithm is based on the homogeneous network and ignores the difference information between nodes. To address this deficiency, the authors improve the original LeaderRank algorithm and propose a type-sensitive LeaderRank algorithm for the richness and semantic complexity of content-associated heterogeneous graph nodes. Firstly, the importance of nodes is divided into the importance of each node type, and then the importance of each type is combined into a global importance by using the distance rule.

The overall structure of this paper is as follows: In section 2, we introduce the basic concepts of content-associated heterogeneous graphs, LeaderRank algorithm and distance rule. In Section 3, we build the model and provide a detailed description of the type-sensitive algorithm. In section 4, we give the process of mining the key nodes of writer influence network to verify the validity of the model. Finally, summarizing discussions and conclusions of the paper will be in the last Section 5.

2. Preliminaries

In this section, we introduce the concepts of content-associated heterogeneous graphs, Pagerank algorithm, and distance rule that will be used in this paper.

2.1. Content-associated heterogeneous graphs

Basic definitions and concepts of content-associated heterogeneous graphs have been illustrated by Zhang at el. [13] in 2019.

Definition 1. A content-associated heterogeneous graph is defined as a graph $G = (V, E, O_V, R_E)$, The specific definition is as follows:

- V: A set of nodes,
- E: A set of links between nodes,
- O_V: A set of object types of nodes,
- R_E: A set of relation types of links

Each node or link belongs to a specific type in O_V or R_E.

Figure 1 shows an academic graph, which is a typical content-associated heterogeneous graph. The node types O_V include paper, place, and author, and the link types include *write*, *cite*, and *publish*. Since a paper can be written by a group of authors and published on a venue, the relationship between the paper and the author is one-to-many, and the relationship between the paper and the site is one-to-one.

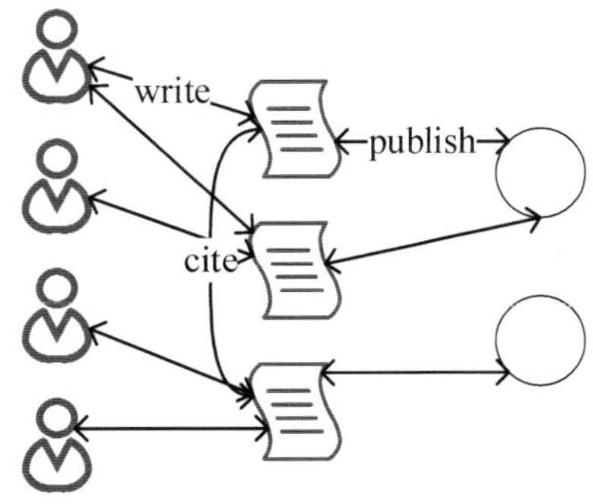

Figure 1. An example of content-associated heterogeneous graph on academic graph

2.2. LeaderRank algorithm

LeaderRank algorithm is a node ranking algorithm proposed by improving PageRank algorithm, which is mainly used to rank nodes of online social networks. LeaderRank algorithm proposes a solution to the problem of inconsistent ranking of disconnected subnet pages in PageRank algorithm. A ground node g is added into the whole network, connected with each node in the network bidirectionaly. The original network becomes a network of N+1 nodes and M+2N links. The addition of ground node makes the original network become a strongly connected network, which solves the problems of node ranking and probabilistic jump in PageRank and speeds up the convergence of algorithm.

The calculation process of the LeaderRank algorithm is an iterative process, which is as follows:

- Allocate one unit of resource to each node except the ground node.
- Each node allocates resources equally to the nodes it points to according to the output.
- Iterates until a steady state is attained.

2.3. Distance rule

Distance rule[14] is a multidimensional utility mergence method. When evaluating an object, it is necessary to combine multiple utilities. Take two-dimensional utility as an example, when the two utilities reach the maximum value at the same time, the combined utility reaches the maximum value. When both utilities reach the minimum value at the same time, the combined utility is zero. The distance rule is defined as follows:

Definition 2. Assume that the utility of n evaluation criteria are $u_1, u_2, \dots, u_n$, and the value range is [0,1]. The calculation formula of the distance rule is as follows:

$$W(u_1, u_2, \dots, u_n) = 1 - \left(\frac{1}{n}\sum_{i=1}^{n}(1-u_i)^2\right)^{\frac{1}{2}} \tag{1}$$

3. Model and method

In this section, we improve the LeaderRank algorithm based on the characteristics of content-associated heterogeneous graphs, proposed a type sensitive LeaderRank algorithm, and combine it with distance rule to establish a model to identify important nodes of content-associated heterogeneous graphs.

Definition 3. Let $G = (V, E, O_V, R_E)$ be a content-associated heterogeneous graph with a ground node. The adjacent matrix A of G can be expressed as:

$$a_{ij} = \begin{cases} 1, if \ (v_i, v_j) \in E \\ 0, if \ (v_i, v_j) \notin E \end{cases} \tag{2}$$

Then the state transition matrix P can be expressed as:

$$p_{ij} = a_{ij}\left(\sum_{V_i \in V} a_{ij}\right)^{-1} \tag{3}$$

For general homogeneous graphs, the importance of each node can be directly calculated by constructing the adjacency matrix and state transition matrix, but this is not applicable to content-associated heterogeneous graphs. To address this problem, a set of type vector t is introduced to distinguish the type differences between nodes.

Definition 4. Let $G = (V, E, O_V, R_E)$ be a content-associated heterogeneous graph with a ground node. For node type T_i, type state vector t_{ij} can be expressed as:

$$t_{ij} = \begin{cases} 1, if \ \ O_{V_j} = T_i \\ 0, if \ \ O_{V_j} \neq T_i \end{cases} \tag{4}$$

As mentioned above, after continuous iteration, the importance of each type of node will converge to a fixed value and then be integrated. For T_i nodes, the calculation process is as follows:

Algorithm 1. Type-sensitive LeaderRank for node type T_i

```
Input: The set of nodes V , the set of neighbor nodes Pi of each node, threshold ε
Output:influence value R_i
1: for each node i,initialize Q_i ← 1/|P_i|
2: for each node i in V
3:      T_i ← 0
4:      for each node j in Pi
5:           T_i ← T_i + Q_j
6: while True do
7:      for each node i in V
8:           R_i ← 0
9:           for each node j in Pi
10:               R_i ← R_i + T_j × Q_j
11:     if the total amount change < ε, then
12:          end
```

Through the above calculation process. We calculated the matrix R containing the importance of each node, where r_{ij} represents the importance of node v_j in node type T_i. We then use min-max scaling to normalize the importance of each node in type T_i to the interval $[0,1]$:

$$u_{ij} = \frac{r_{ij} - min(r_i)}{max(r_i) - min(r_i)} \tag{5}$$

Where $\max(R_i)$ and $\min(R_i)$ are maximum and minimum importance of type T_i nodes.

The normalized importance of each node in each node type is calculated and the importance matrix U is obtained, where u_{ij} represents the importance of node v_j in node type T_i.

Then, the distance rule is used to calculate the global importance of node v_j as follows:

$$w_j = 1 - \left(\frac{1}{n}\sum_{i=1}^{n}\left(1 - u_{ij}\right)^2\right)^{\frac{1}{2}} \tag{6}$$

Finally, the importance order of the content-associated heterogeneous nodes can be obtained by ranking the nodes by w_j.

4. Experimental analysis

In this section, in order to further prove the validity of the importance ranking model for content-associated heterogeneous graphs, an experimental analysis is conducted with the writer influence dataset.

4.1. Data declaration

The data set comes from Kaggle.com, which records the influence relationships of more than 4,000 writers since 1800. A writer may be influenced by more than one other writers, and writers can be divided into five genres. Therefore, with writer as node, writer genre as node type, and influence relationship between writers as edge, the content-associated heterogeneous graphs of writer's influence are constructed. Some statistics of writer content-associated heterogeneous graph are shown in Table 1:

Tale 1. Statistics of writer content-associated heterogeneous graph

	Type	Quantity
Nodes	Novelist	2531
	Poet	904
	Short Story Writer	1038
	Playwright	371
	Essayist	294
Edges	Influence	29411

4.2. Experiment process

According to the formula (3), we can calculate the state transition matrix P. P is a sparse matrix, from which it can be seen that there is an obvious community division phenomenon in this graph.

Let $\varepsilon_1 = \varepsilon_2 = \cdots = \varepsilon_5 = 1 \times 10^{-3}$ and $q_1 = q_2 = \cdots = q_5 = 0.2$. Then, importance matrix U can be calculated. According to the importance matrix U, the top three writers in novelist, poet and playwright are listed in Table 2:

Table 2. Top three writers in Novelist, Poet and Playwright

Genre	Writer	Importance
Novelist	J. K. Rowling	1.00
	Ernest Hemingway	0.85
	Charles Dickens	0.81
Poet	J. R. R. Tolkien	1.00
	Edgar Allan Poe	0.87
	Rabindranath Tagore	0.79
Playwright	William Shakespeare	1.00
	Oscar Wilde	0.86
	Arthur Miller	0.83

Next, we use the formula to calculate the combined importance of each writer node. For example, in the importance matrix U, Oscar Wilde's column values are $[0.87,0.96,0.81,0.89,0.82]^T$. Thus, the global importance for Oscar Wilde is 0.959. Some statistics of global importance calculation results are shown in Table 3:

Table 3. Experimental result

Importance interval	Nodes	Description
1	J. K. Rowling (Novelist), J. R. R. Tolkien (Poet), William Shakespeare (Playwright)... (5 in total)	The most influential writer in a particular field
[0.6,1)	Ernest Hemingway (Novelist), Edgar Allan Poe (Poet), Oscar Wilde (Playwright)... (517 in total)	Writers with great influence in a particular field
[0.3,0.6)	Charles Rhoton (Novelist), Denny Langon (Essayist)... (1928 in total)	Writers with moderate influence in a particular field
[0,0.3)	Coy Pack (Short Story Writer), Nathanial Mankoski (Essayist), (2688 in total)	Writers of little influence in any field

In order to more clearly show the effect of the model on node importance classification, we draw the distribution histogram of node importance values, as shown in Figure 2 below. As can be seen from the figure, the importance of nodes is mainly concentrated in two intervals [0.1,0.25] and [0.6,0.7]. Very few nodes have very high ([0.95,1]) or very low ([0,0.05]) importance. Therefore, this experiment successfully distinguishes important nodes from non-important nodes.

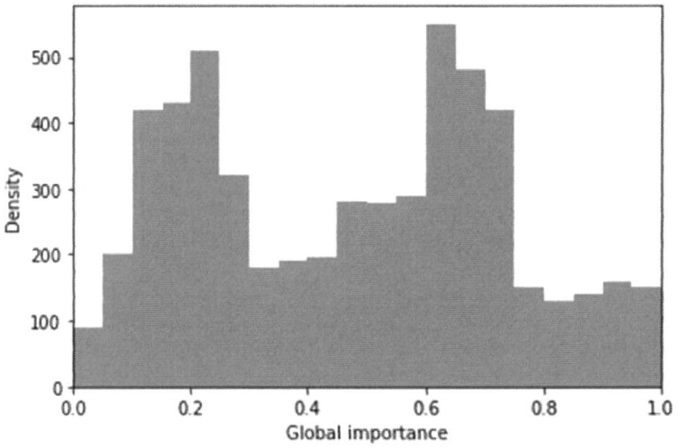

Figure 2. Distribution histogram of node importance values

5. Conclusions

In this paper, we introduced the ranking of node importance in content-associated heterogeneous graphs. At present, most research objects are homogenous networks, and the differential information in the network is ignored, resulting in incomplete or missing information. In addition, the traditional Pagerank algorithm has the problem of inconsistent node importance ranking for disconnected subnets.

In order to address these problems, the authors propose a novel type-sensitive LeaderRank algorithm, and combine it with the distance rule to build the model, which makes full use of the difference information between nodes and improves the calculation efficiency. It provides an efficient, accurate and concise solution to the problem. The experimental results show that this model has a good effect in distinguishing important nodes and non-important nodes in Content-associated heterogeneous graphs.

However, this model does not make full use of the differential information of links, nor does it work as well in heterogeneous networks in other fields as it does in social networks. Thus, in future work, the full use of edge difference information will become the vital aspect of improvement, followed by the study of utility combination rules.

Acknowledgments
We are thankful to the Editor and the reviewers for their valuable comments and detailed suggestions to improve the presentation of the paper. Further, we also acknowledge the support in part by the National Natural Science Foundation of China under grant No.71690233 and No.71671186.

References

[1] HAO PENG, JIANXIN LI, YANGQIU SONG, et al. Streaming Social Event Detection and Evolution Discovery in Heterogeneous Information Networks. ACM Transactions on Knowledge Discovery from Data. 2021;15(5):1-33. doi:10.1145/3447585
[2] Bedru HD, Zhao W, Alrashoud M, Tolba A, Guo H, Xia F. TOSNet: A Topic-Based Optimal Subnetwork Identification in Academic Networks. IEEE Access, Access, IEEE. 2020;8:201015-201027. doi:10.1109/ACCESS.2020.3034997
[3] Jahromi, D. A. (2020). A Quantitative Study of the Perceived Impact of Social Media Networks on Bahraini Users' English Language Learning. Teaching English with Technology, 20(4), 23–40.
[4] Li Y, Chen C, Duan M, Zeng Z, Li K. Attention-Aware Encoder–Decoder Neural Networks for Heterogeneous Graphs of Things. IEEE Transactions on Industrial Informatics, Industrial Informatics, IEEE Transactions on, IEEE Trans Ind Inf. 2021;17(4):2890-2898. doi:10.1109/TII.2020.3025592
[5] Jiung Lee, Mariana Olvera-Cravioto, PageRank on inhomogeneous random digraphs, Stochastic Processes and their Applications, Volume 130, Issue 4, 2020, Pages 2312-2348, ISSN 0304-4149.
[6] Kleinberg JM. Authoritative Sources in a Hyperlinked Environment. Journal of the ACM. 1999;46(5):604. doi:10.1145/324133.324140
[7] Kitsak M, Gallos L K, Havlin S, et al. Identification of Influential Spreaders in Complex Networks[J]. Nature Physics, 2012, 6(11): 888-893.
[8] Han L, Zhou Q, Tang J, Yang X, Huang H. Identifying Top-k Influential Nodes Based on Discrete Particle Swarm Optimization With Local Neighborhood Degree Centrality. IEEE Access, Access, IEEE. 2021;9:21345-21356. doi:10.1109/ACCESS.2021.3056087
[9] C. Shi, Y. Li, J. Zhang, Y. Sun, and P. S. Yu, "A survey of heterogeneous information network analysis," arXiv:1511.04854, 2015.
[10] Mohammad Mehdi Keikha, Maseud Rahgozar, Masoud Asadpour, Mohammad Faghih Abdollahi, Influence maximization across heterogeneous interconnected networks based on deep learning, Expert Systems with Applications, Volume 140, 2020, 112905, ISSN 0957-4174.
[11] Soheila Molaei, Reza Farahbakhsh, Mostafa Salehi. Noel Crespi, Identifying influential nodes in heterogeneous networks[J]. Expert Systems with Applications, Volume 160, 2020, 113580, ISSN 0957-4174.
[12] Lü L, Zhang Y-C, Yeung CH, Zhou T. Leaders in Social Networks, the Delicious Case. 2011. doi:10.1371/journal.pone.0021202.
[13] Chuxu Zhang, Dongjin Song, Chao Huang, Ananthram Swami, Nitesh V. Chawla. Heterogeneous Graph Neural Network. KDD 2019.
[14] Mohammad Mehdi Keikha, Maseud Rahgozar, Masoud Asadpour, Mohammad Faghih Abdollahi, Influence maximization across heterogeneous interconnected networks based on deep learning, Expert Systems with Applications, Volume 140, 2020, 112905, ISSN 0957-4174.

HHL Analysis and Simulation Verification Based on Origin Quantum Platform

Xiaonan Liu[1], Lina Jing[2*], Lin Han[2], and Jie Gao[1]

[1]State Key Laboratory of Mathematical Engineering and Advanced Computing, Zhengzhou, 450001, China

[2]Henan Supercomputing Center, Zhengzhou University, Zhengzhou, 450001, China

[*]Corresponding author's e-mail: liyinxinllx@gs.zzu.edu.cn

Abstract Solving large-scale linear equations is of great significance in many engineering fields, such as weather forecasting and bioengineering. The classical computer solves the linear equations, no matter adopting the elimination method or Kramer's rule, the time required for solving is in a polynomial relationship with the scale of the equation system. With the advent of the era of big data, the integration of transistors is getting higher and higher. When the size of transistors is close to the order of electron diameter, quantum tunneling will occur, and Moore's Law will not be followed. Therefore, the traditional computing model will not be able to meet the demand. In this paper, through an in-depth study of the classic HHL algorithm, a small-scale quantum circuit model is proposed to solve a 2×2 linear equations, and the circuit diagram and programming are used to simulate and verify on the Origin Quantum Platform. The fidelity under different parameter values reaches more than 90%. For the case where the matrix to be solved is a sparse matrix, the quantum algorithm has an exponential speed improvement over the best known classical algorithm.

1. Introduction

The quantum era has arrived. Linear equations, as one of the most important content in linear algebra, play an important role in many fields, so it is very important to solve large-scale linear equations. However, regardless of whether Gaussian elimination or Kramer's rule is used to solve linear equations on a traditional classical computer, the solution time is in a polynomial relationship with the dimension of the equation. With the advent of the era of big data, the growth rate of data has reached an unprecedented level[1], and traditional computers have certain difficulties in processing large-scale data. Therefore, the current situation puts forward higher requirements for us.

In response to this situation, quantum computing bears the brunt, it is currently one of the hottest and best solutions. As we all know, quantum computing is naturally capable of parallel computing due to its superposition, entanglement and reversibility. Compared with the parallel computing capability of classical computers, it has incomparable advantages in multi-machine parallel, multi-thread parallel and instruction level parallel. An N-bit classic register can only store a certain number of bits from 0 to 2^N-1, while in a quantum computer, the number of N-bit quantum registers in the super imposed state is all the numbers from 0 to 2^N-1, which exist at the same time with certain probability[2]. Therefore, one N-bit quantum register can simultaneously store 2^N N-bit binaries. When individuals operate on N qubits, in principle, it means that operate on 2^N states simultaneously[2]. The linear increase in the number of bits in the quantum register causes the storage space to increase exponentially. This is the basic feature of quantum computing memory cells and the premise that quantum computing speed can greatly exceed the classic computing speed[3].

According to the erasure theorem, in an irreversible process, the heat is closely related to the scale of the irreversible operation: The higher the integration, the more heat is generated per unit area[4]. The quantum computer based on the basic principles of quantum mechanics, and the information processing process is unitary transformation, so the reversibility of unitary transformation makes quantum information processing process does not bring heat. Therefore, quantum computing can theoretically solve the problem of high energy consumption, another key technology of modern information processing.

The structure of this paper is as follows: The first part introduces the background and significance of quantum computing. In the second part, the HHL quantum algorithm for solving linear equations is discussed in detail, and its solution idea, main algorithm framework and complexity advantage over classical algorithm are analyzed. In the third part, combining the practical characteristics of the Origin Quantum Platform, the specific case in this HHL algorithm is simulated by means of circuit diagram and programming. The last part is the summary of this paper and the prospect of the follow-up work.

2. HHL quantum algorithm principle

Aram W. Harrow, Avinatan Hassidim and Seth Lloyd explicitly proposed a quantum algorithm for solving linear equations[5], which solves such a problem: In Ax = b, if an Hermite N×N matrix A and a unit vector b are given, one solve for x satisfying the above equation. Equivalently, one can convert the problem to solve $x = A^{-1}b$. Since matrix A is an invertible matrix, this equation must have a unique set of solutions.

First, the N×N matrix A has to be an Hermitian matrix (the conjugate transposed matrix of A is equal to itself). Secondly, the input vector b is a normalized quantum state $|b\rangle$. When the matrix A is not an Hermitian matrix, individuals convert the matrix A into an Hermitian matrix by construction as follows:

$$\hat{A} = \begin{bmatrix} 0 & A \\ A^{+} & 0 \end{bmatrix}$$

At this time, the constructed matrix $\hat{A}$ will be changed from a general matrix to an Hermitian matrix, and subsequent paper will only consider the case where matrix A is an Hermitian matrix. Correspondingly, when $\hat{A}|x\rangle = |b\rangle$ is solved, $|x\rangle = (0 \quad x)^T$ will be found, and x at this time is the solution of the linear equations. Since the quantum state used to store the linear equation vector information must be normalized, the input information of the quantum linear equations algorithm is not a strict vector b, but a normalized quantum state $|b\rangle$ proportional to the vector b. Therefore, the vector b of the linear equations should be converted to $|b\rangle$, which can be converted through $|b\rangle = \sum_{i=1}^{N} b_i |i\rangle$, where b = K ($b_1, \ldots, b_N$) and K are normalization constants, and $\sum_{i=1}^{N} |b_i|^2 = 1$. After the transformation, the information of vector b has been stored in the probability amplitude of $|b\rangle$ in the quantum state. Then, the quantum version is $A|x\rangle = |b\rangle$.

It is worth noted that the HHL quantum algorithm finally solved a quantum state containing x information. That is, the HHL quantum algorithm could not directly solve the size of x, but could only obtain a global characteristic about x. Therefore, the result $|x\rangle$ differs from the solution of the original problem by a normalized constant. However, for many real-world applications, individuals don't need to get the exact value of x directly, just the overall properties. In addition, the normalized constant can also be obtained by means of amplitude estimation in the case that exact value is required. Therefore, HHL quantum algorithm is worth considered when it is necessary to solve large-scale linear equations without requiring the exact value of output directly.

In order to make the problem more clear, the process of HHL quantum algorithm is first analyzed in detail. For the Hermitian matrix A, one express it as the spectral decomposition of the N×N Hermitian matrix:

$$A=\sum_{j}\lambda_j|\mu_j\rangle\langle\mu_j| \tag{1}$$

Where $|\mu_j\rangle$ is the eigenvector of matrix A , and λ_j is the eigenvalue corresponding to matrix A. From $A|\mu_j\rangle=\lambda_j|\mu_j\rangle$ theorem one can get such an equation:

$$A^{-1}|\mu_j\rangle=\frac{1}{\lambda_j}|\mu_j\rangle \tag{2}$$

For the input quantum state $|b\rangle$, if expand it on the eigenvector $|\mu_j\rangle$ of the matrix A, one can get:

$$|b\rangle=\sum_{j}^{N}\beta_j|\mu_j\rangle \tag{3}$$

The equation of $|x\rangle$ can be obtained from the above equations:

$$|x\rangle=A^{-1}|b\rangle=A^{-1}\sum_{j}^{N}\beta_j|\mu_j\rangle=\sum_{j}^{N}\frac{\beta_j}{\lambda_j}|\mu_j\rangle \tag{4}$$

It can be found that the required output vector $|x\rangle$ is actually the input vector $|b\rangle$ with a corresponding reciprocal of the eigenvalue of the matrix A as a coefficient before each of its eigenvectors. Therefore, the purpose of solving linear equations can be transformed into how to obtain the information of these eigenvalues and add them to the corresponding eigenvectors[6]. The specific implementation of the algorithm can be divided into the following three subroutines: Phase Estimation(PE), Controlled Rotation(CR), and Inverse Phase Estimation(PE^{-1})[7]. The quantum circuit of this algorithm is shown in Fig.1, with the first, second and third registers from top to bottom.

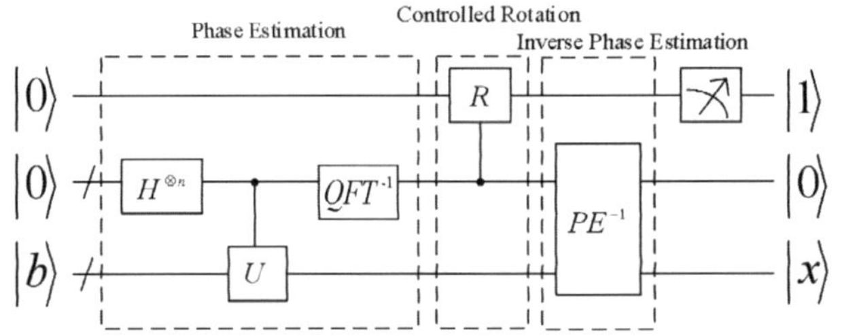

Fig. 1 Quantum circuit of HHL algorithm

The quantum algorithm includes three registers: an auxiliary qubit register with the initial state of $|0\rangle$, a register containing n qubits with the initial state of $|0\rangle$, and a register with the initial state of $|b\rangle$.

The first subroutine, Phase Estimation, is analyzed. Step 1 is the preparation of the initial state. The first register is the auxiliary register and its initial state is $|0\rangle$. The second register is a working register, and its initial state is $|0\rangle$ as well, and it contains n qubits. Here, n is the number of unknowns in linear equations. The third register is initialized to $|b\rangle$. Then, the coefficient matrix A of linear equations is prepared into unitary matrix by e^{-iAt0} . At this point, the joint initial state of the entire system is (excluding the auxiliary register) $|0\rangle^{\otimes n}|b\rangle$. Step 2 is the Phase Estimation operation. The purpose of this operation is to extract the eigenvalues of the coefficient matrix A. One express the eigenvector of the coefficient matrix A as $|\mu_j\rangle$, then λ_j represents the eigenvalue corresponding to the matrix A. Firstly, the eigenvalues of matrix A are extracted into the probability amplitude of the quantum state of the second register by H gate and controlled unitary matrix. And then the phases of the probability amplitude are extracted by the inverse Quantum Fourier Transform (QFT^{-1}) and placed into the ground state of the quantum state of the second register. At this time, the state of the second register is $|\lambda_j\rangle$, where $|\lambda_j\rangle$ is binary storage. The expansion of $|b\rangle$ in the third register is $\sum_j^N \beta_j|\mu_j\rangle$ based on the characteristic vector $|\mu_j\rangle$, and the joint initial state of the whole system is $\sum_j \beta_j|\lambda_j\rangle|\mu_j\rangle$. This step is the acceleration of the whole algorithm, because when calculating the eigenvalues of the coefficient matrix A, it is processed for all the eigenvalues simultaneously, that is, parallel computing.

Although the information of the eigenvalues into the second register in the Phase Estimation has been extracted, the eigenvalues here are stored in the quantum state. According to Formula (4), the eigenvalue λ_j must be stored in the form of coefficient and in the form of reciprocal. The results of the second register do not currently exist in the form individuals expect. Therefore, these can be converted through Controlled Rotation.

Controlled Rotation is mainly implemented in the first register and the second register, which is divided into three steps. First, one change the state of the second register from $|\lambda_j\rangle$ to $\left|\frac{1}{\lambda_j}\right\rangle$ temporarily. Then, through the auxiliary qubit introduced in the first register, under the operation of the controlled gate, the state of the first register is mapped from $|0\rangle$ to the superposition state of $|0\rangle$ and $|1\rangle$ to get $\sqrt{1-\frac{c^2}{\lambda_j^2}}|0\rangle+\frac{c}{\lambda_j}|1\rangle$ (where c is constant). Finally, the state of the second register is restored to $|\lambda_j\rangle$. At this time, the state of the whole system is $\sum_j\left(\sqrt{1-\frac{c^2}{\lambda_j^2}}|0\rangle+\frac{c}{\lambda_j}|1\rangle\right)\beta_j|\lambda_j\rangle|\mu_j\rangle$ (including the first register). In short, the function of Controlled Rotation is to extract the reciprocal of the ground state value proportionally to the corresponding probability amplitude of the ground state by an auxiliary qubit.

When the Controlled Rotation is completed, the state of the entire system is close to the final output. The output of the third register is useful. At this time, the second register and the third register are in an entangled state, which will affect the result. Therefore, the operation purpose of the Reverse Phase Estimation is to de-entangle the second register and the third register, and restore the result of the second register to the initial state. Only the entanglement result of the first register and the third register is left. The entire Inverse Phase Estimation is actually the inverse process of the previous Phase Estimation. After execution, the second register untangles and returns to the initial state. Meanwhile, the result of the entire system is $\sum_j\left(\sqrt{1-\frac{c^2}{\lambda_j^2}}|0\rangle+\frac{c}{\lambda_j}|1\rangle\right)\beta_j|\mu_j\rangle$. Because the first register is in the superimposed state, and what individuals need is when the state of the first register is observed as $|1\rangle$, then the result of the whole system becomes $\frac{c}{\lambda_j}\beta_j|\mu_j\rangle|1\rangle$. This result is proportional to the solution of the linear equations. Therefore, individuals only need to perform the measurement operation on the first register, and discard all the measurement results of $|0\rangle$. The remaining results at this time is the solution.

The complexity analysis of this algorithm mainly comes from Phase Estimation and Controlled Rotation. We define s as the sparsity of the matrix, that is, there are no more than s non-zero elements in each row. And is the conditional number, that is, the absolute value of the ratio between the maximum eigenvalue of matrix A and the minimum eigenvalue. In the third register, the resource required when the coefficient matrix A of the linear equations is prepared into a unitary matrix by e^{-iAt0} is $O(\log(N)s^2t)$. When performing the Phase Estimation, let $t = O(\kappa/\varepsilon)$, where ε is the acceptable error rate of the output vector. Finally, in order to measure the result of $|1\rangle$ in the first register, one also need $O(\kappa)$ iterations. Therefore, for a sparse N×N matrix A, the complexity achieved by the entire HHL quantum algorithm is $O(\log(N)s^2\kappa^2/\varepsilon)$[6][7]. At present, for solving the problem of linear equations, the fastest classic algorithm is based on the conjugate gradient descent method, and its complexity is $O(Ns\kappa/\varepsilon^2)$. It can be seen that in addition to the lack of accuracy, the HHL quantum algorithm has a huge advantage over the classical algorithm in other respects. Under certain conditions, it has an exponential time acceleration.

3. Simulation implementation

In order to verify the feasibility of HHL quantum algorithm, 2×2 linear equations will be simulated and verified. In this paper, the Origin Quantum Platform is selected as the experimental platform.

Founded on September 11, 2017, the Origin Quantum is the first domestic start-up company whose main business is the development and application of quantum computer. The Origin Quantum Platform provides quantum-virtual machines for users to choose, in which 32-bit quantum-virtual machine is free to use and 64-bit application is required. Virtual machine adopts visual programming learning mode: legend + quantum language. Users can easily drag and place legend for quantum simulation operation, and can convert the designed operation into quantum language mode for in-depth learning. QPanda SDK is a C++ host language toolkit for writing quantum programs and applications, which enables users to easily connect and execute quantum programs. For quantum learners, it greatly reduces the difficulty of learning quantum computation.

3.1. Circuit implementation

In order to solve the 2×2 linear equations, four qubits are needed to achieve the system. One qubit as the input $|b\rangle$, one as an auxiliary qubit for Controlled Rotation and measurement, and two qubits for Phase Estimation. For 2×2 linear equations, its coefficient matrix A is selected as:

$$A = \frac{1}{2}\begin{pmatrix} 3 & 1 \\ 1 & 3 \end{pmatrix}$$

Through calculation, individuals can get the eigenvector of matrix A as follows:

$$|\mu_1\rangle = \frac{1}{\sqrt{2}}\begin{pmatrix} 1 \\ -1 \end{pmatrix}, |\mu_2\rangle = \frac{1}{\sqrt{2}}\begin{pmatrix} 1 \\ 1 \end{pmatrix}$$

In this demonstration experiment, for input $|b\rangle$ of the third register, the paper selects two sets of data as a demonstration:

$$|b_A\rangle = \begin{pmatrix} 1 \\ 0 \end{pmatrix}, |b_B\rangle = \frac{1}{\sqrt{2}}\begin{pmatrix} 1 \\ -1 \end{pmatrix}$$

The circuit for solving linear equations is shown in Fig.2. Assuming $|b_0|^2 + |b_1|^2 = 1$, the vector b can be encoded into this state: $|b\rangle = b_0|0\rangle + b_1|1\rangle$. The eigenvalues of matrix A are $\lambda_1=1$ and $\lambda_2=2$, and the corresponding eigenvectors are $|\mu_1\rangle$ and $|\mu_2\rangle$. At this time, λ_1 and λ_2 can be accurately encoded by $|q_1q_2\rangle = |01\rangle$ and $|q_1q_2\rangle = |10\rangle$, respectively. Therefore, after the Phase Estimation, the state $|q_1q_2q_3\rangle$ of the 3-qubit system will be $\beta_1|01\rangle|\mu_1\rangle + \beta_2|10\rangle|\mu_2\rangle$.

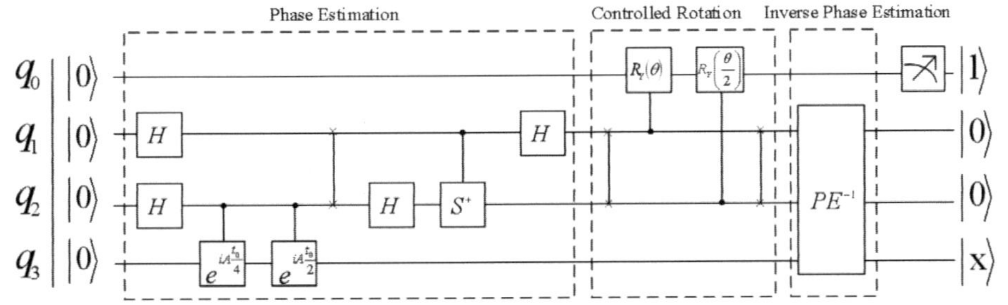

Fig. 2 4-qubits quantum circuit

After the Phase Estimation, in order to temporarily change the state of the second register from $|\lambda_j\rangle$ to $\left|\frac{1}{\lambda_j}\right\rangle$, this section uses a special method that does not require any auxiliary qubits. An essential step is to apply the SWAP gate between $|q_1\rangle$ and $|q_2\rangle$, so the 3-qubit system is converted to $\beta_1|10\rangle|\mu_1\rangle+\beta_2|01\rangle|\mu_2\rangle$ state. This can interpret $|q_1q_2\rangle=|10\rangle$ as the state where the reciprocal of the coded eigenvalue $2\lambda_1^{-1}=2$, and interpret $|q_1q_2\rangle=|01\rangle$ as the reciprocal of the coded eigenvalue $2\lambda_2^{-1}=1$ similarly. In other words, after the SWAP gate, the system state becomes $|q_1q_2q_3\rangle=\sum_{j=1}^{2}\beta_j\left|2\lambda_j^{-1}\right\rangle|\mu_j\rangle$.

When performing a Controlled Rotation, in general, the mapping from $\sum_j\beta_j|\lambda_j\rangle|\mu_j\rangle$ to $\sum_j\beta_j\lambda_j^{-1}|\lambda_j\rangle|\mu_j\rangle\propto|x\rangle$ should in principle use a controlled R operation. The original target operation R is $R|0\rangle\left|\frac{1}{\lambda_j}\right\rangle=\sqrt{1-\left(\frac{1}{\lambda_j}\right)^2}|0\rangle\left|\frac{1}{\lambda_j}\right\rangle+\frac{1}{\lambda_j}|1\rangle\left|\frac{1}{\lambda_j}\right\rangle$.

In this case, R_Y is used to approximate R. R_Y uses a revolving door (Pauli Y) around the Y axis of the Bloch ball, acting on $|0\rangle$ to obtain $R_Y|0\rangle\left|\frac{1}{\lambda_j}\right\rangle=\cos(\frac{1}{\lambda_j})|0\rangle\left|\frac{1}{\lambda_j}\right\rangle+\sin(\frac{1}{\lambda_j})|1\rangle\left|\frac{1}{\lambda_j}\right\rangle$.

At this time, the system status is $|q_0q_1q_2q_3\rangle=\sum_{j=1}^{2}(\cos(\frac{1}{\lambda_j})|0\rangle+\sin(\frac{1}{\lambda_j})|1\rangle)\beta_j\left|2\lambda_j^{-1}\right\rangle|\mu_j\rangle$. When the result of measuring the first quantum register $|q_0\rangle$ is 1, the ideal $|q_3\rangle$ should be $\frac{\beta_1}{\lambda_1}|\mu_1\rangle+\frac{\beta_2}{\lambda_2}|\mu_2\rangle$, but the actual $|q_3\rangle$ is $\sin(\frac{1}{\lambda_1})\beta_1|\mu_1\rangle+\sin(\frac{1}{\lambda_2})\beta_2|\mu_2\rangle$. Therefore, when $\frac{1}{\lambda_1}$ is approximately $\sin\left(\frac{1}{\lambda_j}\right)$, that is, the value of $\sin\left(\frac{1}{\lambda_j}\right)$ is very small, the probability of outputting $|x\rangle$ will be very low, but high fidelity of output can be guaranteed.

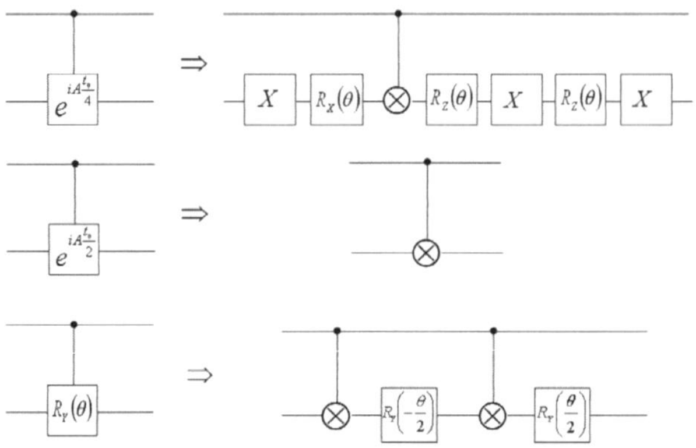

Fig. 3 Operation conversion

This paper choose to draw and simulate through the quantum circuit diagram mode on the Origin Quantum Platform. Quantum gates supported by quantum circuit design[8] just including H gate, NOT gate, phase π gate, Pauli X gate, Pauli Y gate, Pauli Z gate, revolving gate, CNOT gate, SWAP gate, Toffoli gate, and CR gate. For this reason, in the quantum circuit of Fig.2, some operations need to be converted. The conversion scheme is shown in Fig.3, which are two unitary matrices and a controlled rotating Y gate in Phase Estimation. The final quantum circuit diagram implemented on the Origin Quantum Platform is shown in Fig.4. The *init* operation is the initialization of $|q_3\rangle$, that is, the assignment of vector *b*. For $|b_A\rangle$, it is equivalent to not execute any operation. For $|b_B\rangle$ is equivalent to NOT gate + H gate.

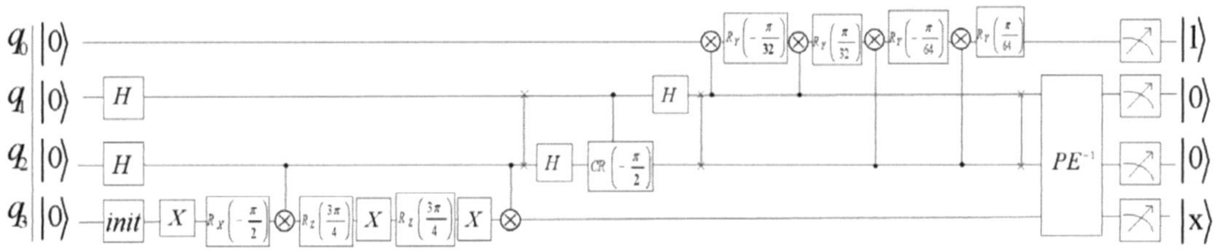

Fig. 4 Quantum circuit diagram on the cloud platform

By analyzing the quantum circuit diagram, it is found that $|q_3\rangle$ is used for the input vector *b* and also used to store the output $|x\rangle$. Therefore, the required $|x\rangle$ is the magnitude of the probability amplitude when $|q_3\rangle$ is $|0\rangle$ and $|1\rangle$ respectively, if the measurement result of the additional qubit $|q_0\rangle$ is 1. Special emphasis is placed on the need to include signs[9][10]. That is, what ultimately need to extract is the probability amplitude of $|0001\rangle$ and $|1001\rangle$. The theoretical value of $|x\rangle$ is calculated through Python's numpy package. For different input $|b\rangle$, the corresponding theoretical values of $|x\rangle$ are:

$$|b_A\rangle = \begin{pmatrix} 1 \\ 0 \end{pmatrix} \rightarrow |x_{A\_theory}\rangle = \begin{pmatrix} 0.75 \\ -0.25 \end{pmatrix}$$

$$|b_B\rangle = \frac{1}{\sqrt{2}} \begin{pmatrix} 1 \\ -1 \end{pmatrix} \rightarrow |x_{B\_theory}\rangle = \begin{pmatrix} 0.7071 \\ -0.7071 \end{pmatrix}$$

In the completed quantum circuit, the different input $|b\rangle$ were measured in turn. Each test iteration is 8192 times (the optional number range is 1-8192 times), and the bar chart of the run results is obtained. Fig. 56 shows the run results.

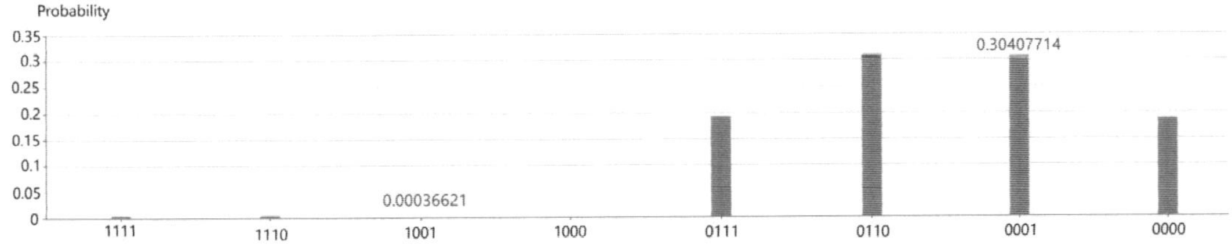

Fig. 5 The operation results with value $|b_A\rangle$

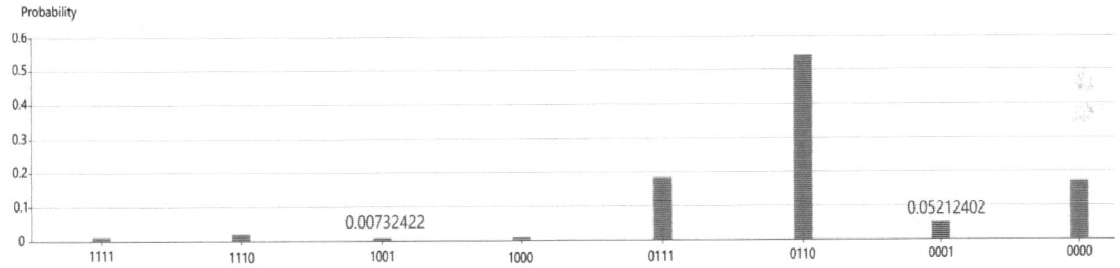

Fig. 6 The operation results with value $|b_B\rangle$

Through the above analysis, it can be clearly known that the useful value is the probability amplitude of $|0001\rangle$ and $|1001\rangle$. The result marks are respectively $|x_{A\_test}\rangle$ and $|x_{B\_test}\rangle$, then

$$|x_{A\_test}\rangle = \begin{pmatrix} \sqrt{0.30407714} \\ -\sqrt{0.00036621} \end{pmatrix} = \begin{pmatrix} 0.551432 \\ -0.019137 \end{pmatrix}$$

$$|x_{B\_test}\rangle = \begin{pmatrix} \sqrt{0.05212402} \\ -\sqrt{0.00732422} \end{pmatrix} = \begin{pmatrix} 0.228307 \\ -0.085582 \end{pmatrix}$$

By normalizing the theoretical value and the measured value, the inner product of the two value is calculated and the fidelity is approximately 0.959081 and 0.910324, respectively.

3.2. Programming implementation

QPanda2 SDK is a fully functional and efficient quantum software development toolkit. It is an open source quantum computing framework developed by the Origin Quantum Platform. It supports mainstream quantum logic gate operations, and can be optimized for quantum programs on different platforms, and can be adapted to a variety of quantum chips. QPanda2 uses C++ as the classic host language and supports quantum languages written in QRunes and QASM. This section provides a quantum programming based on QPanda2 development environment to verify the 2×2 linear equations mentioned in the previous section.

Taking the input $|b_A\rangle$ as an example to show the code, the code is as follows:

Journal of Physics: Conference Series **2113** (2021) 012083 doi:10.1088/1742-6596/2113/1/012083

```
#include <QPanda . h>
USING QPANDA
QCircuit   PE( Qubit*qubit1, Qubit* qubit2, Qubit* qubit 3 )
{
QCircuit cir ;
cir << H( qubit1 ) << H( qubit2 ) << X( qubit3 )<< RX( qubit3 ,−PI / 2 ) << CNOT( qubit2 , qubit3 )
    << RZ( qubit3 , 3 *PI / 4 ) << X( qubit3 )<< RZ( qubit3 , 3 *PI / 4 ) << X( qubit3 )
    << CNOT( qubit1 , qubit3 )<< SWAP( qubit1 , qubit2 )<< H( qubit2 )
    << CR( qubit1 , qubit2 ,−PI / 2 )<< H( qubit1 ) ;
return cir ;
}

QCircuit   ¯PE_ ( Qubit*qubit1, Qubit*qubit2, Qubit* qubit3 )
{
QCircuit cir ;
cir << H( qubit1 ) << CR( qubit1 , qubit2 , PI / 2 )<< H( qubit2 ) << SWAP( qubit1 , qubit2 )
    << CNOT( qubit1 , qubit3 ) << X( qubit3 )<< RZ( qubit3 , −3*PI / 4 ) << X( qubit3 )
    <<RZ( qubit3 , −3*PI / 4 ) << CNOT( qubit2 , qubit3 )<< RX( qubit3 ,−PI / 2 ) << X( qubit3 )
    << H( qubit2 ) << H( qubit1 ) ;
return cir ;
}

QCircuit   CR( Qubit*qubit1, Qubit* qubit2, Qubit*qubit 3 )
{
QCircuit cir ;
cir << SWAP( qubit2 , qubit3 )<< RY( qubit1 , PI / 16 ).control ( qubit2 )
    << RY( qubit1 , PI / 32 ).control ( qubit3 )<< SWAP( qubit2 , qubit3 ) ;
return cir ;
}

int main ( void )
{
init ( ) ;
auto qubits=qAllocMany ( 4 ) ;
auto cbits=cAllocMany ( 4 ) ;
auto prog=CreateEmptyQProg ( ) ;
auto circuit=CreateEmptyCircuit ( ) ;
QCircuit   PE( Qubit* qubit1, Qubit* qubit2, Qubit* qubit 3 ) ;
QCircuit   PE _( Qubit* qubit1, Qubit* qubit2, Qubit* qubit 3 ) ;
QCircuit   CR( Qubit* qubit1, Qubit* qubit2, Qubit* qubit 3 ) ;
circuit << PE( qubits [1] , qubits [2] , qubits [3] )
        << CR( qubits [0] , qubits [1] , qubits [2] )
        << PE_ ( qubits [1] , qubits [2] , qubits [3] ) ;
```

```
prog << circuit<< Measure ( qubits [0] , cbits [0] )<< Measure ( qubits [1] , cbits [1] )
        << Measure ( qubits [2] , cbits [2] )<< Measure ( qubits [3] , cbits [3] ) ;
auto result=runWithConfiguration ( prog, cbits, 10000 ) ;
for ( auto &val : result )
{
std : : cout << val.first << " ," << val . second << std : : endl ;
}
finalize ( ) ;
system (" pause " ) ;
return 0 ;
}
```

By calling *runWithConfiguration*() to get the measurement result of the quantum program, the test iteration is set to 10000 times. For input $|b_A\rangle$, the times of getting $|0001\rangle$ and $|1001\rangle$ are 9920 and 34 respectively. The times were converted into probability and normalized, and the inner product was carried out with the normalized theoretical value to obtain the fidelity of 0.965542. Similarly, for input $|b_B\rangle$, the fidelity is about 0.933564.

4. Conclusion

Regardless of whether it is implemented by the circuit diagram or programming, for different input $|b\rangle$, the output of this HHL quantum circuit is maintained above 90% accuracy, basically meeting the requirements of solving linear equations. As far as the error is concerned, the error of this quantum circuit is mainly caused by the inherent error caused by the Phase Estimation and the choice of the rotation angle when the R_Y approximates the controlled R operation in the Controlled Rotation. In addition, there is random error caused by repeated measurements. The accuracy has reached the required standard, but the success rate is not high. The overall probability of $|0001\rangle$ and $|1001\rangle$ is low, and further research is needed. However, this paper has carried out a complete study of the HHL algorithm from theory to simulation implementation, and the decomposition of advanced logic gates to basic logic gates, which provides a clear reference for the understanding of the algorithm itself and the realization of quantum computing for the next step. It has laid a good foundation for the application research of the broader Shor algorithm and Grover algorithm[11][12][13].

References

[1] Cai X.D. Optical Quantum Computing and Algorithm Implementation[D]. University of Science and Technology of China, 2015.
[2] Guo G.C. Ten Questions of Quantum [J]. Physics, 2019, 48 (3): 189-192.
[3] Wang K.N. The Parallelism of Quantum Computing and Its Philosophical Significance [C]. Shanxi University 2008 National Doctoral Academic Forum (Philosophy of Science and Technology), 2008:9.
[4] Han Y.J, Li C.F, Guo G.C. Principles and Research Progress of Quantum Computing [J]. Science and Technology Review, 2017, 35(23): 70-75.
[5] Harrow A W, Hassidim A, Lloyd S. Quantum Algorithm for Linear Systems of Equations[J]. Physical Review Letters, 2009, 103(15):150502.1-150502.4.
[6] Cai X.D, Weedbrook, C, Su Z.E, et al. Experimental Quantum Computing to Solve Systems of Linear Equations[J]. Physical Review Letters, 110(23):230501.
[7] Pan J, Cao Y, Yao X, et al. Experimental realization of quantum algorithm for solving linear systems of equations[J]. Physical Review A, 2014, 89(2):022313.
[8] Deng X, Deng Y.X. Qsimulation: A Quantum Computing Simulator Tool[J]. Computer Engineering and Science, 2019, 41(05): 843-850.

[9] Huang H.Y, Bharti K, Rebentrost P. Near-term Quantum Algorithms for Linear Systems of Equations[J]. 2019.
[10] Wei J, Ni M, Zhou M, et al. Quantum Algorithm Research Based on IBM Q platform [J]. Computer Engineering, 2018 (12): 6-12.
[11] Long G.L. Research and Development Progress and Future Prospects of Quantum Computers[J]. People's Forum·Academic Frontiers, 2021(07):44-56.
[12] Lu Y, Feng G.R, Li Y.S, et al. Experimental Digital Quantum Simulation of Interaction Fermion System Dynamics in Real Space[J]. Science Bulletin, 2015, 60(2):241- 248.
[13] Harrow A W. Review of Quantum Algorithms for Systems of Linear Equations[J]. Computer Science, 2014.

A simplified algorithm for 3D mesh model considering the influence of edge features

Rumeng Lv[1], Xiaobing Chen[1,2*] and Bingying Zhang[3]

[1] Huaiyin Institute of Technology, Jiangsu Province, China

[1,2*] Huaiyin Institute of Technology, Jiangsu Province, China

[3] Huaiyin Institute of Technology, Jiangsu Province, China

*Corresponding author's e-mail: chenxiaobing@hyit.edu.cn

Abstract. Aiming at the problem that most of the existing grid simplification algorithms for 3D models can not deal with a large number of boundary or non-popular grid models, this paper proposes a grid simplification algorithm for 3D models based on traditional algorithms. The algorithm mainly studies the geometric features of the model, considering the calculation methods and characteristics of edge shrinkage, and introduces the edge feature factors on the basis of the traditional algorithm, that is, the triangular area and side length factors of local area are introduced in the calculation of folding cost. In addition, the gaussian curvature characteristics of the 3D model are also included. Experimental results show that the proposed algorithm can keep the detail features of the mesh model well, and greatly reflect the quality and effect of mesh simplification after simplification.

1. Introduction

In the process of 3d mesh model simplification, under certain constraints, relatively detailed mesh model is sometimes unnecessary to be generated in specific scenes. In order to improve the efficiency of mesh model loading and reduce the simplified mesh model of vertices and faces, a model with fewer faces is used to replace the original one. In recent years, many mesh simplification algorithms are mostly based on the more important algorithm -- edge folding algorithm, and most mesh simplification algorithms are derived and developed based on edge folding algorithm.

In order to simplify the mesh, the algorithm takes the square sum of the distance between the new vertex and the associated plane of the two vertices to be folded as the quadratic error measure, so as to generate the simplified mesh model. The edge folding algorithm proposed by Garland and Heckbert[1][2] in 1997 and 1999 has good quality after simplification. However, the characteristics of model key are not clarified. Zhang Shixue[3] et al. classified vertices and edges of the 3D model and allocated different coefficients, which effectively reduced the generation of long and narrow triangles, but reduced the efficiency of grid simplification. Meng Jun[4] proposed a fast mesh simplification algorithm, which used vertex weights to indicate the importance of vertices and used priority queues storing vertex weights to control the order of edge contraction. Compared with Zhang Shixue's algorithm, the simplification speed of the algorithm was improved, but the feature problems of the model were ignored. Therefore, Li Long[5] et al. introduced absolute curvature of the vertices of the grid model when calculating the folding cost, and the algorithm can effectively preserve the geometric features of the model, but the simplification quality of the algorithm needs to be improved. Hou[6] et al. redesigned the weight calculation method of QEM algorithm, and introduced the square of the maximum deviation found in the triangle adjacent to the vertex as the importance of the vertex into the error measure. The contour information of the model generated by the algorithm was well preserved, but the details of the model were not obvious. Wang Jun[7] introduced the concept of approximate measure of edge curvature, and the algorithm ran efficiently and effectively retained the details of the grid, but the simplified grid model was of low quality.

On the theoretical basis of the above research, aiming at the problems of insufficient detail feature retention and insufficient grid optimization after simplification, this paper proposes a grid simplification algorithm, which utilizes the influence of geometric features of the model on quadratic error to simplify relatively flat regions. At the same time, the detailed parts and key feature areas in the model are better reserved. Experimental results show that the proposed method retains more features of the model than previous methods.

2. Measure of quadratic error

Quadratic error measurement is proposed by Garland and Heckbert. This algorithm is mainly divided into two parts: one is initialization of grid vertices, that is, assigning a quadratic error matrix Q to each vertex in the grid; the other is the iterative process of opposite edge contraction operation. Using the sum of squares of the distance between the new vertex generated by edge folding operation and the associated plane of the two vertices of the folded edge as the error measure, the folding cost of the model is calculated as $\Delta v = v^T Q v$. For vertices in three dimensions $v = \left[v_x, v_y, v_z, 1\right]^T$, The set of planes associated with it is $plane(v_i)$, Let the plane equation P in three-dimensional space be $ax+by+cz+d=0$, Where $a^2+b^2+c^2=1$, d is a constant. This gives us the distance between our new vertex v' and the plane P squared:

$$d^2 = \left(p^T v\right)^2 = v^T\left(pp^T\right)v = v^T K_p v \tag{1}$$

Where Kp is a 4*4 symmetric matrix:

$$K_p = pp^T = \begin{vmatrix} a^2 & ab & ac & ad \\ ab & b^2 & bc & bd \\ ac & bc & c^2 & cd \\ ad & bd & cd & d^2 \end{vmatrix} \tag{2}$$

Let $Q(v) = \sum_{p \in plane(v)} K_p$ be the quadratic error matrix of vertex v. in the edge folding operation, the folding cost is $\Delta(v') = v^T\left(Q_i + Q_j\right)v$. use $Q_i + Q_j$ to represent the quadratic error matrix of new vertex v ', where $\Delta(v')$ is the error cost of new vertex v.

3. Simplified algorithm description

The mesh density of the simplified model generated by QEM algorithm is relatively uniform, and the feature retention effect of the large-scale simplified model generated is general, which cannot be well applied in the field with obvious model details. Therefore, this paper considers factors affecting model features to achieve the purpose of feature retention.

3.1. The area of a triangle's local area

For example, the literature [9] takes edge curvature as a new quadratic error measure, which can improve the simplification quality of the model, but cannot well reflect the detailed features of the model. Therefore, the shape of the triangle in the grid model is determined by using the average area of the local area of the vertex. The larger the average area of the local area of the vertex is, the flatter the feature of the local area will be; otherwise, the triangle shape around the vertex is more dense . In this paper, the local average area of vertex triangle is used to measure the distribution of mesh density in the model, so as to reflect the change of local surface.

There are many ways to calculate the triangular area, such as the cross product of triangular vectors or the formula of triangular apex points is $S(Tr)$, In this paper, the coordinates of three vertices of a triangle are set by using the first calculation method are $v_0(x_0, y_0, z_0)$, $v_1(x_1, y_1, z_1)$, $v_2(x_2, y_2, z_2)$. Let the product of the faces of the triangle be $S(Tr) = \frac{1}{2}\left\|\overline{v_0v_2} \times \overline{v_0v_1}\right\|$, Considering the influence of the area size of the adjacent triangle at a point in the local area and the mesh density on the model feature loss, the average area of the adjacent triangle at vertex IV is calculated as follows:

$$\overline{S(v_i)} = \frac{\sum_{i=1}^{n} S(Tr_i)}{n} \tag{3}$$

Where n is the number of adjacent triangles of vertex v_i and $S(Tr)$ is the area of the i-th triangle in the adjacent triangles of vertex v_i.

3.2. side length

On vertex characteristics were measured after the local area, single use vertex local average size will remain simplified model part of the detailed features, but only triangular mesh morphological characteristics described in the model, will lead to a decline in the quality of grid model, they may also make the uneven distribution of grid, makes the model simplified general effect. In addition, in the calculation of Angle method, if the internal angles of a triangle are equal, the shape of the triangle tends to 1.If the inner Angle of a triangle is close to 0, the triangle shape ranges from 0 to 1, but the calculation is relatively complicated. Based on the above considerations, this paper chooses to calculate the length of the connection between two vertices in the grid model, and the length of side length determines the importance of model features.

The Euclidean distance method is used to calculate the length of the connection between the two vertices in the model. The folded edge length between edges (v_i, v_j) is calculated as follows:

$$l(v_i, v_j) = \sqrt{(x_j - x_i)^2 + (y_j - y_i)^2 + (z_j - z_i)^2} \tag{4}$$

Where the coordinates of v_i and v_j are $v_i = (x_i, y_i, z_i)$, $v_j = (x_j, y_j, z_j)$.

3.3. Gaussian curvature

Literature [1] proposed a second error algorithm in surface has a good effect on simplified, but there is no good to keep the details of the simplified model, this is because the models in the process of simplifying the details are ignored, and can reflect the characteristics of the model surface curvature, to smooth the degree of curvature of curve, the greater the curvature, the greater the degree of curvature of curve, Moreover, the surface of the grid model is a second-order non-differentiable surface, which cannot be calculated by the traditional curvature calculation method. Therefore, gaussian curvature of the model surface is adopted to increase the calculation cost, so that the edges of the non-boundary vertices are processed later.

The vertex curvature is calculated by the sum of the angles and areas of triangles in the grid. Let α be the Angle of the i-th triangle in the vertex neighborhood, and approximate by gauss-Bonnet theorem:

$$\iint_N G dS_N + \sum_{i=0}^{j-1} \alpha_i = 2\pi \tag{5}$$

The Gaussian curvature G is solved by formula (6). Assuming that G is constant in the local region of the vertex, it can be approximated as the surrounding center of gravity region, where D is the sum of angles in the local neighborhood of the vertex, DSN is the area element of the surface, and D is the area of the i-th triangular surface in the neighborhood n of the vertex center of gravity, as shown in Figure 1:

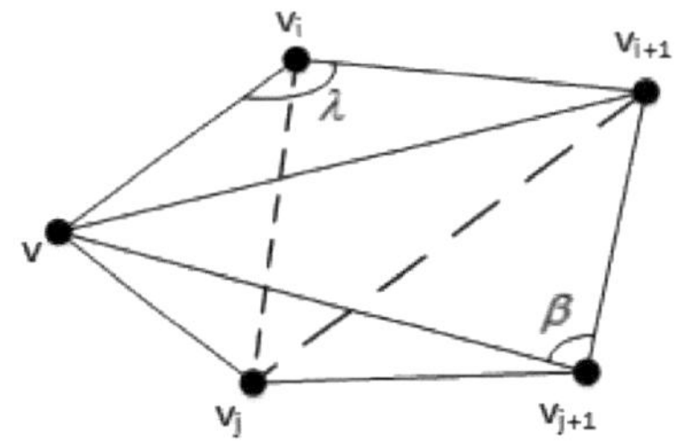

Figure 1. Schematic diagram of solving G

3.4. Error measure and new point position

The purpose of this paper is to reduce the error between the original model and the simplified model and retain the details of the model under the given simplification rate. Since the calculation of error measure and the position of new points are both important factors affecting the simplification effect of the model, the simplified error measure method has a large amount of calculation of geometric measure, so it is not suitable for the grid simplification application with high efficiency requirements, so this paper calculates a new error measure based on the quadratic error measure method of Garland.

In the previous algorithm, the second error calculation is fold after vertex peace in the distance, and price for folding, in this study, the introduction of local regional average size, length and gaussian curvature calculation, the triangle vertex area, the smaller the average area of $\overline{S_{v_i}}$ vertex gaussian curvature G_{v_i} change more intense, thus establishing relation, Denoted by φ, the boundary treatment and vertex characteristic factors of the model are considered when calculating the error, and the distance between the vertex and the relevant plane is calculated. The following formula is used to calculate the quadratic error:

$$\varphi_l = l(v_i, v_j)G(v_i, v_j)\left(\overline{S_{v_i}} + \overline{S_{v_j}}\right)^{-1} \tag{6}$$

Add φ_l as a weight to the edge cost calculation. When the edge (v_i, v_j) shrinks to the new vertex $v = \left[x\ y\ z\ 1\right]^T$, let $G(v_i, v_j) = |G_{v_i}| + |G_{v_j}|$, and map $G(v_i, v_j)$ to [0,1], the total error matrix of the contracted edge is as follows:

$$\begin{aligned} Q(v) &= l(v_i, v_j)G(v_i, v_j)\left(\overline{S_{v_i}} + \overline{S_{v_j}}\right)^{-1}\left(Q(v_i) + Q(v_j)\right) \\ &= \varphi_l\left(Q(v_i) + Q(v_j)\right) \end{aligned} \tag{7}$$

In 3d mesh model, QEM algorithm is generally used to determine the position of new vertices on triangular edges of mesh. In this paper, mesh simplification is carried out based on edge folding method, and the schematic diagram of boundary vertices is shown in Figure 2:

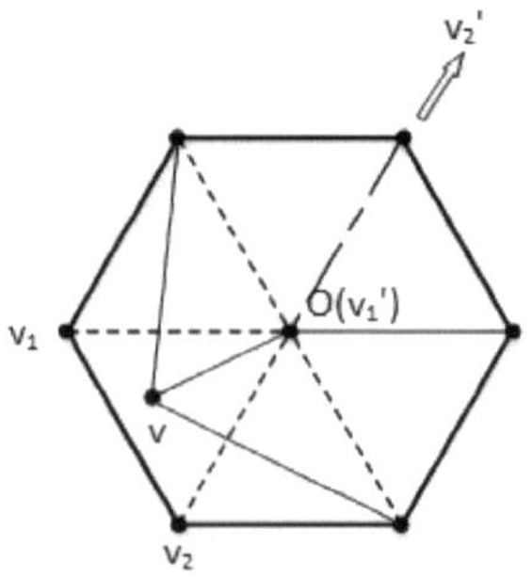

Figure 2. Schematic diagram of boundary vertices

For the edge shrink operation (v_i, v_j)→v, the position of the new vertex v needs to be determined. It is known from Section 3.1 that the edge contraction cost function equation is a quadric surface equation. The new vertex position can be obtained by using the reversibility of the obtained matrix. That is, during each edge contraction, the error is derived and minimized, so that the calculated mesh is more likely to be close to the original mesh model. As a quadratic error, the optimal position of the new vertex x, y and Z coefficients can be easily found through solution.

In this paper, the model grid vertices are classified, namely in the border of vertex to the border, the border edges on the vertex is a border, the border edge is only one adjacent surface, while the other side there are two adjacent surface, grid vertex classification, and adopt different methods to determine the location of the new vertex v, can save computation efficiency, into and keep the model boundary, The non-boundary edge is treated as follows:

If the matrix is irreversible, then $v=\begin{bmatrix} x & y & z & 1 \end{bmatrix}^T$ is the new vertex of the folded edge; If the matrix r is reversible, since the last line of the r matrix is empty, the unique solution of the homogeneous vector v is obtained:

$$v = \begin{bmatrix} q_{11} & q_{12} & q_{13} & q_{14} \\ q_{12} & q_{22} & q_{23} & q_{24} \\ q_{13} & q_{23} & q_{33} & q_{34} \\ 0 & 0 & 0 & 1 \end{bmatrix}^{-1} \begin{bmatrix} 0 \\ 0 \\ 0 \\ 1 \end{bmatrix} \tag{8}$$

The traditional quadratic error measurement method based on feature degree weighting does not take into account the boundary problem of the model. In this paper, the boundary vertices or edges are not simplified, but reserved and restricted. However, if the simplified folding operation of boundary edge is restricted blindly, the simplification can only simplify other edges outside the boundary edge, but the effect of model simplification will be reduced. The key to solve the boundary edge problem is to set an appropriate weight for the boundary edge to increase the folding error of the edge, so as to preserve the boundary characteristics of the grid.

3.5. Algorithm steps

Based on the QEM algorithm, this algorithm introduces the local area, side length and Gaussian curvature of the triangle. According to the relationship between the three, the error cost function is obtained, and the new vertex positions are calculated for different feature vertices of the mesh model, which effectively deals with the problem of preserving the model boundary edges and features in the process of mesh simplification. The algorithm execution process is as follows:

- The user input iteration times and simplified targets were read, topological relationships of vertices in the grid were established, and different vertices in the model were classified, including boundary vertices.
- The quadratic error matrix, neighborhood triangular mean area and gaussian curvature of each vertex are calculated.
- The edge length, edge shrinkage error matrix, position of new vertices and edge shrinkage error of each edge are calculated.
- Sort the calculated edge shrinkage error.
- Each time the smallest error edge is selected from the edge shrinking error to perform the edge shrinking operation.
- The feature factors affected by edge contraction are updated, such as vertex position and edge contraction error.
- If the simplification conditions are not met, skip to Step 6; otherwise, the simplification ends.

4. Experimental results and analysis

In order to evaluate the effectiveness of the method, we will prepare several simplified models for verification, which are mainly compared with Garland method, triangular area weighting and literature [5].Microsoft Visual Studio 2017 platform is mainly used in computers with Windows 10(64-bit) system,

Intel(R) Core(TM) I5-9300H CPU @ 2.40ghz processor and 8G RAM. OpenGL rendering engine is used to render and display the results.

Using the proposed method in this paper to the simplification of 3 d model, and under the condition of the same simplified rate of ratio, in figure 3, the original is given for 3 k simplify the 3 d model to 1 k, in the fish's eye and tail with the local characteristics of the present a good details, and the 3 d model is not obvious characteristics of the local grid density is small. In the case of high simplification rate, the important features of the model are maintained, as shown in the figure below:

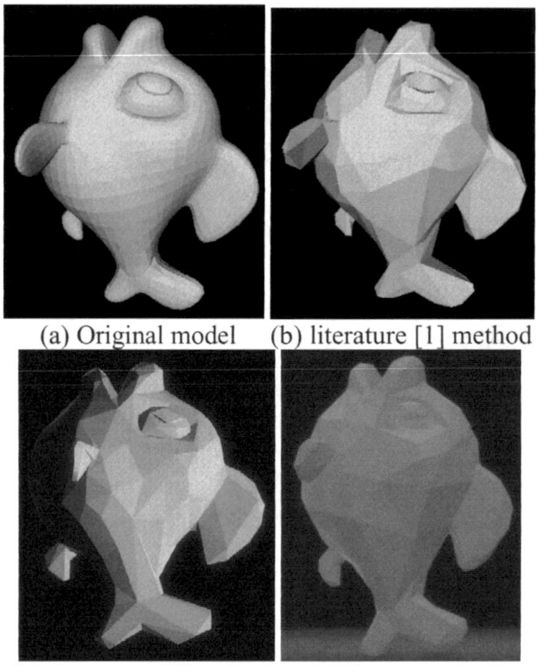

(a) Original model (b) literature [1] method
(c) document [5]method (d) Paper method
Figure 3. 90% effect of fish model simplification

In figure 4, the original simplified to 1 k to 2 k of 3 d model, and use different methods to simplify to 90%, in the cow's stomach, head, etc with the local characteristics of the present a good details, and the 3 d model rate larger piece of characteristics, performance than other methods to keep the method presented in this paper the important features of the model, the diagram below:

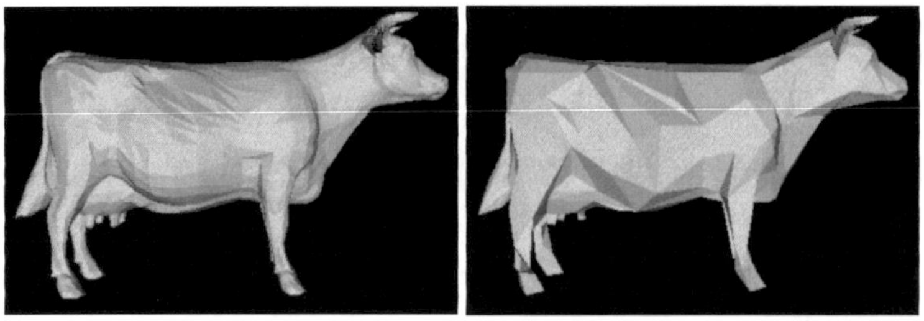

(a) Original model (b) literature [1] method

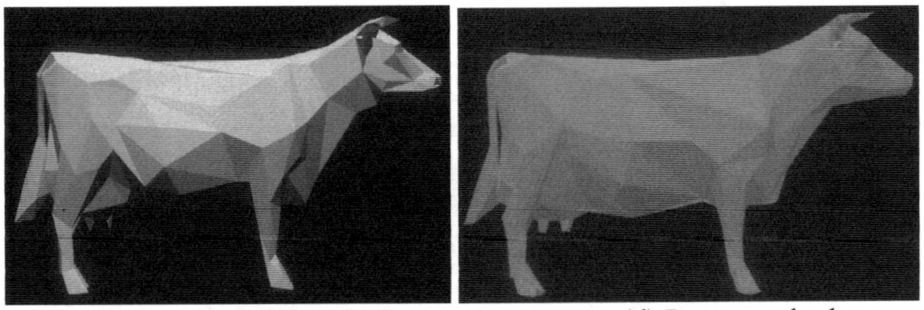

(c) document [5]method (d) Paper method

Figure 4. Cow model simplified 90% effect

Table 1. Comparison of loading time of simplified model.

Model	(a)	(b)	(c)	(d)
Cow	1.63s	1.35s	0.84s	0.63s
Fish	1.15s	1.01s	0.92s	0.62s

According to the experimental results, in the case of unweighted 3D model with high simplification rate, there are more inferior triangular shapes, and the details of the model cannot be maintained. In addition, by considering the edge feature factors, the visual effect of the model is increased, the overall feature retention is improved, and more details of the model are retained.

5. Conclusion

In this paper, a new mesh simplification algorithm based on QEM simplification algorithm is proposed, which weights edge feature factors to folding cost. Experiments show that the folding cost can not only maintain the overall feature variation of the triangular mesh model, but also retain more details, thus improving the overall visual effect of the triangular mesh model. However, the simplification effect of the algorithm is not obvious for areas with low sharpness. Therefore, more factors affecting the simplification effect are considered in the following work to improve the universality of the application of the model.

Acknowledgments

The completion of this paper is inseparable from the guidance of teachers and the help of students. Thank them for accompanying me through this unforgettable time. Finally, I would like to express my sincere thanks to the reviewing teachers and instructors for your precious time in this paper.

References

[1] Garland M, Heckbert P S. Surface simplification using quadric error metrics [J] . Computer Graphics, 1997, 31 (3):209-216.

[2] Paul S. Heckbert and Michael Garland. Optimal triangulation and quadric-based surface simplification [J]. Computational Geometry: Theory and Applications, 1999, 14(1):49-65.

[3] Zhang Shixue, Wu Enhua. Journal of Changchun University of Science and Technology, 2005(02): 52-56 +66.

[4] Meng Jun, Song Lei. Fast Mesh Simplification Algorithm Based on Edge Shrinkage [J]. Computer Engineering and Applications, 2007(20):62-64+116.

[5] Li Long, He Mingyi. Journal of northwest university (natural science edition), 2010, 40(03):419-422.

[6] Hou Bao Ming, Liu Xue Na. A Modified Generating Algorithm of Progressive Mesh Model Based on the Importance Degree of Vertex [J]. Advanced Materials Research, 2014, 3470.

[7] Wang jun, wang xiuhui. Simplification algorithm of grid model based on edge curvature [J]. Journal of China jiliang university, 2016, 27(01):73-79+85.

Research on recommendation algorithm of Graph attention Network based on Knowledge graph

Yiping Zeng[1], Shumin Liu[2]

[1] School of Software Engineering, Jiangxi University of Science and Technology, Nanchang, Jiangxi, 330013, China

[2] School of Software Engineering, Jiangxi University of Science and Technology, Nanchang, Jiangxi, 330013, China

*Corresponding author's e-mail: 6720190972@mail.jxust.edu.cn

Abstract. The introduction of knowledge graph as the auxiliary information of recommendation system provides a new research idea for personalized intelligent recommendation. However, most of the existing knowledge graph recommendation algorithms fail to effectively solve the problem of unrelated entities, leading to inaccurate prediction of potential preferences of users. To solve this problem, this paper proposes a KG-IGAT model combining knowledge graph and graph attention network, and adds an interest evolution module to graph attention network to capture user interest changes and generate top-N recommendations. Finally, experimental comparison between the proposed model and other algorithms using public data sets shows that KG-IGAT has better recommendation performance.

1.Introduction

As more and more data in the Internet can be perceived and acquired, multi-source heterogeneous data including images, texts and labels contain rich user behavior information and personalized demand information[1], and the hybrid recommendation method integrating multi-heterogeneous auxiliary information has attracted more attention from researchers. Knowledge Graph (KG) aims to describe various entities or concepts and their relationships in the real world, and constitutes a huge semantic network Graph, which facilitates the effective integration of various auxiliary information into the recommendation algorithm, thus alleviating the problems of cold start and data sparsity.

Wu X Y et al. [2] use KG feature learning embedding semantic data into low dimension space, by calculating the semantic similarity between projects itself into semantic information of the collaborative filtering recommendation. Wang et al. [3] proposed a RippleNet algorithm that applies KG to the recommendation system for predicting web page click rate. As the graph attention network GAT proposed by literature [4] et al. uses the attention mechanism to replace the static normalized convolution operation, which avoid the noise influence caused by excessive information of neighbor nodes. Wang et al. [5] proposed a new heterogeneous graph attention network based on attention mechanism, providing a new idea for the application of graph attention network to heterogeneous graph. Based on literature [6] KGAT algorithm is put forward, in the end-to-end way explicitly modeling the relationship of higher order connectedness in KG, but this method is computationally expensive. Subsequently, KG was combined with graph convolution network [7] and KGCN[8] was proposed. Literature [9] proposed to use user preference tree to solve the problem of irrelevant entities when exploring user potential preferences along the relationship in KG. However, for large-scale KG, the construction of user preference tree is too complex.

The main contributions of this paper are as follows:

(1) KG merged into the field of collaborative filtering recommendation framework, designed an end-to-end collaborative knowledge map network.

(2) Based on the drawing attention network, points for attention to update the threshold cycle unit of user interest model. So as to obtain user preferences more accurately and achieve personalized recommendations.

2.KG- IGAT algorithm

The overall framework of KG-IGAT model is shown in Figure 1 below.

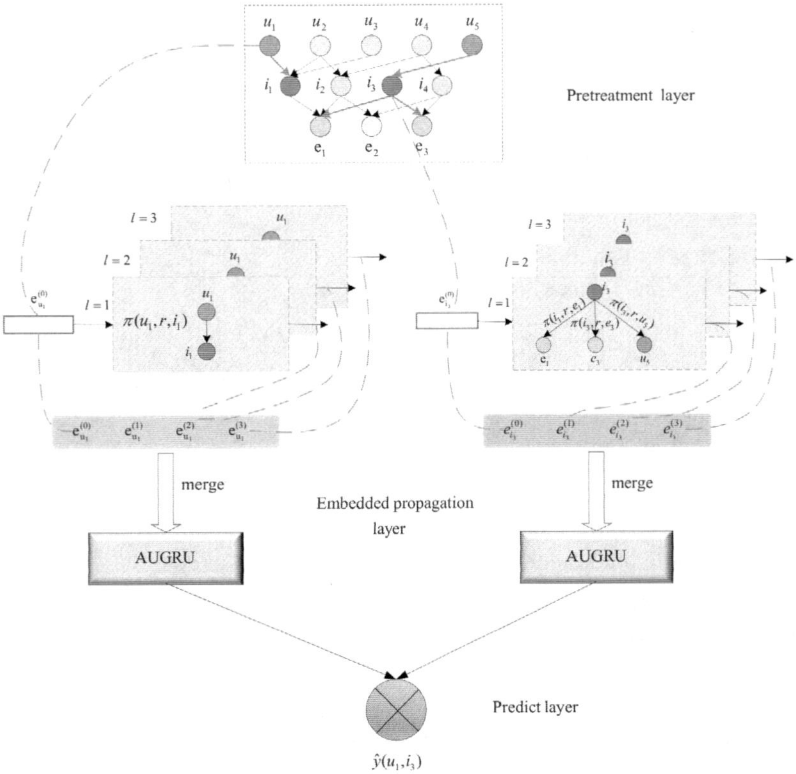

Figure 1. General framework of KG-IGAT

2.1 Pretreatment Layer

In this paper, TransR[11] method is used to preprocess collaborative knowledge graph G. For any given triplet (h,r,t), whether or not it actually exists in G, the following distance formula is used to describe its authenticity:

$$g(h,r,t)=\| M_r e_h + e_r - M_r e_t \|_2^2 \tag{1}$$

Where $M_r, \in \mathbb{R}^{k\times d}$ is the transformation matrix of the relation r, which is used to project the entity from the d-dimensional entity space to the K-dimensional relation space; $\|\cdot\|_2$ represents L2 regularization to prevent overfitting. Here, a lower score means that a given triple $g(h,r,t)$ is more likely to be real in G.

2.2 Attention embedding communication layer

2.2.1 Information dissemination

After the information of G is embedded into the low-dimensional space, the corresponding embedding vector e is generated, and then the information dissemination stage.

Given an entity h, $N_h = \{(h,r,t) \mid (h,r,t) \in G\}$ represents a collection of triples, where h is the header entity. Then, the representation vector e_{N_h} of N_h can be calculated by:

$$e_{N_h} = \sum\nolimits_{(h,r,t)\in N_h} \pi(h,r,t) e_t \tag{2}$$

Where, $\pi(h,r,t)$ is the attenuation factor controlling each transmission of the triplet, indicating how much information is transmitted from t to h through r.

2.2.2 Based on attention mechanism

The second stage of embedding the propagation layer is based on the attentional mechanism. The calculation formula of attention mechanism is as follows:

$$\pi(h,r,t) = (M_r e_t)^T \tanh((M_r e_h + e_r)) \tag{3}$$

Where, tanh is nonlinear activation function; M_r is the transformation matrix on the relation r generated in the embedding layer. The size of $\pi(h,r,t)$ depends on the distance between node h and neighbor node t in relation r in the embedded layer. Then, softmax normalized the attention score calculated for all triples (h,r,t) as follows:

$$\pi(h,r,t) = \frac{\exp(\pi(h,r,t))}{\sum\nolimits_{(h,r',t')\in N_h} \exp(\pi(h,r',t'))} \tag{4}$$

2.2.3 Information aggregation

The entity representation vector e_h of h and the neighborhood representation vector e_{N_h} obtained from its neighbors in the process of information collection are aggregated as the representation form of the first-order neighbor information aggregated by entity h, denoted as $e_h^{(1)}$, $e_h^{(1)} = f(e_h, e_{N_h})$, $f(\cdot)$ is the aggregator function. A graph aggregator is used to connect two vectors for nonlinear transformation, $f_{GraphSage} = Leaky\,\mathrm{Re}LU(e_h \,\|\, e_{N_h})$.

2.2.4 High-order propagation

Next, update e_h iteratively to obtain higher-order information of h. In this paper, the embedding of l -order information of entity h is denoted as e_h^l, and the recursive updating formula is as follows:

$$e_h^l = f(e_h^{(l-1)}, e_{N_h}^{(l-1)}), \tag{5}$$

$$e_{N_h}^{(l-1)} = \sum\nolimits_{(h,r,t)\in N_h} \pi(h,r,t) e_t^{(l-1)} \tag{6}$$

$e_t^{(l-1)}$ is the $(l-1)$ order representation of entity t, storing the neighborhood information of t.

2.2.5 Evolution of interest

In order to deeply explore how user Interest evolution module is added in this stage, this module adopts a new structure-- AUGRU[10] to model user interest, as shown in Figure 2.

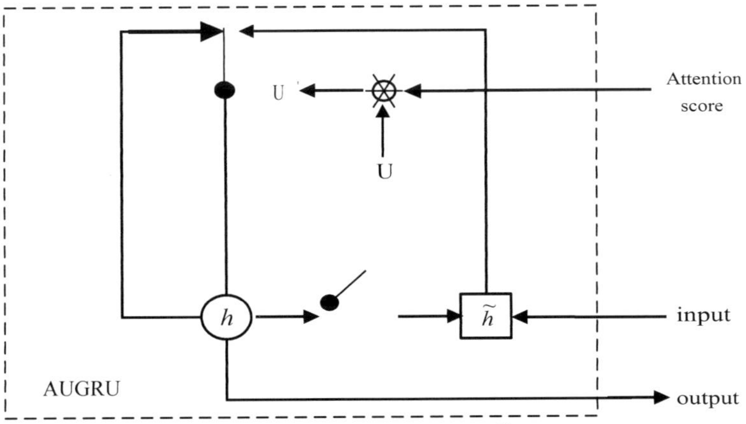

Figure 2. AUGRU internal structure diagram

AUGRU can be expressed by the formula:

$$\tilde{u}_t^{'} = \pi_t \otimes u_t^{'} \tag{7}$$

$$h_t^{'} = (1-\tilde{u}_t^{'})h_{t-1}^{'} + \tilde{u}_t^{'}\tilde{h}_t^{'} \tag{8}$$

Where $u_t^{'}$ is the initial state of updating gate in AUGRU, $\tilde{u}_t^{'}$ is the state of updating gate with attention score π_t added, and $h_t^{'}$, $h_{t-1}^{'}$ and $\tilde{h}_t^{'}$ are the states of hidden layer at AUGRU t moment.

In this interest evolution module, AUGRU uses attention score π_t to influence the update gate in GRU. After the addition of π_t, users' interests are focused according to π_t, and there are fewer attributes corresponding to users' interests in the items to be recommended.

2.3 Prediction layer of model

After the L-layer superposition, we can obtain the multi-scale representation of user node u and item node i as $\{e_u^{(1)},...,e_u^{(L)}\}$ and $\{e_i^{(1)},...,e_i^{(L)}\}$ respectively. Inspired by the multi-scale fusion in the image field[11], the hierarchical aggregation mechanism is adopted to connect the representation of each step into a vector representation, such as: $e_u^{*} = e_u^{0} \| \cdots \| e_u^{L}, e_i^{*} = e_i^{0} \| \cdots \| e_i^{L}$. The above operations can not only adjust the initial embedding parameters, but also control the scope of information aggregation and strengthen the propagation by adjusting L. Finally, this paper predicts the matching score of user and project through the inner product of their representation: $\hat{y}(u,i) = (e_u^{*})^T e_i^{*}$, sort $\hat{y}(u,i)$ in descending order to generate a top-N list.

3.Experimental research

3.1 Experimental Settings

In this paper, movielens-1M data set is selected to conduct comparative experiments between the model KG-IGAT in this paper and existing recommendation algorithms, including embedding based recommendation algorithm CKE[12], path-based recommendation algorithm RippleNet [3] and KGCN [8]. MovieLens is a public, standardized collection of data about how people rate movies. In this paper, DBpedia ontology knowledge base [13] was used to build a knowledge map for movielens-1M dataset, and users, movies and their interactive information were mapped and linked to corresponding entities, and the attributes in the dataset were all regarded as relations. In addition, a user triplet (user, gender, age) is constructed, corresponding to the user's ID, gender, age and other attribute information. The statistics of the dataset are shown in Table 1 below. In this paper, 80% of the interactive information in the data set was randomly selected as the training set and the remaining 20% as the test set.

Table 1. Data set statistics

Data set		MovieLens-1M
User-item Interaction	Users	6040
	Movies	3883
	Interaction	100209
Item KG	Entity	13593
	Relation	3
	Triple	25462

Table 2. Performance comparison of depthindexes of different models

indicators / depth	recall@20	NDCG@20
1	0.2433	0.6694
2	0.2461	0.6718
3	0.2487	0.6735
4	0.2475	0.6729

For the generated top-N list, that is, the most consistent recommendations of the first N items are given. In this paper, NDCG and recall are used to evaluate the model recommendation performance.

NDCG evaluates the sorting performance through the position of the item in the top-N list, so as to obtain the gap between the top-N list and the user's real interaction list, $NDCG@N = \frac{1}{IDCG}\sum_{i=1}^{N}\frac{2^{rel_i}-1}{\log_2(i+1)}$. Where, rel_i represents the probability of the project at the i position, and IDCG is the maximum DCG value under the ideal state, $IDCG@N = \sum_{i=1}^{|REL|}\frac{2^{rel_i}-1}{\log_2(i+1)}$. Where, $|REL|$ is the set of the first N terms of the recommended results arranged according to the highest to lowest probability. It can be seen that the higher the item is in the list, the larger the NDCG is.

recall is an indicator to measure the proportion of items in the top-N list in the test set, $recall@N = \frac{\sum_{u\in U}|R(u)\cap T(u)|}{\sum_{u\in U}|T(u)|}$.Where $R(u)$ is the list of recommendations generated for user u and $T(u)$ is the list of items that the user actually interacts with in the test set.The more accurate the recommendation, the larger the proportion of $T(u)$ in $R(u)$, and the larger the recall.

The initial learning rate of KG-IGAT was set to 0.001, the L2 regularization coefficient was set to 1E-5, and the embedding dimension was set to 64. Adam optimizer was used for optimization, and all parameters in the model parameter set Θ were updated and adjusted.

3.2 Experimental results and analysis

In this section, L is selected as 1, 2, 3 and 4 respectively to determine the most appropriate model depth, and N=20. The experimental results are shown in Table 2:

As can be seen from the table, with the increase of L, recall and NDCG also increase. When L=4, it is found that the values of recall and NDCG decrease somewhat, indicating that the larger L is, the better is not. In addition, when L=3, recall and NDCG values are the maximum, indicating the best model performance. When the threshold is exceeded, the performance degradation is due to the effect of noise on the model, so L=3 is selected in this paper for subsequent experiments.

Taking L=3 as the optimal model depth, the performance of each model at N=20 is compared respectively in the following Table 3:

Table 3. Performance comparison of each model index

indicators / model	recall@20	NDCG@20
CKE	0.2323	0.6478
RippleNet	0.1441	0.5235
KGCN	0.1572	0.5532
KG-IGAT	0.2441	0.6697

Table 3 shows that when N=20, recall@: KG-IGAT> CKE> KGCN> RippleNet and NDCG@: KG-IGAT > CKE> KGCN> RippleNet. RippleNet and KGCN may be superior to CKE in other aspects, but from the two evaluation indexes in this paper, CKE is superior to RippleNet and KGCN. In general, the model KG-IGAT in this paper has the best performance.

In addition, Figure 3 and Figure 4 respectively show the comparison of recall index and NDCG index performance of each model.

The below four models are all based on KG. With the increase of N, the performance of each model shows an overall upward trend. In terms of recall and NDCG, the performance of the proposed model is superior to the other three models. When N=20, recall value: KG-IGAT is 1 percentage point, 10 percentage point and 9 percentage point higher than CKE, RippleNet and KGCN, respectively. NDCG value: KG-IGAT is 2, 14 and 11 percentage points higher than CKE, RippleNet and KGCN.

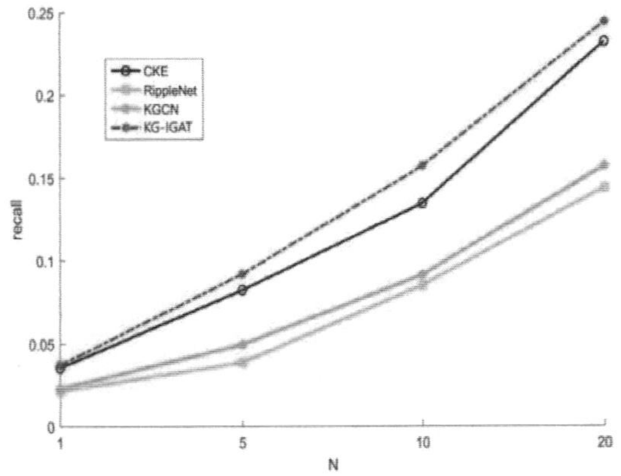

Figure 3. Comparison of recall index performance among models

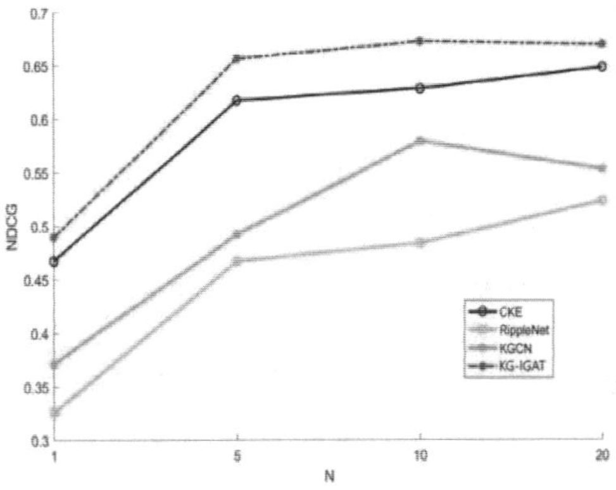

Figure 4. Comparison of NDCG index performance among models

4. Conclusion

This paper proposes a model, KG-IGAT, which makes full use of the information of the central node and the neighbor node in the process of embedded propagation to model, and then disseminates the information aggregation to a higher order. Meanwhile, the evolution of user interest is integrated into the attention mechanism to model, so as to capture the changes of user interest more accurately. A large number of experimental data show that when N=20, the recall rate and NDCG of this model are 24.4% and 66.9%, respectively. Therefore, this model is superior to the other three models and has good robustness.

For the research of intelligent recommendation algorithm, it can combine various algorithms to mix recommendation in the future in order to achieve better results.

Acknowledgement
This work was supported in part by National Natural Science Foundation of China (No. 61761021), Natural Science Foundation of Jiangxi Province (Grant No. 20181bab202018, 20202BAB212003), Special Fund for Postgraduate Innovation of Jiangxi Province(No. YC2020-S481).

References

[1] HUANG L W, JIANG B T, LYU Y B, et al. Survey on deep learning based recommender systems[J]. Chinese Journal of Computers: 2018, 41(7): 1619-1647.
[2] Wu X, Chen Q, Liu H, et al. Collaborative filtering recommendation algorithm based on KG representation learning[J]. Computer engineering, 2018, 44(2): 226-232, 263.
[3] Wang H W, Zhang F Z, Wang J L, et al. Ripple Net: propagating user preferences on the knowledge graph for recommender systems[C] // Proc of the 27th ACM International Conference on Information and Knowledge Management. New York: ACM Press, 2018: 417-426.
[4] CUCURULL G, CASANOVA A, et al. Graph attention networks [J]. ar Xiv: 1710.10903, 2017.
[5] Wang X, Ji H Y, Shi C, et al. Heterogeneous graph attention network[C] // Proc of World Wide Web Conference.New York: ACM Press, 2019: 2022-2032.
[6] Wang X, He X N, Cao Y X, et al.KGAT: knowledge graph attention network for recommendation[C] // Proc of the 25th ACM SIGKDD International Conference on Knowledge Discovery and Data Mining. New York: ACM Press, 2019: 950-958.
[7] KIPF T N, WELLING M. Semi-supervised classification with graph convolutional networks[J]. ar Xiv: 1609. 2907, 2016.
[8] Wang H W, Zhao M, Xie X, et al. Knowledge graph convolutional networks for recommender systems[C] // Proc of World Wide Web Conference. New York: ACM Press, 2019: 3307-3313.
[9] Li Q Y, Tang X L, Wang T Y, et al. Unifying task oriented knowledge graph learning and recommendation[J]. IEEEAccess, 2019, 7: 115816-115828.
[10] Zhou G, Mou N, Fan Y, et al. Deep Interest Evolution Network for Click-Through Rate Prediction[J]. In Proceedings of the AAAI Conference on Artificial Intelligence, 2019: 5941-5948.
[11] Zhao J, Qian Y, Nan F, et al. CNN multi-layer feature fusion and ELM for breast disease diagnosis[J]. Computer engineering and applications, 2020, 56(4): 122-127.
[12] Zhang F Z, Yuan N J, Lian D F, et al. Collaborative knowledge base embedding for recommender systems[C] // Proc of the 22nd ACM SIGKDD International Conference on Knowledge Discovery and Data Mining. New York: ACM Press, 2016: 353-362.
[13] Bizer C, Lehmann J, Kobilarov G, et al. DBpedia: a crystallization point for the Web of data[J]. Journal of Web Semantics, 2009, 7(3): 154-165.

MRDRL-ROS: a Multi Robot Deep Reinforcement Learning Platform based on Robot Operating System

Zhongxuan Cai, Zhen Liang and Jing Ren*

Institute for Quantum Information & State Key Laboratory of High Performance Computing, College of Computer, National University of Defense Technology, Changsha, Hunan, 410073, China

*Corresponding author's e-mail: renjing@nudt.edu.cn

Abstract. Deep reinforcement learning (DRL) has greatly improved the intelligence of AI in recent years and the community has proposed several common software to facilitate the development of DRL. However, in robotics the utility of common DRL software is limited and the development is time-consuming due to the complexity of various robot software. In this paper, we propose a software engineering approach leveraging modularity to facilitate robot DRL development. The platform decouples learning environment into task, simulator and hierarchical robot modules, which in turn enables diverse environment generation using existing modules as building blocks, regardless of the underlying robot software details. Experimental results show that our platform provides composable environment building, introduces high module reuse and efficiently facilitates robot DRL.

1. Introduction

Deep reinforcement learning (DRL) has been one of the most important breakthroughs of artificial intelligence in recent years. The first hallmark was the Deep Q-Network (DQN) achieving superhuman performance on Atari 2600 video games in 2013 [1]. Apart from the field of video games, DRL is also widely applied in robotics, where robot control tasks can be naturally modeled as sequential decision problems, i.e. to determine an action sequence based on sensor states. The feature extraction ability of deep neural networks enables end-to-end control from robot sensors to actuators, developing multi-joint motion control [2], mobile robot navigation [3], 3D manipulation [4], etc. DRL common software is developed in the community, providing benchmark environments [5], DRL platforms [6] and parallel learning architectures [7].

Although the software infrastructure for general reinforcement learning is increasingly sophisticated, robot reinforcement learning faces additional problems arising from the unique characteristics of the robotics field, i.e. the software complexity of environment building. Distinct from general DRL development, robot DRL aims to accomplish a variety of robot tasks with various robot hardware rather than solving a limited set of benchmarks. The environment building requires customization. It usually involves third party robot simulators and communicates with multitudinous robot sensors and actuators like GPS, laser, sonar, camera, mechanical arms, mobile bases, rotors and so on [8], which makes the program extremely complex and produces a huge workload, especially for multi-robot cases.The software complexity makes it rather inefficient and time consuming to develop robot DRL.

In this paper, we propose a robot DRL platform design as the approach to robot DRL software engineering. Following modularity, we abstract robot DRL environment into three reusable modules without loss of generality: the simulation, robot and task modules. The simulation module adapts to different robot simulators and provides easy transfer through module switching. The robot module encapsulates communication details with various robot sensors and actuators, by which custom robot groups can be built without re-engineering underlying details. The task module implements a specific robot task. By modularity, custom learning environments with arbitrary combination of simulators, robot groups and tasks can be built like building blocks.

The major contributions of this paper are:

- We propose a software engineering approach to building robot deep reinforcement learning programs, which addresses the software complexity of robot DRL through modularity from software infrastructure.
- The modularity of proposed model abstracts the environment into three high-level reusable modules and decouples the engineering of robot simulators, robot groups and tasks, which supports customizing environment like building blocks, regardless of underlying details.
- Experiments with different robot type, different simulator and different tasks are conducted to verify the effectiveness of the proposed platform.

2. Related Work

With the development of deep reinforcement learning, several software platforms are proposed to improve the efficiency of DRL development and free researchers from tedious works unrelated to DRL algorithm design, e.g. the Arcade learning environment provided an interface to hundreds of Atari 2600 games as learning environment [5]. Besides out-of-box benchmark environments, the community further considered efficient programming in DRL software engineering. OpenAI gym [6] proposed unified interacting interfaces between DRL algorithms and environments, allowing cross-testing of different algorithms and environments as long as they followed the gym interface specification. Gym's modular design has built a powerful algorithm and environment software ecosystem, making it the de facto standard for DRL research and development.

While the common software for general DRL has developed a lot, its usage and utility in robotics is limited. In robot DRL, the ultimate goal is to achieve multitudinous robot tasks given certain robot hardware, rather than to merely achieve higher performance on fixed benchmark. The environment building needs to be customized and third party robot simulators are usually involved for physics simulation. Gym-gazebo and parallel gym gazebo [7] extend openAI gym with gazebo, a powerful 3D robot simulator. However, they extend fixed environments instead of providing a mechanism to efficiently build diverse robot tasks.

3. The MRDRL Platform

To improve the efficiency of building environments for multitudinous robot DRL tasks, we propose a modular environment model which can be built from abstraction with the above three parts. The modularity provides high code reuse, convenient transfer and ease of environment modification.

3.1. Model architecture

The architecture decouples the environment model for robot DRL into three parts, including simulation, robots and task modules, as shown in Fig. 1.

The environment model bridges communication between DRL algorithms and robot simulators or hardware drivers. On the top of the architecture, the environment interfaces of openAI gym (reset and step) are preserved between the DRL algorithm and the environment to be compatible with gym-based DRL algorithm implementations. At the bottom of the architecture, the environment model integrates rich robot simulators or hardware drivers through ROS communication.

The main idea of our design is the building blocks philosophy. With a set of implemented simulation, robots and task modules, an environment can be efficiently constructed or modified by combing these modules. A set of robots can perform a different learning task, e.g. from exploring a maze to formation, by simply modifying the task module rather than rewriting a completely new environment.

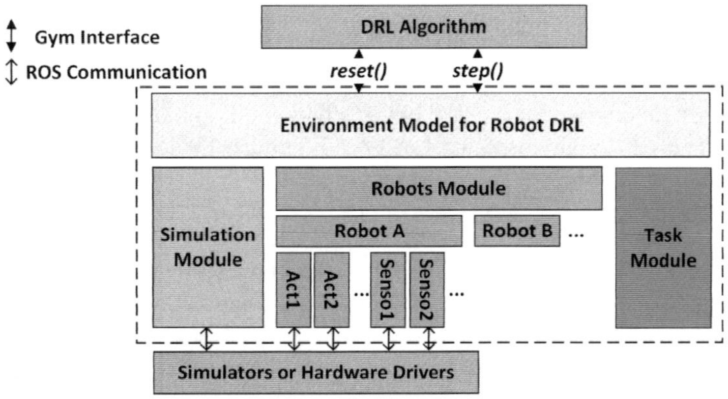

Figure 1. The Robot DRL platform architecture.

3.2. Modularized learning procedure

In this subsection we describe the function of each module in our environment architecture through a top-down approach. Since the architecture keeps the unified environment interfaces of openAI gym, from the perspective of DRL, all the environment model needs to accomplish is to reset and step the simulation world and return feedback through ROS communication.

Rather than directly implementing the reset and step details in a flat process-oriented way for each robot learning task, our model resets and steps the environment through three abstract modules as shown in Fig. 2. To reset the environment, the simulation module is called to reset and unpause the underlying robot simulator. Then the robots module is utilized to wait for sensor messages of each robot from ROS as the raw observation. Once these data received, the simulation is paused by the simulation module, and the raw sensor data is managed by the task module to produce observation in float lists directly available to DRL neural networks.

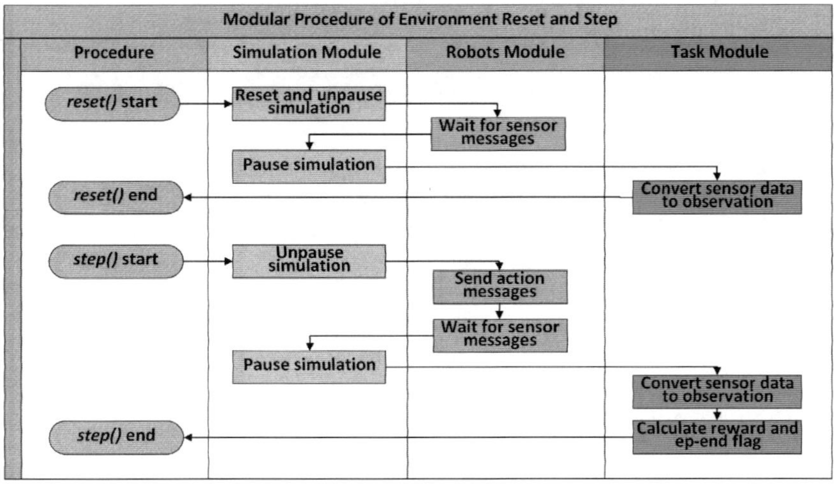

Figure 2. Modularization of environment interaction interface.

Similarly, the step procedure is also achieved with the three proposed modules in Fig. 2. The additional parts includes: (1) sending action command messages by the robots module, which converts the action calculated by DRL algorithms to ROS communication messages and publishes them to corresponding robot actuators, and (2) the task module calculating the reward and the episode-end flag as the step feedback. Given the required functions of each module to accomplish environment reset and step, we design the modules in the following subsections.

3.3. *Simulation module for transfer*

During environment reset and step, the major function of the simulation module is to reset, pause and unpause the underlying robot simulator. The reset function is utilized to restart a new simulation world for a new episode. The pause and unpause function is needed to accomplish frame simulation and learning step by step. The simulation module encapsulates all the communication details of simulator control. When building environment for new robot tasks with the same simulator, the implemented module can be directly reused and no underlying detail of simulator control needs to be taken care of. After several mainstream simulator implemented, transfer between these simulators can be enabled by simply changing to corresponding modules, where all communication details are handled by these modules and hidden from developers.

3.4. *Hierarchical robots module for communications*

The robots module is the core of communication between DRL algorithms and robots. It inputs DRL results as action commands and sends action command messages to each robot actuator, then waits for sensor messages from all robots and returns observations. The major function of the robots module is the conversion of raw action/observation data and messaging with actuators and sensors.

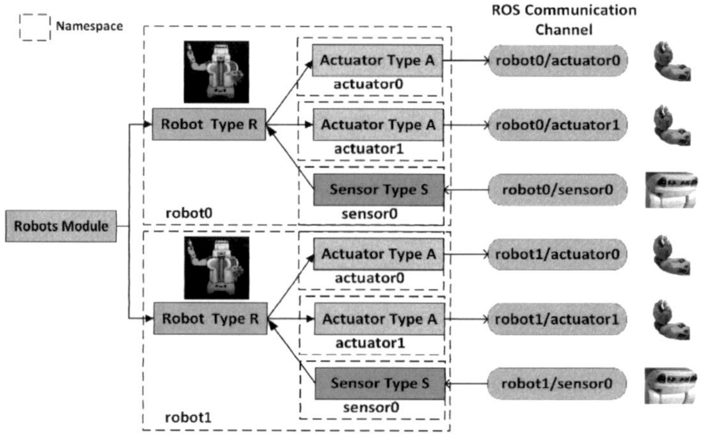

Figure 3. Reusing implemented models through namespace mechanism.

It is natural to organize the group from the group level to robot level to actuator/sensor level as in Fig. 1. The operation of the robots module can be achieved by iteratively querying each robot as submodules to operate. Each robot submodule also iteratively queries its actuators and sensors to accomplish the query of the robot. With the query structure, the upper robot modules can be generated by listing child modules like actuators and sensors. In ROS different robot sensors and actuators communicate on separate topics but the message format is the same for each type. We develop a namespace mechanism by which the communication detail of a specific type of sensor or actuator can be reused and redirected to different channels. As shown by the same colored submodules in Fig. 3, when managing different actuators or sensors of the same type, we can assign namespace to communicate on different channels and reuse existing implementations. The higher robot module of the same type can also be reused through the namespace mechanism.

After implementing all needed types of actuators and sensors once, we can build robot groups by simply combing these modules and assigning namespaces, regardless of any communication details.

The complexity of the underlying communication management only increases with the types of sensors and actuators

3.5. Task module for learning

Distinct from the simulation module and the robots module which handle communication with the simulators and robot drivers, the task module encapsulates the reward signal for robots to accomplish a specific learning task. It processes raw sensor data to task-defined environment observations that match the input of DRL neural network. Besides, the task module also determines the terminate condition for an episode and returns the episode-end flag for further environment reset.

With the proposed reusable task, simulation and robots modules, environment building of robot deep reinforcement learning is decoupled and modularized. After implementing these modules for once, developers are free to build arbitrary learning environment of any task on any group of robots in any simulator by simply combining these high-level modules like building blocks.

4. Experiments

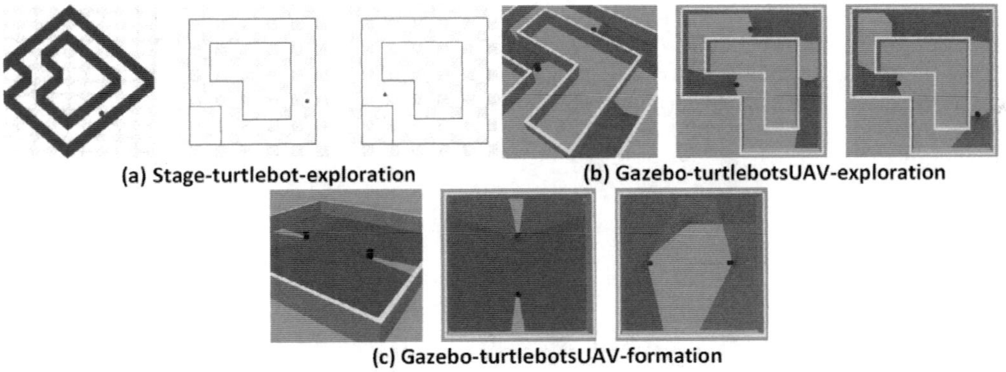

Figure 4. Environment engineering with modularity.

To validate our work, we conduct several robot DRL experiments with a widely used DRL algorithm, proximal policy optimization (PPO), which performs well on both continuous and discrete problems [9]. We build modular environment models and demonstrate modifying simulators, robots and task modules to efficiently switch to new learning environment. The basic environment is the standard benchmark of gym-gazebo, which contains a turtlebot 2 mobile robot which learns to explore in a maze without colliding walls, as shown in Fig. 4.

To build a modular environment model between the DRL algorithm and the gazebo robot simulator, a gazebo simulation module contains communication details to pause, unpause and reset the gazebo simulator. Then we build a hierarchical robots module containing one single turtlebot-laser model which is made up of one actuator model and two sensor models. The actuator model implements communication details with the robot base driver. The laser sensor model and position sensor model implements communication details with the sensor drivers. The task module calculates exploration reward [7]. Since the current learning task only requires laser data as observation, the task module removes the position sensor data from the original sensor data as observation.

To conduct the same learning task through a different simulator, switching to a new simulation module is all that needs in our modular environment model. We demonstrate the modification in Fig. 4(a), where a turtlebot mobile robot learns to explore a maze in Stage robot simulator rather than the gazebo simulator. To build an environment model between the simulator and the DRL algorithm, we load a new simulation module, keeping the existing robots module and task module invariant. The new simulation module for Stage implements communication details for controlling Stage simulator.

Besides modifying simulator, the modularity also enables switching to a new robot group to perform the robot task by modifying the robots module. To build a new robots module, we can further reuse existing robot models, or build new robot models with existing actuator and sensor models. We change the robot group from one single turtlebot to two turtlebots as shown in Fig. 4(b) by simply adding more turtlebot models and assigning distinct namespace. We further add a multi-rotor UAV model consisting of a flight control actuator, laser and position sensors. The simulator we use is the gazebo simulator and the task is still exploration, therefore the two corresponding modules are reused.

Finally, we build a new environment with the same robot simulator and robot group to conduct a new learning task. As shown by Fig. 4(c), we still use two turtlebots and one multi-rotor UAV in the gazebo robot simulator therefore we reuse the existing robots and simulation modules. But rather than exploration, we create a new formation task where each robot aims to navigate and form a triangle formation. The new task module coverts position data to relative polar position and concatenate the laser as observation. It also calculates rewards based on Euclidean distance to the target position [10].

5. Conclusion

In this paper, we propose a common software approach for engineering robot DRL programs. The modularity platform enables reuse of modules to avoid inefficient re-engineering of environments and provides high level abstraction to free algorithm designers from underlying robot software details. Experiments are conducted to verify the effectiveness the proposed approach. Future work is to consider the learning procedure improvement of multi-robot cooperative learning.

Acknowledgments

This work was partially supported by the National Key Research and Development Program of China (No. 2017YF-B1301104) and the National Natural Science Foundation of China (No. 91648204 and 61803375).

References

[1] Mnih, V., Kavukcuoglu, K., Silver, D., Rusu, A.A., Veness, J., Bellemare, M.G., Graves, A., Riedmiller, M.A., Fidjeland, A., Ostrovski, G., Petersen, S., Beattie, C., Sadik, A., Antonoglou, I., King, H., Kumaran, D., Wierstra, D., Legg, S., Hassabis, D.: Human-level control through deep reinforcement learning. Nature 518(7540), 529–533 (2015).

[2] Schulman, J., Levine, S., Abbeel, P., Jordan, M.I., Moritz, P.: Trust region policy optimization. In: Proceedings of the 32nd International Conference on Machine Learning, ICML 2015, Lille, France, 6-11 July 2015, vol. 37, pp. 1889–1897.

[3] Hsu, S., Chan, S., Wu, P., Xiao, K., Fu, L.: Distributed deep reinforcement learning based indoor visual navigation. In: 2018 IEEE/RSJ International Conference on Intelligent Robots and Systems, IROS 2018, Madrid, Spain, October 1-5, 2018, pp. 2532–2537.

[4] Gu, S., Holly, E., Lillicrap, T.P., Levine, S.: Deep reinforcement learning for robotic manipulation with asynchronous off-policy updates. In: 2017 IEEE International Conference on Robotics and Automation, ICRA 2017, Singapore, May 29 - June 3, 2017, pp. 3389–3396.

[5] Bellemare, M.G., Naddaf, Y., Veness, J., Bowling, M.: The arcade learning environment: An evaluation platform for general agents. Journal of Artificial Intelligence Research 47, 253–279 (2013).

[6] Brockman, G., Cheung, V., Pettersson, L., Schneider, J., Schulman, J., Tang, J., Zaremba, W.: Openai gym. ArXiv abs/1606.01540 (2016).

[7] Liang, Z., Cai, Z., Li, M., Yang, W.: Parallel gym gazebo: a scalable parallel robot deep reinforcement learning platform. In: 31st IEEE International Conference on Tools with Artificial Intelligence, ICTAI 2019, Portland, OR, USA, November 4-6, 2019, pp. 206–213.

[8] Kober, J., Bagnell, J.A., Peters, J.: Reinforcement learning in robotics: A survey. The International Journal of Robotics Research 32(11), 1238–1274 (2013).

[9] Schulman, J., Wolski, F., Dhariwal, P., Radford, A., Klimov, O.: Proximal policy optimization algorithms. ArXiv abs/1707.06347 (2017).

[10] Long, P., Fan, T., Liao, X., Liu, W., Zhang, H., Pan, J.: Towards optimally decentralized multi-robot collision avoidance via deep reinforcement learning. In: 2018 IEEE International Conference on Robotics and Automation, ICRA 2018, Brisbane, Australia, May

AUTHOR INDEX